Role for Concrete in Global Development

Proceedings of the International Conference
held at the University of Dundee, Scotland, UK
on 8-9 July 2008

Edited by

Ravindra K. Dhir
Director, Concrete Technology Unit
University of Dundee

Peter C. Hewlett
Group Technical Advisor,
John Doyle Group

Laszlo J. Csetenyi
Research/Teaching Fellow, Concrete Technology Unit
University of Dundee

Moray D. Newlands
Lecturer, Concrete Technology Unit
University of Dundee

 bre press

DUNDEE

ABOUT THE EDITORS

Professor Ravindra K Dhir is founding Director of the Concrete Technology Unit at the University of Dundee. A renowned practitioner in the field of concrete technology, he is a member of numerous national and international technical committees and has published extensively on many aspects of concrete technology, cement science, durability and construction methods.

Professor Peter C Hewlett is a visiting industrial professor to the Concrete Technology Unit, University of Dundee. He is currently Principal Consultant to the British Board of Agrément, Group Technical Advisor to John Doyle Group plc and Chairman of the Editorial Board of Magazine of Concrete Research. An authority on concrete, he was awarded the UK Concrete Society's gold medal in November 2006.

Dr Laszlo J Csetenyi is a Research/Teaching Fellow in the Concrete Technology Unit, University of Dundee. He is a chemical engineer specialising in cement chemistry. His area of expertise includes confinement of wastes (both radioactive and non-radioactive) in cementitious matrices, and investigation into the properties, performance and potential utilisation of waste materials and industrial by-products in cement and concrete.

Dr Moray D Newlands is a Lecturer in the Concrete Technology Unit, University of Dundee. His research involves use of innovative cement combinations for engineering and durability performance, carbonation and chloride resistance of concrete and early age concrete properties. In addition he undertakes technology transfer to the concrete construction industry at a national and international level.

Cover photo: Singapore's overground / underground train system. (George Clerk)

Details of all publications from IHS BRE Press are available from:
www.ihsbrepress.com or
IHS BRE Press, Willoughby Road, Bracknell RG12 8FB, UK
Tel: 01344 328038, Fax: 01344 328005, brepress@ihs.com

Published by IHS BRE Press, 2008

Requests to copy any part of this publication should be made to:
IHS BRE Press, Garston, Watford WD25 9XX, UK
Tel: 01923 664761, Fax: 01923 6642477

EP86
ISBN-13: 978-1-84806-037-1
© IHS BRE Press, 2008

[*Concrete: construction's sustainable option*, Volumes 1–6, EP92, 978-1-84806-043-2]

The views expressed in the papers in this volume of proceedings are those of the authors. The editors and IHS BRE Press do not accept any responsibility for the contents of the papers or for any loss or damage which might occur as a result of following or using data or advice given in the papers.

PREFACE

Concrete and its constituent parts are available and used globally. It has been, is, and will continue to be the major construction material for mankind. As a consequence, we all have a responsibility for concrete's effective design, construction and efficient use. Future resources, energy consumption, performance, durability, environmental and societal impacts as well as economics are all matters on which concrete's sustainability will be judged - and this has a global significance. Increasingly, at both political and practical levels, construction has to address and implement sustainability and towards this goal, concrete engineering has the capability to drive the agenda.

The Concrete Technology Unit (CTU) of the University of Dundee organised this Congress to address these challenges, continuing its established series of events, namely, Global Construction: Ultimate Concrete Opportunities in 2005, Challenges of Concrete Construction in 2002, Creating with Concrete in 1999, Concrete in the Service of Mankind in 1996, Economic and Durable Concrete Construction Through Excellence in 1993 and Protection of Concrete in 1990.

The event was organised in collaboration with three of the world's most recognised institutions: the Institution of Civil Engineers, UK, the American Concrete Institute and the Japan Society of Civil Engineers. Under the theme of Concrete: Construction's Sustainable Option, the Congress consisted of six Events: *(i) Role for Concrete in Global Development, (ii) Precast Concrete: Towards Lean Construction, (iii) Concrete Durability: Achievement and Enhancement, (iv) Designing Concrete for the Visual Environment, (v) Concrete for Fire Engineering, (vi) Harnessing Fibres for Concrete Construction.* In all, a total of 308 papers were presented from 72 countries.

The Opening Addresses were given by Sir Alan Langlands, Principal and Vice-Chancellor of the University of Dundee, Mr John Letford, Lord Provost, City of Dundee, and senior representatives of the Institution of Civil Engineers, the Japan Society of Civil Engineers and the American Concrete Institute. The Event Opening Papers were presented by Dr Andrew Minson, The Concrete Centre, UK, Professor M Samarai, University of Sharjah, United Arab Emirates, Mr Martin Clarke, British Precast, UK, Dr Habib M Zein-Al-Abideen, Ministry of Municipal and Rural Affairs, Saudi Arabia, Mr Erik Larsen, COWI A/S, Denmark, Professor Tom Harrison Quarry Products Association, UK, Professor Carmen Andrade, Institute of Construction Science, Spain, Ms Linda Patterson, Linda Patterson Design, USA, Mr Michel Levy, Setec TPI, France and Dr Ellis Gartner, Lafarge Central Research, France. The Closing Papers for each event were presented by Professor Peter Hewlett, John Doyle Group, UK, Dr Bernd Hans Wolschner, SW Umwelttechnik, Austria, Dr Oladis Troconis, Zulia University, Venezuela, Mr Harry Moats, L M Scofield, USA, Dr Lars Bostrom, Swedish National Testing and Research Institute, Sweden, and Professor Sergio Carmona, University Tecnica Federico Santa Maria, Chile.

The support of over 43 International Professional Institutions and over 20 Sponsoring Organisations was a major contribution to the success of the Congress. An extensive Trade Fair formed an integral part of the event. The work of the Congress was an immense undertaking and all of those involved are gratefully acknowledged, in particular, the members of the Organising Committee for managing the event from start to finish; members of the Scientific and Technical Committees for advising on the selection and reviewing of papers; the Authors and the Chairmen of Technical Sessions for their invaluable contributions to the proceedings.

All of the proceedings have been prepared directly from the camera-ready manuscripts submitted by the authors and editing has been restricted to minor changes, only where it was considered absolutely necessary.

Dundee
July 2008

Ravindra K Dhir
Chairman, Congress Organising Committee

INTRODUCTION

Concrete is the second largest commodity used after water on a global scale with almost one tonne of concrete being used for each person every year. This figure is increasing steadily with the demand for infrastructure to be built in conjunction with industrial production gradually being transposed from Europe and North America to other parts of the world. The response to the effects of climate change such as strengthening sea defence lines incurs extensive construction activity everywhere and the first and foremost material for such development is concrete.

However, cement production heavily relies on mineral resources and can generate carbon dioxide by burning fossil fuels and decomposing limestone. This has significant environmental consequences despite remarkable efforts all over the world including significant developments by cement manufacturers themselves.

Perhaps the only way to mitigate the drawbacks of global development is by improving the performance of concrete whilst lowering its carbon footprint. Advanced use of admixtures and improved cements for example by better utilising the potential of pozzolanic additions is essential in achieving lean construction without compromising, or rather enhancing, quality.

Cement and concrete research is gaining new impetus to resolve the technical challenges of reducing energy and resource demands. Improving kiln and milling technologies, harnessing clean energy resources, utilising wastes and by-products, recycling aggregates, and exploiting local materials and techniques are all making substantial contributions to the development of concrete.

The sustainability agenda also infiltrates architectural design, puts the client/professional relationship into new dimensions, adopts whole life costing and resource efficiency strategies, and provides show-cases in creating low carbon buildings for example by utilising the thermal mass of concrete.

It has to be realised that the information technology revolution fuelled an even more rapid societal change world-wide, creating radically expanding demands for infrastructure and housing, especially in the developing world. There is a global responsibility in tackling this situation, so that the demands and the resulting environmental changes can be balanced.

The Proceedings *'Role for Concrete in Global Development'* dealt with all these subject areas and the issues raised four clearly defined themes: (i) Cements and Admixtures: Future Directions and Performance, (ii) Energy and Resources: Where Next?, (iii) Architecture and Engineering: Appropriate Design, and (iv) Developing World: Responsibilities in Changing Environments and Demands. Each theme started with a Keynote Paper presented by the foremost exponents in their respective fields. There were a total of 90 papers presented during the International Conference which are compiled into these Proceedings.

Dundee
July 2008

Ravindra K Dhir
Peter C Hewlett
Laszlo J Csetenyi
Moray D Newlands

ORGANISING COMMITTEE

Concrete Technology Unit

Professor R K Dhir OBE (Chairman)

Dr M D Newlands (Secretary)

Professor P C Hewlett
John Doyle Group

Professor T A Harrison
Quarry Products Association

Professor P Chana
British Cement Association

Professor V K Rigopoulou
National Technical University of Athens, Greece

Dr S Y N Chan
Hong Kong Polytechnic University

Dr N Y Ho
L & M Structural Systems, Singapore

Dr K A Paine
University of Bath

Professor S P Singh
NIT Jalandhar, India

Dr M R Jones

Dr M J McCarthy

Dr T D Dyer

Dr J E Halliday

Dr L J Csetenyi

Dr L Zheng

Dr S Kandasami

Mr R Lavingia

Dr S Caliskan

Dr M C Tang

Miss A M D Scott (Congress Assistant)

Mr S R Scott (Unit Assistant)

v

SCIENTIFIC AND TECHNICAL COMMITTEE

Dr Chanakya Arya, *Senior Lecturer,*
University College London, United Kingdom

Dr John L Clarke, *Advisor - Technical Services*
The Concrete Society, United Kingdom

Mr Martin A Clarke, *Chief Executive*
British Precast, United Kingdom

Professor David J Cleland, *Head of Planning, Architecture and Civil Engineering*
Queen's University Belfast, United Kingdom

Dr Per Fidjestol, *ACI Chair 234*
Elkem ASA Materials, Norway

Dr Stephen L Garvin, *Construction Director*
Building Research Establishment, United Kingdom

Mr Chris Hendy, *Head of Bridge Design and Technology*
Atkins Highways and Transportation, United Kingdom

Professor M Saeed Mirza, *Professor of Civil Engineering*
McGill University, Canada

Professor S S J Moy *Professor of Structures & Construction*
University of Southampton, United Kingdom

Professor Tarun R Naik, *Professor of Structural Engineering*
University of Wisconsin-Milwaukee, United States of America

Professor Byung Oh, *Professor of Civil Engineering*
Seoul National University, South Korea

Professor Khim Chye Ong *Deputy Head (Infrastructure and Resources)*
National University of Singapore, Singapore

Dr Donald Pearson-Kirk, *Technical Director*
AccordMP, United Kingdom

Professor Subba A Reddi, *Deputy Managing Director (Retd)*
Gammon India Limited, India

Professor Ahmed Sidky, *Professor*
Public Authority for Applied Education and Training (PAAET), Kuwait

Dr Howard P J Taylor, *Technical Director*
Tarmac Precast Concrete Ltd, United Kingdom

Dr Johan Vyncke, *Director*
Belgium Building Research Institute, Belgium

Professor Joost C Walraven, *Professor of Concrete Structures*
Delft University of Technology, Netherlands

Professor Habib M Zein-Al-Abideen, *Deputy Minister*
Ministry of Municipal and Rural Affairs, Saudi Arabia

COLLABORATING INSTITUTIONS

Institution of Civil Engineers, UK
American Concrete Institute
Japan Society of Civil Engineers

SPONSORING ORGANISATIONS WITH EXHIBITION

Aalborg Portland, Denmark

Aggregate Industries

Arcelor Mittal

Arup

BASF Construction Chemicals

British Board of Agrément

British Cement Association

British Precast

Building Research Establishment

Castle Cement Limited

CEMBUREAU

CNS Farnell Ltd

Danish Technological Institute

Dundee & Angus
Convention Bureau

Dundee City Council

Elkem Materials Ltd

Germann Instruments A/S

Grace Construction Products

Halcrow

Holcim Ltd

Heidelberg Cement

John Doyle Group

Lafarge Cement UK

OMYA UK

Oscrete Construction
Products

PANalytical Ltd

Propex Concrete Systems

Quantachrome Instruments

Singapore Concrete Institute

STATS Ltd

The Concrete Centre

United Kingdom Quality Ash
Association

Zwick Testing Machines Ltd

SUPPORTING INSTITUTIONS

American Society of Civil Engineers Construction Institute, USA

Asociación de Ingenieros de Caminos, Canales y Puertos, Spain

Asociación de Ingenieros del Uruguay, Uruguay

Association of Slovak Scientific & Technological Societies, Slovakia

Associazione Italiana di Ingegneria dei Materiali, Italy

Bahrain Society of Engineers, Bahrain

Belgian Concrete Society (BBG), Belgium

Brazilian Concrete Institute (IBRACON), Brazil

Canadian Society for Civil Engineering, Canada

China Civil Engineering Society, China

Chinese Institute of Engineers, Taiwan

Concrete Institute of Australia

Concrete Society of Southern Africa

Consiglio Nazionale degli Ingegneri, Italy

Czech Concrete Society, Czech Republic

Danish Concrete Society (DBF), Denmark

European Concrete Societies Network (ECSN)

Federation of Scientific Technical Unions in Bulgaria (FNTS)

German Society for Concrete and Construction Technology

Hong Kong Institution of Engineers

SUPPORTING INSTITUTIONS

Hungarian Cement Association (MCS)

Indian Concrete Institute

Institute of Concrete Technology, UK

Institution of Engineers, India

Institution of Engineers (IEB), Bangladesh

Institution of Engineers, Malaysia

Institution of Engineers (IESL), Sri Lanka

Institution of Engineers, Tanzania

Institution of Structural Engineers, UK

Instituto Mexicano del Cemento y del Concreto AC, Mexico

Irish Concrete Society, Ireland

Japan Concrete Institute, Japan

Korea Concrete Institute (KCI), South Korea

Netherlands Concrete Society, Netherlands

New Zealand Concrete Society, New Zealand

Nigerian Society of Engineers, Nigeria

Singapore Concrete Institute, Singapore

Slovenian Chamber of Engineers, Slovenia

The Concrete Society, UK

Turkish Chamber of Civil Engineers (TCCE), Turkey

Zimbabwe Institution of Engineers, Zimbabwe

CONTENTS

THEME 2: ENERGY AND RESOURCES: WHERE NEXT?

Keynote Paper

xiv

THEME 3: ARCHITECTURE AND ENGINEERING: APPROPRIATE DESIGN

Keynote Paper

THEME 4: DEVELOPING WORLD: RESPONSIBILITIES IN CHANGING ENVIRONMENTS AND DEMANDS

Keynote Paper

OPENING PAPERS

SUSTAINABLE DESIGN AND WHAT CONCRETE CAN OFFER

A J Minson

The Concrete Centre

United Kingdom

ABSTRACT. Sustainability has risen to the top of the political agenda and is influencing all sectors of society, including the construction sector. Sustainability has many strands to it and the construction sector can contribute to many of these either positively or negatively. Tools and regulations have been developed for design teams to guide them towards or dictate to them sustainable design. However these tools are often limited because they require weighting of different impacts and often concentrate on environmental impacts to the detriment of social and economic sustainability. To understand how the construction sector in general and concrete in particular can contribute to sustainability it is useful to consider the role of those involved in design and where opportunities exist to exploit inherent properties of concrete. Concrete has a very strong sustainability story but it is not a single issue story and is therefore more difficult to communicate. The message must be communicated to the different professions primarily to ensure that operational aspects are given due weighting and also to guard against simplistic checks that favour materials with a single sustainability credential.

Keywords: Sustainability, Concrete, Design, Performance, Embodied.

Dr A J Minson, Director, Technical Services & Head of Structural Engineering. Andrew joined The Concrete Centre in 2004 as Head of Structural Engineering and was promoted to Director in 2007. He and his team seek to provide engineers contractors, contractors, architects and clients with the information and tools required to design buildings and structures more effectively and efficiently using concrete. Sustainability is an ever increasing part of this role and The Concrete Centre seeks to pro-actively take a lead in the construction sector. Prior to his current position, Andrew was a multi-disciplinary design leader in Arup's London office. Before then he completed his doctorate as a Rhodes Scholar at Balliol College, Oxford having begun his career in his home country of Australia.

INTRODUCTION

To drive the sustainability agenda in the construction sector, various voluntary schemes and design tools have been developed over the past two decades. These are now increasingly supported by regulations. In order to respond to our developing understanding of sustainability, these schemes, design tools and regulations are becoming more detailed.

Of the three aspects of sustainability – social, economic and environmental – the latter is more easily measured and quantified. However, even for this aspect there is a difficulty. Different environmental impacts are measured in different units and how to compare and weight these is a matter of debate. Therefore, even before the less measurable social and economic indicators are incorporated into design tools, schemes or regulations, developing these for the environmental aspect alone is problematic.

Therefore, in addressing the question of what is sustainable design, even after twenty years of development of tools and regulations, it is useful to consider it more fundamentally, free of the impulse to weight and score different impacts. To do this, the paper firstly identifies what different members of the traditional design team bring in terms of sustainability to the multi disciplinary design team for a client who seeks a sustainable project, and secondly suggests what concrete can offer.

CONTEXT

The Brundtland report, "Our Common Future", published in 1987 and the UN Conference on Environment and Development ("Earth Summit") in Rio de Janeiro five years later delivered a common language for sustainability. For example: "Sustainable development is development that meets the needs of the present without compromising the ability of future generations to meet their own needs" [1].

In the subsequent 15 years there has been an acceleration in activity in the area of sustainability. In the last two years it has entered centre stage of the political arena and is an issue of concern for the general public.

The UN Department of Economic & Social Affairs, Division for Sustainable Development released the third, revised set of indicators for the UN Commission on Sustainable Development in August 2007. These serve as a reference for countries to develop national indicators and have been developed since previous editions in 1996 and 2001. The fifty core indicators and forty six other indicators are grouped under fourteen themes. The themes and sub-themes are shown in Table 1.

To some extent, construction in general and concrete in particular impact on most of the themes and sub-themes. More significantly, without concrete many of the sub-themes cannot be delivered. These have been indicated with dark shading.

Concrete is virtually essential to ensure "health care delivery" through medical centres and hospitals, "education level" through schools, "transportation" through roads, railways and ports and protection from "natural hazards". Similarly, it enables engineers to address "poverty" through power generation and distribution, water infrastructure and decent housing.

Table 1 Commission for Sustainable Development indicators of sustainable development [2]

THEME	SUB-THEME	THEME	SUB-THEME
Poverty	Income poverty	Consumption and production patterns	Material consumption
	Income inequality		
	Sanitation		Energy use
	Drinking water		Waste generation
	Access to energy		and management
	Living Conditions		Transportation
Governance	Corruption	Natural hazards	Vulnerability to
	Crime		natural hazards
Health	Mortality		Disaster
	Health care delivery		preparedness and response
	Nutritional status	Atmosphere	Climate change
	Health status and risks		Ozone layer depletion
Education	Education Level		Air quality
	Literacy	Land	Land use and status
Demographics	Population		
	Tourism		Desertification
Economic Development	Macro-economic performance		Agriculture
			Forests
	Sustainable public finance	Oceans, seas and coasts	Coastal zone
			Fisheries
	Employment		Marine environment
	Information and communication technologies	Freshwater	Water quantity
			Water quality
	Research and development	Biodiversity	Ecosystem
			Species
	Tourism	**KEY**	
Global economic partnership	Trade	Concrete is required to deliver this sub theme	
	External financing	Construction Adversely Impacts this sub theme	

On the negative side construction, and to a lesser extent concrete, has an adverse impact on other sub-themes and much of designers', contractors' and material suppliers' efforts in sustainability aim to mitigate these impacts. These have been indicated with light shading.

In the delivery of all sub-themes, some may falsely want to minimise concrete's use and use alternative construction materials where possible. However, by considering the construction industry in more detail, and what concrete can offer, it will be seen that concrete is not an unfortunate necessity, but is the sustainable material of choice.

CONSTRUCTION SECTOR'S OPPORTUNITIES FOR POSITIVE CONTRIBUTION TO SUSTAINABLE DEVELOPMENT

The contribution of the construction sector and how this can be improved, can be considered under different headline categories, for example; sustainability impacts (waste, pollution CO_2, etc), sustainable solutions (material innovation, new products, green design etc) or sectors (transport, housing, commercial etc).

For the purpose of this paper, I have chosen to consider the aspects which different professions are responsible for and the opportunities they have for delivering more sustainable construction. This has the benefit of identifying with whom the industry must engage and with what messages.

Civil engineering design and civil engineers

Notwithstanding the work of professional civil engineering institutions to lobby governments for sustainable development, the opportunities for civil engineers to make a positive (or negative) contribution to sustainable development primarily lie in the purpose and function of their projects.

The Three Gorges civil engineering feat in China has successfully broken its own record for annual use of concrete. Whether this project is a sustainability flagship or an environmental disaster that cannot be mitigated by any long-term economic and social benefit, is subjective and primarily based on the project concept. Even if all design decisions and construction processes that bring this project to fruition are the most sustainable possible, it is minor consolation if in concept it is not a sustainable contribution to development.

It is the responsibility of policy decision makers, often advised by civil engineers, to ensure that major civils projects make a positive contribution to sustainable development. The current political climate and awareness amongst electorates now demands it. However, defining what the most sustainable choice is can be very difficult and controversial when the different strands of sustainability are balanced against one another.

Here in the UK we only need to look at the capital to find projects that have strong sustainability credentials: the well known Thames Barrier, providing flood protection, the London ring main providing water to the capital, the proposed sewage overflow tunnel beneath the length of the Thames through London, the Channel Tunnel rail link providing high speed connections to mainland Europe and the Thames Array of wind turbines to help meet renewable energy targets.

Over and above the sustainability credentials of a civil engineering project concept, the additional sustainability benefits arise through the civil engineering design. The design (a subset of which is material specification) can minimise energy, transport and use of materials; maximise re-use and recycling; deliver efficient design; and specify appropriate durability and longevity.

Building design

A range of professionals have input into building projects and can contribute either positively or negatively to sustainable development. Many of the sub-themes presented in Section 'Context' and Table 1 require buildings. Delivery of the buildings themselves is the most significant contribution to sustainability: for example, schools to deliver "education level".

How the building is designed will affect how well it functions, but is of lesser importance than the fact a school is built. Of even less importance, in terms of overall sustainability, is the material used for its structure, unless once again the functionality is compromised. These statements may seem harsh to material producers, but do put the sustainability impacts into context. They are analogous to the 200 : 5 : 1 ratios attributed to a building for relative costs of business (activity in building), building operation and building construction [3].

A summary of the role of key contributors to the building design team is presented in Table 2. There is an overlap between roles, but for the purposes of considering sustainability, the different responsibilities have been allocated where possible. Some sustainability design decisions are truly integrated across several disciplines and this is indicated at the foot of Table 2.

Table 2   Opportunities to Impact Sustainable Development

CLIENT	ARCHITECT	STRUCTURAL ENGINEER	SERVICES ENGINEER
To develop, reuse or maintain status quo	Concept design Orientation Massing Cladding Internal finishes	Structural materials specification Efficiency of design	Energy/CO_2 Thermal comfort Air quality Equipment specification
Scale of development	Fire* Acoustics* Lighting*		Operation manual
Functional requirements	Thermal comfort [†] Air quality[†] Minimum water use		
Location	Landscaping Flooding		
	New sustainable architecture	New structural concepts	New servicing strategies
New integrated design solutions			
Long life/loose fit or Deconstruct/reconstruct			

*   Specialist consultancy (not tabled) may be used on larger/complex projects
[†]   Responsibility is another discipline listed in this table

This paper focuses on design and any input to design by contractors can be attributed to an architectural, structural or services engineering role. The process of construction influences design and to enable the most sustainable construction processes the right design decisions need to have been made.

Client

The client, with the design team, develop the brief for a project and, just as for civils projects, this is the primary determinant of the sustainable impact of a project. To develop or not, to reuse an existing site, to develop on brownfield land, to locate near public transport and decisions on required functionality are very influential on the ultimate sustainability credentials of the development.

Architect

The lead designer can make a wide contribution to sustainable design. The massing, orientation and cladding affect building energy and thermal performance as well as overall visual impact, amenity of spaces and social interactions. These final three effects, together with the function of the building, heavily influence the contribution to a sustainable community.

The architect's responsibilities for the internal finishes gives opportunities to minimise material use and wastage, ensure longevity and enhance performance with respect to affect on acoustics, lighting and thermal mass. These performance issues affect occupant comfort and this is a key determinant of well being and productivity. For example, worker output in the business environment, learning potential in educational establishments, and the sustainability of communities in dense housing developments are all affected by finishes.

Acoustic performance within spaces, between spaces in a development and noise break-in from external sources are all important for occupants. Similarly lighting, not only in terms of energy requirements, is a part of sustainability since glare, uneven lighting or dull lighting all affect comfort. Similar arguments apply for air quality and thermal comfort (see Section 'Services engineers').

Fire is a specialist field of building engineering and warrants the full day of this conference that it has been allocated. It has become a subset of sustainability because of its impact on material usage, environmental pollution through air and water, life safety, economic hardship, loss of livelihood and affect on local economies. Compliance with fire building regulations alone may not deliver the most sustainable project with respect to fire. In the UK, and elsewhere, property safety and climate change is not included in the fire regulations, meaning much of the negative sustainability impact of fire is overlooked.

Flooding has similarly broad sustainability implications. The architect has responsibility for providing flood resistance (not getting wet) and flood resilience (minimising damage/cost/time if wet). If unsuccessful, there is creation of waste and a requirement for new finishes. Also there are economic implications through flood damage, lost business time and business failure. Finally the personal suffering should not be underestimated.

Minimisation of water use is a sustainability goal accepted by the general public and there are many opportunities to use rain water, grey water and minimise usage. Rain water can be collected from building roofs and landscaping. In landscaping the use of SUDS is becoming increasingly common. Even if water is not collected for reuse the peak flows are minimised by attenuation through the collection system.

Structural engineers

The opportunities for structural engineers to influence sustainability lie in material specifications, efficiency of design and new structural concepts. Working with the rest of the design team they can also influence long life/loose fit or deconstructability and respond to and contribute to integrated sustainability design solutions, such as exposed concrete floor soffits for thermal mass.

The minimisation of embodied impacts and maximising longevity/durability is an opportunity. However, availability of comparable data and the fact that the weighting of different environmental impacts is subjective makes consideration of embodied impacts difficult. Durability is a more familiar concept and more easily quantified.

Efficiency of structural design is part of a structural engineer's raison d'être, but this takes nothing away from its importance in terms of sustainability: after all it results in minimising material usage. Efficiency is also achieved when the structure serves other functions such as fire compartmentation, noise separation or provision of thermal mass.

A rediscovered structural concept, driven by the need to meet solar gain targets, is the use of solid facades as structural load bearing elements, avoiding the need for perimeter columns. The vernacular in many of the world's cities and business parks of floor to ceiling glass is being challenged by the need to reduce air conditioning energy consumption. The architect and services engineer can have the discussions regarding energy gains within the space and conclude that solidity is required in the façade to minimise solar gains. A structural engineer can develop this conclusion to propose that this cladding be formed of a structural material which can also support the floor slabs. This is not new, it was common practice for masonry structures. However, for developers, tenants and clients it will be a new style of building.

Services engineers

The prime opportunity for services engineers to influence sustainability is through their responsibility for the interlinked issues of energy/CO_2, thermal comfort and air quality. Lesser opportunities lie in equipment specification (assuming like for like energy/CO_2 performance), provision of building management systems and reporting of building operation strategies. The implementation of minimisation of water use can also fall into the services engineer's remit.

In considering energy/CO_2, there are opportunities to investigate site renewable power generation and use of the ground or groundwater as a heat sink. There is also the opportunity to utilise thermal mass.

WHAT CONCRETE CAN OFFER TO THE DESIGN TEAM

Concrete is generally cheap, available, durable and versatile, and as a result, is often the only practical material option. These very attributes leading to its selection also lead to it offering sustainable benefits. If it is cheap and available, then presumably it is not scarce nor must it be transported some distance. Even though our economics is not dictated by sustainability (yet), the concrete industry is already carrying a burden of environmental regulation and tax on transport and fuel is driven by the green agenda, so cost and availability are already influenced by sustainability.

Concrete's durability results in better whole life performance than alternatives and often longer life. As an example, whilst concrete safety barriers are common around the world, in the UK only recently have they been mandated by the Government for all motorways over 25000 vehicles per day. The change is because the material and the design are robust and durable and offer a more sustainable solution: reduced deaths (cross-over accidents and maintenance of weaker steel barriers), reduced congestion (arising from cross-over accidents and repair work) and reduced maintenance costs.

Concrete's versatility means it can be formed in factory or on site, in whatever shape and with a whole spectrum of performance. As an example, it is difficult to imagine an alternative material for bus-ways, such as in Adelaide and Cambridge. Also, because it is versatile it permits the civil engineer to design it efficiently, hence minimising waste.

Concrete's availability implies that it is local, hence the environmental impacts of extraction are not exported elsewhere. Also social benefits of employment and general economic benefits are enjoyed by the local and regional community.

Concrete can often be the most sustainable material choice – this is even more the case if the concrete mix is designed with sustainability in mind, and the constituent parts are sourced from suppliers who have minimised the impacts of production. Table 3 summarizes key sustainability messages for each constituent part of reinforced concrete that need to be communicated to ensure a correct understanding of concrete's material credentials, even before performance benefits are taken into account. These messages have been compiled for a UK audience.

Table 3 Embodied Impacts of Concrete: Key Messages for UK market

Cement	CO$_2$	29% reduction in direct CO$_2$ emissions compared to 1990, giving an annual CO$_2$ saving of over 3.9 million tonnes. 2010 energy efficiency target met four years early.
	Recycling	In 2006, over one million tonnes of waste derived material was recycled productively, replacing 15% of kiln fuels and 6% of virgin raw materials.
	Biodiversity	Site restoration of quarried land is beyond the requirements of planning permissions.
	Resource Depletion	In 2006, fossil fuel consumption per tonne of cement was reduced by over 23%.
	Waste	The volumes of cement kiln dust going to landfill have been reduced by some 75 per cent since 1998, falling from 289,000 tonnes to 70,000 tonnes in 2006. The industry is a net user of waste, it recycles more waste than it produces.
	Pollution	Industry's total environmental burden to air had reduced by 45% by the end of 2006, compared to 1998.
	H&S	A 32% improvement in lost-time incidents between 2003 and the end of 2006.
	Local	Work with local liaison committees which represents local communities and the local regulators.
Additional cementitious	CO$_2$	As an example: the use of 50% ggbs in some concrete mixes can reduce embodied CO$_2$ by over 40% compared with a CEMI concrete.
	Recycling	ggbs and fly ash are by-products of other industrial processes.
	Resource Depletion	Every tonne of ggbs or fly ash that is used saves about 1.4 tonnes of raw materials and fossil fuels being extracted for cement production.
	Waste	Use of ggbs and fly ash in concrete largely prevents those products being land-filled.
Aggregates	CO$_2$	Mining accounts for 0.5% of the total UK CO$_2$ emissions in 2005.
	Recycling	Recycled and secondary aggregates account for 25% of the total market. There is little evidence that any hard demolition and construction waste is landfilled. i.e. it is reused.
	Biodiversity	In 2005, 754 hectares of land were restored with 131,663 trees planted, 15.5 km of hedgerows planted and 1.25 km of stone walling built. 700 of the 7000 SSSIs were sites of quarries.
	Resource Depletion	Aggregates are abundant and reserves will last for thousands of years. *
	H&S	HSE reportable injuries in 2005 was 293 compared to 367 for the previous year

Table 3   Embodied Impacts of Concrete : Key Messages for UK market, cont'd

Reinforcement	CO_2	Manufacture by the Electric Arc Furnace process is up to six times better than the Basic Oxygen Steel making system used for UK structural steel.
	Recycling	UK produced reinforcement is manufactured from 100% recycled UK scrap steel. Scrap reinforcement is recycled to manufacture new reinforcement.
	Resource Depletion	see recycling above.
Readymix (example of industry provided by Aggregate Industries)	CO_2	20% reduction in CO_2 per tonne of production by 2012 (2003 baseline).
	Recycling	100% virgin aggregate replacement concrete available – up to 50% cement replacement available.
	Biodiversity	Company Biodiversity Action Plan in place.
	Resource Depletion	All cured concrete waste is reused as construction materials.
	Waste	Waste is minimised through recycling techniques.
	Water	All plants recycle water where physically possible.
	Pollution	New equipment purchased is ahead of compliance requirements.
	H&S	30% year on year reduction in LTIFR targeted.
	Local	Community Action Plans in place at all sites by end 2008.
Precast Products	CO_2	Primary contributors are cement and transport. Commitment to reduce cement consumption and use alternative cementitious materials where performance requirements permit. Transport distances are less than 50 miles for majority of products.
	Recycling	The majority of concrete waste produced is reused in the production process. A high proportion of other waste streams are recycled or reused.
	Biodiversity	Companies with factories in more rural areas are increasingly committed to protecting and enhancing the natural environment.
	Resource Depletion	Use of by-products from other industries (fly ash, ggbs, and recycled concrete aggregates) minimise raw material extraction.
	Waste	See recycling above
	Water	Dependency on mains water supplies is being drastically reduced across the industry as companies adopt recycling systems and alternative water sources such as rainwater harvesting.
	Pollution	The industry is closely regulated by the Environment Agency.
	H&S	BPCF operate an industry wide Health & Safety improvement scheme covering 15448 employees: 60% overall improvement since the project start in 2000.
	Local	Many factories operate local community liaison schemes and the industry as a whole helps support local communities by providing much needed employment.

References:

- British Cement Association: www.cementindustry.co.uk/the_industry/performance.aspx
- Quarry Products Association: www.qpa.org/index.htm
- Cement Admixtures Association: www.admixtures.org.uk/publications.asp and personal communication John Dransfield, Secretary , Cement Admixtures Association
- Aggregate Industries Ltd: personal communication Dr Miles Watkins, Director of Group Environmental and Corporate Social Responsibility
- British Precast Concrete Federation (BPCF): personal communication Martin Clarke, Chief Executive BPCF
- *McLaren D, Bullock S, Yousef N, Tomorrow's World: Britain's Share in a Sustainable Future, Friends of the Earth, Earthscan, 1999

What concrete can offer building design

In addition to the items described in Section 'WHAT CONCRETE CAN OFFER TO THE DESIGN TEAM', which apply to both civil engineering and building design, concrete enables clients, architects, structural engineers and services engineers to exploit sustainability opportunities within their areas of responsibility.

What concrete offers the client

Re-use of an existing facility is more likely if it is concrete because of its adaptability and durability and use of brownfield land is more achievable in certain cases through the use of cement for soil stabilisation and solidification.

Decisions on long-life/loose fit or deconstructability are usually made by the client because of cost implications though the engineers need to advise and ultimately to deliver whichever is required. Appropriate design of concrete structures can ensure the former. The latter, deconstructability, is possible with precast structures, but this requirement in isolation does lend itself to lighter materials, albeit with resultant question marks over performance issues (refer Table 4).

Architects

Table 4 summarises the in-use performance benefits of concrete. For the architect the main areas of fire, flooding, acoustics and robustness/security of both internal finishes and cladding are compelling reasons to choose concrete. Thermal mass is as well and is discussed further in Section 'What concrete can offer to services engineers'.

Life safety is a component of sustainability, but the risk to life posed by fire is not included in Table 4 as all materials must meet this basic requirement. It is not something that concrete uniquely offers, but it could be reasonably argued that other materials are more reliant on workmanship to ensure this life safety. The unique advantages of concrete, with respect to fire, are tabled and arise from its inherent property of not combusting, not emitting gases or smoke and its low heat conductance.

Flood resistance (not getting wet) is a function of location, landscaping and massing. These are independent of construction material.

Flood resilience benefits of concrete arise from its dimensional stability when wet, resistance to rot and not suffering corrosion. In addition, concrete (whether it be masonry or reinforced concrete) can be used with minimum finishes, removing the need to use finishes with low flood resilience. The sustainability benefits that concrete offers in terms of flooding are equally significant across environmental, social and economic categories.

Acoustic properties of concrete, both high damping and high mass, minimise sound transmission. Impact transmission from footfall above a slab to below must also be controlled by a resilient layer, but overall, concrete's inherent properties permit fewer finishes to be specified to meet required performance levels. This has sustainable (environmental) advantages, but arguably the most significant advantage is in the social benefit in areas of high density living. These areas also derive much benefit from robust party walls between dwellings – concrete uniquely offers these. Robust partitions are also of benefit in heavily trafficked areas, such as school corridors and, once again, there are concrete solutions available.

Table 4   Summary of Performance Benefits of Concrete

	ENVIRONMENTAL	SOCIAL	ECONOMIC
Fire Resistance	Reduce emissions from fires and wastage of materials.	Reduce loss of livelihood, homes.	Property safety is inherently provided over and above regulated life safety requirement.
Acoustic Isolation Performance	Minimise use of finishes materials which have low lifespan compared with robust concrete surfaces.	Ensures quality of life, particularly in high density living, is not affected adversely by acoustic break-in.	Provide required acoustic separation with minimum finishes, hence minimum cost and maintenance.
Flooding Resilience	Flood resilience of concrete results in minimum wastage of materials following a flood event.	Inconvenience and disruption following flood event minimised.	Loss of businesses and cost of refurbishment minimised.
Robustness/Security	Less risk of damage to finishes, hence less use of materials through whole life cycle.	Solid party walls provide safe, secure housing preventing intruders. Helps build safer communities.	Minimise damage and hence repair and replacement.
Thermal Mass	Reduce heating and cooling energy of buildings.	Reduce overheating.	Lower running costs Lower maintenance costs.

What concrete can offer to structural engineers

Efficiency of design can often be achieved with concrete because of its versatility: in-situ, hybrid concrete or precast; traditionally reinforced, pre-tensioned or post tensioned; linear, planar or three-dimensional elements. This enables the structural engineer to use the material in a manner best suited for each project, thus achieving efficiency. It can also serve multiple functions. A load-bearing wall panel can provide acoustic separation, fire compartmentation, a secure party wall, air tightness and thermal mass. This contrasts with steel, for example, which can do lightweight long spanning skeletal elements very effectively, but other solutions have been less successful to date.

In terms of the "new concept" of load-bearing façades, concrete offers the structural engineer the opportunity to replace perimeter columns with a 2-D panel that supports slab edges. This panel may, or may not, be incorporated with the decorative cladding and insulation. The versatility of concrete makes it better placed than other materials to be used for this new concept. A more conventional way to deliver this 'new concept', and appropriate in low rise situations, is to simply use traditional masonry.

What concrete offers services engineers

Thermal mass benefits are, in theory, no longer unique to heavyweight materials as its positive benefits can now also be achieved through the use of phase change material incorporated into partitions. However, heavyweight materials and concrete in particular offer this benefit in a robust, cheap, readily available manner. They are also tried, tested and non flammable.

The sustainability benefits of thermal mass are presented in Table 4. The social benefit of avoidance of overheating is likely to become apparent when lightweight social housing in a warming climate is in need of air conditioning, but due to energy poverty, the residents are unable to install and operate it.

A key piece of work here in the UK has been to demonstrate that the CO_2 savings arising from thermal mass more than compensate for any additional embodied CO_2. This work was done in detail for a low rise dwelling [4] and more generally office accommodation [5].

My personal view is that the most likely change in building design in the UK, in the near future, will be the wider adoption of exposed concrete soffits in different sectors and by different types of client. This requires an integrated design approach from the architect, structural engineer and services engineer. In the UK, notable examples are in the owner occupier flagship HQ buildings and in the education sector. However, there has recently been the first signs of the beginnings of acceptance of exposed concrete soffits in the speculative office market. Building regulations are likely to be amended to make wider adoption in the housing sector.

ACTIONS FOR THE INDUSTRY

The alternative structural materials of timber and steel have single messages of "low carbon" and "re usable". These are very strong messages because the two most widely understood measures of sustainability are CO_2 footprint and recycling, and re-use is even better than recycling . Concrete is only strong on the CO_2 credential, if thermal mass is considered and re-usability is limited to some precast elements. Therefore, if the discussion remains brief and at headline level, concrete risks coming off second best. However, in the (currently) lesser aspects of embodied impacts concrete performs well and, of far greater significance, is that concrete performs better than other materials during the operational phase of the building.

Concrete has a very strong sustainability story but it is not a single issue story and is therefore more difficult to communicate. The message must be communicated to the different professions and to the parts of government who are developing regulations and tools that are designed for these professionals. It is vital to influence how these regulations and tools are structured and what they contain in detail; primarily to ensure that operational aspects are given due weighting but also to guard against simplistic checks that favour single message materials.

The supply side of our industry must continue to ensure that all that can be done is being done to minimise embodied impacts. A significant part of this work is communicating what is already being done. To ensure continued license to operate, the importance and nature of the messages for the design professionals must be understood by material suppliers in their communication with planners and governments.

As an industry we need to be on guard for future critical sub-issues within sustainability. Currently the highest concern is over CO_2 but in future biodiversity and water are widely identified as becoming the most important sub issues. More broadly, I wonder if the environmental issues may become so critical that the social and economic aspects of sustainability are no longer considered. Therefore, as an industry we must guard against relying too heavily on the social and economic aspects and maintain our efforts in reducing environmental burdens and maximising environmental performance.

REFERENCES

1. U.N. World Commission on Environment and Development, Our Common Future, 1987.

2. UN Department of Economic and Social Affairs, CSD indicators of sustainable development, 3rd edition, August 2007.
   (refer www.un.org/esa/sustdev/natlinfo/indicators/factsheet.pdf)

3. EVANS R, HARYOTT R, HASTE N AND JONES A, The long term cost of owning and using buildings, 1998. London, Royal Academy of Engineering.

4. HACKER J N, DE SAULLES T P, MINSON A J AND HOLMES M S, Embodied and Operational Carbon Dioxide Emissions from Housing: A case study on the Effects of Thermal Mass and Climate Change; Energy and Buildings; Vol. 40, 2008, pp. 375-384.

5. The Concrete Centre, Embodied and Operational CO_2 Emissions, Concrete Structures 6, 2006, pp. 4-5.

DURABILITY AND QUALITY OF CONSTRUCTION: CHALLENGES FACING UAE

M A Samarai

University of Sharjah

L M Qudah

Dubai Accreditation Center

United Arab Emirates

ABSTRACT: This paper highlights major construction projects in Dubai, durability issues, environment challenges and quality control efforts in the region with emphasis on the role of the Dubai Municipality Accreditation Center, academia and industry in assessing and strengthening the resources to implement quality control in Dubai.

Keywords: Durability, Environment, High performance concrete, Construction, Quality control, Testing laboratories.

Mufid A Samarai is a professor of civil engineering, Collage of Engineering at the University of Sharjah, United Arab Emirates and director of The Central Laboratories. He obtained his Ph.D. from the University College London in 1976, and has been engaged in teaching, research and consulting work. He was involved in the repair of over 100 structures and bridges in Iraq, Jordan and UAE. A former vice-president of CIB (Holland) and Bureau member of RILEM (France). His research interests include quality control, low cost housing, non-destructive testing, durability, energy conservation, damage assessment and repair of structures. He has over 120 publications, a book and 4 registered patents.

Lina M Qudah is the Acting director of the Dubai Accreditation Center (DAC) at Dubai Municipality, United Arab Emirates and the head of the Accreditation Decisions Section at DAC. She holds a Masters Degree in Quality Management which she obtained from the University of Wollongong in 2004, and has been engaged in assessments of and training in quality management systems. She was involved in the assessment and accreditation of over 50 testing laboratories and inspection bodies in the field of construction materials in Jordan and UAE. She is the vice chair of the DAC Technical Committee for the Accreditation of Laboratories and the DAC Technical Committee for the Accreditation of Inspection Bodies. She is also a member of the Accreditation Committee (AIC) of the International Laboratory Accreditation Cooperation (ILAC) and a member of the Joint Development Support Committee (JDSC) of ILAC and IAF (International Accreditation Forum). She has 4 publications in quality management and accreditation.

INTRODUCTION

The Gulf region is witnessing an extensive urbanization development and billions of dollars are spent on reinforced concrete construction projects. By 2020 six of the tallest ten buildings in the world are going to be in Dubai. All over the region there are many concrete structures which are in outstandingly good condition compared to their age, but we have also examples of structures that within a few years deteriorated to a poor condition, yet if well maintained could have lasted for many more years. This information can be used for setting up the requirements not only to the concrete but also to the structural design itself, since today we want not only high performance concrete, but also high performance structures. Many reinforced concrete buildings exhibited some signs of distress early in their service life. Awareness of the importance of concrete durability has risen in the region along with the need to stretch concrete to its uttermost limits, considering the hot and arid weather in the region and the need to use high strength concrete. In many parts of the Middle East there are important factors such as unskilled workmanship, deleterious substances, funds and speed of construction that influence the durability of structures and bridges to a great extend. There is also the need to educate the industry and society to the importance of quality control and to realize that the additional cost and effort spent on quality is a greater saving in the long run [1]. This brings out the importance of the assessment of material quality and homogeneity in the site which lead to the increasing demand for in-situ non-destructive testing methods. However, the problem of high strength concrete durability in hot weather is very complex. It can be seen that a concrete structure in the hot, arid environments tends to deteriorate more rapidly than those in temperate regions of the world, unless particular precautions are taken. Maintenance in the Middle East has consistently been treated as the poor relation of the construction industry, attracting only a tacit recognition of its importance, both within the industry and amongst concerned institutes.

There is growing acceptance in the region for living in high-rise buildings, which are architectural forms that make efficient use of land and meet the social need to live in city centres. One building of 30 to 40 floors can have up to about 500 residential units, effectively making it a small town. As towns need parks and roadways, super high-rise apartment buildings also need "space" to make them a community. Until the19th century, buildings of over six stories were rare. It was impractical to have people walk up so many flights of stairs. Also, water pressure could only provide running water to about 15 m. City building has always served human ends, whether of the governments that have vision or of the ordinary people who want to create something that they dreamed and wished to locate their homes in. The shapes of the cities are mirrors of people's ambition and of the civilizations that created them. Cities in UAE are planned environmental revolution changing the desert into a garden with a superb architectural layout of greens and buildings. An environment can be considered 'valued' if its users can show recognition of an empathy with their local social and physical environment, a milieu that in turn serves to buttress their won preferred life-styles. When planning the massive number of high rise buildings and skyscrapers, plans should not only organize space, but also arrange important facets of human lives: their use of time, which they are likely to communicate with, and the meanings they find in their built environment.

The development of durable and high strength steel and concrete enabled the construction of extremely tall buildings, some of which are over 300 meters tall. The other development essential to practical skyscraper development was the invention of the elevator and the development of high performance concrete and high performance structures. As a matter of fact the increased density of human population all over the world, and the lack of land to accommodate these people justify thinking vertically. However, creating a vertical city where 60,000 people can be housed, and where high velocity trains running horizontally and vertically are a necessity, is charming but at the same time alarming. Increasing the complexity of these buildings by making them taller and more importantly increasing the risk of hazards associated with all their operations is neither sensible nor practical [2].

This paper highlights major construction projects in UAE, durability issues, environment challenges and quality control efforts in the region with emphasis on the role of Dubai Municipality Accreditation Center, academia and industry in assessing and strengthening the resources to implement quality control in UAE.

THE ENVIRONMENT AND THE SUSTAINABILITY ISSUE

By far, the most talked-about topic in the architecture universe now is how to reduce the environmental impact of everything from summer cottages to skyscrapers, and there are some remarkable examples – some would say exceptions – of progress. The question is, will high energy prices turn those exceptions into the rule? The challenge is clear. Buildings devour 39% of energy in the US alone, and according to the Department of Energy (DoE), more than factories and automobiles.

Adopting an environmentally responsible attitude towards the specification, design and construction of our building, infrastructure and civil engineering projects, we can choose materials for all the right reasons — and our choice can have beneficial effects all the way through the life-cycle of a building or structure. Construction is considered one of the main consumers of resources and energy. Although it generates work and the need for more labour, cement as the major construction material generates large amounts of CO_2 and hence causes the greatest damage to our environment. As a matter of fact it is responsible for 7% of the world emission of CO_2 to the atmosphere. With the increase in the value of land, real estate agents demand more utilization of the land. The architects and planners should not bend to the demands of restricting the designs to aesthetic values and speed of construction only, but should consider and insist on the social and environmental aspects of their design and planning. With the fast increase in the local economies, there is demand for faster construction, and little concern for quality and durability.

Environmental issues are getting attention worldwide. These issues include not only updating the rules and regulations that enforce sustainable development, but also the economic incentives to incorporate sustainable development designs. The natural ecosystem has become a focal point for urban planning within the past generation, planners of earlier periods tended to regard the environment as a rural or "wilderness" concern, but not applicable to city choices except perhaps in the layout of parks. The news media began to publicize environmental concerns nationally, and the resulting public awareness spurred major legislation, which empowered planners to incorporate ecological values.

Environmental policy now has several dimensions with distinct challenges. The first is that a city, like a human body, has a "metabolism" by which it takes in means of life support and disposes of unneeded and harmful products. In practical terms, this points to the supply of air, water, food, energy and raw materials for manufacturing and the safe removal or reprocessing of wastes. While much of this exchange is conducted by the private market, it impacts how land is used and so falls under public regulations. In the century just past, we somewhat recklessly kept using resources, especially energy. The environmental impact of ever-increasing usage is rapidly becoming a major factor, and if we do not change our energy usage, the costs for future generations could be enormous. Current construction methods result in our buildings being a major user of energy, not only in their construction, but also in their day-to-day operation. Continuing future growth of energy usage will have to stop if we hope to obtain a sustainable habitat on our spaceship, planet Earth [3]. Building activity, the production, maintenance and repair of the built environment, has a significant impact on the environment representing half the total energy consumed in high consuming countries. Notwithstanding the acknowledged need to conserve resources, the rate of consumption is increasing. It could be argued that the increase is a direct consequence of the political agenda of all governments to increase the affluence of their peoples. Affluence and consumption are at the core of the sustainability agenda. That agenda is concerned with understanding the impact of human activity on our environment and from a position of knowledge move towards a position where human activity is sustainable [4].

The UAE is getting stricter on the environment-friendly side to development. More so, because of the rapid scale of development in so short a time, just over 30 years. Development which sometimes has been at the cost of the environment, which is every country's every region's natural heritage. A fast growing population, immense financial resources and an ambitious development program have combined to place strong pressures on the environment in the UAE. The growing consensus from major industry analysts is that technological innovation, if engineered in an environmentally-conscious way, brings both reduced costs and increased efficiencies to their operations. Institutions across the Middle East are recognizing that environmental protection laws can actually help to drive positive economic growth.

However the practical meaning of sustainability is open to much debate. At minimum, it is a call for steady reduction of air and water pollution, reduction of and recycling of wastes, more efficient use of energy and a shift to renewable energy sources, and restraint in the development of rural land. A dedicated view of sustainability demands major shifts in our life styles toward smaller homes, use of mass transit instead of automobiles, and reorientation of marketing toward frugality rather than consumption. While some of these policy choices are made in local communities and households, any large-scale implementation depends on choices by the national government and major corporations that compete in the global market. Yet planers recognize that even small scale choices have incremental effects on their environments, and there is growing interest in alternatives that promote sustainability while not violating the freedom to choose our life styles. Therefore we should aim to achieve a balance between environmental protection, social progress and economic growth, while working in harmony with Earth's ecosystems. Hence, sustainable development means:

- Reduction of emissions
- More efficient use of resources
- Better re-use of waste
- Increased consideration for people's health and safety

Fuelling the environmental building revolution are new materials. Chief among them is glass. It is stronger, safer and more energy-efficient than ever. Low-emissive glass lets visible light in, but keeps heat out. That enables architects to add windows — and light — without having to double up on expensive air conditioning. Smart-Wrap is a transparent building membrane that adjusts its properties to changes in light and temperature to keep buildings climate-controlled and well-lighted.

CONSTRUCTION AND DURABILITY ISSUE

Until the 19th century, buildings of over six stories were rare. It was impractical to have people walk up so many flights of stairs. Also, water pressure could only provide running water to about 50 feet (15 m). The development steel, reinforced concrete, and water pumps have made possible the construction of extremely tall buildings, some of which are over 300 meters tall. The other development essential to practical skyscraper development was the invention of the elevator. Considerable effort in developing improved design and construction procedures to address typical problems encountered in high-rise residential buildings. They found ways to upgrade parking garages enhance envelope durability and improve a multi-unit building's thermal envelope, HVAC systems, accessibility and environmental performance. But one of the most important achievements is the enhancement of steel and reinforced concrete to cope with the ever increasing demand for fast and lasting construction.

Perhaps no other building material was analyzed, studied and modified as concrete. The need for higher performance concrete, which is more durable and stronger, required greater quality control and enhancement of bond between the concrete constituents and materials overcoming the weakness of concrete in tension. All these require higher quality control and higher technology in the manufacture of concrete. In many parts of the Middle East there are important factors such as hot weather, sulfates and chlorides, unskilled workmanship, deleterious substances, funds and speed of construction that influences the durability of structures and bridges to a great extent [6]. There is also the need to educate the industry and society to the importance of quality control and to realize that the additional cost and effort spent on quality is a greater saving in the long. This brings out the importance of the assessment of concrete quality and homogeneity in the site which lead to the increasing demand for in-situ non-destructive testing methods. These methods will close the gap between the long time cubes and cylinders, as indicators of strength and quality and actual conditions of concrete in the structural elements.

UAE is witnessing an extensive urbanization development and billions of dollars are spent on reinforced concrete construction projects. Many reinforced concrete buildings exhibited some signs of distress early in their service life. Awareness of the importance of concrete durability has risen in the region, however, repair techniques are essentially those involving short-term control of the situation and use of imported systems which are not fully tested [7]. What is meant by a durable structure is in practice very subjective and difficult to define precisely. It should refer to a structure maintaining a satisfactory performance over a predetermined period of time without requiring unexpected high costs for maintenance. Therefore, the term service life design (SLD) has replaced the term durability, being a quantifiable and measurable quantity (years). In principle two basically different design strategies to ensure a required service life can be followed: Strategy A, avoid the degradation threatening the structure due to the type and aggressivity of the environment; Strategy B, select an optimal material composition and structural detailing to resist, for a specified period of use, the degradation threatening the structure. Schafer [8] emphasized this line of thought.

There have been many advances in concrete technology during the past 25 years. Full advantage will not be derived from these advances unless they can be shown to be appropriate to the environment in which they are to be used and the application and use of basic concrete design and construction techniques is sound. Good performance can only be achieved when designs allow for build ability, ease of construction, and if high standards of supervision and workmanship are used and imposed. Performance is also dependent on the use of appropriate technologies for specific environments, both macro and micro. In the past there has been a tendency for concrete technology developed in Europe and North America to be imported into other parts of the world, with markedly different environments. The resulting poor performance has led to the realization that existing technologies need to be checked out for their effectiveness in specific locations before use.

The problem of concrete durability in hot weather is very complex. It can be seen that concrete structure in the hot, arid environments tends to deteriorate more rapidly than those in temperate regions of the world, unless particular precautions are taken. Increase in temperature also increases the cracks, including plastic shrinkage cracking and drying shrinkage cracking, facilitate the ingress of salt-laden water and moisture causing disintegration of concrete due to sulfates attack and corrosion. Carbonation, which reduces the passive effect of concrete on reinforcement corrosion, proceeds at a faster rate at the higher temperatures. Even good concrete, where chlorides have been included at the mixing stage or have entered from an external source, will deteriorate more quickly than it would in a temperate climate. It has been reported that chemical agents, e.g., water-reducers retard carbonation, as they would promote densification of concrete. On the other hand, mineral admixtures such as fly ash, silica fume and blast furnace slag has been reported to increase the rate of carbonation [9].

Since high-strength concrete is characterized by a low porosity and a more uniform microstructure compared to that of normal-strength concrete, this indicates a high resistance to penetration of carbon dioxide and chloride ions into the concrete. However, during production of high-strength concrete, both macro cracking due to plastic shrinkage and micro cracking due to self-desiccation may represent potential problems from a corrosion protection point of view. Even for normal-strength concrete, production of concrete with a good and stable air-void system is normally a problem, but in the presence of a high dosage of super plasticizer, the establishment of a good and stable air-void system is even a bigger problem. By increasing the concrete strength from 50 up to 100 MPa the abrasion of the concrete was reduced by roughly 50%. At 150 MPa the abrasion of the concrete was reduced to the same low level as that of high quality massive granite. Very often, it is not the improved strength, which is the primary objective, but rather the overall improved performance. Therefore, the term "high-performance concrete" covers "high-strength concrete".

In many areas, large quantities of resources are being spent on maintenance and rehabilitation of concrete structures due to lack of durability. There is a great challenge, therefore, for the engineering profession to utilize and further develop the technology of high-strength concrete or high-performance concrete for the benefit of the society. The use of admixtures and additives in the gulf has gained greater interest and acceptance leading to an extensive increase in the amount of concrete which contains admixtures. The climatic conditions and the need to stretch concrete to its ultimate capabilities make the use of admixtures not only necessary but a must. However, engineers in the region are finding themselves grappling with products, standards and specifications which emanate from other countries, and working in climatic conditions which the admixtures have not been tested in. The assumption that an admixture that retards at 21°C will do so at 33°C is not valid. Concrete mixtures incorporating fly ash, silica fume, or fine cements frequently have a low to negligible bleeding rate, making such mixtures highly sensitive to surface drying and plastic shrinkage, even under moderately evaporative conditions. Certain admixtures increase the time of initial setting or reduce the amount of water needed for a given initial slump or both, but such concretes may stiffen faster, sometimes too fast even for a cement and an admixture that separately meet all specifications.

It is imperative that government institutes responsible for the construction industry should join hands with the universities to expand the education and implementation of quality control. The cost of row materials and manufacture of concrete is the same for good and bad concrete, the difference of an additional 5% cost of quality assurance could lead to a saving of more than 70% in life cycle of structures. We should never settle in the middle east of building to last 30 year but must target for 70 years and more bridges should last more than 120–150 years and important building should reach 200–250 years. This is not for fetched at all, now there are structures that are being designed for 250 and 500 years. The British library is designed for 1000 years and a temple in China is designed for 5000 years.

Other ways in which concrete construction techniques can be durable and environmentally considerate are:

- Self-compacting concrete (SCC): this requires no vibration for compaction and so reduces noise on site. It also increases construction efficiency and minimizes labour.
- Formwork: this can be designed to maximize re-use and minimize waste.
- Safer chemicals: these include biodegradable, vegetable oil-based release agents.
- Admixtures: for example, many admixtures used in the UK to improve performance are derived from natural products such as lignin, a by-product of the paper industry.
- Specialist techniques such as pre-stressing are becoming more common; these make it possible to use less material and to create longer spans — these lead to enhanced flexibility of buildings in use.
- Use of stainless steel, will not only enhance our construction but it play an important role in having a more durable and sustainable structure.

STEEL VS REINFORCED CONCRETE STRUCTURES

For high rise buildings details of design should be very precise and take into consideration the possibility of distribution of stresses to other elements of the structure and accuracy of joints and connections. This is to avoid concentration of excessive stresses in an element or part of the structure that might lead to over stressing any of the elements.

Reinforced Concrete structures have greater ability to distribute the stresses and forces between different elements and sections due to the continuity of the joints and connecting parts. They are more durable, fire resistant, better sound and heat insulation and longer lasting construction. It has a better resistant to impact and this is an important issue nowadays. This is not the case for steel framed structures unless proper attention given to detailing of joints, connectors, welding and rivets and bolts. Steel structures however, have the ability to absorb high stresses and in a uniform way in three directions. They have the ease of disassemble and replacement which facilitates the changing of strengthening of the sections when reconstructing [10].

A 50% increase in European steel prices during 2004 has left many in the construction industry reviewing design solutions that have a heavy reliance on steel. A study by leading construction economies Franklin and Andrews [11], examining the impact of the steel price rises has found that that full fit out whole project costs for concrete framed buildings are marginally less than for steel framed buildings. Costs are for the 2nd quarter of 2004.

Table 1   Full fit out cost

	CONCRETE	STEEL
3-storey	£5,107,845	£5,190,067
7-storey	£10,796,986	£10,962,115

Concrete's range of inherent benefits – fabric energy storage, fire resistance, sound insulation – means that concrete buildings tend to have lower operating costs and lower maintenance requirements. Recently more rapid development in the field of concrete technology has taken place. Increasing construction challenges in combination with new innovations in materials and construction techniques have strengthened the position of concrete as a construction material. Having been an empirical technology, the concrete technology is now entering into a high-technology profession, with a potential for utilization far beyond the traditional construction industry. An increasing interest for better scientific knowledge in concrete technology has correspondingly taken place.

Along with the rapid development of concrete technology, the definition of high-strength concrete has also been changing. Thus, in the 1950s concrete with a compressive strength of 35 MPa was considered high strength. In the 1960s concrete with 40 to 50 MPa compressive strength was used commercially. In the early 1970s, 60 MPa concrete was being produced. Since the compressive strength depends on both type of specimen and density of concrete, concretes with a compressive strength of more than 50–60 MPa are now normally considered high. High strength concrete possesses compressive strength far beyond what can be utilized by the current structural design practice. Thus, for high-quality natural mineral aggregates compressive strengths is replaced by high quality ceramic aggregate, compressive strength of up to 460 MPa can be achieved. Even with lightweight aggregate, compressive strength of more than 100 MPa, with a density of less than 1900 kg/m^3, can be obtained. Very often it is not the improved strength which is the primary objective but rather the overall improved performance. In the present paper, a brief summary of the most recent development on the high-strength concrete based on mineral aggregates is presented.

The need for higher performance concrete which is more durable and stronger requires greater quality control and enhancement of bond between concrete constituents and reinforcing steel. Also there is a need for better corrosion protection of steel with enhanced details and welding quality. In many parts of the third world there are important factors such as unskilled workmanship, deleterious substances, funds and speed of construction that influences the durability of structures and bridges to a great extend. There is also the need to educate the industry and society to the importance of quality control and to realize that the additional cost and effort spent on quality is a greater saving in the long run. Eco-designers believe the way to sustainable building lies in long-life, adaptable, low-energy design. The earth's resources are best conserved if the service life of a building is prolonged, so the durability and longevity of concrete make it an ideal choice. Anticipating and designing out maintenance and repair, and designing in flexibility for change of use can also extend service life. Concrete construction offers designers a better long-term way of achieving these design aims.

Research into the durability and environmental aspects of concrete elements shows there are often initial cost, life-cycle cost, and environmental advantages to be gained from using high-strength concrete. This can produce slender structures, making savings in materials due to their thinner sections, reduced volume and reduced cover to reinforcement.

MEGA PROJECTS IN UAE: AN OVERVIEW

Throughout the history of architecture, there has been a continual quest for height. Thousands of workers toiled on the pyramids of ancient Egypt, the cathedrals of Europe and countless other towers, all striving to create something awe-inspiring. People build skyscrapers primarily because these are convenient – one can create a lot of real estate out of a relatively small ground area. But ego and grandeur do sometimes play a significant role in the scope of the construction, just as it did in earlier civilizations. Until relatively recently, we could only go so high. After a certain point, it just was not feasible to keep building up. In the late 1800s, new technology redefined these limits. Suddenly, it was possible to live and work in colossal towers, hundreds of feet above the ground.

There have been many advances in concrete technology during the past 25 years. Full advantage will not be derived from these advances unless they can be shown to be appropriate to the environment in which they are to be used, and the application and use of basic concrete design and construction techniques is sound. This required considerable effort in developing improved design and construction procedures to address typical problems encountered in high-rise residential buildings. They found ways to upgrade parking garages enhance envelope durability and improve a multi-unit building's thermal envelope, HVAC systems, accessibility and environmental performance. But one of the most important achievements is the enhancement of steel and reinforced concrete to cope with the ever increasing demand for fast and lasting construction.

The current tallest building is Taipei 101 in Taiwan, which is approximately 530 m tall. However, three projects on three continents have laid claim to be the tallest in the race toward the clouds, recently, these are: The residential building Burj Dubai in Dubai , the New York World Trade Center's 542 m tall Freedom tower and the Shanghai World Financial Center which has broken ground after nearly a decade of delays.

The gulf region is witnessing an extensive urbanization development and billions of dollars are spent on reinforced concrete construction projects. It is estimated that investment in projects in the Gulf states will reach 750 billion dollars in the next 10 years. The oil revenues for the Gulf states might be reaching 300 billion dollar in 2007. By 2020 six of the tallest ten buildings in the world are going to be in Dubai. At present about 30,000, or 24 per cent of the world's 125,000 construction cranes, are currently operating in Dubai. These mega projects will not only exert pressure on resources but also require high performance materials, top project management and greater quality control.

Construction of the one billion dollar Burj Dubai, which will be the world's tallest tower, started in early 2005 and estimated to end in 2008. Over 3000 workers are working on the site where more than 160,000 cubic meters of high-quality concrete and 25,000 metric tones of steel rebar have been used. The height of Burj Dubai is being kept private, and will not be told until it is complete. According to the developers, it will be the tallest building on earth. The height is expected to range between 725 and 800 meters.

The tower's 192 piles have been constructed to depths of more than 50 meters and are bound together by a 3.7-meter thick concrete raft across 8000 square meters, encompassing the tower's entire footprint. Nearly 18000 cubic meters of cement concrete was poured for the tower piles while 15,000 cubic meters of concrete was used for the podium piles. The raft, in addition, comprise of 12,300 cubic meters of concrete, bringing the total concrete poured into the foundation to over 45,000 cubic meters weighing more than 110,000 tonnes. The high performance exterior cladding system will be employed to withstand the extreme temperatures during the summer months in Dubai. Primary materials include reflective glazing, aluminium and textured stainless steel spandrel panels and the vertical stainless tubular fins accentuating the height and the slenderness of the tower. When completed Burj Dubai will hold the record in all four categories as recognized by the New York-based global authority – Council on Tall Building and Urban Habitat – highest structure, roof, antenna, and occupied floor.

The Tower is the central core of the magnificent US$20 billion Burj Dubai district and the new emerging Downtown Dubai. More than 500 international consultants around the globe are currently finalizing the design. At its construction peak, a workforce of more than 20,000 construction workers will be on site to create 45 million square feet of liveable space to accommodate 30,000 homes, making the Burj Dubai site the biggest single construction site in the world.

Dubai airport will have the capacity of 60 million passengers and 3 million tons cargo. The Jabel Ali airport under construction will be of the same capacity as Heathrow and of Chicago with 120 million passenger capacity and 12 million tons of cargo capacity with over site of growth till 2050. The construction was initiated on the Three Palm islands which are the largest man made islands in the world .The first two Jumaira and Jabal Ali will add 120 km of beach to Dubai. The third Palm in Deira with an area of 80 square kilometres will be the largest and equal to the area of Paris and London using one billion cubic madder of rock and sand. The City of Arabia is one of the Mega cities has being built in a unique design that cost about 20 billion dollars in total.

The 10 billion dollar Emirates City in Ajman is a unique housing complex with 70 high-rise buildings surrounding artificial lakes and surrounded by a golf course and entertainment centre. Abu Dhabi the biggest of emirates and the capital is undergoing projects according to the financial news publication reaching 30 billion dollars. The investment in the projects in the Gulf States will reach 750 billion dollars in the next 10 years. The oil revenues for the Gulf States might be reaching 300 billion dollars in 2007.

QUALITY CONTROL AND ACCREDITATION IN UAE

Dubai Municipality is regarded as one of the largest establishments in UAE in terms of the number of people it employs, the volume of services it provides to the public and the projects it carries out. Dubai Municipality is the major driving force behind the development process of Dubai City as a whole. The Municipality was established in the 1940s with total staff strength of three persons who were operating from a single room. However, it was in 1965 that Dubai Municipality came into being officially. The Municipality kept up its steady growth since its inception and has now become a large organization with more than 10,710 staff working in 31 organizational units. In short, the Municipality's growth reflects the growth of Dubai City in general. Two of the main quality control establishments in the Municipality are the Central Laboratories and the Accreditation Center. The international corporation UKAS (United Kingdom Accreditation Services) granted inspections rights to Dubai central laboratory according to the international criteria 17025 ISO and the ISO 90001. Dubai Accreditation Center is the sole authorized accreditation body within the Emirates of Dubai, who is providing accreditation services to CABs using international standards, guidelines and best practices for enhancing confidence in conformity assessment which ensures free movement of goods and services within and across the borders. Dubai Accreditation Center, have adopted the following standards as their criteria for accreditation:

- ISO/IEC 17025: General Requirements for the Operation of Calibration and Testing Laboratories for accreditation of laboratories,
- ISO 15189: General Requirements for accreditation of the Medical Laboratories,
- ISO/IEC 17020: General Criteria for the Operation of Various Types of Bodies Performing Inspections for accreditation of inspection bodies,
- ISO/IEC Guide 62 for the accreditation of quality system certifiers,
- ISO/IEC Guide 65 for the accreditation of product certifiers,
- ISO/IEC 66 for the accreditation of environmental management system certifiers, and
- ISO/IEC 17024 for the Personnel certifiers.

Accreditation by DAC is formal recognition that a Conformity Assessment Body (CAB) is competent to carry out specific tasks, i.e. specific types of tests, calibration methods, inspections, or certification activities. The specific areas of competence are specified in the scope of accreditation (e.g. a laboratory may be accredited to undertake specific chemical tests but not to test electrical products).

DAC accreditation is granted after detailed assessment of the competence of the CAB staff against defined technical and management system criteria. This assessment is conducted by DAC permanent staff in association with external experts, referred to as Technical Assessors, being assessed as per the Technical Standards. External Technical Assessors can be found in the particular field of technology or industry; in CABs, manufacturing companies, universities, research centres and some governmental departments.

Accredited organizations are subject to regular surveillance visits by DAC. DAC uses international standards and criteria for accreditation of CABs; e.g. ISO/IEC 17000 series, ILAC and IAF publications, etc. This uniform approach allows Accreditation Bodies in different countries to establish agreements among themselves, based on mutual evaluation and acceptance of each other's accreditation systems. Such International agreements, called Mutual Recognition Arrangements (MRAs) or (MLAs), are crucial in enabling conformity assessment data to be accepted between countries. In effect, each partner in such an MRA/ MLA recognizes the other partner's accredited CABs and this facilitates trade across borders.

OBSERVATIONS

It is imperative that government institutes responsible for the construction industry join hands with the universities to educate the community and the industry to the importance of quality control and preventive maintenance. The cost of raw materials and manufacture of concrete is the same for good and bad concrete, the difference of an additional 5% cost of quality assurance could lead to a saving of more than 70% in life cycle of structures. We should never settle in the Middle East for buildings to last 30 year but must target for 70 years and more. Bridges should last more than 120–150 years and important building should reach 200–250 years. This is not far fetched at all, now bridges are designed for more than 100 years and there are structures that are being designed for 250 and 500 years. The British Library is designed for 1000 years.

REFERENCES

1. SAMARAI M, Sustainable Development: Environment Challenges Facing Urban Planners And Designers In UAE, Sixth Sharjah Urban Planning Symposium, Sharjah Municipality and South Bank University London, 1-2 June 2003, Sharjah, UAE.

2. ESPER P AND KEANE W, Major Incident and Disaster Management (MIM), Building Magazine, Issue 15 April 2005, published in the IStructE issue of 7 June 2005, London, UK.

3. KAY E A AND WALKER M J, Guide to evaluation and repair of concrete structures in the Arabian Peninsula, The Concrete Society in collaboration with the Bahrain Society of Engineers, The Concrete Society, Crowthorne, 2002, UK. ISBN 0 946691 94 0

4. GOLD J R AND BURGESS J, On the significance of valued environments, John R. Gold & Jacquelin Burgess, ed. Valued Environments, London, George Allen & Unwin, 1982, pp. 1-9.

5. ESPER P, KHALIL P AND WILSON P, The Role of FEM in the Innovative Design of Unconventional Buildings, NAFEMS World Congress on Innovative Engineering Simulation Technology, Orlando, USA, 2003

6. FOOKES PG, POLLOCK D J AND KAY E A, Concrete in the Middle East – Rates of Deterioration, Concrete, London, Vol. 15, No. 9, Sept. 1981, pp. 12-19.

7. WALKER M J, Problems of Working With Reinforced Concrete in Hot and Aggressive Environments - Materials and Techniques Currently in Use to Reduce and Avoid These Difficulties, Annual Concrete technology & Corrosion Protection Conference 2004, 22-23 November, 2004, Dubai, U.A.E.

8. SCHAFER B, How to obtain sustainable environment using performance based regulations and appraisals, 5th World Congress & Exhibition "Building Towards the Third Millennium", UAE, 26-29 March, 2000.

9. NEVILLE A M, Properties of Concrete, 2nd Edition, Pitman, London, 1977.

10. FRANKLIN AND ANDREWS, Economic Bulletin, Vol. 7, 2 July 2004.

11. SAMARAI M, Challenges Facing Repair of Structures in the Middle East, The 4th International Operation and Maintenance Conference in the Arabic Countries: Maintenance under limited resources, Beirut, Lebanon, 20-23 June, 2005.

THEME ONE:

CEMENTS AND ADMIXTURES: FUTURE DIRECTIONS AND PERFORMANCE

SUSTAINABILITY IN CEMENT AND CONCRETE CONSTRUCTION

M Gilbert

British Cement Association

United Kingdom

ABSTRACT. This paper reports the activities of the British Cement Association (BCA) in developing and implementing a sustainable development strategy, in which it has been actively engaged since 2003. Recent achievements (2007/8) are reported and future activities are presented in relation to broadening the strategy to more fully consider the whole-life performance of cement. This includes an overview of the UK Concrete Platform Sustainable Construction Task Group, and the way in which its role of allowing closer working on common issues in the cement and concrete sector will help in the development of a strong cross-sector sustainability message

Keywords: Business case, Carbon strategy, CEMBUREAU, Cement, Cement makers' code, Climate change agreement, EUETS, Responsible sourcing, Sustainable construction task group, Sustainability, UK Concrete Platform, UKGBC, Waste substitute.

Mr M Gilbert, is a qualified architect, fellow of the RSA and Chief Executive of the British Cement Association. He worked as an architect in private practice in Hampshire and London, before joining IBM in diverse roles including Design and Construction Manager, UK Consulting Architect, and in Quality and Business Process Management. Seconded to the British Standards Institution, he led the development of the world's first environmental management systems, now known as ISO 14001, and is the author of Achieving Environmental Management Standards, published by Pitman. Joining BSI he became a Board member with responsibility for Commercial Services in the UK, Europe and the Middle East. Mike is past chairman of the CPA Sustainable Construction Committee and is a member of the Government's Project Board for the Sustainable Construction Strategy.

INTRODUCTION

Climate change is reshaping business practice at remarkable speed in the construction sector, with carbon budgets become increasingly tight for manufactures, and environmental considerations moving closer to the top of developers' design requirements. The UK Government's policies on sustainable construction have been progressively developed over the last few years, culminating in 2007 with the Code for Sustainable Homes, and a draft strategy for sustainable construction in the public sector, which brings together a range of programmes, and departmental initiatives. A key driver for change is the Stern Report [1], which highlights the importance of taking early action to combat climate change. There is also an underlying recognition that more sustainable construction requires action from all parties including the public sector, which is responsible for 40% of non-domestic construction [2]. Other drivers include initiatives such as the Government's Sustainable Procurement Action Plan [3] and the Energy White Paper [4].

For manufacturers of construction materials, a key challenge has, and will continue to be, the environmental performance targets imposed by schemes such as the UK Climate Change Agreement and membership of the European Emissions Trading Scheme (EUETS). However, it is also becoming increasingly important to provide much more attention to downstream sustainability issues, which have become the focus for competing materials.

Alongside tough regulatory requirements, sector organisations and trade associations have been effectively challenged by government to develop and implement their own sustainability strategies.

In the Cement and Concrete sector, this process started in 2003 when progress towards sustainable development [5] was reported by the British Cement Association (BCA). Helping compile this report were the trade bodies: British Precast, Quarry Products Association and The Concrete Centre, all of whom recognised the need for action. Since that time, the BCA and the other trade bodies have each developed a sustainably strategy suited to their respective products and member needs.

Trade Association Activity

Collective progress towards sustainable development has been significant within the cement and concrete sector, with a general move towards setting key performance indicators (KPI), and an on-going programme of measurement. This was a key component of the sustainability white paper "Towards a More Sustainable Precast Industry" [6] published by British Precast in May 2007. It details a programme of measures to improve performance across the precast industry on sustainability, including developing a set of sustainability KPIs. These will allow British Precast to measure and report progress by stakeholders and will also enable performance benchmarking. The initial data set to be collected includes production, quality, resource use, environmental protection, employment, health and safety, transport and community relations. The Quarry Products Association (QPA) has also introduced a scheme for monitoring environmental performance and published its first report in March 2006. The QPA has been refining its approach to sustainable development since 2003, and in 2006 won the Trade Association Forum's annual Environmental Initiative Award for the implementation of its sustainable development strategy.

British Cement Association Activity

During 2004, members of the British cement Association (BCA), working with Forum for the Future, created a task force to define and put in place a plan to improve the UK cement industry's role in delivering sustainable construction. In November 2005, the task force published Working Towards Sustainability, a report launched by Jonathan Porritt, Chairman of the Sustainable Development Commission, setting the strategy for the industry's response to sustainable development. The task force focused on five principal areas: improving communications; the Business Case for cement; a Carbon Strategy to take the industry to 2050; a Cement Makers' Code; and an annual BCA publication of a corporate responsibility report entitled 'Performance'. These activities represent the initial work developed by the task force towards the achievement of the following objectives:

- To maximise cement's contribution towards a more sustainable built environment.
- To monitor progress towards targets for improvement in environmental, economic and social performance.
- To optimise the role that the cement industry can play in assisting the UK with delivery of best practicable environmental options for waste recovery.
- To integrate sustainable development into all UK cement industry strategies, activities and communications.
- To create a framework which will allow the industry to maximise its contribution to the well-being of its employees, neighbours and society.
- To extend our constructive, proactive and sustainable relations with stakeholders.

With regard to environmental performance (objectives (1 to 3), progress has been good, and 2007 figures show a reduction in direct emissions of carbon dioxide from UK cement manufacture of 29% between 1990 and 2006 giving a CO_2 saving of over 3.9 million tonnes. Between 1998 and 2006, the use of waste-derived alternatives has lead to reduction in fossil fuel consumption of 23% per tonne of cement. In respect of waste recovery, over 1 million tonnes was used in 2006, replacing 15% of kiln fuel and 6% of virgin raw materials. Another significant milestone was also reached in 2006, when each of the BCA member companies reached targets set under the UK Government Climate Change Agreement. Since the start of the agreement in 2002, the sector has achieved an improvement in specific energy consumption of 27.5% relative to the base year of 1990. This surpasses the overall sector plan target of a 26.6% improvement by 2010.

Looking at broader sustainable development objectives (4 to 6), the cement industry continues to improve its health and safety performance, and has introduced a zero-tolerance policy towards incidents, and can report a 32% improvement in lost-time incidents between 2003 and the end of 2006. In the same year, BCA members signed up to a Cement Makers' Code that sets out the values ethics and standards they should follow. Supporting this is a regular opportunity for stakeholder engagement with the industry to provide an opportunity for feedback, helping facilitate continuous improvement and better communications with interested parties.

Much has been accomplished towards improving the environmental footprint of cement, but sustainable development is based on a whole-life approach that considers impacts from cradle to grave. Consequently, the in-use performance of cement also forms a part of the BCA sustainability strategy, and the organisation works closely with The Concrete Centre to research and promote the sustainable qualities of concrete such as fire resistance and thermal mass. This type of work is also strongly supported by the BCA at a European level through work with CEMBUREAU, the association for European cement manufacturers. Collaboration is through numerous task groups set up to address specific topics including sustainability, energy use and fire. Examples of recent outputs include two publications in 2007 entitled 'Concrete for Energy-Efficient Buildings [7]' and 'Comprehensive Fire Protection and Safety with Concrete [8].

At the heart of the cement industry's sustainable development strategy is the business case [9], which Forum for the Future helped develop in 2004. This identifies the net costs and benefits to society arising from the manufacture of cement, using proxies where necessary to account for specific impacts or benefits. The overall case can be represented in a simple chart, which graphically demonstrates the overall positive contribution cement makes to society through areas such as employment, use of waste fuels/materials and the important contribution it makes to the built environment (see Figure 1). These can be shown to effectively outweigh negative impacts such as energy use and material extraction. The business case may undergo further development in the future, but will remain the foundation upon which sustainable development in the cement industry is based, and the means by which manufacturing impacts can be set against the beneficial properties of concrete such as durability, fire resistance and thermal mass.

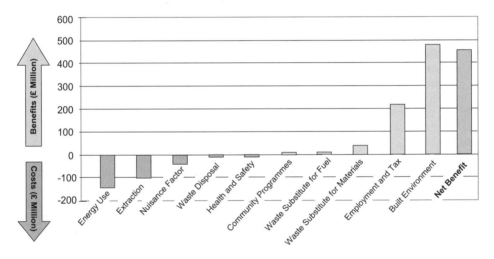

Figure 1 Graphical representation of the business case for UK cement production

Moving forward, there is much more to be accomplished in terms of closing the loop on the manufacturing impacts and whole-life benefits of cement and concrete. This has already started to occur at a more conspicuous level through both national and European level initiatives such as the CEMBUREAU Sustainability Task Force, which the BCA chairs, and the new Sustainable Construction Task Group formed under the UK Concrete Platform.

The UK Concrete Platform Sustainable Construction Task Group

Following publication of the Chapman Report within the cement and concrete sector, in 2006 five leading organisations agreed to form the UK Concrete Platform (UKCP). In addition to the BCA, other organisations joining the Platform are The Concrete Centre, British Precast, Quarry Products Association and the Concrete Society. Its objective is to improve alignment of these organisations and help facilitate joint working on activities of common interest within the member organisations. The UKCP is not a legal entity, but a forum where issues and opportunities can be discussed and solutions identified. The five chief executives comprise the UKCP Industry Management Group which meets quarterly (see Figure 2). Practical work is carried out by five task groups, each of which has agreed terms of reference and deliverables. The Sustainable Construction Task Group has a number of objectives including:

- Understanding current issues, policies and practices within in the sustainable construction field
- Responding to these issues and public consultations affecting the industry
- Reporting progress and promoting solutions to both the internal and external markets.
- Ensuring linkages and partnerships with construction industry initiatives

In 2006 the decision was taken by the Task Group to hold a master class for senior industry practitioners to review the future direction of the sector in respect of sustainable development.

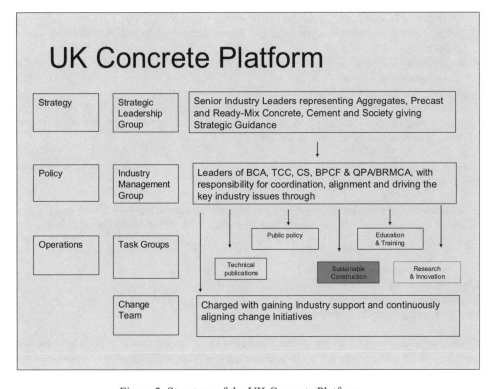

Figure 2 Structure of the UK Concrete Platform

Sustainability Master Class

A sustainability master class took place in May 2007, facilitated by Forum for the Future with The Concrete Centre and led by Jonathon Porritt. Seventeen companies joined with eight trade bodies to discuss where the industry is headed in relation to sustainable development.

The aim of the master class was to develop a shared understanding of how the concrete industry could collectively move forward on sustainability issues. The event helped to crystallize a broad commitment to fundamental change in the way the sector as a whole responds on sustainability and a keenness to accelerate this response.

The programme coming out of the master class has helped bring together the existing initiatives within the cement and concrete sector into a coherent whole that can be communicated internally and externally in a collective strategy. The strategic outputs arising from the master class can be summarised as:

- Compile a report detailing the sector's current position; what is being done and achieved.
- Establish an internal and external communications plan.
- Develop an overarching industry strategy for improvement i.e. a route map for change, including the end vision.
- As part of that strategy, develop a programme of continuous improvement maintaining international competitiveness.
- Establish KPIs and quantification of where we are currently.
- Set specific performance targets for the future (approximately nine objectives to be included).
- Establish a verifiable certification scheme, which will include responsible sourcing for companies and products.

Realising these outputs is currently a key activity for the Task Group, with some of the activities being tackled internally and much of the work relating to the strategy development work being undertaken by an external consultant.

Progress Towards a Sustainable Construction Strategy

Initial work towards the development of a strategy has already achieved some consensus as a result of a preliminary review, which suggests that it should be built around four main objectives:

1. Establish a common framework
2. Improve 'eco-points' and profiles
3. Stakeholder collaboration
4. Communicate targeted messages

These objectives focus on aligning sector action with the expectations of stakeholders, and providing better quality information to decision makers. They will also help communicate the sector's increasing contribution towards sustainable development, and facilitate work with others to help achieve the wider objective of sustainable construction.

Work on the development of a strategy for the concrete industry will have been concluded by spring 2008. In parallel to this, work will also continue of the development of a responsible sourcing scheme, including work on a gap analysis between data requirements and available data. The BCA along with the other sector trade associations are fully engaged with work to develop a responsible sourcing scheme, and are also engaged in a broader context with work on the suitable construction strategy. In the autumn of 2008, the completed strategy will be launched at a high profile event.

Looking forward, the formation of groups such as the UK Green Building Council (UKGBC) [10] and the CPA Sustainable Construction Committee [11] provides an opportunity to align the overall sustainability agenda coming out of the Sustainable Construction Task Group with that of the wider construction industry. This should strengthen the Groups objectives and help ensure its standing with government and the broader construction industry.

CONCLUSIONS

Over the last five years, the BCA and member companies have been actively developing and implementing a sustainable development strategy. The overriding objective to maximise cement's contribution towards a more sustainable built environment is an on-going goal, as is the drive for improvement in environmental, economic and social performance. Having focused largely on sustainable production, the next challenge is to build on the sector's achievements by broadening the strategy to more fully encompass sustainable construction issues. In this way the complete whole-life environmental performance of cement and cementitious materials can be properly presented, leading to a construction industry that is better informed when assessing the sustainability credentials of the materials it uses.

Achieving this objective is being assisted by the Sustainable Construction Task Group formed under the UK Concrete Platform, which is helping unite the various organisations that comprise the cement and concrete sector. Activity within the Group to develop a sustainable construction strategy has been productive, and a launch date for the strategy has been set for autumn 2008.

Close involvement at a European level is also an essential element in assisting the BCA to achieve its broader sustainability objectives and help communicate key messages. Both at a national and European level, there is increased activity related to sustainability and the sector is moving closer towards positioning cement and concrete as the sustainable construction material of choice.

REFERENCES

1. HM TREASURY. Stern Review on the Economics of Climate Change. 2007.

2. Draft Strategy for sustainable Construction – A Consultation Paper, July 2007.

3. DEPARTMENT FOR ENVIRONMENT, FOOD AND RURAL AFFAIRS. Procuring the Future Sustainable Procurement National Action Plan: Recommendations from the Sustainable Procurement Task Force 2006.

4. DEPARTMENT OF TRADE AND INDUSTRY. Meeting the Energy Challenge, a White Paper on Energy. May 2007.

5. THE BRITISH CEMENT ASSOCIATION. Sustainable Development in the Cement and Concrete Sector - Project Summary, 2003.

6. www.britishprecast.org/documents/WhitePaper_A5_8pp_LR.pdf

7. EUROPEAN CONCRETE PLATFORM. Concrete for Energy Efficient Buildings – the Benefits of Thermal Mass. March 2007.

8. EUROPEAN CONCRETE PLATFORM. Comprehensive Fire Protection and Safety with Concrete. April 2007.

9. FORUM FOR THE FUTURE. The UK Cement Industry, Benefit and Cost Analysis. September 2005.

10. www.ukgbc.org

11. www.constructionproducts.org.uk

EVALUATION OF THE OPTIMAL DURATION OF THE TREATMENT OF SYNTHESIZED CEMENTLESS BINDER MIXTURE USING MATHEMATICAL MODELING

S N Kalashnikov

S I Pavlenko   A V Aksenov

Siberian State University of Industry

Russia

ABSTRACT. A group of the SSUI scientists have developed and patented a new cementless binder from industrial by-products (secondary mineral resources) with a compressive strength of 50 to 60 MPa (patent No 2196749). The analysis of the properties new synthesized cementless binder (the free CaO content, the degree of the mixture dispersity D and the compressive strength R_{com}) versus duration of the mixture treatment in planetary mills has been made. The compressive strength of the binder versus duration of the mixture treatment was expressed by mathematical models in the form of polynomials of the second, third and fourth power and as the set of interpolation cubic splines approximating the experimental data. The value of the optimal duration of the mixture treatment has been estimated.

Keywords: New cementless binder, Planetary mill, Mixture, Treatment, Compressive strength, Mathematical model, Cubic spline evaluation, Optimal duration.

Dr S N Kalashnikov works at the Department of Information Technologies at SSUI, Novokuznetsk, Russia.

Dr S I Pavlenko is Professor and Head of Department of Civil Engineering, SSUI, Novokuznetsk, Russia.

Dr A V Aksenov is the Senior Science Collaborator at SSUI, Novokuznetsk, Russia.

INTRODUCTION

A new cementless composite binder made of secondary mineral resources (patent No 2196749) [1] has been obtained at the laboratories of SEE HPE "SibSUI" and ICS&M, Siberian branch RAS. The materials used were as follows: 60 to 80% high-calcium ash from thermal power stations, 10 to 30% used moulding sand from foundry production ("burnt" sand) and 5 to 10% high alumina product (wastes from abrasive works). The mixture of the components was treated at high-speed planetary mills of a new generation designed by the ICS&M, SB RAS [2].

The main parameters of the new binder are dispersity of a mixture, the presence in it of free CaO and compressive strength, all of them being related to the duration of a mixture treatment at the mills.

Evaluation of the optimum duration of a synthesized binder mixture treatment in new mills to increase its compressive strength is given in the paper.

THE OBJECTIVE OF STUDY AND THE ANALYSIS OF EXPERIMENTAL DATA

The Objective of the Study

Formulation of the objective and the methods of investigation depend to a great extent on the characteristics of the aim function and the information about it which is available in the process of solving as well as the one which is known a priori and as a result of the experiments conducted. Then, the methods of mathematical analysis may be used for the investigation [3, 4, 5].

The results of the experiments carried out to investigate the affect of the mixture treatment duration on the mixture dispersity D, the free CaO content in it and the compressive strength of the composite binder are given in Table 1.

Table 1   Experimental Results

DURATION OF MIXTURE TREATMENT, min	DISPERSITY OF MIXTURE, m^2/kg	FREE CaO CONTENT IN A MIXTURE, %	COMPRESSIVE STRENGTH, MPa
0	200	12.60	27.82
1	400	6.33	35.63
3	500	4.82	41.35
6	600	3.92	52.48
10	750	2.80	56.76
15	1000	2.02	43.54

Diagrams of correlation between the above parameters and the duration of treatment are given in Figure 1.

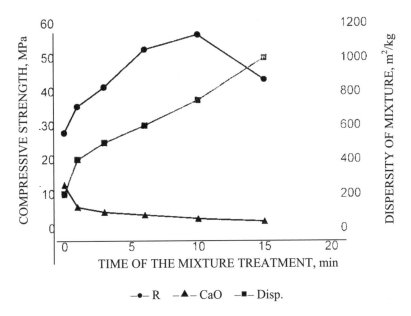

Figure 1 Degree of dispersity, free CaO content and compressive strength of a synthesized binder versus duration of the mixture treatment

Ground for the Choice of the Experimental Points

During the mixture treatment, the dispersity of the mixture D, the free CaO content in it and the compressive strength of the composite binder R are the main parameters, and they depend on the duration of treatment t. The main criterion for the optimization problem solution is the increase in the compressive strength of the composite binder. Parameters D and free CaO as well as the character of the correlation between them and the duration can serve as the source of information to be used for the choice of the experimental points by means of which the mathematical model of the correlation between the parameters R and the duration will be created. The analysis of the experimental data shows that the character of the correlation between the D and CaO parameters and the duration of treatment at the five last experimental points is close to a linear one. If analyzing all six experimental points using the experimental point corresponding to the absence of treatment one can notice the correlation between D and free CaO and the duration of treatment differs greatly from the linear one.

It is attributed to the fact that at the last five experimental points the mixture is in one regime – the dynamic regime of treatment irrespective of its duration. The absence of treatment may be considered to be a different regime when the duration of the treatment is equal to zero corresponding to the first point in the experimental data which disturbs the linear correlation between dispersity and free CaO content and the duration of the treatment. Therefore, when evaluating the optimum duration of the mixture treatment with regard to the increase in compressive strength of a composite binder, the first point in the experimental data should not be taken into consideration as it gives information about the mixture which is beyond the regime of treatment. The maximum compressive strength of a composite binder in the experimental data corresponds to the fifth point.

PLOTTING THE MATHEMATICAL MODELS

Evaluations of the optimum treatment duration were made using three, four and five last points of the experimental data which contained the point corresponding to the maximum compressive strength. Using these points the mathematical models of correlation between compressive strength and the duration of the treatment were plotted as interpolation polynomials in the form of Lagrange of the second, third and fourth power and in the form of the interpolation cubic splines set approximating the experimental data [6, 7, 8].

Evaluation of the Mixture Treatment Optimum Duration by Means of Interpolation Polynomials

Determining the interpolation polynomial of the second power by means of the last three experimental points

Let $t_5 = 10$, $t_6 = 15$; $R_4 = 52.48$, $R_5 = 56.76$, $R_6 = 43.54$

$$R_{4-6}(t) = 52.48\frac{(t-10)(t-15)}{(6-10)(6-15)} + 56.76\frac{(t-6)(t-15)}{(10-6)(10-15)} + 43.54\frac{(t-6)(t-15)}{(15-6)(15-10)} =$$

$$= 52.48\frac{1}{36}(t-10)(t-15) + 56.76\left(-\frac{1}{20}\right)(t-6)(t-15) + 43.54(t-6)(t-10).$$

As a result, the mathematical model "compressive strength versus time" in the form of the polynomial of the second power has been obtained:

$$R_{4-6}(t) = -0.41t^2 + 7.67t + 21.30 \tag{1}$$

According to the evaluation of the mathematical model (1), the maximum compressive strength is achieved at the moment of time t_1^* at which the derivative $R_{4-6}(t)$ is equal to zero and which will serve as the evaluation of the optimum duration of the mixture treatment by this mathematical model.

The derivative of the polynomial (1) is expressed as:

$$R_{4-6}(t) = -0.82t + 7.67 \tag{2}$$

After solving the equation $R'_{4-6}(t_1^*) = 0$, the value $t_1^* = 9.30$ min was obtained with evaluation of the optimum compressive strength having the value R* = 56.96 MPa.

Determining the interpolation polynomial of the third power by means of the last four experimental points

Let $t_3 = 3$, $t_4 = 6$, $t_5 = 10$, $t_6 = 15$; $R_3 = 41.35$, $R_4 = 52.48$, $R_5 = 56.76$, $R_6 = 43.54$

$$R_{3-6}(t) = 41.35\frac{(t-6)(t-10)(t-15)}{(3-6)(3-10)(3-15)} + 52.48\frac{(t-3)(t-10)(t-15)}{(6-3)(6-10)(6-15)} +$$

$$+ 56.76\frac{(t-3)(t-6)(t-15)}{(10-3)(10-6)(10-15)} + 43.54\frac{(t-3)(t-6)(t-15)}{(15-3)(15-6)(15-10)} =$$

$$= 41.35\left(-\frac{1}{252}\right)(t-6)(t-10)(t-15) + 52.48\frac{1}{108}(t-3)(t-10)(t-15) +$$

$$+ 56.76\left(-\frac{1}{140}\right)(t-3)(t-6)(t-15) + 43.54(t-3)(t-6)(t-10).$$

As a result, the mathematical model of the correlation between compressive strength and time in the of the polynomial of the third power has been obtained:

$$R_{3-6}(t) = -0.003t^3 - 0.32t^2 + 6.79t + 23.96. \tag{3}$$

According to the evaluation of the mathematical model (3), the maximum of the compressive strength is achieved at the moment of time t_2^* at which the $R_{3-6}(t)$ derivative is equal to zero and which will serve as the estimation of the optimum duration of the mixture treatment by means of this mathematical model.

The derivative of the polynomial (3) is expressed as:

$$R_{3-6}(t) = -0.009t^2 - 0.64t + 6.79. \tag{4}$$

After solving the equation $R'_{3-6}(t_2^*) = 0$, the value t_2^*–9.34 min has been obtained with the evaluation of the optimum compressive strength having the value $R^* = 56.89$ MPa.

Determining the interpolation polynomial of the fourth power by means of the last five experimental points

Let $t_2 = 1$, $t_3 = 3$, $t_4 = 6$, $t_5 = 10$, $t_6 = 15$;
$R_2 = 35.63$, $R_3 = 41.35$, $R_4 = 52.48$, $R_5 = 56.76$, $R_6 = 43.54$

$$R_{2-6}(t) = 35.63\frac{(t-3)(t-6)(t-10)(t-15)}{(1-3)(1-6)(1-10)(1-15)} + 41.35\frac{(t-1)(t-6)(t-10)(t-15)}{(3-1)(3-6)(3-10)(3-15)} +$$

$$+ 52.48\frac{(t-1)(t-3)(t-10)(t-15)}{(6-1)(6-3)(6-10)(6-15)} + +56.76\frac{(t-1)(t-3)(t-6)(t-15)}{(10-1)(10-3)(10-6)(10-15)} +$$

$$+ 43.54\frac{(t-1)(t-3)(t-6)(t-10)}{(15-1)(15-3)(15-6)(15-10)} = 35.63\frac{1}{1260}(t-3)(t-6)(t-10)(t-15) +$$

$$+ 41.35\left(-\frac{1}{504}\right)(t-1)(t-6)(t-10)(t-15) + 52.48\frac{1}{540}(t-1)(t-3)(t-10)(t-15) +$$

$$+ 56.76\left(-\frac{1}{1260}\right)(t-1)(t-3)(t-6)(t-15) + 43.54\frac{1}{7560}(t-1)(t-3)(t-6)(t-10).$$

As a result, the mathematical model of correlation between compressive strength and time in the form of the polynomial of the forth power has been obtained:

$$R_{2-6}(t) = 0.004t^4 - 0.14t^3 + 1.31t^2 - 0.65t + 35.12. \tag{5}$$

According to the evaluation of the mathematical model (5), the maximum compressive strength is achieved at the moment of time t_1^* at which the derivative $R_{2-6}(t)$ is equal to zero and which will serve as the estimation of the optimum duration of the mixture treatment by means of this mathematical model.

The derivative of the polynomial (5) is expressed as:

$$R_{2-6}(t) = 0.016t^3 - 0.42t^2 + 2.62t - 0.65. \tag{6}$$

After solving the equation $R'_{2-6}(t_3^*) = 0$, the value $t_3^* = 8.92$ min has been obtained with the evaluation of compressive strength having the value $R^* = 57.12$ MPa.

Evaluation of the Mixture Treatment Optimal Duration by Means of Cubic Spline – Interpolation

A cubic spline-interpolation was plotted for the output optimized parameter – compressive strength – at the 1 to 15 min time interval which corresponded to the treatment regime at various durations of the mixture treatment.

With the cubic – spline – interpolation, a coincidence of the neighbouring splines values at the experimental points takes place as well as the coincidence of the first and second derivatives values of these splines at the points indicated.

This time interval is broken up into intervals where $j = 1, ..., n$ where $n = 4$. At each $[t_{j-1}, t_j]$ interval, the $R(t)$ function is expressed by a cubic spline

$$R_j(t) = A_j(t - t_{j-1})^3 + B_j(t - t_j)^3 + C_j(t - t_{j-1}) + D_j(t - t_j), \tag{7}$$

where $A_j = \dfrac{S_j}{6h_j}$; $B_j = -\dfrac{S_{j-1}}{6h_j}$; $C_j = \dfrac{F(t_j) - \dfrac{S_j h_j^2}{6}}{h_j}$; $D_j = -\dfrac{F(t_{j-1}) - \dfrac{S_{j-1} h_j^2}{6}}{h_j}$.

where $h_j = t_j - t_{j-1}$.

For S, the system of linear equations with a three-diagonal matrix was obtained:

$$j = 1, ..., n-1 : \quad S_{j-1} + 4S_j + S_{j-1} = \frac{6(R(t_{j+1}) - 2R(t_j) + R(6t_{j-1}))}{h_j^2};$$

$$j = 0 : \quad 2S_0 + S_1 = 6\left(\frac{R(t_1) - R(t_0)}{h_1^2} - \frac{R(t_0)}{h_1}\right); \tag{8}$$

$$j = n : \quad 2S_n + S_{n-1} = -6\left(\frac{R(t_n) - R(t_{n-1})}{h_n^2} - \frac{R(t_n)}{h_n}\right)$$

As each $[t_{j-1}, t_j]$ intervals has a length different from others, a two-point approximation has been used for the derivatives at the ends of all interval from the first experimental point (with $t = t_0$) of the treatment regime to the last one (with $t = t_n$).

$$R(t_0) = -\frac{R(t_1) - R(t_0)}{h_1}. \tag{9}$$

$$R(t_n) = -\frac{R(t_n) - R(t_{n-1})}{h_n}. \tag{10}$$

The system of equations (8) was solved by the method of running over.

As a result, at each interval, the following cubic splines have been obtained:

$R_1(t) = 0.07(t-1)^3 + 0.04(t-3)^3 + 20.38(t-1) - 17.96(t-3);$
$R_2(t) = -0.05(t-3)^3 - 0.05(t-6)^3 + 17.99(t-3) - 13.34(t-6);$
$R_3(t) = -0.05(t-6)^3 - 0.04(t-10)^3 + 14.98(t-6) - 13.78(t-10);$
$R_4(t) = 0.02(t-10)^3 + 0.04(t-15)^3 + 8.21(t-10) - 12.34(t-15).$

Diagrams of the splines obtained are given in Figure 2.

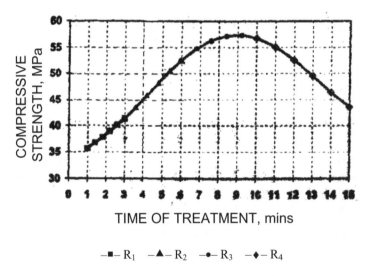

TIME OF TREATMENT, mins

$-\blacksquare- R_1 \quad -\blacktriangle- R_2 \quad -\bullet- R_3 \quad -\blacklozenge- R_4$

Figure 2   Cubic splines $R_1(t)$, $R_2(t)$, $R_3(t)$, $R_4(t)$

The maximum of the final index is achieved at a cubic spline which has a zero derivative at the moment of time t* and which will be the evaluation of the optimum treatment duration.

In the first and second intervals, the derivatives of cubic splines are positive everywhere while in the fourth interval they are negative everywhere.

In the third 5 to 10 min interval, the derivative of the cubic spline is equal to zero.

The derivative of the cubic spline in the third interval has the following form:

$$R_3(t) = 0.15(t-6)^2 + 0.12(t-10)^2 + 1.20. \tag{11}$$

On solving the equation $R_3'(t_4^*) = 0$, the value $t_4^* = 8.99$ min has been obtained.

The evaluation of the maximum compressive strength has the value R* = 57.34 MPa. On receiving the average data of three polynomials in the form of Lagrange and using cubic spline-interpolation, the averaged evaluation has been obtained:

$$t_{opt} = \frac{t_1^* + t_2^* + t_3^* + t_4^*}{4} = \frac{9.30 + 9.34 + 8.92 + 8.99}{4} = 9.14 \text{ min.}$$

The optimum duration for the mixture treatment at the activator mills is 9.14 min.

CONCLUSIONS

1. Mathematical models of correlation between the compressive strength of a new cementless binder and the duration of a mixture treatment in planetary mills have been plotted in the form of polynomials of the second, third and forth power and as a set of interpolation cubic splines approximating the experimental data.

2. By means of the mathematical models "compressive strength of a new binder versus time of the mixture treatment", the value of the optimum treatment duration has been estimated. The averaged evaluation of the optimum treatment duration was 9.14 min which was corroborated by previous experimental investigations (up to 10 min).

REFERENCES

1. PATENT 2196749 RUSSIA, 6C04 and 28/08, Cementless binder. S.I. Pavlenko, S.I. Merculova, A.V. Aksyonov, et al, 2003.

2. AKSYONOV, A V, PAVLENKO, S I, AVVAKUMOV, YE G. Mechanochemical synthesis of a new composite binder from secondary mineral resources, Science edition, ICS&M SB RAS. Monograph / A.V. Aksyonov, S.I. Pavlenko, Ye.G. Avvakumov; editor-in-chief, corresp. member RAS N.Z. Lyakhov, Novokuznetsk, 2002, 62 p.

3. KHIMMELBLAU, D. Applied non-linear programming, Moscow: publishing house "Mir", 1982.

4. VASILJEV, PH P. Numerical methods for the solution of extreme problems, Moscow: Publishing house "Nauka", 1981.

5. FLAKKO, A ET AL, Non-linear programming: methods of successive unconditioned minimization. Moscow: Publishing house "Mir", 1972.

6. MARCHUK, G I. Methods of computing mathematics, Moscow: Publishing house "Nauka", 1976.

7. STECHKIN, S B, SUBBOTIN, YU N. Splines in computing mathematics, Moscow: Publishing house "Nauka", 1976.

8. ALBERG, J, NILSON, E, WALSH, J: translated from English by Yu N Subbotin, Moscow: publishing house "Mir", 1972, 318 p.

STUDIES ON A BLENDED CEMENT FOR SUSTAINABLE DEVELOPMENT

M T G Barbosa   T M de Oliveira   W J dos Santos

Federal University of Juiz de Fora

F A Darwish   S Kitamura   C G Coura

Federal University Fluminense

Brazil

ABSTRACT: Sustainable development is growing considerably because of the increased production of rejects by the industries exhausting the areas in industrial plants. It is evident that there is an increased concern about the preservation of the environment, the reduction of natural resources for the fabrication of traditional construction materials and new initiatives to assist in governmental programmes. This research work aims at investigating an alternative cement for the building industry. On the whole, three different combinations of cement + steel slag + blast furnace slag were found. The chemical and physical tests carried out were the ones prescribed for Portland cement by Brazilian standards, since there was no reference about its use as building material in the literature. Such tests intend to characterize the mixtures in terms of fineness modulus, hardening time, volumetric expansion, compressive strength at 28, 90 and 120 days, and durability in different situations at humidities of 0%, environmental and 100% for an environmental temperature of 27°C.

Keywords: Alternative cement, Sustainable environment, Performance.

Thaís Mayra de Oliveira is a Professor of the Federal University of Juiz de Fora, Brazil,

Sergio Kitamura and **Claudia Gávio Coura** are post-graduate students at the Federal University of Fluminense.

Dr Maria Teresa Gomes Barbosa, is a Professor of the Federal University of Juiz de Fora, Brazil, and member of ACI and IBRACON.

Dr Fathi Aref Darwish, Professor of the Federal University of Fluminense, Brazil.

White José dos Santos Graduate of Civil Engineering of the Federal University of Juiz de Fora, Brazil.

INTRODUCTION

All human activities result in some degree of environmental degradation. The challenge is to minimize this degradation to a level consistent with sustainable development. Sustainable development is integration of environmental, economic and social considerations when deciding whether and how developments should be executed.

For the civil engineering community, the concept of sustainable development involves the use of high performance materials produced at a reasonable cost and with the lowest possible environmental impact. The ways to reduce environmental impact include reduction of waste and emission of harmful greenhouse gases, more efficient use of mineral resources, and increased use of recycled materials such as aggregates, gypsum, plastics and others.

Portland cement concrete is a major construction material worldwide. Its production releases large amounts to CO_2 into the atmosphere. In view of global sustainable development, it is imperative that supplementary cementing materials be used to replace large proportions of cement in the concrete industry.

The most available supplementary cementing material worldwide is fly ash, a by-product of thermal power stations. There is a potential for use of fly ash in concrete, and therefore, for significant reductions in cement production resulting in considerable environmental benefits [1-5].

The search for alternative binders or cement replacement materials has continued for past decades. Research has been carried out on the use of volcanic ash, volcanic pumice, fly ash, blast-furnace slag, and silica fume as cement replacement materials. There are pozzolanic materials, which can improve the durability of concrete and the rate of strength gain and can also reduce the rate of heat liberation, which is beneficial to mass concrete [1-5].

Over the recent decades, Portland cement containing fly ash and or silica fume has gained increasing acceptance while Portland cement containing natural rice husk ash and burnt oil shale is common in regions where these materials are available [3-7].

The tests presented in this article are based on a project investigating alternative cements for building industry. On the whole, three different combinations of cement, steel slag and blast furnace slag were found to be optimal. The results illustrate how different combinations affect the hydration behaviour of Portland cement.

The chemical and physical tests carried out were the ones prescribed for Portland cement by Brazilian standards from ABNT, since there was no reference about the use of these alternative materials as building material in the literature.

Such tests to characterize the mixtures include: fineness modulus, hardening time, volumetric expansion, compressive strength at 28, 90 and 120 days, and durability in different situations, at humidities of 0%, environmental and 100% at 27°C.

EXPERIMENTAL PROGRAMME

Table 1 shows the experimental program and the number of concrete specimens tested for compressive strength.

Table 1   Experimental Programme

| | MIXTURE AND COMPOSITION | | | | COMPRESSIVE STRENGTH, $n°$ ST's | | | | | |
| | | | | | Brazilian Standard | | | Environmental temperature: | | |
	% SS	% BS	% TS	% CPV	28 days	90 days	120 days	H=0%	EH	H=100 %
MR	0	0	0	100	4	4	4	4	4	4
M1	3.0	3.0	6.0	94.0	4	4	4	4	4	4
M2	12.5	12.5	25.0	75.0	4	4	4	4	4	4
M3	17.0	17.0	34.0	66.0	4	4	4	4	4	4

SS – steel slag (see Table 2);
BS – blast furnace slag (see Table 2);
TS – total slag;
CPV – Portland cement, type V – ARI;
ST – specimens tested;
H – humidity;
EH – environmental humidity;
Environmental temperature of 27°C at 28 days

A concrete mixture of 1 : 2.00 : 2.50 : 0.65 was used in order to obtain a compressive strength of approximately 27 MPa at 28 days. Substitution of slag for cement was accomplished observing the maximum and minimum quantities of slag prescribed in the Brazilian standard for the cement type CPII-E. Hence, the substitution of cement involved slags at 6% (3% steel slag + 3% blast furnace slag); 25% (12.5% steel slag + 12.5% blast furnace slag); 34% (17% steel slag + 17% blast furnace slag). For the verification of durability of the mixtures, the specimens were separated into groups of 0%, environmental and 100% humidity.

Table.2   Composition of steel slag and blast furnace slag

COMPOSITION, %	STEEL SLAG	BLAST FURNACE SLAG
CaO	41.97	31.80
FeO	-	24.72
SiO_2	35.47	24.30
MgO	5.69	5.34
Mn_2O_3	0.84	4.78
Al_2O_3	13.20	4.40
TiO_2	0.53	0.46
SO_3	0.07	0.93
K_2O	0.28	x
Na_2O	0.10	x
Fe_2O_3	0.83	x
S	0.87	x
P_2O_5	x	0.61
TiO_2	x	0.46
Cr_2O_3	x	0.43

RESULTS AND ANALYSIS

Table 3 shows the fineness modulus, hardening time and volumetric expansion in the mixture proportions for the alternative cement (MR – reference mixture, M1, M2 and M3). The volumetric expansion indicates the presence of components causing expansion during cement hydration. Table 4 shows the average compressive strength of concrete at 28, 90 and 120 days. Table 5 and Figure 1 show compressive strength of concrete at 28 days at environmental temperature and humidities: 0%, environmental and 100%.

Analysing the results obtained, see Table 5 and Figure 1, it is observed that with the increase of the slag percentage the compressive strength of concrete decreases at the three ages studied. The results shown in Table 5 demonstrate a decrease of the compressive strength of concrete for the ends of the humidity range in relation to environmental humidity. It is also verified that for an environmental humidity, there is a decrease of the strength of concrete with increasing slag percentage, as already observed previously. However, the same is not verified in the case of humidities of 0% and 100%, nor is there a linear decrease of the concrete strength with increasing slag percentage.

Table 3   Fineness modulus, hardening time and volumetric expansion

MIXTURE	FINENESS MODULUS, GPa	HARDENING TIME, hours		VOLUMETRIC EXPANSION, mm	
		BT	FT	cold	hot
MR	24.77	02:37	08:22	0.25	0.00
M1	21.65	03:18	06:03	4.00	0.50
M2	22.10	03:30	05:30	1.75	0.00
M3	25.86	03:00	05:30	1.75	0.00

BT – BEGIN TIME
FT – FINAL TIME

Table 4   Compressive Strength of Concrete

MIXTURE	STRENGTH OF CONCRETE, MPa		
	28 days	90 days	120 days
MR	31.82	35.33	36.50
M1	31.75	33.60	35.65
M2	30.25	30.58	31.44
M3	29.66	30.45	31.16

Table 5   Compressive strength of concrete at different humidities (H)

MIXTURE	STRENGTH OF THE CONCRETE, MPa		
	H = 0%	H = environmental	H = 100%
MR	35.83	38.18	30.47
M1	27.30	32.62	26.52
M2	29.60	32.36	27.26
M3	29.52	29.23	29.31

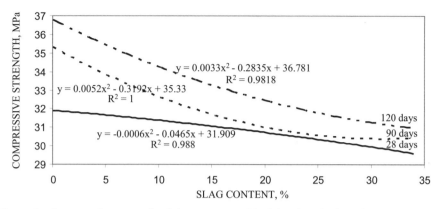

Figure 1   Compressive strength of the concrete x % of the slags in the mixture proportions

To obtain a better analysis of the results it is necessary to do statistical analyses that to obtain a correlation through a regression among compressive strength of concrete, humidity and slag percentage.

The expression (1) and Figure 2 present the result:

$$f_c = 30.253 S^{-0.0116} H^{0.0002} \qquad \text{(error} = 1.09 \text{ MPa)} \qquad (1)$$

Where:

f_c = compressive strength of concrete at 28 days of age, MPa;

S = slag percentage (% steel slag + % blast furnace slag);

H = humidity of air, %.

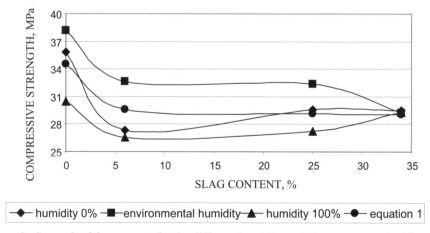

Figure 2   Strength of the concrete for the different humidity and slag percentage in this study

CONCLUSIONS

The results of this experimental program allow affirming that:

1) Compressive strength of concrete decreases at the ends of the humidity range compared to environmental humidity.

2) It is verified that the additions are beneficial at high proportions for low and high humidities (0% and 100%), but not having the same behaviour at environmental humidity.

3) The result obtained through the statistical analysis demonstrates that there are correlations among compressive strength of concrete, humidity and slag amount.

REFERENCES

1. BILODEAU A AND MALHOTRA V M, High Volume Fly Ash System: concrete solution for sustainable development, ACI Materials Journal, Vol. 97, No. 1, 2000, pp. 41-48.

2. BOUZOUBAÂ N, FOURNIER B, MALHOTRA V M AND GOLDEN D M, Mechanical Properties and Durability of Concrete made with High Volume Fly Ash Blended Cement Produced in Cement Plant, ACI Materials Journal, Vol. 99, No. 6, 2002, pp. 560-567.

3. HOSSAIN, K M A; LACHEMI, M. Development of Volcanic Ash Concrete: Strength, Durability and Microstructural Investigations. ACI Materials Journal. Vol. 103, No. 1, 2006, pp 11-17.

4. JOHN V M, Cimentos de escória ativada com silicatos de sódio, Tese de Doutorado, Universidade de São Paulo, Escola Politécnica, São Paulo, SP, 1995, 189 p.

5. MACHADO A T, Estudo comparativo dos métodos de ensaio para avaliação da expansibilidade das escórias de aciaria, Dissertação de Mestrado, Universidade de São Paulo, Escola Politécnica, São Paulo, SP, 2000, 135 p.

PHYSICO-MECHANICAL AND DEFORMATION PROPERTIES OF A NEW CEMENTLESS FINE COMPOSITE CONCRETE OVER A 5-YEAR PERIOD

S I Pavlenko

Siberian State University of Industry

Russia

ABSTRACT. A new cementless binder (patent N°2196749) and a fine concrete on its basis composed of secondary mineral resources (industrial wastes) only have been created by the Siberian State University of Industry (SSUI). The paper presents the data on its physico-mechanical and deformation properties over a period of 5 years. The mixture proportions of the concrete were as follows: 900 to 1000 kg/m^3 new cementless binder; 850 to 1000 kg/m^3 slag sand of 0 to 5 mm grading fractions; 250 to 300 kg/m^3 water. The main physico-mechanical (strength) characteristics of a new fine cementless concrete enhanced during a five-year period monitored since at the age of 28 days by 20 to 30%, with the modulus of elasticity increasing by 54%. At the same time, shrinkage and creep strains ceased at the age of 2 years. The investigations showed that the developed fine cementless concrete performed better than the traditional cement sand concrete and had a higher crack resistance. It meets the requirements of the building regulations and its cost is reduced by a factor of 1.5 to 2 compared with that of the conventional concrete.

Keywords: Cementless concrete, Shrinkage, Creep, 5-year period, Frost resistance, Waterproofness, Water sorptivity.

Dr S I Pavlenko is a head of Department of Civil Engineering at the Siberian State University of Industry. Professor, "Outstanding Scientist of the 21st Century Laureate"

INTRODUCTION

The objective of the investigation was to obtain a new cementless fine concrete on the basis of the created cementless binder patented in 2003 with improved characteristics as compared to those for conventional concretes made of natural components and the concretes developed by us earlier in the 20th century (certificate) of the Government of Russia [1-3].

EXPERIMENTAL INVESTIGATIONS ON THE DEVELOPMENT OF A NEW COMPOSITION FOR CEMENTLESS FINE CONCRETE

The investigations made in this section of work included:

1. Study of the initial materials.
2. Development of the optimum mixture proportions.

Materials

The following materials were used:

– a new cementless binder synthesized by us;
– slag sand (furnace bottom ash) from the Abakan thermal power station (TPS).

Cementless binder (patent No 2196749) has been developed by mechanochemical synthesis at planetary mills designed by the ICS&M SB RAS using the following materials: 60 to 80% high-calcium fly ash from the Abakan thermal power station, 10 to 30% used moulding sand ("burnt sand") from the foundry production of the "Abakanvagonmash" (HAP), wastes of abrasive works.

The methods of its production and properties are given elsewhere [4-7]. Table 1 shows the chemical (oxide) analysis of a binder used in this work. After treatment at a planetary current mill, the binder was a uniform fine material; the degree of dispersity (specific surface) and an X-ray amorphous phase were 600 to 750 m^2/kg and 30%, respectively. The concentration of heavy metals in the mixture was low (up to 100 g/t).

Table 1 Chemical Analysis of Cementless Binder and Slag Sand from the Abakan TPS

MATERIALS	COMPOUND										
	SiO_2	CaO	MgO	Al_2O_3	Fe_2O_3	Na_2O	K_2O	TiO_2	MnO	Ba	LOI
Cementless composite binder	28.62	28.89	14.27	9.92	9.94	0.81	0.37	0.39	0.14	0.61	5.43
Slag sand from the Abakan TPS	56.47	29.92	3.50	8.16	9.63	–	–	–	0.17	–	–

Granulated Furnace Bottom Ash (Slag) up to 5 mm Size Fraction from the Abakan TPS

The bulk density and the absolute density of the slag are 1580 and 2430 kg/m^3, respectively. It is referred to as a dense slag. The chemical analysis of the slag sand is given in Table 1. It is characterized by the absence of free CaO and loss on ignition; according to the State Standard 25589-91 [8], it can be used in concrete as replacement aggregate for natural rubble and sand.

Development of the Optimum Mixture Proportions of Concrete

First, proportions of components for the concrete to be developed have been determined experimentally, namely:

- the optimum water-to-binder ratio;
- the maximum possible proportion of aggregate (slag sand) in concrete;
- the effect of the heat-moisture treatment regime on the compressive strength of concrete.

The optimum water-to-binder ratio was determined by casting 2 x 2 x 2 cm specimens with water at room temperature. The results of the experiment are given in Figure 1.

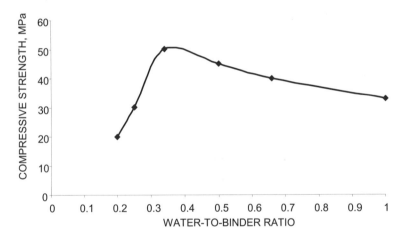

Figure 1   Compressive strength of cementless binder versus water-to-binder ratio

As can be seen from Figure 1, the best water-to-binder ratio was 1/3. Then, the correlation between the compressive strength of the binder and the temperature of mixing water was determined. The best results were obtained at the water temperature of 20–40°C.

The maximum possible proportion of a binder was determined using $7 \times 7 \times 7$ cm specimens cast at a water temperature of 30°C with further hardening for 28 days. The test results are given in Figure 2.

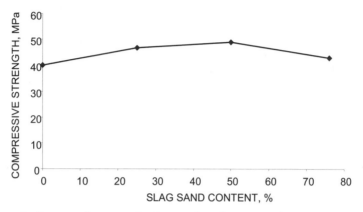

Figure 2   Compressive strength of cementless fine composite concrete versus
the amount of the slag incorporated at 28 days

Analysis of the data in Figure 2 shows that to increase the compressive strength of a fine
concrete it is necessary to incorporate 50% slag sand. As a result of the investigations, the
optimum mixture proportions for a fine concrete on the basis of the created cementless binder
have been determined; they are given in Table 2:

Table 2   Properties of concrete

PROPERTY	VALUE
Binder content, kg/m^3	900–1000
Slag sand content, kg/m^3	850–1000
Water content, kg/m^3	250–300
Slump, cm	7–10
Average density of concrete, kg/m^3	2085
Compressive strength of concrete, MPa	45–55

PHYSICO-MECHANICAL AND DEFORMATION PROPERTIES

Drying Shrinkage and Creep for a 5-Year Period

For the determination of the drying shrinkage and creep strains of cementless ash-slag concrete,
$10 \times 10 \times 40$ cm prisms were tested using the methods of NIIZhB [9]. Prisms were cured at
18 to 22°C and 50 to 75% relative humidity. Shrinkage strains were measured by a dial type
indicator with a scale of 0.001 mm.

The shrinkage strains were 0.4 and up to 0.6 mm at 24 hours and 28 days, respectively.
Between 28 and 180 days, the shrinkage strain increased to 0.65 mm/m and ceased beyond
this period. At 5 years it was 0.07 mm/m versus 0.75 mm/m according to the Building
Regulations 2.03.01–84 for fine concretes.

Besides, the drying shrinkage of the concrete developed was smoother than that of a sand concrete (Figure 3) which was attributable to a lower bleeding of the former. Due to a lower bleeding and higher strength gain as compared to the cement sand concrete, there obtained a higher crack resistance.

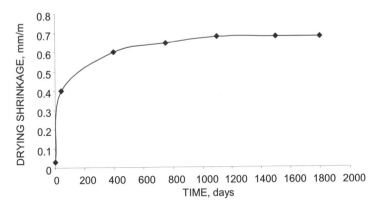

Figure 3   Relative shrinkage of fine cementless concrete

Creep tests were performed on spring devices beginning at the age 28 days. Prisms, $10 \times 10 \times 10$ cm in size, were subjected to a long-term loading at the applied stress constituting 50% of the prism strength (0.5R prism) (Figure 4).

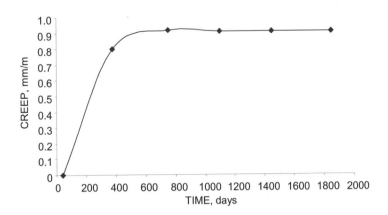

Figure 4   Creep of fine cementless concrete

The creep strain developed noticeably in the early period of loading and then slowed down. As can be seen from the diagram, the relative creep strain of the fine cementless ash-slag concrete achieved 0.8 mm/m at 20 days, and the magnitude of the creep increased insignificantly between 20 and 180 days (up to 0.94 mm/m) ceasing beyond this period. The concrete showed 12 to 15% less creep than the data reported for a fine sand concrete.

The main physico-mechanical and deformation properties of a new concrete have also been studied for 5 years (Table 3). It is evident from the data that concrete increased its compressive and prism strengths by 53 and 58%, respectively, over a 5-year period, the strength gain decreasing with time. The maximum compressive strength and prism strength gain was achieved at 28 days (18 and 26%, respectively) with only 5 and less percent gain at later periods. For comparison: the conventional concrete increases its strength up to 10 and 15–18% at 1-year and 5-years, respectively.

Table 3 Physico-mechanical and deformation characteristics of cementless concrete

PROPERTIES OF CONCRETE	AGE OF TESTING, days						
	1	28	365	730	1095	1460	1825
Compressive strength, MPa	45.9	51.5	57.3	58.9	59.4	61.3	62.1
Prism strength, MPa	34.4	40.7	45.8	48.3	49.9	51.5	52.2
Prism-to-Cube strength ratio	0.75	0.79	0.80	0.82	0.84	0.84	0.84
Modulus of elasticity, MPa	15.5	19.7	28.7	28.9	29.3	29.4	30.4
Compressibility, mm/m	0.84	0.86	1.02	1.03	1.04	1.05	1.07
Extensibility, mm/m	0.08	0.1	0.14	0.15	0.16	0.18	0.18
Relative shrinkage, mm/m	0.04	0.4	0.6	0.65	0.7	0.7	0.7
Relative creep, mm/m	–	–	0.8	0.94	0.94	0.94	0.94

Modulus of elasticity of the concrete was determined using the NIIZhB methods [10]. It increased by 54% over a 5-year period versus 15 to 20% for fine cement concretes according to the Building Regulations 2.03.01.–84. Compressibility and extensibility increased by 23 and 50%, respectively, which exceeded the Building Regulations by 10 to 12%. During the last three years, these properties stabilized.

Frost Resistance of a New Composite Concrete

10-cm cube specimens were tested for frost resistance in the corrosion laboratory of NIIZhB according to the accelerated mode developed by the NIIZhB [11]. Frost resistance was investigated in the altitude chamber "Nema TBV 800" (Figures 5 and 6) according to the 3-d method of the State Standard 10060-95 [12]. Specimens previously saturated with a 5% NaCl solution were placed in a bath at 20°C for 4 days. Two hours after extraction from the bath, the control specimens were tested for compressive strength while the others were weighed and placed into a refrigerating chamber where they were frozen to -50°C. After freezing, the specimens were exposed to thawing in the same baths. Charging the specimens into the refrigerating chamber, their freezing and thawing took 24 hours.

At the end of a definite number of cycles, the specimens were weighed again and tested for compressive strength. Testing was terminated when the strength of the specimens was reduced by more than 15% of the initial strength. After 5 cycles, which correspond to 30 cycles according to State Standard (F50), the compressive strength of specimens was reduced by 6%. After 10 cycles (F100), the surface of the specimens began to scale off, ribs crumbled but the strength of concrete increased by 12%, the coefficient of frost resistance being 1.2. Up to 20 cycles (F200), the increase in the weight of specimens was observed which was attributable to the continuation of the reaction of hydration. After 20 cycles, concrete began to fail.

Figure 5   Thermal altitude chamber "Nema"

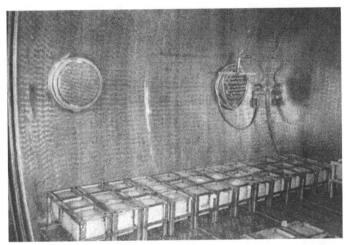

Figure 6   The interior of the thermal altitude chamber "Nema"

Waterproofness of Composite Fine Concrete

The waterproofness tests of cylinder specimens were performed on a standard installation (Figures 7 and 8). Waterproofness was determined in accordance with the State Standard 12730.5-84 [13] using the method of "wet spot". Water was supplied to a lower base of cylinders under the pressure of 2 atmospheres (W2) and kept there for 6 hours. Then the pressure increased by 2 atm. The process continued until the signs of water filtration in the form of drops or wet spot appeared on the upper base of specimens. Waterproofness for the series of specimens was evaluated by a maximum water pressure at which on four out of six specimens infiltration was not observed. Sides of the specimens were previously painted with three layers of paint so that water penetrated through the cylinder bases only. The waterproofness grade of concrete was referred to as W10.

Figure 7   The installation for testing specimens for waterproofness

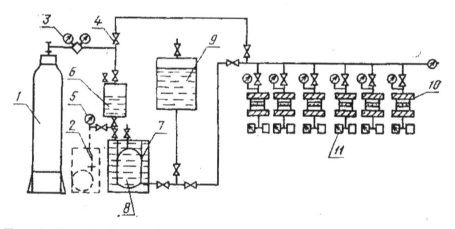

Figure 8   Thematic diagram of the installation for testing specimens for waterproofness

1 – gas bag; 2 – pump; 3 – reducer; 4 – valve; 5 – manometer; 6 – pressure transmitter;
7 – water reservoir; 8 – elastic reservoir with deaerated water; 9 – spare reservoir with
deaerated water; 10 – test pocket; 11 – filtrate weight measurer

Water Sorptivity

The water sorptivity tests were carried out on 70-mm cubes. They were placed in water at 20°C and weighed every 24 h until the difference between the last and previous weighing was 0.1%. It was found that the water sorptivity of this new fine concrete according to mass and according to volume was 9.3% ($W_m = 9.3\%$) and 19.8% ($W_v = 19.8\%$), respectively. These values meet the requirements of the Building Regulations for fine concretes.

CONCLUSIONS

As a result of the 5-year investigation, a new cementless binder and a fine concrete on its basis have been created from secondary mineral resources only.

A new composite fine cementless concrete incorporating a cementless binder [5] and slag sand of 0 to 5 mm size fraction as an aggregate exhibited a higher performance than the concrete developed earlier [2, 14, 15].

Both furnace bottom ash sand from the burning of coal and granulated slags from the metallurgy production meet the standard requirements for use in concretes.

REFERENCES

1. PAVLENKO S I, AKSENOV A V AND MERKULOVA S I, Cementless fine mixture ready for use, "The experience of enterprises in utilization of industrial and domestic wastes", Proceeding of the 2nd scientific-practical seminar, Novokuznetsk, 23–25 November, 1999. Administration of Kemerovo region and Novokuznetsk, State committee for protection of the environment. Novokuznetsk: publishing house "Kuzbass Fair", 1999, pp. 68–71.

2. TECHNICAL CERTIFICATE "On fitness of the product for use in construction at the territory of the Russian Federation" No TC-07-0175-99 (resolution of the Russian Federation government, 27.12.1977 No 1636) "Cementless ash-slag concrete fine mixture ready for use", Register of the Gosstroy Russia, 08.06.1995.

3. PAVLENKO S I AND SHMELKOV M A, Physico-Mechanical and Deformation Properties of Cementless Fine Ash-Slag Concrete over 5-year Period, Proceeding of "Fourth Canmet/ACI International Conference on Durability of Concrete", Sydney, Australia, 1997, Supplementary papers. Published ACI, Farmington Hills, Michigan, USA, 1997, pp. 331–340.

4. PAVLENKO S I, BAZHENOV YU M, AVVAKUMOV E G AND AKSENOV A V, Mechanochemical Synthesis of New Composite Binder from Secondary Mineral Resources, 11th International Congress of the "Chemistry of Cement" Cement's Contribution to Development in the 21st Century (ICCC 2003) Durban, South Africa, 11–16 May 2–3, Congress Proceeding, Volume 3. Alpha (Pty) Ltd. The Al Choice, Holcim Group, Volume 3, 2003, pp. 1217–1226.

5. PATENT 2196749 RUSSIA, C27C04D7/28 Cementless binder (S.I. Pavlenko, S.I. Merkulova, A.V. Aksenov et al.). Inventions Bulletin No 2, 2003.

6. DIPLOMA AND SILVER MEDAL of the VI International Saloon of Industrial Property "ARKHIMED" (Inventions, industrial samples, trade marks), Russia, Moscow, 18–21 March, 2003.

7. PAVLENKO S I AND AKSENOV A V, A new composite binder and fine concrete with secondary mineral resources on its basis. Scientific edition-monograph. Moscow: Publishing house of International Association of the Construction higher education establishments, 2005, 140 pp.

8.   STATE STANDARD 25589-91.

9.   NIIZhB, Recommendations for Study of Creep and Shrinkage of Concrete, MR-1-75-Moscow, Stroyizdat, 1976.

10. NIIZhB, Recommendations for Determining Strength and Structural Characteristics of Concrete Subjected to Short-Story-Long-Term Loading, pp. 10–76. Moscow, Stroyizdat, 1976.

11. ROMANOVA N A AND IVANOVA O S, Hardening of the ash-containing concretes in winter conditions and their frost resistance, Proceedings of the All-Union conference "Concretes on the basis of ash and slag from TPS and their use in construction", Editor, S.I. Pavlenko, 1990. State education USSR, Minenergo USSR, Gosstroy USSR, SMI, Novokuznetsk, Publishing house, 1990, Volume 2, pp. 84.

12. STATE STANDARD 10069-95.

13. STATE STANDARD 12730-84.

14. PAVLENKO S I, Patent of Russia No 2065420. Concrete mixture. M. Inventions Bulletin, 1996, No 23.

15. PAVLENKO S I, MALYSHKIN V I AND BAZHENOV YU M, Cementless fine composite concrete from secondary mineral resources, Scientific edition-monograph. Novokuznetsk, Publishing house of the Siberian branch of the Russian Academy of Sciences, 2000, 142 pp.

GENESIS OF STRUCTURE AND STRENGTH OF AN ARTIFICIAL STONE WITH HIGH CONTENT OF FLUIDIZED FLY ASH

E K Pushkarova   V I Gots   O A Gonchar

S G Guzij   M Stepanuga

Kiev National University of Construction and Architecture

Ukraine

ABSTRACT. Efficiency of the decision of ecological, economic and technical problems is connected to recycling of waste products of a power industry, including fluidized fly ash. From all existing kinds of waste products the power fluidized fly ash differs not only in being pozzolanic, but also having high hydraulic activity. It allows receiving on its basis (at limited use ordinary Portland cement) composite cements and concrete with outstanding physico-mechanical and special properties. In this work composite cements based on fluidized fly ash are investigated at various quantities of Portland cement (5-90%). It is shown, that maximal strength characterizes the compositions containing 45-55% fluidized fly ash that is connected to the formation of products such as hardening hydrogarnets, hydrogehlenite and small quantity of calcite and scawtite. Development of strength of the above-named compositions was investigated and stable increase of physico-mechanical characteristics during 7 years was affirmed. Concrete of class B40 on the basis of different composition binding systems can be obtained and used in road construction as various types of road coverings.

Keywords: Fluidized fly ash, Amorphous aluminosilicates, Composite cements, Hydrogarnets, Hydrogehlenite, Ettringite, Road coverings.

Professor, DrSc E K Pushkarova, is Head of the Building Materials Department, Kiev National University of Construction and Architecture. Since 1980 she specializes in technologies of producing composite materials based on alkaline cementitious materials.

Professor, Dr V I Gots, is Head of the Building Technology and Construction Products Department, Kiev National University of Construction and Architecture. He specializes in technologies of producing composite materials with the use of waste products.

Dr O A Gonchar, is a Researcher in the Scientific Research Institute of Binders and Materials, Kiev National University of Construction and Architecture. Her research topic is to investigate preparation of composite cements and concretes with the use of waste products.

Dr S G Guzij, is a Researcher in Scientific Research Institute of Binders and Materials, Kiev National University of Construction and Architecture. His research topic is to investigate composite materials based on alkaline cementitious materials for special uses.

Ms M Stepanuga is a postgraduate research student at the Kiev National University of Construction and Architecture. Her research topic is to investigate preparation of composite cements and concretes based on waste products.

INTRODUCTION

According to the concept of sustainable development and the Kyoto protocol for achievement of stability of economic development, the waste of a power station should be considered not only as the factor which leads to environmental contamination but also as a source of additional resources for a wide scale of building materials of various purpose [1-5]. The decision of this problem demands development of new conceptual solutions directed on transformation of waste to a commodity, promoting preservation of natural resources and the integrity of the environment.

In this work the opportunity of application of fluidized fly ash as active component of ash-cement mixtures to obtain high-strength concrete is presented. Within the mineralogical composition of such ash, alongside with amorphous aluminium silicates, there is calcium oxide CaO, anhydrite $CaSO_4$ and quartz SiO_2. Fluidized fly ash is characterized not only by high pozzolanic properties, but also high hydraulic activity.

When designing the composition of binding materials, which contain maximum quantity of ashes, one should provide the directed synthesis of new hardening products. These formations provide stability of properties of a material in time. When using high calcium fluidized ashes to obtain binding substances, synthesis of stable structure new formations is achieved due to the sulfate-alkaline activation and the directed synthesis of solid solutions on the basis of ettringite, modified by silicate-anions [6]. Formation of the latter not only stabilizes strength properties of the obtained artificial stone in time, but also interferes with the formation of secondary ettringite and to the development of destructive processes in the structure of the hardened cement stone.

RESEARCH TECHNIQUES

As raw material, the fluidized ash of power station "Zeran" was used. It had the following chemical composition, % by mass, SiO_2 – 42.1; Fe_2O_3 – 7.4; Al_2O_3 – 21.2; Mn_3O_4 – 0.12; TiO_2 – 0.81; CaO – 13.9; MgO – 3.02; SO_2 – 7.2; P_2O_5 – 0.23; Na_2O – 1.47; K_2O – 2.13. According to X-ray diffraction analysis (see Figure 1, curve 1) the fly ash of fluidized bed combustion comprises quartz β-SiO_2, anhydrite β-$CaSO_4$, calcium oxide CaO, magnetite Fe_2O_3 and also an X-ray amorphous alumonosilicate phase obtained at the dehydration of minerals such as illite, kaolinite, and montmorillonite.

Alite was introduced by using Portland cement (see Figure 1, curve 2). On the X-ray diffraction pattern, the basic reflections of alite C_3S are captured. The silica fume of the metallurgical plant (as the modifying component in the composite binder) was introduced by amorphous silicon dioxide (see Figure 1, curve 3). As plasticizing component superplasticizer C-3 was used, which was a sodium salt of the condensation product of naphthalene monosulfate acid and formaldehyde.

River sand with the fineness modulus M=1.4...1.45 was used fine aggregate for the preparation of test mortars. The physico-mechanical properties of binder compositions based on fluidized ashes were defined using high-plasticity paste prepared by vibration. The compaction time was 180 s, vibration frequency 50 Hz, and oscillation amplitude 0.35 mm. The physico-chemical investigations of binder compositions were carried out by using X-ray diffraction (XRD) and scanning electron microscopy (SEM) analyses.

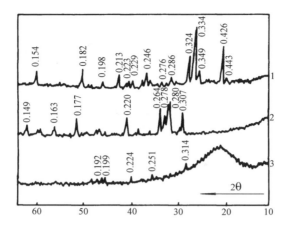

Figure 1   X-ray diffractograms of initial components
1: fluidized bed coal combustion ash; 2: Portland cement; 3: silica fume

RESULTS AND DISCUSSION

Complex research of chemical-mineralogical composition of fluidized ashes has been carried out, which simultaneously shows pozzolanic properties (owing to the presence of aluminosilicate component) and hydraulic properties (due to the presence of anhydrite). It has enabled to put forward a hypothesis regarding preparation of an artificial stone, whose composition included the presented hydrations products except for low basic hydrosilicates of calcium, and also solid solutions on the basis of ettringite.

In order to identify the greatest possible quantity of fluidized ashes, systems containing different quantities of ashes — from 0 up to 100% — were examined (see Table 1). The use of fluidized ashes allows obtaining binder compositions whose engineering properties do not differ from properties of an artificial stone based on Portland cement, and are characterized by stability in time. The technical characteristics offered by these compositions (see Table 1) and the physico-mechanical characteristics of an artificial stone produced on their basis testify (see Figures 2 and 3) that the optimum structure contains 35-85% Portland cement.

However, at later stages (1 year, 3 and 7 years) recession of strength was observed in systems with greater contents of a technogenic component (about 60%) possibly caused by insufficient density of the created artificial stone and gradual destruction of its structure owing to destructive processes. These processes are connected with the recrystallization of hydrates of the new formation in time, and with gradual hydration of fluidized ashes which contain ß-CaSO₄, with the formation of monosulfatic hydrosulfoaluminium of calcium at late stages of hydration (in the case of not sufficient quantity of Portland cement).

At the same time, for fluidized ash contents of 45-65% the structure is characterized by a stable set of strength at all stages of hardening. It is caused by features of processes of structurization and phase composition of new formations, as well as by the presence of an optimum quantity of Portland cement. Thus hydration of ashes takes place with primary creation of low lime hydrosilicates of calcium, ettringite and solid solutions, and is accelerated in time and shows volume stability.

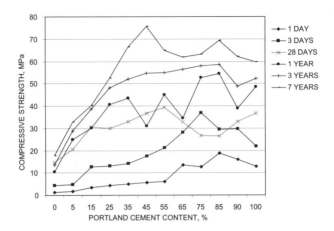

Figure 2 Dependence of compressive strength for ash-cement compositions
after hardening in standard conditions for 7 years

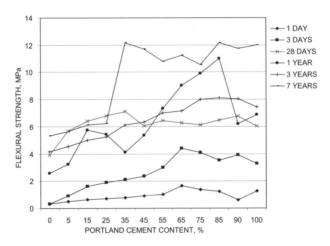

Figure 3 Dependence of flexural strength for ash-cement compositions
after hardening in standard conditions for 7 years

For prevention of the recrystallization processes of hydrosulfo-aluminium phases in the monosulfatic form and of the development of destructive processes in the generated cement stone, the specified systems have been modified by addition of 5% silica fume, used for replacement of ashes [6]. Strength characteristics of binder compositions and respective concretes are presented in Tables 2 and 3.

The results collected for a sufficiently long period (7 years) confirm the formation of a high-strength artificial stone on the basis of ash-cement compositions, modified by silica additives, and that its strength remained stable in time.

Table 1   Composition and technical characteristics of ash-cement mixes investigated

MIX COMPOSITION, mass %	1	2	3	4	5	6	7	8	9	10	11	12
PC M350	-	5	15	25	35	45	55	65	75	85	90	100
Fly ash	100	95	85	75	65	55	45	35	25	15	10	-
					Time of set, hours-min.							
Initial	1-21	0-30	0-25	0-29	0-25	0-30	0-36	0-44	0-40	4-50	3-55	3-20
Final	2-34	1-05	0-42	0-49	0-44	1-05	1-16	1-39	1-20	5-47	5-00	4-15
Cone flow, mm	160	160	160	160	160	160	160	160	160	160	160	160

Table 2   Strength development of ash-cement binder substances obtained with the use of fluidized bed combustion fly ash

No	INITIAL RAW COMPONENTS			COMPRESSIVE STRENGTH, MPa, BENDING STRENGTH, MPa					
	cement	ash	silica fume	1 day	3 days	7 days	28 days	365 days	7 years
1	55	45	-	6.08 / 1.02	21.18 / 2.99	41.25 / 3.33	39.25 / 6.44	45.0 / 7.32	65.52 / 11.72
2	55	40	5	4.88 / 2.49	7.93 / 2.45	39.5 / 7.47	53.75 / 8.78	45.0 / 8.57	68.02 / 12.72

Table 3 Strength development of concretes based on the developed binder compositions

No	BINDER COMPOSITION, %	STRENGTH OF CONCRETE BASED ON THE DEVELOPED BINDER COMPOSITIONS, MPa						
		1 day	7 days	14 days	28 days	360 days	3 years	7 years
1	Portland cement: 55, ashes: 40, silica fume: 5	24.50	60.25	61.50	62.38	64.75	71.20	78.40
2	Ash: 100	1.75	13.50	15.75	28.18	33.25	30.00	27.18
3	Portland cement: 100	16.75	31.25	43.75	47.75	42.25	47.00	52.00

Structural features of hydration products which provide stability for the material in time are given in Figures 4-8.

In accordance with X-ray and thermal analysis data at ages of hardening (7 days) the composition of new the formation includes portlandite $Ca(OH)_2$ (d = 0.193; 0.262; 0.311; 0.493 nm), ettringite (d = 0.498; 0.324; 0.277; 0.261; 0.194 nm), and C-S-H (1) gel (d = 0.307 nm), (see Figure 4, curve 1).

After 1 year hardening of the specimens (according to XRD data), the composition of the new formations is represented by hydrogehlenite C_2ASH_8 (d = 0.261; 0.249; 0,.244; 0.238; 0.236; 0.212; 0.207; 0.161 nm) and calcite $CaCO_3$ (d = 0.385; 0.303; 0.249; 0.227; 0.208; 0.191 nm). Also, the presence of scawtite $Ca_4Si_4O_{14}·2H_2O·CaCO_3$ (d = 0.599 0.355; 0.303; 0.200 nm) and low lime hydrogarnets such as $C_3AS_{0.7}H_{4.6}$ (d = 0.355; 0.311; 0.301; 0.245; 0.228 nm) and $C_3AS_{1.2-1.4}H_{3.6-3.2}$ (d = 0.272 – 0.273 nm) are possible. On the X-ray diffractogram, portlandite $Ca(OH)_2$ (d = 0.263 0.193; 0.169 nm) and ettringite (d = 0.973; 0.561; 0.479 nm)) also are detected (see Figure 4, curve 1). These products are based on the possible formation of a hydrosulphosilicate like hydroellestadite $Ca_{10}(SiO_4)\cdot3(SO_4)\cdot4(OH)_2$ (d = 0.424; 0.331; 0.302; 0.276; 0.269; 0.189; 0.180; 0.169 nm), similar in composition to calcium chondrodite, including $(SO_4)_4^{2-}$ ions. In the composition of new formations, the existence of a hydrosulphosilicate like phase $Ca_6(Si(OH)_6)_3\cdot(SO_4)_3\cdot24H_2O$ is also possible.

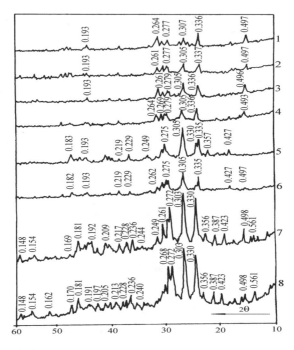

Figure 4 X-ray diffractograms of compositions containing 45% ash and 55% Portland cement (curves 1, 3, 5, 7); 40% ash, 55% Portland cement and 5% silica fume (curves 2, 4, 6, 8) after hardening for 7 (1, 2), 28 (3, 4), 90 (5, 6) and 360 days (7, 8) in standard conditions

As the obtained data testify, the introduction of highly-dispersive silica additives in the composition of binder systems promotes the stabilization of hydrosulfo-aluminate phases in time. Due to introduction of the latter in products of hardening binding substances, ettringite and hydrogarnet type solid solutions $3CaOAl_2O_3\cdot1.6SiO_2\cdot2.8H_2O$ (d=0.314; 0.270; 0.233; 0.197; 0.167; 0.161 nm) are formed, whose presence is proven by microanalysis results (see Figure 6-8) and explains the stable strength characteristics of artificial stone in time. Synthesis of hydrogarnets in hydration products of ash-cement-sulfate binder systems modified by silica fume, is also accompanied by the formation of hydrosilicate, which contains silica anions, similar to epistilbite $(Ca_6(Si(OH)_6)_3\cdot(SO_4)_3\cdot24H_2O)$ (d=0.584; 0.399; 0.369; 0.354 nm). According to electron microscopy results (see Figures 6-8), strength of artificial stone, authentically, is attributed to the crystallization of low lime hydrosilicate of calcium on primarily formed needle crystals of ettringite.

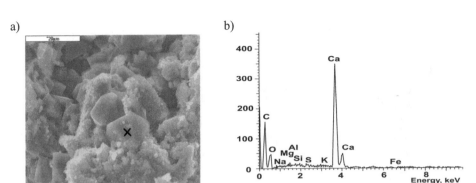

Figure 5   Electron photomicrograph (a) of specimen and results of microanalysis for
(b) separate phases (after 7 days of storage in standard conditions) on the basis of
cement-ash binder system of composition 40% fluidized bed combustion fly ash,
55% Portland cement and 5% silica fume

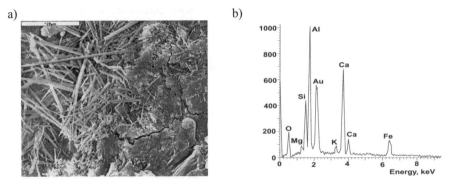

Figure 6   Electron photomicrograph (a) of specimen and results of microanalysis for
(b) separate phases (after 90 days of storage in standard conditions) on the basis of
cement-ash binder system of composition 40% fluidized bed combustion fly ash,
55% Portland cement and 5% silica fume

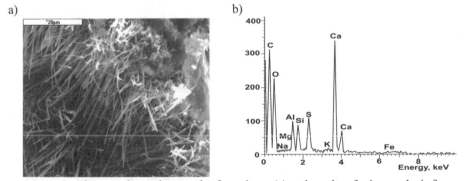

Figure 7   Electron photomicrograph of specimen (a) and results of microanalysis for
(b) separate phases (after 1 year of storage in standard conditions) on the basis of
cement-ash binder system of composition 40% fluidized bed combustion fly ash,
55% Portland cement and 5% silica fume

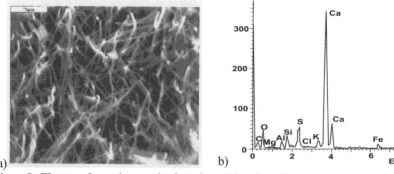

Figure 8   Electron photomicrograph of specimen (a) and results of microanalysis for (b) separate phases (after 3 years of storage in standard conditions) on the basis of cement-ash binder system of composition 40% fluidized bed combustion fly ash, 55% Portland cement and 5% silica fume

On the other hand, some changes take place in the chemical composition of ettringite and related solid solution formations which contain silica-anion complexes. It testifies the ability of the formed artificial stone to structurally adapt to different service conditions [7] that the formation of solid solutions opens an opportunity of some phase changes and a chemical compound of the new formations, depending on service conditions, without deterioration of physical and mechanical characteristics of the obtained artificial stone.

CONCLUSIONS

The developed concretes are expedient for using in underground parts of buildings and constructions where sulfate aggression is present, in the bottom layers of road covering at construction of roads, for hydraulic engineering construction and to prepare special concrete.

REFERENCES

1. GLAVIND M, MATHIESEN D AND NIELSEN C V, Sustainable Concrete Structures: A Win-Win Situation for Industry and Society, Achieving Sustainability in Construction: Proc. International Congress 'Global Construction: Ultimate Concrete Opportunities', Dundee, UK, 2005, pp. 1-14.

2. HARRISON J, The Role of Concrete Post Kyoto: The Unrecognized Sequestration Solution, Achieving Sustainability in Construction: Proceedings of the International Congress 'Global Construction: Ultimate Concrete Opportunities', Dundee, UK, 2005. pp. 426-438.

3. NAIK T R, Sustainability of Cement and Concrete Industries, Proceedings of the International Congress 'Global Construction: Ultimate Concrete Opportunities', Dundee, UK, 2005, pp. 141-150.

4. REINER M, RENS K L AND RAMASWAMI A, Green Buildings and Fly Ash Concrete – The Commerce City, Colorado Project, Proceedings of the International Congress 'Global Construction: Ultimate Concrete Opportunities', Dundee, UK, 2005, pp. 111-118.

5. MEHTA P K, Role of Fly Ash in Sustainable Development, Concrete, Fly Ash and the Environment Proceedings, 1998, pp. 13-25.

6. PUSHKAROVA K K AND DOMOSLAWSKY W, Features of Processes Hydration and Hardening of Binding Compositions based on fluidized fly ash, Proc. of Seventh NCB International Seminar on Cement and Building Materials, New Delhi, India 2000, XL, pp. 125-134.

7. CHERNYAVSKIJ V L, Concrete adaptation, Dnepropetrovsk, Novaya idealogia, 2002, 116 p.

THE INFLUENCE OF CEMENT TYPE AND QUANTITY ON EARLY AUTOGENOUS DEFORMATION OF HSC

D Saje

University of Ljubljana

B M Saje

Ministry of Defence

Slovenia

ABSTRACT. In the paper the time development of shrinkage of nine different high strength concretes, from the beginning of concrete hardening until the age of 2 days, has been presented and analysed. The autogenous shrinkage of these concretes was measured under isothermal ambient conditions. In order to be able to measure early autogenous concrete shrinkage, a special measuring procedure was developed, which makes it possible to measure changes in the specimen length from the very beginning of concrete hardening. Concretes made using different cement types and quantities were compared. Based on an analysis of the experimental results of the shrinkage measurements, it was determined that both the types and the quantities of the different cements significantly influenced the time development of high strength concrete shrinkage during initial concrete hardening.

Keywords: High-strength concrete, High-performance concrete, Silica fume, Cement, Water-binder ratio, Autogenous shrinkage.

Drago Saje, Ph.D., is teaching assistant at the Faculty of Civil and Geodetic Engineering, University of Ljubljana. His main research activities are the mechanical and rheological properties of high performance concrete and the design of concrete and timber structures.

Barbara Mihaela Saje, M.Sc., is building supervisor at the Ministry of Defence of the Republic of Slovenia. Her research activities mainly concern the design and safety of concrete structures, mechanical and rheological properties of ordinary and high-performance concrete at normal and elevated temperatures, nonlinear analysis of reinforced concrete structures, as well as fire resistance of concrete structures.

INTRODUCTION

After concrete has been mixed, the cement hydration process begins. This is followed, depending on the curing conditions, by volumetric changes in the form of shrinkage and swelling. If an uninterrupted water supply is available to all pores of the cement paste during the hydration process, concrete swelling occurs. When water movement into or out of the cement paste is prevented, the concrete shrinks.

If concrete shrinkage due to the carbonisation, plastic shrinkage and temperature shrinkage of concrete is excluded, then concrete shrinkage can be split into chemical and autogenous shrinkage, as well as shrinkage due to drying. Concrete shrinkage due to drying is the consequence of water evaporating from the concrete into the surroundings, whereas autogenous concrete shrinkage, also known as hydration shrinkage, is the result of a self-desiccation process within the pores of the hardened cement paste, i.e. due to the consumption of water during the process of cement hydration [1]. High contents of cement or cementitious materials in high-strength concretes increase the relative volume of the paste in the concrete mix, which leads to increased concrete shrinkage. In the case of high strength concretes, finely ground Portland cement with a large specific surface area (500 m^2/kg) is also frequently used, as well as ordinary Portland cement. The consequence of such a high specific surface area is increased reactivity, and a more intense chemical reaction between the cement and the water. Autogenous shrinkage, which is the consequence of self-desiccation in the pores of the hardened cement paste, is in fact the shrinkage of a sealed concrete specimen, where an impermeable foil or a surface coating prevents moisture exchange between the specimen and the surroundings.

The conditions of concrete shrinkage inside thick concrete elements and very massive structures, such as river dams, where there is practically no humidity exchange between the concrete and the surroundings, are simulated by means of such specimens. Chemical shrinkage of the cement paste is the result of a decrease in the cement paste volume, which occurs due to the chemical binding of water during the cement hydration process. The resulting volume of the products formed in the reaction between the water and the cement is smaller than the volume of the input cement and water together. The main reason for the decreased volume of the cement paste, compared with the volume of the water and the cement, is the higher density of the chemically bound water compared to the free water of the mix before binding [2]. Chemical shrinkage of the cement paste is the reason for autogenous concrete shrinkage.

As hydration proceeds, water is taken out of the system to form hydration products. The pore volume generated during the chemical shrinkage is filled with air. The shrinkage-induced pore volume further increases as hydration continues. An increase in the pore volume is associated with a decrease in the gas pressure. For a closed system, the pressure in the empty pore space will decrease in order to establish thermodynamic equilibrium. A reduction in this pressure implicitly affects the relative humidity in the pore space. A lower relative humidity also affects the thickness of the adsorption layer in the pore wall area. The thermodynamic equilibrium in the pore system requires an increase in the surface tension of the adsorption layer. This process continues as long as hydration takes place. The stress in the adsorption layer causes deformations which are restrained by the effective stiffness of the microstructure. In the initial period of the hardening process, when the modulus of elasticity is still fairly low, the stresses in the adsorption layer may cause large external deformations, called autogenous shrinkage.

In the chemical reaction between the cement and the water, heat is released in the concrete, which raises its temperature and causes it to deform. The size of the concrete temperature change depends on the conditions of the surroundings and on the concrete's composition. The difference between the temperature inside the concrete and the temperature at its surface causes the differential stresses that can cause cracks in the concrete.

The degree of concrete deformation due to temperature changes in the concrete depends on the coefficient of thermal expansion. In the initial hardening phase the concrete is still viscous. Fresh concrete, where the water phase prevails, has a larger coefficient of thermal expansion than that of later hardened concrete. Since it is quite difficult to define the coefficient of thermal expansion of fresh high-strength concrete due to the simultaneous concrete deformation caused by temperature changes and due to autogenous shrinkage. When defining the size of temperature deformations of fresh concrete, researchers take into account a co-efficient of thermal expansion of hardened concrete amounting to $\alpha_T = 10^{-5}/°C$ [3, 4].

EXPERIMENTAL

Materials and mix proportions

When preparing the experimental concrete mixes, coarse stone aggregate, with a maximum size of 16 mm, was used, together with fine sand. The aggregate consisted of washed crushed limestone aggregate from the separation plant at Kresnice, and the fine sand came from a sand-pit at Moravče.

When preparing the concrete mixes, three different types of cement were used: the normal clinker cements CEM II/A-S 42.5R and CEM I 52.5R, and a cement with a low heat of hydration: CEM I 42.5LH, all three of which are produced at the cement-works at Anhovo. The composition of the clinker in these cements is shown in Table 1.

In order to achieve higher strengths, part of the cement was replaced with the mineral additive silica fume, in the dry state. In the case of concrete mixes with low water-binder ratios, adequate workability was achieved by adding a superplasticizer. The new-generation liquid polycarboxylate superplasticizer "Cementol Zeta S" was used, as well as the superplasticizer "Antikorodin". According to its chemical composition Cementol Zeta S is a polycarboxylate, whereas Antikorodin is a powder mixture of silica fume and a superplasticizer which, according to its chemical composition, is a sulphonated naphthalene-formaldehyde condensate. Both of these additives are manufactured by the factory of chemical products, TKK, of Srpenica. The compositions of the experimental concrete mixes are presented in Table 2.

Table 1   Miineral composition of the types of cement clinker used (Bogue)

SHARE OF CLINKER MINERALS					
		Clinker minerals			
Type of cement	Type of cement clinker	C_3S	C_2S	C_3A	C_4AF
CEM II/A-S 42.5R CEM I 52.5R	Normal clinker	64%	15%	9%	9%
CEM I 42.5LH	Clinker with a low heat of hydration	34%	46%	1.5%	15%

Table 2   Properties of the investigated concretes

Mix designation	Type of superplasticizer	Fine aggregate 0-4, kg/m³	Coarse aggregate 4-16, kg/m³	Type of cement	Quantity of binder, kg/m³	Silica fume content, % of binder	Water/binder ratio	$f_{cm,28day}$, MPa
312	(1)	1080	719	(X)	500	10	0.28	89.0
731	(1)	1023	682	(X)	600	10	0.23	90.3
1600	(1)	1133	755	(X)	400	10	0.36	81.4
1620	(2)	1134	755	(X)	400	-	0.40	68.9
1620-1	(2)	1135	756	(Y)	400	-	0.40	74.9
1620-3	(2)	1144	762	(Z)	400	-	0.40	56.9
16210	(2)	1130	752	(X)	400	10	0.40	80.6
16210-1	(2)	1130	752	(Y)	400	10	0.40	85.0
16210-3	(2)	1138	758	(Z)	400	10	0.40	74.0

Key:
(1)   sulphonated naphthalene-formaldehyde condensate
(2)   polycarboxylate
(X)   CEM II/A-S 42.5R (Blaine = 355 m²/kg, containing 15% blast furnance slag)
(Y)   CEM I 52.5R (Blaine = 440 m²/kg)
(Z)   CEM I 42.5LH (Blaine = 367 m²/kg)

The common denominator of all the concrete mixes was the same workability of the mix, based on the slump of the fresh concrete mix. According to the European standard EN 206 [5], the selected workability has the designation S4, which corresponds to a slump within the range 160 - 210 mm, with a permissible deviation of ± 30 mm.

Measuring high-strength concrete shrinkage

Autogenous concrete shrinkage was measured using sealed specimens with the dimensions 100 mm × 100 mm × 400 mm. The specimens used for measuring autogenous shrinkage were sealed by means of polyethylene sheets from the very beginning of the casting of the concrete. A Teflon sheet was inserted between the specimen and the mould in such a way that friction between the specimen and the base of the mould was reduced to a minimum. After the specimen had been prepared in the laboratory in the above described way, it was put in a chamber with a constant humidity of 70% and a temperature of 20°C. The temperature measurements were then begun.

The concrete temperature was measured in the middle of the specimen, using a thermo-couple. As soon as the concrete hardened sufficiently for it to be possible to open the sides of the mould without damaging the concrete, i.e. when the temperature in the concrete specimen started to increase, the electronic measuring of the deformations started. The beginning of the temperature rise in the concrete was assumed to occur at the same time as the beginning of the autogenous concrete shrinkage.

The longitudinal strain in the prism in the initial phase of the concrete shrinkage, i.e. during the period of rapid change, was measured by two inductive strain transducers with a resolution of 0.0001 mm.

For each type of concrete at least three specimens were used to measure the concrete shrinkage.

RESULTS AND ANALYSIS OF THE MEASUREMENTS OF THE AUTOGENOUS SHRINKAGE OF HIGH-STRENGTH CONCRETES

Depending on the type of concrete mix, the process of shrinkage of the investigated concretes started 2.5 to 10 hours after mixing. At this age the concrete is normally not yet exposed to drying, since it is put in a mould and cured, or else covered by foil that prevents water evaporation. It can therefore be stated with certainty that the shrinkage or deformation of concrete at this early age consists of autogenous shrinkage and temperature-related deformations only.

The electronically measured deformations represent the total concrete deformations. The difference between the total deformations and the temperature deformations is the autogenous concrete shrinkage. Since it was not possible to find values of the coefficient of linear thermal expansion of fresh and hardening concrete in the literature, and it is known that, due to the prevailing water phase, this coefficient is larger than the coefficient of linear thermal expansion of hardened concrete [6], the size of the temperature deformations was estimated on the basis of the measured temperature of the specimen by means of the author's measured coefficient of linear thermal expansion of fresh concrete $\alpha_{T1} = 1.48 \cdot 10^{-5}$ /°C [7], and the well-known coefficient of linear thermal expansion of hardened concrete $\alpha_{T2} = 1.0 \cdot 10^{-5}$ /°C. From the time when the concrete temperature started to increase, and up until the time when the specimen temperature under isothermal conditions of the surroundings once again achieved the same temperature as that of the surroundings, a linear distribution of the coefficient of thermal expansion of concrete, from the initial coefficient value for fresh concrete up to the coefficient value for hardened concrete, was assumed.

In the text which follows, an analysis of the time distribution of the shrinkage of nine different concrete mixes is presented. This analysis includes the influence of the cement type and quantity.

Influence of the binder quantity on the time distribution of concrete shrinkage

Figure 1 shows the distribution of the total deformations, the distribution of temperature, and the autogenous shrinkage of the high-strength concrete mixes 731, 312 in 1600. The concretes 731, 312 and 1600 contained 600, 500 and 400 kg of binder per m^3 of concrete, and their water-binder ratios were 0.23, 0.28 and 0.36. In all these concretes 10% of the binder was made up of silica fume. The concrete temperature increased fastest in the case of the concrete 731, followed by the concretes 312 and 1600. The concrete 731 also achieved the highest temperature, and the time at this highest achieved temperature value was, in the case of this concrete, the shortest. The temperature increase in the concrete due to the cement hydration process in the concrete 731 was 6.5°C, whereas in the case of the concrete 312 it amounted to 4.8°C, and in the case of the concrete 1600 to 4.4°C.

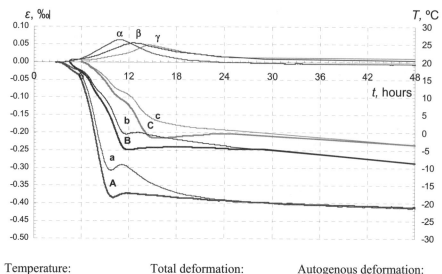

Temperature:

α … concrete 731

β … concrete 312

γ … concrete 1600

Total deformation:

a … concrete 731

b … concrete 312

c … concrete 1600

Autogenous deformation:

A … concrete 731

B … concrete 312

C … concrete 1600

Figure 1   Shrinkage of high-strength concretes 731, 312 and 1600 in the first 2 days of ageing

In the concrete 731 autogenous shrinkage started just as the temperature began to rise, sooner than in the case of the concretes 312 and 1600, and the shrinkage of this concrete was also the largest. At a concrete age of 24 hours the autogenous shrinkage of the concrete 312 was 1.22 times greater then that observed in the case of the concrete 1600. When compared to the shrinkage of concrete 1600 the shrinkage of the concrete 731 was, at the same concrete age, as much as 1.94 times larger.

A larger quantity of binder in the concrete accelerates the cement hydration process. This process is accelerated due to the larger quantity of released heat per unit of concrete than in the case of concrete with a smaller quantity of binder, and at the same time the rapidity of the chemical reaction between the cement and the water is also affected, from the heat transfer aspect, by the quantity of aggregate in the concrete. Less aggregate in the concrete means that, in order to ensure thermal equilibrium, less heat has to be transferred from the cement paste, which is the source of heat in the concrete, into the aggregate. As more heat remains in the cement paste, the latter has a higher temperature, so that the cement hydration process is more intense. Since, in concrete mixes containing larger quantities of cement, the cement hydration is more intense, the consumption of water in the concrete is faster, leading to larger autogenous shrinkage. However, since the concrete 731 had, besides a larger quantity of binder also the smallest water-binder ratio (i.e. it had the least quantity of water per quantity of cement), the process of self-desiccation in the pores of the hardened cement paste was the most intense, causing large autogenous shrinkage.

The larger autogenous shrinkage in the concretes containing larger quantities of binder can also be attributed to the smaller quantity of aggregate in these concretes. Less aggregate in the concrete means a smaller resistance to the shrinkage of the cement paste.

Influence of cement type on the time distribution of concrete shrinkage

Figures 2 and 3 show the influence of cement type on the time distribution of shrinkage in the concretes 16210, 16210-1 and 16210-3, as well as in the concretes 1620, 1620-1 and 1620-3, during the first two days of their ageing. All these concretes had a water-binder ratio of 0.40 and 400 kg of binder per m³ of concrete. The concretes 1620, 1620-1 and 1620-3 did not contain silica fume, whereas the remaining concretes had 10 % of silica fume in the binder.

The concretes 1620 and 16210, as well as the concretes 1620-1 and 16210-1, were made of cement with the same proportions of minerals in the cement clinker. The used cements differ mainly according to the cement grain size. The cement CEM II/A-S 42.5R in the concretes 1620 and 16210 included slag. Since autogenous concrete shrinkage is, in the early stages, the consequence of cement hydration, the autogenous shrinkage of the concretes 1620-1 and 16210-1 at an age of 1 day was 1.71 and 1.19-times larger than the autogenous shrinkage of the concretes 1620 and 16210, mainly because of the smaller grain size of the cement CEM I 52.5R. Fine grained cement densifies the microstructure of the hardened cement paste more than coarser cement. The pores in the developing hardened cement paste are thus smaller, and the autogenous shrinkage is larger. The reason for the slower hydration of the coarser cement CEM II/A-S 42.5R is, as can be seen from the temperature distribution in the tested concretes in Figures 2 and 3, also the presence of slag in this cement. Due to the slag there is less cement clinker than in the cement CEM I 52.5R, without pozzolanic additions. Since there is less cement, during the cement hydration less heat is released, the cement paste temperature is lower, and the cement hydration process is slower.

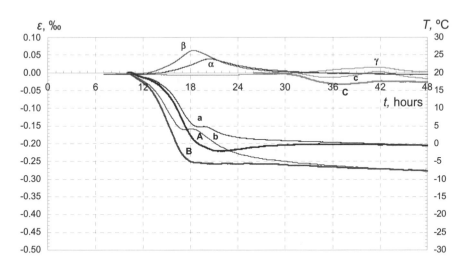

Temperature:	Total deformation:	Autogenous deformation:
α ... concrete 16210	a ... concrete 16210	A ... concrete 16210
β ... concrete 16210-1	b ... concrete 16210-1	B ... concrete 16210-1
γ ... concrete 16210-3	c ... concrete 16210-3	C ... concrete 16210-3

Figure 2   Shrinkage of high-strength concretes 16210, 16210-1 and 16210-3
in the first 2 days of ageing

The cement clinker of the cement CEM I 42.5LH in the concretes 1620-3 and 16210-3 had a different ratio of minerals compared to the cements CEM II/A-S 42.5R and CEM I 52.5R, which were used in the concretes 1620 and 16210, as well as 1620-1 and 16210-1 (Table 1).

The smaller proportion of alite in the slowly binding cement CEM I 42.5LH (1620-3 and 16210-3) compared to the rapidly binding cements (1620 and 16210, as well as 1620-1 and 16210-1) slows down the cement hydration process, and at the same time reduces the autogenous shrinkage of the concrete.

For alite hydration more water is consumed than for belite hydration [8]. Due to the smaller water consumption during the hydration of cement containing a smaller share of alite and a larger share of belite, autogenous shrinkage is considerably less than in the case of the concretes 1620 and 16210, as well as the concretes 1620-1 and 16210-1, which contain cement with a larger share of alite and a smaller share of belite.

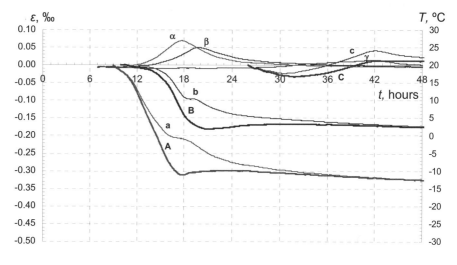

Temperature:
α ... concrete 1620-1
β ... concrete 1620
γ ... concrete 1620-3

Total deformation:
a ... concrete 1620-1
b ... concrete 1620
c ... concrete 1620-3

Autogenous deformation:
A ... concrete 1620-1
B ... concrete 1620
C ... concrete 1620-3

Figure 3   Shrinkage of high-strength concretes 1620-1, 1620 and 1620-3
in the first 2 days of ageing

CONCLUSIONS

Autogenous concrete shrinkage is the consequence of self-desiccation inside the pores of the hardened cement paste, and in high-strength concretes it represents an important part of the total concrete shrinkage. A large part of this shrinkage occurs in the initial stage of rapid concrete hardening.

The results of the analysis of the influence of binder quantity on autogenous concrete deformations showed that when the quantity of binder is increased the amount of autogenous shrinkage increases.

The amount of autogenous concrete shrinkage is also influenced by the type of cement used in the concrete. Concretes made using cement with a smaller share of alite and a larger share of belite, such as concretes made using cement with a lower heat of hydration, show less autogenous shrinkage then that which occurs in the case of concretes made of cement containing ordinary clinker, with more alite and less belite.

REFERENCES

1. PERSSON B, Self–desiccation and its importance in concrete technology. Materials and Structures. Vol. 30, 1997. pp. 293–305.

2. FÉDÉRATION INTERNATIONALE DU BÉTON (FIB): Structural concrete, Volume 1, Textbook on Behaviour, Design and Performace, Updated knowledge of the CEB/FIP Model Code 1990, 1999.

3. WEISS W J, BORICHEVSKY B B, SHAH S P, The influence of a shrinkage reducing admixture on the early-age shrinkage behaviour of high performance concrete, Utilization of High Strength/High Performance Concrete, Proceedings of the 5th International Symposium, Vol. 2, Sandefjord, 1999, pp. 1339–1350.

4. YANG Y AND SATO R, Separation of autogenous shrinkage from shrinkage of high strength concrete under drying, Utilization of High Strength/High Performance Concrete, Proceedings of the 5th International Symposium, Vol. 2, Sandefjord, 1999, pp. 1351-1360.

5. EUROPEAN COMMITTEE FOR STANDARDIZATION: Concrete — Part 1: Specification, performance, production and conformity, European Standard, EN 206-1, 2000.

6. SELLEVOLD E J AND BJONTEGAARD Ø, Thermal dilation – autogenous shrinkage: How to separate?, Autogenous Shrinkage of Concrete, E&FN Spon, 1999, pp. 245–256.

7. SAJE D, Compressive Strength and Shrinkage of High Strength Concrete, Ph.D. Thesis, Faculty of Civil and Geodetic Engineering, University of Ljubljana, 2001.

8. NEVILLE A M, Properties of concrete, Longman, London, 1995, 844 p.

IMPROVING PACKING DENSITY OF CEMENT MATRIX BY ADDING MINERAL ADMIXTURES

A K H Kwan

H H C Wong

University of Hong Kong

China

ABSTRACT. Mineral admixtures are essential ingredients for the production of high-performance concrete. Regarding the contribution of mineral admixtures, there is a common postulation that their addition could improve the packing density of the cement matrix. However, due to the lack of an acceptable method of measuring the packing density of cementitious materials, there has been little experimental evidence supporting this postulation and so far there is no direct method for maximising the packing density of cementitious materials. Herein, a new method, called the wet packing method, is proposed and using this method, the packing densities of double- and triple-blended cementitious materials incorporating different amounts of pulverised fuel ash (PFA) and condensed silica fume (CSF) have been measured. The results show that the packing density of the cementitious materials could be increased by blending cement with either PFA or CSF, and further increased or even maximised by triple blending cement, PFA and CSF at appropriate mix proportions. Based on these results, a ternary packing density diagram has been produced for direct determination of the optimum mix proportions giving maximum packing density.

Keywords: Cementitious materials, Mineral admixtures, Packing density.

Professor Albert K H Kwan is a professor in the Department of Civil Engineering and an associate dean of engineering of The University of Hong Kong. He is also an independent consultant on concrete materials and concrete structures. His specialities include concrete materials and repairs, reinforced and prestressed concrete structures, and tall concrete buildings.

Mr Henry H C Wong is currently a research student in the Department of Civil Engineering, The University of Hong Kong. His research interests are in the areas of packing density of cementitious materials, rheology of cement paste and mortar, self-consolidating concrete and drying shrinkage of concrete.

INTRODUCTION

High-performance concrete (HPC) is a new generation concrete possessing all-round high performance in terms of strength, workability, dimensional stability and durability. As pointed out by Neville [1], "what makes the concrete a high performance one is a very low water/cement ratio". However, since a minimum volume of water must be added to completely fill up the voids between the particles of the ordinary Portland cement (OPC), there exists a lowest limit of water/cement ratio that the cement paste can attain. To further lower the water/cement ratio, it is necessary to reduce the volume of voids between the OPC particles, and to achieve this, mineral admixtures are usually added.

The effect of adding mineral admixtures is similar to that of blending aggregates of different sizes. In concrete mix design, a coarser aggregate is always blended with a finer aggregate. The rationale behind is to reduce the volume of voids between the coarser aggregate by filling the voids with the finer aggregate. Consequently, the amount of cement paste required to fill up the voids of the total aggregate could be reduced. Drawing analogy to blending of aggregates, the voids between the OPC particles could also be filled by adding some finer mineral admixtures, thus resulting in an increase of packing density. As the packing density is increased, less water is required to fill up the voids between the cementitious materials (inclusive of the OPC and any mineral admixtures added). In this way, the water/cementitious materials (W/CM) ratio could be rendered lower than the lowest achievable using OPC alone and a HPC with a very low W/CM ratio could be produced.

Adopting this strategy, some researchers have produced concrete with very high strength. For example, de Larrard and Sedran [2] have maximised the packing density of the cementitious materials by following a certain packing model and successfully lowered the W/CM ratio to as low as 0.14 by weight, thereby achieving concrete strengths of 165 to 236 MPa. Richard and Cheyrezy [3] have even achieved concrete strengths in the order of 200 to 800 MPa by maximising the packing density of all the granular materials in the concrete mix.

Apart from high strength, the benefits of adding mineral admixtures to increase the packing density of the cementitious materials are also reflected in the form of improved workability. For instance, Lange et al. [4] have shown that blending ordinary cement with a finer blast furnace slag based cement could significantly reduce the water demand. Kwan [5] has demonstrated that at a W/CM ratio lower than 0.28 by weight, the addition of condensed silica fume (CSF) could significantly improve the workability of the concrete mix, due to the improvement in packing density of the cementitious materials rendered by the filling effect of CSF. On the other hand, Lee et al. [6] have found that adding pulverised fuel ash (PFA) could improve the fluidity of a cement paste and adding PFA of higher fineness could lead to a larger improvement in fluidity. Similarly, Xie et al. [7] have utilised ultrafine PFA, which has a much higher fineness than ordinary PFA, to produce high-strength, self-consolidating concrete with excellent slump, slump flow and flow velocity; such improved workability was attributed to the higher packing density of the cementitious materials achieved by adding the ultrafine PFA.

Due to the lack of an appropriate test method, currently, it has remained a difficult task to directly measure the increase in packing density rendered by blending OPC with finer mineral admixtures. Therefore, although the above researches have demonstrated the positive effect of adding mineral admixtures on workability, the saying that the enhanced workability may be attributed to the improved packing density resulted from the addition of mineral admixtures is so far just a postulation without any direct experimental evidence. Without a test method for measuring the packing density, it is also difficult to experimentally optimise the mix proportions of cementitious materials so as to maximise the packing density of the cementitious materials for the production of HPC. To resolve this problem, the authors have recently developed a new method, called the wet packing method, to measure the packing density of cementitious materials. The method has been applied to measure the increase in packing density and the maximum packing density that can be achieved by blending OPC with PFA and/or CSF. In this paper, the wet packing method is introduced and the results of packing density measurement using this method are presented so as to provide experimental evidence to support the postulation that the addition of mineral admixtures can significantly increase the packing density. With these results, the optimum mix proportions yielding the maximum packing density can now be determined directly.

MEASURING PACKING DENSITY OF CEMENTITIOUS MATERIALS

Although there exists many test methods for measuring the packing density of solid particles, they have their own limitations when applied to measure the packing density of cementitious materials. In this section, the existing test methods are reviewed and the wet packing method, which can overcome their limitations, is introduced.

Existing Methods for Measuring Packing Density

Existing methods that have been commonly adopted for measuring the packing density of solid particles can be divided into two categories: the dry bulk density method [8] and the standard consistence method [4, 9, 10]. As the name implies, the dry bulk density method measures the bulk density of solid particles under dry condition. Since the cementitious materials are mixed with water in fresh concrete, measuring the packing density under dry condition would lead to negligence of the effects of water and chemical admixtures on the packing behaviour of the cementitious materials. Therefore, employing the dry bulk density method may not yield the true packing density of the cementitious materials in concrete. On the other hand, in the standard consistence method, the water content of the cement paste at a certain arbitrarily chosen consistence level is often assumed to be the same as the minimum voids content of the cementitious materials from which the packing density is calculated. However, such assumption has not been proven so far and as a result the packing density calculated based on the water content may not be correct. In addition, both water voids and air voids may be present between the cementitious particles in the cement paste but only the former type of voids is measured in the standard consistence method. The packing density calculated based on the water content excludes the volume of air voids and thus the packing density tends to be overestimated.

New Method: The Wet Packing Method

With the aim of overcoming the limitations of the existing measurement methods, a new method, called the wet packing method, has been developed. The fundamental principle of the wet packing method is to find the maximum solid concentration of cement paste by directly measuring the wet bulk density of cement paste at various W/CM ratios. At a relatively high W/CM ratio, the cement paste is formed as a suspension with a low solid concentration, whereas at a relatively low W/CM ratio, the water is not enough to fill the voids between the cementitious particles, leading to entrapment of a high volume of air voids and consequently a low solid concentration. In between these two extreme conditions, there exists an optimum W/CM ratio at which the solid concentration is maximum and the volume of air voids is kept at a low level. This maximum solid concentration, which is the maximum volume of cementitious particles that can be held in a unit volume of cement paste, is the packing density of the cementitious materials.

To measure the packing density of cementitious materials, firstly, a cement paste having a high W/CM ratio is produced and the wet bulk density of the paste is measured. Afterwards, the W/CM ratio of the paste is successively reduced and the measurement of wet bulk density is continued. When the W/CM ratio is low enough such that the maximum solid concentration is attained, the packing density, which is equal to the maximum solid concentration, is obtained and the test is completed. To ensure a thorough mixing of cement paste, a special mixing sequence has been developed. Detailed procedures of the wet packing method are as follows (note: all the equipment used comply with BS EN 196: Parts 1-3):

i) Set the W/CM ratio at which the wet packing test is to be carried out. Weigh the required quantities of water, cementitious materials and chemical admixtures (if any), e.g. superplasticiser, and dose each ingredient into a separate container.

ii) If the cementitious materials consist of several different materials blended together, pre-mix the materials in dry for 2 minutes.

iii) Add all the water, half of the cementitious materials and half of the superplasticiser into the mixing bowl and run the mixer at low speed for 3 minutes.

iv) Divide the remaining cementitious materials and superplasticiser into four equal portions. Add the remaining cementitious materials and superplasticiser into the mixing bowl one portion at a time and after each addition run the mixer at low speed for 3 minutes.

v) After mixing, transfer the mixture to a cylindrical mould and fill the mould to excess. Remove the excess with a straight edge and weigh the amount of paste in the mould. The wet bulk density of the paste can be readily calculated through dividing the mass of cement paste inside the mould by the volume of the mould.

vi) Repeat steps (i) to (v) at successively lower W/CM ratios until the maximum solid concentration, i.e. the packing density, is found.

From the test results obtained, the solid concentration may be determined as follows. Let the mass and volume of the paste in the mould be M and V, respectively (the mould used by the authors is of 62 mm diameter $\times$ 60 mm height but any other mould of similar size may also be used). The wet bulk density of the paste is equal to M/V. If the cementitious materials consist of several different materials denoted by α, β, γ and so forth, the solid concentration of the cementitious materials ϕ in the mould may be worked out from the following equation:

$$\phi = \frac{M/V}{\rho_w u_w + \rho_\alpha R_\alpha + \rho_\beta R_\beta + \rho_\gamma R_\gamma} \qquad (1)$$

in which ρ_w is the density of water, ρ_α, ρ_β and ρ_γ are the solid densities of α, β and γ. The packing density of the cementitious materials is the maximum solid concentration obtained by varying the W/CM ratio of the paste.

The wet packing method does not have the limitations of the afore-mentioned existing methods. As the name of the method indicates, in the wet packing method, the packing density is measured under wet condition so that the effects of water and chemical admixtures (if any) on the packing density can be taken into account. Also, instead of determining the water content at a certain arbitrarily chosen consistence level, which does not necessarily correspond to the minimum voids content, employing the wet packing method can directly measure the packing density of cementitious materials based on the results of wet bulk density. Furthermore, since the wet bulk density is measured, both the volumes of water and air voids in cement paste are automatically taken into account. Overestimation of the packing density is therefore prevented.

MATERIALS

Three types of cementitious materials, namely OPC, PFA and CSF, were used in this study. The OPC was a commonly used cement, which had been tested to comply with BS EN 197: 2000. The PFA was a classified ash, which had been tested to comply with BS 3892: Part 1: 1982. The CSF was imported from Norway and according to the supplier, it complied with ASTM C 1240-03. The particle densities of OPC, PFA and CSF had been measured in accordance with BS EN 1097: Part 7: 1999 as 3110 kg/m^3, 2329 kg/m^3 and 2202 kg/m^3, respectively. Their particle size distributions, which were measured using a laser diffraction particle size analyser, are shown in Figure 1.

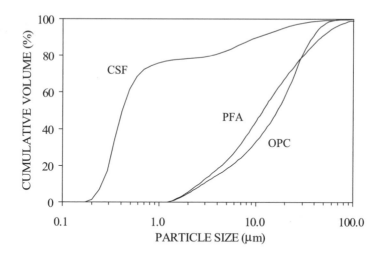

Figure 1 Particle size distributions of OPC, PFA and CSF

The superplasticiser (SP) used was a polycarboxylate ether-based one. It is a liquid-type SP having an active ingredients content (by mass) and a relative density of 20% and 1.03, respectively. The normal dosage of the SP recommended by the supplier, measured in terms of liquid mass, was 0.5 to 3.0 % by mass of the cementitious materials.

Table 1 Mix proportions and measured packing densities of series A, B and C

SERIES	SPECIMEN NUMBER	CEMENTITIOUS MATERIALS (% BY VOLUME)			MEASURED PACKING DENSITY
		OPC	PFA	CSF	
A	A1	100	-	-	0.622
	A2	-	100	-	0.646
	A3	-	-	100	0.397
B	B1	85	15	-	0.637
	B2	70	30	-	0.641
	B3	55	45	-	0.644
	B4	40	60	-	0.643
	B5	25	75	-	0.645
	B6	85	-	15	0.703
	B7	70	-	30	0.726
	B8	55	-	45	0.671
	B9	40	-	60	0.645
	B10	25	-	75	0.590
	B11	-	85	15	0.748
	B12	-	70	30	0.745
	B13	-	55	45	0.689
	B14	-	40	60	0.575
	B15	-	25	75	0.410
C	C1	70	15	15	0.718
	C2	55	30	15	0.731
	C3	40	45	15	0.730
	C4	25	60	15	0.742
	C5	55	15	30	0.736
	C6	40	30	30	0.735
	C7	25	45	30	0.752
	C8	40	15	45	0.665
	C9	25	30	45	0.691
	C10	25	15	60	0.639

MIX PROPORTIONS

In the present study, three comprehensive series of packing density tests (named as Series A, Series B and Series C) were carried out using the newly developed wet packing method. In Series A, all mixes of cementitious materials tested were non-blended, i.e. pure OPC, PFA and CSF. The aim of this series was to find out the reference packing densities of OPC, PFA and CSF under non-blended condition. In Series B, all mixes of cementitious materials tested were double-blended, i.e. OPC + PFA, OPC + CSF and PFA + CSF. These mixes were designed by successively increasing the content of the blending material from 15% to 75% at increments of 15% (by volume). Lastly, in Series C, all mixes of cementitious materials tested were triple-blended, i.e. OPC + PFA + CSF. These mixes were designed by successively increasing the content of each blending material, i.e. either PFA or CSF, at increments of 15% (by volume) until the total content of blending materials, PFA + CSF, reached 75%. The aims of Series B and Series C were to investigate the effects of double and triple blending on the packing density of the cementitious materials and to determine the optimum mix proportions yielding the highest packing density. Altogether, 28 packing density measurements had been carried out. Details of the mix proportions of the three series are listed in Table 1.

Since the cementitious materials have different densities and it is the solid volume of each cementitious material added rather than the mass that is more important, all mixes of cementitious materials were designed on a volumetric basis and the content of each cementitious material in the mix was expressed in terms of its percentage solid volume content, i.e. the ratio of the solid volume of the cementitious material to the total solid volume of cementitious materials in the mix expressed as a percentage. Likewise, the SP dosage was calculated in terms of the mass of SP per total solid volume of cementitious materials in the mix. The dosage of SP was set as the maximum dosage recommended by the supplier, which was 3% by mass of the cement content. Based on this, the dosage of SP in terms of the mass of SP per total solid volume of cementitious materials was calculated as 93.3 kg/m^3. This dosage of SP was added to all the mixes in the three series.

RESULTS AND DICUSSIONS

Packing Densities of Non-Blended Cementitious Materials

By employing the wet packing method, the packing densities of pure OPC, pure PFA and pure CSF were obtained, as tabulated in the last column of Table 1. It is seen that the packing densities of the OPC and PFA were respectively 0.622 and 0.646, each of which is approximately the same as the packing density of mono-size spheres packed in random condition [11]. However, the packing density of the CSF was only 0.397, which is an unexpected low result.

Comparing the results of the OPC and the PFA, it is noted that the packing density of the PFA was higher than that of the OPC. This result is attributed mainly to the different shapes of the OPC and PFA particles. In general, OPC particles are angular and PFA particles are spherical. Since angular particles pack to a lower density and spherical particles pack to a higher density [12, 13], it is likely that the higher angularity of the OPC particles has caused the OPC to have a lower packing density.

Regarding the extremely low packing density of the CSF, it is believed that the low value was due to flocculation of the CSF particles arising from the formation of bonding bridge between the surfaces of the CSF particles and the ether in the copolymer side-chains of SP [14]. In addition, due to the very fines size of the CSF particles, the inter-particle forces were probably strong enough to cause the CSF particles to pack into a loose configuration [15]. Owing to the interaction with the SP and the strong inter-particle forces, the CSF only manifested a very low packing density.

Packing Densities of Double-Blended Mixes

After measuring the packing densities of pure OPC, pure PFA and pure CSF, the effect of double blending on packing density can be evaluated. Using the wet packing method, the packing densities of the double-blended mixes were measured, as presented in the last column of Table 1. Based on these results, the variation of the packing density with the mix proportions of the two cementitious materials in the mix is plotted in Figure 2. From the figure, it can be seen that blending OPC with mineral admixtures can lead to a significant increase in packing density. These results, therefore, provide experimental evidence to verify the postulation that the packing density can be increased by adding mineral admixtures.

On adding PFA to OPC, the packing density of the mix increased more or less linearly with the PFA content. The increase in packing density when the PFA content was increased from 0% to 100% was about 4%. As can be seen from Figure 1, the particle size distributions of both the OPC and the PFA are very much the same. Under such circumstance, the PFA particles may not be able to exhibit a significant filling effect so as to increase the packing density. The improvement in packing density when blending OPC with PFA, therefore, is mainly attributed to the beneficial particle shape effect arising from the spherical shape of the PFA particles.

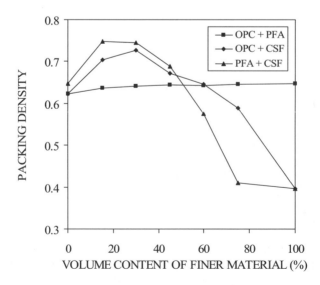

Figure 2   Packing densities of double-blended mixes

On the other hand, the results showed that blending OPC with CSF could markedly improve the packing density. With just 15% CSF added, the packing density already increased from 0.622 to 0.703. With 30% CSF added, the packing density further increased to 0.726, which is 17% higher than the packing density of pure OPC. As both CSF and PFA are spherical in shape, the CSF should also be contributing to the increase in packing density through its beneficial particle shape effect. However, the packing density improvement on adding CSF was much larger than that on adding PFA. This may be attributed to the better filling effect exhibited by the ultrafine CSF particles. As the CSF is about 100 times finer than the OPC, the CSF particles should be able to fit themselves very well inside the voids of the OPC particles, leading to a large increase in particle packing. Based on these results, 30% CSF content should be the optimum dosage yielding the highest packing density. When the CSF content was increased to beyond 30%, the packing density started to decrease until it reached a fairly low value of 0.397 at 100% CSF. This result may be caused by the overfilling of the ultrafine CSF particles between the voids of OPC particles, which pushed the OPC particles apart and destroyed the optimal packing configuration.

Like the blending of OPC with CSF, the blending of PFA with CSF could also markedly improve the packing density of the mix. As both the CSF and PFA are spherical in shape, the marked improvement in packing density should be due solely to the filling effect of the ultrafine CSF. It should be noted that these mixes are not for real application as a mixture containing only mineral admixtures is not practicable. These results are for the purposes of comparing with the case of blending OPC with CSF and for constructing the ternary packing density diagram presented in the next section.

Packing Densities of Triple-Blended Mixes

To further illustrate the effect of blending OPC with the mineral admixtures, packing density measurements have been carried out on the triple-blended mixes. The measured results are presented in the last column of Table 1. Based on these results, the variation of the packing density with the mix proportions of the three cementitious materials is plotted in the form of a ternary packing density diagram in Figure 3.

The results in the ternary packing density diagram illustrate very clearly that regardless of double blending or triple blending, the addition of mineral admixtures to OPC could lead to a very significant increase in packing density. It further provides experimental evidence to verify the common postulation that the packing density of the cement matrix can be increased by adding mineral admixtures.

The results showed that at a fixed CSF content of up to 45%, the packing density increased slightly with the PFA content. This trend was similar to the case of OPC + PFA with no CSF added, indicating that the PFA was contributing a similar particle shape effect in the triple-blended mixes as in the case of double-blended mixes. On the other hand, at a fixed PFA content of up to 45%, the packing density increased markedly with the CSF content until a maximum packing density was reached at about 30% CSF, after which the packing density started decreasing. Hence, regardless of the PFA content, the optimum CSF content for maximum packing density was around 30%. This variation of the packing density with the CSF content was also similar to the case of OPC + CSF, implying that the filling effect and particle shape effect of the CSF in the triple-blended mixes were the same as in the case of double-blended mixes.

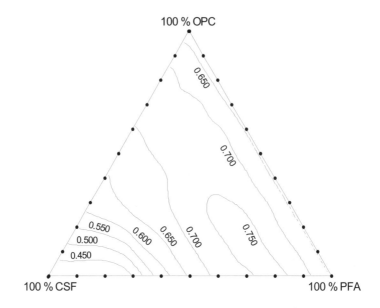

Figure 3   Ternary packing density diagram for triple-blended mixes

Comparing the packing densities of the double-blended and triple-blended mixes, it can be seen that triple blending was superior to double blending in terms of packing density improvement. At the lower right side of the central region in the ternary packing density diagram, there existed a plateau over which the packing density would be higher than 0.750, indicating that the packing density could be increased by more than 20% through triple blending at appropriate mix proportions. The highest packing density achieved was 0.752. Such a high packing density could not be achieved if the OPC was only double blended with either PFA or CSF. Thus, in order to yield a higher packing density, it is more beneficial to have a triple-blended mix than a double-blended mix.

Packing Density, Excess Water and Water Film Thickness

The long-term performance of concrete, such as strength, dimensional stability and durability, are largely influenced by the W/CM ratio of the concrete mix. As mentioned in the introduction, to enhance the performance of concrete, it is necessary to lower the W/CM ratio of the concrete mix. However, there is a minimum W/CM ratio the concrete mix could attain so that the water content is enough to fill up the voids between the solid particles in the concrete mix and to render a desirable workability level for concreting. In this sense, workability is the crucial factor determining the W/CM ratio, which in turn influences the long-term performance of the concrete.

To lower the W/CM ratio but at the same time maintain a sufficient workability level, the packing density of the cementitious materials has to be increased. As demonstrated in the present study, the packing density of cementitious materials can be increased by blending OPC with mineral admixtures. The benefit of so doing is to provide room for lowering the

W/CM ratio without scarifying the workability, which can be explained by using the concepts of excess water and water film thickness. By increasing the packing density of the cementitious materials through adding finer mineral admixtures to the coarser OPC, the former particles would fill into the gaps between the latter particles, thereby displacing the water originally trapped inside the gaps. The displaced water, which is called the excess water in the present context, would be squeezed out and act as water films coating the particle surfaces of the cementitious materials. Consequently, the average thickness of water films on the particle surfaces would be increased, thus dilating the particle skeleton and increasing the separations between the particles. As the separations between the particles are increased, the particle interactions that hinder the mixture from flowing, e.g. inter-particle attractions and frictions, would also be reduced and the workability would be increased. This increased workability provides room for lowering the W/CM ratio while maintaining the workability level. In this way, workability is unaffected but the W/CM ratio can be lowered so that the long-term performance of concrete can be enhanced.

As explained above, the lowering of W/CM ratio achieved by improving the packing density is governed by the amount of excess water and the thickness of water film on the surfaces of cementitious materials. Hence, the excess water and the water film are the fundamental factors governing the W/CM ratio and affecting the long-term performance of concrete. Packing density is only one of the major parameters affecting the amount of excess water and thickness of water film. It is obvious that other parameters, such as specific surface areas and particle sizes of the cementitious materials, would also influence the amount of excess water and thickness of water film. Further research in this direction is recommended.

CONCLUSIONS

A new method for measuring the packing density of cementitious materials, called the wet packing method, has been developed. The method has been successfully applied to measure the packing density of pure OPC, pure PFA and pure CSF as well as their double and triple-blended mixtures. It was found that due to the better spherical shape of the PFA particles, the PFA has a higher packing density than the OPC and that due to the special interaction with the SP and large inter-particle forces, the CSF has a very low packing density.

The packing density results of the blended cementitious materials have provided solid proof that blending OPC with PFA and/or CSF could significantly increase the packing density. Owing to the beneficial spherical shape effect of PFA, the addition of PFA could increase the packing density of OPC at constant CSF content. On the other hand, adding optimum amount of CSF into OPC at constant PFA content would markedly increase the packing density of OPC; such results may be attributed to the strong filling effect and beneficial spherical shape effect exhibited by CSF particles. From the ternary packing density diagram constructed based on the test results, it has been found that triple blending is superior to double blending for increasing the packing density of the cement matrix.

Finally, the influence of packing density on the performance of concrete is discussed. It is believed that the packing density would indirectly govern the long-term performance of concrete through determining the W/CM ratio required to achieve certain workability level. Based on the mechanism of lowering the W/CM ratio through increasing the packing density, the concepts of excess water and water film are introduced. These concepts could be the basis of a new direction for future research.

REFERENCES

1. NEVILLE, A M. Properties of Concrete, Longman, London, 1995, 844pp.

2. DE LARRARD, F AND SEDRAN, T. Optimisation of ultra-high-performance concrete by the use of a packing model. Cement and Concrete Research. 1994, Vol. 24. pp 997-1009.

3. RICHARD, P AND CHEYREZY, M. Composition of reactive powder concretes. Cement and Concrete Research. 1995, Vol. 25. pp 1501-1511.

4. LANGE, F, MÖRTEL, H AND RUDERT, V. Dense packing of cement pastes and resulting consequences on mortar properties. Cement and Concrete Research. 1997, Vol. 27. pp 1481-1488.

5. KWAN, A K H. Use of condensed silica fume for making high-strength, self-consolidating concrete. Canadian Journal of Civil Engineering. 2000, Vol. 27. pp 620-627.

6. LEE, S H, KIM, H J, SAKAI, E AND DAIMON, M. Effect of particle size distribution of fly ash-cement system on the fluidity of cement pastes. Cement and Concrete Research. 2003, Vol. 33. pp 763-768.

7. XIE, X J, LIU, B J, YIN, J AND ZHOU, S Q. Optimum mix parameters of high-strength self-compacting concrete with ultrapulverized fly ash. Cement and Concrete Research. 2002, Vol. 32. pp 477-480.

8. BRITISH STANDARDS INSTITUTION. BS EN 1097 Tests for Mechanical and Physical Properties of Aggregates Part 4: Determination of the Voids of Dry Compacted Filler, BSI, London, 1999, 11pp.

9. BRITISH STANDARDS INSTITUTION. BS EN 196 Methods of Testing Cement Part 3: Determination of Setting Time and Soundness, BSI, London, 1995, 8pp.

10. JONES, M R, ZHENG, L AND NEWLANDS, M D. Estimation of the filler content required to minimise voids ratio in concrete. Magazine of Concrete Research. 2003, Vol. 55. pp 193-202.

11. REED, J S. Principles of Ceramics Processing, John Wiley & Sons, New York, 1995, 658pp.

12. ZOU, R P AND YU, A B. Evaluation of the packing characteristics of mono-sized non-spherical particles. Powder Technology. 1996, Vol. 88. pp 71-79.

13. KWAN, A K H AND MORA, C F. Effects of various shape parameters on packing of aggregate particles. Magazine of Concrete Research. 2001, Vol. 53. pp 91-100.

14. SOMASUNDARAN, P AND ZHANG, L. Modification of silica-water interfacial behavior by adsorption of surfactants, polymers, and their mixtures, in: E. PAPIRER (Ed.), Surfactant Science Series Volume 90: Adsorption on Silica Surfaces, Marcel Dekker, New York, 2000, pp 441-462.

15. YU, A B, BRIDGWATER, J AND BURBIDGE, A. On the modeling of the packing of fine particles. Powder Technology. 1997, Vol. 92. pp 185-194.

FLY ASH IN CONCRETE FOR
ENHANCED DURABILITY AND SUSTAINABILITY

L K A Sear

United Kingdom Quality Ash Association

United Kingdom

ABSTRACT. Fly ash has been used for many years in concrete because it is able to offer many technical advantages such as enhanced durability and performance. For example its ability to improve the sulfate resistance, reduce chloride diffusion, prevent alkali silica reaction, give long term strength gain properties and reduce heat generation in cementitious applications is well known. These benefits have been researched by many people with published papers totalling several thousands. However, it is only in recent years it is increasingly recognised that using fly ash in concrete also results in significant environmental and sustainability benefits, simply by replacing virgin aggregates. For example in foamed concrete it acts as a cementitious binder as used in road sub-base hydraulically bound mixtures, by enhancing a structure durability extending its working life, etc. In this way it is able to significantly reduce overall environmental impact and greenhouse gas emissions. This paper will consider the sustainability and environmental benefits of greater utilisation of the material in cementitious applications. It will review how industry has moved towards reducing environmental impacts using fly ash, the standards and new specifications that have enabled recent changes and look at the future for this important and readily available material.

Keywords: Fly ash, Pulverised fuel ash, PFA, Durability, Sustainability.

Dr L K A Sear is the Technical Director of the United Kingdom Quality Ash Association representing the interests of the UK coal fired power station operators. He represents the members of the UKQAA on a number of British and European Standard Committees. He is involved in the steering committees of many research projects ranging from the environmental aspects of PFA/fly ash through to the thaumasite form of sulfate attack. In addition, Lindon gives many presentations and attends exhibitions and conferences relating to the applications of fly ash. Consequently he has a broad knowledge of the use of fly ash in concrete, fill applications, grouting etc.

INTRODUCTION

Fly ash, whether it is coal fired power station ash or volcanic ash, has been used for many years to produce hydraulic cement. In general concrete structures properly made with these pozzolanic materials have proven to be durable without the need for continual remediation and/or replacement. In addition pozzolanic materials reduce the overall environmental impacts of making concrete by reducing the amount of Portland cement required with a given concrete mix. Fly ash from coal fired power stations is a readily available pozzolanic binder, which is not subject to any environmentally expensive processing. But as well as acting as a binder, it is quite possible to use fly ash as an aggregate, replacing natural materials. In whatever way fly ash is added to concrete, in most circumstances as well as environmental and sustainability benefits, there are often significant technical benefits, for example reduced chloride permeability, resistance to alkali silica reaction, enhanced sulfate resistance, etc. In this manner the use of fly ash in concrete and cementitious systems is more sustainable. However, in the future ash quality issues will have to be fully addressed and the introduction of beneficiation processes will become increasingly necessary.

THE PRODUCTION OF FLY ASH

Fly ash is a by-product of the combustion of coal in a power station. Coal contains minerals that were laid down when the coal measures were deposited, many millions of years ago. It is these minerals when burned that form the fly ash. The coal is ground to a fine powder, similar to talcum powder, and burned in the power station furnace at in excess of 1250°C within 2 to 4 seconds. The high temperature coupled with the ash being in a gas stream results in rounded particles of glassy material being formed. These fly ash particles are extracted from the gas stream using electrostatic precipitators as shown in Figure 1 and may be used in a number of ways, primarily within the construction industry.

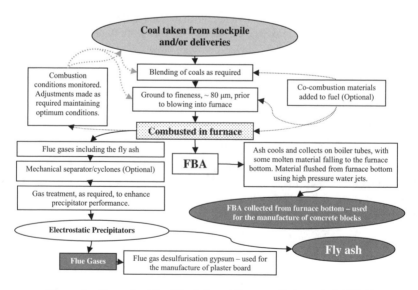

Figure 1   Flow chart for Fly Ash and Furnace Bottom Ash (FBA)
Production at a coal fired power station.

The resulting dry ash will be tested for compliance with the appropriate standard. For example fly ash production to EN450-1:2005, when it is used as a Type II addition is shown in Figure 2. Fineness and Loss On Ignition (LOI) are the common parameters used for control purposes, though some selection based on other criteria such as colour may be used at some power stations. Depending on the requirements of the fly ash, the material may be tested and either accepted for direct sale to the customer, processed to bring the material within the specification or rejected, where it will be sent to the mono fill disposal site found at or near to most power stations. Normally for use in concrete the fly ash would be supplied dry, though conditioned ash may be supplied as filler aggregate to cement, precast concrete and grouting companies.

The use of fly ash in concrete, in whatever form it is supplied, is beneficial to the environment as it reduces the amount of Portland cement and/or replaces virgin aggregates. Fly ash will continue to be produced as pulverised coal fired electricity generation will continue for the foreseeable future [1], though this is felt by many to be dependant on development of clean coal technologies and carbon capture [2]. Coal fired power generation, irrespective of the CO_2 issues, forms the predominate backbone of UK power generation as shown in Figure 3. Even though the stock of coal fired power stations is rather old, modernisation and retrofitting of equipment has kept them operational in many cases well beyond their original design life. With an ever expanding UK population and the apparent lack of an energy policy for a number of years, coal fired generation has proven to be reliable, cost effective and readily available. For this reason ash production has risen in recent years to ~6,000,000 tonnes per annum. However, there are ash quality issues to consider.

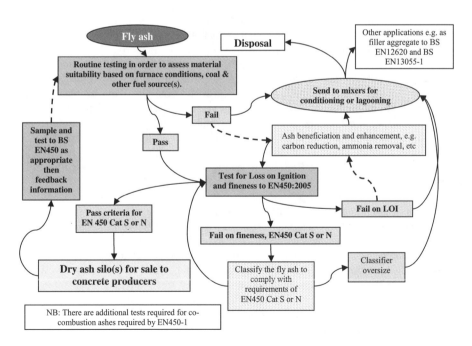

Figure 2   Flow chart for production of Fly ash for use in concrete
as a Type II addition complying with BS EN450:2005.

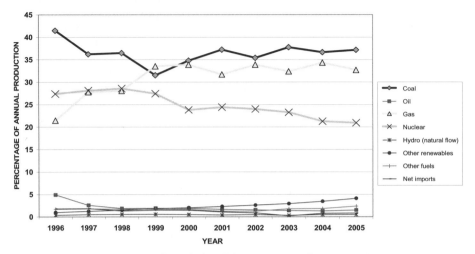

Figure 3   Proportion of Electricity per Annum by Energy Source
Source: DTI Energy Statistics [3] (Dukes 5.4).

Ash Quality Issues

Ash quality is seen by some to have declined in recent years. Loss On Ignition (LOI) of <5% was common place 10 years ago, but in recent years a mean of 6% is more common. For use in concrete, loss on ignition (LOI) is seen as a major issue particularly when air entrainment is required. The UK standard for concrete, BS8500 [4], limits LOI of fly ash as an addition to a maximum of 7%. However, in order to reduce the emissions of nitrous oxides from power stations, the industry has been fitting so called low NOx burners. This has resulted in an increase in loss on ignition within fly ash in recent years [5]. Eventually all coal fired stations will have to fit these burners and this trend has resulted in carbon reduction systems being fitted to some power stations. Within the UK the STI electrostatic system [6] has been fitted to three power stations, which is capable of producing ash with LOI's as low as 2.5% from 10% or higher LOI ash. Other stations are investing in the Rocktron [7] approach for full ash beneficiation.

The injection of ammonia into the furnace gas stream has been another issue in recent years. This is designed to increase precipitator efficiency when using low sulfur coals. Though the amounts of ammonia are extremely small, there have been incidents where concrete made with ammoniated ash has released noticeable amounts of ammonia [8]. The alkali environment in cementitious systems releases ammonia gas and as ammonia levels above 10ppm are easily detectable by man, the issues are one of smell and health and safety rather than having a detrimental effect on the concrete. This problem is not insurmountable and the ammonia can be removed efficiently with the appropriate equipment. One power station in the UK is fitting this equipment to overcome this problem.

It is envisaged in the future that such beneficiation equipment will become more prevalent in order that the fly ash industry can continue to supply the cementitious and concrete markets for years to come.

Where does all the ash go to?

Products from coal fired power stations are used in a variety of construction applications. Many of these applications are cementitious; the largest proportion of the UK fly ash produced going towards making aerated concrete blocks, with concrete, cement manufacture, fill and grouting being the others. In all these applications ash is used because it acts both as inert filler and as a pozzolanic material, enhancing strength and durability.

The market breakdown [9] for 2005 is shown in Figure 4, but note that it excludes the material that is landfilled. Some ~50% of the fly ash produced within the UK is currently being landfilled, which should be considered a wasted resource rather than a waste. Of the fly ash that is disposed most is conditioned, which is where it is mixed with water, typically ~15% and disposed of in mono landfill site, being transported as an aggregate would be. Some is lagooned, i.e. mixed with copious amounts of water and pumped as slurry to settlement lagoons. In either case the fly ash could be recovered and used in concrete, though there is an apparent reduction in ash reactivity with time.

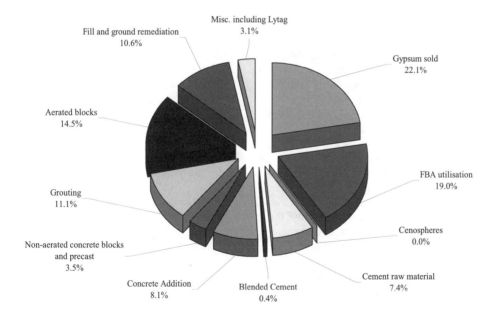

Figure 4   The markets for coal fired power station products.
The proportions of sold material by application – Jan to Dec 2005.

Why do concrete producers take fly ash?

There are many reasons why fly ash is used within the cement and concrete industry. While we will mainly consider the environmental and sustainability aspects in this paper, there are numerous technical benefits imparted to concrete containing fly ash, it is economical, it is available in larger quantities, etc. These benefits have been well documented over the years and the current standards for concrete reflect the technical performance of fly ash concrete.

ENVIRONMENTAL ASPECTS OF FLY ASH IN CONCRETE

Let us consider that fly ash collected from the electrostatic precipitators can either be used or disposed of. Disposal involves the ash being sent to large mixers, similar to concrete pan mixers, where water is added. This is known as conditioning. The resulting earth dry/conditioned ash is loaded on trucks or conveyors to be taken to the disposal site. Mixing the ash with water and transporting it to the landfill site in itself requires energy and has an environmental impact. However, if ash is to be used in concrete it is normally supplied as a dry material in a cement tanker. As there is no energy and water use involved for dry ash the overall environmental impacts are reduced.

The cement industry has become a big user of fly ash both as a kiln feed and for blending into cement. The ash may be added as a minor addition constituent, which can be up to 5% of cement, or within blended cements. These additions comply with BS EN197-1 'Common Cements' and are an effective way for a cement supplier to reduce the overall emissions associated with Portland cement manufacture.

Fly ash for use as a concrete addition at the mixer is normally supplied complying with EN450:2005, where it can be counted as being part of the cement content of the concrete, see BS8500. EN450 describes two basic types of fly ash, Category N and Category S. The differences are described below.

EN450:2005 Category N fly ash

Category N, which we shall call 'normal' fly ash, is fly ash that is taken dry direct from the power station. EN450 imposes a series of quality control requirements on the ash, such as fineness, chemical properties, etc but this material would normally be controlled by a process of selection and rejection based on the various control parameters. As a result Category N fly ash is generally sourced straight from the power station silos without any processing. This could be considered as zero environmental impact at the factory gate from production and even possibly a negative impact if the energy for disposal is taken into account.

EN450:2005 Category S fly ash

Category S, which we shall call 'special' fly ash, is again dry ash from the station. In the majority of cases this is processed to remove the coarser ash particles within the material. Typically this is done with air swept classifiers and the process reduces the water demand and increases the strength of the resulting concrete by removing the misshapen and generally coarser fraction. This requires energy, typically 9.75 kW/h per tonne of Category S fly ash, which equates to ~4.2 kg of CO_2 per tonne of product.

However, in relation to the improvements in reactivity and water demand of the resulting concretes, classification is a positive environmental benefit.

What are the relative impacts of using fly ash?

As we have established, the environmental impacts of producing fly ash are minimal at the factory gate with little or no processing being required. Some users take fly ash as conditioned material for use within concrete. Conditioning is the addition of typically ~15% water to the ash to prevent dust problems. This is carried out in large mixers similar to those for mixing concrete. The resulting damp ash is either sold for other applications or disposed of in mono landfills. It could be said that the act of conditioning the ash adds to the environmental costs of disposal whereas using dry ash to EN450 Category N fly ash in concrete further reduces the overall impact.

Table 1 shows the calculations inherent in comparing a 40 MPa CEM I concrete with mixes containing both 30% and 50% EN450 fly ash additions. A figure of 960 kg/tonne of CO_2 was used for CEM I within these computations. The calculations would be similar if CEM II/B-V (30%) or a CEM IV/B (50%) blended cements as opposed to mixer additions were to be specified. No admixtures are used and a typical UK average figure of 0.43 kg of CO_2 is produced to generate 1 kWh of electricity. It is clear that the classification of fly ash has no significant effect on the overall reduction in CO_2 emissions, as the power consumption is relatively minimal. It is clear the use of fly ash results in significant reductions in overall CO_2 emissions.

Table 1   The impacts of producing a 40 MPa design strength @ 28 days concrete.

ENVIRONMENTAL IMPACT	CEM I ONLY	CAT N	CAT S	CAT N	CAT S
Normal replacement level	0%	30% fly ash		50% fly ash	
Total Cementitious 40 MPa concrete kg/m^3	280	320	310	410	395
Normal extra over of total cementitious required to maintain strength @ 28 days	0%	+15%	+10%	+45%	+40%
Typical CEM I reduction in kg/m^3 for 40 MPa concrete @ 28 days	N/A	-56 kg/m^3	-63 kg/m^3	-75 kg/m^3	-83 kg/m^3
CO_2 reduction achieved	N/A	- 53.76 kg/m^3	- 60.5 kg/m^3	- 72.0 kg/m^3	-79.7 kg/m^3
Electrical energy – CO_2 produced to process material @ 0.430 kg per kW/h	N/A	NIL	9.75/1000 x 93 x 0.43 = 0.39 kg/m^3	NIL	9.75/1000 x 197.5 x 0.43 = 0.83 kg/m^3
Overall reduction in CO_2 emissions per m^3 of concrete produced	N/A	- 54 kg/m^3	- 60 kg/m^3	- 72 kg/m^3	- 79 kg/m^3
Percentage reduction in comparison with CEM I only concrete	0%	-20%	-22%	-27%	-29%

As fly ash is pozzolanic, the reactions are relatively slow in comparison with modern CEM I. There are benefits in specifying strength at 56 days, by which time the pozzolanic reaction will have had some significant contribution to the measured strength. Table 2 repeats the 40 MPa concrete calculation, but based on a 56 day concrete strength.

Table 2   The impacts of producing a 40MPa design strength @ 56 days concrete.

ENVIRONMENTAL IMPACT	CEM I ONLY	CAT N	CAT S	CAT N	CAT S
Normal replacement level	0%	30% fly ash		50% fly ash	
Total Cementitious 40 MPa concrete kg/m^3	265	280	270	375	360
Normal extra over of total cementitious required to maintain strength @ 56 days	0%	+6%	+2%	+40%	+36%
Typical CEM I reduction in kg/m^3 for 40 MPa concrete @ 56 days	N/A	-69 kg/m^3	-76 kg/m^3	-78 kg/m^3	-85 kg/m^3
Overall reduction in CO_2 emissions per m^3 of concrete produced	N/A	- 66 kg/m^3	- 73 kg/m^3	- 75 kg/m^3	- 81 kg/m^3
Percentage reduction in comparison with CEM I only concrete	0%	-26%	-29%	-29%	-32%

Are these the only environmental benefits are there others?

In addition to the obvious benefits of reducing overall CO_2 emissions when replacing Portland cement, fly ash can act as an aggregate, the so called inert filler. To produce 1 tonne of virgin aggregate takes ~21 kg of CO_2, whereas fly ash is CO_2 neutral. Fly ash can comply with both BS EN12620 Aggregates for Concrete [10] and BS EN13055-1 Lightweight Aggregates for Concrete, Mortar and Grout [11] as filler aggregate and therefore it may be used within concrete and grouts to replace virgin aggregate. In concrete mixes this only becomes practicable when either very coarsely graded fine aggregates and/or very low cement contents are being used. For example it is quite possible to design a 10 MPa concrete mix for pumping with a low cementitious content, by adding fly ash as filler aggregate. With such an application it is important to remember that the particle density of fly ash is typically ~2.3 where naturally aggregate is higher ~2.6, i.e. a 13% increase in volume per unit mass. The increased volume has to be allowed for in the mix design.

In grouting, fly ash has proven to be far superior to virgin aggregates for most applications. The inherent round particle shape of fly ash in comparison with many virgin materials reduces the required water content for a given workability and makes the grout easy to pump. The pozzolanic reaction, coupled with the lower water content, gives better strengths with fly ash than the virgin aggregate equivalent. This leads to a reduction in Portland cement content and, due to the reduced particle density of fly ash, less material being required. These differences can be very substantial, with one grouting contract [12] reporting that using fly ash grouts reduces vehicle movements by 40% and material cost by ~50% in comparison with those for Portland cement and virgin sand grout.

Other factors

The above calculations do not take into account transportation of the material to the user. Many environmental profiles use simple assumptions for transport using average travel distances, often ignoring the impacts of shipping, handling etc for imported materials. In the real world transport is a major environmental impact and whether it is fly ash, cement, aggregates, etc a comprehensive environmental assessment has to take transport into account. For example Parrott [13] concluded that transporting the raw materials from the source to the concrete plant and the concrete to the site accounted for ~10% of the environmental impacts of producing the concrete on average. If imported materials are used the additional transport and handling would increase the environmental impacts significantly.

What is clear is that the overall environmental impacts are different depending on circumstances. Issues such as transport distances, imported materials, methods of transportation, application for the product, exposure conditions, etc all have to be assessed on a case by case basis in order to draw sensible and accurate conclusions.

ENHANCED DURABILITY

Fly ash imparts many technical benefits to concrete. These include resistance to the penetration of chlorides reducing corrosion of reinforcing, preventing alkali silica reaction, reducing the heat of hydration and reducing the risk of cracking, etc. These benefits, when fly ash concrete is used in the appropriate applications, can extend the working life of a structure.

Chloride Ingress

The ability of fly ash concrete to reduce the permeability in respect of chlorides is well known with in excess of 480 papers published on the properties of fly ash concretes. The improved performance of fly ash concrete is reflected within BS8500, the UK National specification for concrete. Table 3 is a small extract of the 100 year design life tables for XD3 exposure. This shows the technical benefits of using fly ash, which are cement and combination types IIB-V and IVB-V, and are reflected within the specification as significantly lower strengths, minimum cement contents and maximum water/cement ratios are required when using fly ash in comparison with CEM I.

Table 3   Extract of BS8500 UK National Specification for Concrete.

NOMINAL COVER mm	$55 + \Delta C$	$60 + \Delta C$	$65 + \Delta C$	CEMENT/COMBINATION TYPES
XD3 for 100 year design life	$C45/55^{E)}$ $0.35^{F)}$ 380	$C40/50^{E)}$ 0.40 380	$C35/45^{E)}$ 0.45 360	CEM I, IIA, IIB-S, SRPC
	$C32/40^{E)}$ 0.45 360	C28/35 0.50 340	C25/30 0.50 340	IIB-V, IIIA
	C25/30 0.50 340	C25/30 0.55 320	C25/30 0.55 320	IIIB, IVB-V

The reduction in cementitious content when using fly ash can result in very significant additional reductions in environment impact in some exposure classifications. For example using Table 3 with 60 mm nominal cover using the IIB-V mixes results in a reduction of CO_2 of ~39% and for IVB-V a reduction of 45% in comparison with CEM I.

Alkali Silica Reaction

Similarly with Alkali Silica Reaction (ASR), the addition of fly ash to the concrete significantly reduces the risks of this deleterious reaction occurring. This is recognised within BRE Digest 330 [14] and BS8500, with no distinction being drawn between Category S or N material. At least 25% fly ash is required to enhance the resistance to ASR and higher proportions are required for the most reactive aggregate/cement combinations.

Sulfate Resistance

Fly ash concrete gives increasing sulfate resistance with increasing ash content. In respect of the formation of ettringite, a minimum of 25% fly ash of the total cementitious content is required to give sulfate resistance to concrete as per BRE Special Digest 1 [15] and BS8500. However, for the thaumasite form of attack, recent research shows there are considerable advantages in using 50% fly ash of the total cementitious content with the higher classes of sulfate exposure.

Carbonation

It is continually levelled against PFA concrete that carbonation is greater than for Portland cement, especially with higher proportions of ash, e.g. > 30-55%. In most research organisations it is normal to assess the carbonation of concrete using accelerated testing regimes, by increasing the proportion of CO_2 to which the concrete samples are exposed. While this accelerates the ingress of CO_2, it doesn't reflect the true performance of materials such as pozzolanas as it fails to accelerate the hydration characteristics and the pore blocking of PFA that lower permeability and reduce the accessibility of CO_2 to the concrete. It is generally accepted that concrete of equal 28 day strength has similar carbonation performance irrespective of cementitious type, including fly ash based concrete. It is this premise that is used within BS8500 in respect of the carbonation exposure classes, XC 1 to 3.

REDUCING THE CARBON FOOTPRINT OF CONCRETE FURTHER

The substitution of Portland cement with fly ash reduces overall CO_2 emissions. However, if a structure could be designed and constructed to last for a longer period of time and/or use fewer materials for the same performance criteria, this would automatically reduce the carbon footprint. As well as the above obvious comparisons, fly ash can be used to significantly reduce the carbon footprint of a concrete construction by taking advantage of the durability enhancement possible using such pozzolanic materials. As just one example of best environmental practice, Heathrow Terminal 5 required a 7MPa Tensile Strength concrete for runway construction [16]. They used fly ash to reduce the overall carbon footprint and by judicious design and modern admixtures managed to increase the flexural strength sufficiently to reduce the runway thickness. They have also experimented with 40% fly ash contents, to further reduce environmental impacts. The result is a more durable structure using less material, producing less overall emissions and a reduced carbon footprint.

SUSTAINABILITY

The UK power industry has produced fly ash since the 1950's and a considerable amount of fly ash has been produced over the years. Many hundreds of millions of tonnes of fly ash are no longer accessible as the number of stations has reduced from in excess of 100 to only 18 coal fired stations. The closed power station sites being developed and the ash disposal sites have been reclaimed or developed, usually for industrial purposes and occasionally for housing. However, on the remaining coal fired power station there is some 55,000,000 tonnes of fly ash readily available and a further 60,000,000 tonnes may be accessible if required. Barlow Mound [17], see Figure 5, is an example of a large fly ash stockpile. In addition the combustion of pulverised coal is unlikely to cease in the foreseeable future, for even with carbon sequestration, fly ash will still be produced. The stockpiles of fly ash form a readily available mineral resource for future generations. They would need extracting, screening, drying and possibly grinding or classification for use in concrete, but they could be put to beneficial use. All these technologies already exist and unlike some other secondary materials, there is no need to import fly ash as supply outstrips demand and large quantities of material are available on stock.

Figure 5   Barlow Mound, Drax Power Station.

CONCLUSIONS

Using fly ash in concrete and other cementitious applications can significantly reduce the overall environmental impact by substituting for Portland cement and virgin aggregates. Depending on the application and the exposure conditions, very significant reductions are possible and the enhanced durability and extended lifetime of the resulting structures can lead a further reduced overall environmental impact.

Fly ash is sustainable for the foreseeable future as there are significant amounts of both freshly produced and stockpile fly ash available within the UK that could be beneficially used. There is no need to resort to imported fly ash with appropriate beneficiation.

REFERENCES

1   DTI, Energy White Paper, Our energy future - creating a low carbon economy, The Stationery Office (TSO), Norwich, February 2003.

2   Analysis of Responses to the Energy Review Consultation, DTI Publication 06/1565, June 2006.

3   Energy statistics are available from DTI at www.dti.gov.uk/energy/statistics/index.html.

4   BS 8500-1:2002 Concrete. Complementary British Standard to BS EN 206-1. Method of specifying and guidance for the specifier, BSI, Chiswick, London
BS 8500-2:2002 Concrete. Complementary British Standard to BS EN 206-1. Specification for constituent materials and concrete, BSI, Chiswick, London.

5   VAN DEN BERG J W, Effect of low NOx Technologies on fly ash quality, Vliegasunie b.v., The Netherlands, ECOBA conference, 1998.

6   GASIOROWSKI, S A & BITTNER J D, High Quality Fly Ash Production and Fuel Recovery by Separation Technologies' Beneficiation Processes, AshTech 2006 Proceedings, Birmingham, UK, May 2006.

7   SMALLEY N, MICHAEL P & WATT J H, Implementation of a RockTron PFA Beneficiation Process Plant, AshTech 2006 Proceedings, Birmingham, UK, May 2006.

8   Investigation of ammonia adsorption on fly ash, EPRI Interim Report, December 1998, EPRI, Palo Alto, CA, USA.

9   UKQAA annual production statistics, available from www.ukqaa.org.uk.

10 BS EN 12620:2002 Aggregates for concrete, BSI, London.

11 BS EN 13055-1:2002 Lightweight aggregates. Lightweight aggregates for concrete, mortar and grout.

12 UKQAA report on Bosty Lane, Walsall grouting contract, January 2004.

13 PARROTT L, Environmental impacts of transport relative to those of concrete, CIA Environmental Factsheet, UK Concrete Industry Alliance, November 1999.

14 Alkali silica reaction in concrete, BRE Digest 330, 2004 Edition, BRE, Watford, UK.

15 Concrete in aggressive ground, BRE Special Digest 1, 3rd edition, 2005, BRE, Watford, UK.

16 SPARKS A, New mix reduces concrete on T5, New Civil Engineer 03/04/2003.

17 Further information about Barlow Mound is available on www.draxpower.com.

THE PROPERTIES OF HYDRAULICALLY CLASSIFIED ULTRA FINE ASH IN CONCRETE

T L Robl

R F Rathbone

J G Groppo

University of Kentucky

United States of America

ABSTRACT. An improved Class F fly ash was produced from an ash pond (lagoon) at a Kentucky (USA) power plant where it had been in storage for a minimum of 10 years. This ash was beneficiated via hydraulic classification which removed most of the coarse ash, carbon, magnetite and cenospheres and produced an ultra fine (D_{50} of 4 to 7.4 μm) ash (UFA). This material was used at a substitution level of 5% to 35% for Type I ordinary Portland cement (OPC) in concrete and compared with the performance of commercially available Class F fly ash. Water reduction was found to be 6% to 7% at substitution levels of 20-35%. The development of compressive strength was more rapid than with commercially available Class F materials, reaching 100% of OPC control at 14 to 21 days, versus 56 to 112 days for commercial Class F ash. Rapid chloride permeability testing suggested that the UFA concrete would exhibit a superior resistance to chloride penetration compared to OPC concrete or standard Class F ash concrete. Although the UFA addition required a higher dosage of air entraining admixture to achieve a target air content, the increase was caused by the particle fineness rather than the presence of unburned carbon. These data indicate that controlling the particle size distribution can result in consistent air entrainment performance.

Keywords: Ultra fine fly ash, Compressive strength, Concrete, Ponded ash, Hydraulic classification.

Dr Thomas L Robl, is an Associate Director of the University of Kentucky, Center for Applied Energy Research. He directs the Environmental and Coal Technologies Research group.

Mr Robert Rathbone, is a Senior Scientist at the University of Kentucky, Center for Applied Energy Research. He is in charge of the Center's concrete and pozzolan laboratories.

Dr Jack Groppo, is a Senior Engineer at the University of Kentucky, Center for Applied Energy Research. He is in charge of the Center's mineral processing research.

INTRODUCTION

Sustainable approaches to concrete construction include the use of recycled material, the minimization of energy and the extension of the life of the structure. This makes the utilization of high performance fly ash particularly attractive. One common approach to producing a high performance ash is to enhance its fineness by selectively removing its coarser sizes. This has commonly been done commercially using air classifiers.

Another method is to use a hydraulic classifier or a hydro-cyclone for this purpose. This has the drawback of producing a wet product. However, this approach is ideal for processing ash that has been landfilled, either in abandoned settling basins (ponds or lagoons) or in controlled fills. This ash is abundant, has no in situ value and under the right circumstance can even be a nuisance.

Technology has been developed at the University of Kentucky, Center for Applied Energy Research [1] to hydraulically process low calcium (Class F) landfilled ash. The technology has been demonstrated at a pilot scale at the Ghent Power Station in Carroll County, Kentucky [2] and Coleman Station in Hancock County, Kentucky. As described in previous papers at this symposium [3-5], the technology can produce a low loss on ignition (LOI) pozzolan of exceptional fineness. One of the products of the technology is ultra fine ash (UFA). This material has an average particle diameter (D_{50}) of 3.5 to 7 μm and has almost no material retained on a 325 mesh (45 μm) screen. This may be compared to run of plant Class F ash with D_{50} more typically in the 20 μm to 25 μm or even higher and 15% or more retained on a 325 mesh screen.

MATERIALS AND TESTS

Materials

The materials for this study were produced in a pilot plant located at the University of Kentucky and in the field demonstration unit described above. These materials were produced from an ash pond at the Ghent Power Station and Coleman Power Station. The ponded Ghent ash used in this study has been stored for at least 10 years, whereas the Coleman ash was collected from an active pond.

Our approach to the recovery of the UFA consisted of a primary classification step to remove most of the coarse, (i.e. +100 mesh) materials, followed by the addition of a dispersant and then secondary classification to remove most of the ash greater than about 10 μm in diameter. These materials were compared with ASTM Class F fly ash that is commercially available in the area.

Methods

Chemical characterization

Major elemental analysis of fly ash and Portland cement was completed by X-ray fluorescence. Crystalline phases on ground samples of cement and ash were identified by X-ray powder diffraction. Carbon content was determined using a Leco CHN-2000, whilst loss on ignition (LOI) was determined by heating the samples in air at 750°C for 4 hours. Type I ordinary Portland cement (OPC) was used in the study and was sealed in mylar bags to minimize hydration during storage.

Aggregates characterization

Concrete aggregates were analyzed for gradation (ASTM C 136), specific gravity and absorption (C 127 and C 128), and organic impurities (C 40) [6].

Mortar testing

The evaluation of fly ash performance in mortar was completed by following guidelines described in ASTM C 311, with the exception of the mortar air test in which Vinsol resin was not used as the air entraining admixture (AEA). The AEAs used were: abietic acid-sodium salt (5.9% solution), dodecyl sulfate (5% solution), Master Builders MicroAir, and Grace Darex II. For brevity, only the MicroAir data is reported herein.

Concrete testing

ASTM protocol was employed for concrete testing, in both the fresh and hardened state. The following standard methods were used [7]:

C 39: Compressive strength of cylindrical concrete specimens

C 138: Unit weight, yield, and air content of concrete

C 143: Slump of concrete

C 403: Time of setting

C 192: Making and curing concrete test specimens

C 231: Air content of concrete by the pressure method

C 1202: Rapid chloride permeability

Pozzolanic reactivity

The ability of fly ash to react with soluble calcium hydroxide and form supplementary cementitious material was determined in solution using a British Standard test (BS EN 196-5) to evaluate pozzolana cements [8]. This method uses 20 g of cement and 80 ml of deionized water. However, for fly ash pastes, 20 g of cement and 5 g of fly ash was mixed with 80 ml water and the solution analyzed after 8 days.

The calcium hydroxide content of solid cement/fly ash paste was determined using thermogravimetric analysis. During heating of cement paste there is a progressive loss of surface and bound water as the sample is heated to 700°C (at 20°C/min). Superimposed on this weight loss profile is a deflection that is caused by the relatively rapid loss of water from $Ca(OH)_2$. The concentration of solid $Ca(OH)_2$ was calculated from these weight loss data [9].

RESULTS AND DISCUSSION

Cement and Fly Ash Analysis

The major oxide compositions of the Portland cement (CEMEX Type I) and fly ash samples are presented in Table 1.

Table 1   Chemical composition of cement and fly ash, values in mass % on a dry basis

	SiO$_2$	Al$_2$O$_3$	Fe$_2$O$_3$	CaO	MgO	Na$_2$O	K$_2$O	LOI	C
OPC	21.9	5.4	2.5	59.9	2.4	0.2	0.7	2.00	Nd
Class F fly ash	45.1	19.1	18.8	3.3	0.9	0.4	2.4	1.63	0.64
Ghent UFA[1]	50.6	26.9	4.9	1.2	0.9	0.4	2.7	2.65	1.50
Coleman UFA	52.5	25.6	10.0	1.4	0.8	0.2	2.4	3.85	Nd

[1]UFA is the notation for the ultrafine fly ash products

Pozzolanic Reactivity

Calcium hydroxide in solution

Based on the initial mortar data it was tentatively concluded that the ultrafine fly ash exhibited a very high degree of pozzolanic reactivity and produced supplementary cementitious material at a rapid rate. However, the concrete strength data indicated that, while the UFA produced higher compressive strengths than the standard Class F fly ash, it may not be as chemically reactive as the mortar data indicated. The ability of the ultrafine ash to consume Ca(OH)$_2$ from solution as part of the pozzolanic reaction was therefore determined using the British Standard test described in the Experimental Section. The Ghent ultrafine ash paste (cement + fly ash + water) was found to plot in the unsaturated field and substantially lowered the calcium content of the solution when compared with the control paste (cement + water) or the standard Class F fly ash paste. The data indicate that the ultrafine fly ash was indeed performing as a highly reactive pozzolan.

Calcium hydroxide in hardened cement paste

Although the British Standard test is very useful for comparison of different pozzolans, its high water : cement ratio (i.e. 4.0) precludes it from giving quantitative Ca(OH)$_2$ concentrations that exist in mortar and concrete, which typically have water:cement ratios on the order 0.5 and lower. The determination of Ca(OH)$_2$ in paste, mortar and concrete can be attempted in two ways: high pressure extraction of pore solution followed by determination of Ca^{2+} and OH$^-$ concentration of the solution, and direct determination of solid Ca(OH)$_2$ concentration. As described in the Experimental Section, the latter method was selected for this study. Data were obtained for cement pastes containing 20% Class F fly ash and Ghent UFA fly ash. A separate paste was prepared that contained 20% finely ground inert quartz powder as an additional control sample. Buy subtracting the Ca(OH)$_2$ of the reactive materials from the quartz and control a measure of the Ca(OH)$_2$ reacted is obtained. From these data the concentration of Ca(OH)$_2$ was quantified over the course of curing the paste samples. Within the first day, the Ca(OH)$_2$ concentration increased rapidly, whereupon it increased at a much slower rate for the control paste, and ostensibly reached a plateau for the sample containing fly ash and quartz powder. This trend has been observed in previous studies [10]. Of interest is the overall lower concentration of Ca(OH)$_2$ in the Class F and Ghent UFA pastes, with the latter containing the lowest concentrations. During the first three days of the tests the UFA paste absorbed Ca(OH)$_2$ at a rate of about 4 times that of the Class F ash. At the end of 22 days, a total of 0.23 g/g Ca(OH)$_2$ per gram of UFA versus 0.09 g/g of Ca(OH)$_2$ per gram of Class F ash. These data provide additional evidence that the ultrafine ash is behaving as a reactive pozzolan.

Mortar Testing

The addition of fly ash, as expected, resulted in a reduction in the quantity of water required to achieve constant flow (i.e. workability). The addition of Ghent and Coleman UFA gave higher water reductions than standard Class F fly ash (Table 2). This "filler effect" is caused by the small size of the spherical particles [11].

Table 2   Water demand in mortar to achieve constant flow

	MEDIAN DIAMETER, μm	WATER DEMAND, %
Control	18	100.0
20% Class F Ash	20	97.0
20% Coleman UFA	8	94.0
20% Ghent UFA	4	91.1
40% Ghent UFA	5	86.7

[1] Underflow ash is an intermediate-sized material from the ash classifier

The ultrafine classified ash products required substantially greater dosages of air entraining admixture (AEA) to achieve 18% air than did the control mortar or even mortar prepared with "standard" fly ash; approximately twice as much AEA was required for the Ghent UFA and Coleman UFA mortars than for the control or standard fly ash mortars (Figures 1 and 2). The higher demand for an AEA to achieve a constant air content is typical when fly ash is used as a cement replacement. This is generally believed to be a result of adsorption of the AEA from solution by unburned carbon in the fly ash. To test this, the Coleman UFA and Ghent UFA were ignited in air at 750°C to remove the carbon, and the mortar air test conducted on the ignited product. The results shown in Figure 3 indicate that for the Ghent UFA, removing the carbon had no impact on the air entraining properties. This suggests that the high degree of fineness probably caused the lower entrained air content. However, for the Coleman product, removal of the carbon caused a significant increase in entrained air (Figure 3).

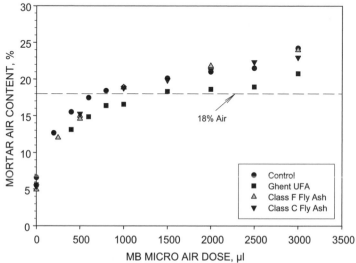

Figure 1   Entrained air content of mortar prepared with standard fly ash and Ghent UFA

The reason for the different behaviour is probably related to the length of storage of the fly ash in the impoundment. The Ghent ash has been stored in the impoundment for greater than 10 years, whereas the Coleman UFA was obtained from a portion of the ash impoundment where ash was being disposed by the power plant. As a consequence, carbon in the Ghent impounded ash has oxidized to a greater extent than the Coleman ash carbon. It has been known for many years that storage of activated carbons can lower their adsorption capacity for organic molecules due to formation of oxygenated functional groups on the carbon surface [12]. Thus, it is reasonable to infer that fly ash carbon can also become "deactivated" during prolonged storage in an impoundment.

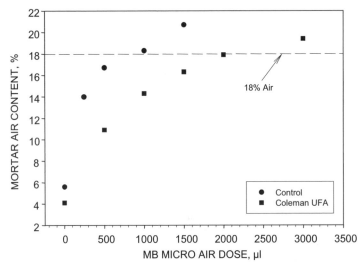

Figure 2   Entrained air content of mortar prepared with Coleman classified product ash

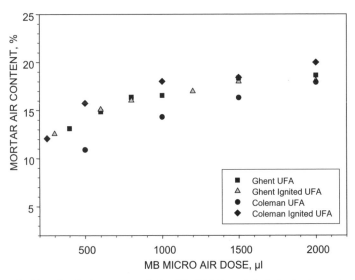

Figure 3   Entrained air content of classified UFA and their ignited counterparts

Compressive Strength

The Ghent UFA that was produced in the pilot plant at the Ghent power plant performed well in mortar as regards its compressive strength. As shown in Figure 4, at a 20% replacement rate for Portland cement the Ghent and Coleman UFA mortars exceeded the strength of the control mortar within 7-10 days, mainly due to the water reduction, and achieved approximately 120-125% of control by 56 days. Compared with standard Class F ("F ash") the UFA exhibited superior performance at all stages of curing. These data support earlier work at the CAER which showed that the ultrafine ash is quite reactive in mortar and produces additional cementitious material within the mortar at a rapid rate, although ash from the pilot plant did not perform as well as that from the laboratory process. Some of the difference is explained by the lower water reduction produced by the pilot plant ash (7% versus 9% for the laboratory UFA), which had a significant influence on early strength. This is probably a consequence of an overall larger particle size of the pilot plant UFA which slows the pozzolanic reaction rate.

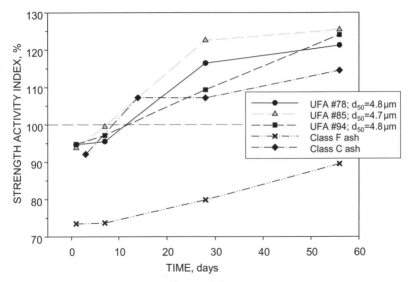

Figure 4   Strength activity index of fly ash mortar.
Fly ash replacement was 20% of Portland cement by weight

Concrete Testing

In this phase of the project there were two series of concrete tests. The first series examined the performance of the Ghent UFA in three different concrete mixes: low strength (21 MPa), moderate strength (41 MPa) and high strength (55 MPa). Fly ash was used at a 20% cement replacement level. The proportions of these concrete mixes are provided in Table 3. The second series of tests examined the use of Ghent UFA in addition to standard F ash; that is, using the UFA as a blend with the F ash. The mix proportions for the F/UFA blend concretes are shown in Table 4. The concrete was prepared at constant slump and no chemical admixtures were used.

Table 3   Concrete mix proportions

MIX, kg/m^3	AGGREGATE		CEMENT	FLY ASH	WATER	W:CM
	COARSE	FINE				
21	1061	881	282	0	170	0.604
21 F	1062	881	226	56	165	0.586
21 UFA	1058	879	226	56	167	0.591
41	1008	837	360	0	182	0.505
41 F	1007	836	288	72	176	0.490
41 UFA	1099	744	288	72	171	0.475
55	1047	866	415	0	133	0.330
55 F	1046	865	332	83	127	0.314
55 UFA	1050	868	332	83	124	0.307

Table 4   Concrete mix proportions for Class F/UFA blends

MIX, kg/m^3	AGGREGATE		CEMENT	FLY ASH	WATER	W:CM
	COARSE	FINE				
Control	1008	837	360	0	182	0.505
20% F	1007	836	288	72	177	0.491
20% F, 5% UFA	1019	846	270	90	174	0.482
25% F	1005	834	270	90	176	0.490
25% F, 5% UFA	1011	838	252	108	171	0.475

The water reduction achieved in concrete was not as high as in mortar: the highest water reduction obtained with Ghent UFA was approximately 6.5%. This is approximately ½ the water reduction reported in the literature for UFA [13] and only marginally better than that obtained with F ash (3-4% reduction). The reason for the lower water reduction is probably the presence of numerous non-spherical, irregular-shaped fly ash particles as well as spherical particles that were "welded" together. Unlike other high performance fly ash, the Ghent UFA contains a larger amount of the non-spherical and welded particles (Figure 5) that would probably not provide the same degree of water reduction as spherical dispersed particles.

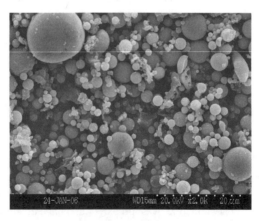

Figure 5   SEM photograph of Ghent UFA

Blending the UFA with F ash resulted in a significant reduction in water required to achieve a constant slump (Table 5), which typically will improve concrete compressive strength. The added water reduction is caused by the UFA improving the grading of the fly ash in the smaller size range, thus providing an additional or extended filler effect in the concrete paste. Set time measurements indicated that the replacement of Portland cement with 20% Ghent UFA delayed the concrete set by about 13 minutes.

Table 5   Water reduction in concrete

CONCRETE MIX	% REPLACEMENT OF CEMENT		WATER REDUCTION, % OF CONTROL
	Class F FA	UFA	
Control	0	0	0.0
20F	20	0	2.8
20F/5UFA	20	5	4.5
25F	25	0	3.0
25F/5UFA	25	5	5.9

Compressive Strength

Low strength (21 MPa) mix

The Ghent UFA exhibited similar early strengths (i.e. 1 and 7 days) as the F ash in the low strength concrete mix, due to the similar water reduction (2-3% for both fly ashes). However, the UFA concrete gained strength at a faster rate after 7 days curing than the F ash concrete (Figure 6) due to the former's smaller particle size distribution. The UFA also produced a significantly higher long-term strength.

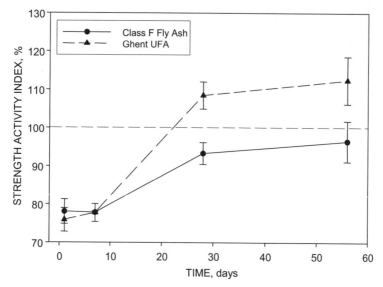

Figure 6   Strength activity index of 21 MPa fly ash concrete

Moderate strength (41 MPa) mix

The UFA performance, relative to the standard F ash, in the moderate strength mix was similar to the low strength mix (Figure 7), except that the UFA concrete achieved a long term SAI of approximately 120% instead of 110%. The early age SAI was higher for the UFA concrete because the water reduction with UFA was 6.3%, while the F ash produced a water reduction of only 2.8%.

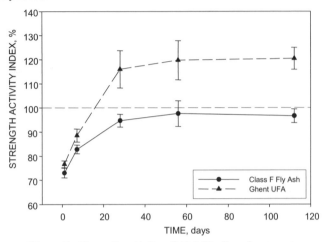

Figure 7   Strength activity of 41 MPa fly ash concrete

High strength (55 MPa) mix

In the high strength concrete, the Ghent UFA produced a water reduction of 7%, while the Class F ash produced a water reduction of 4.8%. Despite the higher water reduction, the Ghent UFA concrete had lower early strengths than the F ash concrete. However, after approximately 7 days curing the Ghent UFA concrete strength surpassed that of the F ash concrete (Figure 8) although the difference in strength between the two was not as substantial as with the lower strength mixes.

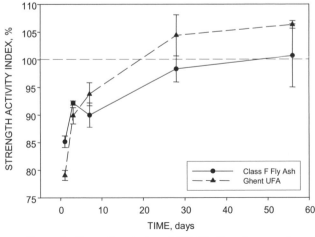

Figure 8   Strength activity of 55 MPa fly ash concrete

Compressive strength of F ash/UFA blends

An overarching goal of this project from its inception has been to increase the use of fly ash in concrete. This goal may not always be realized if the UFA product is used instead of standard F ash. However, because the UFA produces superior water reduction and compressive strength it could potentially replace an additional 5-10% of Portland cement in concrete that already contains Class F fly ash at typical replacement rates of 20-25%, but without lowering the early compressive strengths. To test this, the concrete mixes listed in Table 3 were prepared and tested, and the results are shown in Table 6. It is evident that an addition of 5% UFA (as a % of Portland cement replaced) resulted in a significant decrease in early strengths, despite a greater water reduction in the F/UFA blends. In fact, the additional 5% of UFA resulted in a lower SAI compared with the F ash concrete until the 56 day test, after which the strengths of the F/UFA blend concretes were higher.

Table 6   Strength Activity Index of F Ash/UFA blend concretes

CONCRETE MIX	STRENGTH ACTIVITY INDEX, % OF CONTROL				
	1 day	7 day	28 day	56 day	112 day
Control	100	100	100	100	100
20F	71.8	81.6	94.0	97.9	98.6
20F/5UFA	65.5	75.3	88.6	105	102
25F	60.1	73.8	89.3	94.0	94.0
25F/5UFA	53.7	69.1	87.5	94.4	98.0

Chloride Permeability

Although unconfined compressive strength is the most often reported property of concrete, durability of the product is of equal, or even sometimes greater, importance. The rapid chloride permeability (RCP) test provides a fairly rapid qualitative means of assessing the permeability of concrete. Table 7 provides the results of the RCP test for a variety of concretes after 60 days curing. The data indicate that the ultrafine ash substantially lowered the total charge passed and thus, by inference, would produce a less permeable, more durable concrete. Unlike the compressive strength results, addition of 5% UFA to an F ash concrete substantially lowered the chloride permeability. It is inferred from the data that even small additions of Ghent UFA could result in improvements in concrete durability.

Table 7. Rapid Chloride Permeability

CONCRETE MIX	RCP, coulombs	ASTM CHLORIDE PERMEABILITY RATING
6000	2418	Moderate
6000 F	1200	Low
6000 UFA 20%*	426	Very Low
6000 UFA 40%*	150	Very Low
3000 Control	3423	Moderate
3000 F Ash	2237	Moderate
3000 UFA	1194	Low
20F	2194	Moderate
20F/5UFA	1580	Low
25F	1627	Low
25F/5UFA	1040	Low

* UFA 20% refers to a cement replacement of 20%; UFA 40% is a cement replacement of 40%

REFERENCES

1. ROBL T L, GROPPO J G, TAPP K AND McCABE M L, Advanced Multi-Product Coal Ash Processing Technology, Supplementary Papers, Eight CANMET/ACI International Conference on Fly Ash, Silica Fume and Natural Pozzolans in Concrete, 2004, pp. 1-8.

2. ROBL T L, GROPPO J G, JACKURA A AND TAPP K, Field Testing of an Advanced Mulit-Product Coal By-Product Processing Plant at Kentucky Utilities Ghent Power Plant., Conference Proceedings, AshTech 2006, United Kingdom Quality Ash Association, May 15-17, 2006, Birmingham, UK, 2006, 7 p.

3. RATHBONE R, ROBL T L, McCABE M L AND TAPP K, The Effect of Hydraulically Classified Ultra Fine Ash on Mortar Properties, *Cement Combinations for Durable Concrete*, R.K. Dhir, T.A. Harrison and M.D. Newlands (Eds.), Thomas Telford, London, 2005, pp. 65-70.

4. ROBL T L AND GROPPO J G, The Production of Superpozzolan from Coal Fired Utility Ash Ponds, Sustainable Concrete Construction, R.K. Dhir, T.D. Dyer and J.E. Halliday, (Eds.), Thomas Telford, London, 2002, pp. 221-229.

5. ROBL T L, GROPPO J G AND HOBBS A, The Pozzolanic Nature of Ponded Fly Ash, Exploiting Wastes in Concrete, R.K. Dhir, and T.G. Jappy, (Eds.), Thomas Telford, London, 1999, pp. 57-66.

6. ASTM International, Annual Book of ASTM Standards, Section 04.01- Cement; Lime; Gypsum, 2000.

7. ASTM International, Annual Book of ASTM Standards, Section 04.02- Concrete and Aggregates, 2000.

8. British Standards International, Methods of Testing Cement; Part 5. Pozzolanicity Test for Pozzolanic Cements, BS EN 196-5, 1995.

9. MARSH B K AND DAY R L, Cement and Concrete Research, Vol. 18, 1988, pp. 301-310.

10. Lea's Chemistry of Cement and Concrete (edited by Peter C. Hewlett), Butterworth Heinemann, 2001, 1057 p.

11. DODSON V, Concrete Admixtures, Van Nostrand Reinhold, 1990, 211 p.

12. MENENDEZ J A, PHILLIPS J, XIA B AND RADOVIC LR, Langmuir, Vol. 12, 1996, pp. 4404-4410.

13. SUBRAMAMIAM K V, GROMOTKA R, SHAH S P, OBLA K AND HILL R, Journal of Materials in Civil Engineering, Jan-Feb 2005, pp. 45-53.

INFLUENCE OF BLENDING OF GGBFS ON BOND BETWEEN REINFORCING STEEL BARS AND CONCRETE

D K Jain

J Prasad

A K Ahuja

Indian Institute of Technology Roorkee

India

ABSTRACT. For concrete and steel to work together in a structural element it is necessary that stresses to be transmitted between the two materials. Bonds stresses are in effect longitudinal shearing stresses acting on the surface between the steel and concrete. Factors that contribute to bond strength are chemical adhesion, friction and bearing of bar projections against the concrete. Ground Granulated Blast Furnace Slag, a by-product of iron industries that have latent hydraulic property, has been used in concrete because it improves the durability, especially in structures exposed to aggressive environments. This is attributed to improvements in the microstructure and reduction of voids in hydrated concrete. Present study examines weather such attributes could also improve the structural requirements such as compressive, tensile, or bond strength. Three mixes with low, medium and high compressive strength were observed and cement was replaced at 0 to 70 % by equal weight in all mixes. The observed strengths of blended concrete were lower than that of cement concrete at 28 days. On the basis of equal compressive strength, blended concrete showed higher split tensile, flexural tensile and bond strengths. The bond strength of concrete was closer to its tensile strength than to its compressive strength. Better geometry of ribs in larger diameter bars improved bonding.

Keywords: Bond, Mineral admixture, Blast furnace slag, Pull-out test, Tensile strength.

Dr J Prasad is an Associate Professor in the faculty of Civil Engineering Department of the Indian Institute of Technology, Roorkee. He has 30 years experience of research and teaching in the IIT. His research interest includes the design and behaviour of reinforced concrete structures in aggressive environments and concrete technology.

Dr A K Ahuja is an Associate Professor in the faculty of Civil Engineering Department of the Indian Institute of Technology, Roorkee. He has 30 years experience of research and teaching in the IIT. His research interest includes the use of GGBFS in concrete, durability of concrete and wind engineering.

D K Jain is currently a research student in the Civil Engineering Department of the Indian Institute of Technology, Roorkee. He has 15 years experience of teaching at graduate level. His research interest includes the performance of GGBFS blended concrete in marine environment and design of RC structures.

INTRODUCTION

Cement concrete is the most widely used construction material in structures, but it has shortcomings such as environmental pollution due to the generation of large quantities of carbon dioxide, high energy consumption of the cement production process and reduced service life in aggressive environments. To overcome these difficulties, the use of mineral admixtures such as Ground Granulated Blast furnace slag (GGBFS), Fly ash, Silica fume, Metakaolin, Rice husk etc. in concrete is advisable. In India, three are large quantities of Blast furnace slag and fly ash available and their availability is increasing day to day. Blast furnace slag is a by-product of iron industries that has latent hydraulic property. India has reasonable experience in using slag cement, but the use of GGBFS as an addition in concrete at the time of mixing is uncommon [1].

Bond is used to describe the means by which the slip between concrete and steel is prevented or minimized. The bond between concrete and reinforcement is necessary to ensure a composite interaction of the two materials that requires to transferring the stresses from one to another. The importance of the bond strength between concrete and reinforcing bar on cracking, deflection, and anchorage of steel in reinforced concrete has been realized for a long time. In 1913, Abrams performed pull-out tests using plain bars to determine the effect of settlement and shrinkage of concrete on the bond property [2].

The bond between steel and concrete is developed by chemical adhesion, friction and bearing of bar projections against the concrete. These properties depend on the physical and chemical characteristics of hydrated concrete. The phase and structure of hydration products of slag are somewhat different from those of cement [3]. Slag hydration products are poorly crystalline Calcium silicate hydrate (C-S-H) with a lower Ca/Si ratio [4]. Slag hydrates are generally found to be more gel-like than the products of cement hydration, hence add denseness to the cement paste [5]. Therefore, with equal mix proportion replacement of cement by GGBFS, the bond behaviour to reinforcement can be affected. Lot of research data available on the bond between cement concrete and steel bars, yet very few studies are available in India on the bond strength of GGBFS blended concrete with reinforcing steel [6].

OBJECTIVE OF WORK

GGBFS has been used in concrete because it improves the durability, especially in structures exposed to aggressive environments. This improvement is attributed to improvements in the microstructure and reduction in voids in hydrated concrete. Present study is undertaken to examine whether such attributes could also improve the structural requirements, especially bond, between the reinforcement and the surrounding concrete. In addition to bond, the effects on compressive and tensile strength of concrete are also studied.

BOND IN REINFORCED CONCRETE

Bond stresses are in effect longitudinal shear stresses acting on the surface between steel and concrete in structural elements. Factors contributing to bond strength are chemical adhesion, friction and bearing of the bar ribs against the surrounding concrete. The share of each of these components in bond resistance depends on the properties of concrete, surface conditions of reinforcing steel and also varies with the level of stress in the reinforcement. Bond strength initially comes from the weak chemical bond between steel and hardened concrete, but it is not a significant component of bond. Rich concrete mixes have greater adhesion hence higher bond strength [7]. After the break-up of adhesion, friction and bearing of reinforcement ribs contributes to bond strength. Friction is the major component of bond strength in the case of plain rounded bars, whereas in ribbed bars bond is mostly due to bearing or mechanical interaction between ribs and the surrounding concrete. Bond strength also depends on the diameter of the reinforcing bars [8] and deformed bars develop higher bond strengths compared to plain bars due to ribs.

Bond failure in structures may be initiated either by break-up of adhesion, by splitting of concrete around the reinforcement, by crushing of concrete in front of ribs or by shearing of concrete between the ribs. Bond failure may be of longitudinal splitting type or pull-out type, depending on the depth of cover to reinforcement. If both the cover over the bars and spacing between bars in structure are large, than pull-out failure will occur.

The likelihood of pull-out failure is increased if the concrete is weak or porous. In the case of reinforced beams, splitting type failure is the most common. The ribs of reinforced bars introduce redial forces causing circumferential tensile stresses in the concrete surrounding the bars and tending to split the concrete along the weakest plane. The direction of spitting crack depends upon the relative value of bottom and side cover and on bar spacing.

BLAST FURNACE SLAG

Blast furnace slag is a by-product obtained in the manufacturing of pig iron in the blast furnace. The quenching process of molten slag by water converts it into a granulated slag of whitish colour. This slag, when finely ground and combined with cement, has been found to exhibit excellent cementitious properties. Hydration of GGBFS largely depends on the breakdown and dissolution of the glassy structure by hydroxyl ions released during the hydration of cement. GGBFS reacts more slowly with water than cement, but it can be activated chemically, mechanically, or thermally.

Blending of GGBFS increases the workability, finish-ability, cohesiveness and setting time in concrete due to the increase in paste volume. Segregation and chances for bleeding in blended concrete is lower. Partial replacement of cement by GGBFS reduces the hydration temperature and also prolongs the time for peak temperature of concrete. Drying shrinkage of blended concrete is higher than that of cement concrete and increases with the increase in GGBFS content [4]. Due to the prolonged hydration process, later age strength of GGBFS blended concrete is higher.

There exists a certain optimum GGBFS content at which the maximum strength of blended concrete can be achieved. With all other factors remaining constant, chloride penetration in GGBFS concrete is lower due to its lower permeability and chemistry of pore solution. Sulphate resistance of concrete is increases by GGBFS blending because of lesser C_3A and $Ca(OH)_2$ in the paste of concrete. GGBFS is very effective in preventing expansion due to Aggregate-Silica Reaction, because of the higher alkali binding capacity of hydration products of slag. Frost resistance and carbonation resistance of GGBFS blended concrete is as good as of cement concrete.

EXPERIMENTAL WORK

OPC grade-43 and a commercially available GGBFS having physical and chemical properties listed in Tables 1 and 2 were used. Yamuna river sand with 2.52 fineness modulus, 2.70 bulk specific gravity and confirming to zone II of IS: 383 [9] was used as fine aggregate. Coarse aggregate was a crushed stone with 16 mm maximum size and 2.66 bulk specific gravity. Potable water was used in mixing and curing of specimens. A superplasticizer based on modified sulphonated naphthalene formaldehyde and specific gravity equal to 1.2 was used in Mix 1. Thermomechanically treated deformed steel bars of Fe 415 grade in 8, 10 and 12 mm nominal diameter were used. The key properties of steel bars are given in Table 3.

On the basis of trial mixes and by using linear regression analysis to 28-day compressive strength of mixes, three mix proportions were obtained for low, medium and high strength concrete, as given in Table 4. In each mix proportion, cement was replaced by GGBFS on equal weight basis, at 0, 10, 30, 40, 50 and 70% levels (CS00, CS10, CS30, CS40, CS50 and CS70). The water/binder ratio was kept constant for all replacement levels in the mixes.

Table 1   Physical properties of binder

PROPERTY	CEMENT	GGBFS
Blain fineness, m^2/kg	245	340
Specific gravity, g/cm^3	3.15	2.86
Initial setting time, minutes	126	-
Final setting time, minutes	173	-
Compressive strength, MPa		
at 3d	18.4	10.6 [*]
at 7d	31.8	24.4 [*]
at 28d	44.1	41.9 [*]
Slag Activity Index (SAI)	-	Confirming Grade 100
Glass content in percent	-	91.00

[*]With 50% CEMENT + 50% GGBFS

Table 2   Chemical composition of binder

OXIDES, %	SiO_2	CaO	Al_2O_3	Fe_2O_3	SO_3	Na_2O	K_2O	MgO	LOI	IR
Cement	19.55	65.02	5.48	3.60	2.20	0.21	0.34	1.70	1.52	0.30
GGBFS	33.50	35.30	12.50	0.80	1.30	0.08	0.40	11.00	1.00	1.25

Table 3   Properties of reinforcing steel

	NOMINAL DIAMETER, mm		
	8	10	12
Unit weight, kg/m	0.398	0.601	0.896
Cross sectional area, mm^2	50.67	76.57	114.33
Yield stress, MPa	447	459	454
Ultimate tensile stress, MPa	562	585	578
Elongation, %	28	29	29
Bend test	Pass	Pass	Pass

Table 4   Concrete mix proportions

COMPOSITION	MIX 1	MIX 2	MIX 3
Binder (cement + ggbfs), kg	445.5	403.3	312.1
Water, l	155.9	173.4	181.0
Fine aggregate, kg	695.0	689.6	711.6
Coarse aggregate, kg	1131.5	1125.2	1161.0
Superplasticizer, l	4.45	Nil	Nil
Water/Binder ratio	0.35	0.43	0.58

The test specimens consisted of cubes of 150 mm for compressive strength test, cylinders of $\varnothing 150 \times 300$ mm for split tensile test, prisms of $100 \times 100 \times 500$ mm for flexural tension test, cubes of 100 mm with centrally placed steel bar for pull-out test and reinforced concrete beams of $100 \times 150 \times 1200$ mm with one steel bar with splice in tension to be used for flexural bond test. A helix of 6 mm diameter plain steel bar at 25 mm pitch was provided symmetrically in the cubes of pull-out test to prevent any possible failure of concrete under compression. The beam of flexural bond test consisted of one bar of 8, 10, or 12 mm diameter with 200 mm long splice at the centre. To ensure splitting mode failure, the length of splice was selected in such a way that the stress developed in steel would be less than its yield stress. Minimum three specimens were used for each parameter at any instant, except in the case of RC beam where this was two only.

Compressive, flexural tensile and split tensile strength tests were carried out as per specifications given in IS: 516 and IS: 5816 [10 and 11]. The pull-out test was carried out as per specification of IS: 2770 [12] on a universal testing machine of 10-ton capacity (Figure 1). The load at failure of specimen in pull was measured. The average bond stress (U_a) was computed using a uniform bond stress distribution along the embedded length of the bar, although the actual bond stress distribution was not uniform [13].

$$U_a = \frac{P_{ua}}{\pi D_b L_a} \qquad \text{Eq. (1)}$$

Where, P_{ua} is the failure load, D_b is the nominal diameter of steel bar and L_a is the anchorage length that was equal to 100 mm in present study.

Beams were tested under a four-point loading arrangement with a simple supported span of 1000 mm on a testing machine of 5-ton capacity (Figure 1). Load was applied in the form of two symmetrically concentrated loads spaced at 200 mm centre to centre. The value of load at the appearance of the first crack and at failure was recorded and the location of the neutral axis was observed from the crack patterns. The contribution of adhesion and friction parts in bonding was calculated on the basis of the load at first crack. Splitting mode of failure indicates that the splice reached its maximum capacity; therefore bond strength could be determined directly from the stress developed in the steel bar [14]. The stress in the steel (f_s) was calculated based on elastic cracked section analysis. The analysis ignored the tensile stresses in the concrete below the neutral axis and assumed a parabolic-linear stress-strain behaviour for concrete. The bond stress (U_f) at failure was calculated from Equation 2.

$$U_f = \frac{A_{st} f_s}{\pi D_b L_s} \qquad \text{Eq. (2)}$$

Where, A_{st} is the area of tension steel, f_s is the stress in steel, D_b is the nominal diameter of steel bar and L_s is the length of splice that was equal to 200 mm in present study.

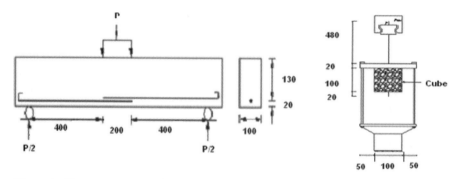

Figure 1   Test arrangements for beam and cube in bond (dimensions indicated in mm)

RESULTS AND DISCUSSION

Compressive, Split and Flexural Strength

The compressive, split tensile and flexural tensile strength at 28 days for all mixes with different contents of GGBFS are given in Table 5. Strengths of blended concrete at 28 days were lower than that of the cement concrete. This was expected as the cement clinker content decreased in the total binder content and hydration of GGBFS is slower than that of cement.

Table 5   Strengths of concrete mixes at 28 days

BINDER	STRENGTH, MPa			$f_{sp}/\sqrt{f_c}$	$f_r/\sqrt{f_c}$
	Compressive, f_c	Splitting, f_{sp}	Flexural, f_r		
Mix 1					
CS00	59.5	5.6	6.9	0.726	0.895
CS10	57.0	5.4	7.0	0.715	0.927
CS30	49.4	5.0	6.8	0.711	0.967
CS40	50.6	5.0	6.9	0.703	0.970
CS50	43.0	4.7	6.4	0.717	0.976
CS70	36.0	4.4	5.8	0.733	0.967
Mix 2					
CS00	45.2	4.2	5.9	0.625	0.878
CS10	42.7	4.0	5.6	0.612	0.857
CS30	35.2	3.8	5.7	0.640	0.961
CS40	37.0	3.9	5.5	0.641	0.904
CS50	34.0	3.7	5.3	0.635	0.909
CS70	28.4	3.3	4.8	0.619	0.901
Mix 3					
CS00	30.1	2.8	3.7	0.510	0.674
CS10	28.9	2.6	3.8	0.484	0.707
CS30	26.2	2.7	4.8	0.527	0.938
CS40	26.2	2.3	3.9	0.449	0.762
CS50	24.4	2.1	3.6	0.425	0.729
CS70	21.1	1.6	3.1	0.348	0.675

By comparing the split tensile and flexural tensile strengths with the compressive strength of concretes, we found better results for blended concrete in tensile strength. The values of the ratio of split and flexural tensile strengths to the square root of the compressive strength of the same concrete are given in Table 5. The ratio between flexural and compressive strength was clearly higher for blended concrete and its optimum was at 30 to 40% replacement level for all mixes. This means if we design a concrete of the same compressive strength with or without GGBFS, the tensile strength of the blended concrete would be higher.

This increase in the tensile strength of GGBFS blended concrete is attributed to the better bonding between paste and aggregate and within the paste itself. This is due to the characteristics of hydration products of slag, which is more compact and denser, compared to those of cement.

Bond Strength by Pull-Out Test

Results of the pull-out test and calculated bond strength are given in Tables 6 to 8. The bond strength of blended concrete was lower compared to cement concrete with the same reinforcement. This was due to the lower cement clinker content and slower rate of hydration of GGBFS in blended concrete. The bond strength of any concrete with or without GGBFS increased with the diameter of the reinforcement. Reasons for this are associated with the manufacturing process of ribs on steel bars. Ribs on larger diameter bars are prominent and sharper in comparison to the smaller diameter bars; hence the bond due to mechanical bearing of ribs on concrete is greater.

Table 6   Results of pull-out test on cube specimens of mix 1

BINDER	BAR DIA., mm	FAILURE LOAD, t	BOND STR., MPa	BOND TO COMP. STR. RATIO	BOND TO SPLIT STR. RATIO	BOND TO FLEX.STR. RATIO
	8	3.77	15.25	0.256	2.723	2.210
CS00	10	5.33	17.23	0.290	3.077	2.497
	12	6.65	17.93	0.301	3.202	2.599
	8	3.59	14.51	0.255	2.687	2.073
CS10	10	5.26	17.03	0.299	3.154	2.433
	12	6.48	17.46	0.306	3.233	2.494
	8	3.35	13.53	0.274	2.706	1.990
CS30	10	4.81	15.56	0.315	3.112	2.288
	12	6.09	16.42	0.332	3.284	2.415
	8	3.26	13.20	0.261	2.640	1.913
CS40	10	4.67	15.10	0.298	3.020	2.188
	12	6.01	16.20	0.320	3.240	2.348
	8	2.93	11.86	0.276	2.523	1.853
CS50	10	4.08	13.21	0.307	2.811	2.064
	12	5.32	14.34	0.333	3.051	2.241
	8	2.47	10.01	0.278	2.275	1.726
CS70	10	3.57	11.56	0.321	2.627	1.993
	12	4.49	12.10	0.336	2.750	2.086

Table 7   Results of pull-out test on cube specimens of mix 2

BINDER	BAR DIA., mm	FAILURE LOAD, t	BOND STR., MPa	BOND TO COMP. STR. RATIO	BOND TO SPLIT STR. RATIO	BOND TO FLEX.STR. RATIO
	8	3.40	13.75	0.304	3.274	2.331
CS00	10	4.92	15.92	0.352	3.790	2.698
	12	6.04	16.29	0.360	3.878	2.760
	8	3.30	13.35	0.313	3.337	2.383
CS10	10	4.92	15.92	0.373	3.980	2.843
	12	5.96	16.07	0.376	4.018	2.870
	8	3.10	12.54	0.356	3.300	2.200
CS30	10	4.60	14.88	0.423	3.917	2.611
	12	5.60	15.10	0.429	3.974	2.649
	8	3.29	13.31	0.360	3.412	2.419
CS40	10	4.45	14.40	0.389	3.692	2.618
	12	5.64	15.21	0.411	3.899	2.765
	8	3.10	12.54	0.369	3.389	2.366
CS50	10	4.24	13.72	0.404	3.708	2.589
	12	5.35	14.43	0.424	3.899	2.722
	8	2.67	10.80	0.380	3.272	2.250
CS70	10	3.69	11.94	0.420	3.618	2.487
	12	4.51	12.16	0.428	3.685	2.533

Table 8   Results of pull-out test on cube specimens of mix 3

BINDER	BAR DIA., mm	FAILURE LOAD, t	BOND STR., MPa	BOND TO COMP. STR. RATIO	BOND TO SPLIT STR. RATIO	BOND TO FLEX.STR. RATIO
	8	2.74	11.10	0.369	3.828	3.000
CS00	10	4.02	13.00	0.432	4.483	3.514
	12	5.50	14.83	0.493	5.114	4.008
	8	2.75	11.12	0.385	3.972	2.927
CS10	10	4.09	13.23	0.458	4.726	3.483
	12	5.44	14.67	0.508	5.239	3.861
	8	2.78	11.25	0.429	3.879	2.744
CS30	10	4.18	13.51	0.516	4.659	3.295
	12	5.34	14.40	0.550	4.966	3.512
	8	2.78	11.25	0.431	4.018	2.885
CS40	10	4.22	13.66	0.523	4.879	3.503
	12	5.36	14.44	0.553	5.157	3.703
	8	2.62	10.60	0.434	4.417	2.944
CS50	10	3.98	12.87	0.527	5.363	3.575
	12	5.07	13.67	0.560	5.696	3.797
	8	2.46	9.95	0.472	4.523	3.210
CS70	10	3.50	11.34	0.537	5.155	3.658
	12	4.66	12.56	0.595	5.709	4.052

An increase in the ratio of bond strength to compressive strength was found with increasing GGBFS content in concrete, while the ratios of bond strength to split or flexural strength were nearly the same for all concretes (Tables 6 to 8). These values led to two conclusions, namely, that bond strengths were directly proportional to the tensile strength of concrete and, for the same compressive strength, blended concretes developed a better bond with the reinforcement. This meant that bond failure occurred when the tensile strength of concrete reached its limit. Higher bond strengths of blended concretes were attributed to the higher tensile strength compared to cement concrete of the same compressive strength. This higher tensile or bond strength of blended concrete may be due to the more gel-like hydrates of slag.

The variations of bond strength with GGBFS content in different mixes are shown in Figure 2. These curves show the reduction in bond strength with an increase in GGBFS content for all mixes, but the slope of reduction was reduced with lowering the grade of concrete. This indicates that GGBFS blending is more advantageous in lean concretes with lower binder contents and higher water to binder ratios. These curves show that there is an increase in bond strength with an increase in the diameter of bars for all mixes.

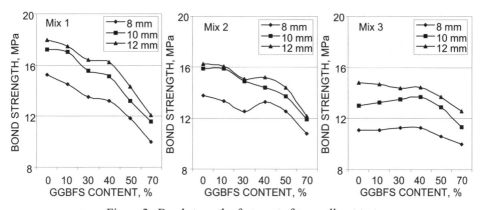

Figure 2   Bond strength of concrete from pull-out test

Bond Strength by Beam Test

The loads at observing the first crack and at failure in flexural bond test of beam, with other calculated values are given in Table 9. The calculated stress in steel was less than its yield strength, hence all beams failed due to failure in bond between the steel and the concrete, as it was our objective. No shear or flexural failures were observed. The results indicated that the bond strength also got reduced with the addition of GGBFS in concrete, but if we compare it on the same compressive strength basis, the GGBFS blended concrete had a higher ratio of bond strength to compressive strength. The bond strength observed by this test also increased with the diameter of bars. Possible reasons for this were already discussed in the previous section.

Table 9   Observations and calculated values of beam tests

BINDER	BAR DIA., mm	FC LOAD, t	FAILURE LOAD, t	STRESS IN BAR, MPa	BOND STRESS (U_f), MPa	AV. BOND STRESS, MPa	FC / FAIL LOAD, %	COMP. STR. (f_c), MPa	$U_f / \sqrt{f_c}$
				Mix1					
	1-8	2.1	2.6	412.71	8.25		80.8	59.5	1.129
CS00	1-10	2.6	3.4	353.45	8.84	8.71	76.5		
	1-12	3.0	4	300.95	9.03		75.0		
	1-8	2.0	2.5	399.35	7.99		80.0	50.6	1.215
CS40	1-10	2.6	3.3	355.04	8.88	8.64	78.8		
	1-12	3.0	3.9	301.92	9.06		76.9		
				Mix2					
	1-8	1.8	2.4	380.96	7.62		75.0	45.2	1.142
CS00	1-10	2.0	2.9	301.48	7.54	7.68	69.0		
	1-12	2.5	3.5	262.41	7.87		71.4		
	1-8	1.6	2.1	349.36	6.99		76.2	37.0	1.210
CS40	1-10	2.0	2.7	299.91	7.50	7.36	74.1		
	1-12	2.2	3.2	253.23	7.60		68.8		
				Mix3					
	1-8	1.5	2.2	340.14	6.80		68.2	30.1	1.172
CS00	1-10	1.8	2.7	268.89	6.72	6.43	66.7		
	1-12	2.0	3.1	216.49	6.49		64.5		
	1-8	1.4	1.9	301.59	6.03		73.7	26.1	1.250
CS40	1-10	1.7	2.5	259.89	6.50	6.38	68.0		
	1-12	1.9	2.8	220.76	6.62		67.9		

The ratio of load at first crack to the one at failure indicated the share of the adhesion component in bond strength. A higher value of this ratio was observed for GGBFS blended concrete (Table 9 and Figure 3), hence blended concrete had stronger adhesion to the reinforcement. The better adhesion of GGBFS blended concrete may be due to the denser and more gel-like hydrates of slag in comparison to those of cement clinker.

The values of bond strength obtained by the beam test were lower than those obtained by the pull-out test, because of the failure pattern. Cubes had a failure of the pull-out type, while all beams failed in split. The pull-out test measured the anchorage bond stress, while the beam test measured the flexural bond stress. Flexural bond failure is more realistic in RC structures. The values obtained by the beam test were closer to the specifications given in standards such as the Eurocode [15], and relevant Indian standards [16]. The Indian standard [12] and ASTM specifications [17] recommend the pull-out test to compare the bond strength. The Indian standard mentions that in the case of deformed bars, the pull-out test gives a larger ultimate failure load corresponding to a large slip. The Indian standard mentions the value of slip (0.025 mm and 0.25 mm) corresponding to the load taken, in calculation of the permissible bond stress. In the present study, our objective was to compare the performance of cement and blended concrete in bond, hence we were more concerned with the relative value in the test method.

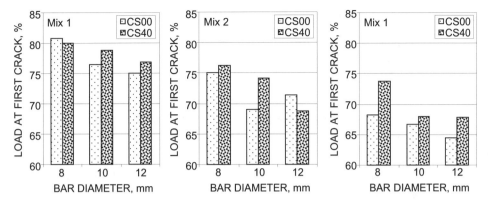

Figure 3   Contribution of adhesion in bond strength

CONCLUSIONS

On the basis of the experimental results presented, the following conclusions can be drawn

- GGBFS blended concrete shows lower strength in compression, tension and in bond at 28 days if cement is replaced by equal weight and the water to binder ratio is kept constant.
- Split and flexural tensile strengths of GGBFS blended concrete are higher than that of cement concrete if both concretes have the same compressive strength.
- Bond strength of GGBFS blended concrete is also higher than that of cement concrete if both concretes have the same compressive strength.
- Bond strength of concrete increases with the bar diameter due to the more prominent and sharper ribs on larger diameter bars.
- Blending with GGBFS in lean mixes is more beneficial for improvement in strength properties of concrete.
- Contribution of adhesion to bond strength is slightly higher in the case of GGBFS blended concrete than in cement concrete.
- Beam test under flexural bond failure gives a more realistic value of bond strength between reinforcement and concrete.

REFERENCES

1. SARASWATI S AND BASU P C, Concrete composites with ground granulated blast furnace slag, The Indian Concrete Journal, June 2006. pp. 29-40.

2. GJØRV O E, MONTEIRO J M AND MEHTA P K, Effect of condensed silica fume on the steel-concrete bond, ACI Material Journal, Vol. 87, 1990, pp. 573-580.

3. GARCIA J I E, FUENTES A F, GOROKHOVSKY A, FRAIRE-LUNA P E AND MENDOZA-SUAREZ G, Hydration products and reactivity of Blast-furnace slag activated by various alkalis, Journal of American Ceramic Society, Vol. 86, No. 12, 2003, pp. 2148-2153.

4. MALHOTRA V M, Supplementary Cementing Materials for Concrete, CANMET, Canada, 1987, 428 p.

5. AMERICAN CONCRETE INSTITUTE, GGBFS as a cementitious constituent in concrete, ACI Manual of Concrete Practice, Committee Report No. ACI 233R-95, 1997, Part I

6. BALASUBRAMANIAN K, KRISHNAMOORTHY T S, GOPALAKRISHNAN S, BHARATKUMAR B H AND KUMAR G, Bond characteristics of slag based concrete, The Indian Concrete Journal, August, 2004, pp. 39-43.

7. SOROUSHIAN P S, CHOI K B, PARK G H AND ASLANI F, Bond of deformed bars to concrete: Effects of confinement and strength of concrete, ACI Materials Journal, Vol. 88, 1991, pp. 227-232.

8. LARRARD F D, SCHALLER I AND FUCHS J, Effect of bar diameter on the bond strength of passive reinforcement in high-performance concrete, ACI Materials Journal, Vol. 90, 1993, pp. 333-39.

9. BUREAU OF INDIAN STANDARDS, Specifications for coarse and fine aggregates from natural sources for concrete, IS: 383 - 1970.

10. BUREAU OF INDIAN STANDARDS, Methods of tests for strength of concrete, IS: 516-1999.

11. BUREAU OF INDIAN STANDARDS, Method of test for split tensile strength of concrete, IS: 5816-1970.

12. BUREAU OF INDIAN STANDARDS, Methods of testing bond in reinforced concrete: Pull-out test, IS: 2770 (Part I) – 1967.

13. EZELDIN A S AND BALAGURU P N, Bond behavior of normal and high-strength fiber reinforced concrete, ACI Materials Journal, Vol. 86, 1989, pp. 515-523.

14. HAMAD B S AND ITANI M S, Bond strength of reinforcement in high-performance concrete: The role of silica fume, casting position, and superplasticizer dosage, ACI Material Journal, Vol. 95, 1998, pp. 499-511.

15. BEEBY A W AND NARAYANAN R S, Designers handbook to Eurocode 2, Part 1.1: Design of concrete structures.

16. BUREAU OF INDIAN STANDARDS, Plain and reinforced concrete – Code of practice, IS: 456 - 2000.

17. AMERICAN SOCIETY OF TESTING MATERIALS, Standard test method for pullout strength of hardened concrete, ASTM C900-06.

UTILISATION OF DEPOSITED CLAY IN SOUTH LIBYA AS A POZZOLANIC MATERIAL

A M Akasha
H M Abdulsalam
Sebha University
Libya

ABSTRACT. Pozzolanic materials be it natural or artificial have the potential of use in the manufacture of mortars and concrete. Nowadays many concrete producers worldwide recognize the value of pozzolans as enhancements to their products. Most pozzolans used in the world today are industrial by products such as condensed silica fume, blast furnace slag, fly ash, burnt clay.....etc. The aim of this work is to study the suitability of south Libya clay as replacement for Portland cement in mortar and concrete. Samples of clay were collected from different places in south Libya. The samples were ground into fine powder, thermally treated at different temperatures and taken for chemical and mineralogical tests. To evaluate their pozzolanic activity, clays were applied as composites, with different percents of cement replacement. The investigation of the pozzolanic reaction and the hydration kinetics of the various compositions included quantitative mineralogical and chemical analyses of the corresponding hydrated specimens, and determination of the chemically combined water, total water free lime contents, and bulk density. The primary result found up to now has been presented in the first part of this paper, more extensive work results will be presented in the second part.

Keywords: Pozzolanic materials, Durability, Cement replacement materials.

A M Akasha, Associated Professor, Head of Civil Engineering Department, Sebha University – Libya.

H M Abdulsalam, Postgraduate Student, Civil Engineering Department, Sebha University – Libya.

INTRODUCTION

An extensive research work has been carried out in the past year by the industrial research center on the natural raw materials in the south region of Libya. The investigation shows that there are many raw materials that can be used in the building material industry, one of that materials was natural pozzolan [1]. From that point we start thinking to carry out a wide range investigation for the possibility of using the natural pozzolans as replacement cement material in concrete. Pozzolana is a siliceous material which whilst itself possesses no cementitious properties, either processed or unprocessed and in finely divided form, reacts in the presence of water with lime at normal temperatures to form compounds of low solubility having cementitious properties. Pozzolanas may be natural or artificial, fly ash being the best known in the latter category. These were used with lime to make concrete before the advent of cement. Currently their principal use is to replace a proportion of cement when making concrete. The advantages gained are economy, improvement in workability of the mix with reduction of bleeding and segregation. Other advantages are greater imperviousness, to freezing and thawing and to attack by sulphates and natural waters. In addition, the disruptive effects of alkali-aggregate reaction and heat of hydration are reduced. It is generally held that the addition of natural pozzolanas reduce the leaching of soluble compounds from concrete and contribute to the permeability of the concrete at later ages [2].

The main justification for using pozzolanas is the possibility of reducing costs. If they are to reduce costs, they must be obtained locally and it is for this reason that they have not so far been much in use.

When mixed with cement the silica of the pozzolana combines with the free lime released during the hydration. Silicas of amorphous from react with lime readily compared to those of crystalline form and this constitutes the difference between active pozzolanas and materials of similar chemical composition which exhibit little pozzolanic activity. It is commonly thought that lime-silica reaction is the main or the only one that takes place, but recent information indicates that alumina and iron if present also take part in the chemical reaction [2].

Figure 1 Profile section in Askeda and Brack places

WORK STEPS AND RESULT ANALYSIS

The clay used in this research was collected from different chosen pleases in the Sebha city reign in South Libya, according to the result of the research work carried out in the past year by the Libyan industrial research center on the natural raw materials in the South reign of Libya. These pleases have been given numbers and symbols as shown in Table 1.

Table 1   Sample numbers and places of collection

NO.	SYMBOL	PLACE
1	A	Sebha
2	B	Tamenhint
3	C	Brack
4	D	Agar
5	E	Taroot
6	F	Askeda

Representative samples have been taken from each place, ground and sieved so that the sample would to pass the sieve 200μm and retained on the sieve 150μm. The loss in weight for all the samples has been taken at 400, 500, 600, 800, and 1000 degree centigrade as shown in Figure 2. According to the heat treatment, we chose the low heating degree rate with high losses in weight samples. The PH has been tested for the chosen samples after 5 min and 30 min as shown in Figure 3.

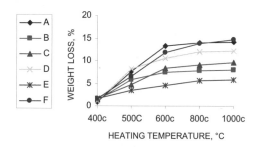

Figure 2   Weight loss for different tempretures

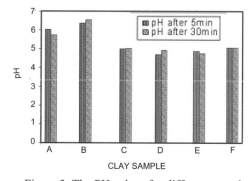

Figure 3 The PH values for different samples

The cement replacement percentages are shown in Table 2. For all samples with different cement replacement percentages, the bulk density, chemically combined water and total water content has been evaluated, the summarized results are shown in Table 3. From the results it is clear that the bulk density of all samples increased with time which means that hydration carried on.

Table 2 Percentage replacement of clay in mixes

MIX	I	II	III	IV	V
Cement %	90	80	70	60	50
Burnt clay %	10	20	30	40	50

Table 3a Total water content, chemically combined water and bulk density at different periods

Sample	Mix	TOTAL WATER CONTENT (%)				CHEMICALLY COMBINED WATER (%)				BULK DENSITY (kg/m³)			
		1 day	7 day	28 day	60 day	1 day	7 day	28 day	60 day	1 day	7 day	28 day	60 day
A	I	22.28	26.19	26.00	26.87	11.30	18.72	18.38	22.19	2174	2192	2199	2050
	II	21.61	23.53	24.00	24.35	10.42	17.94	16.42	19.87	2147	2180	2198	2198
	III	23.86	26.22	28.62	28.93	11.96	17.47	18.41	21.90	2098	2122	2166	2147
	IV	24.3	28.63	30.34	32.22	12.34	18.34	18.64	22.11	2032	2078	2095	2080
	V	24.54	28.28	32.64	31.86	12.26	16.67	17.66	20.91	1991	2022	2091	2022
B	I	21.93	23.16	23.08	23.24	13.38	17.68	17.47	19.62	2217	2248	2248	2270
	II	22.49	24.22	24.15	23.72	12.24	16.61	17.11	19.32	2171	2226	2200	2242
	III	21.71	25.24	25.43	25.56	7.27	16.20	16.54	18.14	2120	2141	2148	2181
	IV	23.51	26.29	25.90	26.67	11.06	16.03	16.12	16.90	2085	2121	2126	2150
	V	23.23	25.59	27.50	24.66	9.91	14.23	15.74	15.40	2028	2068	2084	2120
C	I	16.23	23.58	24.36	23.70	10.88	16.20	17.47	19.80	2218	2223	2241	2279
	II	16.3	25.00	26.27	26.26	9.22	16.62	17.06	18.78	2163	2189	2196	2231
	III	20.16	25.00	27.24	26.67	9.58	16.58	16.81	19.17	2122	2141	2146	2174
	IV	19.87	25.93	28.83	27.23	9.24	15.91	15.93	18.75	2057	2079	2086	2105
	V	18.24	25.64	29.78	27.27	7.32	13.81	16.57	18.00	2023	2037	2032	2061

Table 3b   Total water content, chemically combined water and
bulk density at different periods

Sample	Mix	TOTAL WATER CONTENT (%)				CHEMICALLY COMBINED WATER (%)				BULK DENSITY (kg/m³)			
		1 day	7 day	28 day	60 day	1 day	7 day	28 day	60 day	1 day	7 day	28 day	60 day
A	I	16.11	24.01	23.51	27.54	10.45	16.67	18.11	20.89	2208	2210	2212	2240
	II	16.5	24.69	26.75	31.09	8.09	17.37	17.72	22.12	2131	2147	2158	2183
	III	18.85	26.67	29.00	34.10	6.96	16.59	19.25	22.15	2060	2073	2083	2099
	IV	17.41	27.40	30.77	38.59	7.96	16.86	19.31	22.78	2025	2057	2050	2054
	V	20.09	26.32	33.12	40.41	7.04	16.22	19.02	21.20	1965	1991	2008	2020
B	I	15.82	23.62	24.64	23.21	11.56	15.87	18.79	20.95	2192	2225	2230	2231
	II	17.71	25.30	26.98	26.47	10.26	16.24	19.70	21.75	2154	2190	2176	2172
	III	17.60	25.26	25.33	27.70	10.49	17.11	17.83	20.27	2089	2142	2148	2126
	IV	18.48	25.87	26.90	27.71	9.95	15.69	16.78	20.23	2045	2050	2095	2078
	V	19.15	26.79	28.88	30.40	10.26	14.61	16.50	18.07	2008	2030	2042	2028
C	I	23.87	19.22	24.78	23.94	13.01	17.58	18.09	21.21	2177	2062	2190	2198
	II	22.8	29.38	27.94	28.51	13.69	19.21	19.22	20.75	2116	2124	2134	2116
	III	23.53	30.80	32.00	29.01	11.81	18.45	19.70	21.40	2066	2073	2056	2056
	IV	23.92	31.77	35.24	31.19	12.4	18.38	19.86	20.98	2018	2006	1995	1993
	V	25.31	31.18	37.28	29.68	11.11	19.55	19.67	20	1953	1966	1958	1943

CONCLUSIONS AND RECOMMENDATIONS

Conclusions

The following conclusions were drawn from the work:

1. Weight loss occurred for most clay samples at 600°C, which is small compared to that of Portland cement clinker manufacture.

2. The increase in the bulk density with time shows that the lime-silica reaction takes place.

Recommendations

Due to the short time given for preparing this paper, the following work has to be published in the second part of the paper.

1. More chemical analysis is needed for the different samples and their physico-chemical and mineralogical changes due the treatments followed with using XRD, DTG, and DTA.

2. Quantitative mineralogical and chemical analysis of the corresponding hydrated specimens. XRD, DTG, TG, chemically combined water, free lime contents, and compressive strength are needed.

REFERENCES

1. Industrial Research Journal. "Raw materials in Libya South region". Vol., No.2, 1996.

2. S.K. DUGGAL. "Building Materials". Second Edition 2003.

3. JOHN NEWMAN. "Advanced Concrete Technology" First Edition 2003.

EVALUATION OF STRENGTH RETROGRESSION AND COMPRESSIVE STRENGTH OF CONCRETES WITH GRANITE AND KAOLIN

U T Bezerra

Federal Centre of Technology Education of Paraíba

G de A Neves   D G Fuzari   H C Ferreira   H L Lira

Federal University of Campina Grande

Brazil

ABSTRACT. The increase in the production of industrial residues motivated the accomplishment of this work. Two of these residues, granite with crystalline silica and kaolin with high alumina concentration, had been mixed with sand and Portland cement, composing a shotcrete that presents good characteristics of combat to the phenomenon of the strength retrogression. The concretes had been divided in four groups: slurry of cement and sand; slurry of cement, sand and granite; slurry of cement, sand and kaolin; and slurry of cement, sand, granite and kaolin. All the results had been evaluated by a factorial planning using three variables: environment temperature of rupture in the test of compressive strength (30°C, 100°C and 180°C); granite concentration (0%, 50% and 100% by substitution of sand) and kaolin concentration (0%, 50% and 100% by substitution of sand). For each combination three specimens were prepared. The results show that the presence of the granite and kaolin, separately or together on the concretes, decreases the compressive strength for the low temperatures. However, for the biggest adopted temperatures, the granite and the kaolin increases the compressive strength and reduce the strength retrogression phenomenon. In conclusion, it can say that the addition of the residues in shotcrete can combat the phenomenon of strength retrogression and reduces its presence in the environment, mainly with regard to the granite that, for containing crystalline silica, is potential causer of silicoses.

Keywords: Granite, Kaolin, Portland cement, Strength retrogression, Refractory castables.

Dr U T Bezerra, Graduated in Civil Engineering, MSc in Production Engineering, Dr in Materials Science and Engineering, Federal Centre of Technology Education of Paraíba, Brazil.

Dr G de A Neves, Graduated in Mine Engineering, MSc in Chemical Engineering, Dr in Process Engineering, Federal University of Campina Grande, Brazil.

Msc D G Fuzari, Graduated in Civil Construction, MSc Student in Science and Materials Engineering, Federal University of Campina Grande, Brazil.

Dr H C Ferreira, Graduated in Civil Engineering, MSc in Chemical Engineering and Dr in Chemical Engineering, Federal University of Campina Grande, Brazil.

Dr H L Lira, Grad in Industrial Chemistry, MSc in Chemical Engineering and Dr in Chemistry, Federal University of Campina Grande, Brazil.

INTRODUCTION

Several research institutes all over the world have employed efforts to solve serious environmental problems on earth. This is not different with researches developed in the engineering area, where several studies are addressed to the incorporation of residues in mortar and concrete, as a way to reduce the environmental impact. The incorporation of mineral industry residue in concrete is a good alternative to the economy of natural resources, raw materials diversification and to optimize the concrete properties. The addition of pozzolanic material to Portland cement reduces the consumption of clinker, produces more C-S-H and reduces the cost, with consequent increase in durability [1, 2].

In this way, the research group from UFCG/CEFET-PB is using residues, mainly form mining industries, in the preparation of mortar and concrete with the aim to improve the refractory properties. The preference for the industrial residue originating mainly from ore refining, in contrast to other residues, is due to the physical and mineralogical properties fit to be used as raw materials to several applications, i.e. refractory, ceramic tile, mortar and concrete to civil construction.

The northeast region of Brazil presents rocky formations, predominantly igneous with granite occurrence and great variety [3]. These rock formations are commercially explored in the shape of ornamental rocks to civil construction and can generate a significant amount of residue that can be applied to other uses.

Beside granite, other mineral that is extracted in the northeast of Brazil is kaolin whose purification technology generates thousands of tons of solid residues per year, which is discharged in the environment. Despite the kaolin is very important to several industries, such as rubber, plastics, ink, paper, etc., it can create a great environmental impact [4, 5].

In the last decade, in addition to good mechanical and durability properties, some concretes has been submitted to thermal treatment [6], as the case of oil well cementation submitted to vapour injection, ceramic industries, etc. These needs directed this research to refractory materials incorporated in concrete. So, the aim of this work is to incorporate in concrete, the two materials, granite and kaolin residues, in a way to increase the refractory properties and to decrease the strength retrogression phenomenon.

Both the physical and mechanical properties of Portland cement are characterized by scattered values as a function of different service conditions. This is especially true at elevated temperatures, responsible for phase transformations that alter the behaviour of the cement [2, 6]. Above 110°C, the cement depicts significant reduction in compressive strength. This phenomenon is known since the '50s as strength retrogression [1, 2, 8], however, its causes and consequences have still to be fully unveiled.

The main crystalline constituents responsible for the strength of Portland cement are C-S-H and portlandite [1, 2, 8-10]. When heated, C-S-H undergoes phase changes which modify the properties of the cement. Between 110°C and 120°C, C-S-H is converted to crystalline hydrated α-C_2S, characterized by high density, high permeability and low compressive strength. At 202°C, C_3S is formed. Its properties are similar to α-C_2S, nevertheless, as the temperature increases, the cement becomes dehydrated, thus favouring the formation of phases similar to those encountered in the clinkered material.

To minimize this effect, crystalline silica has been added to Portland cement, especially for oil well cements [2, 8]. Additions of up to 40% silica increase the pozzolanic reaction and reduce the expansion and cracking of the cement. When crystalline silica is added as fine silica powder or silica sand, the natural sequence of phase formation of C-S-H is affected. At 120°C it is converted to tobermorite, which has low permeability and contributes to high compressive strength. At 150°C, tobermorite changes to xonotlite and/or gyrolite, whose permeability and compressive strength are similar to those of tobermorite. By further increasing the temperature to 250°C, gyrolite transforms into truscotite, with higher permeability and lower compressive strength compared to tobermorite. Other phases may also develop, such as pectolite and scawtite. The content of Portlandite not involved in the pozzolanic reaction is converted to lime and water [6]. It should be also considered that the sequence of reactions mentioned earlier is not unique, i.e. tobermorite is not necessarily transformed into xonotlite, gyrolite and truscotite. Other studies [2] have also reported the conversion of truscotite into xonotlite, contrary to what is commonly observed. For that, different temperature and curing conditions are involved. The presence of free CaO, instead of silica, initially favours the formation of xonotlite, whereas free silica first assists the formation of truscotite [11].

In the case of silica-free Portland cement, two exothermic phase transformations are reported by Lankard [11], corresponding to the conversion of C-S-H into tobermorite at 105°C, and the conversion of Portlandite into CaO at 550°C. The chemical composition of all such phases has not been yet fully established. Tobermorite, for instance, may include the presence of aluminium and sodium in its crystalline network, in addition to interstitial defects [2]. Nevertheless, all of them show similar strength and permeability. On the other hand, Cao and Detwiler [12] also reported that the network of interstitial pores grows as the curing temperature increases. Curing at lower temperatures prevents the fast migration of excess water and the consequent formation of deleterious pores. Up to 22 other secondary compounds may also be formed, including foshagite, hillebrandite, jaffeite, afwillite, calcium-chondrodite, reyerite, kilchoanite, and rankinite.

The reduction of strength retrogression may be obtained by reducing the CaO/SiO_2 ratio [8, 9]. This is accomplished by the addition of silica, i.e. the formation of tobermorite, which consists of 6 SiO_2 and 5 CaO, instead of the original 2 SiO_2 and 3 CaO. Therefore, tobermorite always requires more SiO_2 (3 times) than CaO (1.67 times). Large silica particles react with C-S-H to form tobermorite, however, smaller particles along with C-S-H are converted to gyrolite and truscotite without forming tobermorite. This reduces the phase variations and preserves the microstructure of the cement resulting in a mechanically stable material.

MATERIALS AND METHODS

The residues incorporated to the concrete were granite and kaolin. The granite is used due to the significant amount of silica, that can avoid the strength retrogression, and kaolin residue was used due to the amount of alumina, that can contribute to the refractory properties. Three samples of granite sawing and kaolin residues were selected and collected from processing industries located at the Paraiba States. The residues from industries were:

- granite residue: from FUJI S/A. Mármores e Granitos; located in the Industrial Centre of Campina Grande-PB, Brazil.
- kaolin residue: from Caulisa Indústria S/A, located in the city of Juazeirinho-PB, Brazil.

Beside these materials, CPIV RS Portland cement was used, conforming to the Brazilian norm NBR 5737 and being equivalent with cement type IS (MS) of the ASTM C 595 norm, i.e. Portland Blastfurnace Slag - Moderate Sulphate Resistance.

The sand applied passed the 2.000 mm (10 mesh) sieve and retained on the 0.075 mm (200 mesh) one. Fresh water was used for the tests.

Characterization of the kaolin and granite residues included particle size distribution (Cilas, model 1064), phase identification, performed by X-ray diffraction (Shimadzu, model XRD 6000, with $Cu_{k\alpha}$ radiation (30kV/40 mA), chemical composition, analyzed by X-ray fluorescence (Shimadzu, model EDX-900).

Compressive and retrogression strengths were measured on cylindrical samples with 5.0×10.0 cm of dimensions. Three samples were used as reference, three with granite residue addition and three with kaolin residue addition, for each adopted temperature conditions.

Retrogression strength is known to occur at temperature values that present discrete change as a function of cure conditions of Portland cement. So, the following temperatures were used, according to the literature [2, 13]: 30°C, 100°C, 120°C, 180°C and 230°C. The water/cement ratio was 0.45. Following the processing, the samples were demoulded after 24 hours and immersed in water at room temperature until the seventh day of moulding. The mechanical tests were carried out on samples after heating at each adopted temperatures for two hours, using a universal machine press with 100 kN capacity.

The composition of each sample is showed on Table 1. Each residue was used at 50% substitution for sand.

Table 1   Composition of concretes

MATERIAL, g	CONCRETE REFERENCE	CONCRETE WITH GRANITE	CONCRETE WITH KAOLIN
cement CP IV	600.0	600.0	600.0
sand	1200.0	600.0	600.0
granite	-	600.0	-
kaolin	-	-	600.0
water	240.0	240.0	240.0
w/c ratio	0.45	0.45	0.45

RESULTS AND DISCUSSIONS

The chemical compositions of granite and kaolin residues (mass %, by X-ray fluorescence) are shown in Table 2. It can be observed that the kaolin residue has great amount of silica (>55%) and Al_2O_3 (36%), and contains 1% of Fe_2O_3. These oxides are very important in the refractory properties of the residue. The granite residue has high amount of SiO_2, more than 60%, and more than 14% of Al_2O_3. The Fe_2O_3 and CaO contents are close to 6%. The presence of Fe_2O_3 and CaO found in the granite residue is due to the use of metallic dust and lime as abrasive and lubricant, respectively. Na_2O and K_2O are from feldspar and the mica in the granite.

Table 2   Chemical composition of the residues

RESIDUE	SiO_2, %	Al_2O_3, %	Fe_2O_3, %	K_2O, %	TiO_2, %	CaO, %	MgO, %	Na_2O, %
granite	62.77	14.38	6.56	3.78	-	6.28	-	3.52
kaolin	56.50	36.00	1.00	6.14	0.13	-	-	-

Figures 1 and 2 show the particle size distribution of the granite and kaolin residues. It can be verified that the granite residue presented a curve with modal behaviour with peak at 7 μm and wide distribution of particle sizes, with D_{10} of 0.9 μm, D_{50} of 5.6 μm, D_{90} of 31.3 μm, with average value of 11.3 μm. Kaolin residue presented a curve with bimodal behaviour, a wide distribution of particle sizes, with D_{50} of 53.0 μm, D_{10} of 5.0 μm and D_{90} of 135.0 μm, with average value of 54.4 μm.

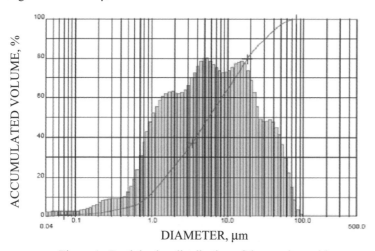

Figure 1   Particle size distribution of the granite residue

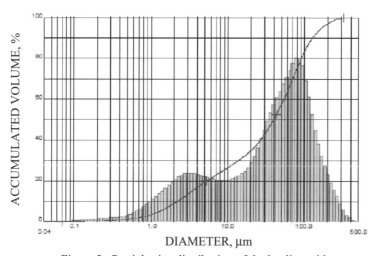

Figure 2   Particle size distribution of the kaolin residue

Figure 3 shows the X-ray diffraction pattern of the kaolin and granite residues. The crystalline phases of the kaolin residue are muscovite mica (interplanar spacing 1.004 nm), kaolinite (0.732 nm) and quartz (0.334 nm). For the granite residue, it is observed crystalline phases of quartz (0.334 nm), feldspar albite (0.319 nm), muscovite mica (1.004 nm) and calcite (0.303 nm).

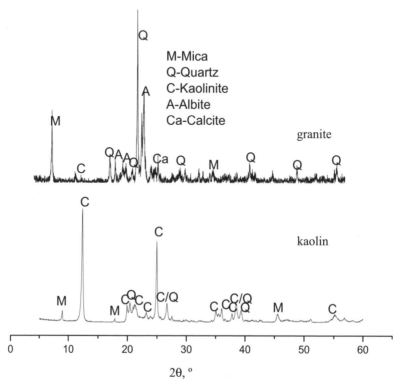

Figure 3   X-ray diffraction of the granite and kaolin residues

After submitted to the high temperatures during six days, the concretes had significant loss of water, which contributed for the limitation of the compressive strength. At this rate, the concretes had been submitted to the severe conditions of assay soon at the beginning of the hydration. These results can be seen in Table 3, as average losses for three specimens each.

Table 3   Loss of water

TEMPERATURE °C	REFERENCE g	%	GRANITE g	%	KAOLIN g	%	GRANITE+KAOLIN g	%
30	5.37	1.30	6.74	1.97	8.89	2.58	4.54	1.32
100	36.06	8.48	33.37	9.22	35.66	9.18	36.80	9.90
180	39.77	9.47	36.89	10.48	35.07	9.56	34.90	9.74

As the temperature increases, the loss of water also increases and this compromises the hydration process. However, the function of temperature is to speed up the hydration reactions, and this effect can be observed in the compressive strength of the specimens for the highest temperatures. A concern related to the heating mentions the possibility of generalized fractures due to water exiting under pressure. However, this can be repaired by using a pozzolanic cement type that presents low concentration of C_3A, which is the most expansive compound of Portland cement.

From 100°C to 180°C, a stabilization of loss of water is observed, indicating probably a limit. At the highest temperatures, this limit approaches to 40 g, representing 50% of the amount of water used in each specimen, i.e. the water/cement ratio tends to stabilize around 0.225. This is very important, because some studies show that the amount of water strictly necessary for the hydration of Portland cement varies of 0.20 to 0.25 [1, 13]. In addition, these studies show that the lower the water/cement ratio, the better the mechanical properties – a fact evidenced from the results of this research.

The average compressive strength of the cement samples is shown in Table 4, and the results obtained from the factorial experimental planning is listed in Table 5.

Table 4   Compressive strength, N/mm²

GRANITE, %	KAOLIN (%)					
	0			100		
	30°C	100°C	180°C	30°C	80°C	180°C
0	18.46	18.46	12.73	4.33	12.48	15.86
0	12.61	15.92	12.73	2.93	11.46	14.00
0	10.19	17.83	11.20	3.31	7.00	14.00
100	4.97	3.95	8.53	2.93	6.62	8.53
100	5.47	6.49	6.75	3.82	8.40	12.73
100	3.82	4.84	7.00	2.80	5.86	10.31

Table 5   Results of the factorial experimental planning (F distribution)

EFFECT	F0 CALCULATED	F0 LISTED	ANALYSIS
Granite, %	84.93	4.26	S
Kaolin, %	9.81	4.26	S
Temperature, °C	22.96	3.40	S
interaction granite vs. kaolin	24.78	4.26	S
interaction granite vs. temperature	3.18	3.40	Not S
interaction kaolin vs. temperature	16.91	3.40	S

S = significance at 95 % reliability.

Figure 4 shows the results of compressive strength versus temperature. The increment of the compressive strength is observed with the increase of the temperature for the specimens that had received residues and the decrease from this resistance for the specimens without residues. It is interesting to observe that, although the reference concrete has superior resistance for the temperature of 30°C, at the same time as the temperature develops, the phenomenon of strength retrogression starts (after 120°C). However, the presence of the pozzolanic material in great amount in the specimens with residues, favours the consumption of the excess of portlandite, and the consequent formation of C-S-H. A characteristic that favours this fact is the size of the residue particles (<#200), therefore, high specific surfaces contribute to enhanced interaction between particles of the cement and the residues.

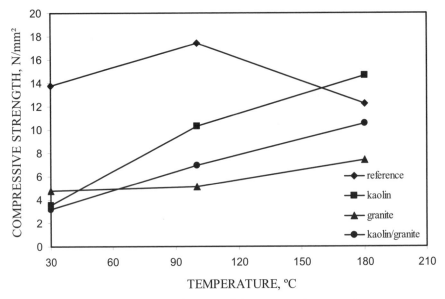

Figure 4   The development of the compressive strength

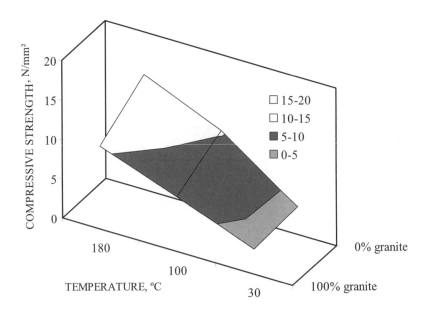

Figure 5   Specimens with 100% of kaolin

In Figure 5 this effect is better observed, mainly for the concrete prepared with kaolin residue. The presence of the granite, in this case, reduces the compressive strength, but it indicates the possibility of improved resistance at high temperatures already shown in this research.

The effect of the granite can also be explained in Table 5. In this table, all the effects have been significant, i.e. improved variation of results, because the calculated values of F0 have been superior to the listed F0 in accordance with F distribution of probabilities. The effect of the interaction between the residues also revealed to be significant (24.78 > 4.26).

CONCLUSIONS

The addition of granite and kaolin residues contributed to prevent the strength retrogression phenomenon, because the Portland cement itself does not show typical anti-retrogression characteristics. Optimum contents and performance of the residues were approximately 75% for kaolin and 25% for granite. Higher contents are probably necessary for temperatures up to 180°C. In this case the residues react with the portlandite to form a pozzolanic reaction (C-S-H formation). Also when the residues are being used in the preparation of concretes and mortars to be applied in civil construction they are eliminated from the environment, in an adequate way.

The next step of this investigation involves an XRD study of the formation of crystalline phases as a function of the testing temperature.

REFERENCES

1.   MEHTA P K AND MONTEIRO P J M, Concrete: microstructure, properties and materials, McGraw-Hill Professional, New York, 2005, 659 pp.

2.   TAYLOR H F W, Cement chemistry, Thomas Telford, New York, 2003, 459 pp.

3.   CHIOSSI N J, Geologia aplicada à engenharia, Editora da USP, São Paulo, 1987, 427 pp.

4.   MENEZES R R et al., Use of granite and kaolin sawing wastes in the production of ceramic bricks and tiles, Journal of the European Ceramic Society, Vol. 25, 2005, pp. 1149-1158.

5.   MENEZES R R et al., Utilização do resíduo do beneficiamento do caulim na produção de blocos e telhas cerâmicos, Revista Brasileira de Cerâmica, Vol. 12, No. 1, 2007, pp. 226-236.

6.   NEVES G A, FERREIRA H C AND SILVA M C, Aproveitamento de resíduos da serragem de granito para confecção de tijolos e telhas cerâmicas, 45° Congresso Brasileiro de Cerâmica, 2001, Florianópolis, pp. 2701-2713.

7.   BAZĂNT P Z AND KAPLAN M F, Concrete at high temperatures: material properties and mathematical models, Longman Group, Essex, 1996.

8. NELSON E B, Well cementing, Dowell Schlumberger Educational Services, Houston, 1990.

9. HEWLETT P C *et al.*, Lea's chemistry of cement and concrete, Elsevier/Butterworth-Heinemann, Burlington, 2004.

10. NEVILLE A M, Properties of Concrete, Longman, London, 1995, 844 pp.

11. LUKE K *et al.*, Some factors affecting formation of truscottite and xonotlite at 300°C-350°C, Cement and Concrete Research, Vol. 11, 1981, pp. 197-203.

12. CAO Y AND DETWILER R K, Backscattered electron imaging of cement pastes cured at elevated temperatures, Cement and Concrete Research, Vol. 25, 1995, pp. 627-638.

13. BEZERRA U T, Composites Portland-biopolymer for oilwell cementing, Doctorate thesis, Natal, UFRN, 2006, 269 pp.

USING ACTIVE ADDITIVES WITH PORTLAND CEMENT

A Brahma

Saâd Dahleb University

Algeria

ABSTRACT. There is a clear interest to use additions to Portland cement due to the benefits, which need not be demonstrated anymore. Additions permit, for example, to lower the costs and adjust some properties of Portland cements. This survey deals with selected possibilities and limits of utilization of additions in Portland cement. Physical and chemical properties of Portland cement with additions of sand, limestone and tuff were considered. Finally, the objective of this study was to show the interest that one must grant to active additions such as volcanic tuff, which is a very abundant material on the surface of terrestrial globe. By appropriate modification of the Portland cement manufacturing technology, inclusion of active additives, volcanic tuff in present case, can be a profitable venture.

Keywords: Ordinary Portland cement, Addition, Grindability, Hydrothermal regime, Strength

Dr A Brahma is lecturer at Civil Engineering Department, Saâd Dahleb University, Algeria. His main research interests include the spontaneous deformation of hydraulic materials, the prediction and modelling of the properties of concrete.

INTRODUCTION

Additives include all natural or artificial matter that could contribute to the formation of cement. In general, additions are introduced at the final phase of cement manufacture and do not necessitate additional expenses. Their importance is significant as they allow to reduce product price and adjust physical and chemical properties of cement, especially in the case of an active addition.

In the production of cement, additions are classified as inert or active. First, chemically inert additions, principally affect only the product cost. Active ones, however, are more important in this respect since they entail additional properties to cement. Examples for both are presented in this paper. The effect of inert and active additions are shown for sand and limestone, and tuff, respectively.

THE INFLUENCE OF INERT ADDITIONS, SAND AND LIMESTONE, ON PORTLAND CEMENT

Sand is known to be inert in ordinary conditions. It was used here for comparative purposes, i.e. to establish specific effects achieved by limestone addition. For these two types of additions, similar suitability tests in terms of setting time and consistency measurements are prescribed by the relevant standards. Mechanical performance and some other factors determining the use of additions are also considered.

Materials

Clinker: originated from the cement plant of Sour El Ghozlane (South East of Algeria) and was obtained after the cooling stage.

Limestone: was sourced from the quarry of the cement plant of Sour El Ghozlane.

Sand: originated from Boussaâda (North of the Sahara). Results of the chemical analyses of clinker, limestone, sand and gypsum are given in Table 1.

Preparation of mixtures

Separate grinding of cement and some additions was opted for in order to show the effect of the fineness of the additions. The fineness of cement (clinker + gypsum) was then maintained constant to a value of approximately 3300 cm^2/g.

Addition of sand or limestone in small proportions (5; 10 and 15% of weight of cement) was repeated for several fineness values.

During this survey three fineness values of limestone additions (3400, 4100 and 6300 cm^2/g) were used. These were designated as limestone I, II and III, respectively.

For sand, three specific surfaces (1000, 2700 and 4600 cm^2/g) were retained and designated as sand I, II and III, respectively.

Table 1   Chemical compositions of the used materials, %

OXIDES	CLINKER	GYPSUM	LIMESTONE	SAND
SiO_2	21.79	3.31	4.37	94.94
Al_2O_3	5.15	0.41	0.96	1.14
Fe_2O_3	3.01	0.23	0.51	0.39
CaO	66.24	29.75	51.27	1.14
MgO	2.04	2.00	0.96	0.34
SO_3	0.09	37.32	0.10	0.03
K_2O	0.37	0.06	0.17	0.32
Na_2O	0.14	0.07	0.08	0.09
Loss to fire	0.39	23.57	41.33	1.45
Insoluble residue	7.1	7.1		
Free CaO	0.56			

Grindability of limestone and sand additions

The specific surface values of limestone and sand, as a function of the duration of grinding, are shown in Figure 1. One can note that, in the case of limestone, grinding becomes less efficient between 25 and 50 minutes. This is presumably due to the bonding of the matter. This last phenomenon does not appear with sand, which is however much harder to grind.

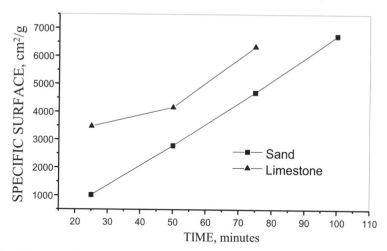

Figure 1   Variations of the specific surfaces of additions as a function of the duration of grinding

Mechanical strength

The mechanical strength in compression and by bending is the main property of cement that can be measured qualitatively. The results of the mechanical tests obtained with limestone addition are shown in Figure 2 and those with the use of sand are shown in Figure 3. In both cases, there is a clear reduction to note in the compressive strength with increased percentage of the additions. On the other hand, flexural strength appears to be less influenced by the fineness, the quantity and the type of addition. In contrast to some studies, limestone can be considered basically as inert material. In the course of our tests, there was a moderate improvement observed for the fresh properties of paste, which can be attributed to the improved granularity of the anhydrous mixture.

Hydrothermal regime

Selected samples were subjected to hydrothermal treatment at a temperature of 190°C and pressure of 12 bars for two hours. Table 2 shows the obtained results. It can be noticed, by comparison to Figure 3, that there is an increase in the strength with the addition of sand that can be explained by the formation of calcium hydro-silicates.

On the other hand, with the addition of limestone, there is a reduction of strength, which is presumably due to the decomposition of the latter. Cement with limestone is therefore not to be recommended for pre-fabrication. However, as a remedy, sand can be added to compensate for this, as shown by the results in Table 2.

Table 2   Results of mechanical tests

VARIANTS	Rc, N/mm^2	R$_f$, N/mm^2
V_1 = 15% limestone addition III	34.2	6
V_2 = 10% sand addition III	46.7	7.6
V_3 = 15% limestone III + 16% sand	40.4	5.8
V_4 = 5% limestone III + 5% sand III	42.4	6.4

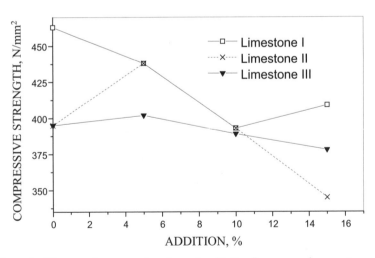

Figure 2   Change of compressive strength with the fineness and percentage of limestone addition

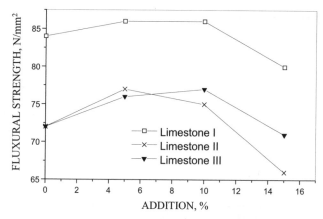

Figure 3   Change of flexural strength with the fineness and percentage of limestone addition

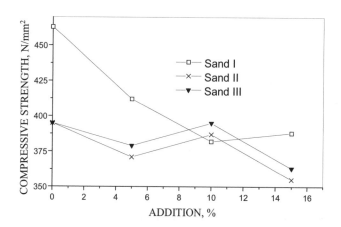

Figure 4   Change of compressive strength with the fineness and percentage of sand addition

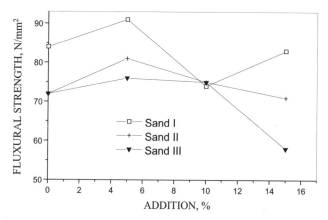

Figure 5   Change of flexural strength with the fineness and percentage of sand addition

THE INFLUENCE OF THE ACTIVE ADDITION: TUFF

Materials

Clinker: the clinker used was obtained from the cement plant of Meftah (Algeria).

Tuff: The activity of the tuff was determined by consumption of CaO, determined as 103 mg of CaO/g. Chemical compositions of clinker, tuff and gypsum are given in Table 3.

Table 3   Chemical composition of materials

COMPOSITION, %	MATERIAL		
	Clinker	Tuff	Gypsum
SiO_2	22.1	67.5	11.75
Al_2O_3	5.38	13.8	3.25
Fe_2O_3	2.9	2.32	1.82
CaO	64.9	4.1	31.87
MgO	0.7	0.85	2.25
SO_3	0.19	0.05	27.35
K_2O	1.09	2.85	0.18
Na_2O	0.1	1.81	0.08
Loss on Ignition	2.28	6.3	21.38
Free CaO	0.13	-	-
Insoluble residue	0.15	-	-

Grindability

The evolution of specific surfaces as a function of the duration of grinding of clinker and tuff are given in Table 4. It appears clearly that tuff grinds better than clinker. The relative results of simultaneous grinding of the two materials are regrouped in Table 5. It is to note that the addition of tuff can be mostly accounted for the improvement noted in the grindability. This improvement appears to be acceptable given the high percentages of the addition used.

Table 4   Grindability of clinker and tuff

Time of grinding, minutes	5	10	20	30	40	50
SSB of clinker, cm^2/g	2820	3620	4230	4680	5680	6120
SSB of tuff, cm^2/g	3180	4580	5390	6240	7300	8000

Mechanical tests

Results of the mechanical tests are summarised in Table 6. The strength remained high with increased percentage of tuff addition. The strength was also observed to grow when the fineness of the mixtures increased.

Table 5   Evolution of the specific surface of mixtures as a function of percentage of addition

TIME OF GRINDING, minutes Additions, %	5	10	20 SSB, cm^2/g	30	40
0	2950	4100	4680	5240	6100
10	3060	4560	4980	5710	6370
15	3250	4800	5160	5790	6490
20	3560	4670	6130	6950	7780
25	3590	4570	6180	7050	7410
30	3670	4790	5870	6180	7130
35	3820	4500	5960	6310	7540

Table 6   Results of mechanical tests

ADDITION, %	SSB, cm^2/g	R_f, N/mm^2			R_c, N/mm^2		
		2 days	7 days	28 days	2 days	7days	28days
	2950	4.7	6.8	7.5	22.6	38.2	40.3
	4100	5.1	6.9	7.2	27.0	46.3	52.2
5	4680	5.4	7.0	7.4	29.4	47.9	57.7
	5240	5.4	7.1	7.9	31.1	48.8	59.7
	6120	5.6	7.1	8.2	34.0	50.0	59.4
	3060	4.6	6.5	7.0	19.8	34.2	45.8
	4560	5.0	7.0	7.5	25.5	42.3	52.9
10	4980	5.5	7.0	8.0	27.2	43.5	56.4
	5710	5.7	7.1	8.2	28.5	47.3	58.0
	6370	5.9	7.2	8.5	30.8	47.9	60.3
	3250	4.1	5.9	7.0	18.6	31.7	44.0
	4800	4.2	6.4	8.1	22.1	38.2	49.0
15	5160	4.7	6.6	8.0	26.2	41.3	55.4
	5790	4.8	6.7	7.0	27.1	42.1	53.8
	6490	4.9	6.9	7.7	27.7	42.8	52.2
	3460	3.9	5.4	6.5	17.3	31.0	43.7
	4670	4.0	5.5	6.8	20.4	36.0	45.0
20	6130	4.2	5.9	7.6	24.8	38.4	46.9
	6950	4.4	6.0	6.9	25.2	38.7	50.9
	7780	4.7	6.3	6.8	25.6	39.0	52.9
	3590	3.6	5.1	6.4	16.4	29.8	42.0
	4570	3.8	5.2	6.8	17.0	31.8	43.3
25	6180	4.0	6.1	7.5	20.0	34.0	44.6
	7050	4.4	6.2	6.8	23.2	37.5	47.8
	7410	4.7	6.3	7.1	25.0	38.2	50.3
	3670	3.5	5.0	6.1	15.5	29.0	39.6
	4790	3.6	5.7	6.8	16.3	31.2	41.0
30	5870	4.0	5.9	7.5	18.5	32.5	43.1
	6180	4.4	6.2	7.5	18.5	33.7	43.9
	7230	4.5	6.3	7.2	25.0	39.7	51.0
	3820	3.0	4.8	5.9	13.0	26.5	38.0
	4500	3.4	5.1	6.4	13.8	28.6	41.2
35	5960	3.9	6.1	7.1	16.3	31.8	43.0
	6310	4.1	6.2	7.5	17.3	32.5	43.8
	7540	4.3	6.3	6.9	19.2	35.7	46.0

The influence of the fineness of tuff addition

In the course of this test, the specific surface of cement (clinker + gypsum) was maintained constant at a value, determined to be 3200 cm^2/g. The results of the evolution of strength as a function of the fineness of the addition are given in Table 7. One can indeed note a clear improvement of the strength, when the fineness of the addition grows. This constitutes an important difference with regard to the case where inert additions are used.

Table 7   Influence of the fineness of the addition

TUFF ADDITION, %	SSB, cm^2/g	R_f, N/mm^2			Rc, N/mm^2		
		2 days	7 days	28 days	2 days	7 days	28 days
	3400	3.1	4.6	6.1	13.0	23.0	29.4
25	4000	3.2	4.8	6.5	13.4	22.3	32.6
	4600	3.2	4.9	6.6	12.4	22.0	33.2

CONCLUSIONS

In general, the inert addition decreased cement strength. Hence, these are not to be used if there is a requirement for an extensive improvement compared to the strength of its class. From this point of view, inert additions present an interest for purely economical reasons, which is not relevant in such cases given the quantity that may be possible to add to cement. Nevertheless, the strength of cement is variable and using inert additions could lead to fatal mistakes without continuous control on site.

Active additions present a particular interest. These allow, in addition to reducing production cost, for providing some important properties to cement.

This survey intended to prove that it would be possible to include substantial quantities of active additions in cement. This quantity can further increase if the addition is ground finer. A modification to the manufacturing technology of Portland cement can then be implemented with the use of active additions: one can consider utilizing the improved fineness when using a high percentage of very active addition.

REFERENCES

1. MOUISSI O, Etude comparative des ajouts sable et calcaire au ciment Portland, Mémoire de fin d'études, Université de Blida, 1996.

2. RAPPORT U RE G, Ciments aux ajouts, Alger, 1989.

3. BASTIDE J C, La fabrication du ciment, Eyrolles, 1993.

4. GOUDJIL S AND MEDJEBEUR N, Choix et optimisation de l'ajout au ciment Portland, Mémoire d'ingénieur, Université d'Oran, Algeria, 1992.

PERFORMANCE OF ENERGETICALLY MODIFIED CEMENT (EMC) AND ENERGETICALLY MODIFIED FLY ASH (EMFA) AS A POZZOLAN

H Justnes
SINTEF
Norway

V Ronin
Luleå University of Technology
Sweden

ABSTRACT The Energetically Modified Cement (EMC) technology consists of processing a blend of ordinary Portland Cement (OPC) and a pozzolan through multiple high intensity grinding mills to impart increased surface activation of the OPC and pozzolan particles. Fly ash (FA) of low reactivity as such was selected as pozzolan. FA may be ground together with all cement forming ready cement, or alternatively ground with a small amount of cement ($\approx$5%) and used as a pozzolan added to the concrete mixer. Environmental benefits are 1) activation of FA that otherwise is not suitable for addition to concrete, 2) reduce the CO_2 emission to the atmosphere (40% using 50% FA) and 3) reduce the energy consumption in the binder production. Performance of EMC with 50% FA is compared to OPC and simple 50:50 blends of OPC and fly ash as binder in mortar and concrete in terms of setting time, strength development, drying shrinkage, sulphate resistance and alkali-silica reactivity. The strength of EMC with 50% FA is comparable to OPC, setting time is similar, and alkali-silica reactivity is lowered as well as drying shrinkage. Activated FA as pozzolan may replace up to 70% of cement in concrete. Improved performance in spite of small increase in fineness is explained by formation of agglomerates of OPC/FA grains of high inner surface. Furthermore, breaking of closed glassy aluminosilicate spheres in the fly ash will increase its reactivity. Full scale industrial projects (parking lot and highway pavements) in Texas, USA, with concrete based on EMC with 50% FA are presented, showing low cracking tendencies.

Keywords: Carbon dioxide reduction, High volume fly ash, Blended cement, Setting time, Compressive strength, Durability.

Professor H Justnes, Chief Scientist at SINTEF Building and Infrastructure, Concrete, in Trondheim, Norway. He is also Adjunct Professor in Cement and Concrete chemistry at the Norwegian University of Science and Technology.

Professor V Ronin, Adjunct Professor at Luleå University of Technology, Luleå, Sweden. His research fields are cement & concrete technology, innovative blended cements and recycling of industrial by-products in concrete.

INTRODUCTION

The energetically modified cement (EMC) technology was developed at Luleå University of Technology in Luleå, Sweden, by Dr. Vladimir Ronin *et al.* in the early 1990's. The EMC technology employs a high intensity mechanical activation process to increase the reactivity of Ordinary Portland cement (OPC) with high filler and/or pozzolan replacements. The EMC technology consists of processing a blend of OPC and filler/pozzolan through multiple high intensity grinding mills to impart increased surface activation of the OPC and pozzolan particles. The high intensity grinding is typically accomplished by multiple stages of vibratory or stirred ball mills. The grinding circuit and type of grinding mills are typically custom designed for the raw materials to produce EMC low in Portland clinker with performance characteristics equivalent to parent OPC, or to make EMC with similar clinker content and superior properties. The process can also be used to activate pozzolans of low reactivity (like certain fly ashes) and used them as addition to the concrete mixer later on.

A number of EMCs, and concrete based on them, has been tested at Luleå University of Technology (e.g. [1-9]) as well as at SINTEF (e.g. [10, 11]) both for performance and microstructural changes to understand mechanism.

A recently developed energetically modified product is fly ash with ≈5% cement treated in the vibration mill that can be added together with Portland cement in the production of a concrete in a conventional mixer. It has been shown that the amount of fly ash can be increased from about twenty percent with untreated fly ash to the level of seventy percent with modified fly ash maintaining the required strength level. A commercial product with energetically modified fly ash (EMFA) has been introduced in Texas, USA, under the trademark CemPozz, and it is this product that has been used in the present investigation under the abbreviation EMFA. One interesting field observation using concrete produced with EMFA was that there seems to be significantly less appearance of cracks when producing slabs on ground and highway paving in comparison with the general experience using traditional concretes.

EXPERIMENTAL

The material used in the major part of this study is a 50:50 inter-grind (EMC) of TXI-Ordinary Portland cement (OPC) and Reliant Energy's Limestone Power Station fly ash (FA). The physical and chemical characteristics of the EMC are compared to that of OPC, FA and a conventional 50:50 blend of OPC and FA, while the EMC performance in mortar and concrete is compared to that of neat OPC and OPC with 50% FA replacement.

Chemical analyses have been performed according to ASTM D-4326 and ASTM C-114 while the particle size distributions of EMC cement and the constituent raw materials (OPC and FA) have been performed with the use of Hariba laser scattering particle size analysis.

Time of setting of EMC paste were compared to that of reference OPC paste using the Gilmore apparatus according to ASTM C-266. The paste consistency was verified using the Vicat needle per ASTM C-187. Evaluation of water demand and compressive strength development of mortar and concrete has been made in accordance with ASTM C-109, ASTM C-311 and ASTM C-192. Sulfate resistance was evaluated according to ASTM C-1012, while alkali silica reactivity (ASR) was tested per ASTM C-441.

Paste samples for DTA/TG (Differential Thermal Analysis/Thermo Gravimetry) were crushed to a fine powder and dried at 105°C (i.e. to remove physically adsorbed water). The DTA/TG experiments were carried out by a NETZSCH 409 STA with a heating rate of 10°C/min until 1000°C and nitrogen as a carrier gas. The sample ($\approx$ 150 mg) was contained in an alumina crucible and alumina powder was used as a reference. The accuracy of the temperature determined for phase transitions was within ±2°C, while the accuracy of the mass losses was within ±0.3 mg.

Paste samples for MIP/HeP (Mercury Intrusion Porosimetry/Helium Pyknometry) were bits of about 5 mm size. The MIP experiments were carried out with the Carlo Erba Porosimeter (Model 2000) that records the pore size (radii) distribution of the sample between 5 and 50,000 nm, assuming cylindrical pores. The density of solid materials, ρ_s, was determined by Micrometrics AccuPyc 1330 He-pyknometer, while the particle density, ρ_p, was determined by Carlo Erba Macropores Unit 120. The accuracy of total porosity is within ±0.5 and density within ±0.01 units.

RESULTS AND DISCUSSION

The chemical analysis of EMC and its constituents are listed in Table 1 and corresponding particle size distributions in Table 2. The chemical analysis corresponds to an ASTM Class F fly ash. The EMC grinding process was effective in reducing the coarse fraction of the fly ash. The percentage of the simple blend retained on 325 Mesh was decreased from 12% to 5% by EMC method. This specific type of fly ash is relatively coarse and has significantly lower pozzolanic activity as compared to the other ashes in the area. Another study of EMC using 50% ASTM Class F fly ash (noted FAP for chemical analysis in Table 1) replacement [4] up to 28 days curing revealed that fine particles of fly ash and cement formed agglomerates of outer size comparable to cement grains but with a considerable inner surface explaining increased reactivity. The results for these samples [4] including analyses after 2.5 years (50/50 sealed/wet cured) are listed in Tables 3 and 4 for TG and porosimetry, respectively. EFAP denotes energetically modified 50/50 OPC/FAP paste, while BFAP is 50/50 OPC/FAP blended paste.

Table 1   Chemical composition of EMC (50/50 OPC/FA) and its constituents, as well as fly ash for TG and porosity test (FAP)

COMPOUND	OPC, %	FA, %	EMC, %	FAP [4], %
CaO	62.4	15.0	40.9	2.5
SiO_2	17.8	49.4	33.2	53.0
Al_2O_3	4.0	19.6	6.3	25.0
Fe_2O_3	3.9	5.2	4.1	9.5
SO_3	3.2	0.8	1.6	-
Na_2O	<0.1	0.3	0.1	-
K_2O	0.3	1.2	1.2	-
Insolubles	0.5	51.3	21.6	-

Table 2   Particle size distribution

PARAMETER	OPC	FA	50/50 BLEND	EMC
Median size, μm	16.0	14.3	14.3	11.8
Min. size, μm	1.5	1.3	1.3	1.5
Max. size, μm	50	100	100	70
Specific surface, cm^2/cm^3	5624	6624	6075	7220
< 10 μm, %	61	38	52	63
> 325 Mesh, %	5	20	12	5

Table 3   Features from thermal analysis for EFAP/BFAP paste as a function of time

AGE	TOTAL MASS LOSS, %	DEGREE OF HYDRATION, %	CH, %	CH/MASS LOSS, %
6 h / 12 h	6.89 / 7.30	28 / 29	4.11 / 3.90	60 / 53
1 day	8.64 / 7.96	69 / 64	8.65 / 6.92	100 / 87
3 days	9.93 / 9.73	79 / 78	9.28 / 8.67	93 / 89
7 days	10.38 / 10.20	83 / 82	9.18 / 8.62	88 / 85
28 days	11.15 / 10.89	89 / 87	7.37 / 8.25	66 / 76
910 days	13.46 / 14.27	108 / 114	3.99 / 5.78	30 / 41

For the EFAP and BFAP pastes in Table 3, the total mass loss is only marginally higher for EFAP than for BFAP at the different termini, except for 910 days when it actually lower probably due to a denser matrix halting reaction. The CH content reaches a higher level in EFAP than in BFAP at 1 day, but decreases faster as a function of time and reaches a lower level at 28 days. This indicates that the pozzolanic reaction of fly ash is faster in EFAP (already between 1 and 3 days) than in BFAP (mostly between 7 and 910 days), which is understandable considering that spherical shells in the fly ash are crushed in the milling process allowing simultaneous reaction on two sides of the glassy fly ash wall. The consumption of CH in EFAP paste is significantly more than in BFAP paste after 28 and 910 days, being +13% and +27%, respectively.

The general usual trends in Table 4 are that the porosity decreases as a function of time and the specific surface increases as a function of time as the pores becomes smaller in size but higher in numbers (e.g. gel pores). The average density of solids decreases as a function of time due to increasing amount of crystal water as hydration proceeds. The porosity of EFAP paste is smaller than the blended BFAP paste from about 7 days due to higher degree of hydration/pozzolan reaction, and is particularly much lower after 910 days. SEM images of 7 days paste in Figure 1 show that the EFAP paste appears much denser than BFAP paste. The pore size distribution of the two samples plotted in Figure 2 reveals a substantial pore refinement with the average pore openings of EFAP and BFAP being 11 and 22 nm, respectively. The pore size distribution is very different; whereas BFAP has a bimodal size distribution of pore openings with a considerable amount around 600 nm and the rest below 100 nm. EFAP, on the other hand, has only a small amount of pores with openings above 40 nm. The reason why $\varepsilon_{Hg} > \varepsilon_{He}$ in particular at 910 days (see Table 4) is probably that highly pressurized mercury are crushing delicate structures and opens otherwise inaccessible pores.

Table 4   Specific surface, S_g, particle density (ρ_p), solid density (ρ_s), Hg accessible porosity (ε_{Hg}) and He accessible porosity (ε_{He}) of EFAP / BFAP pastes as a function of curing time

AGE	S_g, m²/g	ρ_p, kg/m³	ρ_s, kg/m³	ε_{Hg}, % (v/v)	ε_{He}, % (v/v)
6 h /12 h	8.4/9.7	1300/1231	2588/2519	48.2/47.7	49.8/51.1
1 day	20.0/15.5	1302/1243	2373/2359	43.7/44.7	45.2/47.3
3 days	32.8/22.7	1349/1313	2264/2260	39.3/38.4	40.4/41.9
7 days	30.6/20.7	1377/1383	2235/2248	37.6/35.9	38.4/38.5
28 days	40.2/27.2	1349/1371	1931/2102	31.7/34.7	30.1/34.8
910 days	35.7/44.7	1324/1180	1609/1856	23.2/40.5	17.7/36.4

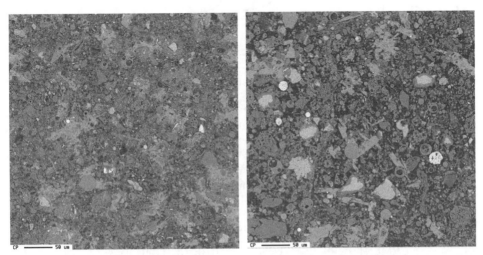

Figure 1   Backscattered electron images (250×) of EFAP (left) and BFAP (right) pastes cured for 7 days at 20°C. The matrix of BFAP appears more porous than the matrix of the EFAP paste, and there are more unreacted cement grains in the BFAP paste. The light gray areas of calcium hydroxide seems to be mass like in EFAP paste, and a mix of mass like and larger individual crystals in the BFAP paste (probably a matter of available space)

The setting behavior of EMC paste is very similar to that of the reference OPC, as can be seen from Table 5. Conventional high volume fly ash (HVFA) Portland-pozzolan blended cements, on the other hand, have typically longer set time; 3-5 hours for initial set and 5-7 hours for final set.

Table 5   Setting time of Paste of OPC and EMC (50/50 OPC/FA)

PROPERTY	OPC	EMC
w/cm	0.24	0.22
Initial setting time, hours:minutes	2:29	2:26
Final setting time, hours:minutes	3:33	3:41

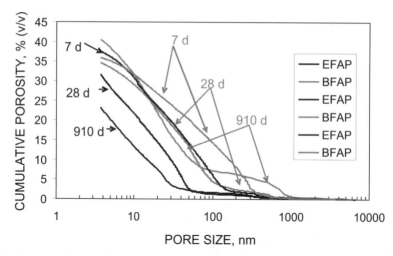

Figure 2   Pore size distribution in EFAP and BFAP pastes (50/50 OPC/FA) at 7, 28 and 910 d

Table 6 represents the data for water demand and the compressive strength development of mortars based on EMC cement (30, 50 and 70% FA) in comparison with ordinary Portland cement and ordinary Portland cement with 20 and 40% of replacement with FA that has not been subjected to the EMC process (reference blends). EMC with 30% fly ash and water-to-cementitious material ratio (w/cm) 0.40 can be considered as high strength / high performance alternative for the newly developed blended cements, while 70% FA replacement may consider a high performance HVFA (high volume fly ash) cement.

Table 6   Compressive strength development, MPa

CEMENT TYPE	w/cm	CURING TIME, days			
		1	3	7	28
OPC[1]	0.48	10.3	26.6	30.0	38.6
EMC (50% FA)[1]	0.43	14.7	22.9	27.2	41.1
EMC (30% FA)[1]	0.40	36.7	-	51.6	62.4
EMC (70% FA)[2]	0.42	11.0	23.0	28.0	-
EMC (70% FA)[2]	0.38	12.5	23.4	30.0	-
80% OPC[1]+20% FA	0.46	6.5	20.4	23.6	35.8
60% OPC[1]+40% FA	0.44	3.8	15.1	17.7	29.6

[1]OPC from Texas, USA. [2]OPC (CEM I 42.5) from Sweden

According to Table 6, the EMC cement made by 50% OPC and 50% FA gave about 40% higher strength after 24 hours than the reference OPC. This EMC mortar had slightly lower compressive strength than OPC mortar at 7 days, but was superior to OPC mortar after 28 days. The EMC (50% FA) performed significantly better than Portland-pozzolan blended cements with 20% and 40% fly ash replacements. The workability of this EMC appears better than the OPC. The high fly ash content in combination with optimized particle size distribution allows 10% reduction in w/cm, which along with the increased reactivity of FA contributes to higher long-term strength. The EMC based on 30% FA and 70% of OPC had compressive strength evolution in line with rapid hardening Portland cement and gave a 28 days flexural strength of 9.6 MPa.

The setting time of HVFA EMC (70% FA) is in line with OPC; initial and final set 2 h 40 min and 3 h 50 min, respectively. No water reducing agent has been used for the mortar with w/cm = 0.42 in Table 6. It had 10-12% higher flow than OPC mortar with same water content. HVFA EMC is very sensitive to addition of superplasticizer. Only 0.1% by mass of cementitious material (cm = OPC + FA) leads to a reduction of w/cm from 0.42 to 0.38 while maintaining flow. Mortars with HVFA exhibit excellent surfaces without flaws. According to Table 6, HVFA EMC shows improved 1 day strength and comparable 7 days strength to OPC mortar of comparable flow.

The EMC strength development relative to OPC was also evaluated using concrete cylinders as shown for the recipes in Table 7 and compressive strengths in Table 8. Even though the earlier compressive strength of EMC concrete was 10-14% lower than OPC concrete, the 28 days strength was 11% higher.

Table 7   Concrete mix design parameters

PARAMETER	OPC	EMC (50% FA)
Cement, % of mass	13	13
Sand, % of mass	38	38
Coarse aggregate, % of mass	45	45
w/cm	0.67	0.66
Slump, mm	50	50
Air content, % (v/v)	2.0	1.5
Unit weight, kg/m^3	2357	2447

Table 8   Concrete strength development

CEMENT TYPE	w/cm	COMPRESSIVE STRENGTH, MPa			
		1 d	7 d	14 d	28 d
OPC	0.67	16.0	22.1	26.4	29.0
EMC-50% FA	0.66	13.8	19.4	23.7	32.2
% of reference	-	86	88	90	111

Table 9 represents the change in length of mortar bars exposed to sodium sulfate solution and the maximum permissible values for specimens. Total six specimens for each type of cement have been tested. The mortar bars made with EMC cement (50% FA) have slightly improved sulfate resistance over reference OPC. The expansion after 4 weeks was roughly one fourth of the maximum permissible level for blended cement. Table 9 also shows that mortar bars made with EMC cement have considerably better resistance (92% improvement) with respect to alkali-silica reactivity (ASR) than OPC mortar bars.

Table 9   Expansion of mortar due to sulfate exposure and ASR

SULFATE RESISTANCE PER ASTM C 1012, % Δ-length		
Cement	OPC	EMC (50% FA)
Max. limit (at 4 weeks)	0.012	0.041
Exposure for		
1 week	0.006	0.006
2 weeks	0.012	0.011
3 weeks	0.013	0.011
4 weeks	0.013	0.011
ASR per ASTM C 441, % Δ-length		
Cement	OPC	EMC (50% FA)
After 14 days	0.026	0.002 (-92%)

Table 10   Slump and compressive strengths of concrete mixtures using EMFA

EMFA, %	35	50	50
OPC+EMFA, kg/m^3	256	273	249
EMFA, kg/m^3	93	136	125
Water, kg/m^3	106	191	137
25 mm aggregate, kg/m^3	1127	1097	1038
Fine aggregate, kg/m^3	827	742	919
Air-entrainer, ml/m^3	155	0	155
Water reducer, ml/m^3	580	0	657
w/cm	0.40	0.70	0.55
Slump, mm	44	216	152
7 days σ_c, MPa	25.6	9.8	14.6
28 days σ_c, MPa	33.8	19.9	26.2
56 days σ_c, MPa	36.8	24.9	31.2

In Texas, USA, a truck stop was made in October 2004 using concrete with 250 kg/m^3 cementitious material (40% OPC and 60% EMFA). The 28 day strength was 20 MPa and the surface excellent without cracks as indicated by the photos in Figure 3.

Laboratory experiments showing reduced drying shrinkage is reported elsewhere [12]. Later a part of Highway 69 in Texas was made by concrete with 50% OPC replacement by EMFA. The total cementitious material was 280-300 kg/m^3 and it achieved 15-20% higher 28 day strength than specified and 50% reduction in cracking compared to traditional pavement according to Texas Department of Transportation. Surface finish was excellent and reduced labour requirements. Some photos are shown in Figure 4.

Table 11   Slump and compressive strengths of concrete mixtures using EMFA

EMFA, %	55	55	60	60
OPC+EMFA, kg/m^3	273	273	249	243
EMFA, kg/m^3	150	150	149	146
Water, kg/m^3	136	158	132	148
25 mm aggregate, kg/m^3	1097	1097	1068	1038
Fine aggregate, kg/m^3	823	848	854	825
Air-entrainer, ml/m^3	0	0	155	116
Water reducer, ml/m^3	696	1005	657	464
w/cm	0.50	0.58	0.53	0.61
Slump, mm	140	165	133	171
7 days σ_c, MPa	15.9	12.7	15.2	12.7
28 days σ_c, MPa	27.6	24.7	26.0	23.6
56 days σ_c, MPa	34.4	30.4	31.4	29.5

σ_c = compressive strength

Figure 3   Pouring concrete for the truck stop (left) and concrete surface of the truck stop after setting (right)

Figure 4   Fresh (left) and hardened (right) concrete surface for Highway 69

Regarding the energy consumption of producing EMC (50% FA) versus OPC, the following statements can be made: The manufacturing process of OPC consists primarily of quarrying or blasting of raw materials (limestone, clay), crushing, grinding, blending and conveying of the said raw meal to cement kilns where at high temperatures (about 1450°C) the formation of Portland clinker takes place. The obtained clinker is further ground with gypsum to produce the final product Portland cement. EMC cement contains typically 50% of OPC and 50% of fly ash (FA). Production of such cement includes primarily grinding of FA to obtain fraction < 250 microns (if required), blending of the ground FA with OPC and processing of the said blend through EMC vibrating milling system to obtain the product with the similar size distribution as commercially available Portland cements (fraction < 150 microns). A comparison of the energies involved in the two processes is shown in Table 12, and one can see that the EMC with 50% fly ash only require 54% energy compared to OPC production. 50% OPC replacement should account for 50% less CO_2, but since EMC require somewhat more electrical energy in grinding the saving may be about 40% (providing that the energy production involves burning of fossil fuel).

Table 12   Energy consumption OPC vs. EMC

CEMENT	OPC	EMC (50% FA)
Clinker production energy:	3.16 GJ/ton (878 kWh/ton)	1.58 GJ/ton (50% clinker)
Burning energy:	*100%*	*50%*
Electrical power grinding cement:	100 kWh/ton	(50% OPC) 50 kWh/ton
Electrical power EMC process:	0	38 kWh/ton
Electrical power:	*100%*	*88% (38+50 = 88)*
Total energy	(100%·878+100%·100) /(878+100) = 100%	(50%·878+88%·100) /(878+100) = 54%

CONCLUSIONS

EMC cement based on 50% of ASTM Class F fly ash (FA) and 50% of ordinary Portland cement (ASTM Type I)) showed about 40% higher compressive strength after 24 hours than the reference Portland cement. Compressive strength development for EMC at 7 and 28 days are in line with that of the pure ASTM Type I Portland cement.

The EMC (50% FA) gave respectively about two times and three times the 1-day compressive strength of conventional blends of Portland cement with 20 and 40% replacement by ASTM class F fly ash. The 7 and 28 days compressive strength were also significantly higher.

EMC cement based on 30% of FA and 70% of Portland cement showed strength development in line with rapid hardening Portland cement, which enables production of high performance/ high strength FA concretes.

EMC (50% FA) had less water requirements (increased workability) compared to simple blends, which contributes to higher strength along with the increased fineness.

The mortar samples produced with EMC (50% FA) had improved sulfate resistance. The change of length values stood at just over 1/4 of the permitted level after 4 weeks and 1/10th of the permitted level after 15 weeks.

The mortar samples produced with EMC (50% FA) had considerably lower alkali-silica reactivity (up to 92% lower change in length) in comparison with ordinary Portland cement.

Fly ash of low pozzolanic activity can be activated by the energetic modification technique (EMFA) together with a small amount of OPC ($\approx$5%) and be used as a pozzolanic additive to mortar and concrete replacing cement. The addition of such a pozzolan reduces shrinkage as shown by laboratory experiments and through practice.

The generated data revealed very promising areas for further research in the field of HVFA high performance cements and concretes with significantly improved environmental profile enabling 46% savings in energy and at least 40% less CO_2 emissions.

REFERENCES

1. HEDLUND H, RONIN V AND JONASSON J-E, 5th International Symposium on Utilization of High Strength/High Performance Cement, Sandefjord, Norway, 20-24 June 1999, ed. by I. Holand and E. J. Sellevold, Norwegian Concrete Association, Oslo 1999, pp. 1144-1153, ISBN 82-91341-25-7.

2. JOHANSSON K, LARSSON C, ANTZUKIN O N, FORSLING W, RAO K H AND RONIN V, Cement and Concrete Research, Vol. 29, 1999, pp. 1575-1581.

3. JONASSON J-E, RONIN V AND HEDLUND H, Proceedings of the 4th International Symposium on the Utilization of High Strength/High Performance Concrete, Paris, France, August 1996, Presses Pont et Chaussees, Paris, 1996, pp. 245-254.

4. JUSTNES H, ELFGREN L AND RONIN V, Cement and Concrete Research, Vol. 35, 2005, pp. 315-323.

5. JUSTNES H, DAHL P A, RONIN V AND ELFGREN L, Proceedings of the 6th CANMET/ACI International Conference on Recent Advances in Concrete Technology, June 2003, Bucharest, Romania, pp. 15-29.

6. RAO K H, RONIN V AND FORSBERG K S E, Proceedings of the 10th International Congress of the Chemistry of Cement (Ed. by H. Justnes), Gothenburg, Sweden, June 1997. Inform Trycket AB, Gothenburg, 3ii104, 9 pp. (ISBN 91-630-5497-5).

7. RONIN V, US Patent Nr. 6,936,098 B2 (2005).

8. RONIN V, US Patent Nr. 6,818,058 B2 (2004).

9. RONIN V AND JONASSON J-E, Report 1994:03, Division of Structural Engineering, Luleå University of Technology, Luleå, Sweden, 1994, 24 pp.

10. RONIN, V AND JONASSON, J-E, Proceedings of International Conference on Concrete under Severe Conditions, Vol. 2, K. Sakai, N. Banthia and O.E. Gjørv (Eds.), E.&F.N. Spon (ISBN 0419 19860 2), Sapporo, Japan, August 2-4, 1995, pp. 898-906.

11. RONIN, V, JONASSON, J-E AND HEDLUND, H, Proceedings of the 10th International Congress on the Chemistry of Cement (Ed. by H. Justnes), Gothenburg, Sweden, June 1997. Inform Trycket AB, Gothenburg, 2ii077, 8pp. (ISBN 91-630-5496-5).

12. JUSTNES H, RONIN V AND JONASSON J-E, Proceedings of the Sixth International Symposium on Cement & Concrete by the Chinese Ceramic Society, Xi'an, P.R. China, September 19-22, 2006, 15 p.

CHALLENGING 75 MICRON LIMITS
FOR PAVEMENT CONCRETE

C Hazaree

Iowa State University
United States of America

ABSTRACT. This paper studies the influence of the aggregates fraction passing 75 micron (FP75) in fine aggregates on the properties of pavement concrete. Two series of trials were conducted with 35% cement replacement by fly ash. Natural sand (NS) and crushed stone sand (CS) were used with FP75 ranging from 0 to 20%. The mechanical strengths, static modulus of elasticity and abrasion resistance of various mixes were also evaluated. The results indicate that for a constant consistency (compaction factor) as the FP75 content increases, the water demand increases; setting times increase for NS; while they decrease for CS; mechanical strength, static modulus of elasticity (SME) and abrasion resistance decrease. The concepts of specific abrasion and specific abrasion ratio can be utilized for characterizing and comparing the abrasion resistances of concrete mixes. Ratios of SME at 90 and 28 days are obtained to evaluate the temporal evolution of SME for different mixes. Finally recommendations regarding the material and construction specifications involving higher FP75 content are made.

Keywords: Microfines, Mechanical Strength, Static modulus of elasticity, Specific abrasion

Chetan Hazaree, obtained his BS degree from the University of Pune in 2002, where he was awarded Prof. P.K. Mehta fellowship for carrying out research on concrete incorporating high volumes of fly ash and lagoon ash. Subsequently he worked for three years as Materials Engineer on various Highway and infrastructure construction projects. These include BOT and World Bank funded projects. He has also earned the Advanced Concrete Technology certificate from the City and Guilds of London Institute. He has published several papers on mechanical aspects, abrasion resistance of roller compacted concrete, and concrete with marginal materials, gap graded concrete and high volume coal ash concrete. Currently he is pursuing his graduate studies in Geotechnical and Materials engineering at the Iowa State University, USA. He is working on the freeze/thaw, permeability and sorptivity aspects of concrete as a part of research under Dwight David Eisenhower Research Fellowship awarded by the National Highways Institute, USA. He is also serving on two Transportation Research Board committees.

INTRODUCTION

There has been a long-term growing demand for aggregates in India in order to produce concrete engendering increased gap between demand and supply of good quality virgin aggregates. This problem is further exaggerated with the aggressive introduction and implementation of infrastructure projects all over India. Reasonable quality aggregates are becoming scarce and uneconomical to acquire. With the need to conserve depleting reserves of good quality aggregates; obligations to use industrial wastes and by-products; and with continually escalating economic pressure in time bound projects; engineers are increasingly driven to use locally available marginal or inferior materials in concrete. It is becoming essentially critical for civil engineers to use durable, high-performance and sustainable materials, produced, at a reasonable cost and with the lowest possible environmental impact. This need is driving engineers towards making innovative sustainable technological developments in concrete. This paper primarily reports the findings of one such research that looked into the applicability of higher percentages of microfines (fraction passing the 75 micron sieve, FP75) in pavement concrete.

LITERATURE REVIEW

Even though the problem regarding aggregates is widely faced and tackled in practice, limited research has been done on this topic. Ahmad and El-Kourd [1] conducted research on natural and crushed stone very fine sand in Riyadh, Saudi Arabia. They observed that for a constant slump the water demand increased with increase in the very fine sand (VFS) content; compressive strength decreased linearly with increase in the VFS content; bond strength remained unaffected by the increase in VFS and drying shrinkage strains increased with increase in the VFS content [1]. Celik and Marar conducted similar research with FP75 content up to 30% of sand replacement at a constant water-cement ratio. They observed that the slump and air content decreased as the dust content increased; water permeability decreased as the FP75 increased; increasing FP75 up to 10% improved the compressive and flexural strength while reducing the drying shrinkage strains, beyond 10% compressive and flexural strength decreased whilst the drying shrinkage strains decreased; impact resistance improved up to 5%, beyond which it decreased and minimum water absorption value was obtained at 15% [2].

RESEARCH SIGNIFICANCE

Sustainability is becoming increasingly important in today's world. Judicious usage of available natural resources with sufficient recycling and reuse are becoming the key drivers of a balanced growth. This becomes more important in the developing countries like India, where the population explosion and rapid urbanization are calling for a substantial infrastructural demand. With mammoth infrastructure development projects undertaken all over India, good quality, virgin resources are becoming scarce driving civil engineers towards the usage of marginal materials. With timely research and technical development, the problem could be acutely analysed and resolved. This research was undertaken with the motivation of utilizing very fine aggregates in concrete for pavements. In the literature there is no such consolidated study available that has looked into critical material properties for pavement concrete. This paper presents the results of the effects of excessive quantities of fraction passing 75 micron (FP75) on the mechanical strengths, elastic modulus and abrasion resistance of pavement concrete mixes. Since the chemical admixtures are costly in developing countries, the primary objective of this study was to generate a data inventory for further research, therefore no chemical admixtures were used. Both natural sand (NS) and crushed stone sand (CS) were used with FP75 content ranging from 0 to 20%. It is anticipated that this research would help specification writers, engineers and contractors while selecting very fine aggregates or fine aggregates with microfines suitably. Table 1 presents the specified limits of aggregate fraction passing by various standards.

Table 1   Specifications for fraction passing 75 micron

STANDARD	STD. REF.	AGGREGATE FRACTIONS, %					REMARK
		Coarse aggregate		Fine aggregate		All-in aggregate	
		uncrushed	crushed	uncrushed	crushed		
ASTM	C 33-01	1	1	3	3	NA	Note A
		1	1	5	4	NA	Note B
BSI	882:1992	2	4	4	16	11	
IS	383:1970	3	3	3	15	NA	

Note A: Abrading surface, different tolerances allowed
Note B: Non-abrading surface

MATERIALS AND METHODOLOGIES

Binders

Ordinary Portland cement (OPC) complying with IS 8112:1989 (Equivalent to BS 12: 1991) was used in the investigations. Class F fly ash (FA) from one of the local Thermal Power Station meeting IS 3812:2003(BS 3892: 1997) was used along with OPC in cement replaced mixtures. The physical characteristics and chemical composition of OPC and fly ash are summarized in Table 2. The initial and final setting times of cement were 192 and 240 minutes, respectively and the lime reactivity of fly ash was 4.5 MPa at 28 days.

Table 2   Physical properties and chemical constitution of binders

TESTS		UNIT	BINDERS	
			cement	fly ash
Physical tests				
Blaine fineness		m^2/kg	297	365
Specific gravity		g/cm^3	3.15	2.31
Soundness (Autoclave)		%	1.10	0.90
Lime reactivity		MPa	NA	4.80
Compressive strength at	3d	MPa	34.5	NA
	7d	MPa	42.2	NA
	28d	MPa	54.5	NA
Chemical composition				
Silicon dioxide		%	21.00	61.30
Aluminium oxide		%	5.20	32.30
Ferric oxide		%	3.80	4.20
Calcium oxide		%	61.20	0.90
Sulphur as SO_3		%	2.00	NA
LOI		%	1.25	1.10
Alkalis		%	0.50	0.70

Aggregates

The coarse aggregates were crushed Deccan trap basalt aggregates with nominal maximum size of 20 mm. These were processed from the integrated crusher having a vertical shaft impactor (VSI). All the coarse aggregates were angular and had combined shape indices value (includes flakiness and elongation) in the range of 16-30% depending on the size fractions. Two types of fine aggregates were used in the trials. Natural sand was uncrushed-clean, siliceous and natural from the Krishna River; while crushed stone sand was crushed basalt and was obtained from the same crusher from which the coarse aggregates were obtained. No mica was found in the sand. It was ensured that the particle size distribution of both the fine aggregates were almost identical All these aggregates complied with the requirements of IS 383-1970 (BS 882:1992). The geological origin, physical characteristics and particle size distribution of individual fractions of concern to the concrete mix proportioning are summarized in Table 3.

Table 3   Aggregate properties

PROPERTY	AGGREGATE TYPE			
Size fractions, mm	4.75-25.00	4.75-12.5	0-4.75	0-4.75
Parent rock	Basalt	Basalt	Siliceous	Basalt
Type	Crushed	Crushed	River sand	Crushed
Specific gravity (bulk), g/cm^3	2.98	2.965	2.74	2.87
Water absorption, %	1.15	1.15	1.68	2.4
Sieve size, mm	Cumulative fractions passing, %			
20.000	98.0	100.0	100.0	100.0
10.000	10.6	98.2	100.0	100.0
4.750	0	0	100.0	100.0
2.360	0	0	86.9	86.6
1.180	0	0	67.6	66.5
0.600	0	0	42.8	42.5
0.300	0	0	19.9	19.8
0.150	0	0	6.8	6.9
0.075	0	0	0	0

Table 4 represents the 24 h water absorption of various fractions of aggregates. These values are important in appreciating the instantaneous water absorption as it will decide the water demand and slump retention of the mix; while the 24-hour water absorption will also have some influence on drying shrinkage. It was observed that the instantaneous (first 15 min) water demand of the finer aggregate fractions reached values close to 24 h water absorption.

Table 4   Water absorption of different fine aggregate fractions

DETAILS	24 h WATER ABSORPTION, %						
Fraction, mm	4.75-2.36	2.36-1.18	1.18-0.6	0.6-0.3	0.3-0.15	0.15-0.75	0.75-0.0
Natural sand	1.2	1.2	1.4	1.5	1.8	1.9	2.8
Crushed sand	1.4	1.4	1.7	2.2	2.8	3.1	4.3

Fraction Passing 75 μm

This size fraction was obtained individually from either type of fine aggregates by sieving. Final fine aggregate gradings were obtained by mixing weighted proportions of aggregate passing 75 μm with the individual gradings. All the percentages passing 75 μm in subsequent discussions represent the cumulative percentage passing.

Water and Admixtures

Well-water confirming to IS456: 2000 and used in ready mix concrete (RMC) plant supplying concrete for pavements was used for the study. No admixture was used.

MIXTURE PROPORTIONING

The control mixes were proportioned for a compaction factor of 0.85 ± 0.02, as it is widely used in pavement construction specifications in India. Cement content of 370 kg/m^3 was used in the control trials; whilst a cement content of 275 kg/m^3 and fly ash content of 145 kg/m^3 was used in rest of the trials. A cement substitution factor of 1.5 was used in replacing cement with fly ash (values were subsequently rounded off to nearest 5 kg).

The percentage combination of coarse aggregate fractions was kept constant for all the trials, to ensure that the aggregate grading for optimum flexural strength in retained. FP75 was directly substituted for an equal weight of fine aggregate. Table 5 shows the summary of the concrete mixture proportions used in these investigations.

Table 5   Mixture proportions and fresh properties

MIX DESIGNATION		w/b	MIXTURE PROPORTIONS, kg/m^3				CF	SETTING TIME, h:min.		
			water	CA	FA	FP75		IST	FST	FST/IST
Control	NSC0	0.378	140	1250	830	0	0.870	4:20	6:00	1.38
	CSC0	0.411	152	1250	830	0	0.860	3:45	5:15	1.40
NS	NS0	0.338	142	1295	695	0	0.880	5:05	7:10	1.41
	NS4	0.338	142	1295	670	28	0.890	5:15	6:30	1.24
	NS8	0.360	151	1279	635	55	0.875	5:45	6:55	1.20
	NS12	0.393	165	1255	593	81	0.845	6:10	7:10	1.16
	NS16	0.429	180	1226	554	106	0.830	6:00	7:30	1.25
	NS20	0.457	192	1205	518	130	0.820	6:40	7:30	1.13
CS	CS0	0.362	152	1295	695	0	0.875	4:30	6:15	1.39
	CS4	0.374	157	1285	664	28	0.860	4:10	5:40	1.36
	CS8	0.393	165	1272	630	55	0.855	3:45	5:00	1.33
	CS12	0.421	177	1249	591	81	0.835	3:10	4:05	1.29
	CS16	0.452	190	1225	554	106	0.820	2:55	3:35	1.23
	CS20	0.488	205	1198	516	129	0.810	2:20	2:40	1.14

SPECIMEN FORMULATION AND TESTING SCHEDULE

In the present series of trials, mixing sequence was tailored as per the sit needs and with an objective of simulating the real life workability obtained from the twin shaft mixing plant. Full amount of coarse aggregates and half of the fine aggregates were mixed for 2 min followed by addition of ¾ water and admixture and mixing for another 1.5 min; followed by addition of binders and remaining fine aggregate and mixing for 1 min. This was followed by addition of reaming ¼ water and mixing for 2 min; subsequent to which the mixer was covered and the mix allowed to rest for 2 min followed by 2 min of mixing.

Table 6   Concrete testing schedule

TEST	SPECIMEN TYPE	SPECIMEN GEOMETRY	DAYS	REF. STD.
Setting time	Mortar	600 × 125 × 125	NA	IS 8142-1976
UUCS	Cube	150	7, 28, 90	IS 516-1959
MOR	Beam	700 × 150 × 150	7, 28, 90	IS 516-1959
CST	Cylinder	∅ 300 × 150	7, 28, 90	IS 5816-1999
SME	Cylinder	∅ 300 × 150	28, 90	ASTM C469-94
AR	Rectangular prism	65 × 65 × 60	28, 90	IS 1237-1980†

Note: All mechanical strengths were checked at 7, 28, 56 and 90 days;
   † Concrete sieved over 16 mm sieve

Table 6 gives the specimen description and testing schedule adopted for the testing program. The test ages of 28 and 90 days were selected based on probable traffic opening timings for fly ash mixes. The Table also provides information about the testing standards followed for various tests.

RESULTS AND DISCUSSIONS

Water demand, Consistency and Setting Behaviour

The water demand, consistency and setting times of the mixes are shown in Table 5. It was observed that the water demand for the cement replaced mix with 0% FP75 remained almost similar to that of the control mixes. With the increase in the FP75 there was a consequent increase in the water demands of the mixes. The increase ranged between 1.43 to 37% and 0 to 34% respectively for natural and crushed stone mixes. The increase in the water demand could be attributed to the increased clay and silt content in case of NS; while the increased water absorption rates in the finer fractions of CS.

Contradictory behaviours were observed for the setting times of these mixes. NS on one hand showed extended setting increasing with FP75 content. This increase ranged from 0:45 to 2:20 hr: min for initial setting time and from 0:30 to 1:30 hr: min for final setting time. This could again be attributed to the increased clay and silt content, which in turn might have provided lower mechanical strength during the initial hours after casting of concrete specimens. CS on the other hand showed a drastic reduction in the setting times of the mixes. For lower percentage of FP75, there was a small increase in the setting time, which could probably be attributed to the fly ash effect; while at higher percentage of FP75, reduction in the setting times were observed. This could be attributed to the increase in the water absorption of the aggregates. With an increase in the FP75, the fraction with relatively higher water absorption increased resulting into higher early age water demand and higher quantity of water absorption by the finer fractions in the initial hours.

Mechanical Strength

Table 7 shows the mechanical strengths of all the mixes. It was observed that the temporal evolution of strength of the fly ash mixes was substantially affected by the fly ash/pozzolanic effect. This was observed at the test age of 90 days, when the strength gains over 28 days strength were significantly larger than the corresponding control mixes. This was irrespective of the FP75 content, but the increments were affected by the FP75 content in the mixes.

As the FP75 increased, the contribution rate of pozzolanic effects decreased. Further to this, there was an increase in the strength drop with the increase in the FP75 content. This could be attributed to the increased water demand and the consequent increase in the water-binder ratios.

Figures 1 and 2 show the decrease in the strength with FP75 content. It was observed that mixes incorporating NS were more sensitive to changes in FP75; while CS mixes showed relatively less sensitive behaviour. It was also observed that the ratio of tensile strength to compressive strength increased with age, thus illustrating a better evolution of the tensile strength as compared to the compressive strength. Further to this the percentage tensile strength of NS mixes was relatively lower than the corresponding CS-mixes. Moreover, at a constant cement factor, even though the water-binder ratios of CS mixes were relatively higher than the corresponding NS mixes, the CS mixes showed a better mechanical behaviour. This illustrates the importance of better aggregate-paste interlock obtained in the crushed rock aggregates.

Table 7   Mechanical strength of concrete

MIX	UUCS, MPa			MOR, MPa			CST, MPa		
	7 d	28 d	90 d	7 d	28 d	90 d	7 d	28 d	90 d
NSC0	47.1	56.5	58.2	5.2	6.3	6.5	2.6	3.2	3.3
CSC0	49.2	58.2	62.1	5.5	6.5	7.0	2.7	3.3	3.4
NS0	39.6	53.0	67.7	4.2	5.7	7.4	2.2	3.0	3.6
NS4	36.6	50.2	62.2	3.5	5.0	6.3	1.8	2.5	3.2
NS8	34.2	44.2	55.2	3.3	4.3	5.4	1.6	2.2	2.7
NS12	29.2	37.7	45.9	2.7	3.4	4.2	1.3	1.7	2.1
NS16	27.2	35.5	43.1	2.4	3.3	3.9	1.3	1.7	2.0
NS20	18.6	25.2	31.2	1.6	2.3	2.6	0.9	1.2	1.3
CS0	43.6	57.2	73.5	5.0	6.5	8.2	2.5	3.3	4.2
CS4	42.2	55.2	69.2	4.8	6.0	7.7	2.4	3.1	3.8
CS8	39.3	49.2	62.2	4.2	5.3	6.8	2.2	2.7	3.4
CS12	35.6	45.3	56.2	3.7	4.8	6.0	1.9	2.3	3.1
CS16	33.9	43.2	53.1	3.3	4.6	5.5	1.6	2.2	2.7
CS20	29.2	37.7	42.2	2.8	4.0	4.7	1.4	1.9	2.3

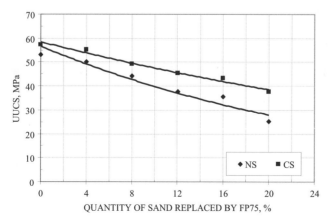

Figure 1   Variation of UUCS with FP75

Static Modulus of Elasticity

Figure 3 shows the temporal evolution of SME of all the mixes between 28 and 90 days. Since all the concrete mixtures were cured isothermally, hence these modulus values represent isothermal static modulus of elasticity. It was observed that the stress-strain behaviour of concrete mixes differs depending on age and FP75 content. Further to this the ratio of (E_{90}/ E_{28}) was calculated for all the mixes and is represented in Figure 3. It was observed that the SME decreased with increasing FP75 content, but not in the same proportion as the UUCS. This might be attributed partially to the aggregates and their behaviour and to the pozzolanic effects produced in individual mixes. Furthermore, it was observed that this ratio depends on the fine aggregate type (and he resulting aggregate-paste interlock), pozzolanic effects and age of curing. Similar to the UUCS sensitivity to the type of fine aggregates (and hence FP75), even SME values of NS mixes are relatively more sensitive to the change in the FP75 content than the corresponding CS mixes. An in depth analysis and discussion of the stress-strain curves of concrete mixtures is beyond the scope of this paper.

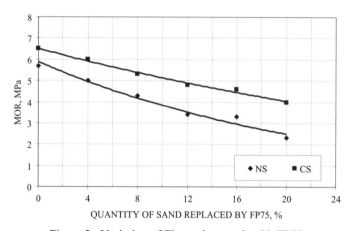

Figure 2   Variation of Flexural strength with FP75

Abrasion Resistance

Important notions

The following notions are defined on the basis of the following two assumptions:

1. The strength of the skin of concrete (one aggregate deep) is having the same compressive strength as the core of the concrete;
2. The time rate of abrasion of concrete remains same throughout the testing time i.e. up to 2 mm depth of wear.

Specific Abrasion (SP_{abr}) is defined as the ratio of time required to reach a depth of wear of 2 mm to the compressive strength of concrete at that age.

Specific Abrasion ratio (κ) is defined as the ratio of specific abrasion at a particular age to the specific abrasion at 28 days.

The abrasion resistances of various mixes were evaluated using the time required in minutes to reach a wear depth of 2 mm. It was observed during the physical testing of concrete that the wear due to abrasion is a function of mortar and surface quality. The abrasion resistances were not found to be solely dependent on the compressive strength of concrete, but were dictated by the curing period and binder constitution. Figure 4 represents the time required for 2 mm abrasion wear, the specific abrasion values for all the mixes. In general, it was observed that the specific abrasion increased with curing and age thus illustrating an improvement in the strength of the skin concrete. Further to this, the SP_{abr} decreased with increase in the FP75 content. The decrease in NS mixes was more than the corresponding CS mixes, thus illustrating the significance of a better aggregate-paste interlock. Additionally, the specific abrasion ratios at 90 days decreased with increase in the FP75 content, but with no specific trend. This illustrates that even for the same strength of concrete; the strength of skin concrete may be different leading to differences in the SP_{abr}. The concept of SP_{abr} can be applied in estimating the compressive strength requirements for a particular value of abrasion resistance, provided sufficient database is available.

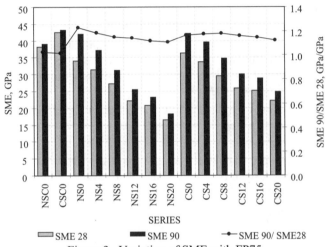

Figure 3   Variation of SME with FP75

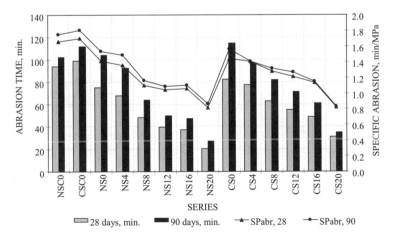

Figure 4   Abrasion resistance

CONCLUSIONS AND RECOMMENDATIONS

The following conclusions and recommendations could be made based on the presented study. For a given slump:

1. The water demand, depending on the fine aggregate type increases as the FP75 replacing equivalent weight of fine aggregate increases. In NS clay and silt content and in crushed stone sand higher instantaneous water absorption besides the increased surface area due to FP75 leads to such an increase. Consistency is affected in similar manner.

2. Contradictory behaviour is observed in setting times of these mixes. Increasing percentage of FP75 in NS leads to increase in the setting times; while increasing percentage of FP75 in CS leads to reduction in setting times with respect to the corresponding control mixes.

3. Mechanical strength decrease as the FP75 increases. These follow non-linear (power) function. NS is relatively more sensitive than CS to change in strengths with changes in the FP75 content.

4. SME also shows behaviour similar to the mechanical strength, but follows slightly different trend.

5. The concept of specific abrasion and specific abrasion ratio can be applied in studying the relative strength of the core and skin of the concrete. This study shows that the abrasion resistance is a function of parameters beyond compressive strength. The abrasion resistance of concrete mixes decreased with the increase in FP75 content.

6. For given cement factor, even though the CS mixes had a higher water-binder ratio when compared to the corresponding NS mixes, the CS mixes outperformed NS mixes because of the inherent strength and better aggregate-paste interlock.

7. The mechanical strengths should not be used alone as criteria for specifying concrete mixes for pavements. Even if some mixes show god mechanical strength behaviour, they might fail in complying with the elastic modulus or abrasion resistance tests.

8. In NS FP75 content can be cautiously allowed up to 8%, provided no other deleterious material is present; while in CS FP75 content may be limited to 12%. Caution must be exercised in hauling, compacting and finishing time vis-à-vis the setting behaviour of these mixes. With the usage of high range water reducers, these percentages could be increased when backed up by sufficient lab and field experience.

9. The usage of fly ash or lagoon ash with higher C content or higher levels of cement replacements must be very carefully executed.

REFERENCES

1. AHMED A E AND EL-KOURD A A, Properties of Concrete Incorporating Natural and Crushed Stone very fine sand, ACI Materials Journal, Vol. 86, No. 2, 1989, pp. 417-424.

2. CELIK T AND MARAR K, Effects of crushed stone dust on some properties of concrete, Cement and Concrete Research, Vol. 26, No. 7, 1996, pp. 1121-1130.

EFFECTS OF CALCIUM NITRATE AND ALKANOLAMINES ON THE SETTING AND STRENGTH EVOLUTION OF PORTLAND CEMENT PASTES

N Chikh **M C Zouaoui**

University of Constantine   University of Oum El-Bouaghi

Algeria

S Aggoun   R Duval

University of Cergy-Pontoise

France

ABSTRACT. The purpose of this work is to investigate the effect of using some admixtures such as calcium nitrate, triethanolamine and tri-isopropanolamine, on the setting and hardening process of cement pastes at 20°C temperature. Tests were carried out on specimens cast considering various added rates of admixtures and types of cements. It is established that the calcium nitrate alone acts as a setting accelerator, but has relatively little beneficial effect on the long term development of mechanical resistances. The results obtained indicated that both alkanolamines used alone performed well as a hardening accelerator at all ages regardless of the cement type used. The combined additions produced at very early age significant and promising results with respect to both setting and hardening acceleration, and to a continuous compressive strength increase with time.

Keywords: Cement paste, Hardening accelerator, Setting accelerator, Early age, Compressive strength, Setting time.

Pr N Chikh, is Professor, University of Constantine, Algeria. His main research fields is early age concrete and strengthening of concrete elements with FRP materials.

Dr M Cheikh-Zouaoui, is Maître de Conférences, University of Oum El-Bouaghi, Algeria. His main research fields is cement and concrete properties.

Dr S Aggoun, is Maître de Conférences, University of Cergy-Pontoise, France. His main research fields is cement and concrete properties.

Pr R Duval, is Professor, University of Cergy-Pontoise, France His main research fields is use of chemical admixtures for concrete.

INTRODUCTION

Various admixtures have been used to produce a concrete with sufficient strength at a very early age. Calcium chloride has been in the past the most widely used set accelerator for this purpose. However, the presence of chloride causes serious problems regarding corrosion of reinforcing bars embedded in concrete members. This has renewed interest to develop a number of chloride-free admixtures. Dodson [1] presented a review of non-chloride, non-corrosive set accelerating salts. His investigation, started in 1962, found the calcium formate, $Ca(CHO_2)_2$, can be used as a set accelerator [2]. The second salt to fulfill requirements was calcium nitrate, $Ca(NO_3)_2$, which was patented in 1969[3]. After an intensive research, it was established that the calcium nitrate was also a very effective corrosion inhibitor for metal imbedded in concrete.

The effect of calcium nitrate additions on the setting characteristics of cements and the steel corrosion has been studied by Justnes [4,5]. It was initially assumed that the accelerating effect was dependent on the aluminate content of the cement. This assumption was not verified for cements at low temperature (5°C). The results obtained showed that calcium nitrate does function as a set accelerator for cement between 7°C and 20°C, but that the set accelerating efficiency of calcium nitrate depends very much on the cement type used. It was also found that the set accelerating efficiency of calcium nitrate seems to increase with increasing belite content in cement or other cement characteristic promoting belite formation in the clinker process.

Triethanolamine (TEA), a low tertiary alkanolamine, is used as a grinding aid in cement manufacture and as a constituent in certain admixture formulations in concrete practice. Depending on the cement type and addition rate, TEA can produce either set acceleration or retardation. An addition rate of 0.02% to the type I Portland cement, TEA acts as a set accelerator, at 0.25% as a mild set retarder and at 0.5% a severe retarder and at 1% a very strong accelerator [6,7]. The effect on the strength development in cement pastes is also dependent on the added amount of TEA [8,9]. The addition of small amounts of higher tertiary alkanolamines such as tri-isopropanolamine (TIPA) resulted in interesting increases in the strength of cement pastes at different ages [10,11]. A recent study carried out on the strength enhancing mechanism of tri-isopropanolamine presented compressive strength data for 10 Portland cements tested as cement paste after 28 days of hydration [12]. The average strength improvement with 200 ppm TIPA added to the mix water was 10%.

The purpose of the present study is to find combinations of calcium nitrate with other compounds so as to fulfill the criteria for both a setting and a hardening accelerator. In this respect, the choice of alkanolamines presenting a developed spatial molecular structure like TEA and TIPA seems to be appropriate to verify previously indicated criteria.

EXPERIMENTAL STUDY

The physical properties and the chemical analyses of the two cements (C1 and C2) used in this investigation are shown in Table 1. Both cements present similar specific surface and tricalcium silicate (C_3S) content but their tricalcium aluminate (C_3A) content are different, with C2 cement containing a lower amount in C_3A. Each of the cement type used is manufactured by separate cement plant situated in east Algeria.

Table 1 Physical properties and chemical analyses of the cements

CHEMICAL COMPOSITION	CEMENT 1 (C1) %	CEMENT 2 (C2) %
CaO	64.36	63.91
SiO$_2$	22.00	21.62
Al$_2$O$_3$	5.02	4.49
Fe$_2$O$_3$	2.94	5.37
SO$_3$	1.94	1.92
MgO	2.07	1.66
K$_2$O	0.47	0.25
Na$_2$O	0.26	0.08
Mineralogical Composition	%	%
C$_3$S	51.28	52.48
C$_2$S	24.68	22.69
C$_3$A	8.33	2.82
C$_4$AF	8.94	16.32
Physical Properties		
Ignition Loss, %	0.64	0.81
Blaine surface, m^2/kg	352	332

The general formula of the calcium nitrate used in the present study is: XNH$_4$NO$_3$ YCa(NO$_3$)$_2$ ZH$_2$O. The coefficients X, Y and Z values are respectively X = 0.092, Y = 0.500 and Z = 0.826, which correspond to 19.00% Ca^{2+}, 1.57% NH$_4^+$, 64.68% NO$_3^-$ and 14.10% H$_2$O. The chemical formulae of the triethanolamine and the tri-isopropanolamine added was respectively N(CH$_2$CH$_2$OH)$_3$ and N(CH$_2$CHOHCH$_3$)$_3$. These additions were used for the purpose of providing the hardening performance to the blend admixture. The w/c ratio for the various mixtures is 0.3. The characteristics of theses mixes are shown in Table 2.

The setting test on the cement pastes was performed using a Vicat apparatus and following EN196-3 procedure. The hardening criteria were determined through simple compressive strength tests on cubic specimen 50 x 50 x 50mm^3. These latter were cast in metallic moulds and kept for 24 hours at 20±1°C and at 55±5% RH. They were then demoulded and conserved in water at 20°C until required for testing. Using a hydraulic testing machine, compressive strength tests were carried out according to the code NF18-406. Each compressive strength value represents the average of the results from 3 specimens tested.

RESULTS AND DISCUSSIONS

Setting Times and Compressive Strength

Tests were carried out on specimen at the age of 1, 3, 7 and 28 days. The various results obtained, regarding setting times and compressive strength, are given in Table 3 and Figures 1 and 2. Both C1 and C2 cements present higher dicalcium silicate content but a different comparing content in aluminate and aluminate-ferrite.

Table 2   Mixes investigated

CEMENT TYPE	MIXES DESIGNATION	% CN	% TEA	% TIPA
	M10	0	0	0
	M11	1.00	0	0
	M12	0	0.05	0
	M13	0	0.10	0
C1	M14	0	0	0.05
	M15	0	0	0.10
	M16	1.00	0.05	0
	M17	1.00	0.10	0
	M18	1.00	0	0.05
	M19	1.00	0	0.10
	M20	0	0	0
	M21	1.00	0	0
	M22	0	0.05	0
	M23	0	0.10	0
C2	M24	0	0	0.05
	M25	0	0	0.10
	M26	1.00	0.05	0
	M27	1.00	0.10	0
	M28	1.00	0	0.05
	M29	1.00	0	0.10

Table 3   Experimental setting times and compressive strengths results

MIXES INVESTIGATED	SETTING TIMES min		COMPRESSIVE STRENGTH N/mm^2			
	Initial	Final	1 day	3 days	7 days	28 days
M10 (Control)	137	205	18	25	38	60
M11 (1%CN)	125	195	27	36	41	67
M12 (0.05%TEA)	150	235	28	36	45	70
M13 (0.10%TEA)	150	225	36	64	76	96
M14 (0.05%TIPA)	140	210	32	48	80	96
M15 (0.10%TIPA)	130	210	40	52	88	112
M16 (1%CN+0.05%TEA)	75	130	25	32	40	72
M17 (1%CN+0.10%TEA)	80	120	28	60	72	80
M18 (1%CN+0.05%TIPA)	75	110	36	48	80	88
M19 (1%CN+0.10%TIPA)	75	100	44	52	82	96
M20 (Control)	188	330	11	17	35	50
M21 (1%CN)	150	260	9	15	29	60
M22 (0.05%TEA)	190	240	17	25	40	62
M23 (0.10%TEA)	180	230	32	56	76	100
M24 (0.05%TIPA)	175	210	32	48	80	100
M25 (0.10%TIPA)	170	200	36	52	88	120
M26 (1%CN+0.05%TEA)	130	195	14	26	50	70
M27 (1%CN+0.10%TEA)	120	180	36	60	72	88
M28 (1%CN+0.05%TIPA)	120	175	28	48	88	104
M29 (1%CN+0.10%TIPA)	110	165	36	56	92	120

Effect of calcium nitrate

In regards to C1 cement, with normal tricalcium aluminate content (8.33%), the calcium nitrate affected significantly the evolution in the early age compressive strengths, which increased by 50% at 1 and 3 days compared to the control mix. However, the improvements noticed on the setting times were very moderate. Considering C2 cement, which has a low tricalcium aluminate content (2.82%), the obtained results indicate that calcium nitrate acts as a good setting accelerator. Compared with control mix initial and final setting times, the presence of calcium nitrate reduces them by 20%. However, as the early age compressive strength was not improved, calcium nitrate can not be classified as a hardening accelerator though it improved the 28 days strength by almost 20%.

Thus the efficiency of the calcium nitrate as a set accelerator seems to be dependent on the cement chemical composition, confirming thereby the results obtained by Justnes [5].

Effect of triethanolamine

The addition of 0.05% and 0.10% TEA with C1 cement produced minor effect on M12 and M13 mix setting times. However, the improvement was significant on the compressive strength. With 0.05% TEA, the increase in strength, compared to control mix M10, was about 55% at 1day, 44% at 3 days, 18% at 7 days and 15% at 28 days. The influence was further enhanced with 0.10% TEA as it led to an increase of about 100% at 1day, 156% at 3 days, 100% at 7 days and 60% at 28 days.

Regarding C2 cement rich in C_4AF, the addition rate of 0.05% and 0.10% TEA reduced the final setting times to about 72% of the control mix M20. It also improved the compressive strength at all ages of M22 mix, about 55% at 1day, 47% at 3 days, 14% at 7 days and 24% at 28 days. More effect was experienced with 0.10% TEA as the recorded increase was about 190% at 1day, 230% at 3 days, 117% at 7 days and 100% at 28 days.

These results show that the triethanolamine performed well as a hardening accelerator regardless of the cement type used. In addition, the strength increase was more significant with TEA content passing from 0.05% to 0.10%.

Effect of tri-isopropanolamine

The use of respectively 0.05% and 0.1% TIPA with C1 cement had a small effect on the M14 and M15 mix setting times. However, it produced a major improvement on the compressive strength at all ages. With 0.05% TIPA, the increase in strength for M14 mix compared to control mix M10 was about 80% at 1 day, 90% at 3 days, 110% at 7 days and 60% at 28 days. Further strength increase was recorded with 0.10% TIPA, about 120% at 1 day, 110% at 3 days, 130% at 7 days and 85% at 28 days.

In the case of C2 cement, the effect of using 0.05% TIPA reduced the final setting time by about 40%. It also improved considerably the compressive strength at all ages of M24 mix, about 185% at 1 and 3 days, 130% at 7 days and 100% at 28 days. A added rate of 0.10% TIPA produced results very similar to the previous final setting times, whereas the compressive strength increased by about 225% at 1 day, 205% at 3 days, 150% at 7 days and 140% at 28 days compared with control mix M20.

Regardless of the cement type used, the performance of tri-isopropanolamine as a hardening accelerator was greater than that of triethanolamine. Besides, the strength increase produced by TIPA in the rate range 0.0% to 0.05% was more significant than that in the rate range 0.05% to 0.10%.

Effect of calcium nitrate combined with triethanolamine

The use of calcium nitrate in combination with triethanolamine resulted in their joining effects. The initial and final setting times were reduced respectively to 55% and 62% for C1 cement and to 66% and 58% for C2 cement of their respective control mix values.

Regarding C1 cement, the compressive strength increase for short and long term ages was respectively more than 35% and 20% for M16 mix and more than 60% and 30% for M17 mix. Further improvement was obtained with C2 cement which contains high aluminate-ferrite content. In comparison with control mix, the strength increase recorded respectively for short and long term ages was more than 30% and 40% for M26 mix and more than 230% and 75% for M27 mix.

These results clearly indicate that calcium nitrate combined to triethanolamine leads at the early ages to some interesting results in terms of accelerating both setting and hardening.

It is worth indicating that the performance in strength produced by TEA in the rate range 0.05% to 0.10% was far greater than that produced in the rate range 0.0% to 0.05%.

Effect of calcium nitrate combined with tri-isopropanolamine

Such combination of admixtures resulted in their joining effects, thus initial and final setting times were reduced respectively to 45% and 50% for C1 cement and to 40% and 50% for C2 cement of their respective control mix values. The mechanical resistances were also enhanced at all ages.

With respect to control mix M10, the compressive strength raised respectively for short and long term ages by more than 80% and 60% for M14 mix and more than 110% and 85% for M15 mix. Such increase was further improved with M28 (M29) mix as it recorded more than 155% (225%) and 110% (140%) respectively for short and long term ages. Thus better admixture reactivity took place with C2 cement.

Thus calcium nitrate combined to tri-isopropanolamine resulted at early ages in a very interesting performance in terms of accelerating both setting and hardening.

It should be emphasised that the increase in strength produced TIPA in the rate range 0.0% to 0.05% was far more important than that in the rate range 0.05% to 0.10%. The dosage of 0.05% TIPA seems to be close to optimum for theses cements.

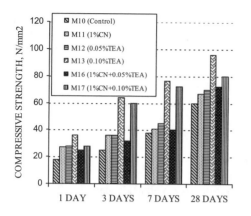
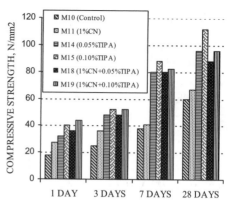

Figure 2(a)   Compressive strength for C1 mixes

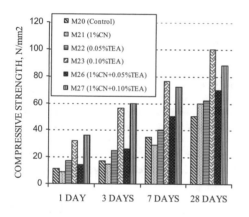
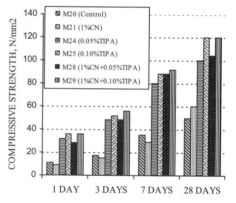

Figure 2(b)   Compressive strength for C2 mixes

CONCLUSIONS

Tests have been carried out on cement pastes specimens to investigate the effect of using some admixtures such as calcium nitrate, triethanolamine and tri-isopropanolamine, on the setting time and compressive strength evolution. In view of the admixture rates added, the main results show that:

- The calcium nitrate acted mainly as a setting accelerator with efficiency depending on the cement chemical composition. In the long term, the effect on the strength increase was insufficient for it to be considered as a hardening accelerator.

- Both triethanolamine and tri-isopropanolamine performed well as a hardening accelerator at all ages and regardless of the cement type used.

- Used with the same rate, tri-isopropanolamine is in most cases far more efficient in terms of strength increase than triethanolamine.

- The combination of calcium nitrate with either triethanolamine or tri-isopropanolamine resulted in their synergistic effects with time, translated by a reduction in the initial and final setting times and a significant strength enhancement at all ages of the cement pastes, particularly at early ages.

- It may be inferred that calcium nitrate acted as a setting accelerator then either triethanolamine or tri-isopropanolamine took over by accelerating mainly the hardening phase.

REFERENCES

1. DODSON V H, Concrete Admixture, V N Reinhold, New York, 1990, pp. 73-102.

2. DODSON V H, FARKAS E AND ROSENBERG A M, U S Patent No. 3,210,207, Oct 5, 1965.

3. ANGTADT R L AND HURLEY F R, U S Patent No. 3,427,175, Feb. 11, 1969.

4. JUSTNES H AND NYGUARD E C, The influence of technical calcium nitrate additions on the chloride binding capacity of cement and the rate of chloride induced corrosion of steel embedded in mortars. International Conference on Corrosion and Corrosion Protection of Steel in Concrete, Sheffield, UK, July1994, pp.491-502.

5. JUSTNES H AND NYGUARD E C, Technical calcium nitrate as set accelerator for cement at low temperatures. Cement and Concrete Research. Vol. 25, 1995. pp.1766–1774.

6. RAMACHANDRAN V S, Action of triethanolamine on the hydration characteristic of tricalcium silicate. Journal of Applied Chemistry and Biotechnology. Vol. 22, 1972. pp.1125-1138.

7. RIXOM R AND MAILVAGANAM N, Chemical Admixtures for Concrete, E & FN Spon, London, 1999, 437 pp.

8. HEREN Z AND ÖLMEZ H, The influence of ethanolamines on the hydration and mechanical properties of Portland cement. Cement and Concrete Research. Vol. 26, 1996, pp.701-705.

9. AIAD I AND ABO-EL-ENEIN S A, Rheological properties of cement pastes admixed with some alkanolamines. Cement and Concrete Research, Vol. 33, 2003, pp.9-13.

10. JUSTNES H, Accelerator blends for Portland cement. Second International Symposium on Cement and Concrete Technology in the 2000s, Istanbul, Turkey, Sept 2000, pp.433-442.

11. GARTNER E AND MYERS D, Influence of tertiary alkanolamines on Portland cement hydration. Journal of the American Ceramic Society. Vol. 76, 1993, pp.1521-1530.

12. SANDBERG P J AND DONCASTER F, On the mechanism of strength enhancement of cement paste and mortar with tri-isopropanolamine. Cement and Concrete Research. Vol. 34, 2004, pp.973-976.

THE EFFECT OF CRYSTALLIZED SLAG ON THE MECHANICAL BEHAVIOUR OF ALKALI ACTIVATED SLAG CONCRETE

L Zeghichi

B Mezghiche

M'sila University

Algeria

ABSTRACT. This paper reports the results of an investigation on the activation of blast furnace slag. It examines the mechanical strength behaviour and the elastic properties of slag cement concrete activated with Na_2CO_3. The main parameter studied was the effect of crystallized slag aggregates on the evolution of compressive and tensile strengths, modulus of elasticity and deformations of alkali activated slag concrete. The results showed that crystallized slag plays an important role in slag activation for the development of mechanical strength and modulus of elasticity of concrete (an increase of 10% compared to ordinary concrete).

Keywords: Alkali activated slag, Crystallized slag strength, Modulus of elasticity, Deformation.

Dr L Zeghichi is a senior scientist at the civil engineering laboratory, Biskra, Algeria. Her field of specialization is cement and valorisation of by-products. She has published several research papers in reputed national and international journals.

Dr B Mezghiche is a scientist at the civil engineering laboratory, Biskra, Algeria. His field of specialization is cement and additions.

INTRODUCTION

Alkali activation of slag is a method to produce concrete with slag (ground granulated blast furnace slag) without the use of any Portland cement by substituting Portland cement with 100% alkali activated slag [1, 2].

Binding of basic slag (activated slag) originates from the reaction of two components in which the silico-aluminous part is represented by the finely ground granulated slag, and the alkaline component is composed of basic metals, giving an alkaline reaction in water [3].

The mechanical strength of alkali activated slag concrete depends on the type of binding, the fineness of grinding, the nature of basic metals and the density of slag [4].

The choice of aggregates is important, their quality plays a great role, they cannot only limit the strength of concrete, but due to their characteristics, they affect the durability and performance of concrete.

In this paper, we studied the effect of crystallized aggregates on the mechanical behaviour of alkali activated slag concrete.

Blast furnace slag is a by-product obtained in the manufacture of pig iron in the blast furnace, by the reaction of the earthly constituents of iron ore with the limestone flux. Slag originating from a molten stream may be treated by various methods, following the desired type:

- When the slag is allowed to cool slowly, it solidifies into a grey crystalline stony material, know as air cooled or dense slag. This forms the material used as concrete aggregate.

- When the molten slag is cooled rapidly by being poured into a large excess of water, known as quenching, the material breaks up into small particles and solidifies as glass. This product is called granulated slag and is used as cement.

EXPERIMENTAL

Materials

Granulated blast furnace slag

The granulated slag originated from the metallurgic industry (Annaba, Algeria) and was obtained by rapid cooling in water. Its chemical composition is given in Table 1. The density of the granulated slag was 2900 kg/m^3 and its Blaine fineness was 3200 cm²/g.

Table 1   Chemical composition of slag

CONSTITUENTS, %								
SiO_2	Al_2O_3	Fe_2O_3	CaO	MgO	SO_3	K_2O	Na_2O	Cl^-
40.8	5.2	0.53	43.01	6.4	0.80	3.02	0.01	0.007

Activator

The activator used for this experimental work was Na_2CO_3 with a density of 1250 kg/m^3, prepared by adding a corresponding quantity of water.

Bulky Aggregates and Fines

Crushed stone (crystallized slag)

The crushed crystallized slag used was classified into three fractions: 3 to 8 mm, 8 to 16 mm and 16 to 25 mm. The material density was 2550 kg/m^3, the apparent bulk density was 1270 kg/m^3.

Sand

The dune sand used was clean, siliceous and fine sand, whose 0 to 3 mm fraction was taken from the M'sila region (Algeria). The sand density was 2540 kg/m^3, the apparent bulk density was 1540 kg/m^3, its sand equivalent was 75%.

Mixture Proportions

The concrete mixture proportions used were determined by the "Dreux Gorisse" method [5]:

Ordinary concrete
Binder: 350 kg/m^3, sand: 590 kg/m^3, gravel: 1163 kg/m^3, water: 188 l/m^3.

Basic concrete with natural aggregates
Binder (granulated slag): 390 kg/m^3, sand: 565 kg/m^3, gravel: 1115 kg/m^3, basic solution: 196 l/m^3.

Basic concrete with crystallized slag
Binder (granulated slag): 390 kg/m^3, sand: 565 kg/m^3, crystallized slag: 1085 kg/m^3, basic solution: 196 l/m^3.

For the study of concrete, triplicate samples were prepared: series of 10 cm cubes and prisms of 10 × 10 × 40 cm dimension.

RESULTS AND DISCUSSION

Basic slag is an artificial stone used in concrete with cement, basic solution and aggregates.

Study of mechanical strength

The tests carried out on cubes of basic concrete showed the effect of crystallized slag on the development of compressive strength.

Compressive strength test result at 7, 28, 180 and 365 days of hardening are presented in Figure 1.

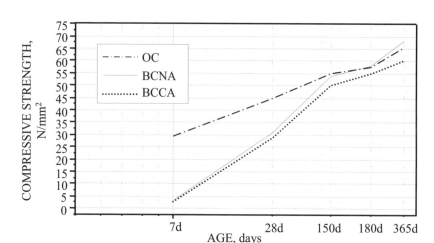

Figure 1   Compressive strength as a function of concrete type

Legend: O.C: Ordinary concrete.
B.C.N.A.: Basic concrete with natural aggregate.
B.C.C.A.: Basic concrete with crystallized slag.

We noted an improvement of strength for basic concrete before 28 days, this meant that the hydration of slag at early ages was slow, the use of waterglass as activator accelerated the hydration of slag and increased the compressive strength considerably.

Tensile strengths for the same types of concrete at 28 and 365 days are given in Figure 2.

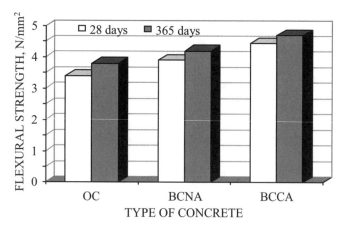

Figure 2   Tensile strength as a function of concrete type

Notable strength improvement was observed when concrete contained crystallized slag. This improvement was due to the good adhesion between the aggregate and hydrated paste. This phenomenon can be explained by the superficial rugosity of slag aggregates.

Study for elastic properties

In this part of work we evaluated the modulus of elasticity for the same types of concrete. The equipment used is shown in Figure 3.

Figure 3   Measuring deformations on compression

The obtained results are presented in Figure 4.

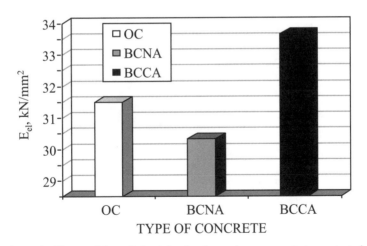

Figure 4   The modulus of elasticity for the various concrete types tested

The results showed that basic concrete with slag aggregates had a modulus of deformation as high as for ordinary concrete, and an elastic modulus which was 7% greater then for ordinary concrete. Crystallized slag may play an active role in the hardening of basic concrete.

CONCLUSIONS

It has been established that it is possible to prepare a basic concrete with high mechanical strength using crystallized slag.

It has been shown that crystallized slag plays an active role in the hardening of basic concrete, it is recommended for many applications such as bridges, marine structures, etc.

REFERENCES

1. COLLINS F AND SANJAYAN J G, Early age strength and workability of slag pastes activated by sodium silicates, Magazine of Concrete Research, Vol. 53, No. 5, October 2001, pp. 321–326.

2. PUERTAS F AND FERNANDEZ F, Mineralogical and microstructural characterisation of alkali-activated fly-ash/slag pastes, Cement and Concrete Composites, 2003, pp. 287-292.

3. MEZGHICHE B, Propriétés mécaniques du béton basique, Séminaire international, Ghardaia (Algérie), 1994, pp. 323–335.

4. ZEGHICHI L, MEZGHICHE B AND CHEBILI R, Study of the effect of alkalis on the slag cement systems, Canadian Journal of Civil Engineering, Vol. 32, No. 5, October 2005, pp. 934–393.

5. DAUPAIN R, LANCHON R AND SAINT-AROMAN J C, Granulats, sols, ciments et bétons, Edition Casteilla, Paris, 2000, 235 p.

INFLUENCE OF VISCOSITY MODIFYING ADMIXTURE ON THE COMPOSITION OF SCC

A Borsoi

M Collepardi

S Collepardi   E N Croce

Enco

A Passuelo

University of Ancona

Italy

ABSTRACT: The role played by the mineral additions is important to obtain a reliable SCC without using an excessive amount of cement, just to compensate the absence of mineral additions. However, there are cases where these mineral additions are not available due to their shortage. In such cases, a higher dosage of viscosity modifying admixture could be employed to compensate for the shortage of solid powder without using an excessive amount of cement. The purpose of the work is to study whether or not this target can be achieved. A special VMA was adopted based on welan gum. Different SCCs were manufactured with cement content in the range of 350-400 kg/m^3 with and without 200 kg/m^3 of ground limestone filler. The VMA was not needed when the volume of cement + ground limestone was about 190 l/m^3. When the ground limestone is not used a reliable SCC can be obtained by increasing the dosage of VMA up to 7 l/m^3. In such a case, the VMA acts as "liquid" filler to compensate for the absence of solid filler.

Keywords: Segregation, Self-Compacting Concrete, Superplasticizer, Slump Flow Viscosity Modifying Agent.

Antonio Borsoi is a laboratory technician of Enco. He is active in the area of concrete mixture design.

Mario Collepardi is author or co-author of numerous papers on concrete technology and cement chemistry.

Silvia Collepardi is a research civil engineer and director of the Enco Laboratory, Ponzano Veneto, Italy.

Emanuela Nunzia Croce is Dr. in Geological Sciences and is responsible for the petrography and mineralogical studies.

Alexandra Passuelo is a researcher from Brazil working in the University of Ancona on concrete properties. She has published some papers in the area of concrete technology.

INTRODUCTION

One of the key points [1] in the manufacture of self-compacting concrete (SCC) is the volume of the fine 'powder' (V_f), including cement and mineral addition in the form of fly ash, silica fume, blast furnace granulated ground slag (BFGGS), ground limestone and the finest particles of sand (< 75μm). This volume should be in the range of 170-200 l/m^3 of concrete (Figure 1). If V_f is > 200 l/m^3, there could be difficulties in transporting this concrete because of its excessive viscosity. On the other hand, if V_f is too low (< 170 l/m^3) there could be some risk of segregation, specially when the volume of coarse aggregate V_g is too high (> 340 l/m^3) [2]. In the absence of any mineral addition, V_f would coincide with the volume of portland cement (V_c) and then the cement content should be in the range of 535-630 kg/m^3 if a specific gravity of 3.15 g/cm^3 is assumed for Portland cement. This excessive cement content would cause cracks promoted by thermal and drying shrinkage which would severely jeopardize the durability of the concrete structure.

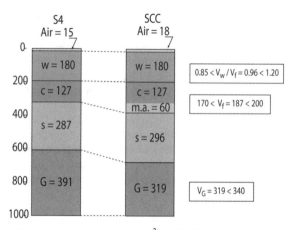

Figure 1   Typical volume composition (l/m^3) of SCC Vs. S4. V_f = powder volume + cement (c) + mineral addition (m.a.); w = water; s = sand volume; G = gravel

Therefore, the availability of fine mineral filler to replace a significant volume of portland cement is absolutely needed to manufacture SCC. However, the SCC technology is widespread in the area of precast concrete (about 30%) but not yet in that of the ready mixed concrete (only 1%) .This means that on the batching plant of a ready-mixed concrete there a silo should be available for the mineral addition of the filler only for some intermittent and sporadic production of SCC. Moreover, in Europe with the advent of the new cement norms (EN 197-1) blended cements rather than pure Portland cement are manufactured, so that there is a shortage of fly ash, BFGGS and ground limestone to be used as mineral additions on the concrete batching plant.

SCOPE

The purpose of the present work is to assess whether the shortage or the absence of mineral additions can be compensated for the use of a viscosity modifying agent (VMA) in manufacturing stable SCC. In general, this admixture is considered to be an optional ingredient for the SCC manufacture to counterbalance an accidental excess of mixing water.

This can happen when wet aggregate is used or when excess water is surreptitiously used in the absence of control on the free water of the aggregate. Indeed, another key point of the SCC technology is the *water-fine powder* ratio in volume (V_w/V_f) which must be in the range of 0.85-1.20 (Figure 1) [2]. If V_w/V_f is lower than 0.85 the concrete would become too viscous and cannot be easily transported by pump or chute; if V_w/V_f is too high (>1.20) again there would be a risk of segregation. However, this drawback can be managed through an adequate check of the water in the aggregates with respect to their absorption, without using VMA.

In the present work, VMA should be a compulsory admixture in the SCC manufacture and play an important role in counterbalancing the shortage or even the absence of any fine mineral addition except that already present in the blended cement. When SCC is sporadically produced in ready mix batching plant the intermittent use of a liquid-based VMA could be a cheaper and more effective solution to face the shortage of fine fillers. In other words, VMA could act as a 'liquid filler' instead of the powder filler.

MATERIALS AND TEST METHODS

Different SCCs were manufactured and assessed in terms of segregation and fluidity by using the following ingredients:

- CEM II B/L 32.5 R according to EN 197-1 was used as cement with a specific gravity of 3.09 in which 25% of the portland clinker is replaced by limestone in the grinding mill;

- natural sand (0-4 mm) and gravel (4-16 mm) both with a specific gravity of 2.7;

- ground limestone (max. size of 63 μm) with a specific gravity of 2.70 was used in some SCCs to attain a V_w/V_f of about 1;

- a polycarboxylic polymer with 20% active matter and a specific gravity of 1.1 was used to reach a slump flow level of at least 650 mm;

- in some SCCs a VMA, based on the welan gum [3] with a specific gravity of 1, was used to mitigate bleeding and segregation (Figure 2);

- an adequate amount of mixing water was adopted to keep the water-cement ratio (w/c) in the range of 0.55-0.50.

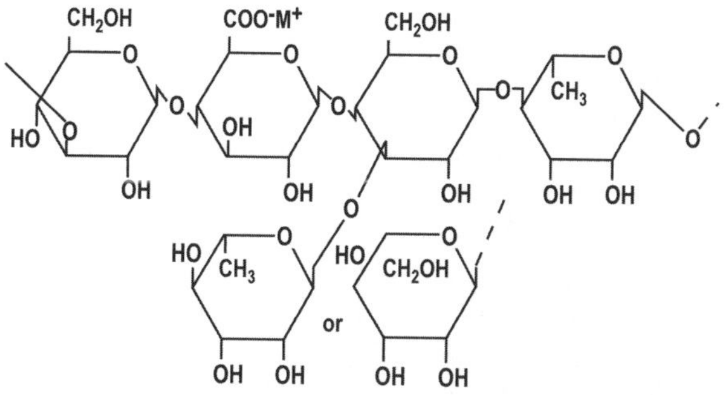

Figure 2 Chemical composition of the polymer welan-gum

Slump flow and L-box test were used to characterize the properties of the concrete mixtures in the fresh state: in particular the slump flow was used to determine the mobility of the concrete (it must be at least 550 mm), whereas the H_2/H_1 ratio in Figure 3 was adopted as parameter to assess the segregation (H_2/H_1 must be at least 0.80 to ensure high passing ability SCC) [2].

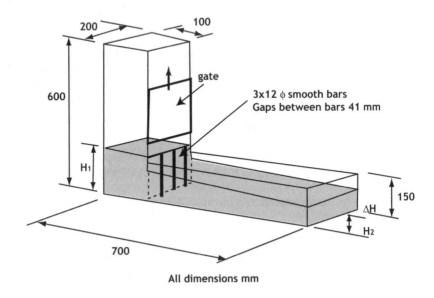

Figure 3   General assembly of the L-box

Compressive strength was determined on cube specimens cured at 20°C with RH > 95% from 1 to 28 days.

RESULTS AND DISCUSSION

Table 1 shows the content of the ingredients (in kg/m³) and their volume (in l/m³) in SCC mixtures A, B and C. Table 1 also presents the concrete properties in the fresh state in terms of slump flow and segregation assessed by visual rating and measurement of H_2/H_1 from the L-box test.

Mixture A behaves as an acceptable SCC for the slump flow (650 mm) as well for the absence of segregation. The volume (V_f) of fine materials (cement and ground limestone) is set to 187 l/m³ which is in the middle of the recommended range of 170-200 l/m³. The *water-fine powder* volume ratio (V_w/ V_f) is at 1.02, which is in the middle of the allowed range of 0.85-1.20 for an acceptable SCC.

However, if the superplasticizer dosage is increased from 5.5 kg/m³ of the mixture A to 6 kg/m³ of the mixture B, in order to increase the slump flow of the mixture from 650 in the mixture A to 750 mm of the mixture B, H_2/H_1 decreases from 0.82 (mixture A) to 0.75 (mixture B) indicating that there is some segregation; this was confirmed by visual observation.

Table 1   Composition and properties of fresh SCC mixtures A, B and C

COMPOSITION	A		B		C	
	by mass	by volume	by mass	by volume	by mass	by volume
Cement, kg/m³	350	113	347	112	347	112
Ground limestone, kg/m³	200	74	197	73	197	73
V_f, l/m³	---	187	---	185	---	185
Sand (0-4 mm), kg/m³	805	298	777	295	797	295
Gravel (4-16 mm), kg/m³	810	300	805	298	805	298
Water, l/m³	190	190	189	189	189	189
Superplasticizer, kg/m³	5.5	5.0	6.0	5.5	6.0	5.5
VMA, kg/m³	---	---	---	----	2.0	2.0
Air, %	---	20.0	---	26.5	---	25.5
w/c		0.55		0.55		0.55
V_w/V_f		1.02		1.02		1.02
PROPERTY						
Slump Flow, mm		650		750		750
H_2/H_1 (L Box Test)		0.82		0.75		0.83
Segregation (Visual Rating)		No		Slight		No

On the other hand, if the increase in superplasticizer content from 5.5 to 6.0 kg/m³ is accompanied by the addition of a small amount (2 kg/m³) of the VMA, the slump flow increases from 650 mm (mixture A) to 750 mm (mixture C) without any segregation at all as confirmed by its H_2/H_1 (0.83) which is higher than the minimum value (0.80) needed for unsegregable SCC. This is the typical role played by VMA expected in SCC technology: to counterbalance an excess of water or superplasticizer that could increase fluidity and favour the transformation of an un-segregable acceptable SCC (mixture A) to a segregable un-acceptable mixture (mixture B).

A new unexpected role played by the VMA admixture will be illustrated in the present paper: how to remove from an acceptable SCC, like mixture of C of the present work, all the mineral filler without any segregation outcome. This, for instance, could be advantageously used to eliminate the filler silo in a concrete batching plant where SCC is produced intermittently. Table 2 shows the compositions and the properties in the fresh state of mixtures C, D and E. The elimination of about 200 kg/m³ of ground limestone of the C mixture was replaced by an increase of 172 kg/m³ of sand content in mixture D; moreover, to increase the slump loss from 700 mm (in mixture C) to 750 mm in mixture D, there was a slight increase in superplasticizer amount (from 6 to 8 kg/m³); finally the dosage of VMA was significantly increased from 2 to 7 kg/m³ in order to counteract the segregation caused by the reduction in the volume of fine material ($V_f/V_c = 112$ l/m³ which is out of the recommended range of 170-200 l/m³ for an unsegregable SCC). In spite of the strong increase in VMA dosage to eliminate the segregation of mixture D, this attempt was unsuccessful, as confirmed by the low H_2/H_1 value of 0.70.

In order to be successful in facing this problem, the cement content of the mixture was tentatively increased from about 350 (in mixture C) up to 400 kg/m³ in mixture E, corresponding to a V_f value of 129 l/m³. Moreover, the amount of mixing water was increased from about 190 kg/m³ in mixture C to 200 kg/m³, so that the workability in terms of slump flow increased from 700 mm (mixture C) to 750 mm in mixture E. The most significant results for the E mixture consist in obtaining unsegregable SCC, even if V_f and

V_w/V_f are out of the usual recommended ranges of 170-200 l/m³ and 0.85-1.20, respectively. So, the elimination of mineral filler, such as the ground limestone, can be compensated for a significant increase of VMA from 2 to 7 kg/m³ and a slight increase of blended cement (CEM II L-B 32.5 R) from 350 to 400 kg/m³ equivalent to only 300 kg/m³ of portland cement.

Table 2 Composition and properties of fresh SCC mixtures C, D and E

COMPOSITION	C		D		E	
	by mass	by volume	by mass	by volume	by mass	by volume
Cement, kg/m³	347	112	347	112	400	129
Ground limestone, kg/m³	197	73	---	---	---	---
V_f, l/m³	---	185	---	112	---	129
Sand (0-4 mm), kg/m³	797	295	969	359	910	337
Gravel (4-16 mm), kg/m³	805	298	805	298	810	300
Water, l/m³	189	189	191	191	200	200
Superplasticizer, kg/m³	6.0	5.5	8.0	7.3	7.7	7.0
VMA, kg/m³	2.0	2.0	7.0	7.0	7.0	7.0
Air, %	---	25.5	---	25.7	---	20.0
w/c	0.55		0.55		0.50	
V_w/V_f	1.02		1.67		1.55	
PROPERTY						
Slump Flow, mm	700		750		750	
H_2/H_1 (L Box Test)	0.81		0.70		0.82	
Segregation (Visual Rating)	No		Yes		No	

The results of compressive strength are shown in Figure 4. In all the concrete, the strength development is the same (about 8 MPa at 1 day and 35 MPa at 28 days) due to the same w/c value (0.55), except for the E mixture that had higher strength because of the lower water-cement ratio (0.50).

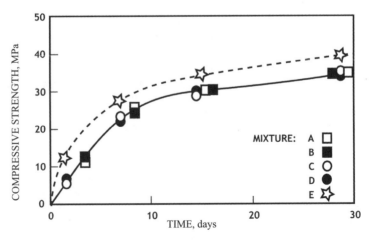

Figure 4 Cube compressive strength from 1 to 28 days for the different mixtures

CONCLUSIONS

The results of the present work confirm that the VMA admixture can be advantageously used to attenuate the consequences of the unavoidable changes in the amount of water of the aggregate or to counteract the segregating effect caused by slight increase of superplasticizer dosage focused to an increase of the slump flow as in the C mixture with respect to mixture A (Table 1).

The incorporation of VMA led to complete elimination of the mineral filler: in such a case a slight increase in cement content must be accompanied by a significant increase in the dosage of VMA (for instance from 2 to 7 kg/m^3) in order to obtain an unsegregable SCC even in the absence of the mineral filler.

The cost increase due to the dosage increase of VMA is compensated by the cost reduction due to the elimination of mineral filler. This means that at least in the first and sporadic production of ready mixed SCC, the silo for the mineral filler is not needed and can be replaced by some drums of 'liquid filler' in form of VMA.

REFERENCES

1. COLLEPARDI M, COLLEPARDI S, OGOUMAH OLAGOT J J AND TROLI R, Laboratory-Tests and Field-Experience of High-Performance SCC, Proceedings of the Third International Symposium on, "Self Compacting Concrete", Rejkjavik, Iceland, 17-20 August 2003, pp. 904 – 912.

2. OKAMURA H AND OZAWA K, Mix Design for Self-Compacting Concrete, Concrete Library of JSCE 25, 1995, pp. 107-120.

3. ROBINSON G, MANNING C E AND MORRIS E R, Conformation and Physical Properties of the Bacterial Polysaccharides Gellan, Welan and Rhamsan, Food Polymers, Gels, Colloids, Special Publication, R. Sol. Vol. 82, 1991, pp. 22-33.

SOME FACTORS INFLUENCING CEMENT AND SUPERPLASTICISER COMPATIBILITY

W Kurdowski

A Garbacik

Center of Building Materials

S Grzeszczyk

Opole University of Technology

Poland

ABSTRACT. In the last years there has been great progress in the understanding of the mechanism of superplasticizer interaction with cement paste. However, many experimental facts are still controversial and it is far from having the possibility of full control of the behaviour of the system cement – superplasticizer. In order to explain some controversial opinions the model of Portland cement which was composed of alite with different aluminates addition was studied. The examined mixtures were the following: alite, alite with addition of gypsum and calcite, alite with C_3A and gypsum, alite with Klein complex and gypsum. The addition of kleinite was adjusted to give approximately the same or grater content of ettringite in the paste. The rheology properties of the paste with sulphonated naphthalene formaldehyde condensate and carboxylic admixture have shown that the replacement of calcium aluminate by kleinite can give very similar rheology of the paste. The acceleration or retardation of hydration of alite also influences the rheology of the paste. Conclusion is taken that the formation of ettringite governs the rheology of the paste, but also the change in the rate of C-S-H formation.

Key words: Alite, C_3A, Kleinite, Rheology, Ettringite, C-S-H, Superplasticizer.

Professor W Kurdowski works at the Center of Building Materials, Krakow, Poland.

Dr A Garbacik works at the Center of Building Materials, Krakow, Poland.

Professor S Grzeszczyk works at Opole University of Technology, Opole, Poland.

INTRODUCTION

The compatibility of the system cement-superplasticizer presents a very important problem within concrete technology [1-3]. It is well know that one cement with admixture gives very good results concerning concrete fluidity and retention of good workability during long period whilst another may presents a very poor response to the same superplasticizer [3]. Many factors have an influence on cement-superplasticizer compatibility including C_3A and alkali sulphates content in cement, calcium sulphate type added as setting retarder, free lime concentration in clinker as well as specific surface of cement.

The reasons of these factors influencing workability were perceived to be adsorption properties of different cement phases (i.e. being particularly high for C_3A and occurs in substantial amounts within a few seconds) [4]. On the other hand Ramachandran [4] states that the adsorption of superplasticizers occurs on the hydrating products rather than on unhydrated phases. Aitcin [3] discusses in details the situation with sulfonated polymers which SO_3^- groups are adsorbed on active sites on C_3A surface, competing in this action with SO_4^{2-} anions from calcium or alkali sulphates.

However, Nawa and Eguchi [5] found that in the system C_3A- gypsum the adsorption of SMF is lower for the prehydrated mixture. Bonen and Sarkar [6] found a very good correlation by plotting adsorption against the product of the C_3A fraction times the cement fineness. Assuming one can recognize that interaction between a PNS superplasticizer and C_3A is crucial when controlling the rheology of Portland cement mortars and concretes. Bonen and Sarkar [6] conclude that the initial flow of cement paste is mainly governed by the fineness and C_3A content of the cement. However, flow loss is best correlated with ionic strength of the pore solution. They conclude that C_3A and ettringite content has a marginal effect on it.

Next Jiang et al. [2] have found that soluble alkali sulphate content is one of the major parameters controlling fluidity and fluidity loss in cement paste containing polynaphthalene-sulfonate (PNS) superplasticizer. Optimum soluble alkali content with respect to fluidity and fluidity loss is 0.4-0.5% Na_2O_e. In cement with this optimum amount of soluble alkalis the C_3A content has practically no effect on fluidity loss. Kim et al. [7] determined the super-plasticizer adsorption capacity of the same cements and found that incompatible cements have higher adsorption capacity of PNS superplasticizer because of a lack of soluble alkali sulphates.

The addition of Na_2SO_4 contributes to increase the slump area by reducing the amount of PNS adsorbed. These findings are in disagreement with commonly accepted opinion that the adsorption of superplasticizer has the positive effect on cement paste rheology by increasing electrostatic repulsive forces between cement particles. The authors explain this disagreement by admitting that the PNS molecules remaining in solution may then act as an additional repulsive barrier between cement particles, further contributing to improve the fluidity of the paste. Additionally as hydration proceeds, the PNS molecules adsorbed on the solid surfaces become ineffective due to incorporation into newly formed hydration products. The PNS remaining in solution is then progressively adsorbed onto the newly hydrated surfaces; the continuous supply of PNS molecules maintains the electrostatic and steric repulsive forces to conserve paste fluidity.

Another approach was presented by Prince et al. [8] which state that ettringite and modification of formation of this phase in the presence of PNS has the decisive influence on rheology of cement paste. Particularly the growth of long needles of ettringite that usually decreases paste flowability is delayed by PNS addition. Prince [8, 9] underlines that the interaction of superplasticizer molecules with ettringite germs could explain also the high consumption of superplasticizer molecules in the early stage of hydration. However, the examined by Prince mixture is relatively far from cement systems.

One can conclude that the understanding of the mechanism of superplasticizer interaction with cement paste has been explored. However, many experimental facts are still controversial and it is far from the possibility to have the full control of the behaviour of the system cement - superplasticizer. In order to obtain some new information of the factors governing the rheology of cement paste with superplasticizer experiments on model system composed of main cement phases i.e. alite, tricalcium aluminate and gypsum was carried out. We also examined the influence of ettringite formation in the paste when the source of aluminate ions is not C_3A, but Klein complex.

MATERIALS AND METHODS

In the experiments classical methods were applied, i.e. X-ray for phase analysis and rotation viscometer Viscotester VT 550 for rheological measurements and semiadiabatic micro-calorimetry for heat of hydration measurements. Rheological properties were established on the basis of flow curves for increasing and decreasing shear rate. Yield value and plastic viscosity were appointed on the basis of Bingham model.

As a Portland cement model alite was used to which several phases were added, principally tricalcium aluminate and gypsum, but also calcite. Cement phases, namely alite and C_3A were prepared in laboratory furnace by heating at 1500°C, three times with intergrinding. Alite contained 1% MgO, 2% of Al_2O_3 and 1% of Fe_2O_3.

One sample of C_3A had no admixture and the second contained 3% of Na_2O. Klein complex were burned at 1300°C. For these samples preparation the pure AR materials i.e. Merck calcium, magnesium, alumina and iron hydroxides as well as calcium sulphate and amorphous silica – Aerosil, Degussa were used. Alite contained 0.3% of free lime and C_3A 0.4%.

$CaSO_42H_2O$ was the by-product of gas desulfurisation and X-ray examination has shown that the only detectable phase was gypsum. Calcite was the natural mineral and no others phases were X-ray detectable. Also anhydrite was natural mineral AII with no lines of other phases on X-ray pattern.

Alite and tricalcium aluminate were X-ray examined using Rietveld program. Alite contained 88% of monoclinic, 8% of rhombic and 4% of triclinic phases. C_3A was cubic phase and contained 0.6% of portlandite. The sample with Na_2O admixture, labelled C_3A_{Na}, was also cubic, but contained 1% of rhombic phase and 2% of $C_{12}A_7$. Klein complex was pure kleinite.

All components used for mixtures preparation were ground to 300 m^2/kg.

In our experiments two superplasticizers were used. One was the sulphonated naphthalene formaldehyde SNF and the second the acrylic copolymer AC formed of three kinds of groups of acid derivative: carboxylic, alkyloesthric and polyetherester. Superplasticizers were applied as 40% water solutions.

Eight mixtures were prepared which composition is given in Table 1. Klein complex additions were so adjusted to ensure the equal and higher quantity of ettringite content formed in comparison to the samples with C_3A. In the mixture with higher addition of Klein complex the later was labelled kleiniteH.

Table 1 Mixture compositions[1]

COMPONENT, %	COMPONENT CONTENT, %
Alite[2]	Alite, 78 + C_3A_{Na}, 13.3 + gypsum, 8.7
Alite, 92.6 + gypsum, 7.4[3]	Alite, 85.9 + kleinite, 4.3 + gypsum, 9.8[5]
Alite, 92.6 + calcite, 7.4	Alite, 78.7 + kleiniteH, 6.5 + gypsum, 14.8[5]
Alite, 78 + C_3A, 13.3 + gypsum, 8.7[4]	Alite, 78 + C_3A, 13.3 + gypsum, 8.7 + KOH[6]

[1] in all pastes w/c = 0.3,
[2] to some samples 2% SNF was added and to all samples 0.5% of carboxylic superplasticizer was added,
[3] one mixture with 5.9% of anhydrite,
[4] adjusted as for monosulphate, approximately as in Portland cement,
[5] paste with kleinite contained 1% of carboxylic superplasticizer because of high viscosity,
[6] 0.5 mole of KOH in mixing water.

RESULTS AND DISCUSSION

The first series of experiments were undertaken with SNF and it was found that to obtain low plastic viscosity a relatively great addition of this admixture must be employed. As we can see in Figure 1 the carboxylic superplasticizer is much more effective and 0.3% addition gives the same result as 2% of SNF. The rheology of the mixture of alite with kleinite is changing quickly with time and after one hour the plastic viscosity became two times greater then after 10 minutes. Simultaneously Newtonian liquid became transformed in Bingham body with a relatively significant yield value (Figure 2).

The rheological behaviour of the mixture of alite with kleinite and gypsum is quite similar to the mixture with C_3A and gypsum. These results will be discussed further for the pastes with carboxylic admixture. It must be underlined that the behaviour of the pastes with SNF is quite analogical to those with carboxylic admixture.

The first series of measurements can be summarized in a following manner. The plastic viscosity depends of the quantity of ettringite formed (higher content of kleinite in mixture). It is seen especially in the increase of plastic viscosity after one hour. The same increase is noted in the case of the mixture of alite with gypsum, but here there is an increase of C-S-H gel content.

The rheological measurements of the mixture with carboxylic admixture have shown that addition of gypsum or anhydrite to alite changes insignificantly the plastic viscosity of this mixture (Figures 4 and 5). Relatively grater changes are caused by addition of calcite. In this case the plastic viscosity is about four times higher. It must be caused by quicker formation of C-S-H gel because, as has shown Nonat [10] calcite presents a heterogeneous germs for this phase.

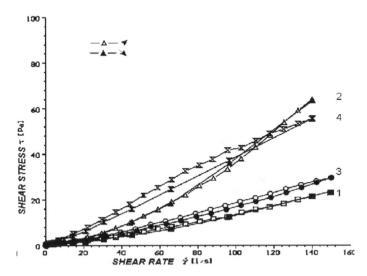

Figure 1   Flow curves of the mixtures after 10 minutes
1 – alite + gypsum + 0.5% AC, 2 – alite + kleinite + gypsum + 0.3% AC,
3 – alite + kleinite + gypsum + 0.5% AC, 4 – alite + kleinite + gypsum + 2%SNF

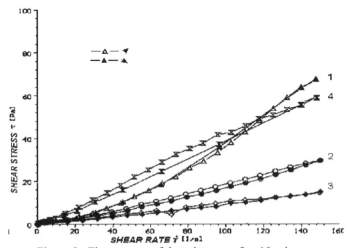

Figure 2   Flow curves of the mixtures, after 10 minutes
1 – alite + kleiniteH + gypsum + 0.3% AC, 2 – alite + kleiniteH + gypsum + 0.5% AC,
3 – alite + kleinite + gypsum + 1% AC, 4 – alite + kleinite + gypsum + 2% SNF

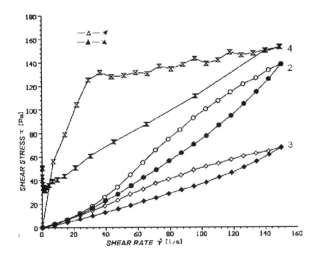

Figure 3   Flow curves after 1 hour
NB It was not possible to make the measurement of mixture 1

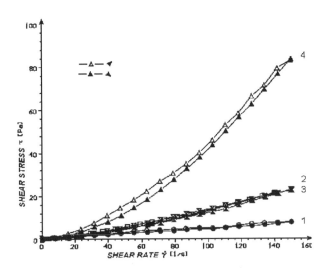

Figure 4   Flow curves of the pastes
1 – alite, 2 – alite + gypsum, 3 – alite + anhydrite, 4 – alite + calcite, after 10 minutes

After one hour flow curve of alite with gypsum shows a relatively great hysteresis (Figure 5) which can be probably caused by temporary thixotropy of the system, may be linked with partial recrystallization of gypsum. There is also a marked hysteresis on the ascending branch of the viscosity curve of mixture 4 (Figure 3) which is probably caused by ettringite needles which during mixing became oriented in parallel to the flow direction of the paste.

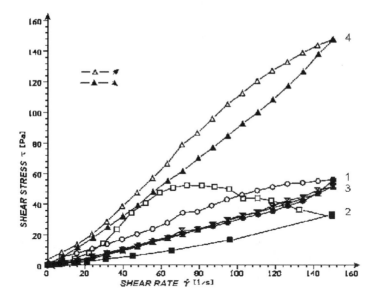

Figure 5   Flow curves of the same pastes as in Figure 1, but after 1 hour

As can be expected the paste of the mixture: alite with C_3A and gypsum has much higher viscosity and the comparison of flow curves after 10 minutes and one hour shows that this viscosity is quickly increasing with time (Figures 6 and 7). After one hour the paste of this mixture presents even a small yield value (Figure 7).

No differences in rheological behaviour of the mixture alite with C_3A_{Na} with comparison the one with C_3A was found. It could be expected because the phase composition of C_3A and C_3A_{Na} were very similar and consequently their reactivity also should be similar.

Very interesting is comparison of flow curves of the mixture alite + C_3A to the curves of alite with Klein complex, which is shown in Figures 8 and 9. The addition of kleinite in quantity to give theoretically the same ettringite content as in mixture with C_3A gives the paste with higher plastic viscosity, so that 1% of superplasticizer must be used. Thus the measurements were repeated with this higher addition of carboxylate also for mixtures containing C_3A and kleiniteH. The results are shown in Figures 8 and 9. The plastic viscosity was the lowest for the paste of the mixture containing smaller quantity of kleinite. However, much higher plasticity has the paste with the higher content of kleinite i.e. kleiniteH. This one stiffened before one hour and the measurement at this time was impossible (Figure 8). The plastic viscosity of two remaining pastes i.e. alite with C_3A and alite with smaller quantity of kleinite is practically the same (Figure 8). However, it is slightly lower for higher shear rate for the mixture with kleinite.

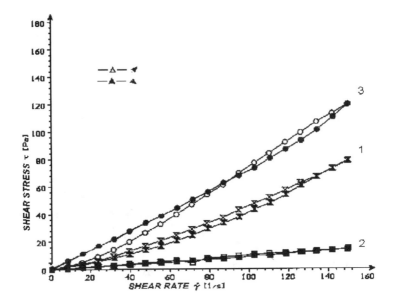

Figure 6   Flow curves of the pastes
1 – alite + C₃A + gypsum, 2 – alite + kleinite + gypsum,
3 – alite + kleiniteH + gypsum, after 10 minutes

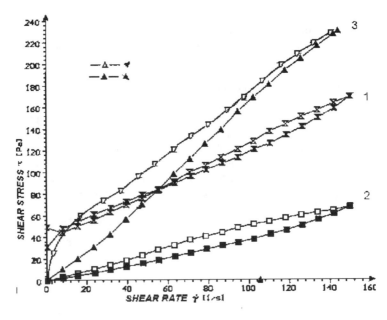

Figure 7   Flow curves of the same pastes as in Figure 6, but after 1 hour

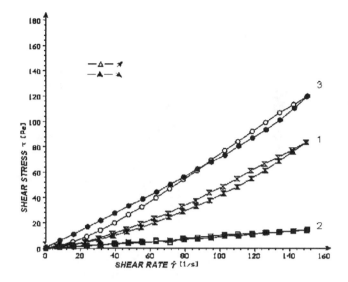

Figure 8   Flow curves of the pastes
1– alite + C$_3$A + gypsum, 2 – alite + kleinite + gypsum,
3 – alite + kleiniteH + gypsum, after 10 minutes. Carboxylate addition 1%

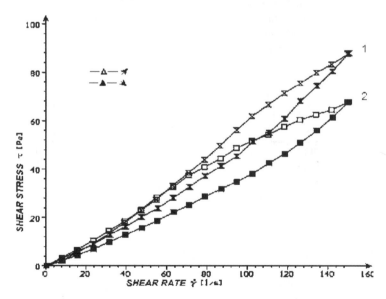

Figure 9   As in Figure 8, but after 1 hour

Additional problem is the kinetic of ettringite, formation which is not the same in case of C$_3$A and kleinite. The X-ray examination of paste after 0.5 hour and 1 and 2 hours has shown that the content of ettringite in the paste is the lowest in case of kleinite and the highest in sample with kleiniteH (Figure 10). In the light of this results the behaviour of the paste are quite understandable: the higher the content of ettringite formed the higher is the plastic viscosity of the paste.

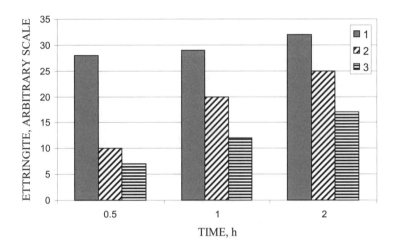

Figure 10   Ettringite content in the mixtures
1 – mixture alite + kleinite H + gypsum, 2 – alite + C$_3$A + gypsum,
3 – alite + kleinite + gypsum

The result of X-ray examination is confirmed by microcalorimetric measurements presented in Table 2. The heat of hydration of the sample alite + C$_3$A and gypsum is much higher than that for the mixture with kleinite after 10 minutes, but after 20 minutes the heat is the same and after 1 hour higher for the second sample.

Table 2   Heat of hydration in cal/g

TIME	ALITE + C$_3$A + GYPSUM	ALITE + C$_3$A + GYPSUM + KOH	ALITE + KLEINITE + GYPSUM
10 min.	7.37	3.34	4.74
20 min.	8.00	3.74	7.96
30 min.	8.46	4.09	10.54
1 h	9.37	4.92	15.43
2 h	10.44	6.04	19.31
6 h	15.96	10.03	35.96
12 h	59.54	59.68	88.44
20 h	112.13	120.01	120.00*

* after 19 hours

The influence of KOH addition on the rheology of some pastes was also investigated. KOH was added with mixing water in quantity ensuring the concentration of 500 µl/l. The results are presented in Figures 11 and 12. As it is seen in these Figures the paste of mixture alite with gypsum has a little lower plastic viscosity after 10 minutes with KOH addition, but evidently higher after 1 hour. It is linked with faster hydration of alite with KOH addition, which is well known accelerator of C$_3$S reaction with water. Quite different situation is observed in C$_3$A containing mixture. The plastic viscosity of this paste is lower after 10 minutes as well as after one hour.

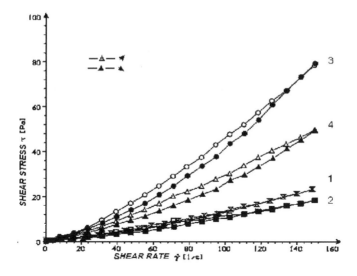

Figure 11   Flow curves of the pastes
1 – alite + gypsum, 2 – alite + gypsum + KOH,
3 – alite + C₃A + gypsum, 4 – alite + C₃A + gypsum + KOH, after 10 minutes

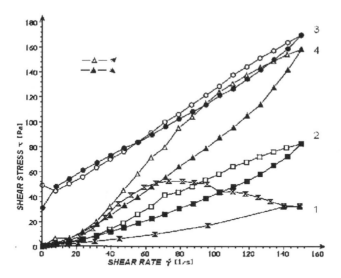

Figure 12   Flow curves of the same pastes as in Figure 6, but after 1 hour

The reason of this behaviour is evident from microcalorimetric measurements (Table 2) from which it is seen that the heat of hydration of this paste with KOH addition is much lower till six hours after mixing with water. Thus the lower is also the degree of hydration and thus lower hydrates content in the paste causes its lower plastic viscosity. However the reasons of such influence of KOH on hydration of this mixture with C_3A should be explained in further researches.

CONCLUSIONS

On the basis of obtained results the following conclusions can be drawn:

- there is a predominant influence of hydration products on rheology of the pastes made of the mixtures of alite with aluminates and are the model of Portland cement,

- independently what phase gives aluminate ions to the solution, ettringite formation controls the rheology of the paste,

- there is no difference in influence of rheology of two admixtures studied (SNF and AC),

- C-S-H formed shortly after mixing the binder with water (alite + calcite) influence the plastic viscosity of the paste.

REFERENCES

1. ERDOGDU S, Cement and Concrete Research, Vol. 30, 2000, pp. 767-773.

2. JIANG S, KIM B G AND AITCIN P C, Cement and Concrete Research, Vol. 29, 1999, pp. 71-78.

3. AITCIN P C, Cement Wapno Beton, 73, pp 217-224, 2006.

4. RAMACHANDRAN V S, MALHOTRA V M, JOLICOEUR C, AND SPIRATOS N, Superplasticizers, Properties and applications in concrete, Ministry of Public Works and Government Services, Canada, 1998.

5. NAWA T AND EGUCHI H, 9th Intern. Cong. Chem. Cem., New Delhi, India, Vol. IV, 1992, pp. 597-603.

6. BONEN D AND SARKAR S L, Cement and Concrete Research, Vol. 25, 1995, pp. 1423-1434.

7. KIM B G, JIANG S, JOLICOEUR C AND AITCIN P C, Cement and Concrete Research, Vol. 30, 2000, pp. 887-893.

8. PRINCE W, EDWARDS-LAJNEF M AND AITCIN P C, Cement and Concrete Research, Vol. 32, 2002, pp. 79-85.

9. PRINCE W, ESPAGNE M AND AITCIN P C, Cement and Concrete Research, Vol. 33, 2003, pp. 635-641.

10. NONAT A, Cement Wapno Beton, Vol. 72, 2005, pp. 65-73.

EFFECTS OF TRIETHANOLAMINE AND TRI-ISOPROPANOLAMINE ON CEMENT PASTES

M Cheikh Zouaoui

A Boudchicha

University of Larbi

N Chikh

University of Constantine

Algeria

ABSTRACT. The purpose of this work is to investigate the effects of tri-isopropanolamine and triethanolamine used in the combination with calcium nitrate on the setting and hardening process of cement pastes at 20°C temperature. Tests were performed on specimens from various mixes considering cement with normal tricalcium silicate content. The results obtained indicate that the calcium nitrate used alone acts mainly as a setting accelerator and may improve the compressive strength at 24 hours. However, in the long term, its strength increase is not significant so as to be considered as a hardening accelerator. In return, the use of calcium nitrate in combination with tri-isopropanolamine or triethanolamine leads at very early age to significant results with respect to both setting and hardening acceleration, and to a continuous compressive strength increase with time, but the results with tri-isopropanolamine were generally better than those with triethanolamine.

Keywords: Calcium nitrate, Tri-isopropanolamine, Triethanolamine, Cements paste, Strength, Setting time.

Dr M Cheikh Zouaoui, Assistant Professor. Dept. of Civil Engineering, University of Larbi Ben Mehidi, Oum Bouaki Algeria, since 1993. Member of the Laboratory of Material and Durability of the Constructions, University of Constantine, Algeria and the laboratory of materials and constructions sciences, University of Cergy Pontoise, France. Main research interest: New materials, durability of constructions, reinforced and pre-stressed Concretes.

Dr A Boudchicha, Assistant Professor. Dept. of Civil Engineering, University of Larbi Ben Mehidi, Oum Bouaki Algeria, since 1993. Member of the laboratory of Modelling, Material and Structure LM^{2}S, University of Cergy Pontoise, France. Main research interest: New Materials, formulation of the concretes, reinforced and pre-stressed Concretes.

Dr N Chikh, Professor. Dept. of Civil Engineering, Campus Mentouri, University of Constantine, since 1987. Member of the Laboratory of Material and Durability of the Constructions, University of Constantine, Algeria. Main research interest: New materials, durability of constructions, FRP, reinforced and pre-stressed Concretes.

INTRODUCTION

Various admixtures have been used to get a concrete with sufficient strength at a very early age, and calcium chloride has been, for some time, the most widely used set accelerator. However, there has been controversy over the use of calcium chloride in concrete containing embedded metal in view of the possibility of corrosion. Many countries have made provision in the relevant codes of practice to prevent or limit its use where steel reinforcement is present [1]. This has renewed interest in chloride-free accelerators as replacements for calcium chloride in reinforced concrete. Dodson [2] presented a review of non-chloride, non-corrosive set accelerating salts. It was established that calcium nitrate can be used as set accelerator and was also found to be very effective corrosion inhibitor for metal imbedded in concrete.

The effect of calcium nitrate additions on the setting characteristics of cements and the steel corrosion has been studied by Justnes [3, 4]. The results obtained showed that calcium nitrate does function as a set accelerator for cement between 7°C and 20°C, but the set accelerating efficiency of calcium nitrate depended very much on the cement type used. It was also found that the set accelerating efficiency of calcium nitrate seems to increase with increasing belite content in cement or other cement characteristic promoting belite formation in the clinker process.

The addition of small amounts of higher tertiary alkanolamines, such as tri-isopropanolamine (TIPA) resulted in interesting increases in the strengths of cement pastes at different ages [5-6]. A recent study carried out on the strength enhancing mechanism of tri-isopropanolamine presented compressive strength data for 10 Portland cements tested as cement paste after 28 days of hydration [7]. The average strength improvement with 200 ppm TIPA added to the mix water was 10%.

On the other hand, triethanolamine (TEA) combined with calcium nitrate was proposed in 1981 like non-corrosive concrete accelerator for concretes. This chemical product accelerates the reaction between tricalcium aluminates and the gypsum [8]; and when it is used with proportions of 0.1 to 0.5% of the weight of cement, it causes the immediate setting of the mixture (from 2 to 6 minutes).

The purpose of the present study is to find combinations of calcium nitrate with other compounds so as to fulfill the criteria for both a setting and a hardening accelerator. In this respect, we will do a comparison of the use of alkanolamines presenting a developed spatial molecular structure like the tri-isopropanolamine (TIPA) and another with less developed molecular structure like the triethanolamine (TEA) to investigate in which proportions these two additives can give interesting results.

EXPERIMENTAL STUDY

Materials

The cement used to carry out this work was supplied by a cement plant in east Algeria. Its chemical and mineralogical composition is given in Tables 1 and 2.

- Designation: Cement CPJ, CEM II/A 42.5.
- Production: Cement factory of Ain Touta, Department of Batna, Algeria.
- Absolute Density = 3100 kg/m³
- Specific Surface (Blaine) = 3200 cm²/g

Table 1   Chemical composition of the clinker

COMPONENTS	CaO	SiO$_2$	Al$_2$O$_3$	Fe$_2$O$_3$	MgO	SO$_3$	K$_2$O	Na$_2$O	L.O.I.
%	64.36	22.0	5.02	2.94	2.07	1.94	0.47	0.26	0.64

Table 2   Mineralogical composition of the clinker (Bogue)

MINERALS	C$_3$S	C$_2$S	C$_3$A	C$_4$AF
%	51.28	24.68	8.33	8.94

The general formula of the calcium nitrate (CN) used in the present study is: X.NH$_4$NO$_3$ Y.Ca(NO$_3$)$_2$ Z.H$_2$O. The coefficients X, Y and Z values are respectively X = 0.092, Y = 0.500 and Z = 0.826, which correspond to 19.00% Ca^{2+}, 1.57% NH$_4^+$, 64.68% NO$_3$ and 14.10% H$_2$O. The chemical formula of the tri-isopropanolamine added was N(CH$_2$CH$_2$CH$_2$OH)$_3$ [C$_9$H$_{21}$O$_3$N], corresponding to a developed spatial molecular structure. The chemical formula of the triethanolamine added was N(CH$_2$ CH$_2$OH), corresponding to a simple spatial molecular structure. These additives were used for the purpose of providing the hardening criteria to the blend admixture. The w/c ratio for the various mixes investigated was 0.3. The characteristics of theses mixes are shown in Table 2.

Table 3   Mixes investigated

ADDITIVE	MIXES STUDIED	% NC	% ADDITIVE
NC	M10	0.00	0.00
	M11	1.00	0.00
NC + TIPA	M12	0.00	0.05
	M13	1.00	0.05
	M14	0.00	0.10
	M15	1.00	0.10
NC + TEA	M20	0.00	0.00
	M21	1.00	0.00
	M22	0.00	0.05
	M23	1.00	0.05
	M24	0.00	0.10
	M25	1.00	0.10

Setting Times

The setting test on the cement pastes was performed using a Vicat apparatus and following EN196-3 procedure. The setting is registered by penetrating a needle of a fixed cross section and with a constant force into the cement paste. The initial setting time is determined to be when full depth intrusion of the needle is not obtainable, while the final setting time is taken when the needle no longer penetrates the cement paste at all.

Compressive Strengths

The hardening criteria were determined through simple compressive strength tests on cubic specimens of 50 x 50 x 50 mm³. These latter were cast in metallic moulds and kept for 24 hours at 20±1°C and 55±5% RH. They were then demoulded and conserved in water at 20°C until required for testing. Using a hydraulic testing machine, compressive strength tests were carried out according to the code NF18-406. Each compressive strength value represents the average of the results from 3 specimens.

Results

The purpose of this work is to investigate the effects of using tri-isopropanolamine (TIPA) and triethanolamine (TEA) combined with calcium nitrate on the cement pastes, setting process and hardening. Tests were carried out on specimens at the age of 1, 3 and 7 days. The results obtained regarding setting times and compressive strength, are given in Table 4 and Figure 2.

Table 4   Experimental setting times and compressive strengths results

MIXES (%NC, %ADDITIVE)		SETTING TIME min		COMPRESSIVE STRENGTH MPa		
		Initial	Final	1 day	3 days	7 days
NC	M10 (0.00, 0.00)	137	205	18	25	38
	M11 (1.00, 0.00)	125	195	27	36	41
NC+TIPA	M12 (0.00, 0.05)	140	210	32	48	80
	M13 (1.00, 0.05)	75	110	36	48	80
	M14 (0.00, 0.10)	130	210	40	52	88
	M15 (1.00, 0.10)	75	100	44	52	82
NC+TEA	M22 (0.00, 0.05)	150	235	28	36	45
	M23 (1.00, 0.05)	75	130	25	32	40
	M24 (0.00, 0.10)	150	225	36	64	76
	M25 (1.00, 0.10)	80	120	28	60	72

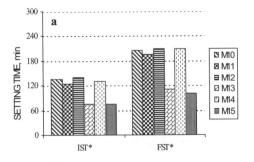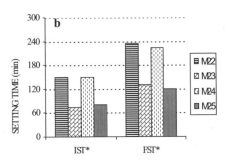

Figure 1   (a) Initial (IST) and final setting times (FST) for TIPA mixes, (b) Initial and final setting times for TEA mixes

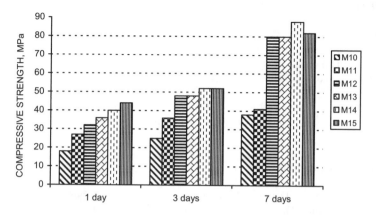

Figure 2-a   Compressive strength for TIPA mixes

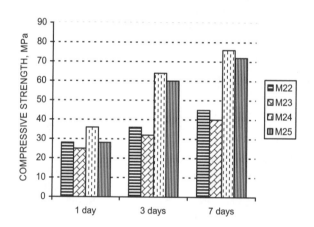

Figure 2-b   Compressive strength for TEA mixes

DISCUSSION

Effect of Calcium Nitrate (CN)

In regards to the cement used, with normal tricalcium silicate content (8.33%), the calcium nitrate affected significantly the evolution of compressive strengths at early age, which increased by 50% at 1 and 3 days compared to the reference. However, the improvement noticed in setting times was very moderate. As the early age compressive strength was not improved, CN cannot be classified as a hardening accelerator. Thus the efficiency of the calcium nitrate as a set accelerator seems to be dependent on the chemical composition of cement, confirming the results obtained by Justnes [4].

Effect of Tri-isopropanolamine (TIPA)

The use of 0.05% and 0.1% TIPA had a small effect on the initial and final setting times. However, it produced significant improvement on the compressive strength at all ages. With 0.05% TIPA, the increase in strength in relation to the reference mix was about 80% at 1 day, 90% at 3 days, and 110% at 7 days. The influence was further enhanced with 0.10% TIPA where the recorded increase was about 120% at 1 day, 110% at 3 days and 130% at 7 days. These results show that the tri-isopropanolamine performed well as a hardening accelerator and the increase in strength was not significant from using 0.05% to 0.10% TIPA.

Effect of Triethanolamine

The use of 0.05% and 0.1% TEA does not involve a remarkably positive effect on the initial and final setting times. However, it produced a significant improvement on the compressive strength at all ages. For 0.05% of TEA we obtain an increase in relation of the reference mix of about 56% at 1 day, 44% at 3 days, and 17% at 7 days. The effect is more accentuated with 0.1% of TEA where the increase in the compressive strength in relation to the reference mix was about 100% at 1 day, 156% at 3 days, and 100% at 7 days. These results show that the triethanolamine performed well as a hardening accelerator and the increase in strength was significant from using 0.05% to 0.10% TEA.

Effect of Calcium Nitrate Combined With Tri-isopropanolamine

The use of calcium nitrate in combination with tri-isopropanolamine resulted in their joint effect. Thus the initial and final setting times compared with the reference mix values were reduced respectively to 45% and 50% for 0.1 % TIPA. The compressive strength increase for early and long term ages was more than 90% and 45% respectively. However, it is worth indicating that the increase in strength produced with 0.05% TIPA compared to 0.0% TIPA was far more important than that produced from using 0.10% TIPA with respect to 0.05% TIPA. The dosage of 0.05% TIPA seems to be close to the optimum dosage. These results clearly show that calcium nitrate combined with tri-isopropanolamine leads at the early ages to very interesting and promising results in terms of accelerating both setting and hardening.

Effect of Calcium Nitrate Combined With Triethanolamine

The use of calcium nitrate in combination with triethanolamine resulted in their joint effect. Thus the initial and final setting times compared with the reference mix values were reduced respectively to 55% and 60% for 0.1 % TEA. The compressive strength increase for using 0.05% of TEA with 1% of NC was about 39% at 1 day, 28% at 3 days, and 5% at 7 days, whereas with 0.1% of TEA we observe clearer increases in compressive strengths, which reaches 56% at 1 day, 140% at 3 days, and 89% at 7 days. These results clearly show that calcium nitrate combined with triethanolamine leads at the early ages to very interesting and promising results in terms of accelerating both setting and hardening. The use of both tri-iso-propanolamine and triethanolamine in combination with calcium nitrate in cement pastes resulted in their joint effect and leads at the early ages to very interesting and promising results in terms of accelerating both setting and hardening.

The analysis of Figure 3 shows that their effects on the initial and final setting times are similar and the reduction of time setting is more accentuated with 0.1 % NC. The analysis of Figure 4 shows that the TIPA mixes generally generate more improvement in the compressive strength than that of the TEA mixes except for mixes with 0.1 % of TEA at 3 days which gives more improvement than the TIPA mixes.

Comparative Analyses

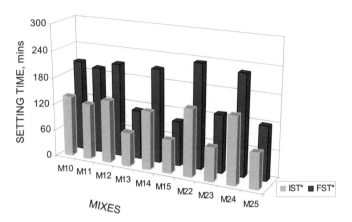

Figure 3   Initial and final setting times for all the mixes

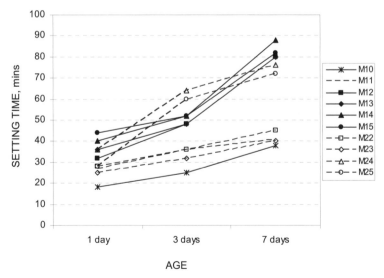

Figure 4   Compressive strength evolution for all the mixes

CONCLUSIONS

Tests have been carried out on cement paste specimens to investigate the effects of using, as additives, calcium nitrate (CN), tri-isopropanolamine (TIPA) and triethanolamine (TEA) on the setting time and compressive strength evolution. The main results show that:

Calcium nitrate acted mainly as a setting accelerator with efficiency depending on the chemical composition of cement. It may improve the one day compressive strength but in the long term, the effect on the strength increase was insufficient for it to be considered as a hardening accelerator. Both of the tri-isopropanolamine and the triethanolamine performed well as hardening accelerators at all ages.

The combination of both additives resulted in their joint effect with time, translated by a reduction in the initial and final setting times by more than 40% and a significant strength enhancement, more than 45%, at all ages of the cement pastes. Initially, calcium nitrate acted as a setting accelerator then tri-isopropanolamine and triethanolamine took over by accelerating mainly the hardening phase.

The combination of tri-isopropanolamine with calcium nitrate gives similar results in setting times with those of triethanolamine, and the reduction of setting time is more accentuated with 1% NC. The combination of tri-isopropanolamine with calcium nitrate gives better results in compressive strength than those with triethanolamine, except for mixes with 0.1 % of TEA at 3 days.

REFERENCES

1. RIXON, R. and MAILVAGANAM, N. Chemical Admixtures for Concrete. E & FN Spon, 1999, London.

2. DODSON, V. Concrete Admixture - Set Accelerating Admixtures. V.N. Reinhold, 1990, New York.

3. JUSTNES, H. and NYGUARD, E.C. The influence of technical calcium nitrate additions on the chloride binding capacity of cement and the rate of chloride induced corrosion of steel embedded in mortars. Proceedings of International Conference on Corrosion and Corrosion Protection of Steel in Concrete, Sheffield, UK, July 1994, pp. 491-502.

4. JUSTNES, H. and NYGUARD, E.C. Technical calcium nitrate as set accelerator for cement at low temperatures. Cement and Concrete Research, 25 (8), 1995, pp. 1766–1774.

5. GARTNER, E. and MYERS, D. Influence of tertiary alkanolamines on Portland cement hydration. Journal of the American Ceramic Society, 76 (6), 1993, pp. 1521-1530.

6. JUSTNES, H. Accelerator blends for Portland cement. Second International Symposium on Cement and Concrete Technology in the 2000s, Istanbul, Turkey, Sept 2000, pp. 433-442.

7. SANDBERG, P.J. and DONCASTER F. On the mechanism of strength enhancement of cement paste and mortar with triisopropanolamine. Cement and Concrete Research, 34, 2004, pp. 973-976.

8. TOKAY V. U.S Patent N° 4, 337, 094, June 1982.

MODIFIED POLYAMINIC ADDITIVES AGAINST THE CARBONATION OF PORTLAND CEMENT CONCRETE AND CORROSION OF REINFORCING STEEL BARS

G Rinaldi

F Medici

University of Roma

Italy

ABSTRACT. In the absence of aggressive environmental media, such as carbon dioxide, a good quality Portland concrete may be realized without chemical admixtures in the mixes; however in severe environmental conditions, even a good quality concrete cannot provide the protection against the carbonation of the cured cement and consequently the corrosion of the steel bars, above all in the presence of chloride ions. In this paper the effect of some organic additives intended for the control and prevention of the carbonation of cement mortars, together with chloride corrosion of steel reinforcement is experimentally investigated. Some polyaminoalkylolic and polyaminophenolic additives, respectively obtained by reacting tetraethylenepentamine (TEPA) with formaldehyde and with formaldehyde and phenol are experimented: their resistance against carbonation was tested on cylindrical specimens of mortars, containing different type and percentage of abovementioned chemical admixtures. The carbonation depth was measured after six weeks in a carbonation room. Weight loss and electrochemical measurements were also carried out on steel specimens submerged in a simulated pore solution containing chloride ions in presence of the additive to be tested. The best results were obtained by using the chemical admixtures which definitely absorb the carbon dioxide from the surrounding environment; at a 'critical' percentage the best additives determine also the protection of the steel bars against the chloride-corrosion.

Keywords: Admixtures, Carbonation, Chloride corrosion, Mortars, Durability

Dr G Rinaldi, Industrial Chemist, Associate Professor of Materials Science and Technology, Dept. of Chemical Engineering and Materials, University of Roma - La Sapienza, Italy

Dr F Medici, Chemical Engineer, Associate Professor of Materials Technology and Applied Chemistry, Dept. of Chemical Engineering and Materials, University of Roma - La Sapienza, Italy.

INTRODUCTION

Portland reinforced concrete is the worldwide most common construction material. In the absence of chemically aggressive media such as high values of atmospheric carbon dioxide and chloride ions in the environment, a good-quality, durable Portland concrete may be obtained by the adoption of a sufficient 'cover' of the steel bars. However, in severe environments (almost everywhere, at present time, due to the steady increase of atmospheric carbon dioxide), even a good-quality Portland concrete is subjected to the progressive carbonation of the free-lime present in the cured matrix, with some lack of durability.

In fact the carbonation determines the fall of the pH-value in the pore solution and the damage (cracking) of the protective oxide layer (lasting only if the pH-value stays over 11) around the steel bars. The layer will be easily destroyed and the steel corroded when the pH falls below the value of 9 or (above all) due to the presence of chloride ions. Both phenomena results in a dangerous reduction of the designed service-life of the structure.

Therefore many corrosion-preventing systems are in use [1]: more resistant stainless steel, coating of the carbon steel with epoxy resins, cathodic protection, surface impregnation of the concrete, addition of corrosion inhibitors into the mixes. Perhaps the easier and cheaper method to face corrosion of steel reinforcements probably is the use of corrosion inhibitors as admixtures in reinforced concrete. The inhibiting activity of poly-amines and of their derivatives against the corrosion of steel in acidic, neutral or alkaline solutions containing aggressive ions such as chlorides, sulphates, nitrates is a well-known property, and various modified poly-aminic products are already employed in industry [2-3].

Different poly-aminic admixtures formerly synthesized [4] were positively employed for the regulation of the carbonation of hydrated lime, both in air-hardening [5] and in hydraulic mortars [6-7-8]; they were tailored to respectively speed up or cut down the reaction. Moreover, similar chemical agents have been experimented as additives in Portland-carbon fly ash mixes [9] to ameliorate the early-strength development of the mortars, with good results.

In this paper different poly-aminic chemical agents, able to cut down or avoid the carbonation of hydrated lime are experimented to enhance the durability of Portland reinforced concrete due to their action in:

- controlling and preventing the carbonation of the free-lime in the cured cement matrix,

- inhibiting the corrosion of steel bars in chloride-containing solutions.

Polyalkylenepolyamino-alkylolic (PAA) or -phenolic (PAP) chemical additives have been added to the water-mix of Portland cement mortars; the resistance against carbonation was tested on cylindrical specimens containing different type and percentage of the abovementioned chemical products. The carbonation depth was measured after six weeks in a carbonation box. Electrochemical measurements were also carried out on carbon steel specimens submerged in chloride-containing solutions in the presence of the 'admixtures' to be tested.

Tests in simulated alkaline pore solution were also carried out; final corrosion tests were executed on chloride-containing mortars with steel rod reinforcement. The best results were obtained by using the chemical admixtures of the PAA-type, which definitely absorb the carbon dioxide from the surrounding environment. At a 'critical' percentage they determine also the protection of the steel against the corrosion even in presence of chlorides.

EXPERIMENTAL AND RESULTS

Materials

The Portland cement was the standard industrially produced material CEM II A-L 42.5R (EN 197/1); siliceous sand was employed for the mortars (see Table 1).

Table 1   Siliceous sand sieve analysis

SIEVE SIZE, mm	UNDERSIZED, %	OVERSIZED, %
4.76	97.2	2.8
2.38	72.6	24.6
1.19	11.6	61.0
0.59	0.4	11.2
0.221	0.0	0.0

Two types of water soluble additives were obtained by modifying the tetra-ethylene-pentamine (TEPA), i.e. a molecule with high content of aminic groups. Type I (PAA) additives were synthesized [9] by reacting TEPA with formaldehyde in molar ratios of 1: 0.1 - 1: 0.2 - 1: 0.4 - 1: 0.6; polyalkylenepolyaminoalkylolic products with increasing content of alkylolic functions ($-CH_2OH$) linked to the aminic groups of the TEPA molecule (Table 2) were obtained.

For the type II (PAP) additives, the reaction of TEPA with formaldehyde and phenol [4] in molar ratios of 1: 1: 0.2 - 1: 1: 0.4 - 1: 1: 0.6 - 1: 1: 0.8 allowed the obtainment of polyalkylene-polyamino-phenolic products with increasing content of phenylic functions ($-CH_2 -C_6 H_4 -OH$) linked to the aminic groups of the TEPA molecule (Table 3).

Carbon AISI 1040 (Table 4) steel rods (diameter 9.95 mm, height 150 mm) were employed for the corrosion experiments (weight loss and electrochemical measurements). Specimens were always mechanically polished, degreased with carbon tetrachloride (Soxhelet), pickled in inhibited 3 N HCl to eliminate rust, rinsed with distilled water, dried and stored in a desiccator.

Table 2   Main properties of the PAA-Type I additives.

ADDITIVE	MEAN MOLECULAR WEIGHT, g	SPECIFIC GRAVITY, kg/m³, 20°C	VISCOSITY, Pa.S, 20°C
PAA 01	195	915	12
PAA 02	220	918	15
PAA 04	240	930	29
PAA 06	310	960	40

Table 3   Main properties of the PAP-Type II additives.

ADDITIVE	MEAN MOLECULAR WEIGHT, g	SPECIFIC GRAVITY, kg/m³, 20°C	VISCOSITY, Pa.S, 20°C
PAP 02	250	980	17
PAP 04	285	1010	1525
PAP 06	310	1012	2934
PAP 08	370	1030	4050

Table 4   Chemical composition of the carbon steel rods (AISI 1040).

ELEMENT	PERCENT, by weight
Carbon	0.395
Manganese	0.75
Silicon	0.38
Sulphur	0.02
Phosphorous	0.003

Mortars

A cement/sand ratio of 1:3 by weight was always adopted; the water/cement ratio was 0.5. The different additives were previously dissolved into the full volume of water-mix; their dosage ranged from 2 to 4% by weight of cement. The resistance against carbonation was experimented on cylinders (diameter 20 mm, height 60 mm), casting 15 specimens for each mixture to be tested. The curing of the mortars was carried out at 20+/-0.5°C, relative humidity > 95%, for two days. The volume mass and water adsorption of the cured mortars are in Table 5.

Table 5   Volumetric mass and water adsorption of the mortar without additive and with different PAA – Type I and PAP – Type II additives.

SAMPLE	VOLUMIC MASS, kg/m³	WATER ADSORPTION, % by weight
Mortar without additives	1706	1.06
PAA 01	1709	1.07
PAA 02	1702	1.11
PAA 04	1701	1.02
PAA 06	1710	1.07
PAP 02	1710	1.04
PAP 04	1715	1.02
PAP 06	1709	1.02
PAP 08	1716	1.05

Carbonation Tests

After curing, the specimens were placed in a carbonation box (20+/-0.5°C, R.H. 60%, mean carbon dioxide concentration 20%) for a period of six weeks. The carbonation depths were measured at 2, 4, 6 weeks: each cylindrical specimen was split and the surfaces of fracture tested by means of the phenolphthalein-spray test (Figure 1). The results are summarized in Table 6; each measure is the mean value of five specimens.

Table 6   Carbonation depth measurements (mm) on fractured specimens of mortars containing different type and percentage of the additives, in comparison with the same mortar of Portland cement-sand without additive.

SAMPLE	% ADDITIVE (w/w of cement)	DEPTH OF CARBONATION, mm		
		2 weeks	4 weeks	6 weeks
Portland-Sand	-	1	3	8
Unmodified TEPA (ref)	3	1	4	6
PAA 01	4	0	1	1
PAA 02	4	0	1	1
PAA 04	2	0	2	2
PAA 04	4	1	2	3.5
PAA 06	4	2	4	6
PAP 02	2	4	4	5
PAP 02	4	3	5.5	6
PAP 04	4	4.5	5	6.5
PAP 06	4	3.	4	5.5
PAP 08	4	1	2.5	5

Figure 1   Carbonation depth after six weeks: mortars without additives (left),
with PAA 02 (right).

Inhibiting Activity

A preliminary evaluation of the inhibiting action of the additives against the corrosion of steel was carried out by means of an 'intensified' test, in a strong acidic environment. The aim was the identification of the additive with the best performance in the worst conditions; therefore the tests were carried out directly on the steel rods submerged (20+/-0.5°C) in 1 N HCl (pH = 0) with gentle stirring. The additive to be experimented was previously dissolved (0.1-1.0 mmole/l) into the acidic solution. In the weight loss experiments (lasting 4 hours) each specimen before and after the test was immersed (5 minutes) in inhibited 3 N HCl solution to eliminate the corrosion products, then carefully rinsed (distilled water), dried in ethyl alcohol and weighed. Duplicate tests were always carried out (Table 7).

Table 7   Weight-loss corrosion tests (4 hours) in 1 N HCl. Additive content
corresponding to the best performance.

SOLUTION	ADDITIVE CONTENT, mmol/l	CALCULATED CORROSION, g/m^2. day
1N HCL	--	104.37
PAA 01	0.3	28.02
PAA 02	0.3	28.60
PAA 04	0.35	28.80
PAA 06	0.4	31.53
PAP 02	0.3	28.53
PAP 04	0.4	37.11
PAP 06	0.3	25.17
PAP 08	0.4	28.22
Unmodified TEPA	0.3	44.31

The polarization measurements [10] were then performed (Corrograph, Amel) in a conventional cell with a Platinum counter-electrode and a saturated calomel electrode (reference). The curves for anodic and cathodic reactions were recorded (sweep rate 0.5 mV/s); the Tafel straight lines were then obtained from the polarization curves. The current density and the corrosion potential were finally determined and the corrosion intensities calculated (Table 8).

In 1 N HCl solution (pH = 0), the best of the PAA-type additives experimented were the less-modified PAA 01 and PAA 02; the best of the PAP-type additives in the same conditions were the more modified PAP 06 and PAP 08.

Table 8   Electrochemical corrosion tests in 1 N HCl: effect of the various PAA (Type I) and PAP (Type II) additives, contents as in Table 7.

	1 N HCL NOT INHIBITED	1 N HCL INHIBITED	
		Type I	Type II
Current corrosion density, mA/m²	1.8	0.5-0.22	0.12-0.18
Corrision potential, V	-0.492	-0.458/-0.470	-0.440/-0.455
Corrosion intensity, g/m². day	64.8	9.8-12.3	8.7-11.5

Corrosion of Steel in Concrete

As preliminarily stated, Type I additives with the higher action against the carbonation of cement in mortars show a contemporary good (even not the best) inhibiting action against the corrosion of steel in a worst (pH = 0) chloride containing solution (1 N HCl). To verify the action of the additives in the 'real' different 'concrete environment', two series of experiments were carried out. In the first set of tests, steel rod specimens were submerged for 24 hours into a saturated $Ca(OH)_2$ solution (artificial pore solution) containing 2% of chloride ions (added as NaCl) and the inhibitor to be tested.

The experimented inhibitor content was the same as in Table 7. Before and after the exposure, each steel specimen was washed with water, immersed (5 minutes) in a solution of inhibited 3N HCl and washed again, then dried (ethyl alcohol); finally each specimen was weighed to determine the weight loss (Table 9).

In the second series of tests, some 'reinforced' mortar prismatic specimens (as in ASTM G109) were prepared with steel rods (cover 5 mm). Chlorides (NaCl, 2% by weight of cement) were build into the mortars as contaminants; the additive (contents as in Table 6) to be tested was previously dissolved into the water-mix. After curing (2 days at 20+/-0.5°C and R.H. > 95%) and 12 weeks in the carbonation box, the specimens were completely broken to examine the state of the steel surfaces (Table 10 and photos in Figure 2).

Table 9   Weight-loss corrosion tests of steel rods exposed to the artificial alkalyne pore solution containing 2% Cl⁻. 24 hours, 20+/-0.5°C.

SOLUTION	ADDITIVE	ADD.CONTENT. mmol/l	WEIGHT LOSS, g/m^2
Not inhibited	-	-	36.12
Inhibited	TEPA	0.3	27.15
	PAA 01	0.3	20.90
	PAA 02	0.3	20.99
	PAA 04	0.35	24.35
	PAA 06	0.4	25.12
	PAP 02	0.3	27.11
	PAP 04	0.4	27.00
	PAP 06	0.3	25.02
	PAP 08	0.4	25.13

Table 10   Surface of the steel rods embedded in mortars containing 2% of NaCl as contaminant after twelve weeks of carbonation.

SPECIMEN	ADDITIVE	% w/w of CEMENT	ASPECT OF THE SURFACE
Not inhibited	-	-	Fully rusted
Inhibited	PAA 01	4	No rust
	PAA 02	4	No rust
	PAP 06	4	No rust
	PAP 08	4	No rust

Figure 2 Aspect of steel rods reinforcement after 12 weeks of carbonation. Mortars without additives (left); with PAA 02 (centre); with PAP 06 (right).

DISCUSSION

The carbonation depths of the mixtures (Table 6) clearly show that both types of the synthesized additives can reduce the carbonation of the Portland cement mortars. Polyamino-alkylolic (PAA) products undoubtedly behave better than the polyamino-phenolic (PAP); the PAA 01 and PAA 02 are the more effective: in comparison with the not additivated mortar there is a reduction of about 87%. The reduced carbonation depth could be attributed to a lower permeability of the additivated mortars, but it must point out that the volumic mass and the water adsorption of the cured matrices containing the different additives were always near the same as those of the not additivated mortar (Table 5).

The higher effectiveness of the PAA products is surely attributable to their stronger absorptive action towards carbon dioxide, respect to the PAP products [4]. The higher alkalinity and higher content of not hindered aminic groups [4] justify the better performance of polyamino-alkylolic molecules, in comparison with the polyamino-phenolic. A sufficient quantity (4% respect to the cement) of the additive is needed to assure a long exposure to the atmospheric CO_2 without the attainment of high carbonation depths. It can be argued that one week in the carbonation box could correspond to about one year in an ordinary environment [11].

In conclusion the PAA type I additives with a low content of alkylolic functions behave better due to their higher content of aminic groups of stronger alkalinity. Unmodified TEPA is effective at a lesser extent due to its excessive polarity, even if TEPA could be employed with satisfying results, from the point of view of the resistance against the carbonation; nevertheless the additives need to exert also an inhibiting action against the corrosion of steel reinforcement, above all in the presence of chloride ions. In all cases the additives, both Type I and Type II determine a protective effect in pure HCl (Table 7-8) and in simulated alkaline pore solution, even in the presence of chlorides (Table 9-10). A corrosion inhibiting effect can be generally explained by the fact that inhibitors displace (due to their strong bonding) ionic species from the oxidised steel surface, forming a durable passivating layer. In fact inhibitors act because they are 'chemisorbed' from aqueous solutions onto the oxidised steel surface [12] and the absorbate-phase is formed even if chloride ions are present [13]. The polar molecules of an inhibitor can change the electrochemical characteristics of the metal surface, blocking the cathodic centres (the inhibitor acts as 'donor' of electrons to the 'acceptor' metal, cathodic, through the bond responsible of the chemisorption); similar blocking action can be exerted on the anodic centres: the electrons of the metal surface migrate preferentially towards the positively charged groups of the inhibitor rather than towards the anodic sites of the surface. Even incomplete 'coverage' of the surface (and the concentration of inhibitors is surely very low) determines the rise of an 'energetic barrier' preventing corrosion.

The PAP products were originally tailored as inhibitors for HCl pickling solutions, and polyamino-phenolic molecules behave a little better than the PAA in 1 N HCl solution (Table 7-8). Their action is effective even in alkaline chloride containing pore-solution, but the polyamino-alkylolic agents become a little more effective (Table 9). This is probably due to the different structure of the molecules in acidic and alkaline environment: the PAA molecule is more flexible even in alkaline medium, whilst the PAP molecule (owing to the presence of polar phenolic ring) results in a more rigid structure [4] that hinders the action of full 'coverage' of the metal surface. The thickness and the structure of the absorbate-phase (which depend on the structure, chemical composition and concentration of the inhibitor) of the PAP Type II additives are evidently less effective in neutral-alkaline environment. Further research is needed in order to explain this phenomenon.

Nevertheless the PAA additives are more effective in simulated alkaline pore solution in the presence of chloride ions, and this behaviour constitutes another advantage, as well as their higher activity against the carbonation of mortars: Type I additives are surely more suitable for a global protection of reinforcing steel even in the presence of chlorides (Table 10).

CONCLUSIONS

The additives obtained by modifying the tetraethylenepentamine (a product containing an higher number of aminic functions in the molecule respect to the more diffused ethanolamines and to their derivatives) are more effective in the action against the carbonation of Portland cement mortars.

The highest reduction of the carbonated layer can be attained by using the polyamino-alkylolic agents; the effect of the polyamino-phenolic products is less effective.

Both types of modified agents exert inhibiting action against the corrosion of steel in HCl solutions and in chloride-contaminated mortars, but in simulated alkaline pore solution containing chloride ions the polyamino-alkylolic additives behave better.

In conclusion, the PAA additives are probably more suitable due to their dual action against the carbonation of cured Portland cement mortars and against the corrosion of steel reinforcement in presence of aggressive chloride ions.

REFERENCES

1. BROOMFIELD, J P. Corrosion of steel in concrete, E and F N Spon Ed., London, 1997. pp. 20-33

2. HERN, Z and OLMEZ, H. The influence of ethanolamines on the surface properties of Portland cement pastes. Cement and Concrete Research. Vol. 27 (6), 1997. pp 805-812

3. BATIS, G, RAKANTA, E, SIDERIS, K K. Corrosion protection investigations with the use of NN'- dimethylaminoethanol corrosion inhibitor. Proc. Int. Conf. Univ. of Dundee, July 6, 2005. Vol 'Admixtures-enhancing concrete performance', R K Dhir, P C Hewlett, M D Newlands Ed., Th.Telford Publ., London, 2005. pp 79-88

4. RINALDI, G. Acid gas absorption by means of aqueous solutions of regenerable modified polyalkylene-polyamine. Industrial and Engineering Chemistry Research. Vol. 36, 2000. pp 3778-3782

5. MEDICI, F and RINALDI, G. Polyaminophenolic additives accelerating the carbonation of hydrated lime in mortars. Environmental Engineering Science. Vol. 19 (4), 2002. pp 271-276

6. MEDICI, F, PIGA, L and RINALDI, G. Behaviour of polyaminophenolic additives in the granulation of lime and fly ash. Waste Management. Vol. 20, 2000. pp 491-498

7. MARRUZZO, G, MEDICI, F, PANEI, L, PIGA, L and RINALDI, G. Characteristics and properties of a mixture containing fly ash, hydrated lime and organic additives. Environmental Engineering Science. Vol. 18 (3), 2001. pp 159-165

8. RINALDI, G and MEDICI, F. Hydrated lime pastes containing coal fly ash and polyaminophenolic additives for the inertization of fly ashes from municipal solid wastes incinerators. Proc. Int. Conf. Univ. of Dundee, Sept. 9, 2002. Vol. 'Sustainable Concrete Construction', R K Dhir, T D Dyer, J E Halliday Ed., Th. Telford Publ., London, 2002. pp 139- 150

9. RINALDI, G. Additivi poliamminoalchilolici per il miglioramento della resistenza di malte contenenti ceneri volanti. La Chimica e l'Industria. Vol. 70, 1988. pp 91-94 and Vol. 71, 1989. pp 34-37

10. SCHASCHL, E. Methods for evaluation and testing of corrosion inhibitors. Corrosion Inhibitors, C C Nathan Ed., National Association of Corrosion Engineering Pub., Houston USA, 1973. pp. 156-172

11. HO, D W S, LEWIS, R K, Carbonation of concrete and its prediction. Cement and Concrete Research, Vol. 17, 1987. pp. 489-504

12. TRABANELLI, G. Corrosion mechanisms. Chemical Industries. Vol. 28. F. Mansfeld Ed., M. Dekker, N.Y. USA, 1987. pp 15-55

13. RIGGS, O L. Theoretical aspects of corrosion inhibitors. C C Nathan Ed., National Association of Corrosion Engineering Publ., Houston USA, 1973. pp 13-17

THEME TWO:

ENERGY AND RESOURCES: WHERE NEXT?

CONCRETE BUILDINGS IN VIEW OF THE EC ENERGY PERFORMANCE OF BUILDINGS DIRECTIVE

M Öberg

Cementa AB

Sweden

J S Damtoft

Aalborg Portland Group

Denmark

ABSTRACT. A heavy material is capable of buffering and utilising a large part of the free heat gains, such as solar radiation and heat from occupants in buildings. Concrete can therefore decrease energy consumption as well as improve thermal comfort. In a joint CEMBUREAU/BIBM/ERMCO effort the energy related benefits of heavy construction were investigated and documented. Energy balance calculations were undertaken for heavy and lightweight buildings in various European climates and for residential as well as office cases. The results show that a solid residential building under normal conditions requires 2-7% less bought energy for heating compared to a similar light building. This has a significant environmental and economical impact. For an office building the advantage is usually larger, which is explained by larger internal gains during office hours that often require mechanical cooling. Comparing thermal comfort, the difference is in the order of 10 to 20%. Energy consuming cooling facilities can be avoided in many heavy buildings. The advantages are even larger if thermal mass is actively taken into account in building design.

Keywords: Energy performance, Thermal mass, Thermal stability.

Dr M Öberg is working at the R&D Department of Cementa AB (Heidelbergcement group). He has previous experience as structural designer and precast concrete project manager and presented his thesis concerning life time optimisation of concrete buildings at Lund University in 2005. He is active in environmental issues concerning concrete and has been chairman of the CEMBUREAU/BIBM/ERMCO task force on energy performance of concrete buildings.

Jesper Sand Damtoft is vice president, R&D of the Aalborg Portland Group and is director of the Group Research and Development Centre. He holds a M.Sc. in geology but has worked with development of cement and concrete as well as use of alternative raw materials and fuels in the cement production since 1985. He has been engaged in several projects dealing with environment and sustainability and is presently chairman of the Nordic network Concrete for the Environment.

INTRODUCTION

Operation of buildings is the largest single source of greenhouse gases in Europe. This was the primary reason for launching the EC-directive on Energy Performance of Buildings, 'EPBD'[1], which was enforced in Member States from January 2006.

The aim of this work is to clarify and position the energy related properties of concrete in buildings, in view of the EPBD. It has been carried out by a CEMBUREAU task force including representatives of the Precast and Ready mix concrete associations BIBM, and ERMCO.

Thermal mass has been known to have a positive influence on energy use and thermal comfort in buildings. This work aims at quantifying the effect of thermal mass. The technical principle is illustrated in Figure 1. Free energy gains are stored instead of being removed from the building and can to be utilised at a later time. In a building this is primarily a 24 hour cycle, which is examined in this work, but also seasonal storage can be the case.

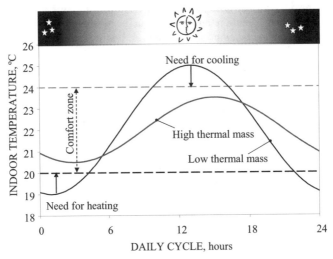

Figure 1 Principle for utilisation of the diurnal thermal storage potential.
After Johannesson et al. [2]

The climate change as such also significantly affects thermal indoor conditions. Arup [3] states that London will be as hot as Marseille in 2080. New buildings need to be adapted accordingly to safeguard health and comfort. Heavy buildings provide good thermal stability which is a robust and environmentally friendly solution to the problem, reducing, or in many cases eliminating, the need, for mechanical cooling.

The EPBD features the following main items of interest with regard to thermal mass:

- A common framework for a methodology of calculation of the <u>integrated</u> energy performance of buildings. This is outlined so that the benefits of heavy materials may be taken into account when calculating energy performance.
- Minimum requirements on the energy performance of buildings including cooling.
- Requirements to check compliance in terms of the measured energy use, as built.

- A CO_2 indicator may be included in the assessment of energy performance. This promotes the use of low exergy[1]) energy sources, which is feasible in heavy buildings because of their lower effect demand.
- Stating that passive cooling concepts should be employed to decrease the problem with peak loads, which increases the electricity cost and disrupts the energy balance of southern Europe during summer.
- Stating that good energy performance must not inflict on the quality of indoor environment, which is easy to fulfil with heavy buildings.
- Energy certifications of buildings are introduced, which strongly increases the general attention to the issue and specifically the market value of energy efficiency.

The common methodology referred to in the EPDB, is the EN ISO 13790 [4] standard. In a simplified method for energy balance calculations given in the standard thermal mass is addressed by the so called utilization factor, cf. Figure 2.

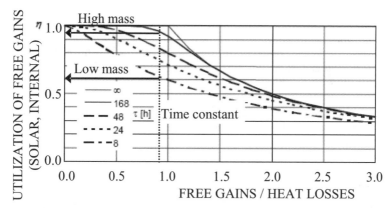

Figure 2   Utilisation of free energy gains according to EN ISO 13790 [4]. Example: Gain/loss ratio 0.90 => gain utilisation; Heavy building 0.95 and light building 0.60

The 'time constant', τ, is defined as the heat capacity divided by heat losses from the building. A heavy building therefore has a higher time constant than a light weight building, with equal heat losses, and consequently obtains a higher gain utilisation factor, η. Free gains is the energy input from persons, electrical appliances etc inside the building and the solar radiation. Losses are the energy lost by transmission and air leaks through the climate shell and by the exhaust ventilation air. Typically the relation between gains and losses is low during mid winter and high during summer because of the solar radiation. Under extreme conditions the difference in utilisation between heavy and light buildings is small. For example during Nordic mid winter conditions with $\gamma < 0.1$, there is full utilisation of free gains irrespective of thermal mass. The same building in March may have $\gamma = 0.9$ such as the example in Figure 2, whereby there is a large difference in utilisation of free gains between heavy and light. The time constant also indicates how fast the temperature changes in a building where heating or cooling is disrupted. With a high time constant the temperature remains stable for a long time (days) while the low time constant implies rapid change.

---

[1] *Low exergy heating and cooling systems use low valued energy which can be supplied by sustainable energy sources such as solar collectors, heat pumps etc. Fossil fuels and electricity are high valued (exergy) sources.*

METHOD

The energy performance was examined by a large number of energy balance calculations using several different computer tools and models including the standard method referred to in the EPDB, EN ISO 13790 [4]. A theoretical building with simple geometry and real existing buildings from different European countries, ranging from Sweden to Portugal, were calculated.

The programmes estimate the required energy input to maintain a desired comfort temperature range by heating or, if relevant, cooling, using the specific climate data for the location of the building. The following energy balance programmes were applied: Be06 [5], Consolis Tool [6], Maxit Energy [7], TCasa [8], and VIP+ [9]. Sensitivity analysis was conducted with regard to thermal insulation, window orientation and operating regimes.

RESULTS

The calculations indicate that a solid residential building under normal conditions requires 2-8% less energy for heating compared to the corresponding lightweight building. The advantage for heating energy, in absolute terms, is more pronounced in a temperate rather than an extreme climate cf. Table 1.

The number of clear days and the window orientation are also of importance, affecting the solar gains. Comparing thermal comfort and risk for overheating, the benefit of thermal mass is even higher than with regard to heating, see Table 2.

Table 1   Results of calculation on a sample residential building with different climates. Consolis Tool calculation

REGION	PREDICTED NET ENERGY USE FOR HEATING kWh/m²·year			MEAN MONTHLY TEMPERATURE October - April	
	Solid	Lightweight	Difference	%	°C
Polar circle	128.7	133.4	4.7	3.7%	-7.9
Northern Europe	66.7	70.7	4	6.0%	1.1
Northern Europe coastal	53.1	57.4	4.3	8.1%	3.4
UK	37.6	43.1	5.5	14.6%	5.9
Benelux	42.2	48.8	6.6	15.6%	5.6
Central Europe	49.2	53.3	4.1	8.3%	3.8
Alpine	60.6	65.9	5.3	8.7%	1.4
Mediterranean	8.0	12.2	4.2	52.5%	12.1

For office buildings thermal storage can be further utilized, because of larger internal gains during office hours and larger fluctuations of internal gains between day and night. Hence, cooling facilities can be avoided in many cases in concrete buildings. In Figure 3 and Table 3 results from calculations of an office building are presented.

Table 2   Results of calculations on theoretical dwelling building with different thermal mass
and window orientations applying different computer programmes

CLIMATE	PROGRAMME	ORIENTATION OF WINDOWS	ENERGY USE FOR HEATING/COOLING, kWh/m²·year		DIFFERENCE LIGHT/SOLID
			Solid	Light	%
Stockholm	Consolis	E/W	66.7	70.7	6.0
Stockholm	Consolis	S	51.5	56.5	9.7
Stockholm	VIP+	E/W	64.5	66.9	3.7
			11.3*	13.2*	16.5*
Stockholm	VIP+	S	54.5	60.1	5.3
			12.4*	15.0*	20.3*
Würzburg	TCasa	E/W	60.3	61.7	2.4
Würzburg	TCasa	S	54.1	56.0	3.5
Denmark	Be06	E/W	47.3	48.0	1.2
			3.4*	4.3*	

* Theoretical cooling to maintain +27°C indoor temperature

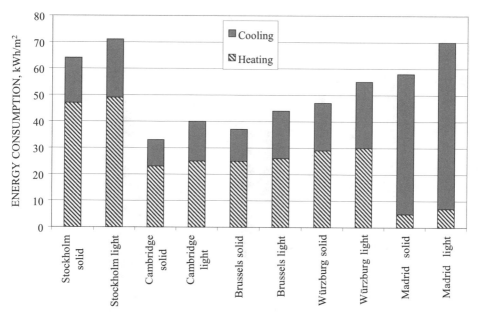

Figure 3   Difference in energy use for heating and cooling for theoretical office building in
different European climates. VIP + calculation

Table 3 Results of calculations on a theoretical office building with different thermal mass and window orientations on applying different computer programmes

CLIMATE	PROGRAMME	ORIENTATION OF WINDOWS	ENERGY USE FOR HEATING/COOLING, kWh/m²·year		DIFFERENCE LIGHT/SOLID
			Solid	Light	%
Stockholm	VIP+	E/W	50.0	54.1	7.3
			13.1*	15.9*	20.7*
Stockholm	VIP+	S	37.4	41.9	12.0
			14.5*	18.3*	26.3*
Denmark	Be06	E/W	38.0	43.6	14.6
			3.4*	4.3*	

* Theoretical cooling to maintain +27°C indoor temperature

Intelligent combinations of heating, ventilation, solar shading and building structure, for instance with increased night cooling, can further improve the utilization of thermal mass. This can be modelled with energy balance programs.

In Figure 4 it is shown how the heavy building is capable of utilizing the increased night ventilation, in this case reducing the cooling demand to 40% instead of 62% for the light building. In many cases, artificial cooling can be avoided in a building with high thermal mass. This provides substantial savings with regard to investment as well as operating costs.

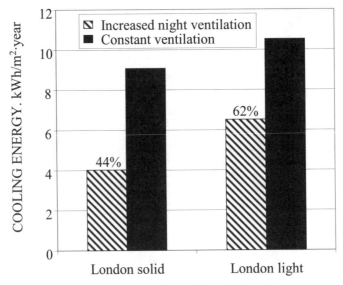

Figure 4 Difference in energy use for cooling of theoretical office building with and without increased night ventilation

DISCUSSION

The **thermal stability** of heavy buildings provides competitive advantages by <u>maintaining an even indoor climate whilst consuming a minimum of bought energy.</u>

If the design of building structure and HVAC systems is done in a holistic and active way, the benefits can be increased. Thus there is a moderate but consistent inherent advantage and a significant potential advantage.

Inherent and potential benefits

The inherent case can be defined as the effect of thermal mass when nothing particular is done to utilize it. Heat is then naturally absorbed by the construction when room temperature is increasing under the influence of solar gains and heat from internal sources such as persons, electrical lights and apparatus. When the room temperature goes down during the night, heat is released again, cf. Figure 1. Either a certain outdoor or indoor temperature fluctuation is required to utilize the storage effect. During steady state conditions, such as Nordic mid winter, the building mass is thus of little benefit. A light building will utilize free gains as much as a heavy one, cf. Figure 2. In the spring, however, the difference is significant.

Thermal mass can also be activated to obtain higher utilisation of the energy storage. In an active solution, energy is transferred to the construction by means of a flowing medium such as water in coils or air in ducts. In this case high thermal conductivity is crucial to distribute the heat from the flowing medium to the room effectively. One example of activated thermal mass is the Swedish 'Termodeck' system [10], using concrete hollow core slabs to distribute and store heating or cooling energy with air as energy carrier. The example in Figure 4 could in principle also be defined as a simple type of activation.

With reference to calculations conducted in this project and literature studies the inherent advantage with regard to energy use on a yearly basis is 2-7%. The effect on indoor temperatures is larger and may provide an option to avoid mechanical cooling in many cases, resulting in substantial savings both with regard to investment and operation. By simply increasing the night ventilation, a heavy structure can be cooled down enough to maintain a comfortable indoor temperature during a summer day.

If the potential advantage is activated the benefits increase. It is difficult to give some general figures. Case studies show up to 30-40% difference. By some applications such as glass facades, a massive building frame is a prerequisite to obtain thermal stability.

Economical consequences

With the current European energy prices the 'inherent' difference in energy use for heating between solid and light, in the dwelling case, responds to less than 1 Euro/m^2, year which may seem insignificant in the perspective of the total annual operating cost of a building.

However, if a whole life cycle costing approach is taken, the difference is very significant. Furthermore, energy prices tend to increase, cf. Figure 5, so that also moderate energy savings will significantly improve the operating economy of buildings in the future.

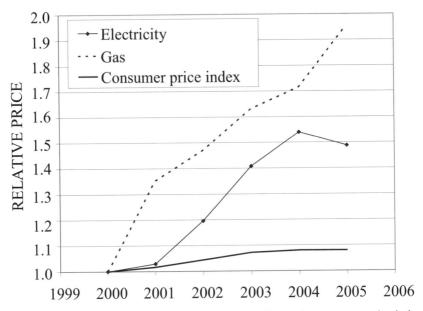

Figure 5 Price increase of energy in Sweden in relation to the consumer price index

In the case of electric heating, the application of low and high tariffs can be utilised by loading the building frame with cold or heat during low tariffs. This can greatly affect operating costs. The low and high tariffs on electricity in the UK are 4.29 and 7.83 p/kWh, respectively.

For the active 'potential' level whereby for example heating is reduced by 10 kWh/m^2 and cooling by 20 kWh/m^2, the saving is 2.5 Euro/m^2 per year or 80-120 Euro/m^2 for a calculation horizon of 50 years.

Substantial savings with regard to investment costs can be obtained if the heating and cooling installations are optimised with regard to the thermal mass.

Environmental context

The primary goal of the EPBD is to reduce the emission of greenhouse gases from the built environment. The reason why the EPBD focuses entirely on operation of buildings is explained in Figure 6, which shows that the energy required to produce buildings is of a secondary order compared to the operation phase. Also moderate differences with regard to energy use during operation will have a significant effect on the life cycle energy use, and thus the environmental load, of a permanent building with a life span of 50 to 100 years.

Heavy construction materials provide inherent benefits with regard to comfort, economy and environment which can further be utilised by holistic design and operation of buildings. The EPBD with related new standards and calculation procedures, as well as climate changes will stimulate the interest in obtaining buildings with good thermal stability.

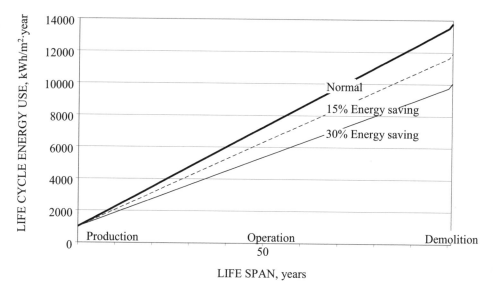

Figure 6   Energy use for life cycle phases of a multi-family residential building, with two heating energy saving scenarios, after [11]

REFERENCES

1. DIRECTIVE 2002/91/EC OF THE EUROPEAN PARLIAMENT AND OF THE COUNCIL of 16 December 2002 on the energy performance of buildings, Official Journal of the European Communities, Brussels, 2003

2. JOHANNESSON G et al., Möjlighet till energieffektiva hus genom helhetssyn och modern beräkningsmetodik, ByggTeknik No. 3, Stockholm, 2006, 66 pp. In Swedish.

3. ARUP, Too hot to handle. Building No. 6, 2004, London

4. ISO DIS 13790:2005 Thermal performance of buildings - Calculation of energy use for space heating, CEN/TC 89, Brussels

5. Be06, Computer programme for energy balance calculations, Danish Building Research Institute, 2005

6. JOHANNESSON G, Consolis Energy+, Spreadsheet tool for energy calculations, (Under development) KTH - Building Technology, Stockholm, 2005

7. Maxit Energy, Computer programme for energy balance calculations, Maxit AB, Stockholm, 2005

8. TCasa, Computer programme for energy balance calculations, Bundesverband Deutscher Zementindustrie, Berlin, 2004

9. VIP+, Computer programme for energy balance calculations VIP+, STRUSOFT, Malmö, Sweden, 2002

10. BUNN R, Termodeck the thermal flywheel, Building Services Journal, No. 5, 1991

11. ÖBERG M, Integrated Life Cycle Design – Application to Swedish Concrete Multi-Dwelling buildings, Lund University. Division of Building Materials, Report TVBM-3103, Lund, Sweden, 2005, 117 p.

PROPERTIES OF SEWAGE SLUDGE ASH AND ITS POTENTIAL USE IN CONCRETE

J E Halliday

University of Dundee

United Kingdom

ABSTRACT. There are currently 1.2 million tonnes of unavoidable waste produced each year from the treatment of domestic wastewater and industrial effluents. Although the incineration process reduces the overall volume of material by approximately 65%, there is a need to find alternative disposal for the resulting ash. Research has shown that materials rich in calcium, silica and alumina, which have been subjected to high temperatures, can produce residues exhibiting cementitious properties, for example, (i) fly ash (a by-product of the generation of electricity using coal) and (ii) ground granulated blast-furnace (a waste product from the manufacture of pig iron) fall into this category. The physical and chemical properties of eight UK sewage sludge ashes (SSAs) were investigated in this study to assess their suitability for use in concrete. The research concluded that the SSAs were reddish in colour and that the particles were porous and irregularly shaped. The maximum particle size, for all SSAs, was found to be <600 μm (all SSAs were found to be much coarser than Portland cement and flyash). The major bulk oxides were found to be calcium, alumina, silica, iron and phosphate. The compositions of the SSAs were such that they lie in the region between latent hydraulic and pozzolanic behavior, suggesting that they may undergo such reactions. The chloride and sulfate contents of all SSAs were found to be low. The trace metal analysis found that the common metals for all SSAs were barium, copper, nickel and zinc.

Keywords: Additions, Concrete, Latent hydraulic, Physical and chemical properties, Pozzolanic, Sewage sludge ash.

Dr Judith E Halliday is a Research Fellow within the Concrete Technology Unit. She is a graduate of Civil Engineering at the University of Dundee and her PhD research investigated the potential use of waste from thermal processes, namely, cement kiln dust (CKD), incinerator fly as (IFA) and sewage sludge ash (SSA), as a cement constituent. She has since been involved in numerous projects, including; (i) Innovative cement combinations for concrete performance, (ii) Facilitating the wider use of coarse and fine RA from washing plants and (iii) Developing the sorting and processing of construction and demolition wastes and create methods for use of recycled aggregates in a range of higher value applications.

INTRODUCTION

There are currently 1.2 million tonnes of unavoidable waste produced each year from the treatment of domestic wastewater and industrial effluents [1]. The cessation of disposal to sea (introduced in 1998 by the Urban Waste Water Treatment Directive [2]), the unsustainability of landfill disposal (EC Directive on Landfill of Waste legislating a decrease in the amount of waste disposed to landfill [3]), and the decrease in the popularity of sewage sludge fertilisation, have all been driving forces leading to an increase sewage sludge incineration, and it is therefore appropriate that research into the ultilisation of waste from incineration is executed now, in order to ensure that the process remains sustainable as its popularity grows.

Incineration of sewage sludge involves combustion to temperatures in excess of 850°C, which reduces the sludge to a residue (sewage sludge ash - SSA). Although the incineration process reduces the overall volume of material by approximately 65% [1], there is a need to find alternative disposal routes for the resulting ash.

Research shows that materials rich in calcium, silica and alumina, which have usually been subjected to high temperatures, can produce residues exhibiting cementitious properties, for example, (i) fly ash (a by-product of the generation of electricity using coal) and (ii) ground granulated blast-furnace (a waste product from the manufacture of pig iron) fall into this category [4].

Previous work carried out at the University of Dundee [5, 6] found that SSA had potential for use as a cement constituent, because of its pozzolanic or latent hydraulic characteristics [7] (Evans, 2001). Moreover, it was found that by using SSA, as cement, properties of concrete including drying shrinkage and freeze/thaw resistance were improved. Work was also carried out on the environmental impact of the material when used in concrete. It was found that, with respect to trace metal and dioxin content and the susceptibility of major trace metals to leach from mortar specimens, UK SSAs were environmentally benign [8].

However, SSAs are derived using different techniques, which can result in variation of the physical and chemical properties. It was decided that it would be beneficial to carry out a study looking at these variations and assess the SSAs suitability for use as an addition in concrete.

EXPERIMENTAL PROGRAMME

The following tests were carried out in order to assess the physical and chemical properties of SSAs, and therefore their potential use in concrete.

Physical Properties

Moisture Content
SSA moisture content was determined in accordance with the procedure described in BS EN 450 [9] and involved oven-drying two samples (of approximately 10 g each), at 105°C for 1 hour. The samples were cooled in a desiccator prior to reweighing. The moisture content was calculated as the percentage ratio of change in mass to mass of dry material.

Loss-on-Ignition

The loss-on-ignition (LOI) of the ashes was determined by igniting approximately 1.0 g of material in a furnace at 975°C for 1 hour [10]. The samples were re-weighed after cooling and the LOI calculated by expressing the change in mass as a percentage of the original mass of material. It is a test which is carried out on flyash to determine the amount of unburnt solids present.

Particle Density

The particle density of the material followed the EN 12620 [11] method for fine aggregate. Approximately 30 g of material was weighed into a 100 ml capacity density bottle and ¾ filled with de-aired water, stoppered, shaken for 1 minute and place in a vacuum for 2 hours. The bottle was then completely filled with de-aired water (using a syringe passing through the stopper). The mass of the density bottle when completely filled with material and water was determined.

Scanning Electron-Microscopy (SEM)

Scanning electron microscopy is a technique based on the principle of using a very fine probe of electrons (ranging from 0.3 to 30kV) focused at the surface of a specimen and scanned across it in parallel lines. An electron detector is then used to detect electrons re-emitted from the specimen surface (known as secondary electrons), the signal from which is then used to build up an image of the surface. The SEM produces images with different magnifications which are brought about by altering the extent of scan on the specimen whilst keeping the size of the display constant. The specimens investigated in this study were mounted on aluminium studs, coated with gold film and analysed in a JEOL microscope. Images were obtained using a standard digital camera.

Particle Size Distribution

A Malvern/E particle size analyser was used to determine the particle size distribution of the sewage sludge ashes. Approximately 1.0 g of the material was mixed in a beaker with 50 ml of tap water and a small quantity of sodium metahexaphosphate (a dispersant) to prevent agglomeration of particles during testing. The water and material were transferred into the test area of the particle size analyser. Particle size distributions were established from the degree of scattering of a collimated, monochromatic, 18mm diameter laser beam passing through the sample, and the method of measurement is based on the principle that the angle of deflection increases proportionally with particle size.

Fineness

Fineness measurements, on dry material, were carried out by wet sieving 1.0 g of material through a calibrated 45 μm sieve under a water pressure of 70-80 N/mm^2, as specified in BS RN 450 [9]. The sieve and residue is oven dried at 105°C, and cooled in a dessicator prior to re-weighing. Fineness was calculated by expressing the sieve retention as a percentage of the original mass of the material.

Specific Surface Area – Blaine Test

The specific surface area of the material was determined using an electronic air permeability tester (ToniPERM). The fineness of cement is measured as specific surface by observing the time taken for a fixed quantity of air to flow through a compacted material bed of specified dimensions and porosity. Under standardized conditions the specific surface of cement is proportional to the square root of t (where t is the time for a given quantity of air to flow through the compacted material bed).

Chemical Properties

Bulk Oxide and Trace Metal Analysis - X-ray Fluorescence Spectrometer (XRFS)
The bulk oxide and trace metal analysis was carried out using a Philips MagiX sequential fluorescence spectrometer with a CuKα source (40 mA, 40 kV) coupled with a Philips auto sampler. The ash was compressed into a standard pellet and analysed using in-house calibrations.

Mineralogy (X-Ray Diffraction)
The mineralogy of SSA was determined using a Hilton Brooks x-ray diffractometer (XRD) with monochromatic CuKx source and curved graphite single crystal chrometer (40 mA, 40 kV). A small amount of previously ground SSA was ground further in a pestle and mortar to ensure random crystal orientation and true peak intensities.

A single test specimen was prepared by firmly compacting the material into a specimen slide. Each sample was analysed over a range of 5 to 60 degrees 2θ at a scan rate of 1 degree/minute on 0.1 degree increments.

RESULTS AND DISCUSSION

Physical Properties

The colour of the ashes ranged from a light brown to a dark reddish brown. The colour is most probably the result of ferric chloride or ferrous sulfate used during chemical conditioning, as there was a relationship between the Fe_2O_3 content (determined using XRF) and colour, with an increase in iron oxide resulting in a redder material.

Figure 1 shows SEM images of four SSAs (A, B, C and D) and it can be seen that the ashes are very similar in both size and particle shape. Each ash is made up of large (mostly <200 μm in size) irregular shaped particles, onto which smaller particles (<5 μm) are fused. This produces a very porous structure which may cause workability problems when used as a constituent in concrete.

Table 1 gives the LOI values for the SSAs. Most values are below 3.0, with the exception of Ash H (which has a value of 9.8%). All ashes exhibit LOI values below those limits (of 12.0%) set by BS EN 450.

The particle densities (results shown in Table 1) range from 1880 to 2280 kg/m^3, with Ash F exhibiting the highest density and Ash E the lowest. The particle density of Ash H is probably misleading, as larger particles could not be included in the test.

Figure 2 show the particle size distribution of Ashes A to G and it can be seen that they are all below 600 μm. Ashes A and B are the finest, with over 70% below 101 μm, whilst Ashes C and G are slightly coarser, with approximately 62% below this size. Ashes D, E and F were found to be the coarsest, with 35 to 43% passing the 101 μm sieve. Ash H was not included, as there were large nodules present, most probably due to the conditioning process.

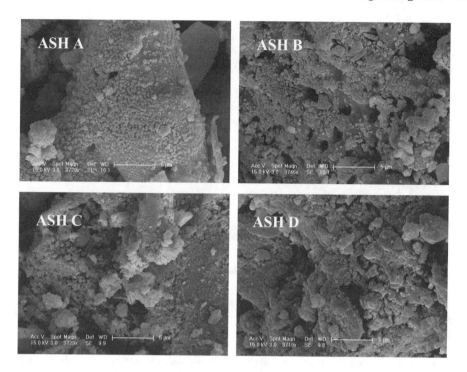

Figure 1   SEM images of SSAs A, B, C and D

Table 1   Physical properties of UK SSAs

PROPERTY	SEWAGE SLUDGE ASH							
	A	B	C	D	E	F	G	H
Moisture content. %	-	12.6	28.8	21.7	26.2	11.5	30.1	27.1
LOI, %	1.8	2.6	2.2	2.6	3.9	2.1	2.3	9.8
Particle density, kg/m³	2250	2190	2160	2060	1880	2280	2070	2060
Fineness, (% ret. 45 µm sieve)	64.5	66.3	71.3	79.8	81.8	75.4	75.1	89.8
Blaine Fineness, g/cm²	2470	2080	2100	1610	1150	1640	1770	850

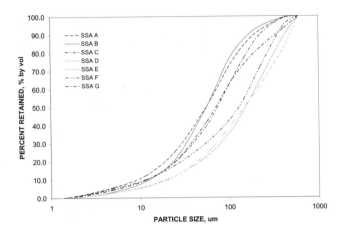

Figure 2   Particle size distribution for UK SSAs

The percentage of material retained in a 45 μm sieve, for each SSA, is given in Table 1. Ash A is the finest material, whilst Ash H is the coarsest. The Blaine test (results also given in Table 1) is a good indicator of the fineness of a material (the higher the Blaine value, the finer the material), and it can be seen that these results agree with those given by the 45 μm sieve test (see Figure 3).

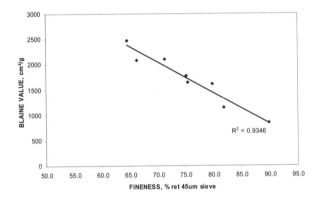

Figure 3   Relationship between fineness and Blaine value

Chemical Properties

Table 2 gives the bulk oxide results for Ashes A to H. Quartz is present in high amounts, which would be attributed to the use of sand in the process. Calcium and alumina are also present in relatively high quantities. The compositions of the ashes are shown on a CaO-SiO_2-Al_2O_3 ternary diagram in Figure 4. Also shown on the diagram are the regions inhabited by hydraulic, latent hydraulic and pozzolanic materials such as Portland cement, ground granulated blastfurnace slag (GGBS) and PFA respectively. It is apparent that all SSAs (with the exception of Ash H) lie in the region between latent hydraulic and pozzolanic behavior, suggesting that they may undergo such reactions. Iron oxide (Fe_2O_3) is most probably a result of the use of ferric chloride or ferrous sulfate during chemical conditioning for secondary sludge, and results in Table 2 show that this is present in quite significant quantities, although they differ from source to source. This would also be the reason for variation colour of the ashes. The high phosphate levels is most probably a result of the composition of the raw material (sludge) coupled with the fact that it is an inorganic compound and will not decompose or volatilize during combustion. The chloride levels are sufficiently low for use in concrete, as are the potassium and sodium levels.

The trace metal analysis results are shown in Table 2 and the common metals (and the most abundant) for all ashes are barium, copper, nickel and zinc. Other elements are found in most ashes, although amounts are small and in some cases insignificant.

Table 2 Chemical properties of UK SSAs

PROPERTY	SEWAGE SLUDGE ASH							
	A	B	C	D	E	F	G	H
Bulk Oxide Analysis, % by mass								
CaO	9.7	13.4	10.6	7.3	11.4	11.6	16.9	33.5
SiO_2	28.8	28.1	27.9	33.4	27.1	29.3	25.2	18.1
Al_2O_3	17.2	14.1	16.7	24.2	26.7	12.3	13.9	10.5
Fe_2O_3	14.5	12.8	18.5	11.2	5.5	21.8	8.6	6.6
MgO	3.4	2.5	3.0	1.9	2.2	2.1	3.2	2.1
MnO	0.3	0.5	0.3	0.2	0.3	0.2	0.4	0.3
TiO_2	1.6	1.5	1.2	1.1	2.0	1.1	1.3	0.9
K_2O	1.6	1.5	1.3	1.4	1.2	1.2	1.4	1.1
Na_2O	1.5	0.8	0.8	0.8	0.7	0.7	0.9	0.6
P_2O_5	16.6	18.0	16.0	13.9	18.3	13.7	21.6	12.3
Cl	0.0	0.0	0.0	0.0	0.0	0.1	0.0	0.5
SO_3	0.8	0.8	0.9	0.6	0.9	1.2	0.9	4.5
Trace Metal Analysis, % by mass								
Barium	0.1605	0.2393	0.1823	0.1612	0.1337	0.1748	0.1946	0.1499
Bromine	0.0009	0.0000	0.0020	0.0002	0.0014	0.0048	0.0000	0.0193
Cerium	0.0000	0.0000	0.0000	0.0000	0.0000	0.0007	0.0000	0.0000
Cobalt	0.0000	0.0042	0.0204	0.0008	0.0000	0.0006	0.0000	0.0000
Chromium	0.0901	0.1248	0.0684	0.0374	0.1183	0.0660	0.2966	0.1870
Copper	0.2215	0.1825	0.0888	0.0900	0.0829	0.0000	0.2789	0.1989
Nickel	0.0252	0.0638	0.0237	0.0194	0.0390	0.0232	0.1633	0.1069
Lead	0.0498	0.0782	0.0567	0.1045	0.0406	0.0791	0.0848	0.0664
Rubidium	0.0123	0.0121	0.0120	0.0167	0.0111	0.0143	0.0088	0.0001
Antimony	0.0146	0.0063	0.0294	0.0163	0.0000	0.0578	0.0253	0.0176
Tin	0.0364	0.0428	0.0285	0.0297	0.0132	0.0346	0.0417	0.0246
Strontium	0.0596	0.0475	0.0450	0.0349	0.0453	0.0432	0.0623	0.0555
Yttrium	0.0041	0.0022	0.0030	0.0040	0.0023	0.0018	0.0235	0.0008
Zinc	0.3252	0.3582	0.2930	0.2431	0.1834	0.2827	0.9321	0.6172
Zirconium	0.0197	0.0487	0.1362	0.0534	0.0405	0.0408	0.0389	0.0195

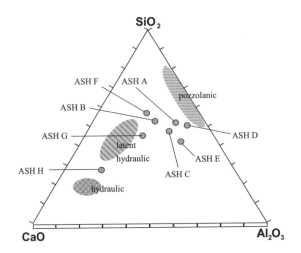

Figure 4   CaO-Si O_2-Al_2O_3 ternary diagram depicting the compositions of SSAs

The mineralogical composition of each ash was determined using XRD techniques, and quantified using Reitveld refinement. Results in Table 3 show that the main compounds are quartz (SiO_2), hematite (Fe_2O_3), albite ($NaAlSi_3O_8$), magerite, $CaAl_2(Si_2Al_2)O_{10}(OH)_2$ and whitlockite ((Ca, Mg)$_3(PO_4)_2$). Ash H contains ettringite ($Ca_6Al_2(SO_4)_3(OH)_{12}.26H_2O$), which is a compound formed when Portland cement undergoes hydration. Traces of iron and aluminium were also found (although <0.5%).The amount of amorphous material present in the ashes ranges from 57.2% to 81.6% which is an indication that it may possess pozzolanic properties.

Table 3   Mineralogical composition of UK SSAs

COMPOUND,% by mass	SEWAGE SLUDGE ASH							
	A	B	C	D	E	F	G	H
Quartz, SiO_2	5.6	5.5	8.2	9.1	10.3	3.7	4.1	2.4
Hematite, Fe_2O_3	2.6	4.4	4.9	1.1	1.6	3.8	0.8	0.8
Albite, $NaAlSi_3O_8$	7.0	1.1	3.7	5.3	1.2	1.6	15.8	2.5
Whitlockite, (Ca, Mg)$_3(PO_4)_2$	7.4	14.9	15.1	4.5	13.5	4.8	11.9	11.0
Magerite, $CaAl_2(Si_2Al_2)O_{10}(OH)_2$	5.3	4.7	7.2	1.7	2.9	2.4	2.7	1.6
Cristobalite, SiO_2	1.2	1.6	1.8	1.3	5.9	0.6	1.2	2.4
Tridymite, SiO_2	2.4	1.9	1.7	0.8	1.5	1.2	0.7	1.5
Iron, Fe	0.4	0.4	0.3	0.1	0.4	0.3	0.1	0.4
Aluminium, Al	0.2	0.2	0.3	0.0	0.2	0.0	0.5	0.4
Calcite, $CaCO_3$	0.0	1.0	0.9	0.0	0.0	0.2	0.0	0.9
Ettringite, $Ca_6Al_2(SO_4)_3(OH)_{12}.26H_2O$	0.0	0.0	0.0	0.0	0.0	0.0	0.0	2.2
Amorphous	68.2	65.6	57.2	76.2	62.6	81.6	62.6	77.5

CONCLUSIONS

The following conclusion could be made:

- The colour of the ashes ranged from a dark reddish brown to a light brown. The colour is most probably the result of the ferric chloride or ferrous sulfate used during chemical conditioning. The particles were found to be irregular in shape and porous in nature.

- The average LOI values were found to be below 3.0, with the exception of Ash H (which has a value of 9.8%). All ashes exhibit LOI values below those limits (of 12.0%) set by BS EN 450 – for using PFA as a Part 2 – Type 1 addition: (nearly inert material in concrete production). The particle densities ranged from 1880 kg/m^3 to 2280 kg/m^3, with Ash F exhibiting the highest density and Ash E the lowest. The maximum particle size of the ashes was found to be 600 μm.

- The bulk oxide composition results showed that quartz is present in high amounts. Calcium and alumina are also present in relatively high quantities. The compositions of the ashes (shown as a CaO-Si O_2-Al_2O_3 ternary) found that all SSAs (with the exception of Ash H) lie in the region between latent hydraulic and pozzolanic behavior, suggesting that they may undergo such reactions.

- The trace metal analysis found that the common metals (and the most abundant) for all ashes are barium, copper, nickel and zinc. Other elements are found in most ashes, although amounts are small and in some cases insignificant.

REFERENCES

1. DEPARTMENT OF THE ENVIRONMENT, Transport and the Regions. Building a better quality of life. A strategy for more sustainable construction, April 2000.

2. EUROPEAN UNION, Urban Waste Treatment, EEC Council Directive, 1991 (91/271/EEC)

3. EUROPEAN UNION Landfill Directive, EU Council Directive, 1999 (99/31/EC)

4. R.C. JOSHI AND R.P. LOHTIA, Fly Ash in Concrete: Production Properties and Uses, Taylor & Francis Ltd (Netherlands), 1997

5. DYER T D, HALLIDAY, J E AND DHIR, R K, Hydration Reaction of Sewage Sludge Ash for Use as a Cement Component in Concrete Production, Recycling and Reuse of Sewage Sludge, International Symposium, 2001.

6. DHIR, R K, DYER, T D, HALLIDAY, J E AND PAINE, K A Value-added Recycling of Incinerator Ash, DETR Ref 39/3/476cc, Sept. 1998

7. EVANS, T The journey to sustainable treatment and use or disposal of waste water biosolids, Recycling and Reuse of Sewage Sludge, University of Dundee, Ed Dhir, Limbachiya and McCarthy, 2001.

8. HAND-SMITH, G, Sludge disposal: the future of high-tech sludge incineration, Waste Management, March 1999 pp 41-42

9. BRITISH STANDARD INSTITUTE, BS EN 450, 1995. Fly ash for use in concrete: Definitions, requirements, and quality control. London.

10. BRITISH STANDARD INSTITUTE, BS EN 196-2, 2005. Methods of testing cement: Chemical analysis of cement, London

11. BRITISH STANDARD INSTITUTE, BS EN 12620, 2002. Aggregates for Concrete, London

RECYCLING OF CONSTRUCTION AND DEMOLITION WASTE AND CRUSHED GLASS CULLET AS CONSTRUCTION MATERIALS

C S Poon   C S Lam

S C Kou   D Chan

Hong Kong Polytechnic University

Hong Kong

ABSTRACT. Recent developments in the recycling of construction and demolition wastes show that recycled coarse aggregates can be used for making new concrete with satisfactory quality. But many research studies show that the use of recycled fine aggregates (< 5 mm) would have adverse effects on the workability and dimensional stability of the concrete due to its high water absorption (>10%) characteristics. However, the disadvantage can be avoided if the recycled fine aggregates are used for making concrete masonry blocks, since only a low workability concrete mixture is required in its manufacturing process. Laboratory test results show that concrete paving blocks prepared with the recycled fine aggregates can achieve satisfactory performance in compressive strength. However, the water absorption and the drying shrinkage of the blocks produced increase with increasing recycled fine aggregate content and limit the applications of the produced blocks. This study aims to optimize the production of concrete masonry blocks using recycled fine aggregate by introducing recycled crushed glass cullet into the concrete mixtures. Glass cullet, being a non porous material, is able to reduce the water absorption and improve the dimensional stability of the blocks. Although the use of glass in concrete materials may induce a chemical reaction between alkali in cement and silica in glass, the so-called "Alkali-Silica reaction (ASR)", which may cause possible expansion and cracking of the blocks. The produced blocks show satisfactory mechanical and durability performance when fly ash is used to control the potential ASR expansion.

Keywords: Construction and demolition waste, Glass cullet, Masonry

C S Poon, is Professor of the Hong Kong Polytechnic University.

C S Lam, is an M Phil Candidate of the Hong Kong Polytechnic University.

S C Kou, is a PhD graduate of the Hong Kong Polytechnic University.

D Chan, is a Research Associate of the Hong Kong Polytechnic University.

INTRODUCTION

As landfill space in Hong Kong is running out, the high waste generation rate becomes a major concern. In 2004, there were about 17,500 tons of waste required to be landfilled per day, of which about 40% was construction and demolition (C&D) waste. In order to reduce the waste, the Hong Kong government has implemented a charging scheme on C&D waste at the end of 2005. The purpose of this policy is to encourage contractors to reduce, reuse and recycle C&D waste. In fact, the reuse of C&D waste is quite common in other countries where this waste can be used as an aggregate replacement in concrete and concrete masonry products.

However, the use of recycled aggregates for masonry production is still relatively new. An early attempt was made by Collins et al. [1] who used recycled aggregates in the manufacture of blocks for a beam-and-block floor system. The blocks were 440 mm long, 215 mm wide and 100 mm high. Recycled aggregates were used to substitute 25 to 75% by weight of both natural coarse and fine aggregates. For blocks with a recycled aggregate replacement level of 75%, a compressive strength of 6.75 MPa and a transverse strength of 1.23 MPa were reported.

On the other hand, Jones et al. [2] suggested that the C&D waste can be used in concrete building blocks. However, they found that high recycled aggregate replacement level had an adverse effect on properties of the blocks, but it was expected that the target strength can be achieved by using a lower replacement level, whilst maintaining an economical cement content.

Furthermore, Poon et al. [3] used recycled fine aggregates as a replacement of natural aggregates in making precast concrete paving blocks which complied with the Hong Kong Standard [4] (a compressive strength of not less than 30 MPa). In fact, the Hong Kong government has published a specification on concrete paving blocks using recycled aggregates. It is specified that 70–100% recycled aggregates have to be used in paving block production in which the mix proportion must contain at least 40% recycled fine aggregates.

However, the water absorption and the drying shrinkage of the produced blocks increase with an increase in the recycled aggregate content, which limits the applications of the resulting blocks [3, 5]. Therefore, there is still a need to further understand the factors affecting the engineering properties of the masonry products using recycled materials in order to optimize the production.

This study therefore aims to optimize the production of concrete paving blocks using recycled fine aggregate by introducing recycled crushed glass into the concrete mixtures. The effects of aggregate properties and aggregate-to-cement (A/C) ratio on the properties of the prepared blocks were determined. Since recycled crushed glass (RCG) was incorporated, the concern for alkali-silica reaction (ASR) was inevitable. The adequate amount of fly ash that needed to be added in order to suppress the potential expansive reaction was determined as well.

MATERIALS

Cementitious Materials
In this study, Ordinary Portland cement and pulverized fuel ash (fly ash), complying with BSEN 197-1:2000 [6] and BS 3892 [7] respectively, were used. The properties of the cementitious materials are shown in Table 1. The cement and fly ash were commercially available in Hong Kong.

Table 1 Chemical composition of cement and fly ash

| | CHEMICAL COMPOSITION | | | | | | | DENSITY kg/m^3 | SPECIFIC SURFACE AREA cm^2/g |
	SiO_2	Fe_2O_3	Al_2O_3	CaO	MgO	SO_3	LOI		
Cement	19.61	3.32	7.33	63.15	2.54	2.13	2.97	3160	3520
Fly ash	56.79	5.31	28.21	<3	5.21	0.68	3.90	2310	3690

Recycled Crushed Glass (RCG)
The recycled crushed glass (RCG) used in this study was mainly from post-consumer beverage bottles, sourced locally. The glass bottles were washed and crushed mechanically. The RCG was sieved in the laboratory to produce the < 5 mm fraction of aggregates. The RCG was a blend of three different types of beverage glass with three different colors (30% Colorless, 40% Green and 30% Brown). The grading of the RCG satisfied the requirement for fine aggregates according to BSEN 12620:2002 [8] after sieving. The properties and sieve analysis of the RCG are shown in Table 2 and Figure 1, respectively.

Recycled Crushed Aggregate (RCA)
In this study, the recycled crushed aggregate (RCA) used was mainly concrete rubbles sourced from a C&D waste recycling facility in Hong Kong. The size fractions of the aggregate used were less than 5 mm (recycled fine aggregate). The properties and sieve analysis of the aggregate are shown in Table 2 and Figure 1 respectively. The 10% fine value of RCG and RCA are presented in Table 2, which was determined according to BSEN 12620:2002 [8].

Table 2 Properties of RCA and RCG

MATERIALS	DENSITY, kg/m^3	WATER ABSORPTION, %	10% FINE VALUE, kN
RCG	2500	~0	107
RCA	2530	10.3	120

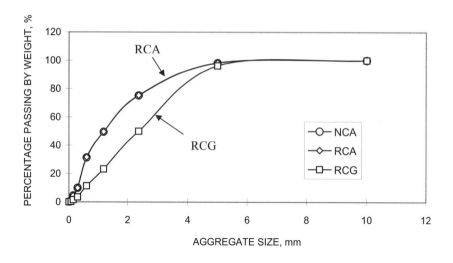

Figure 1   Sieve Analysis of RCG and RCA

MIX PROPORTIONS

The study was divided into two parts. The first part aimed to determine the effects of increasing the RCG content and changing the aggregate-to-cement ratio (A/C) on the properties of the paving blocks. In the second part of the study, mortar bars were prepared to determine the potential ASR expansion and the effectiveness of using fly ash as an ASR suppressant. After finding the adequate fly ash dosage, a single concrete mixture was subsequently prepared to determine if the incorporation of fly ash would significantly alter the properties of the blocks.

In part one, two series of concrete mixtures were prepared. The concrete mixtures in Series I was prepared with a fixed aggregate-to-cement (A/C) ratio of 4, but with varying RCG contents as shown in Table 3. The effects of the RCG content on the properties of the paving blocks were determined. On the other hand, the concrete mixtures in Series II were prepared with 50% RCG but with varying A/C as shown in Table 4. The influences of the change in the A/C ratio on the properties of the paving blocks were then determined.

Table 3 Mix proportions of the concrete mixtures in Series I

NOTATION	RCG kg	RCA kg	CEMENT kg	A/C RATIO	ADDED WATER kg
RCG-4	23	-	5.8	4	2.3
25RCA/RCG-4	17.2	5.8	5.8	4	2.1
50RCA/RCG-4	11.5	11.5	5.8	4	2.1
75RCA/RCG-4	5.8	17.2	5.8	4	2.0
RCA-4	-	23	5.8	4	2.1

Table 4   Mix proportions of the concrete mixtures in Series II

NOTATION	RCG, kg	RCA, kg	CEMENT, kg	A/C RATIO	ADDED WATER, kg
50RCA/RCG-3	11.5	11.5	7.7	3	2.2
50RCA/RCG-4	11.5	11.5	5.8	4	1.9
50RCA/RCG-6	11.5	11.5	3.8	6	1.9

Because the results in part one indicated that 50% RCG content was the optimal for paving block production, four mortar bars mixtures (Table 5) were prepared with 50% RCG and different fly ash additions (i.e. 0, 5, 10 and 15% by weight addition of total aggregates) in accordance with the procedures outlined in ASTM C1260 [9] in the second part of this study.

Since the mortar bar test results revealed that 10% PFA addition could adequately suppress the potential expansive reaction, a single concrete mixture was subsequently prepared with 50% RCG, 50% RCA and 10% PFA (Table 6) and its corresponding properties were measured.

Table 5   Mix proportions of the mortar bar mixtures

NOTATION	RCG, g	RCA, g	CEMENT, g	PFA, g
50RCA/RCG	495	495	440	0
50RCA/RCG5P	495	495	440	49.5
50RCA/RCG10P	495	495	440	99.0
50RCA/RCG15P	495	495	440	148.5

Table 6   Mix proportions of the concrete mixture prepared with fly ash

NOTATION	RCG, kg	RCA, kg	CEMENT, kg	PFA, kg
50RCA10P-B	11.5	11.5	5.8	2.3

BLOCK FABRICATION

The paving blocks were fabricated in steel moulds with internal dimensions of 200 x 100 x 60 mm. After mixing the materials in a pan mixer, about 3 kg of the materials were placed into the mould in three layers. The first two layers were compacted manually by hammering a wooden plank on the surface layer to provide an evenly distributed compaction. The last layer was prepared by slightly overfilling the top of the mould (approximately 5 mm) and the overfilled materials were subjected to a static compaction twice by using a compression machine.

The load was increased at a rate of 600 kN/min until 500 kN was reached for the first compaction. After removing the excessive material with a trowel, a second compaction was applied at the same rate until 600 kN was reached. After that, the blocks were demoulded after 24 hrs and were cured in water for 28 days.

The mortar bars were prepared using a wet-mixed method. The sieved aggregates were mixed with the cementitious materials in different proportions. Then, the mixes were fabricated into steel moulds which had internal dimensions of 285 x 25 x 25 mm.

TEST METHODS

The compressive strength of the paving blocks was measured using a loading machine with a capacity of 3000 kN with a corresponding loading rate of 400 kN/min. On the other hand, the tensile splitting strength, skid resistance and abrasion resistance of paving blocks were determined according to BS 6717:2001 [10]. Cold water absorption and density of the blocks were determined in accordance with AS/NZS 4456 [11] and BSEN 12390-7:2000 [12] using a water displacement method for hardened concrete, respectively. The potential ASR expansion was assessed in accordance with ASTM C1260 using the prepared mortar bars. Although the standard requires a test period of 14 days, the expansion of the mortar bars was measured up to 28 days.

RESULTS AND DISCUSSION

The test results of the paving blocks in Series I and II are summarized in Table 7. It was found that the density of the blocks was not noticeably affected by the change in neither RCG content nor the A/C ratio. However, the increasing use of RCG tended to reduce the compressive strength of the paving blocks as shown in Figure 1. The reduction could be attributed to the lower intrinsic strength of RCG compared to that of RCA as indicated by the ten percent fines value. Similarly, an increase in the A/C ratio also decreased the compressive strength (as shown in Figure 2) as a result of the low cement content.

Table 7 – Test results of the paving blocks in Series I and II

NOTATION	DENSITY kg/m^3	TENSILE SPLITTING STRENGTH, MPa	SKID RESISTANCE, BPN	ABRASION, mm	WATER ABSORPTION %
Series I					
RCG-4	2247	3.5	105	20	3.3
25RCA/RCG-4	2270	3.1	101	21	3.3
50RCA/RCG-4	2260	3.4	106	20	4.3
75RCA/RCG-4	2275	3.9	108	19	4.7
RCA-4	2242	3.8	105	20	6.3
Series II					
50RCA/RCG-3	2250	4.6	95	19.5	3.2
50RCA/RCG-4	2260	3.4	105	20.0	4.3
50RCA/RCG-6	2169	2.8	105	23.5	5.4

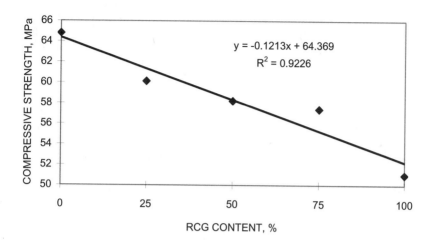

Figure 2   Influence of RCG content on the compressive strength

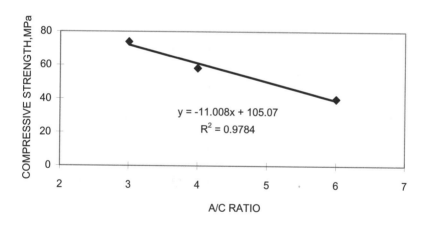

Figure 3   Influence of A/C ratio on the compressive strength

Despite a general trend could not be drawn between RCG content and the tensile splitting strength, a decrease in the A/C ratio certainly increased the tensile strength of the paving blocks. The results in Table 7 show that the skid resistance was not affected by the RCG content or the A/C ratio. On the other hand, it was found that the RCG content did not cause a reduction in the abrasion resistance; however, an increase in the A/C ratio decreased the abrasion resistance.

Due to the negligible water absorption of the RCG, the paving block, which contained a higher RCG content, had a lower water absorption value. Furthermore, due to the higher aggregate content as a result of the higher A/C ratio, the corresponding water absorption of the paving block increased as the A/C ratio increased. After analyzing the results in Figure 2 and Table 7, it is recommended to incorporate 50% RCG in the paving blocks in order to obtain a balance between strength and other durability properties.

The test results of the mortar bars prepared with 50% RCG content and different PFA additions are shown in Figure 4. It is shown that the mortar bar mixture prepared without PFA was susceptible to ASR according to the definition of ASTM C1260 where 0.1% expansion is the limit at 14 days. However, as the addition of PFA increased from 0 to 10% by weight of the total aggregates, the expansion of the mortar bar was suppressed to a very low level up to the measurement at 28 days. The expansion was even negligible when the PFA content was 15%.

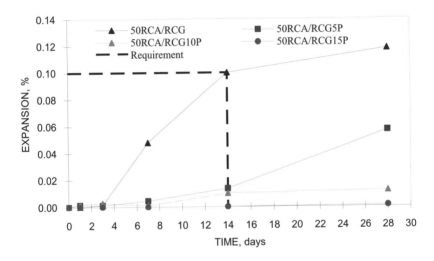

Figure 4   Expansion of the mortar bar mixtures up to 28 days

In order to assess the influence of the use of PFA on the properties of the paving blocks, a single mixture was prepared with 50% RCG, 50% RCA and 10% PFA addition. The test results are summarized in Table 8 and the concrete mixture prepared with the same RCA and RCG contents but without PFA addition was cited as a reference.

The results indicated that the use of PFA as an additional cementitious material slightly decreased the density of the paving blocks. However, due to the pozzolanic reaction of the fly ash particles, the 28-day compressive strength was about 20% higher than that of the reference mixture. The higher compressive strength of the paving block also led to a better abrasion resistance of the paving blocks prepared with fly ash.

Nevertheless, the addition of fly ash did not affect the skid resistance and the water absorption of the paving blocks. Based on the findings, the paving block production can be optimized by using 50% RCG, 50% RCA, 10% PFA by weight of total aggregates and an A/C ratio of 4 in the concrete mixtures.

Table 8 Properties of the paving blocks prepared with PFA

NOTATION	DENSITY, kg/m^3	COMP. STRENGTH MPa	SKID RESIST. BPN	ABRASION RESIST. mm	WATER ABSORPTION %
50RCA/RCG-4	2260	58.2	105	20.5	4.3
50RCG10P-B	2242	70.0	105	19.0	4.2

CONCLUSIONS

Based on the results of this study, the following conclusions can be drawn:

1. The compressive strength of the paving blocks decreased with an increase in the RCG content.

2. The compressive strength is directly affected by the aggregate-to-cement (A/C) ratio.

3. The abrasion resistance decreased as the A/C ratio increased.

4. The increasing use of RCG decreased the water absorption of the paving blocks.

5. The increase in the A/C ratio increases the water absorption of the paving blocks as a result of the higher aggregate content.

6. The use of 10% PFA by weight of the total aggregates was an effective means to minimize the potential expansive reaction due to ASR measured in accordance with ASTM C1260.

7. The addition of PFA increased the compressive strength of the paving blocks while other properties remained unchanged.

REFERENCES

1. COLLINS, R J, HARRIS, D J AND SPARKES, W. Blocks with recycled aggregate: beam-and-block floors. BRE Report IP 14/98. Building Research Establishment. United Kingdom, 1998.

2. JONES, N, SOUTSOS, M N, MILLARD, S G, BUNGEY, J H AND TICKELL, R G. Developing precast concrete products made with recycled construction and demolition waste. In: Limbachiya MC, Roberts JJ, editors. Proceedings of the International Conference on Sustainable Waste Management and Recycling: Construction Demolition Waste. London: Kingston University, 2004, pp. 133-140.

3. POON, C S, KOU, S C AND LAM, L. Use of recycled aggregates in moulded concrete bricks and blocks. Construction and Building Materials. Vol. 16(5), 2002, pp 281 – 289.

4. CIVIL ENGINEERING AND DEVELOPMENT DEPARTMENT (CEDD). General specifications for civil engineering works. HKSAR, 1992.

5. CHAN, D AND POON, C S. Using recycled building site waste as aggregates for paving blocks. (Accepted by Proceedings of ICE: Waste and Resource Management for publication)

6. BSEN 197-1:2000, Specifications for Portland cement. British Standards Institution, 2000.

7. BS 3892, Pulverized fuel ash, Part 1: Specification for pulverized fuel ash for use with Portland cement. British Standards Institution, 1997.

8. BSEN 12620:2002, Specifications for aggregates from natural sources for concrete. British Standards Institution, 2002.

9. ASTM C1260. Standard test method for potential alkali reactivity of aggregates (mortar-bar method). Annual Books of ASTM Standards, 2001.

10. BS 6717:2001, Precast, unreinforced concrete paving blocks-requirements and test methods. British Standards Institution, 2001.

11. AS/NZS 4455. Masonry units and segmental pavers. Australian/New Zealand Standard, 1997.

12. BSEN 12390-7:2000. Testing concrete: methods for determinations of density of hardened concrete. British Standards Institution, 2000.

MECHANICAL PROPERTIES OF CONCRETE CONTAINING HIGH DENSITY POLYETHYLENE AS AGGREGATES

S Gavela

S Kolias

K Kordatos

V K Rigopoulou

National Technical University of Athens

Greece

ABSTRACT. Increasingly plastics become the material of choice by product designers because they provide useful properties and for this reason the worldwide production of plastics is growing year after year. Concrete, the most widely used material, is a potentially very promising market for recycled plastics. The present research deals with the study of the use of thermoplastic wastes as a part of the conventional aggregates in concrete. A widely used thermoplastic polymer was studied, high density polyethylene (HDPE). The polymer was industrial waste and it was used without any treatment taking account of economy in practice. Mixes containing HDPE at percentages 0, 7, 12 and 20% by volume of aggregates were made. Flexural and compressive strength and dynamic and static modulus of elasticity were determined at various ages. A study of the homogeneity of the mix was also performed. The results suggest that HDPE can be used as concrete aggregates for certain applications.

Keywords: Concrete, Plastic waste, High density polyethylene (HDPE)

S Gavela is a PhD Candidate at the National Technical University of Athens. Her thesis focuses on the use of thermoplastic wastes as aggregates in concrete.

S Kolias is an Associate Professor at the Civil Engineering Department of NTUA, Greece.

K Kordatos, is a Lecturer at the Chemical Engineering Department of NTUA, Greece. His teaching and research interest include New Inorganic Materials and Nanomaterials.

V Kasselouri – Rigopoulou is a Professor at the Chemical Engineering Department of NTUA, Greece. Her teaching and research fields include High Temperature Chemistry and Technology (Cement, Ceramics, Glass) and New Inorganic Materials.

INTRODUCTION

The fraction of plastic wastes in household wastes is large and increases with time so it is of great importance to find applications for the reuse of plastic waste. Additionally, industrial wastes also continue to increase. At the present study plastic waste and specifically a widely used polymer, high density polyethylene (HDPE), is proposed as a replacement of a part of the conventional aggregates of concrete.

Several industrial wastes can be recycled as a substitute for various virgin materials in concrete [1, 2]. Koide et al [2] at their study concluded that the use of plastic aggregate in concrete has several advantages. The plastic aggregates were made by crushing lumps of plastic produced by cooling a molten mixture which contained about 85% polyethylene terephthalate (PET) and about 15% polypropylene (PP) and polyethylene (PE). Other alternative aggregates that have been examined are recycled tyre rubber [3, 4], and melamine – formaldehyde thermosetting plastics [5]. Also, in order to improve certain concrete properties, such as toughness and tensile strength, polyolefin fibres are introduced in the concrete mixture [6].

The influence of temperatures (2–250°C) on compressive strength of concrete containing thermoplastic wastes as aggregates has been studied [7]. It was found that significant reduction of strength is associated with the temperature of 250°C and higher and therefore these materials should not be allowed for buildings. However other uses such as pavements, culverts, river banks etc could be found for this kind of concrete.

EXPERIMENTAL PROCEDURE

Concrete included in the research was prepared with cement type CEM II 42.5, fine and coarse crushed limestone aggregates, plastic aggregates and tap water. The plastic aggregates used (Figure 1) are industrial wastes from high density polyethylene (HDPE). They were used without any treatment for practical and economic reasons. The grain size of the polymers did not vary widely and they were used as a replacement of the conventional fine aggregates. A sieve analysis was performed on the gradation of the plastic as well as to the fine and coarse crushed limestone aggregates that were also used (Figure 2), according to ASTM C136.

Figure 1   Industrial wastes from high density polyethylene

The results of the determination of specific gravity and water absorption of the aggregates are shown in Table 1. Four different mixes were made: one reference mix (0% by volume replacement of conventional aggregates) and three mixes containing HDPE aggregates (7, 12 and 20% by volume replacement of conventional aggregates). The water to cement ratio was kept constant, 0.6. Table 2 shows details of the proportions of all mixes.

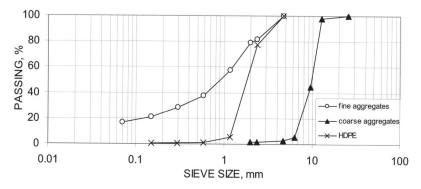

Figure 2   Gradation of plastic, fine and coarse aggregates

Table 1   Physical properties of aggregates

MATERIAL	SPECIFIC GRAVITY, kg/m^3	WATER ABSORPTION, %
Coarse aggregates	2680	0.92
Fine aggregates	2620	1.75
HDPE	910	0.00

Table 2   Proportions for mixes containing HDPE

MIX COMPONENT	VOLUME REPLACEMENT OF CONVENTIONAL AGGREGATES, %			
	0% (RM)	7% (HDPE7)	12% (HDPE12)	20% (HDPE20)
Water, kg/m^3	205	205	205	205
Cement, kg/m^3	350	350	350	350
Fine aggregates, kg/m^3	876	754	666	526
Coarse aggregates, kg/m^3	896	896	896	896
HDPE, kg/m^3	-	43	73	122

$70 \times 70 \times 280$ mm prisms and $100 \times 100 \times 200$ mm prisms were cast. After 24h the specimens were demoulded and put in constant conditions: 20°C and RH>95%.

7 and 28 days after the specimen preparation flexural strength, compressive strength and dynamic modulus of elasticity were determined from the average of three $70 \times 70 \times 280$ mm prisms. The flexural strength of the prisms was measured using one-third point loading as

described in EN 12390-5. The two parts of the specimen after the flexural test were used for compressive strength determination (equivalent cube test) by applying the load through square steel plate of the same size as the cross-section of the prism (70 × 70 mm). Static modulus of elasticity was determined from 100 × 100 × 200 mm prisms 28 days after preparation.

Dynamic modulus of elasticity in GPa is given by Equation (1):

$$E_d = 4 \cdot 10^{-15} \cdot n^2 \cdot L^2 \cdot p \tag{1}$$

where n is the fundamental resonant frequency (Hz), L is the length of the specimen (m) and p is its density (kg/m^3). The fundamental resonant frequency is measured with the test method prescribed in ASTM C215-91. Since it is a non destructive method, it has the advantage that the same specimen can be used for the determination of dynamic modulus of elasticity at various ages.

As it is already known, specific gravity of the mix constituents is one of the primary causes of segregation. The low unit weight of polymers and the lack of fines that can be the result of the use of poorly graded polymers could lead to segregation of the mix. For this reason the study of the scanned digital images of cross sections of the specimens was thought to be necessary, in order to find out if separation of the constituents of the mixture has taken place. 100 × 100 × 200 mm prisms were cut parallel to pouring direction. With optical observation of the cross sections the homogeneity or heterogeneity of the mixes could be concluded, but for better results the cross sections were digital scanned and the images received were edited with image software. The different colour of the polymers especially helped in distinguishing the polymer aggregates and with the software tools that enable the user to automatically select coloured objects the polymer aggregates were outlined. Image analysis software additionally have the possibility to automatically select light coloured objects as limestone aggregate particles so whatever the colour of the polymer is this method can be used [7].

RESULTS AND DISCUSSION

Strength

The results of flexural and equivalent cube compressive strength tests are plotted in Figures 3 and 4 against the percentage of HDPE replacement of aggregates. It can be seen that both compressive and flexural strength decreases as the percentage replacement of the aggregates increases. This is attributed to the lower strength of the plastic wastes and also to the lower bond characteristics of these materials with cement paste.

The reduction of strength expressed as % reduction in relation to the mixes containing no plastic replacement of the aggregates is shown in Figures 5 and 6 plotted against the percentage of replacement. From the coefficients of the regression lines computed to pass through the origin, it can be seen that the rate of reduction is greater for the compressive strength than for the flexural strength for HDPE. This relationship between flexural strength and compressive strength is important as it indicates whether with the use of the plastic wastes the tensile behaviour of the material is improved. The coefficients of the regression lines for the reduction of flexural and compressive strength were calculated as averages of all ages (7, 28 and 90 days).

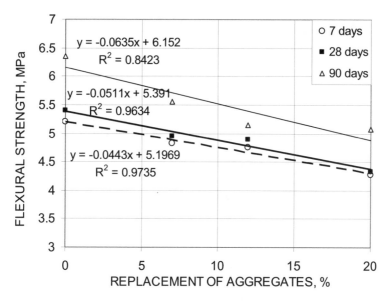

Figure 3   Relationship of flexural strength and % HDPE replacement of aggregates

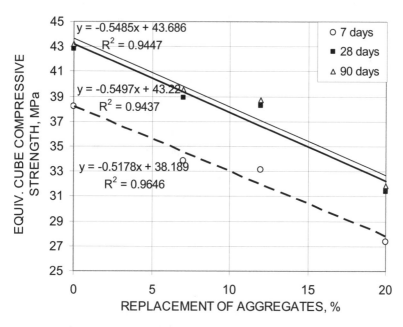

Figure 4   Relationship of compressive strength and % HDPE replacement of aggregates

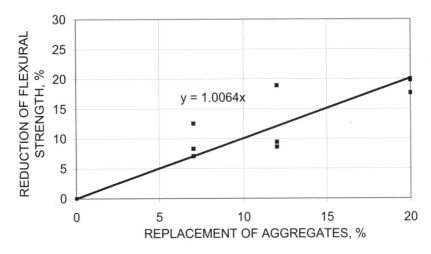

Figure 5   Flexural strength reduction against % HDPE replacement of aggregates

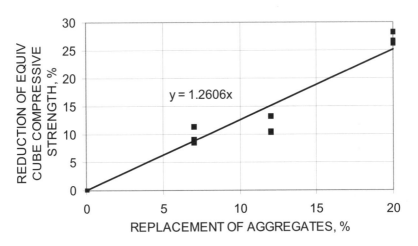

Figure 6   Equivalent cube compressive strength reduction against % HDPE replacement of aggregates

Modulus of elasticity

The dynamic and static modulus of elasticity of concrete containing different percentages of plastic aggregates are plotted against % replacement in Figures 7 and 8. It can be seen that the modulus values decrease rather rapidly as the plastic aggregate content increases. The reduction is attributed to the lower modulus of elasticity of the plastic wastes. Figure 9 shows the % reduction of the static and the dynamic modulus of elasticity in relation to the % replacement of aggregates. It can be inferred that the reduction of the modulus of elasticity is greater in the static modulus values than in the dynamic ones reflecting the fact that the behaviour of plastic materials is very much influenced by the frequency and the magnitude of loading.

The rate of decrease of modulus of elasticity is higher than that of the flexural strength and consequently the load or temperature stresses will be reduced. More important however, is the influence of the % plastic replacement of aggregates to the ratio E/f of the modulus of elasticity to flexural strength. This is shown in Figure 10. It is noted that the ratio E/f is a rough estimation of the ratio σ/f of the load or temperature induced stresses to flexural strength, which defines the safety margin and the fatigue response of the structural element. It can be seen that for the case of HDPE replacement the ratio decreases significantly.

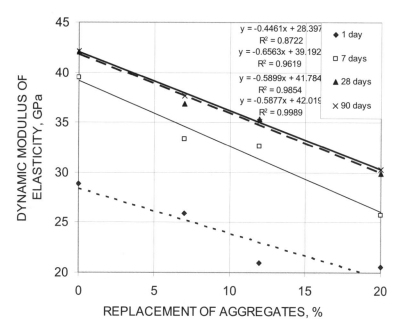

Figure 7   Dynamic modulus of elasticity against % HDPE replacement of aggregates

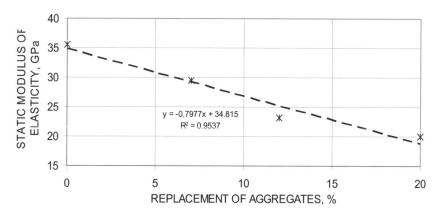

Figure 8   Static modulus of elasticity against % HDPE replacement of aggregates

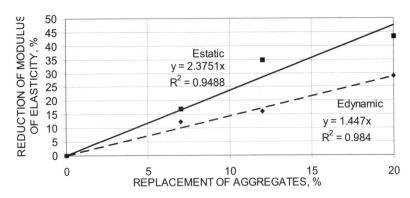

Figure 9   Percentage reduction of static and dynamic modulus of elasticity at the age of 28 days in relation to the % replacement of aggregates

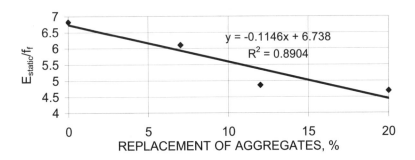

Figure 10   Relationship of static modulus of elasticity to flexural strength ratio and % by volume replacement of conventional aggregates

Homogeneity of the mix

One representative image from each mixture containing HDPE aggregates received after scanning the cross sections of the specimens is shown in Figure 11.

As it can be seen, segregation of the mix due to the lower unit weight of the polymers from the conventional aggregates has not taken place and HDPE is dispersed all over the specimen.

CONCLUSIONS

The replacement of a part of the conventional aggregates by high density polyethylene results in a reduction of strength (compressive and flexural) as well as of dynamic and static modulus of elasticity. This reduction is attributed to the lower strength and modulus of elasticity of the polymers and to the relatively lower bond strength of the interface between the polymer and the mortar. It could be said that there is a strong indication that by increasing the HDPE content in the mix, the material is improved from the point of view of the stress ratio f_f/f_c. Additionally, as the percentage replacement increases, the rate of modulus of elasticity decrease seems to be higher than that of flexural strength, so load or temperature induced stresses will be reduced. Digital imaging of concrete prism specimens containing HDPE aggregates showed that the plastics are well dispersed.

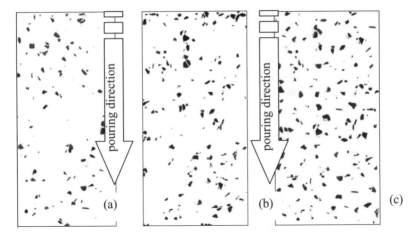

Figure 11   Representative images received after editing the images of cross sections of specimen containing (a) 7% by volume of aggregates HDPE, (b) 12% by volume of aggregates HDPE, (c) 20% by volume of aggregates HDPE

ACKNOWLEDGMENTS

A part of the laboratory work was carried out in the Laboratory of the Hellenic Cement Industry Association and the authors would like to express their appreciation for the facilities made available.

REFERENCES

1. PAPADAKIS V G AND TSIMAS S, Greek supplementary cementing materials and their incorporation in concrete, Cement and Concrete, Vol. 27, 2005, pp. 223-230.

2. KOIDE H, TOMON M AND SASAKI T, Investigation of the use of waste plastics as an aggregate for lightweight concrete, Sustainable concrete construction, Proceedings of the International Conference held at Dundee, Scotland, UK, 2002.

3. RAGHAVAN D, HUYNH H AND FERRARIS C F, Workability, mechanical properties, and chemical stability of a recycled tyre rubber-filled cementitious composite, Journal of materials science, Vol. 33, 1998, pp. 1745-1752.

4. SIDDIQUE R AND NAIK T R, Properties of concrete containing scrap-tire rubber – an overview, Waste Management, Vol.24, 2004, pp. 563-569.

5. ZIARA M M, DWEIK H S AND HADIDOUN M S, Engineering properties of concrete made with ground melamine–formaldehyde thermosetting plastics as sand replacement, Proceedings of the 6th international conference on concrete technology for developing countries, Jordan, 2002, pp. 233-242.

6.   TAGNIT – HAMOU A, VAHHOVE Y AND PETROV N, Microstructural analysis of the bond mechanism between polyolefin fibers and cement pastes, Cement and Concrete Research, Vol. 35, 2005, pp. 364-370.

7.   KASSELOURI – RIGOPOULOU V, KOLIAS S AND GAVELA S, Study of the thermal behavior of concrete containing thermoplastic wastes as aggregates, Proceedings of the 6th international conference Global Construction – Ultimate Concrete Opportunities, Dundee, 2005, pp. 799 – 805.

8.   HUNTER A E, AIREY G D AND COLLOP A C, Effect of asphalt mixture compaction on aggregate orientation and mechanical performance, Proceedings of the 8th Conference on Asphalt Pavement for Southern Africa, South Africa, 2004.

VIABILITY OF ELECTRIC ARC FURNACE SLAG (EAFS) AS NON-CONVENTIONAL MATERIAL IN CONCRETE MIX

K Shavarebi Ali

A R M Ridzuan

University Teknologi MARA

A B M Diah

Kolej Universiti Teknikal Kebangsaan

Malaysia

ABSTRACT. This paper reports the research on the feasibility and viability of Electric Arc Furnace Slag (EAFS) in concrete mix. Research work was carried out in the laboratory to produce concrete with good hardening properties using EAFS as coarse aggregate. Concrete mixes with water / cement ratios ranging from 0.47 to 0.63 were prepared in order to analyse concrete strength incorporating 10%, 50% and 100% of slag aggregate. Physical properties for Natural Aggregate (NA) and Steel Slag Aggregate (SSA) were compared and investigated material. The results showed that compressive strength of concrete specimen incorporating 100% of steel slag was higher compared to that of natural aggregate concrete. The mixture with EAFS aggregate presented the highest strength at 28 days. Similarly, some of the physical properties of the EAFS aggregate were better than those of natural aggregate (granite). With respect to resistance to carbonation, Steel Slag Concrete (SSC) showed comparable performance.

Keywords: Demand on construction materials, Electric Arc Furnace Slag (EAFS) aggregate, Concrete, Strength, Durability, Carbonation.

Kamran Shavarebi Ali MCIOB is a Senior Lecturer in the Building Department of the Faculty of Architecture, Planning and Surveying at Universiti Teknologi MARA, Malaysia. His research focuses on the usage of non conventional materials in concrete production. He is a member of the Chartered Institute of Building, UK. He published a number of international papers in this field.

Dr Ahmad Ruslan Mohd Ridzuan, is a Senior Lecturer, Faculty of Civil Engineering, University Teknologi MARA, Malaysia. His interest includes recycled aggregate concrete, utilization of waste in concrete and self compacting concrete. He published several papers in this area.

Datuk Assoc. Prof. Dr Abu Bakar Mohamad Diah, Associate Professor at Kolej Universiti Teknikal Kebangsaan, Malaysia. His research interest includes blended cement concrete and durability of concrete. He is currently the chairman of the Concrete Society of Malaysia.

INTRODUCTION

Concrete is one of the most important and widely used construction materials. Its consumption rose tremendously with the growth in population and urbanization. However, along with an increased consumption, there is also an increased generation of industrial waste and by-product material.

The disposal of industrial waste and by-product material is also an environmentally sensitive problem facing waste managers throughout the world. As environmental quality standards become more stringent and the volume of waste generated continues to increase, the traditional disposal methods are no longer accepted and there is therefore great pressure to change [11].

The use of Non-Conventional materials increases. These materials can be incorporated in concrete to facilitate several benefits including the modification and improvement of certain material properties, the conservation of non-renewable natural resources and the utilisation of industrial by-products [8].

Using Electric Arc Furnace Slag (EAFS) is a good example. Steel slag, a by–product of steel-making operation, steel scrap is melted in an electric arc furnace along with fluxing agents. Steel Slag chemistry is based upon the fluxing practices and impurities from the selected scrap. The steel slag aggregate, presently produced by a steel plant in the northern Peninsula of Malaysia, is utilized in road construction.

There are two main reasons using by-product (Steel Slag) as aggregate in concrete, firstly it can reduce the environmental problem due to the production of primary aggregate in quarry and conserving our natural resources and secondly it will overcome the issue of waste disposal. Slag is currently temporarily stored or used as landfills in steel plant compound or sent to designated landfills [13].

In Malaysia, very little is known about the use of steel slag in the manufacture of concrete. Therefore, the main thrust of this investigation was to evaluate the viability of steel slag aggregate for use as coarse aggregate in fresh concrete, as such material is likely to provide both environmental and economic advantages.

Most of the previous research regarding slag aggregate has been done in concrete mix and focused on one mix proportion which limits the data on the advantages of steel slag aggregates. However this perception might be changed in Malaysia and if several concrete mix proportions were tested.

In this study, an investigation has been carried out to quantify the properties of concrete made by fully and partially replacing natural coarse aggregate with steel slag aggregate. The properties investigated were aggregate characteristics, compressive strength development and resistance to carbonation.

MATERIALS AND TECHNIQUES

Materials

The materials used in this study were ordinary Portland cement in compliance with MS 522 [6] and BS EN 197-1 [3]. River sand and coarse natural aggregate (granite) complied with BS EN 12620 [2]. Crushed electric arc furnace oxidizing slag used in this study was obtained from a steel plant (located in the north of Peninsular Malaysia) and was of nominal minimum size of 10 mm and maximum size of 20 mm. The 20 mm and 10 mm coarse aggregates were then combined in the ratio of 2:1 respectively for both the natural aggregate and slag aggregate concrete mixes.

Physical Properties of Aggregates

The physical properties of all aggregates in terms of specific gravity, aggregate crushing values and water absorption are presented in Table 2, determined in accordance with BS 1881: Part 3 [5].

Mix Design of Slag Aggregate Concrete (SAC) and Natural Aggregate Concrete (NAC)

The designed water / cement ratios of the mixes ranged from 0.47 to 0.63. Since there is no existing standard method of designing concrete mixes incorporating slag aggregate, slag aggregate concrete mixes were derived simply by replacing the natural coarse aggregate proportion in the natural aggregate concrete mix design developed using conventional mix design method [15], with slag coarse aggregate (air dried). The mix proportions are shown in Table 3.

Compressive Strength Test

For each concrete mix, 48 numbers of 100 mm size cubes were made in standard steel moulds. After 24 hours, the cubes were removed from the mould and cured in water at 23°C until testing. The cube specimens were tested for compressive strength in accordance with BS 1881: Part 116 [1]. The reported observations are the average of three measurements of compressive strength at each curing age.

Measurement of Depth of Carbonation

Since natural atmospheric carbonation is a slow process, an accelerated carbonation chamber was used. During exposure, an atmosphere of 4% CO_2 (Carbon Dioxide) and 55% RH (Relative Humidity) was applied. The process of measuring the carbonation depth was done as recommended by RILEM CPC-18 (1984). Carbonation depth was monitored at intervals of 5, 10 and 15 weeks. Nuruddin, 1997 [12] suggests that exposure under these conditions of one week is approximately equivalent to one year of normal atmospheric carbonation. Prior to placing in the accelerated carbonation chamber, all specimens were dried for 14 days after the 28 days water curing process in the laboratory, to ensure that the specimens were in a surface-dry condition. The depth of carbonation was determined by spraying the surface of the broken concrete, which has been split perpendicular to the exposed faces, with a solution of phenolphthalein (1% phenolphthalein in 70% ethanol).

RESULTS AND DISCUSSION

Chemical Composition

Blast Furnace Slag and Steel Slag have similar chemical constituents, the proportions are quite different. However, the more pronounced difference between the two slags is their difference in their mineralogical composition. Table 1 gives the chemical analysis of the steel slag used, which is compared with Blast Furnace Slag (BFS) and Electric Arc Furnace Slag (EAFS). As it is obvious, as opposed of blast furnace slag, steel slag cannot be used in the production of blended cement; due to its low SiO_2 content which makes it less pozzolanic and FeO content which causes a difficult grinding process [9].

Physical Properties of Steel Slag

Specific gravity

According to the laboratory results shown in Table 2, steel slag has a higher specific gravity. The specific gravity of most natural aggregate ranges between 2.6 and 2.7. Granite also falls in this range [10]. However, the specific gravity for steel slag is between 3.0 and 3.7, which is higher than that of natural aggregate [7]. Result obtained for EAFS or steel slag show a specific gravity of 3.48 which is also within that range and is higher compared to granite. This may occur due to its higher density in contrast to granite. The quality and properties of the aggregate can lay on its specific gravity. The higher hardness of EAFS could improve abrasion resistance of concrete [10].

Absorption

The water absorption of the steel slag aggregate was more than twice of that of granite aggregate due to the porosity of steel slag particles compared with granite. The higher water absorption of steel slag may be attributed to the impervious nature of steel slag aggregate compared to the granite aggregate [7].

Table 1   Chemical analysis of different types of slag

CONSTITUENT	COMPOSITION, %	
	EAF Slag	BFS
CaO	37 – 56	34 – 48
SiO_2	8 – 29	31 – 45
FeO	15 – 45	0.1 – 2.44
MnO	4 – 6	2 – 10
MgO	3 – 8	1 – 15
Al_2O_3	4 – 9	10 – 17
P_2O_5	< 1	1 – 3
S	0.2 - 2	-

Sources: Southern Steel Berhad, Malaysia & M. Shekarchi 2004

Aggregate crushing value (ACV)

Based on Table 2, steel slag has lower crushing value compared to granite. Thus it is expected that steel slag aggregate may produce concretes that are more durable and stronger than granite aggregate concrete. Higher ACV of granite aggregate will lead to low resistance of the aggregate against compressive loads [7], but it is still between of 20 to 25 ACV value that can be accepted as concrete aggregate.

Table 2 Comparison of physical properties of aggregate

PHYSICAL PROPERTIES	AGGREGATE	
	Granite	Steel Slag
SSD, specific gravity, g/cm^3	2.61	3.48
Water absorption, %	0.8	1.8
Aggregate Crushing Value (ACV), %	24.79	19.63

Sources: Obtain from the Laboratory

Table 3 Mix Proportions for SAC and NAC

COMPOSITION, kg/m^3	WATER/CEMENT RATIO		
	0.63	0.57	0.47
OPC	320	355	435
Water	204	204	204
Coarse Aggregate	726	744	802
Sand	1065	1010	860

Compressive Strength Test

The compressive strength development in concrete containing different aggregates and different percentages of slag aggregate are shown in Tables 4 to 6. As expected, the compressive strength of both these mixes increased with the period of curing. Generally the trend is similar to Natural Aggregate Concrete (NAC) and Slag Aggregate Concrete (SAC) mixes that is the higher the water cement ratio the lower the compressive strength. From the results it can be seen that all the mixes managed to attain the 28 day target strength.

For the same design strength, the Slag Aggregate Concrete (SAC) mixes showed higher compressive strength as compared to the corresponding Natural Aggregate Concrete (NAC) mixes. The higher compressive strength of Slag Aggregate Concrete (SAC) may be attributed to high water absorption of steel slag aggregate which reduces the effective water-cement ratio of the SAC mixes, thus resulting in higher compressive strength than the corresponding Natural Aggregate Concrete (NAC) mixes. Also the shape of the steel slag aggregate was more angular with a rough texture, which provided enhanced bonding to the concrete matrix compared to natural aggregates.

Table 4   Compressive strength results of W/C = 0.63 (water cured) concrete mix

COARSE AGGREGATE	STRENGTH, N/mm^2			
	Age, days			
	3	7	28	90
Granite (control)	9	15	22	24
Slag (10%)	10	16	22	24
Slag (50%)	12	18	23	26
Slag (100%)	14	19	25	27

Table 5   Compressive strength results of W/C = 0.57 (water cured) concrete mix

COARSE AGGREGATE	STRENGTH, N/mm^2			
	Age, days			
	3	7	28	90
Granite (control)	14	22	32	35
Slag (10%)	15	24	33	36.0
Slag (50%)	17	26	34	38
Slag (100%)	20	28	37	41

Table 6   Compressive strength results of W/C = 0.47 (water cured) concrete mix

COARSE AGGREGATE	STRENGTH, N/mm^2			
	Age, days			
	3	7	28	90
Granite (control)	18.0	29	43	47
Slag (10%)	19	32	43	48
Slag (50%)	23	35	46	51
Slag (100%)	27	38	50	54

Durability Performance of NAC and SAC with Respect to Carbonation

The result of the depth of carbonation measurement after 5, 10 and 15 week's exposure to 4% CO_2 (Carbon Dioxide) in an accelerated carbonation chamber for the Natural Aggregate Concrete (NAC) and Slag aggregate concrete (SAC) mixes are shown in Figures 1 and 2. The depth of carbonation of the Slag Aggregate Concrete (SAC) mixes was comparable to the corresponding natural aggregate concrete NAC mix with only very small differences in performance with a decreased water / cement ratio. The reason for this is that the higher water absorption of the slag aggregate reduces the effective water / cement ratio, hence it can be deduced that the utilization of slag aggregate does not have a detrimental effect on the durability of Slag Aggregate Concrete (SAC) with respect to resistance to carbonation.

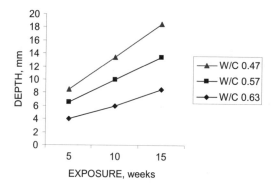

Figure 1   Depth of Carbonation of the control sample

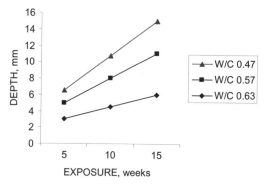

Figure 2   Depth of carbonation of the 100% steel slag aggregate concrete sample

CONCLUSION

No significant differences in physical properties of slag aggregate were observed as compared to granite aggregate. Tests also showed that it was possible to obtain Slag Aggregates (SA) complying with the standard. It must be mentioned here that the crushed Slag Aggregates (SA) were kept in a sealed container and used when required. No adjustment of mixing water content was made to account for the higher water absorption of Slag Aggregate (SA). From a strength point of view, the Slag Aggregate Concrete (SAC) compared well with the corresponding Natural Aggregate Concrete (NAC) and could be considered for various potential applications. The durability performance of Slag Aggregate Concrete (SAC) and Natural Aggregate Concrete (NAC) were comparable. Currently, research on longer strength development (1 year), other mechanical aspects (Tensile Strength, Flexural Strength, Modulus of Elasticity and Drying Shrinkage) and other laboratory tests (Permeation Tests, Air permeability, Sulphate Attack and Near Surface Absorption) are carried out to evaluate the durability of steel slag. Every technical aspect will be identified in using steel slag for the construction industry in Malaysia.

The benefits resulting from the study will provide the construction industry with technical information on a valuable resource that has a key role in meeting the challenges of sustainable construction.

ACKNOWLEDGEMENTS

The authors thankfully acknowledge the research grant provided by the Institute of Research, Development and Commercialization of the University Teknologi MARA (UiTM). The authors are also thankful to RST Teknologi Sdn Bhd, Prai Industrial State, Pulau Pinang, Malaysia for donating the steel slag for this investigation. Authors are also thankful to Miss Norishahaini bt Mohamed Ishak for assisting in this research. Not to be forgotten is Dr. Haji Mohamad Nidzam Rahmat who was always there when I needed him. May god reward them accordingly.

REFERENCES

1. BRITISH STANDARDS INSTITUTION, BS 1881: Part 116, Method for Determination of Compressive Strength, London, 1983.

2. BRITISH STANDARDS INSTITUTION, BS EN 12620:2002, Aggregates for Concrete, 2002, 47 p.

3. BRITISH STANDARDS INSTITUTION, BS EN 197-1:2000, Cement. - Part 1: Composition, specifications and conformity criteria for common cements, 2000, 46 p.

4. BRITISH STARDARDS INSTITUTION, BS EN 12620: 2002, Aggregates for Concrete.

5. BRITISH STANDARDS INSTITUTION, BS EN 12620: 2002, Aggregates for Concrete.

6. MALAYSIAN STANDARDS 522: Part 1, Specification for Portland Cement, SIRIM Malaysia, 1989.

7. MEHTA P K AND MONTERIO P J, Concrete: Structure, Properties and Methods, 2nd edition, 1993.

8. MONTGOMERY D C, Instant Chilled steel slag aggregate in concrete, Cement and Concrete Research, Vol. 21, pp. 1083-1091.

9. SHEKARCHI M et al, Study on the Mechanical Properties of Heavy weight Replaced aggregate concrete using Steel Slag as aggregate, International Conference on Concrete Engineering and Technology, Universiti Melaya, Malaysia, 2004.

10. NEVILLE A M, Properties of Concrete, 3rd Edition, Pitmen Publisher, 1990.

11. NIDZAM R M, Soil Stabilization Utilization Wastepaper Sludge Ash, Unpublished PhD Thesis, University of Glamorgan, Wales, UK, 2004.

12. NURUDDIN M F, Permeation & Carbonation Characteristics of PFA Concrete, MSc Thesis, Dundee University, 1992

13. NEW STRAITS TIMES, Proposal to recycle slag for road construction, 17 March, 1997.

14. RILEM CPC-18, Measurement of hardened concrete carbonation depth, Materials and Structure, Vol. 17, No. 102, 1984, pp. 437-440.

15. TEYCHENNE D C, FRANKLIN R E AND ERNTROY H C, Design of Normal Concrete Mixes, 1988.

EXPERIMENTAL STUDY FOR USING RUBBER POWDER IN CONCRETE

M Khorami

Islamic Azad University

Iran

E Ganjian

Coventry University

United Kingdom

ABSTRACT. Sustainable development towards environmental protection and prevention of pollutant accumulation requires that special attention be paid to the development of a method for recycling and reapplication of waste materials, especially those that are normally not exhaustible. In this context, scrapped rubber tyres may be conceived as a type of non-replenishing waste material. In this research some of the most important features of rubber concrete have been investigated. Keeping coarse and fine aggregate constant in normal concrete, the influencing of different weight percentages (5, 7.5, and 10 %) replacement of powder of rubber tyre with cement in concrete is investigated in this paper. Experimental results show that replacing of powder rubber with cement up to 5% by weight have no important change in prevalent properties of concrete, but more increment of these materials changes these properties significantly.

Keywords: Concrete, Rubber, Sustainable development, Mechanical property.

Morteza Khorami, Islamic Azad University Branch Eslamshahr, Building and Housing Research Center, Tehran, Iran

Esmaiel Ganjian, Coventry University, Department of Built Environment, Coventry, UK.

INTRODUCTION

Waste rubber has received a great deal of attention for disposal or utilization because solid waste disposal is a major environmental issue to cities around the world. There are many ways for waste rubber to be useful. However, to harmonize with our environment, waste rubber should be converted to a sophisticated form for better utilization. Innovative solutions to solve the tyre disposal problem have been long in development. Researchers have found that waste rubber can be disposed of by several methods. The easiest disposal method is just burial in a landfill. However, as waste rubbers discarded in a landfill tend to float on top, mosquito breeding or illegal disposal is causing severe environmental pollution. Also, scrap rubber, as a fuel source is a possible method because incineration of scrap rubber has a high caloric value. Rubber asphalt appears to be a promising method in the future. State highway agencies have been proactive in their efforts to evaluate usable materials and to recycle or incorporate such materials into the highway system whenever possible. The use of waste tyres in the construction of asphalt pavements has been a success to the point where the Intermodal Surface Transportation Efficiency Act (ISTEA) (Intermodal 1991) mandated the use of recycled rubber in any federally assisted asphalt pavement project. The rubber improves asphalt ductility and increases the temperature at which the asphalt softens. Asphalt rubber is also used for waterproofing membrane, crack and joint sealer, hot-mix binder, and roofing material.

Owing to the problems associated with waste tyre modified asphalt, more and more attention has been paid to waste tyre modified Portland cement concrete. As opposed to waste tyre modified asphalt, which needs the wet process, waste tyre modified concrete utilizes the low-cost "dry process", with a portion of aggregates replaced by waste tyre rubbers. Most of researches have shown rubber concrete has very high toughness. This is desirable because conventional concrete is a brittle material.

Eldin and Senouci reported that concrete mixtures with tyre chips and crumb rubber aggregates exhibited lower compressive and splitting tensile strengths than regular Portland cement concrete. There was approximately 85% reduction in compressive strength and 50% reduction in splitting tensile strength when coarse aggregate was fully replaced by coarse crumb rubber chips. However, a reduction of about 65% in compressive strength and up to 50% in splitting tensile strength was observed when fine aggregate was fully replaced by fine crumb rubber. Both of these mixtures demonstrated a ductile failure and had the ability to absorb a large amount of energy under compressive and tensile loads [1].

Khatib and Bayomy used fine crumb rubber and tyre chips to replace a portion of fine or coarse aggregates. They found that the rubber-filled concrete showed a systematic reduction in strength, while its toughness was enhanced. They also proposed a regression equation to estimate the strength of rubber-filled concrete [2]. Topçu investigated the particle size and content of tyre rubbers on the mechanical properties of concrete. He found that, although the strength of rubber-filled concrete was reduced, the plastic capacity was enhanced significantly [3].

Schimizze et al. developed two rubberized concrete mixes in 1994 using fine rubber granules in one mix and coarse rubber granules in the second. While these two mixes were not optimized and their design parameters were selected arbitrarily, their results indicate a reduction in compressive strength of about 50% with respect to the control mixture. The elastic modulus of the mix containing coarse rubber granular was reduced to about 72% of that of the control mixture, whereas the mix containing the fine rubber granular showed a reduction in the elastic modulus to about 47% of that of the control mixture. The reduction in elastic modulus indicates higher flexibility, which may be viewed as a positive gain in rubberized Portland concrete cement mixtures used in stabilized base layers in flexible pavements.

The objective of this study was to evaluate the feasibility and performance of waste tyre modified concrete using larger sized fibre chips. Other aspects that may affect the performance of modified concrete, including characteristic of material, grading of rubber, tyre resources, and percent of replacement, were also investigated.

The past researches are often based on aggregate replacement by tyre rubber which was the main goal for this kind of application to prevent the entrance of waste tyres to environment rather than improvement of concrete properties. In this research, an attempt was made to not only investigate the past trend of research which was the replacement of aggregates by waste tyres but also to investigate the cement replacement by waste tyre. A number of researchers were believed that the replacement of cement by waste tyre even though it will reduce the amount of cementing materials but by filling the holes in the concrete will reduce stiffness.

MATERIALS AND CONCRETE MIX DESIGN

The materials used to develop the concrete mixes in this study were fine aggregate, coarse aggregate, tyre rubber, and cement. The fine aggregate (sand) and the coarse aggregate (gravel) were ordinary from Iranian mine. The type of scrap tyre rubber were used, powder rubber, which are material with gradation close to that of cement. Type I Portland cement was used in all mixes. The properties of fine and coarse aggregates were determined according to Iranian standard test methods which are similar to ASTM standard test methods C 127, C 128, C 29 and C 136. Specific grading for the fine and coarse aggregate and rubber are shown in Figure 1.

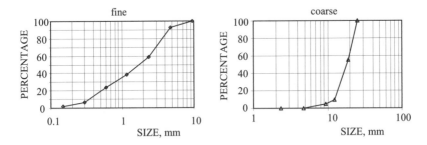

Figure 1   Aggregate grading

To develop the other rubberized concrete mixes, all mix design parameters were kept constant except for the cement constituents. Then rubber powder was used as a replacement for an equal part of cement by weight. Three groups of rubberized mixes were developed with 5, 7.5 and 10% replacement for cement.

The control sample in this research is named CS. Mix design specifications for samples are given in Table 1.

Table 1 Mix design specifications

SAMPLE NAME	CEMENT	POWDER RUBBER, kg/m³	FINE AGGREGATE, kg/m³	COARSE AGGREGATE, kg/m³	REPLACEMENT OF RUBBER POWDER FOR CEMENT, %
CS	380	0.0	858	927	Control
RC5	361	19.0	858	927	5
RC7.5	352	28.0	858	927	7.5
RC10	342	38.0	858	927	10

TEST RESULTS AND ANALYSIS

Cubic specimens of 150 mm were prepared to measure their compressive strength, carried out according to the standard BS 1881: Part 116: 1993. Also the tensile strength and modulus of elasticity of the specimens were determined according to BS 1881: Part 117: 1983 and BS 1881: Part 121: 1983, respectively. Flexural strength test was also carried out according to BS 1881: Part 118: 1983 by making prismatic samples of 100 × 100 × 500 mm. The results of measurement values taken in laboratory, and their analysis are described below.

Compressive strength test

The result of the compressive strength test at 28 day is presented in Figure 3. As it is seen, with 5% rubber replacement, the compressive strength change is about 5%, but addition of rubber at 7.5 and 10% would further reduce the strength by 10 to 45%.

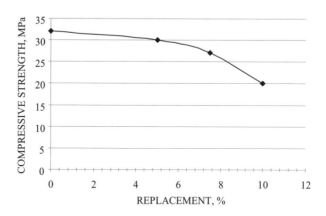

Figure 3 Compressive strength test results at 28 day

The reasons for reduction in compressive strength of concrete are more related to differing properties of rubber particles and cement. These factors include:

1- Since part of the cement is replaced by rubber particles, this causes a reduction in the quantity of cement used as adhesive materials. On the other hand, compressive strength of concrete depends on physical and mechanical properties of these materials (which have considerable superiority over rubber). A reduction in compressive strength of concrete is therefore predictable.

2- Because of the lower specific weight (or specific gravity) of rubber materials and also due to lack of bonding between rubber and the concrete mass, during moulding and vibration, rubber particles tend to move toward the upper surface of the mould, resulting in high concentration of rubber particles at the top layer of the specimens. Non-uniform distribution of rubber particles at the top surface tends to produce non-homogeneous samples and leads to a reduction in concrete strength at those parts, resulting in failure at lower stresses.

The findings of this research and other previous ones reveal that addition of 5 % by weight crumb rubber would not have noticeable impact on concrete strength.

Modulus of elasticity test

Replacing rubber powder for cement will reduce Young's modulus. The results of Young's modulus tests are given in Figure 4. Cement characteristics affect the modulus of elasticity. Considering concrete as a base model of a composite compound consisting of two phases (aggregate and cement), it is observed that the impact on aggregates is due to the modulus of elasticity and to the volumetric ratio of these particles in concrete.

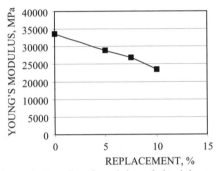

Figure 4   Results of modulus of elasticity tests

Modulus of elasticity of concrete containing rubber particles is defined by the formula given by Li and associates; they applied formula to calculate Young's modulus for normal concrete. To achieve this formula, they assumed concrete as a substance composed of cement paste and aggregates and then the formula was corrected for concrete containing rubber.

$$E_a = \frac{f_1 E_1 + f_2 E_2}{f_1 + f_2}$$

E_a is the equivalent modulus of elasticity in concrete, a part of which is replaced by tyre rubber bits.

E_1 and E_2 are the modulus of elasticity for aggregates and rubber respectively, and f_1 and f_2 are volumetric percentages. ($f_1+f_2+f_3=1$, f_3 is the volumetric percentage of cement).

Modulus of elasticity for the various constituents of concrete are approximately as following:

Modulus of elasticity of coarse aggregates 30,000-120,000 MPa
Modulus of elasticity of cement paste 10,000-30,000 MPa
Modulus of elasticity of tyre rubber 5-1200 MPa

Hence, an increase in rubber replacement for cement in concrete, the equivalent modulus of elasticity and consequently modulus of elasticity of concrete is reduced which is directly related with the volume of rubber added.

Tensile strength

The findings of tensile strength are given in Figure 5. Tensile strength of concrete was reduced with addition of rubber.

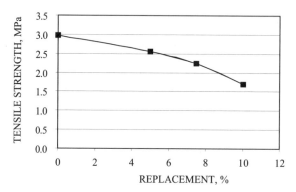

Figure 5   Results of tensile strength tests

Griffith's theory describes the behaviour of materials when exposed to tensile forces. According to the theory, materials contain microcracks and crevices, which act as stress concentration points when exposed to tensile forces and cause surface failure to generation of new surfaces. Thus, energy stored in the materials is freed. If the liberated energy is sufficient to develop cracks, the conditions for immediate failure is provided, but if barriers are emerged, crack expansion is halted so that exerted pressure is increased.

In concrete also, micro crack grows and expands if stress is applied. When crack expansion in the cement paste is confronted with barriers such as a large cavity, an un-hydrated cement particle and or a soft material that requires greater energy to disintegrate, it stops advancing. Tyre rubber as a soft material can act as a barrier against crack growth in concrete. Therefore, tensile strength in concrete containing rubber should be higher than the control sample.

However, the results showed opposite of this hypothesis. The reason for this: the transfer region between rubber and cement may act as a micro crack due to loose bonding between the two materials; the weak transfer region accelerates concrete breakdown. Therefore, it was thought that the adhesion between the matrix rubber and the aggregates was very poor. The more rubber powder added, the more weak boundary layers would be produced at the interface between the matrix rubber and the aggregates. Thus, it resulted in a decrease of tensile strength

Flexural strength

The results of flexural strength tests are shown in Figure 6. Addition of rubber to concrete reduces flexural strength. A reduction of 15% with respect to the control sample was observed.

According to a general principle governing flexure, flexural stresses exerted on concrete produce tensile stress on one side of neutral axis and compressive stress on the other, so that with combination of the coupled tensile and compressive forces, they can neutralize the flexural moment.

Due to low (negligible) tensile strength of concrete as compared to its compressive strength, in lower stresses and before concrete reaches its ultimate strength in the compression region, failure will occur. As a result the most important factor in reducing flexural strength, as well as the compressive strength is lack of good bonding between rubber particles and cement paste.

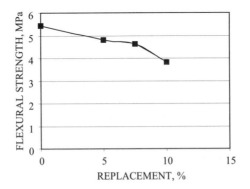

Figure 6   The results of flexural strength testing

CONCLUSIONS

A number of laboratory samples were prepared and tested in this research, in accordance with the prevalent Standards. The following research findings are therefore based on the laboratory studies and may differ with change in conditions such as material characteristics, changes in mix proportion, curing procedure, and also use of additives. The findings of the current research are as follows:

1. Compressive strength of concrete depends on percentage added. In general, compressive strength will be reduced with increasing the percent of rubber addition in concrete, though with 5% replacement of cement by rubber, decrease in compressive strength is low (less than 5%) with no noticeable changes in concrete properties the highest reduction is related to 7.5 and 10% replacement. With a reduction in compressive strength of about 20 to 40% for cement replacement.

2. Modulus of elasticity of concrete was reduced with the replacement of rubber for cement with lower modulus of elasticity. Reduction in modulus of elasticity was 18 to 36% by powdered rubber.

3. Tensile strength of concrete is reduced by addition of rubber particles. The most important reason being lack of proper bonding between rubber and the matrix, as bonding plays the key role in reducing tensile strength. For 5 to 10% cement replacement by rubber powder, the reduction is a bout 15 to 30%.

4. Addition of rubber to concrete caused a reduction in its flexural strength reduction was about 12 to 30% for cement replacement.

The reduction in physical characteristics in rubber concrete may limit its usage in some structural applications, but rubberized concrete may have some desirable characteristics such as lower density, higher impact and toughness resistance, enhanced ductility and better sound insulation into which more research is required. In this regard, some factors are important: size and percentage of rubber, type of cement, use of chemical and mineral admixtures, and methods of pre-treatment of rubber particles on the characteristics of concrete.

ACKNOWLEDGMENTS

This study was partially supported by the Building and Housing Research Center and Azad University. They provided support in the preparation of this paper and for continued research on rubberized concrete.

REFERENCES

1. ELDIN N N AND SENOUCI A B, Rubber tyre particles as concrete aggregates. ASCE Journal of Materials in Civil Engineering, 1993, pp. 478-496.

2. KHATIB Z K AND BAYOMY F M, Rubberized Portland cement concrete, ASCE Journal of Materials in Civil Engineering, 1999, pp. 206-213.

3. TOPÇU I B, The properties of rubberized concrete. Cement and Concrete Research, 1995, pp. 304-310.

4. AMOS A D AND ROBERTS M, Use of ground rubber tyres in Portland cement concrete, Proceedings of the International Conference of Concrete, University of Dundee, Scotland, 2000, pp. 379-390.

5. FATTUHI N I AND CLARK N A, Cement-based materials containing tyre rubber. Journal of Construction and Building Materials, 1996, pp. 229-236.

6. HERNANDEZ-OLIVARES F, BARLUENGA G, BOLLATI M AND WITOSZEK B, Static and dynamic behaviour of recycled tyre rubber-filled concrete, Cement and Concrete Research, 2002, pp. 1587-1596.

7. NAIK T R AND SINGH S S, Utilization of discarded tyres as construction materials for transportation facilities. Report No.CBU-1991-02, UWM Center for By-products Utilization. University of Wisconsin-Milwaukee, Milwaukee, 1991, 16 pp.

8. ROSTAMI H, LEPORE J, SILVERSTRAIM T AND ZUNDI I, Use of recycled rubber tyres in concrete, Proceedings of the International Conference on Concrete, University of Dundee, Scotland, UK, 2000, pp. 391-399.

9. SEGRE N AND JOEKES I, Use of tyre rubber particles as addition to cement paste. Cement and Concrete Research, 2000, pp. 1421-1425.

10. SIDDIQUE R AND NAIK T R, Properties of concrete containing scrap-tire rubber — an overview, Waste Management, 2004, Vol. 24, pp. 563-569

A NEW INSULATING MATERIAL FORMULATED USING CAR INDUSTRY WASTES, FLAX FIBRES AND POLYPROPYLENE

B Dupre   **B Laidoudi**

V Fouemina   **M Queneudec**

University of Picardie Jules Verne

France

ABSTRACT. Reuse of plastic resources for building materials is today a reality and also a very interesting opportunity for plastic board producer with an aim of utilising by-products as raw materials. The L.T.I. "Laboratoires des Technologies Innovantes" use to work with that sort of materials for many years and in particular with car industry wastes from used tyre rubber wastes or hybrid wastes from plastic board composites. Because of the physical characteristics of industrial by-product aggregates, elaborated composites will lead to very interesting mechanical, thermal and hydrous performance. In this paper, we will show the influence of the Aggregate/Cement ratio, on the one hand, on the bulk density of composites and on the other hand on their overall performance. We will also show the influence of the size of aggregates obtained by crushing for different granulometry. We expect the research to lead to obtain high Aggregates/Cement ratio to very low weight concrete composites with a very low thermal conductivity for thermal insulation while preserving mechanical and hydrous properties.

Keywords: Fibrous aggregates, Plastics, Plant wastes, Cement composites, Light concrete, Thermal insulation.

Dr B Dupre is a doctor in Process Engineering at the University of Picardie Jules Verne, Amiens, France. His major field of interest is the valorisation of plant resources in building materials.

Dr B Laidoudi is a doctor in process engineering at the University of Picardie Jules Verne, Amiens, France. His major field of interest is the comprehension of thermal behaviour of building materials.

V Fouemina is research student at the University of Picardie Jules Verne, Amiens, France, and his major field of interest is new ways of valorisation for industry by products.

Professor M Queneudec, is head of the "Laboratoire des Technologies Innovantes", University of Picardie Jules Verne, Amiens, France, she is specialised in the design and characterisation of building materials.

INTRODUCTION

Since many years car industry have developed a new generation of composites with plastic polymers and vegetal fibres or charges. As it was an opportunity to substitute non renewable material with plant resources, there were no good solutions for recycling these composites because of difficulties to separate plastics and plant material.

The LTI work on recycling topics for many years, for example, O Yazoghli Marzouk et al. studied the reuse of plastic waste cementitious concrete composites and show that the use of plastic aggregates can lead to very interesting bearing insulator [1]. The researchers of the LTI also studied the reuse of rubber wastes in cement composites [2, 3] and they also have great interest for the reuse of plant resources in construction material, for example the use of flax wastes to elaborate insulator wood concrete [4, 5].

So we used crushed plastic and flax composites from car industry to obtain a sort of fibrous aggregates with different sizes, and we used this fibrous material to elaborate a wood/cement composite. Some researchers from Romania showed the influence of aggregates size on wood cement composite properties [6], and it was thus interesting to appreciate the influence of the size of that new type of aggregates on the composite's properties. The interest of these sort of materials is their low weight, that's why we studied the influence of the volumetric ratio Aggregate/Cement (A/C) on the physico-mechanical properties.

MATERIALS AND METHODS

Crushed fibrous aggregates

Crushed plastic composites come from plastic board production, flax and polypropylene panel door composites. To obtain the crushed fibrous aggregates we used a crusher with four rotating swords. Crushed material from various size hopper led to obtain different types of fibrous aggregates, 2 to 10 mm. This difference will be important to explain mechanical, hydraulic and thermal properties of the elaborated composites. Crushed fibrous aggregates used in this work are summarised in Table 1.

Table 1 Different aggregate sizes and apparent dry bulk density

Aggregate size, mm	2.0	8.0	10.0
Apparent dry bulk density, kg/m^3	75	67	55

Cement and water

The cement used was Portland cement CEM I 52.5 according to the EN 196-1 standard. Water came from the French local network without any admixture to modify the composition.

Composite production

In order to produce the concrete samples we first used the Agresta formulation intended for cover wood concrete composites using the mineral bearing wood aggregate Agreslith-C [7]: 2l aggregates 0.6 kg cement and a water/cement volumetric ratio. Then we modified the A/C ratio from 4 to 8 to reduce the apparent bulk density of the composite.

The crushed aggregates and cement were mixed beforehand in a mixer and the water was added gradually without stopping the mixing. The mixture was thus been worked for at least 3 minutes. It was then introduced into the mould in two layers, and between each addition, the mixture was placed on a jolting table for 60 hits. Then the mixture was levelled and placed before and after de-moulding to cure at 98% of relative humidity and 25°C for 28 days. De-moulding was after 24 hours.

Measurement of Mechanical performance

Both the compressive and tensile strengths were evaluated on 4 x 4 x 16 cm prismatic samples, in compliance to the method specified in the EN 196-1 standard.

Thermal conductivity

Thermal conductivity was determined by the TPS (Transient Plane Source) method [8] on cylindrical composites (10 cm diameter and 8 cm high). Results are obtained with MATLAB© program calculating thermal conductivity [9].

The results correlated to the final density of the elaborated composites according to aggregates size and A/C ratio.

Maximum dimensional variations

maximum dimensional variations have been measured with an electronic comparator between a completely wet state (putting composites completely in water until saturation) and a dry state after drying composites in an oven at 50°C until constant weight.

RESULTS AND ANALYSIS

The objective of the work was to find a new way to valorise these car industry by-products or wastes. The incorporation of various quantities of crushed fibrous aggregate in the cement matrix enabled us to identify a limit beyond which material performance is not satisfactory.

Composites Apparent bulk density

We initially measured the dry apparent bulk density of the prepared samples in order to show lightening anticipated by the increase in the volumetric A/C ratio. The three following figures show crushed fibrous aggregates used in this study.

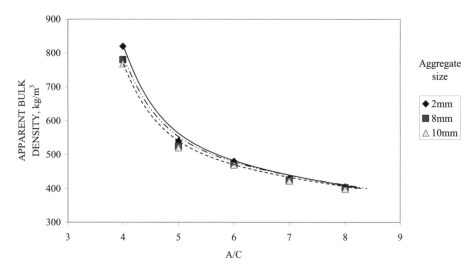

Figure 1   Apparent bulk density of composites according to the A/C volumetric ratio
and aggregate size

The use of a high A/C volumetric ratio involves an important lightening of the composite. Indeed, for a A/C ratio of 4, usually used for the wood concretes, the apparent bulk density of the composites is around 800 kg/m^3 but it great falls with a A/C ratio of 5 and higher, to reach values close to 400 kg/m^3 for a A/C ratio of 8. However, an other important information demonstrated in this graph, is the weak influence of the aggregate size.

Mechanical strength

We evaluated the mechanical resistances of materials obtained with the various aggregates used and for different A/C volumetric ratios. One can thus observe the evolution of the mechanical flexural strength and compressive strength according to A/C volumetric ratio for a given aggregate size.

As one can see it in the following graph (Figure 2), the flexural strength decreases naturally with the increase in the proportion of crushed fibrous aggregate within the cementing matrix. The performance obtained is close to that observed in the case of light concretes with the same bulk density.

It will also be noted that the size of the aggregate has ultimately little influence on the flexural strength, but the coarser crushed aggregates (10 mm) lead all the same to better performance in flexural behaviour.

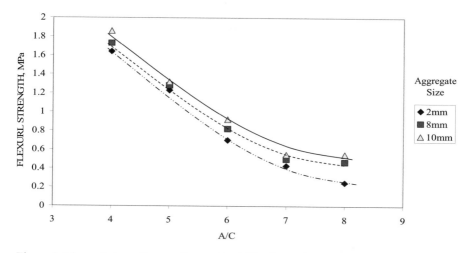

Figure 2 Flexural strength according to the A/C volumetric ratio for each aggregate size

We proceeded in the same way for the study of compressive strength, varying A/C ratios for different fibrous aggregate sizes. Figure 3 shows the obtained results.

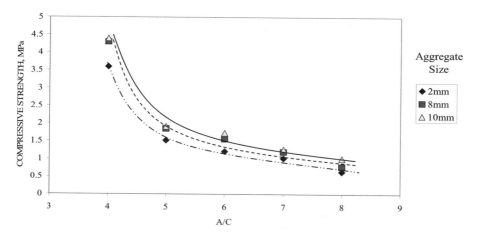

Figure 3 Compressive strength according to the A/C volumetric ratio for each aggregate size

The compressive strength significantly decreases beyond an A/C volumetric ratio of 4. These performance is highly completely interesting for A/C = 4, because there are higher than those obtained for of the same density wood concretes. Strength values are slightly lower than 2 MPa for a ratio A/C = 5. Until a ratio of 8, these compressive strengths remain satisfactory (1 to 1.5 MPa). Few differences are observed when the size of the aggregates varies although as in the case of flexural strength, better performance is observed when the size of the aggregate increases. This can be explained by a greater connection of large particles. Small particles generate a too large specific surface which is harder to cover with cement. However the use of this type of composite does not require excessive mechanical resistance, but a sufficient behaviour for the maintenance of an insulating structure. All in all, one can say that these materials show satisfactory mechanical characteristics for use as filler materials for thermal insulation. Compared to marketed materials, for example agresta's wood concrete [7], for an A/C ratio of 4, the mechanical strength is indeed better for our composites by 3.5 to 4.5 MPa for compressive strength and by 1.6 to 2.0 MPa for flexural strength.

Thermal Conductivity

Thermal conductivities of sample composites prepared with crushed fibrous aggregates of various sizes, and with variable A/C ratios, were gathered on Figure 4.

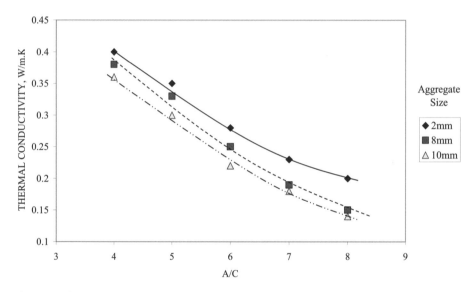

Figure 4 Thermal conductivity according to A/C volumetric ratio for each aggregate size

For an A/C ratio of 4, thermal conductivities have been measured around 0.4 W/m.K, but these values tend to decrease with the increase in A/C ratio leading to very good heat insulators for ratios higher than 6. In accordance with our preceding observations, the use of a finer aggregate does not lead to better performance, indeed the measured thermal conductivities are always higher in the case of a 2 mm fibrous aggregate. The increase in A/C ratio will lead to an increase of the material porosity, promoted by the use of bigger size particles, and which will lead to better thermal performance because of the increase in the air filled spaces within the material.

Maximum Dimensional Variations

The problems encountered during the preparation of this type of materials are often related to the use of a vegetable fraction and the inclusion of certain porosity, these two characteristics involving a considerable increase in the water sensitivity of the elaborated composites. We thus measured the dimensional variations between specimens saturated with water by total immersion until obtaining a constant mass, and in a completely dry state, after drying out at 50°C until constant mass. These measurements were taken with an electronic comparator of a 1 µm sensitivity. The results obtained are presented in the following figure.

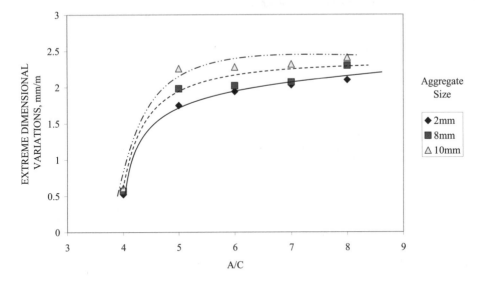

Figure 5 Maximum dimensional variations according to A/C volumetric ratio for each size of aggregates

For an A/C ratio lower or equal to 4, the composites maximum dimensional variations are completely acceptable (< 1 mm). On the other hand, when one increases the proportion of fibrous aggregates, these variations increase appreciably, but remain in the range of acceptable values, around 2 mm/m. The dimensional variation is generally required to be lower than 1 mm/m [8] for wood concretes highly sensible to moisture, but actually, the marketed wood concretes more generally present higher maximum dimensional variations [10]. In our case, the increase in these values is due to an increase in the vegetable fraction, flax fibres, being able to absorb a very high quantity of water. However, the values obtained for the usual proportions for wood concretes show an unquestionable improvement.

CONCLUSIONS

The results obtained at the time of this study showed the possibility of using fibrous aggregates originating from the car industry. The mechanical performance obtained even for high proportions of aggregates is acceptable and ensures a satisfactory behaviour to these materials for use as filler for thermal insulation. Indeed, the use of a high aggregates ratio makes it possible to reach low thermal conductivities, and to obtain considerably lighter materials. The materials produced this way can thus be regarded as remarkable competitors to the existing wood concretes on the market.

This work also made it possible to show that it is not useful to work with a too fine aggregate. It was to show that a 2 mm crushed aggregate would lead to a mechanical performance always being lower than that obtained with crushed aggregates of 8 or 10 mm. Moreover, preparing fine crushed aggregates represent a higher energy cost, it is thus advantageous to use coarser crushed aggregates.

If for higher proportions of fibrous aggregates, we observed an increase of dimensional variation, given that this material is never used without any coating.

Lastly, the current economic data create opportunities for the use of new alternative resources, or the implementation of a recycling die. The use of a hybrid composite originating from the car industry shows the potential of certain matter to be recycled in order to prolong the materials life cycle by maintaining a considerable performance level.

REFERENCES

1. YAZGHLI MARZOUK O., DHEILLY R-M., QUENEUDEC M., Reuse of plastic wastes in cementitious concrete composites, Cement combination for durable concrete, Global construction: Ultimate Concrete Opportunities, Eds R.K. Dhir, T.A. Harrison, M.D. Newlands, Thomas Telford publishing Ltd, Dundee, Scotland, July 2005. pp 817-824.

2. LAIDOUDI B., BENAZZOUK A., MARMORET L., LANGLET T., QUENEUDEC M., recycling of automotive industry rubber waste: Thermal performances od cement rubber composites. Recycling and reuse of waste materials, Eds R.K. Dhir, M.D. Newlands, J.E. Halliday, Thomas Telford publishing Ltd, Dundee, Scotland, Sept 2003. pp 355-364.

3. BENAZZOUK A., MEZREB K., GOULLIEUX A., QUENEUDEC M., Transport of fluids in cement-rubber composites, Cem. Concr. Comp., Vol. 26, 2004; pp 21-29.

4. DUPRE B., GOULLIEUX A., MARMORET L., QUENEUDEC M., Influence of life culture treatments of flax to produce flax waste aggregate cement composites, Cement combination for durable concrete, Global construction: Ultimate Concrete Opportunities, Eds R.K. Dhir, T.A. Harrison, M.D. Newlands, Dundee, Scotland, July 2005. Thomas Telford Ltd, pp 23-30.

5. DUPRE B., GOULLIEUX A., DHEILLY R-M., QUENEUDEC M., A New way of valorisation for oleaginous flax by products: wood concrete composite with flax shaves, elaboration and properties, Revue de Cytologie et Biologie végétales-Le Botaniste, 2005, N°2005-28 (s. i.), p.22–27.

6. EUSTAFIEVICI M., MUNTEAN O., MUNTEAN M., Influence of the wood waste characteristics and its chemical treatments on the composites properties. NOCMAT/3-Vietnam International Conference on Non-Conventional Materials and Technologies, 2002, pp 107-112.

7. MERLET J.D., VERZAT A., BLACHE B., Chape en composite ciment-bois – Agreslith-C , Avis technique du CSTB (Centre Scientifique et Technique du Bâtiment), N 16/01-422, april 2002.

8. GUSTAFSSON S.E., Transient plane source techniques for thermal conductivity and thermal diffusivity measurements of solid materials, Review Scientific Instruments, Vol. 62-3, pp 797-804, 1991.

9. BOUGUERRA A., LAURENT J. P., GOUAL M. S., QUENEUDEC M., The measurement of the thermal conductivity of solid aggregates using the transient plane source technique. J. Phys. D. Applied Physics, Vol. 30, October 1997, pp 2900-4.

10. PIMIENTA P., CHANDELLIER J., RUBAUD M., DUTRUEL F., NICOLE H., Etude de faisabilité des procédés à base de bétons de bois. Cahier du CSTB 2703, January-February 1994.

SULFATE RESISTANCE OF GRADE 30
RICE HUSK ASH (RHA) CONCRETE

K Kartini   M S Hamidah

Universiti Teknologi MARA

H B Mahmud

Universiti Malaya

Malaysia

ABSTRACT. The use of pozzolanic materials in the construction industry has been a common practice all over the world for many years. This is because quality of concrete can be improved by incorporating supplementary cementing material such as rice husk ash (RHA). Concrete undergoes degradation when exposed to sulfate-bearing environments. Dissolved sulfates intrude into the concrete pore structure causing progressive precipitation of sulfate-bearing phases in the internal microstructure of the concrete. This change leads to swelling, spalling and cracking of concrete and causes a significant gain in weight as well as decreasing in strength. This paper reports on the experimental study to investigate the strength and durability properties of concrete containing RHA subjected to 5% magnesium sulfate ($MgSO_4$) solution. The rate of deterioration was assessed by measuring the relative change in weight and compressive strength for 5 cycles, each cycle comprising 30 days of immersion in $MgSO_4$ solution followed by 7 days of air drying.

Keywords: Pozzolan, Rice husk ash, Ordinary Portland cement, Magnesium sulfate, Durability, Compressive strength, Weight change, Length change.

Engr K Kartini is an Associate Professor, Faculty of Civil Engineering, Universiti Teknologi MARA, Malaysia and currently doing her PhD in concrete technology.

Dr M S Hamidah is a Lecturer, Faculty of Civil Engineering, Universiti Teknologi MARA, Malaysia.

Dr H B Mahmud is an Associate Professor, Department of Civil Engineering, Universiti Malaya, Malaysia. His research interests include cement replacement technology and high strength high performance concrete.

INTRODUCTION

Malaysia as a rice growing country, which allocated an average of 734,659 units of area for paddy growing between the years of 1980 to 1990. This, in turn, attracted more attention due to environmental pollution of disposing the rice husk [1]. The Malaysian Environmental Quality Act (EQA) which was established in 1974, and amended in 1984, requires selection of appropriate disposal of solid residue in order to avoid any harm to human health as well as the environment. With that in mind, an attempt is made to predict possible future developments in the utilization of these industrial by-products.

Rice husk ash, which is the product of burning rice husk is a pozzolanic material that can be used as partial cement replacement in concrete [2-5]. Its cement replacement capability not only reduces the demand for Portland cement, but it has been reported that when used as blended cement, it performed better in a sulfate environment [6-8]. The reasons that can be associated to it are: (i) consumption of portlandite [$Ca(OH)_2$] reducing the formation of gypsum due to the pozzolanic reaction; (ii) densification of pore structure thus reducing the permeability of sulfate ions because of the formation of secondary C-S-H; and (iii) reduction of C_3A content because of the replacement of ordinary Portland cement with a pozzolanic material which reduces the aluminate bearing phases. Knowing that sulfate attack plays a major role in the durability of concrete as a result of environmental pollution and industrial products; therefore it is the intention of the present study to investigate the potential use of a local pozzolanic material *i.e.* RHA concrete to resist sulfate attack. In this study, the effect of environmental sulfate salt ($MgSO_4 \cdot 7H_2O$) on RHA concrete grade 30 was assessed in terms of its compressive strength and weight changes.

EXPERIMENTAL PROGRAMME

Materials

The rice husk used in this research was obtained from a local rice mill and later burnt in a ferrocement furnace at the laboratory. At any time about 50 to 60 kg of rice husk was burn slowly on its own for about 24 hours. The ash was left to cool inside the furnace for another 24 hours before taken out for grinding using a Los Angeles machine. Only amorphous ash which lies within the middle third of the furnace was taken for grinding as it is considered as quality ash.

Fineness of RHA and OPC, determined in accordance to BS 3892: Part 1 [9] as that retained on 45 μm sieve was 21.87% and 6.12%, respectively. As specified in BS 3892, and ASTM C430 [10], the prepared RHA conforms to grade A for dry pulverized-fuel ash *i.e.* 12.5% to 34% fineness. The specific surface area of RHA and OPC was determined using the nitrogen absorption method ranged between 10.857 m^2/g to 17.463 m^2/g and 1.757 m^2/g, respectively. The chemical composition of RHA and OPC used in the present investigation is shown in Table 1. Figure 1 shows the X-ray diffraction (XRD) pattern of RHA particles; indicating that the structure of silica present in RHA used was amorphous material with a diffused peak of 140 counts at about $2\theta = 22°$. Other materials used in the concrete mixture were Portland cement-Type II, granite coarse aggregate of 20 mm maximum size and mining sand of 5 mm maximum size as fine aggregate. The fineness modulus for the coarse aggregate and fine aggregate was 2.43 and 4.61, respectively. The superplasticizer (Sp) used was sulphonated naphthalene formaldehyde condensed polymer based admixture.

Table 1   Chemical composition of OPC and RHA

CHEMICAL COMPOSITION, %	OPC	RHA
Silicon dioxide (SiO_2)	21.38	96.70
Aluminium oxide (Al_2O_3)	5.60	1.01
Ferric oxide (Fe_2O_3)	3.36	0.05
Calcium oxide (CaO)	64.64	0.49
Magnesium oxide (MgO)	2.06	0.19
Sodium oxide (Na_2O)	0.05	0.26
Potassium oxide (K_2O)	-	0.91
Phosphorous oxide (P_2O_5)	-	0.01
Titanium oxide (TiO_2)	-	0.16
Sulphur trioxide (SO_3)	2.14	-
Loss on ignition	0.64	4.40

Mix Proportions

The control OPC concrete was designed to achieve a strength of 30 N/mm^2 using the DOE method [11]. Based on this design method, 325 kg/m^3 of cement content was adopted for all mixes. The water to cementitious material ratio (W/B) of the control mix was 0.63 with a slump of about 40 mm to 50 mm. Due to the adsorptive character of cellular RHA particles and its high fineness, an increase in the amount of RHA content resulted in a dry concrete mix, therefore Sp was used to enhance the fluidity of the mixes. The Sp dosage in subsequent mixtures was tailored to achieve slump in the range of 100 mm to 150 mm.

The range of mixes studied was limited to emphasize on the principal replacement additive parameter and hence all mixes had a coarse aggregate content of 940 kg/m^3 and 900 kg/m^3 of fine aggregate. The only variation in the mixes were the percentage content of RHA and the necessarily adjusted W/B and Sp content to achieve the required slump. Table 2 summarises the mix proportions for various RHA concrete mixes, without and with Sp.

For determination of the resistance against sulfate attack, concrete 100 mm cubes were used. They were air dried or water cured for 7 days before immersion in water as control or in 5% $MgSO_4$ solution for 5 cycles. The pH value of $MgSO_4$ solution was monitored to be between 8.5 and 8.8 during the test and the solution was replaced at the beginning of every new cycle.

Table 2   Mixture Proportion of RHA concrete without and with Sp

SERIES	MASS PER UNIT VOLUME OF MATERIALS, kg/m^3					Sp, %	W/B	SLUMP, mm
	Cement	RHA	Water	Aggregate				
				Fine	Coarse			
OPC	325	-	205	900	940	-	0.63	50
RHA20	260	65	221	894	930	-	0.68	45
OPCSp	325	-	205	900	940	0.40	0.63	130
RHA30Sp	228	97	205	900	940	1.61	0.63	120

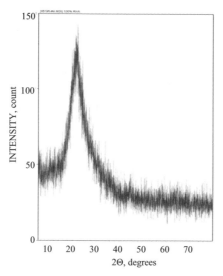

Figure 1   XRD of RHA particles

RESULTS AND DISCUSSION

Compressive Strength

To assess the deterioration of concrete due to sulfate attack, the change in compressive strength of the specimens after a scheduled immersion in 5% $MgSO_4$ for 30 days and 7 days air drying period was taken, and this is shown in Table 3 and presented graphically in Figure 2. The rate of deterioration at the end of each cycle for pre-cured (air dried and water cured) specimens were assessed by comparing with the corresponding compressive strength of specimens cured in water (control). Table 4 shows the relative compressive strength of specimen immersed in sulfate solution to that cured in water.

From Figure 2 it can be seen that the concrete specimens with 7 days initial curing in water (pre-cured in water) before immersion in $MgSO_4$ solution showed higher compressive strengths compared to those with 7 days initial air drying. The highest compressive strength at the end of 5th cycles for specimens pre-cured in water was attained by the OPC concrete followed by OPCSp and RHA20 and lastly RHA30Sp concretes. For specimens with for 7 days pre-cured in air laboratory environment, the highest compressive strength attained after the 5th cycle was OPC concrete followed by OPCSp, RHA30Sp and lastly the RHA20 concretes. The reason is due to insufficient hydration for the air dried specimens, whereas for the initial water cured specimens, densification of the pore structure developed as proper hydration products formed because of the presence of water. Meyer [12] and Dias [13] concluded that for durable concrete, proper curing is important as poor curing primarily affects the surface zone of the concrete.

The compressive strength of concrete specimens immersed in water (control) as shown in Figure 2 demonstrates that all the series have a higher compressive strength. The highest compressive strength was dominant in OPCSp concrete followed by RHA30Sp, OPC and RHA20 concretes. This finding is in line with the results by the authors in regards to the mechanical properties of these concrete [14-16].

Comparing the compressive strength of concrete specimens immersed in water (control) with those immersed in $MgSO_4$ solution as shown in Figure 3, demonstrates that concrete immersed in water (control) have higher compressive strength those immersed in $MgSO_4$ solution. This means that the strength of concrete reduces when the concrete is immersed or exposed to $MgSO_4$. The results confirm this: as the amount of RHA increases, it reduces the C_3A content and this increases the resistance against sulfate.

Table 3   Compressive strength of OPC and RHA concrete subjected to immersion in water or sulfate solution

SERIES	INITIAL EXPOSURE CONDITION	IMMERSION	COMPRESSIVE STRENGTH AFTER EXPOSURE CYCLES, σ_c, N/mm^2				
			1st	2nd	3rd	4th	5th
OPC		Water (Control)	40.9	42.1	43.8	45.1	46.6
RHA20	Water		40.4	40.7	38.5	43.8	45.5
OPCSp			45.5	45.9	46.8	47.6	48.8
RHA30Sp			45.2	44.5	45.5	45.2	47.7
OPC			21.4	37.2	35.6	39.1	42.2
RHA20	Air		34.3	34.2	30.9	37.5	40.6
OPCSp			39.1	38.9	42.5	39.1	41.1
RHA30Sp		5% $MgSO_4$	37.9	37.6	37.8	39.1	40.7
OPC			26.0	42.3	43.2	42.9	45.4
RHA20	Water		36.5	40.6	37.1	45.3	44.3
OPCSp			39.8	40.1	42.1	44.7	44.3
RHA30Sp			43.0	43.5	43.5	44.5	43.7

Table 4 shows the remained compressive strength of concrete subjected to sulfate solution to that cured in water. There is a loss in strength for concrete exposed to $MgSO_4$. Examination of Table 4 indicates that the initial loss of strength of OPC concrete is higher compared to RHA concrete, however in due time the loss of strength between OPC and RHA20 concretes is marginal. This indicated that RHA concrete had high resistance against sulfate attack. Insignificant contribution of Sp in RHA concrete to resist sulfate attack is noticeable.

Table 4   The relative compressive strength with respect to strength under water curing

SERIES	INITIAL EXPOSURE CONDITION	IMMERSION	RELATIVE COMPRESSIVE STRENGTH AFTER EXPOSURE CYCLES, σ_c, %				
			1st	2nd	3rd	4th	5th
OPC			52.3	88.3	81.3	86.7	90.6
RHA20	Air		84.9	84.2	80.5	85.6	89.2
OPCSp			85.9	84.6	90.8	82.2	84.2
RHA30Sp		5% $MgSO_4$	83.9	84.6	82.9	86.4	85.2
OPC			63.7	100.4	98.6	95.3	97.4
RHA20	Water		90.4	99.9	96.3	103.4	97.2
OPCSp			87.4	87.2	89.9	93.9	90.9
RHA30Sp			95.2	97.9	95.5	98.3	91.6

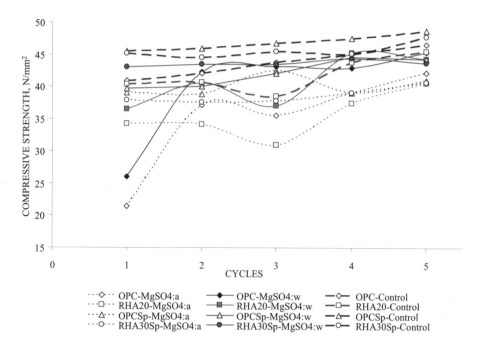

Figure 2   Compressive Strength of OPC and RHA concrete stored in MgSO₄ solution subjected to initial air drying or water curing

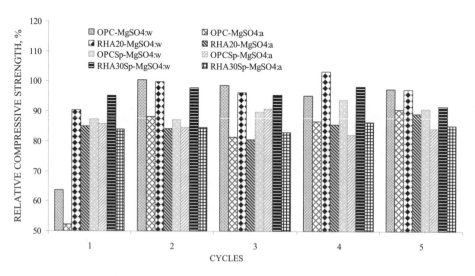

Figure 3   The relative compressive strength of OPC and RHA concrete subjected to sulfate solution to that cured in water

Weight Change

Changes in weight with respect to immersion cycles are depicted in Figure 4. An increase in weight was observed for specimens subjected to 7 days initial air drying prior to immersion in $MgSO_4$ solution, whereas for mixes pre-cured in water showed a weight decrease when subjected to immersion in $MgSO_4$ solution. RHA20 concrete shows a higher weight increase followed by RHA30Sp, OPCSp and OPC concretes for specimens pre-cured in air. However, when initially water cured, RHA30Sp concrete exhibited the lowest weight decrease followed by RHA20, OPCSp and OPC concretes. This again shows that especially for RHA concrete, proper curing is of paramount importance in order to have durable concrete.

Comparison between the 7 days initial water curing followed by immersion in sulfate solution and water (control), there is a smaller weight change for the specimens immersed in $MgSO_4$ solution with 7 days initial water cured. The Figure also shows that the weight loss for RHA30Sp concrete immersed in sulfate solution is lower than for OPC concrete immersed in water (control), which indicates that the RHA concrete was less affected by sulfate solution. The weight loss for RHA concretes immersed in water (control) taken at the end of the 5th cycle were -0.377% and -0.380%, for without and with Sp, respectively, whereas for counterparts with 7 days initial air drying prior to immersion in sulfate solution, the weight gain of was 3.044% and 2.723%, respectively. Similarly, a weight loss of -0.169% for RHA20 concrete and weight gain of 0.255% for RHA30Sp concrete, both subjected to 7 days initial water curing, were observed at the end of the 5th cycle from the experimental study.

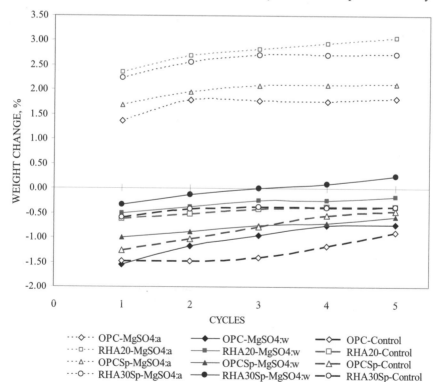

Figure 4   Weight change of OPC and RHA concretes subjected to immersion in water (control) and in $MgSO_4$ solution

CONCLUSIONS

1. The compressive strengths of concrete specimens with 7 days initial air dried prior to immersion in $MgSO_4$ were lower compared to those of 7 days initial water cured. This indicates the importance of curing for RHA concrete when immersed in sulfate solution.

2. Concrete immersed in water (control) had higher compressive strength compared to specimens subjected to $MgSO_4$ solution.

3. For concrete immersed in $MgSO_4$ solution, the compressive strength of OPC concrete at the end of the 1st cycle for both air dried and water cured was lower compared to RHA concrete under the same initial curing conditions. However, in due time the difference in compressive strength between these concretes became marginal.

4. Adding superplasticizer to the RHA concrete did not show much improvement in terms of compressive strength especially towards the end of the 5th cycle.

5. Weight increase was observed for specimens with 7 days initial air drying, whereas the reverse was true for specimens subjected to 7 days initial water curing. The RHA concrete with 7 days initial air drying showed a higher percentage of weight change compared to the same concrete with 7 days initial water curing. Thus, suggested that for RHA concrete exposed to $MgSO_4$ solution, initial water curing must be provided.

6. RHA concrete with 7 days initial water curing prior to immersion in $MgSO_4$ solution exhibited a smaller weight loss compared to OPC concrete immersed in water. It also showed that the weight loss of RHA concrete with 7 days initial water curing in $MgSO_4$ solution was smaller than for the OPC concrete in $MgSO_4$ solution. This indicated that RHA concrete has a high resistance against sulfate attack.

ACKNOWLEDGEMENTS

The authors wish to express their appreciation to the Universiti Teknologi MARA and the Universiti Malaya for their assistance, cooperation and support.

REFERENCES

1   SAID M A, Pengurusan Sumber dan Alam Sekitar, Birotek ITM, ISBN 967-958-108-X, 1999, pp. 202.

2. COOK D J, Rice Husk Ash in Cement Replacement Materials, Concrete Technology and Design, Edited by R N Swamy, Surrey University Press, United Kingdom, 1986, Vol. 3, pp. 171-195.

3. HWANG C L AND WU D S, Properties of Cement Paste Containing Rice Husk Ash, ACI SP-114, Editor: V M Malhotra, 1989, pp. 733-765.

4. MEHTA P K AND MONTEIRO J M, Concrete Structure, Properties and Materials, 2nd edition, Prentice-Hall, Eaglewood Cliffs, NJ, 1993, pp. 263-264.

5. ZHANG M H AND MALHOTRA V M, High Performance Concrete Incorporating RHA as a Supplementary Cementing Material, ACI Material Journal, Vol. 93, 1996, pp. 629-639.

6. MANGAT P S AND KHATIB J M, Influence of fly ash, silica fume and slag on sulfate resistance of concrete, ACI Materials Journal, 1995, No. 5, pp. 542-552.

7. IRASSAR E F, DI MAIO A AND BATIC O R, Sulfate attack on concrete with mineral admixtures, Cement and Concrete Research, 1996, Vol. 26, No. 1, pp. 113-123.

8. WEE T H, SURYAVANSHI A K, WONG S F AND RAHMAN A K M A, Sulfate resistance of concrete containing mineral admixtures, ACI Material Journal, 2000, Vol. 97, No. 5, pp. 536-549.

9. BRITISH STANDARDS INSTITUTION, BS 3892:1982 Part 1, Specification for Pulverized-Fuel Ash for Use with Portland cement, London, 1982.

10. AMERICAN SOCIETY FOR TESTING AND MATERIALS, ASTM C 430: Standard Test Method for Fineness of Hydraulic Cement by the 45 μm (No. 325) sieve, Section 4: Construction, USA, 1993, pp. 223-225.

11. DEPARTMENT OF ENVIRONMENT (DOE). Design of Normal Concrete Mixes, BRE Publication, United Kingdom, 1986.

12. MEYER A, The important of the surface layer for the durability of concrete structures, SP100-Concrete Durability-Katherine and Bryant Mather International Conference, American Concrete Institute, Detroit, 1987, pp. 49-61.

13. DIAS W P S, Influence of curing method and duration of sorptivity of concrete and mortar, Proceeding of the 4th CANMET/ACI International Conference, Editor V M Malhotra, Sydney, Australia, 1997, Vol. 2, SP170-55, pp. 1073-1095.

14. KARTINI K, MAHMUD H B AND HAMIDAH M S, The influence of superplasticizer on the workability and strength of RHA concrete, Proceeding of 7th International Conference on Concrete Technology in Developing Countries, (ICCT) – Sustainable Development in Concrete Technology, Editors Hashem Al-Mattarneh, I Azmi and A Zakiah, ISBN-967-958-162-4, Kuala Lumpur, 2004, pp. 331- 337.

15. KARTINI K, MAHMUD H B AND HAMIDAH M S, Influence of Rice Husk Ash on the strength property of the normal strength concrete, Proceeding of 22nd Conference of Asian Federation of Engineering Organization (22 CAFEO), Vol. 2, Myanmar, 2004, Sect 403-01.

16. KARTINI K, MAHMUD H B AND HAMIDAH M S, Strength properties of Grade 30 Rice husk Ash concrete, Proceeding of the 31st Conference in Our Worlds in Concrete and Structure (OWICS) – Housing for the Nation, ISBN-981-05-5560-1, Vol. XXV, Singapore, 2006, pp. 199-206.

USE OF WASTES GENERATED WITHIN COAL-FIRED POWER STATIONS FOR THE SYNTHESIS OF CALCIUM SULPHOALUMINATE CEMENTS

M Marroccoli

M Nobili

A Telesca

G L Valenti

University of Basilicata

Italy

ABSTRACT. Rapid-hardening and dimensionally stable calcium sulphoaluminate (CSA) cements, when compared with Portland cements, have very interesting properties and are also able to give a significant contribution to the sustainability of the cement manufacturing process, due to relatively low synthesis temperatures, reduced thermal input to the kiln and CO_2 generation, easy grindability and utilization of solid industrial wastes in the technological cycle. By-products coming from a fluidized bed combustion (FBC) plant and a traditional coal-fired power station, both located in the industrial area of Sulcis-Sardinia (which also includes an aluminium plant based on the Bayer manufacturing process from commercial bauxite) were investigated as raw mix components for the synthesis of CSA cements. Mixtures containing limestone, bauxite as well as FBC bottom and/or fly ash (up to 22%), pulverized coal fly ash (up to 24%), flue gas desulphurization waste (up to 6%) were heated for two hours in a laboratory electric oven at temperatures ranging from 1200 to 1300°C. The conversion of reactants was generally complete and high selectivities towards the desired hydraulic phases ($C_4A_3\bar{S}$ and dicalcium silicate) were obtained, particularly at about 1250°C.

Keywords: CSA cement manufacture, Raw mix components, FBC fly and bottom ash, Pulverized coal fly ash, FGD waste.

Prof M Marroccoli is Professor of Material Science at the Department of Environmental Engineering and Physics, University of Basilicata, Italy. Her scientific activity deals with cement technology and utilization of industrial by-products.

M Nobili is doing her doctoral studies in Environmental Engineering at the University of Basilicata, Italy. Her research activity deals with the fluidized bed combustion technology and the reuse of industrial by-products.

Dr A Telesca is Researcher in Materials Science and Technology at the Department of Environmental Engineering and Physics, University of Basilicata, Italy. His main research interests are the development of special cements and the utilization of industrial wastes as sources of raw materials.

Prof G L Valenti is Professor of Materials Technology and Applied Chemistry at the Department of Environmental Engineering and Physics, University of Basilicata, Italy. He is mainly engaged in researches concerning the development of special cements and the utilization of industrial by-products.

INTRODUCTION

The most important properties of CSA cements are related to their ability of generating ettringite ($C_6A\bar{S}_3H_{32}$) upon hydration of calcium sulphoaluminate ($C_4A_3\bar{S}$). This latter can react with water in two ways: *i)* together with lime and calcium sulphate to give only ettringite (reaction i); *ii)* combined with calcium sulphate alone to give ettringite and aluminium hydroxide (reaction ii):

$$C_4A_3\bar{S} + 6C + 8C\bar{S} + 96H \Rightarrow 3C_6A\bar{S}_3H_{32} \qquad (I)$$

$$C_4A_3\bar{S} + 2C\bar{S} + 38H \Rightarrow C_6A\bar{S}_3H_{32} + 2AH_3 \qquad (II)$$

Ettringite produced in the reaction I) shows an expansive behaviour which can be exploited by special binders like shrinkage-resistant and self-stressing cements [1-3].

Other CSA formulations, first developed by the China Building Materials Academy, are characterized by rapid hardening, dimensional stability as well as outstanding chemical and physical durability [1, 4-8]. In these cements $C_4A_3\bar{S}$, according to the reaction II), quickly hydrates to give, in a less alkaline environment (due to the absence of lime), a non-expansive ettringite. Such binders, depending on both proportioning of raw materials and synthesis temperature, may contain, besides $C_4A_3\bar{S}$, calcium aluminates, dicalcium silicate, calcium sulphosilicate ($C_5S_2\bar{S}$), anorthite (CAS$_2$) and gehlenite (C$_2$AS). Among the silica-containing phases, only C$_2$S is able to regulate strength and durability of hydrated cements, especially at medium and long ages, owing to the generation of calcium silicate hydrate.

Moreover CSA cements, compared to Portland cements, give a pronounced environment-friendly character to their manufacturing process [9-16]. In this regard important features are: a) CSA clinkers can be synthesized at temperatures (1250–1350°C) lower than those requested by Portland clinkers (1450-1500°C); b) there is a reduction of thermal input to the kiln and CO$_2$ generation due to the lower limestone concentration in the raw mix, since $C_4A_3\bar{S}$ requires less CaO for its synthesis; c) CSA clinkers are easier to grind; d) hardly recoverable wastes and by-products can be utilised. Among the residues that can be used as raw materials for the manufacture of CSA cements, it has to be mentioned the fluidized bed combustion (FBC) waste whose utilization and disposal are made difficult by the occurrence of exothermal and expansive phenomena due to the relatively high content of lime and calcium sulphate.

The FBC technology is very effective in the burning of solid fuels, ensuring also a low environmental impact [17, 18]. Among the main advantages of this technique, there is the possibility of removing "in situ" sulphur dioxide through injection of calcium-based sorbents [19, 20]. However, the effectiveness of such desulphurization technique is not elevated and a high calcium/sulphur molar ratio is required in the feed. This leads to the generation of a considerable amount of waste. Finally, FBC ash, compared to that generated in a pulverized coal-fired power station, has a considerably lower glass content (due to the lower combustion temperature) and consequently a poor pozzolanic activity. For these reasons FBC waste is generally unsuitable for the use in ordinary cement and concrete and therefore new application fields have to be searched for. FBC waste, mainly composed by exhausted sulphur sorbent and coal ash, contains CaO, SiO$_2$, Al$_2$O$_3$ and SO$_3$ as major oxides, thus representing a potential raw material for the manufacture of calcium sulphoaluminate cements [16].

However, FBC waste has a relatively low Al_2O_3 content and its use in the raw mix generating CSA clinker needs (in order to obtain a significant saving of an expensive natural material like bauxite) additional and cheaper sources of alumina, such as pulverized coal fly ashes.

This paper was aimed at evaluating on laboratory scale the suitability of wastes and by-products, generated within a FBC plant and a traditional power station, as raw mix components for the manufacture of rapid-hardening and dimensionally stable CSA cements. In particular, FBC fly and bottom ash (originated from a circulating fluidized bed combustor), as well as a pulverised coal fly ash (FA) and a flue gas desulphurization (FGD) waste were used.

The peculiar feature of this research lies in the fact that both power stations are placed in the same industrial area (Sulcis, Sardinia), which also includes an aluminium plant based on the Bayer manufacturing process from commercial bauxite. An investigation by differential thermal analysis (DTA), scanning electron microscopy (SEM) and mercury porosimetry on the hydration properties of a CSA cement, obtained by heating natural raw materials in a pilot rotary kiln, was preliminarily performed. Afterwards, conversion and selectivity of waste-based raw mixes, burnt in a laboratory electric oven, were evaluated by means of X-Ray diffraction (XRD) analysis.

EXPERIMENTAL

Hydraulic Behaviour of CSA Cement

Materials and testing procedures

CSA cement was obtained from a sulphoaluminate clinker synthesized in a pilot rotary kiln at 1300°C by using bauxite, limestone and gypsum as raw materials. Its Blaine fineness was 0.500 m²/g. $C_4A_3\bar{S}$, calcium sulphate, calcium aluminates and calcium silicates were, in the order, the main mineralogical phases.

As a reference term, a II-A/LL class 42.5R Portland limestone cement (Blaine fineness, 0.460 m²/g), containing 12% of limestone and 4.5% of natural gypsum, was also investigated. The chemical composition, in terms of major oxides, was evaluated by X-Ray fluorescence (XRF) analysis and is listed in Table 1. CSA cement samples were paste hydrated and submitted to DTA, SEM and mercury porosimetry.

Table 1   Chemical composition of sulphoaluminate cement and Portland cement, mass %

	SULPHOALUMINATE	PORTLAND
CaO	41.4	62.0
Al_2O_3	27.6	4.8
SiO_2	5.1	17.1
Fe_2O_3	1.5	2.4
SO_3	22.3	3.1
MgO	1.0	2.2
Loss on Ignition*	-	5.3
Total	98.9	96.9

* according to EN 196-1 standard

As far as mercury porosimetry is concerned, pastes of CSA and Portland cements (water cement mass ratio, 0.5), shaped as cylindrical discs, were cured in a FALC WBMD24 thermostatic bath at 20°C for times ranging from 2 h to 28 days. Before the SEM observations, at the end of each aging period, the CSA cement samples were broken, treated with acetone and diethyl-ether to stop hydration and stored in a desiccator over silica gel–soda lime ensuring protection against water and carbon dioxide.

For DTA analysis CSA cement samples were treated with the above mentioned procedures except for a final step of grinding after which they were obtained in a pulverized form. Mechanical testing on CSA and Portland cements was carried out according to EN 196-1 standard. The compressive strength measurements, at various curing times, are reported in Table 2.

Table 2   Mortar compressive strength of Sulfoaluminate and Portland cements, MPa

	3 HOURS	8 HOURS	1 DAY	2 DAYS	7 DAYS	28 DAYS
Sulphoaluminate	24.0	38.9	59.1	67.5	68.5	68.7
Portland	-	2.0	20.2	29.1	41.3	46.1

Characterization techniques on hydrated samples

DTA was carried out by means of a Netzsch Tasc 414/3 apparatus, operating between 20°C and 1000°C with a heating rate of 10°C/min, in order to identify ettringite and aluminium hydroxide. SEM apparatus was a JEOL JSM-840 equipped with a high vacuum evaporation chamber (SCD 004 from Balzers Unionliechtenstein) and an energy dispersive X-ray analyser.

Porosity measurements were performed by means of a Thermo Finnigan Pascal 240 Series porosimeter (maximum pressure, 200 MPa) equipped with a low-pressure unit (140 Series) able to generate a high vacuum level (10 Pa) and to operate between 100 and 400 kPa. For each sample, two plots can be obtained from the porosimetric analysis: (a) cumulative and (b) derivative Hg intruded volume vs. pore radius. With increasing pressure, mercury gradually penetrates the bulk sample volume. If the pore system is composed by an interconnected network of capillary pores in communication with the outside of the sample, mercury enters at a pressure value corresponding to the smallest pore neck. If the pore system is discontinuous, mercury may penetrate the sample volume if its pressure is sufficient to break through pore walls. In any case, the pore width related to the highest rate of mercury intrusion per change in pressure is known as the "critical" or "threshold" pore width. Unimodal, bimodal or multimodal distribution of pore sizes can be obtained, depending on the occurrence of one, two or more peaks, respectively, in the derivative volume plot.

Synthesis of CSA Clinkers from Waste-Based Raw Mixes

Raw materials and burning tests

The chemical composition (in terms of major oxides) of the raw materials employed was evaluated by XRF analysis and is reported in Table 3.

Table 3   Chemical composition (dry basis) of limestone, bauxite, FBC fly ash, bauxite, FBC bottom ash, pulverised coal fly ash (FA) and FGD waste, mass %

	LIMESTONE	BAUXITE	FBC FLY ASH	FBC BOTTOM ASH	FA	FGD WASTE
CaO	54.70	1.69	24.20	43.12	4.30	36.04
SO$_3$	-	0.03	12.80	25.89	0.04	51.11
Al$_2$O$_3$	-	55.22	13.71	5.85	22.80	0.08
SiO$_2$	-	6.48	23.23	18.45	35.08	0.10
MgO	0.30	0.00	1.04	1.00	1.13	0.37
SrO	-	0.03	0.00	0.00	0.11	-
P$_2$O$_5$	-	0.01	0.00	0.00	0.10	-
TiO$_2$	-	2.34	0.82	0.48	1.52	-
Fe$_2$O$_3$	-	6.25	6.74	3.15	8.20	-
Mn$_3$O$_4$	-	0.00	0.07	0.08	0.10	-
L.o.I.*	42.61	27.68	16.26	1.39	25.85	12.28
Total	97.61	99.73	98.87	99.41	99.23	99.98

*loss on ignition, according to EN 196-1 standard

Four mixtures (A1, A2, B1, B2), having the composition shown in Table 4, were investigated. Mixtures A1 and A2 were prepared with limestone, bauxite, FBC fly-ash, FA and FGD waste. Both B1 and B2 contained limestone, bauxite, FBC fly (60%) and bottom (40%) ash and FA; moreover FGD waste was added to mixture B2. All the mixtures were heated in a laboratory electric oven for 2 hours at 1200°, 1250° and 1300°C, then submitted to XRD analysis, performed by a Philips PW1710 apparatus operating between 5° and 60° 2θ (Cu Kα radiation). Table 5 shows the potential concentration values of $C_4A_3\bar{S}$ and C$_2$S in the burning products of the four mixtures. They were calculated assuming that SO$_3$ and Al$_2$O$_3$ on the one hand, and SiO$_2$, on the other, react to give only $C_4A_3\bar{S}$ and C$_2$S, respectively; furthermore, solid solution effects were neglected.

Table 4   Composition of raw mixtures, mass %

MIXTURE	A1	A2	B1	B2
limestone	47.04	46.77	48.53	44.13
bauxite	17.65	17.06	18.23	16.22
FBC fly and bottom ash	-	-	21.61	10.08
FBC fly ash	17.08	10.46	-	-
FA	13.97	19.43	11.63	23.52
FGD waste	4.27	6.28	-	6.05

Table 5   Potential composition of $C_4A_3\bar{S}$ and C$_2$S in fired products of raw mixtures, mass %

MIXTURE	A1	A2	B1	B2
$C_4A_3\bar{S}$	39.54	41.61	38.01	41.30
C$_2$S	43.07	43.70	45.53	39.43
$C_4A_3\bar{S}$ +C$_2$S	82.61	85.31	83.54	80.73

RESULTS AND DISCUSSION

Hydration Behaviour of CSA Cement

The peculiar strength development of rapid-hardening and dimensionally stable CSA cement is related to the formation rate of ettringite which is very high at early ages and then rapidly declines, due to lack of water, inasmuch as a high (0.78) stoichiometric water solid mass ratio is required for the full formation of ettringite and aluminium hydroxide [21].

The rapid increase of the ettringite concentration, followed by the attainment of a steady-state, can be observed from the DTA thermograms, shown in Figure 1, where the endothermal peaks at about 160-170°C and 270-280°C are related to ettringite and aluminium hydroxide, respectively.

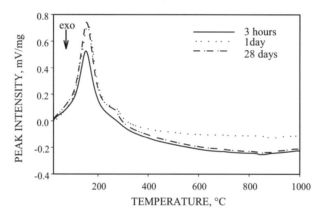

Figure 1   DTA for CSA cement pastes cured at 3 hours, 1day and 28 days

From the SEM observations, shown in Figure2, it is clearly seen the fast growth, within the curing period 3-72h, of ettringite prismatic crystals having an hexagonal cross section.

CSA cements behave differently from Portland cements as far as the development of porosity in the hydrated systems is concerned.

The evolution of the porosimetric curves during Portland cement hydration, as a function of curing time and water/cement ratio, is well established [22-26]. Both increased curing time and decreased w/c ratio give rise to lower values of total porosity and threshold pore width. The differential curves for pastes cured at early ages tend to exhibit a sharply defined initial peak, indicating a unimodal distribution of pore sizes. As curing time increases, a second peak appears at smaller pore sizes thus suggesting a bimodal distribution.

The first peak is related to the lowest size of pore necks connecting a continuous system. The second peak seems to correspond to the pressure required to break through the blockages formed by the hydration products which isolate the interior pore space.

Figures 3-I and 3-II show respectively cumulative and derivative Hg volume for Portland cement pastes cured at 12 hours, 1 day, 7 days and 28 days. As expected, the distribution of pore sizes is unimodal at earlier ages while it is bimodal at longer periods.

Figure 2   SEM (SE) microphotographs of CSA cement pastes
cured at 3h (A), 8h (B), 16h (C) and 3d (D)

Figures 3-III and 3-IV show respectively cumulative and derivative Hg volume for CSA cement pastes cured at 2 hours, 6 hours, 12 hours, 1 day, 7 days and 28 days. It can be noted that a bimodal distribution rapidly occurs. After 2 hours of aging the lower porosity region contributes about 25% of the total intruded volume while at longer curing times the role of the smaller pores becomes dominant.

The development of the pore structure is initially very fast and a prevailing region of lower porosity is established because sufficient hydration products are rapidly formed to reduce and isolate the interior space. At longer curing times the evolution of porosity proceeds very slowly because hydration is almost stopped.

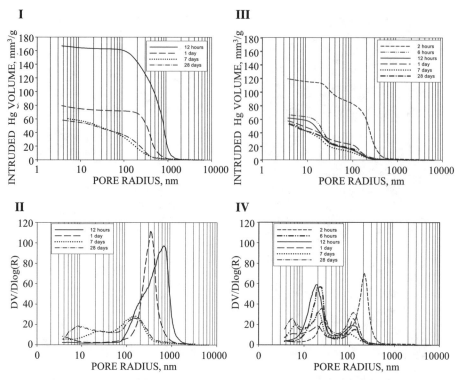

Figure 3   Intruded Hg volume vs. pore radius for cement pastes cured at various ages:
(I) Portland cement, cumulative plot; (II) Portland cement, derivative plot;
(III) sulphoaluminate cement, cumulative plot; (IV) sulphoaluminate cement, derivative plot

The comparison between the results obtained with Portland and CSA cement pastes at the 12 hours of aging clearly shows the strong difference in the porosimetric characteristics. The evolution of porosity for CSA cement pastes cured in the period 1-28 days is also completely different from that observed for Portland cement pastes, even if at 7 and 28 days total intruded volume, first and second threshold pore radius do not vary significantly. It is interesting to note the appearance of a trimodal distribution inasmuch as the lower porosity field is split in two regions with two threshold pore radii ranging from 20 to 25 nm and from 5.5 to 7 nm. The peculiar development of porosity during the hydration of rapid-hardening dimensionally stable and durable CSA cements has a very important role in promoting their technical behaviour.

Synthesis of CSA Clinkers from Waste-Based Raw Mixes

From the examination of the XRD data concerning the burning products of all the investigated mixtures it can be argued that $C_4A_3\bar{S}$ and C_2S are, in the order, the main mineralogical phases. Unreacted compounds were absent in the burning products of mixtures A1, A2 and B2 while mixture B1, upon heating at all the temperatures investigated, showed an almost negligible presence of $CaSO_4$.

Mixtures A1 and B2, when heated at 1200°C, revealed a complete absence of secondary phases; upon burning at 1250°C and 1300°C, they showed the presence of brownmillerite, C_4AF, and calcium sulphosilicate, $C_5S_2\bar{S}$, respectively, in little amounts. As far as mixtures A2 and B1 are concerned, at every heating temperature, respectively weak peaks of $C_5S_2\bar{S}$ and C_4AF were generally detected.

Figures 4 and 5 as well as Figures 6 and 7 indicate, for mixtures A1, A2, B1 and B2, respectively, the XRD intensities of the main peaks of $C_4A_3\bar{S}$ and C_2S, in the order, as a function of the burning temperature. It was generally observed a significant influence of the synthesis temperature on the $C_4A_3\bar{S}$ and C_2S concentrations. However, 1250°C seemed to be the optimum temperature for obtaining the maximum amount of both phases.

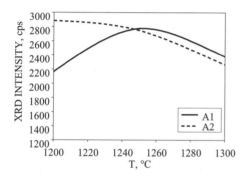

Figure 4 $C_4A_3\bar{S}$-XRD intensity (main peak, counts per second) for the burning products of mixtures A1 (solid curve) and A2 (broken curve) vs. synthesis temperature

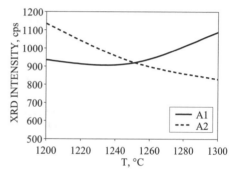

Figure 5 C_2S-XRD intensity (main peak, counts per second) for the burning products of mixtures A1 (solid curve) and A2 (broken curve) vs. synthesis temperature

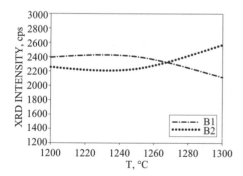

Figure 6 $C_4A_3\bar{S}$-XRD intensity (main peak, counts per second) for the burning products of mixtures B1 (dash-dot curve) and B2 (dotted curve) vs. synthesis temperature

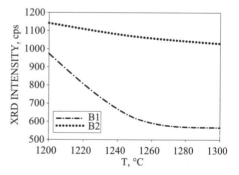

Figure 7 C_2S-XRD intensity (main peak, counts per second) for the burning products of mixtures B1 (dash-dot curve) and B2 (dotted curve) vs. synthesis temperature

CONCLUDING REMARKS AND FUTURE RESEARCH DEVELOPMENTS

The technical behaviour of rapid-hardening and dimensionally stable CSA cements is mainly related to the fast ettringite formation and the peculiar porosity development during the early stages of hydration.

Among the interesting environmentally friendly features of the manufacturing process of CSA cements, there is the possibility of using by-products generated within coal-fired power stations like FBC fly and bottom ashes as well as pulverized coal fly ash and FGD waste.

A laboratory scale investigation on CSA clinker generating raw mixes consisting of limestone, bauxite and up to 40% by-products from coal-fired power stations has shown that high amounts of $C_4A_3\bar{S}$ and dicalcium silicate can be obtained especially at a synthesis temperature around 1250 °C.

The object of future investigation will be the study of the hydration behaviour at early and long curing times of CSA cements consisting of the above mentioned clinkers and FGD.

REFERENCES

1. MUZHEN S U, KURDOWSKI W AND SORRENTINO F P, Development in non-Portland cements, Proc. 9th Int. Congr. Chem. Cem., New Delhi, Vol. I, 1997, pp. 317-356.

2. KOUZNETSOVA T V, Development of special cements, Proc. 10th Int. Congr. Chem. Cem., Göteborg, Vol. 1, 1997, p. 1i001.

3. SCRIVENER K L, Properties, applications and practicalities of special cements, Proc. 11th Int. Congr. Chem. Cem., Durban, Vol. 1, 2003, pp. 84-93.

4. DENG JUN-AN, GE WEN-MIN, SU MUZHEN, LI XIU-YING, Sulfoaluminate cement series, Proc. 7th Int. Congr. Chem. Cem., Paris, Vol. IV, 1980, pp. 381-386.

5. WANG YANMOU, SU MUZHEN, The third series cement in China, Proceedings of the Third International Symposium on the Cement and Concrete. Shanghai, Vol. 3, 1993, pp. 116-121.

6. MUZHEN SU, YANMOU WANG, LIANG ZHANG, DEDONG LI, Preliminary study on the durability of sulfo/ferro-aluminate cements, Proc. 10th Int. Congr. Chem. Cem., Göteborg, Vol. 4, 1997, p. 4iv029.

7. TANGBO SUI, YAN YAO, Recent progress in special cements in China, Proc. 11th Int. Congr. Chem. Cem., Durban, Vol. 4, 2003, pp. 2028-2032.

8. GLASSER F P AND ZHANG L, High-performance cement matrices based on calcium sulfoaluminate–belite compositions, Cement and Concrete Research, Vol. 31, 2001, pp. 1881-1886.

9. MEHTA P K, Investigations on energy-saving cements, World Cem. Technol., Vol. 11, 1980, pp. 166-177.

10. BERETKA J, SANTORO L, SHERMAN N AND VALENTI G L, Synthesis and properties of low energy cements based on $C_4A_3\bar{S}$, Proc. 9th Int. Congr. Chem. Cem., New Delhi, Vol. III, 1992, pp. 195-200.

11. BERETKA J, DE VITO B, SANTORO L, SHERMAN N AND VALENTI G L, Hydraulic behaviour of calcium sulphoaluminate-based cements derived from industrial process wastes, Cement Concrete Research, Vol. 23, 1993, pp. 1205-1214.

12. MAJLING J, SAHU S, VLNA M AND ROY D M, Relationship between raw mixture and mineralogical composition of sulphoaluminate belite clinkers in the system $CaO-SiO_2-Al_2O_3-Fe_2O_3-SO_3$, Cement Concrete Research, Vol. 23, 1993, pp. 1351-1356.

13. BERETKA J, CIOFFI R, MARROCCOLI M AND VALENTI G L, Energy-saving cements obtained from chemical gypsum and other industrial wastes, Waste Management, Vol. 16, 1996, pp. 231-235.

14. IKEDA K, FUKUDA K AND SHIMA H, Calcium sulphoaluminate cements prepared from low-alumina waterworks slime, Proc. 10th Int. Congr. Chem. Cem. Göteborg, Vol. 1, 1997, p. 1i025.

15. ARJUNAN P, SILSBEE M R AND ROY D M, Sulfoaluminate-belite cement from low-calcium fly ash and sulphur-rich and other industrial by-products, Cement and Concrete Research, Vol. 29, 1999, pp. 1305-1311.

16. BERNARDO G, MARROCCOLI M, MONTAGNARO F AND VALENTI G L, Use of fluidized bed combustion wastes for the synthesis of low energy cements, Proc. 11th Int. Congr. Chem. Cem., Durban, Vol. 3, 2003, pp. 1227-1236.

17. GRACE J R, AVIDAN A A AND KNOWLTON T M, Circulating fluidized bed, Chapman & Hall, London, 1997, 585 p.

18. ÅMAND L E AND LECKNER B, Emissions of nitrogen oxide from a circulating fluidized bed boiler-the influence of design parameters, 2nd International Conference on CFBC, Compiègne, 1988, pp. 457-463.

19. LYNGFELT A AND LECKNER B, SO_2 capture in fluidized bed boilers: re-emission of SO_2 due to reduction of $CaSO_4$, Chemical Engineering Science, Vol. 44, 1989, pp. 207-213.

20. MONTAGNARO F, SALATINO P AND SCALA F, The influence of sorbent properties and reaction temperature on sorbent attrition, sulfur uptake, and particle sulfation pattern during fluidized-bed desulfurization, Combustion Science and Technology, Vol. 174, 2002, pp. 151-169.

21. BERNARDO G, TELESCA AND VALENTI G L, A porosimetric study of calcium sulfoaluminate cement pastes cured at early ages, Cement and Concrete Research, Vol. 36, 2006, pp. 1042-1047.

22. BENTUR A, The pore structure of hydrated cementitious compounds of different chemical composition, Journal of American Ceramic Society, Vol. 63, No. 7-8, 1980, pp. 381-386.

23. MIDGLEY H G AND ILLSTON J M, Some comments on the microstructure of hardened cement pastes, Cement and Concrete Research, Vol. 13, 1983, pp. 197-206.

24. GARBOCZI E J, Permeability, diffusivity, and microstructural parameters: A critical review, Cement and Concrete Research, Vol. 20, 1990, pp. 591-601.

25. COOK R A AND HOVER K C, Mercury porosimetry of hardened cement pastes, Cement and Concrete Research, Vol. 29, 1999, pp. 933-943.

26. DIAMOND S, Mercury porosimetry – An inappropriate method for the measurement of pore size distributions in cement-based materials, Cement and Concrete Research, Vol. 30, 2000, pp. 1517-1525.

STUDY ON ALKALI-ACTIVATED PHOSPHOROUS SLAG – FLY ASH CEMENTITIOUS MATERIALS

F Yanghao

W Cheng   C Yikan   R Liyun

Hehai University, Nanjing

China

ABSTRACT. The performance of alkali-activated phosphor slag-fly ash cement with different phosphor slag/fly ash proportion was studied, and the pore and microstructures of the hardened pastes were analyzed. The results show that the alkali-activated phosphor slag-fly ash cement has normal setting performance, which differs from that of alkali-activated blast furnace slag cement. Within the proportion of 0~30% (w/w), adding fly ash into the alkali-phosphor slag system results no significant change in setting time. Compared with ordinary Portland cement, alkali-phosphor slag cement is characterized with higher compressive but lower flexural strengths, especially, the early flexural strength being much lower than that of ordinary Portland cement. Adding fly ash into alkali-phosphor slag system will reduce the compressive strength, while the flexural strength can be improved. The corrosion and frost resistances of alkali-activated cement are much better than those of ordinary Portland cement, while its drying shrinkage is much greater than that of the latter one. The hardened alkali-activated phosphor slag-fly ash cement paste is very dense with very low porosity and small mean pore diameter.

Keywords: Phosphor slag, Fly ash, Alkali-activated cement, Performance, Microstructure

Professor Fang Yonghao is a professor of the Department of Materials Science and Engineering, Hehai University, Nanjing, China. He works mainly in the area of cement based materials.

Mr Wang Cheng is a graduate student in the Department of Materials Science and Engineering, Hehai University, Nanjing, China.

Mr Ceng Yikan is a graduate student in the Department of Materials Science and Engineering, Hehai University, Nanjing, China.

Mrs Ren Liyun is a graduate student in the Department of Materials Science and Engineering, Hehai University, Nanjing, China.

INTRODUCTION

Alkali-activated cement has attracted great interest for its specific desirable properties and for that it can make full use of industrial wastes to save resources and reduce pollution to the environment [1-3]. However, up to the present, only research on alkali-activated blast furnace slag cement has obtained desirable results. Since the resource of blast furnace slag is very limited, and it has been made full use in cement industry and civil engineering, there is a shortage for slag supplement. It is necessary that new raw material for substituting blast furnace slag be investigated to develop alkali-activated cement. Phosphor slag is a type of waste from yellow phosphor production process, of which the main components are CaO and SiO_2. Quenched phosphor slag is composed mainly of glassy phases. Because of its comparability both in terms of chemical and phase compositions to blast furnace slag, it is promising to use phosphor slag as a raw material to prepare alkali- activated cement [4]. Fly ash has been used in investigating on alkali-activated slag–fly ash cement [5, 6], and its efficiency has been recognized. In the present work, the performance of alkali-activated phosphor slag-fly ash cement and microstructure of the hardened paste were studied.

EXPERIMENTAL

Materials

The chemical composition of phosphor slag (PS) and fly ash (FA) used in the experiments are shown in Table 1. The slag and fly ash were grounded to have the specific surface areas of 355 m^2/kg and 452 m^2/kg, respectively. The water glass (WG) used as the alkaline activator was diluted and the SiO_2/Na_2O module was adjusted by NaOH to 1.4, which is the optimum value obtained from a preparatory experiment. P.O 42.5 cement (OPC) manufactured by the Zhongguo Cement Plant was used as the reference, those properties are shown in Table 2.

Table 1   Chemical composition of phosphor slag and fly ash, % (w/w)

MATERIAL	SiO_2	Al_2O_3	Fe_2O_3	CaO	MgO	F^-	P_2O_5	SO_3	IGNITION LOSS
PS	41.23	4.40	4.87	44.06	0.27	2.73	1.44	-	1.70
FA	50.76	30.37	4.04	3.53	1.09	-	-	0.66	3.32

Procedure

Phosphor slag, fly ash and water glass were mixed in the proportion as shown in Table 2. The water requirement and setting time were determined according to GB/1346-2001. Paste specimens were prepared and cured at 20±1°C with RH > 90% for 28 days. The hydration of the specimens was terminated with absolute alcohol and the specimens were vacuum-dried and analyzed by MIP and SEM. The mortar strength was tested according to GB/T17671-1999. The w/(PS+FA) ratio was 0.42, including the water in the water glass to give the mortar fluidity be 160±10 mm, being consistent with that of the reference OPC (160 mm, w/c=0.5). The drying shrinkage and frost resistance tests were conducted according to DL/T 5150-2001. Corrosion resistance test was conducted to the mortar specimens cured for 28 days. The specimens were immersed in different corrosive solutions for 3 months, and then the compressive strength was tested. Corrosion resistance factor F_3 is obtained by dividing the strength of the corroded specimen by that of the specimen cured in water for the same age.

RESULTS AND DISCUSSION

Performance of Alkali-Activated Phosphor Slag-Fly Ash Cement

Strength and setting time

Table 2 shows the water requirement for normal consistency, setting time, compressive and flexural strengths of the alkali-activated cements with different fly ash/phosphor slag ratios. The total amount of water glass and NaOH equivalent to Na_2O was 5% of the sum of the fly ash and phosphor slag. The water requirement in the table includes the water in the water glass. It can be seen from Table 2 that the water requirement of the alkali-activated phosphor slag-fly ash cement is a little lower than that of the reference OPC. Not as that the alkali-activated blast furnace slag cement usually shows, the alkali-activated phosphor slag-fly ash cement sets in a normal time. The reason for it is that the phosphor slag contains a certain amount of soluble phosphor, which will retard the setting of the alkali-activated cement [7]. Changing of fly ash dosage in the range of 0~30% (w/w) results no significant change in the setting time.

Table 2   Properties of alkali-activated phosphorous slag-fly ash cement

SAMPLE	MIX PROPORTION IN MASS		WATER REQUIREMENT %	SETTING TIME, minutes		W/B	STRENGTH /MPa			
							Flexural		Compressive	
	PS	FA		Initial	Final		3 d	28 d	3 d	28 d
PF0	100	0	24.2	97	130	0.42	4.7	6.6	30.9	98.8
PF1	90	10	23.9	110	145	0.42	4.0	7.0	27.8	73.8
PF2	85	15	23.5	117	150	0.42	4.4	9.8	25.8	78.1
PF3	80	20	23.2	120	150	0.42	3.8	8.7	25.5	65.9
PF4	70	30	23.3	125	160	0.42	3.1	5.9	17.2	51.3
OPC			26.0	150	215	0.50	6.5	7.8	24.3	53.1

On the one hand, the fly ash is not as active as the phosphor slag, so that the higher that fly ash proportion, the slower the general reaction in the cement, and the slower the setting process; On the other hand, however, increasing the proportion of fly ash will reduce the content of soluble phosphor in the cement, thus the retarding effect of soluble phosphor is weakened and the setting process enhanced. The influence of fly ash on the setting time of the cement is the sum of the two contrary effects. The compressive strength of the alkali-activated phosphorous slag cement with no fly ash developed very rapidly and the 3 d and 28 d strengths reached as high as 30.9 MPa and 98.8 MPa, respectively, being much higher that those of the reference OPC. While the flexural strengths, both for 3 d and 28 d, are all much lower that those of the reference OPC. They all meet the requirement of GB175-1999 for P.O 42.5 cement. The probable reason for the lower flexural strength is that the main hydration product in the alkali-activated cement is gel-like substance. The lower crystal/gel ratio made the paste very brittle. When fly is added, the compressive strength is significantly decreased with the increase of fly ash, but the strengths at both 3d and 28d all meet the requirement for P.O 42.5 cement, if the mass percentage of fly ash is lower than 30%. Increasing the percentage of fly ash has only little effect on the flexural strength at 3d. However, there is an appreciable trend that the flexural strength at 28d increases with the increase of fly ash up to 20%, then decreases as the addition of fly ash exceeded the optimum amount. The flexural strength at 28d even exceeded that of the reference OPC cement when the addition of fly ash was in the range of 15% to 20%. This influence of fly ash can be ascribed to its micro-aggregate effect.

Corrosion resistance

Table 3 shows the compressive strengths of the specimens after being corroded by acid and sulfate solutions and artificial sea water for 3 months. There is no significant change in the compressive strengths of the PF0 and PF2 specimens after being corroded by all the corrosive solutions, the corrosion resistance factors F_3 are all higher that 0.96. As a contrast, the strengths of the reference OPC cement is significantly reduced, especially that corroded by the HCl solution, of which the compressive strength is reduced by 18%.

The results indicate that the alkali-activated phosphor slag-fly ash cement is much more resistant to acid, sulfate and other composite salt solutions than OPC. One reason for the fact is that the contents of the components such as $Ca(OH)_2$ which can react with the acid, sulfate and other composite salt solutions are much lower in the alkali-activated phosphor slag-fly ash cement pastes than those in the OPC paste, and the other is that the alkali-activated cement pastes are denser than that of OPC, which will hinder the penetration of the corrosive media.

Table 3   Results from corrosion experiment for the mortar specimens

SPECIMEN	COMPRESSIVE STRENGTH CURED IN WATER, MPa	CORROSION SOLUTION					
		5% HCl		3% Na_2SO_4		Artificial sea water*	
		Compressive strength, MPa	F_3	Compressive strength, MPa	F_3	Compressive strength, MPa	F_3
PF0	78.3	75.5	0.96	76.8	0.98	76.3	0.97
PF2	80.6	77.3	0.96	79.6	0.99	79.8	0.99
OPC	60.2	49.6	0.82	52.5	0.87	53.6	0.89

* 2.7% NaCl + 0.32% $MgCl$ + 0.22% $MgSO_4$ + 0.13% $CaSO_4$.

Drying shrinkage

The shrinkages of the mortar specimens with the age are shown in Figure 1. Just like alkali-activated blast furnace slag cement, the shrinkages of the alkali-activated phosphor slag cement (PF0) at all ages are much higher than those of the reference OPC, being twice as high at 28 d and 56 d. Adding fly ash will reduce the shrinkage, the shrinkages of PF2 at 28 d and 56 d are reduced by about 20% compared with those of PF0. The reason of the shrinkage reduction is that the micro-aggregate effect of the fly ash restricted the contraction of the cement paste.

Freeze/thaw resistance

Figure 2 shows the change of the relative dynamic elastics module P_n with freeze/thaw cycles for the mortar specimens. It can be seen that the freeze/thaw resistance of the both alkali-activated cements PF0 and PF2 is much higher than that of the reference OPC. The relative dynamic elastics module of the OPC specimen has reduced by more than 50% after 75 cycles, while those of the alkali-activated cements PF0 and PF2 are only reduced by less than 40% after 100 cycles.

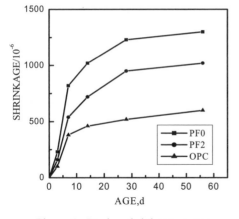

Figure 1   Drying shrinkage vs. age
for mortar specimens

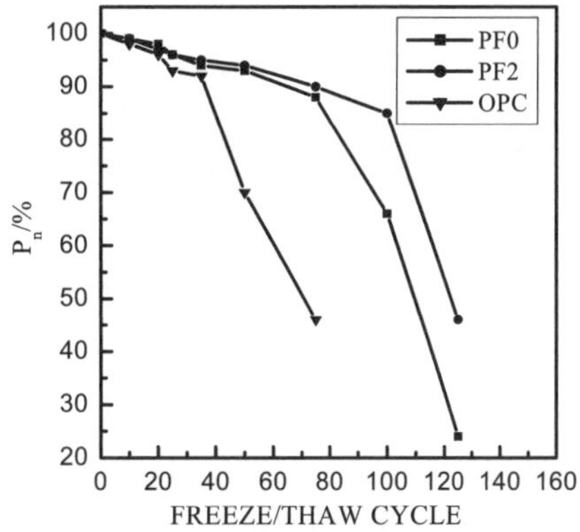

Figure 2   Relative dynamic elastics module
vs. freeze/thaw cycles for mortar specimens

Microstructure

Figures 3a and 3b are the SEM images of the hardened OPC and PF2 pastes, and Figure 3c is the enlarged image of the squared area in Figure 3b. The OPC paste is composed mainly of C-S-H gel, $Ca(OH)_2$ and AFt with very dense structure. The structure of the PF2 paste is even denser. The main components of the paste are C-S-H and C-A-H gels, among which included a large amount of glass particles, which come from the fly ash, and small crystals (Figure 3c). XRD analysis gave no exact information about the mineral composition the crystals. The dense gel structure including glass particles and small crystals can explain the fact that the alkali-activated phosphor slag-fly ash cement has higher compressive and flexural strengths.

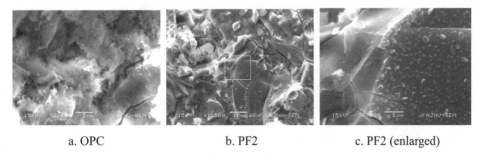

a. OPC   b. PF2   c. PF2 (enlarged)

Figure 3   SEM photos of hardened cement paste cured for 28 d

Pore structure

The porosity and pore size distribution for the hardened cement pastes are shown in Table 4. It can be seen that the porosities of the alkali-activated cement pastes are much lower than that of the OPC paste, especially those pores are fewer which are larger than 100 nm, which will give severe adverse influence on the performance of the pastes. It is the low porosity and higher gel pore content, meaning high gel content that resulted in the high strength and high resistance to freeze/thaw and corrosion of the alkali-activated cement.

Table 4 Porosity and pore distribution for the hardened cement pastes

SAMPLE	POROSITY, $\times 10^{-2}$ cm^2/g	PORE DISTRIBUTION, %			
		<10 nm	10~10^2 nm	10^2~10^3 nm	10^3~5×10^3 nm
PF0	10.12	57.1	29.3	9.2	4.4
PF2	11.28	59.4	26.1	10.5	4.0
PO 42.5	16.02	42.3	30.3	19.0	8.4

CONCLUSIONS

1. Alkali-activated phosphor slag-fly ash cement has normal setting performance. Adding fly ash into the alkali-phosphor slag system results in no significant change in setting time.

2. Compared with ordinary Portland cement, alkali-activated phosphor slag cement is characterized with higher compressive but lower flexural strengths. Adding fly ash into the alkali-activated phosphor slag system will reduce the compressive strength, while the flexural strength can be increased. The optimum proportion of fly ash is 15%~20%.

3. Corrosion and freeze/thaw resistances as well as drying shrinkage of alkali-activated cement are much greater than those of ordinary Portland cement.

4. The hardened alkali-activated phosphor slag-fly ash cement paste is very dense with very low porosity and small mean pore diameter.

REFERENCES

1. GLUKHOVSKY V D, ROSTOVSKAYA G S AND RUMYNA G V, High strength alkali-slag cements, 7th Int. Congress on Cement Chemistry, Paris, Vol. V, 1980, 164 pp.

2. DOUGLA S E, BILODEAU A, AND MALHOTRA V M, Alkali activation of ground granulated blast furnace slag, Cement and Concrete Research, Vol. 21, No. 1, 1991, pp. 101-108.

3. BROUGH A R AND ATKINSON A, Sodium silicate-based, alkali-activated slag mortars - Part I, Strength, hydration and microstructure, Cement and Concrete Research, Vol. 32, No. 6, 2002, pp. 865-879.

4. CHEN L, ZHU C G, AND SHENG G H, Mechanical properties and microstructures of alkali-activated phosphor slag cement, Journal of Chinese Ceramic Society (in Chinese), Vol. 34, No. 5, 2006, pp. 604-609.

5. XIE Z AND XI Y, Hardening mechanisms of an alkaline-activated class F fly ash, Cement and Concrete Research, Vol. 31, No 8, 2001, pp. 1245-1249.

6. SMITH M A AND OSBORNE G J, Slag/fly ash cement, Journal of World Cement Technology, Vol. 8, No. 6, 1977, pp. 223-233.

7. GONG C AND YANG N, Effect of phosphate on the hydration of alkali-activated red mud-slag cementitious materials, Cement and Concrete Research, Vol. 30, No.7, 2000, pp. 1013-1016.

UTILISATION PF GLASS CULLET FOR THE PRODUCTION OF BINDING MATERIALS

A I Puzanov

JCS Perm Production of Foamed Silicates

S I Puzanov   A A Ketov

Perm State Technical University

Russia

ABSTRACT. Glass cullet as a raw material possesses a number of valuable properties: high strength, chemical stability, availability and a relatively low cost. At present, glass cullet can not be completely utilized, and some part of it (up to 25%) is accumulated in disposal dumps which can be harmful to the environment. Utilization of glass cullet is a serious problem for municipalities all over the world. The production of artificial silicate materials implies a significant consumption of material and power resources, and mining and processing mineral resources has an unfavorable effect on the environment. Nevertheless, the anthropogenic source of silicate materials – glass cullet – is used in extremely limited quantities. Fine glass cullet is characterized by high ion-exchange activity, especially for Na^+ ions. Besides, the amorphous structure of the material implies that a number of chemical and phase reactions proceed on the surface of glass particles. Moreover, silicate glass is characterized by high mechanical strength, so glass powder can be utilized for the production of binding materials and concrete. It was discovered that the glass concrete obtained from a chemically modified glass has higher strength than the concrete from an ordinary glass aggregate. Then, optimum modification conditions were suggested. The obtained results prove that glass can be used as an aggregate for producing concrete with relatively high strength. It was shown that glass as an aggregate can be used, with certain precautions, to manufacture a wide range of concrete and binding materials.

Keywords: Binding materials, Glass cullet utilization, Dispersed glass, Ion-exchange modification.

Dr A I Puzanov is a consulting engineer at the JCS Perm Production of Foamed Silicates specializing in the reuse of glass cullet in construction materials.

S I Puzanov is a lecturer assistant at the Chemical Engineering Department of the Perm State Technical University.

Professor A A Ketov is a full Professor in chemical engineering at Perm State Technical University. His research interests include creation of inorganic materials and development of new products. He specializes in the processes of environmental technologies and reuse of wastes.

INTRODUCTION

Glass cullet as a raw material possesses a number of valuable properties: high strength, chemical stability, availability and a relatively low cost. At present, glass cullet cannot be completely utilized, and some part of it (up to 25%) is accumulated in disposal dumps, that can be harmful to the environment. Utilization of glass cullet is a serious problem for municipalities all over the world.

For example, the U.S. Environmental Protection Agency [1] gives data concerning volumes of unutilized glass accumulated in the environment. The data shows that the volume of annually accumulated glass cullet has stabilized at a value of 12.7–13.3 million tons. The volume of utilized glass cullet is about 2.7–2.8 million tons annually. So, there is no uptake for about 9.9–10.7 million tons of glass cullet, therefore it is land-filled and accumulates in the environment. It should be noted, that the ratio of accumulated and utilized secondary glass does not change. Similar tendency is observed in other countries. For example, according to [2], the volume of annually accumulated glass cullet in Germany has stabilized at the value of 3.4–3.6 million tons. About 20–25% of annually produced glass accumulates in the environment.

Thus, the quantity of glass stored and discharged to the environment is comparable to the geological resources. The production of artificial silicate materials implies a significant consumption of material and power resources, and mining and processing mineral resources has an unfavorable effect on the environment. Nevertheless, the anthropogenic source of silicate materials – glass cullet – is used in extremely limited quantities.

The recycling of unsorted waste glass poses major problems because glass producers have no easy way for economical reuse. For example, New York City recycling facilities are able to sell plastics and metals, but there are virtually no takers for the glass [3-6]. Some limited applications have been found in asphalt (glasphalt) [7] and roadway fill.

But for the total utilization of glass cullet it is necessary to recycle about tens of million tons of the material annually. We suppose that such volumes of silicate raw materials can be taken up only for the production of building materials. It is usually advised to use glass cullet as an additive, for example, as aggregate for pozzolan concretes or additive in the process of sintering cement clinker.

As a matter of fact, such approach does not utilize specific properties of glass as a material. Besides, introduction of glass aggregate into the composition of concrete induces the reaction between alkalis of the cement and amorphous silica contained in the aggregate, which results in cracking and destruction of the concrete [8-9].

One of the most perspective ways of glass cullet utilization is using it in the production of binding materials [10]. T.R. Jones, R.D. Pascoe and P.B. Hegarty [11] have shown, that such approach is highly effective. They obtained a binding material called 'casamic', and the technology of its production is based on the cementing properties of dispersed glass realized in the presence of free Ca^{2+} ions enabling hardening of the material.

As it was shown earlier [12], dispersed glass acts as a cationic ion-exchanger. This assumption is confirmed by the high pH values of aqueous dispersions of glass powders. Hydrated pastes of the dispersed glass contain highly reactive particles, which can interact with the components of the solution and promote hardening. Thus, dispersed glass can act as a cementing material. So, our purpose was to find out necessary conditions for producing solid concrete-like materials of highly dispersed glass cullet.

RESULTS AND DISCUSSION

Addition of Water Glass

The pH of aqueous dispersions of glass powders is about 8–10. So, amorphous silicon dioxide contained in the structure of glass can be partially dissolved in this solution. The processes of dissolving and precipitation of silicates in dispersed glass pastes may cause formation of condensing silicate phases between glass particles and hardening of the composition. Addition of water glass encourages the process of hardening, for it already contains dissolved silicates, necessary for the formation of new phases between glass particles.

A standard solution of water glass with the density of 1430 kg/m^3 and the modulus of 3.2 was taken for the experiments. Glass cullet was ground in a vibro-centrifugal mill to obtain powder with the average particle size of 60 micrometers. In Figure 1, the growth of compressive strength of the obtained material with time is shown.

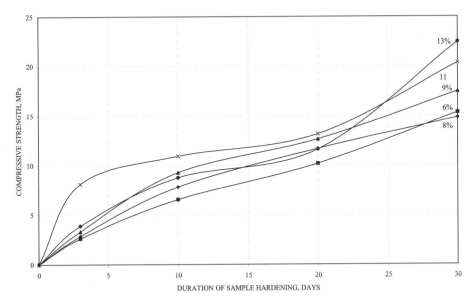

Figure 1   Growth of compressive strength of samples with time at 25°C
(different contents of water glass in the solution).

It can be seen that the increasing water glass content improves the compressive strength of the blocks. With the increase of water glass content from 6 to 13%, the compressive strength grows 1.5 times. The maximum compressive strength achieved in 30 days is 22 MPa.

From the technological point of view, it is necessary to accelerate the process of hardening. For this purpose, it is advisable to increase the processing temperature. The necessary experiments have been carried out. The samples were treated for 5 hours at various temperatures. The samples treated at 100-120°C had the maximum compressive strength. Addition of quartz sand also contributed to the growth of sample strength. This probably can be explained by the fact that the excess free alkali causes corrosion of silicon dioxide, contained in quartz, which is then bound with the formation of silicate compounds.

Thus, binding of extra Na^+ ions favors the strength of the composite materials. However, it is not recommended to use crystalline quartz for this purpose, for it is rather stable in alkalis. One of the methods of eliminating Na^+ ions from the system is a preliminary ion-exchange treatment of glass powder.

Addition of Water Glass into Dispersed Ion-exchange Modified Glass

Dispersed glass is characterized by a high cation-exchange ability, due to the migration of cations (mostly Na^+ ions) into the solution. The process can be described with the following scheme:

$$glass\text{-}Na + H_2O \leftrightarrow glass\text{-}H + NaOH,$$

ionic form:

$$glass\text{-}Na + H^+ \leftrightarrow glass\text{-}H + Na^+,$$

where glass-Na – is a Na^+ ion on the surface of glass attached to the glass matrix able to migrate into the solution. So, dispersed glass can be considered as a non-stoichiometric salt of a weak acid and a strong base. In this case, water acts as a donor of H^+ ions, so, the rate of the process can be essentially increased, if we use acid instead of water for the treatment of dispersed glass. Na^+ ions can be replaced not only by H^+, but also by ions of heavy metals, which form silicates insoluble in water – for example, by ions of Ca.

This process can be described by the following scheme:

$$2\cdot glass\text{-}Na + Ca^{2+} \leftrightarrow glass_2\text{-}Ca + 2\cdot Na^+$$

When dispersed glass is treated by the solutions of acetic acid and calcium chloride, an exchange of Na^+ into H^+ and Ca^{2+}, respectively, takes place. After this process, the concentration of H^+ and Ca^{2+} ions in the solution decreases, and the concentration of Na^+ ions increases. The data obtained are given on Figure 2.

So, dispersed glass is characterized by a high ion-exchange activity. It can be easily transformed into H^+ or Ca^{2+} form, hereinafter called H-glass and Ca-glass.

We have studied the tendencies of hardening the composite materials obtained from the ion-exchange modified glass. The process of hardening was carried out at 25 and 60°C. A solution of water glass was added into the composition as a source of soluble silicates. The results obtained are given in Figure 3.

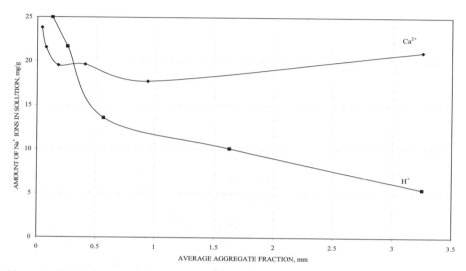

Figure 2   The influence of the average particle size of glass powder (mm) on the quantity of Na^+ ions (mg/g) migrated into the solution in the process of ion exchange.

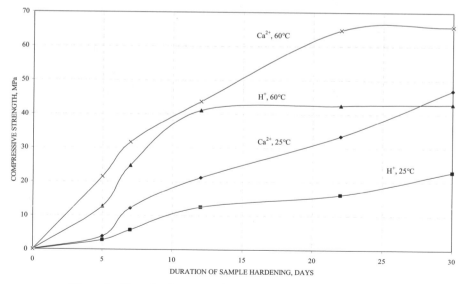

Figure 3   Growth of compressive strength of composite materials on H-glass and Ca-glass with time at 25 and 60°C.

It can be seen, that in four weeks (standard period of time for control concrete strength) the compressive strength of the samples obtained from Ca-glass and H-glass was 21 and 44 MPa, respectively. In 45 days, the compressive strength of the samples obtained from Ca-glass and H-glass was 38 and 74 MPa, respectively. At 60°C, the compressive strength of 21 MPa for Ca-glass is achieved in 6 days, the compressive strength of 44 MPa for H-glass is achieved in 12 days.

So, the obtained composite materials can compete with standard pozzolanic concretes in strength and hardening rate. Besides, the materials can be used for decorative purposes. They are white, and the addition of pigments allows producing the material of any color. Some items, produced according to the suggested technology, are given on Figure 4.

Figure 4   Construction materials, produced according to the suggested technology.

CONCLUSIONS

Dispersed glass possesses cementing properties and can be used for the production of binding materials and concretes. The technology should be based on the cationic ion-exchange activity of the glass surface. The suggested technology allows obtaining glass-containing binding materials, which may harden under various conditions. The obtained concrete-like materials have high strength and good decorative properties. The suggested technology enables us to create new effective binding materials and to utilize glass cullet.

ACKNOWLEDGEMENTS

The authors would like to acknowledge the financial help provided by JCS Perm Production of Foamed Silicates for this project.

REFERENCES

1. UNITED STATES ENVIRONMENTAL PROTECTION AGENCY. Municipal solid waste generation, recycling, and disposal in the United States: Facts and Figures for 2003. Report No. 05-18, Washington, April. 2005.

2. GLUSING, A K, CONRADT, R. Dissolution kinetics of impurities in recycled cullet. Recycling and Reuse of glass Cullet: Proceedings of International Symposium 19-20 March 2001, Dundee UK. pp 29-41.

3. KIRBY, B. Secondary Markets for Post-Consumer Glass. Resource Recycling, June, 1993.

4. TROMBLY, J. Developing Non-Traditional Glass Markets. Resource Recycling, October 1991.

5. Glass Feedstock Evaluation Project. Report No. B6 to the Clean Washington Center, Dames and Moore Inc., Seattle, 1993.

6. REINDL, J. Development of Non-Traditional Glass Markets. Resource Recycling, October 1991.

7. HUGHES, C. S. Feasibility of Using Recycled Glass in Asphalt. Virginia, Transportation Research Council, Charlottesville, Va., Report VTRC 90-R3, March 1990.

8. HOBBS, D. W. Alkali-Silica Reaction in Concrete, Thomas Telford, London, 1988.

9. SWAMY, R. N. (ed.), The Alkali-Silica Reaction in Concrete, Van Nostrand Reinhold, 1992.

10. PATTENGIL, M. Glass as a Pozzolan, Albuquerque Symposium on Utilisation of Waste Glass, Second Prod. 1973

11. JONES, T R, PASCOE, R D, HEGARTY, P B. A novel ceramic (casamic) made from unwashed glass of mixed colour. Recycling and Reuse of Waste Materials: Proceedings of the International Symposium 9-11 September 2003, Dundee UK. pp 577-585.

12. KETOV, A. Peculiar Chemical and Technological Properties of Glass Cullet as the Raw Material for Foamed Insulation. Recycling and Reuse of Waste Materials: Proceedings of the International Symposium 9-11 September 2003, Dundee UK. pp 695-704.

RECYCLING TECHNOLOGY FOR CONCRETE AGGREGATE USING UNDERWATER ELECTRIC PULSE DISCHARGE

S Maeda   **M Shigeishi**

T Namihira   **M Ohtsu**

H Akiyama

Kumamoto University

Japan

ABSTRACT. In this research, the aggregate collection technology from waste concrete mass by pulsed power was examined. In addition, high-voltage-short-input-impulse called pulsed power and aggregate collection by underwater pulsed electrical discharge pulsed power were studied. Pulsed power recycled aggregate is of improved quality, although an increased number of voltage pulses applied leads to aggregate damage. The quality of pulsed power recycled aggregate can be assessed by the fineness modulus. Pulsed power recycled fine aggregate is of low quality attributed to its hardened cement paste and collected fine aggregate contents. Powder arises when the pulsed power applied to concrete is lower than that used at the ready-made aggregate collection technology from waste concrete.

Keywords: Recycle, Pulsed electric power, Recycled coarse aggregate, Powder.

S Maeda, ME, is a graduate student, Graduate School of Science & Technology, Kumamoto University, Japan.

Dr M Shigeishi, Associate Professor, Graduate School of Science & Technology, Kumamoto University, Japan.

Dr T Namihira, Associate Professor, Graduate School of Science & Technology, Kumamoto University, Japan.

Dr M Ohtsu, Professor, Graduate School of Science & Technology, Kumamoto University, Japan.

Dr H Akiyama, Professor, Graduate School of Science & Technology, Kumamoto University, Japan.

INTRODUCTION

The quantity of construction waste is approximately 20% of industrial waste as a whole. However, its disposal amounts to 30% and moreover, 60% of this is illegally dumped. It is necessary to recycle industrial waste to encourage a recycling society and raise awareness of the shortage of disposal sites. Industrial waste consists of concrete mass, construction sludge, waste wood and mixed waste. Concrete accounts for the highest percentage of construction waste.

The recycling rate of concrete mass was 98% in 2002, and kept this high rate in 2005. However, its main use is as a sub-grade material. Whilst the demand for this subgrade material is set to decrease in the future, the amount of concrete mass will increase due to proposed demolition of various urban facilities that were built in times of high economic growth. In order to rebuild these concrete structures, a lot of aggregate will be required. However, the use of natural aggregate will be difficult due to natural aggregate depletion and environmental concerns [1, 2].

It is therefore necessary to use concrete mass as a recycled aggregate. Current concrete technology methods fracture concrete mechanically, which can cause damage to the aggregate within concrete. Also, the surface of the aggregate contains mortar and cement paste. As a result, the aggregate quality diminishes and the use of recycled aggregate within structural concrete can cause difficulties. It is therefore necessary to obtain high quality recycled aggregate, which is not damaged during processing with optimal removal of mortar and cement paste from surface.

This research studies the potential use of recycled aggregate collected from technologies using the control breaking technique from high pressure pulsed power which should decrease the mortar content, limits the damage to aggregates, and minimises the powder content usually produced by fracture treatment. The research looked at coarse recycled aggregate (retained on a 5 mm sieve), and tested the quality for use in concrete.

EXPERIMENTAL SETUP

The high voltage Marx generator used in the study of water breakdown and waste water treatment is shown schematically in Figure 1. The high voltage pulses were obtained from a Marx generator with a variable number of stages. The stage capacitance was 0.22 μF, the stage series resistance was 1 Ω, and the inter-stage inductance was 100 μF. The charging voltage was constant peak output voltage in the range of 120 to 480 kV (positive polarity) with nearly current in a circuit load [3].

The recycled aggregate was examined for a range of impressed pulsed power for quality grade [4, 5].

Density and Absorption

The results of the density and absorption tests are shown in Table 1 and Figures 2 and 3. Surface dry density and absolute dry density was found to increase with the applied impressed pulse shots, and was similar to that of natural aggregate. This was due to the removal of cement/mortar from the aggregate surface by impact pulsed power. However, the difference between 80 times pulse shots density and 100 times pulse shots density was small in comparison with the results. Similar results were found for absorption values. The cause of this phenomenon was thought to be that 80 times pulse power recycled aggregate had little adhered mortar. On the other hand, the removed adherence mortar had a significantly different natural density and absorption, attributed to the damage by the impact of impressed pulse shots within concrete.

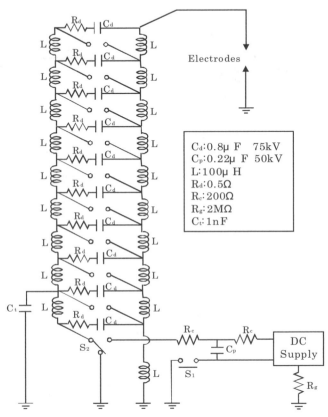

Figure 1   Circuit diagram of a Marx generator

QUALITY OF RECYCLED COARSE AGGREGATE

Table 1   Recycled aggregate for concrete class H

	RAW AGGREGATE	RECYCLED AGGREGATE FOR CONCRETE CLASS H
Surface dryness density, g/cm^3	3.06	-
Absolute dryness density, g/cm^3	3.04	2.5 <
Absorption, %	0.49	3.0 >

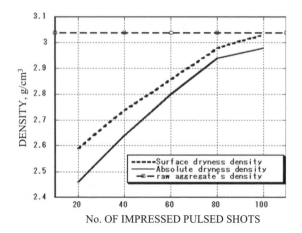

Figure 2　Density test results

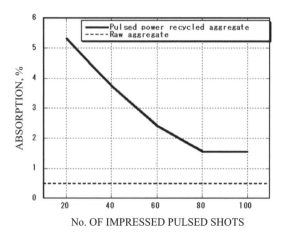

Figure 3　Absorption test results

However, pulsed power recycled aggregate meets the criteria in JIS A 5201 "Recycled aggregate for concrete class H" [6], so it is possible to maintain the aggregate quality by the impressed pulse shots being stopped before the aggregate suffers damage.

Screening Test

The results of the screening test of impressed pulsed shots are shown in Figures 4 and 5. It can be seen that 20 times impressed pulsed shots resulted in large grain size aggregates and 100 times impressed pulsed shots also produced a few large grain size aggregates. It is thought that the earlier stage of impressed pulsed shots leads to a lot of adherence mortar in aggregate, hence it results in an increase of grain size of aggregate. Increasing of impressed pulsed shots leads to removal of mortar from aggregate, decreases grain size and draws closer in terms of grain size to raw aggregate.

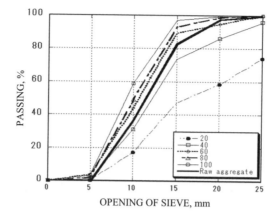

Figure 4   Grain distribution curves

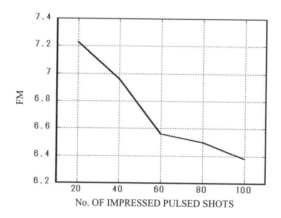

Figure 5   The finite modulus

Nevertheless, the comparison of grain size distribution between 80 times impressed pulsed shots and raw aggregate showed the former becoming reduced in grain size. This is for the impact of pulse shots causes damage and grain refining, similarly to the results of density test and absorption test.

The difference between the 80 times and 100 times impressed pulsed shots was that the former had over 45% retained mass on sieves (5 mm and 10 mm). According to JIS A 5021 "Recycled aggregate for concrete" having over 45% retained mass between adjacent sieves conforms to class H. Therefore, it cannot be used for recycled aggregate for concrete without adjusting the grain size at the time of use. Figure 6 shows the fineness modulus graph of 20, 40, 60, 80, 100 times impressed pulsed shots. This graph has two distinct characters for the primary stage and the late stage of pulsed shots. The line of primary stage is attributed to the separation of mortar the by impact of pulsed shots. The line of the late stage in turn relates to the impact of pulsed shots damaging the aggregate and degrading its quality. Degradation of aggregate is caused by unnecessarily high pulsed shots. Hence the point of intersection of the two lines is thought to be the most efficient for mortar removal by the impact of pulsed shots.

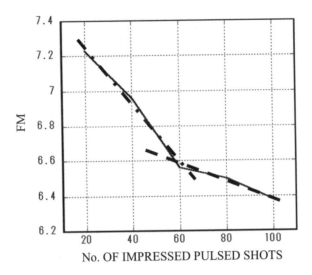

Figure 6 The finite modulus

The most efficient impressed pulsed shots can be selected by checking the fineness modulus prior to starting to collect the required character recycled aggregate.

Aggregate Strength Test

Aggregate strength of pulsed power recycled coarse aggregate was determined according to TS A 0006 [7]. The result of the aggregate strength test is shown in Table 2 and Figure 7.

Table 2 Aggregate strength test results

No. OF IMPRESSED PULSED SHOTS	50 TIMES	75 TIMES	100 TIMES	RAW AGGREGATE
Sample, g	2792	3080	3240	3326
2.5 mm or less, g	87	71	59	44
100 kN crushing value, %	3.1	2.3	1.8	1.3

It was found that aggregate strength of pulsed power recycled coarse aggregate was to increase with the increase of the number of impressed pulsed shots and was close to the value for raw aggregate. However, the 100 kN crushing value of 50 times impressed pulsed shots was more than twice than that of the raw aggregate. This factor was arrived at by visual observation of the recycled aggregate of 50 times for adhered mortar, which fell off from the aggregate by loading, and the separated mortar was accounted as the 100 kN crushing value. Therefore, the strength of aggregate became lower than that of the raw aggregate. From this result it follows that if the aggregate has a lot of adhered mortar at the primary stage of pulsed shots, it is thought that the loading separates adherence mortar from the aggregate before recycled aggregate crushes.

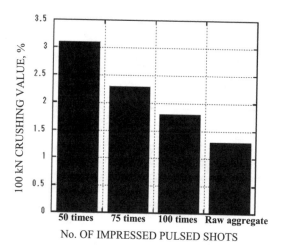

Figure 7 100 kN crushing value

It could be expected that aggregate cracks upon the increase in the number of pulsed shots, but from the comparison at 100 kN crushing value between 50 times, 75 times and 100 times, it was found that 100 kN crushing value of a lower number of impressed pulsed shots was higher than the value with a great number of impressed pulsed shots. This was explained by the effect of mortar separated by the impressed pulsed shots being greater than the effect of weakening of aggregate by the increase of the number of impressed pulsed shots. Therefore, the latter result showed that it was possible to separate only the adherence mortar and to measure the amount of adherence mortar by the control of the loading conditions.

GRADE QUALITY OF POWDER

Now, mass concrete was used as recycled aggregate by the grinding action. However, these methods generate a lot of powder and cost a lot due to the treatment of secondary waste. Hence, the powder produced during the treatment of concrete mass became a key problem in collecting recycled aggregate. In this research, the residue collected from grinding mass concrete by high power pulsed power was used. The quality of the recycled fine aggregate and powder product was also discussed in view of using the treated mass concrete residue as recycled fine aggregate. Impressed pulsed shots enable collecting aggregate by the control-breaking-technique facilitated by high voltage pulsed power and not by grinding action. As a result, it has the potential of not producing powder and can efficiently collect recycled fine aggregate. Also, recycled fine aggregate is defined as retained on the 0.15 mm sieve and completely passing the 10 mm sieve, whereas material finer than this is called powder.

Powder products

Powdered products were collected as the residue of pulsed power recycled aggregate in 25 times and 50 times impressed pulsed shots and the effect of impressed pulsed shots on the powder product was assessed. The results of the test are shown in Table 3 in comparison with the powder product obtained by friction action as reference [8].

Table 3   Aggregate strength results

No. OF PULSED SHOTS	25 TIMES	50 TIMES	FRICTION ACTION
Total mass, kg	60.45	60.40	1000
Powder product, kg	0.81	4.87	340
Powder product, %	1.34	8.06	34

The comparison of powder products between 25 times and 50 times impressed pulsed shots shows that the powder product obtained by the 50 times one is larger and tends to increase with the quantity of adhered mortar in aggregate, crushed cement paste mass and the proportion of aggregate crushed by the impact of pulsed shots. It appears that powder product is more likely to be produced by high-voltage pulsed shots compared to the product obtained by friction action.

Screening tests

The grade of the pulsed power recycled fine aggregate was assessed by the screening test and expressed by the grain size distribution. The results of the screening test are shown in Figures 8 and 9.

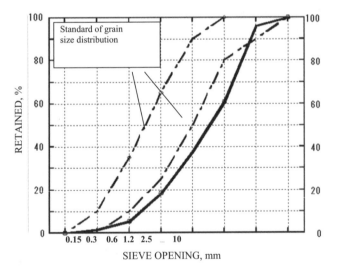

Figure 8   Grain size distribution of powder obtained by 25 times impressed pulsed shots

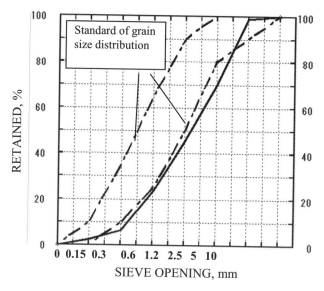

Figure 9   Grain size distribution of powder obtained by 50 times impressed pulsed shots

The comparison of the grain size distribution of recycled fine aggregate between 25 times impressed pulsed shots and 50 times impressed pulsed shots shows that the grain size distribution is similar, but the grain size distribution of the 50 times impressed one tends to shift to finer fractions. This is thought to be influenced by the crushed cement paste mass and the proportion of aggregate crushed by the impact of pulsed shots and can be monitored by the fineness modulus of aggregates obtained by 25 times and 50 times impressed pulsed shots.

These two grain size distributions occurred to exhibit below standard grain size distribution. Therefore, the grading of grains dictated to use this residue as a recycled fine aggregate. The cause of being below standard could be influenced by the mixed cement paste mass by impressed pulsed shots. Consequently, aggregate treated by impressed pulsed shots hardly produces powder, but produces mortar and cement paste as a mass instead. The quantity of mortar adhered to aggregate appears to be influenced by various grain sizes.

The character of pulsed power technology controls the breakage through the control of impressed conditions and the number of impressed pulsed shots. Therefore, pulsed power technology has a potential to produce high-quality fine aggregate by suitable settings of electrical pulses resulting in fine aggregate and adherence mortar.

Density and absorption tests

Density and absorption tests of pulsed power recycled fine aggregate were carried out to measure grade quality. The results are shown in Table 4. It was found for the 50 times impressed pulsed shot aggregate that its absorption increased and its density decreased compared to the results with the 25 times impressed pulsed shot aggregate. This result shows the decrease of quality as a fine aggregate. In the case of recycled coarse aggregate, an increase of quality was confirmed by the increase of the number of pulsed shots that caused the mortar separating from the aggregate and the cement paste mass fraction. However, for recycled fine aggregate, decreased aggregate quality can be explained by the damage of aggregate upon the increased number of pulsed shots, the cement paste mass and the mortar separated from aggregate as collected with the recycled fine aggregate.

On comparing current test results with the standard recycled aggregate classes, the recycled fine aggregate can be classified as recycled aggregate for concrete class H and class M [9]. Within class L of recycled aggregate for concrete [10], the recycled fine aggregate of 25 times impressed pulsed shots conformed in terms of absorption, but this may not mean that high-quality recycled aggregate was collected.

From these results, the collected recycled fine aggregate appears to be but high-quality. This was attributed to mortar and cement paste mass being mixed with the collected fine aggregate. It may be necessary to separate mortar and cement paste mass, and collect fine aggregate for quality recycled fine aggregate to achieve improvement. The effects of pulsed power on the physical properties of mortar, cement paste mass and fine aggregate need to be looked at in the future. Only separate collection of mortar, cement paste and fine aggregate has the potential of obtaining high-quality recycled fine aggregate.

Table 4 Density and absorption results of pulsed power recycled fine aggregate

No. OF IMPRESSED PULSED SHOTS	ABSORPTION, %	SURFACE DRY DENSITY, g/cm^3	ABSOLUTE DRY DENSITY, g/cm^3
25 times	11.78	2.28	2.04
50 times	13.74	2.23	1.96
Recycled aggregate for class H	3.5>	—	2.5<
Recycled aggregate for class M	7.0>	—	2.3<
Recycled aggregate for class L	13.0>	—	—

CONCLUSIONS

This research assessed the possibility of decreasing the amount of mortar in recycled concrete aggregate by using a new technique which in turn would not damage the aggregate. The conclusions are as follows:

(1) From the result of the density and absorption test of every 20 times pulsed shots it was found that the quality of recycled coarse aggregate improved when the number of impressed pulsed shots increased, and the pulsed power recycled coarse aggregate of 60 times impressed pulsed shots met the standard quality.

(2) The grain refining of aggregate caused cracks and subsequent damage to the aggregate when the number of impressed pulsed shots increased.

(3) The fineness modulus of recycled aggregate has a potential to evaluate clear-cut distinction between the stages of mortar separation and crack development by impressed pulsed shots.

(4) From the aggregate strength test result it was found that the strength decreased by the increased number of impressed pulsed shots. The quantity of adhered mortar can be controlled by load conditions, because the adhered mortar is preferentially separated from aggregate by the load.

(5) The quantity of powder product obtained when aggregate was collected by impressed pulsed shots was little compared to existing recycled aggregate collection techniques.

(6) The fineness modulus of recycled fine aggregate of 25 times and 50 times impressed pulsed shots fell outside the standard grain order hence requiring size control.

(7) Recycled fine aggregate was found to be refining similarly to recycled coarse aggregate by the increase of the number of impressed pulsed shots.

(8) The density and absorption of collected recycled fine aggregate of 25 times and 50 times impressed pulsed shots conformed to recycled aggregate for concrete class L, but its quality was low.

In conclusion, under impressed pulsed shots conditions of $0.4\,\mu F$ and $400\,kV$, the recycled coarse aggregate suffered low damage whilst removing the low adherence mortar using 50~60 times impressed pulsed shots. Applying less than 50~60 times, the aggregate retains some mortar, whereas with more than 50~60 times, the collected aggregate becomes damaged and its quality would not improve by increased number of impressed pulsed shots.

Recycled fine aggregate became of low quality by the cement paste mass and mortar mass collected with the fine aggregate.

Therefore, the problem to solve remains as follows:

(1) The aggregate suffers damage and grain refining by the increased number of impressed pulsed shots. It is however possible to protect aggregate from degradation by efficiently impressed conditions and by the number of impressed pulsed shots, therefore it is necessary to stock data for the impressed conditions prior to collection of aggregates.

(2) It is necessary to develop a technique for separating and collecting recycled coarse aggregate and recycled fine aggregate.

In the future, the research aims to generate high-quality recycled aggregate, seek optimized solutions of friction concrete under other impressed conditions, and investigate into the dependence of material characteristics in friction concrete.

REFERENCES

1. MINISTRY OF LAND, INFRASTRUCTURE AND TRANSPORT, Amount of factory waste emission, 2002.

2. MINISTRY OF LAND, INFRASTRUCTURE AND TRANSPORT, Amount of factory waste emission, 2005.

3. LISITSYN V, NOMIYAMA H, KATSUKI S AND AKIYAMA H, "Thermal processes in a Streamer Discharge in Water".

4. JAPANESE INDUSTRIAL STANDARDS COMMITTEE: JIS A 1110 Method of test about density and absorption of coarse aggregate, 2002.

5. JAPANESE INDUSTRIAL STANDARDS COMMITTEE: JIS A 1102 Method of screening test.

6. JAPANESE INDUSTRIAL STANDARDS COMMITTEE: JIS A 5021 Recycled aggregate for concrete class H, 2005.

7. JAPANESE INDUSTRIAL STANDARDS COMMITTEE: TS A 0006 Concrete used as recycled aggregate, 2003.

8. MITSUBISHI MATERIAL GROUP, Development of environment loading reduction type cement from concrete waste.

9. JAPANESE INDUSTRIAL STANDARDS COMMITTEE: JIS A 5022 Recycled aggregate for concrete class M, 2006.

10. JAPANESE INDUSTRIAL STANDARDS COMMITTEE: JIS A 5023 Recycled aggregate for concrete class L, 2006.

POWDERED GLASS CULLET AS LATENT HYDRAULIC ADDITION FROM WASTE GLASS BOTTLES

K Yamada

A Satoh

S Ishiyama

Akita Prefecture University

Japan

ABSTRACT. Wider application of waste glass containers is required. One of the most prospective ways of using glass cullet originated from glass containers is material for some types of addition for concrete. The authors have been conducting many experiments including accelerated test methods for ASR in an attempt of applying recycled glass cullet as addition in concrete. From the research, it was found that ground powder glass under 0.075 mm particle diameter has two opposite effects to strength of hardened mortar specimens; one is an increasing effect to strength by latent hydration of glass, and the other is a decreasing effect to strength from air entrainment by glass powder while mixing. The most effective ratio for substitution of ground glass powder for cement in mortar is 10% of cement.

Keywords: Glass cullet, Ground powder glass, Cement substitute, Alkali-silica reaction, ASR, Pessimum size, Sustainable construction, Recycling.

Dr K Yamada, a Certificated Architectural Engineer and professor at the Department of Architecture and Environmental Engineering at the Akita Prefecture University. He has been in charge of the Materials Laboratory that includes research and development related to fracture and ductility of fibre reinforced concrete, durability of concrete and other building materials, wooden structures, life-cycle assessment and sustainable construction. His research interests cover many aspects of design, construction and assessment of building in an environmentally friendly way towards sustainable development.

Ms A Satoh, an aspiring graduate student whose research field is mechanical and chemical behaviour of concrete including ASR. She is especially interested in the field of interface characteristics of concrete, such as improvement of construction joints in concrete and crack propagation in concrete.

Mr S Ishiyama, a research associate at the Department of Architecture and Environmental Engineering at the Akita Prefecture University. He received his MSc in engineering from the Tohoku Institute of Technology. His research interests include durability of fibre reinforced concrete, permeability of concrete and recycling of waste materials in concrete.

INTRODUCTION

Glass is commonly used for various containers, window panes, video displays and light bulbs. Although recycled glass cullet comprises the major raw material for glass itself, nearly one million tons of glass has been allocated to landfill in Japan alone. Though this waste glass is considered to be good building material after the legislation of recycling law, there have not been good applications in the field of concrete. One of the most prospective ways of using glass cullet is addition in concrete as crushed glass cullet (GC) and ground powder glass (GPG).

The literature reveals the advantages and disadvantages of glass cullet used for concrete aggregate and other applications [1, 2]. It is well known that the most significant problem for glass in concrete is alkali-silica reaction (ASR) that causes detrimental expansion. Nevertheless, it is very interesting and useful to know that the use of powdery glass cullet (PGC) and GPG do not cause detrimental expansion [3-5]. Also GPG is known to have an ability for pozzolanic reaction in combined use of high-alkali cement [3-5].

Authors have been examining many aspects of specimens that contain glass in concrete, many of which projects focus on porosity distribution of the specimens to examine the mechanism which causes expansion or increases (or decreases) the strength of specimens [6-8]. In this research, the authors employed mortar-bar test method (JIS A 1146) for the purpose of finding an effective use of GPG through examining expansion, strength and porosity distribution of the specimens.

EXPERIMENTS

The authors conducted experiments with observing JIS's mortar bar test method (JIS A 1146), in which two sets of experiment were conducted. One is an experiment to detect the precise size of GC that does not cause detrimental expansion (called GC test), and the other is an experiment to reveal the performance of GPG for admixture for concrete (called GPG test). In the former experiment, single-size or mixed-size glass cullet of two colours (green and clear) was used. In the experiment, green GPG under 75 μm was used and clear PGC under 150 μm was used as a reference.

Materials and Mix Proportions

The cement used was ordinary Portland cement in which the measured total alkali was 0.65%. The glass cullet originated from waste glass containers was gathered by Akita municipality, and crushed with impact crusher. The GC was washed, dried and sieved into the following 8 grades; 2.36, 1.18, 0.85, 0.6, 0.425, 0.3, 0.212 and 0.15 mm. The name of GC is called after the mesh size of the sieve on which the GC was retained in this study. Usually 5 grades from 2.36 mm to 0.15 mm were used, but 8 grades were used to reveal the pessimum size in more precise sizes. The part passing through the 0.15 mm sieve is PGC, and then the green PGC was furthermore ground by ball mill. After it was sieved with a 75 μm-sieve, the passing-through part was GPG for specimens. Figure 1 represents the grain size distribution of cement, clear PGC and green GPG measured with a laser diffraction particle size analyser.

Table 1 shows the mix proportions for GC test. The test is based on the mortar bar test method (JIS A 1146), in which the glass content is 100% and the aggregate-to-cement ratio is 2.25. NaOH solution was prepared to meet the requirement of total alkali in specimens, which is 1.2% in Na_2O equivalent to cement weight, making the amount of solution 300g (Na_2O: 0.387mol/l). Along with single-size cullet specimens, mixed-size cullet specimens were also made, observing the mix proportion of different size GC stipulated in JIS. The number of specimens tested was 3 for each mixture, amounting to 54 in total.

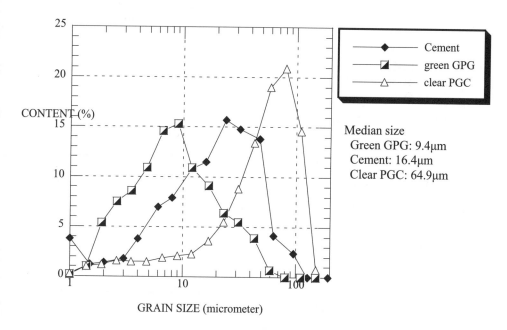

Median size
Green GPG: 9.4μm
Cement: 16.4μm
Clear PGC: 64.9μm

GRAIN SIZE (micrometer)

Figure 1   Grain size of cement and green GPG measured with
laser diffraction particle size analyser

Table 1   Mix proportions of specimens for GC test for evaluation of size-dependent ASR

NAME	AGGREGATE (CLEAR OR GREEN)				CEMENT	NaOH SOLUTION
	Size	Weight	Size	Weight		
	mm	g	mm	g	g	g
0.15	0.15	1350			600	300
0.212	0.212	1350			600	300
0.3	0.3	1350			600	300
0.425	0.425	1350			600	300
0.6	0.6	1350			600	300
0.85	0.85	1350			600	300
1.18	1.18	1350			600	300
2.36	2.36	1350			600	300
Mixed	0.15	202.50	0.6	168.75	600	300
	0.212	168.75	0.85	168.75		
	0.3	168.75	1.18	168.75		
	0.425	168.75	2.36	135.00		

Table 2   Mix proportions of specimens for GPG test for evaluation of cement substitution by green GPG with reference by clear PGC

NAME	SAND	BINDER		CEMENT	NaOH
		PGC	GPG		
		Clear	Green		SOLUTION
	g	g	g	g	g
SR	1350	0	0	600	360
SW1	1350	60		540	360
SW3	1350	180		420	360
SW5	1350	300		300	360
SG1	1350		60	540	360
SG3	1350		180	420	360
SG5	1350		300	300	360

Table 3   Evaluated issues in GC test and GPG test

TEST	CONDITIONS	BEHAVIOUR		STRENGTH		POROSITY
		Expansion	Weight Increase	Compressive	Bending	
GC	Mortar Bar Method	x	x			
GPG	Mortar Bar Method	x	x	x	x	x
	Autoclave Cure			x	x	
	Standard Cure			x	x	

Table 2 shows the mix proportion for GPG test. The test method was also based on JIS A 1146, but the cement was substituted for GPG or PGC from 0% to 50%. The quantity of NaOH solution was increased by 20% to acquire adequate workability for the mixture, because GPG and PGC was so fine that workability worsened, but total alkali was the same as stipulated in JIS. The number of specimens tested was 3 for each mixture (21 in total).

Specimens and Test Procedures

The specimens had a dimension of 40mm x 40mm x 160mm. After being cast in moulds, the specimens were cured in saturated humid condition (95% RH) at 20 degrees C for 1 day and then the first measurement was done. The curing conditions, procedures and methods for the measurement of mortar bar test are strictly stipulated in JIS, which the authors observed. After the first measurement, JIS A 1146 stipulates that the specimens should be placed in a saturated humid storage chamber whose temperature is 40 degrees C for 6 months. The measurement should be conducted at an interval of 1 month after two measurements done at 2 weeks and 1 month since demoulding. To compare the strength in GPG test, autoclave cure at 160 degrees C for 8 hours was conducted, which is a major commercialized process for concrete industry. Also standard water curing at 20 degrees C in water was conducted as reference. Table 3 is a summary of the entire test conducted in this research, in which v indicates the conducted ones.

RESULTS AND DISCUSSIONS

ASR Expansion Ratio from GC Test

Figure 2 and Figure 3 represent expansion ratios as a function of GC size. The pessimum size for clear cullet is 0.6 mm at the early period and then shifting to 0.85 mm at later. The pessimum size for green cullet is one size larger than that of the clear one. This result is not consistent with previous studies [9] which indicate the same pessimum size throughout the whole period, but the previous studies were based on the results by a coarser pitch system of sieve sizes, it means 2.36 - 1.18 - 0.6 - 0.3 - 0.15 mm (5 grades). This study is based on the results by a finer pitch system of sieve sizes with internal sieve sizes like 2.36 - 1.18 - 0.85 - 0.6 - 0.425 - 0.3 - 0.212 - 0.15 mm (8 grades). So precisely speaking, there is a slight shift of pessimum size at the later period, and the pessimum size of clear cullet specimen is different from that of the green one. JIS A 1146 declares that if the expansion ratio is smaller than 0.1%, the aggregate is not detrimental. Then it should be noted that the cullet under 0.3 mm (clear cullet) or under 0.425 mm (green cullet) would not cause harmful expansion in concrete, if such single-size cullet was used.

Expansion Ratio and Weight Increase from GPG Test

Figure 4 represents expansion ratios against the curing duration. As is expected from Figure 2 and 3, the expansion is very small being almost the same as that of SR (reference), and is far from the detrimental value. Though the values are small, the larger the added GPG or PGC amount, the larger the expansion ratio. This experimental result agrees with the results from previous study evaluated with specimens in which GC was mixed [7-9]. Figure 5 represents the weight increase against the curing duration. There are two inconsistent tendencies of weight increase observed here. The clear PGC causes the proportional weight increase to the added PGC, while green GPG produces an opposite result. It means that the larger the added green GPG amount, the smaller the weight increase, proportional only to the mixed weight of cement. The authors do not rule out that green GPG was used for hydration, judging from the results of strength (Figure 6-7) and porosity (Figure 8-9). Then the possible cause of the results is that clear GPG produced water-absorptive gel in proportional to the content, while green GPG did not produced such gel. The reason for it is unknown, and the mechanism should be investigated with more experiments.

Compressive and Bending Strength from GPG test

Figure 6 represents the compressive strength of mortar bar specimens cured for 1 day, 2 weeks, 4 weeks and 90 days. As for clear PGC, strength decreases according to the PGC substitution, which derives from the decrease of cement in the mixture. However, green GPG achieves almost the same strength at 10% substitution (SG1) and maximum strength at 30% substitution (SG3), exceeding that of SR. As we saw in Figure 5, the weight increase ratio was lower than that of SR, but the strength of the substituted ones was the same or even higher than that of SR.

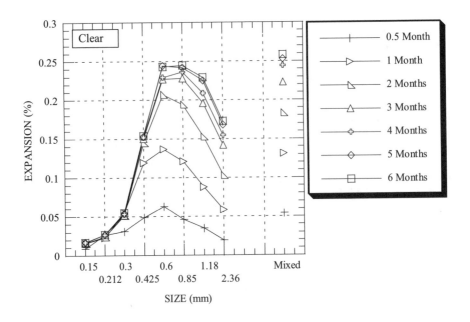

Figure 2   Expansion of mortar bar specimens with different sizes of
clear GC aggregate tested under JIS A1146.

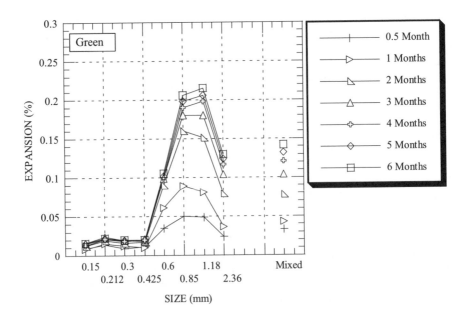

Figure 3   Expansion of mortar bar specimens with different sizes of
green GC aggregate tested under JIS A1146.

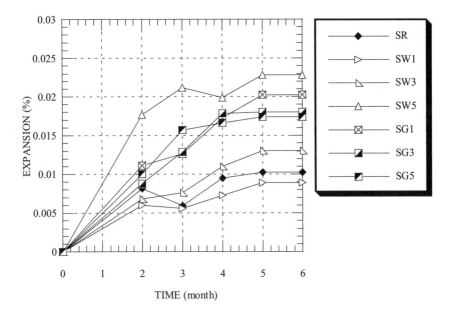

Figure 4   Expansion of mortar bar specimens with green GPG or
clear PGC tested under JIS A1146.

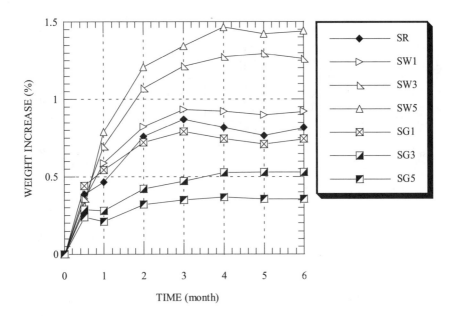

Figure 5   Weight increase of mortar bar specimens with green GPG or clear PGC
cured under JIS A1146 condition.

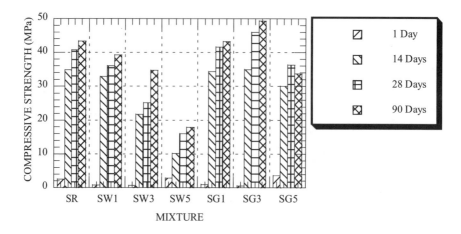

Figure 6   Compressive strength of mortar bar specimens with green GPG or clear PGC cured under JIS A1146 condition for different curing duration.

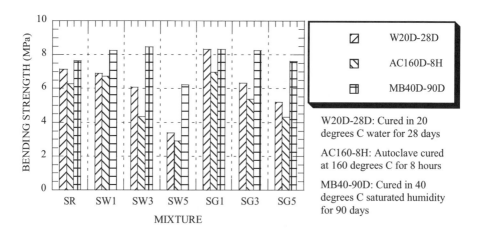

Figure 7   Bending strength of mortar bar specimens with green GPG or clear PGC tested in three different curing conditions.

Figure 7 represents the bending strength of mortar bar specimens whose curing conditions are different. The strength is dependent on maturity, as is expected, and then the strength in the MB40D-90D case is the highest. It is notable that SG1 is the highest in each curing case exceeding SR. In the case of W20D-28D (20 degrees C, 28 days), temperature and duration is not thought to be sufficient for ASR caused by GPG even when the particle size is under 75 μm.

The mechanism with which green GPG has a strengthening effect has not been investigated yet. But it would be possible to say that the small size well under 75 μm (See Figure 1) worked as a latent hydraulic admixture that dissolved in NaOH solution even in the case of W20D-28D. Figure 6 shows that the 1 day compressive strength for SG1 and SG3 is lower than that of SR, but the 14 day strengths for both mixtures are higher than that of SR. This fact also reveals that GPG worked as a latent hydraulic addition that needs more time for hydration than cement.

Porosity Distribution from GPG Test

Figure 8 represents porosity distribution of specimens immediately after demoulding. There is a sharp peak near 1 μm for the SW5 and SG5 mixtures, and a small peak for the SR mixture. This indicates that GPG has a strong air-entraining effect, especially in the region of very small pore sizes. It is well known that chemical air-entraining admixtures (liquid type) bring about larger pores well over 10 μm. Smaller pores are very advantageous because these have little detrimental effect to strength and improving effect for freezing-and-thawing resistance making the pore spacing smaller.

Figure 9 represents the porosity distribution of specimens after 3 months of accelerated curing under JIS A 1146. The sharp peak for the SW5 and SG5 mixtures observed immediately after demoulding diminishes and shifts to 0.1 ~ 0.01 μm. This small pore range is generated from the original large pores (1 μm) after being packed with hydrated CSH gel. Therefore it is obvious that the total porosity (an area between the x-axis and the plotted graph) for SW5 and SG5 mixture after 3 months is also higher than that of SR, judged from the results obtained immediately after demoulding.

Relationship between Strength and Total Porosity

Figure 10 represents the bending and compressive strength of specimens as a function of total porosity after 3 months of accelerated curing stipulated under JIS A 1146. There is a strong relationship between compressive strength and total porosity, whereas very weak relationship between bending strength and porosity.

From Figure 10, it is noted that the compressive strength of SG3 and SG5 is well above the regression line which suggests some positive effect other than latent hydraulic by green GPG on compressive strength. It is also observed that SG1 and SG3 have almost the same porosity as SR, which is compatible with the well known fact that the same porosity causes the same strength.

To summarize, green GPG has two opposite effects to strength of hardened mortar specimens; one is an increasing effect to strength by latent hydration of glass, and the other is a decreasing effect to strength from air entrainment by glass cullet while mixing. As for clear PGC, increasing effect of strength would be very small due to its coarser size than GPG, and then the decreasing effect by air entrainment would have worked.

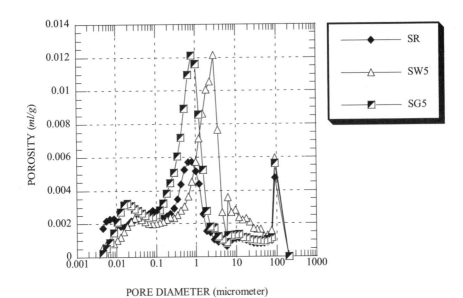

Figure 8   Porosity distribution of specimens immediately after demoulding.

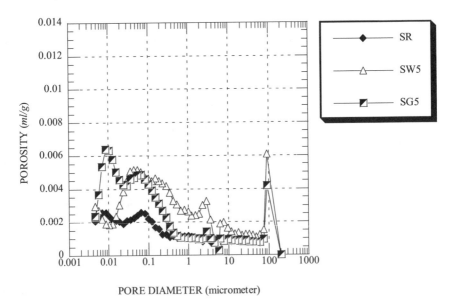

Figure 9   Porosity distribution of specimens after 3 months of accelerated curing under JIS A 1146 mortar bar method.

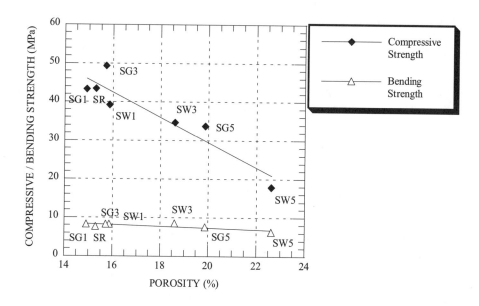

Figure 10   Bending and compressive strength of specimens as a function of total porosity after 3 months of accelerated curing under JIS A 1146.

CONCLUSIONS

The authors conducted experimental research in an attempt of finding new applications for waste glass containers as addition in concrete. The glass used was glass cullet ranging from 2.36 to 0.15 mm, ground powder glass (GPG) whose median diameter was smaller (9.4 μm) than that of ordinary cement, and powdery glass cullet (PGC) whose median diameter was larger (64.9 μm) than that of ordinary cement.

The findings are as follows;

[1] The cullet under 0.3 mm (clear cullet) or under 0.425 mm (green cullet) would not cause detrimental expansion in concrete.

[2] GPG and PGC have two effects on mortar specimens; latent hydration of glass and air entrainment by glass.

[3] Green GPG had large effect of latent hydration and small effect of air entrainment, which enhanced compressive strength, whereas clear PGC had large effect of air entrainment and small effect of latent hydration, which could not enhance compressive strength.

REFERENCES

1. DHIR, R. K., LIMBACHIYA, M. C., AND DYER, T. D. eds. Recycling and Reuse of Glass cullet, Thomas Telford, 2001.3, 292 pp.

2. LIU, T. C. AND MEYER, C. eds. Recycling Concrete and Other Materials for Sustainable Development, ACI, SP-219, 2004, 164 pp.

3. DYER, T. D. AND DHIR, R. K. Use of Glass Cullet as a Cement Component in Concrete, Recycling and Reuse of Glass cullet, Thomas Telford, 2001.3, pp. 157-166.

4. DYER, T. D. AND DHIR, R. K. Chemical Reactions of Glass Cullet Used as Cement Component, Journal of Materials in Civil Engineering, Vol. 13, No. 6, 2001.12, pp. 412-417.

5. SHAO, E., LEFORT, T., MORAS, S. AND RODRIGUEZ, D. Studies on Concrete Containing Ground Waste Glass, Cement and Concrete Research, Vol. 20, 2000. pp. 91-100

6. YAMADA, K. AND ISHIYAMA, S. A discussion on ASR Expansion Characteristics of Mortar Made of Waste Glass Cullet Aggregate, Journal of Structural and Construction Engineering, No. 607, 2006.9, pp. 1-6. (to be printed)

7. YAMADA, K. AND ISHIYAMA, S. Maximum Dosage of Glass Cullet as Fine Aggregate in Mortar, Proceedings of the International Conference "Achieving Sustainability in Concrete", 2005.7, pp. 185-192.

8. YAMADA, K., KATO, E., SATOH, A. AND ISHIYAMA, S. Recycling of Waste Bottle Glass Cullet as Fine Aggregate, Proceedings of the International fib Conference, 2006.7, pp. 1-9 in Session 18 on CD-ROM.

9. YAMADA, K., ABE, Y. AND ISHIYAMA, S. ASR Expansion Caused by Glass Cullet under Rapid Test Method, Proceedings of Sustainable Waste Management and Recycling: Glass Waste, 2004.9, 109-116.

SULFATE RESISTANCE OF COMPOSITE PORTLAND CEMENTS CONTAINING STEEL SLAG AND GRANULATED BLASTFURNACE SLAG

O Özkan

I Yüksel

Zonguldak Karaelmas University

Turkey

ABSTRACT. The objective of this study is to investigate sulphate resistance of composite Portland cements produced with basic oxygen furnace (BOF) slag and granulated blast-furnace slag. An experimental study was conducted for this purpose. Ground granulated blast-furnace slag (GGBFS) and BOF slag were mixed in selected ratios with clinker-gypsum and composite cement was produced. Three types of cement were produced. The first series contains GGBFS, the second contains BOF slag, and the third series contains 60% GGBFS and 40% BOF slag with clinker-gypsum. The slag replacement ratios are selected as 20, 40, and 60% of clinker by weight in all series. Some physical characteristics and sulphate resistance of these series are tested. Test results showed that sulphate resistance of mortar specimens produced by using these cements is gradually increased with respect to reference cement as the replacement ratios are increased. The maximum increase measured in sulphate resistance was 7% for sodium sulphate, and 3% for magnesium sulphate.

Keywords: Blast furnace-slag, Composite cement, BOF steel slag, Sulphate resistance.

Dr Ömer Özkan was born in Ankara, Turkey in 1975. He graduated from the Yıldız Technical University, Department of Civil Engineering (1996), MSc and PhD at Gazi University, Graduate School of Natural Applied Sciences (2003). He is working as an assistant professor at the Zonguldak Karaelmas University, Alaplı Technical Vocational School since 2003. He has published 12 refereed papers in international and national journals, 19 proceedings papers in international and national congresses and symposia. He is working on Construction Materials (Industrial waste, solid waste, cement and concrete).

Dr İsa Yüksel is assistant professor of civil engineering at The Zonguldak Karaelmas University where he has been engaged in structural and construction materials research and teaching for 15 years. His research and teaching have encompassed a range of topics in analysis and design of concrete structures, and utilization of industrial wastes in concrete. He earned his M.Sc. and PhD. degrees in the structures division of civil engineering from Yıldız Technical University after completing his B.Sc. degree at Middle East Technical University (METU).

INTRODUCTION

The environmental regulations, requiring waste disposal minimisation, force the reuse of waste materials. Solid wastes of iron and steel factories can be used as raw material in other industries such as cement and concrete sectors. In Europe, it is estimated that 12 million tons of basic oxygen furnace (BOF) steel slag is generated annually [1]. Also, the European Community (EU) has declared targets to protect the environment and to guarantee a cautious and efficient use of natural resources. Solid wastes should be reused in order to use natural resources efficiently and for sustainable development.

The use of recycled waste cementitious materials is becoming of increasing importance in construction practice. Such materials, including ground granulated blast-furnace slag (GGBS), can be incorporated in cementitious materials to modify and improve certain properties for specific uses, to conserve non-renewable natural resources and to utilize industrial by-products. The possibility of being able to recycle or process materials to use as a partial replacement for cement in mortar or concrete, or to stabilize soils, has great potential economic benefits in all areas of the construction industry [2]. Also, durability characteristics of concrete containing blast furnace slag have been investigated by many researchers. Alkali silica reactivity, resistance to chloride penetration, freezing and de-icing salt resistance can be shown as examples of such characteristics [3, 4, 5, 6, 7]

BOF slag is a by-product produced during the conversion of iron ore or scrap iron to steel. It is composed of silicates and oxides of unwanted elements in the chemical composition of steel. The mineralogical composition of steel slag, changes with its chemical composition. Olivine, merwinite, calcium silicates (C_2S, C_3S), C_4AF, C_2F, CaO–FeO–MnO–MgO in solid solution (RO phase) and free CaO are common minerals in steel slag [8]. Shi and Qian found that the free CaO content increased the basicity of the steel slag that increased the reactivity of the steel slag [9]. However, high free CaO content in steel slag has been shown to produce volume expansion problems [10, 11].

Portland cement-based materials subjected to sulphate attack may suffer from two types of damage: loss of strength of the matrix due to the degradation of calcium-silicate-hydrate (C-S-H), and volumetric expansion due to the formation of gypsum or ettringite that leads to cracking. Loss of strength has been linked to decalcification of the cement paste hydrates upon sulfate ingress, especially C-S-H, while cracking and expansion is attributed to the formation of expansive compounds [12].

In this study, GGBFS and BOF steel slag is used to partially replace clinker to produce composite Portland cement. Some physical and mechanical properties of the composite cement and sulphate resistance of mortars produced with these cement were investigated.

EXPERIMENTAL DETAILS

Materials

Clinker and gypsum used in this study were provided from a local cement factory. BOF slag and GBFS were provided from Ereğli Iron and Steel Works Company in Turkey. Chemical properties of these four materials are shown in Table 1. Standard sand was used in order to manufacture mortar specimens.

Table 1   Chemical compositions of clinker, gypsum, GGBFS and BOF slag (wt. %)

MATERIALS	CaO	SiO_2	Fe_2O_3	Al_2O_3	MgO	SO_3	LOI
GBFS	37.80	35.10	0.70	17.54	5.50	0.70	1.08
BOF slag	58.53	10.72	15.30	1.71	4.27	0.04	1.11
Clinker	66.11	21.57	3.17	5.09	1.74	1.35	0.77
Gypsum	32.57	0.67	0.24	0.21	2.20	46.56	22.98

LOI: Loss On Ignition

Procedure

Ground granulated blast-furnace slag (GGBFS) and ground BOF slag were mixed in selected ratios with clinker-gypsum and composite cement was produced. Three types of cement were produced. The first series contains GGBFS, the second contains BOF slag, and the third series contains 60% GBFS and 40% BOF slag with clinker-gypsum.

Table 2   Composition of cement mixtures

CODE	MATERIALS	CLINKER %	GYPSUM %	GGBFS %	BOF %
C	100 % Clinker (Reference)	95	5	0	0
G1a	80 % Clinker-Gypsum + 20 % GGBFS	76	4	20	0
G1b	60 % Clinker-Gypsum + 40 % GGBFS	57	3	40	0
G1c	40 % Clinker-Gypsum + 60 % GGBFS	38	2	60	0
G2a	80 % Clinker-Gypsum + 20 % BOF slag	76	4	0	20
G2b	60 % Clinker-Gypsum + 40 % BOF slag	57	3	0	40
G2c	40 % Clinker-Gypsum + 60 % BOF slag	38	2	0	60
G3a	80 % Clinker-Gypsum + 12 % GGBFS + 8 % BOF slag	76	4	12	8
G3b	60 % Clinker-Gypsum + 24 % GGBFS + 16 % BOF slag	57	3	24	16
G3c	40 % Clinker-Gypsum + 36 % GGBFS + 24 % BOF slag	38	2	36	24

The slag replacement ratios are selected as 20, 40, and 60% of clinker by weight in all series. Table 2 illustrates the composition ratios of the mixtures used in our study. The series are divided into groups and symbolized by suffixes (a, b, c) with respect to their varying ratios in compositions; for instance the code "S3c" denotes a material that is composed of 40% clinker-gypsum, 36% GGBFS and 24% BOF slag. BOF slag, gypsum, and GBFS were supplied in granulated condition. These materials were first ground in a mill separately in the laboratory. Clinker, GBFS, and BOF slag were ground for 3, 4, and 3 hours respectively. Clinkers, gypsum, GBFS, and BOF slag were ground till the fineness of the material was 2400-2500 cm^2/g. Then they were mixed in proportions as shown in Table 2, and ground again till the fineness of the mix was 3100–3300 cm^2/g.

Mortar specimens were produced in accordance with the Turkish Standard TS EN 196-1 [13]. These were 40x40x160 (width x height x length) mm in dimensions and they were prepared with the cement mixtures given in Table 3, cement, standard sand and tap water with the proportions of 1, 3 and 0.5 respectively. Specimens were first cured in a fog room at 20°C for 24 hours, and then demolded and cured in water at 20±3°C until the 7th day. Some physical properties of the cements produced are examined. Percentages of cement particles remained on the 32 and 90 μm sieves, specific surface values, and specific weights were determined according to the procedure described in the TS-EN 196-6 standard [14].

Mortar specimens are examined for their sulphate resistance. The mortar specimens are kept in water at 20±3°C until the 7th day after being held in the curing room for one day and being taken out of the moulds. At the end of the 7th day, three samples are kept in a solution of 4% Na_2SO_4 and three samples in 4% solution of $MgSO_4$ until the end of the 28th day. The specimens held in the solutions are tested for their compressive strength at the end of the 28th day. For the 28 day mortar specimens, a compression machine controlled by a computer is used to determine resistance to sulphates. The loading speed is selected to be 1 kN/s in the measurements of compressive strength.

RESULTS AND DISCUSSIONS

The physical characteristics of cements produced are shown in Table 3. GBFS is harder to grind than BOF slag. GBFS and BOF slag were ground 4 and 3 hours, respectively, to become of the same specific surface area. There are similarities between specific surface area values and the results of the sieve analysis of the cement produced in the laboratory. As can be seen from Table 3, GGBFS and BOF slag replacement decreases the specific weight of the cement. 5% decrease, the maximum decrease, is observed for the 60% replacement ratio with respect to reference cement. This decrease shows the difference of GBFS and BOF slag from clinker. There is no difference between GGBFS and BOF slag related with the decrease of specific weight. Table 4 lists compressive strength values for the mortar specimens cured in pure water for reference, and cured in sulphate solutions for the resistance of sodium sulphate and magnesium sulphate.

When cement-based materials are exposed to sodium sulphate attack, gypsum and ettringite are produced which can cause expansion in concrete. Formation of gypsum plays an important role in the damage of the material. Gypsum results in softening of the material. There is a close relationship between the $Ca(OH)_2$ content and gypsum formation [15]. Ettringite formation results in cracking and expansion of the material. Expansion is related to the water absorption of crystalline ettringite [16, 17]. The presence of a pozzolanic material results in an increase in the resistance to sodium sulphate attack. On the other hand, effectiveness of pozzolanic material against sulphate attack is dependent on the maximum temperature reached during the producing of the pozzolan. Wild et al. concluded that the calcination temperature for clay below 900°C produces a marked loss in sulphate resistance when the pozzolanic product partially replaces cement in mortar. Pozzolan materials prevent the harmful effect by binding $Ca(OH)_2$. Some results supporting this point of view were obtained in our study and it was observed that all substitution materials increased the durability of concrete exposed to Na_2SO_4 solution [18].

Table 3   Physical characteristics of cements

CODE	FINENESS (wt. %)		SP SURFACE (cm^2/g)	SP GRAVITY
	>32 μm	>90 μm		
C	21.00	0.90	3330	3.12
G1a	21.20	1.10	3115	3.05
G1b	21.90	1.15	3108	3.01
G1c	21.80	1.15	3090	2.95
G2a	21.15	1.18	3214	3.06
G2b	22.10	1.00	3213	3.02
G2c	22.15	1.25	3152	2.97
G3a	22.25	1.25	3138	3.05
G3b	22.50	1.45	3028	3.01
G3c	22.55	1.55	3055	2.96

Table 4   Compressive strength (f_c) of reference and other specimens exposed to sulphate attack

CEMENT	F_C REFERENCE SPECIMEN (MPA)	F_C SPECIMENS EXPOSED TO SULPHATE ATTACK (MPA)	
	Water	Na_2SO_4	$MgSO_4$
C	37.3	33.9	34.7
G1a	37.7	37.3	35.0
G1b	32.4	34.0	30.4
G1c	29.9	32.0	29.3
G2a	28.6	28.0	27.4
G2b	27.9	28.7	27.0
G2c	24.3	25.4	23.8
G3a	31.6	31.9	30.3
G3b	31.3	30.4	30.7
G3c	30.7	32.5	29.4

The mortar specimens affected by sulphate are also tested for their residual compressive strength. In this residual compressive strength measurements the samples cured in sulphate solution are compared to those cured in water. Sulphate resistance is computed by the formula P_s/P_w, here P_s indicates the compressive strength for the sample cured in sulphate solution and P_w for the sample cured in water. Figure 1 gives the sulphate resistance of the specimens.

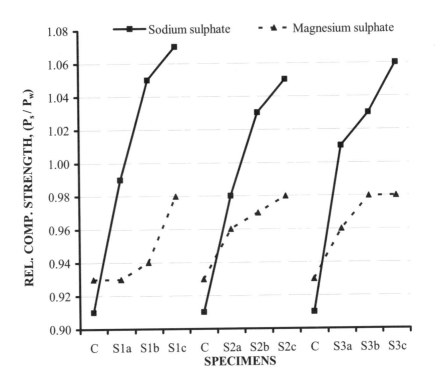

Figure 1   Sulphate resistance of mortar specimens

Much more compressive strength loss was observed in specimens produced with reference cement and cured in corrosive solutions with respect to the specimens in other groups exposed to sulphate attack. 9% and 7% compressive strength decrease was observed in specimens exposed to Na_2SO_4 and $MgSO_4$ attack, respectively. The best results were found from G1c, G2c, and G3c cements. The residual compressive strength of mortar specimens of these series (G1c, G2c, and G3c) cured in Na_2SO_4 solution were higher than that of the reference specimens cured in water. There are 7, 5, and 6 % differences between them, respectively. The same differences are 1, 3, and 3% for specimens cured in $MgSO_4$ solution. These results shows that composite Portland cements shows higher sodium sulphate and magnesium sulphate resistance for all replacement ratios selected in this study. The result that P_S/P_W ratio is greater than 1.0 is more interesting. Therefore it can be said that sulphate resistance of composite cements including GBFS and BOF slag is higher than that of normal Portland cement.

It was concluded that wastes or by-products that have pozzolanic property are resistant to sulphates [19, 20, 21]. Pozzolans react with $Ca(OH)_2$ and C-S-H gel is formed. This formation increases the sodium sulphate resistance of materials. However it does not show the effect with magnesium sulphate solution. Because the C-S-H gel produced after pozzolanic reactions is more sensitive and weak against magnesium sulphate attack [15, 18, 22]. As a result, cements containing GBFS and BOF slag in different ratios are more resistant to sodium sulphate attack which is consistent with the literature.

CONCLUSIONS

New composite cements included GGBFS and BOF slag were produced in the laboratory. Physical characteristics and sulphate resistance of these cements were investigated. The test results showed that cements containing GGBFS and BOF slag are resistant to sulphate attacks. As the substitution ratio increases, sulphate resistance increases. However this increase is not at the same level for sodium sulphate and magnesium sulphate. Sodium sulphate resistance is higher than magnesium sulphate resistance. Also, these cements have lower specific weight than normal cements. Usage of BOF slag in cement production with GGBFS will revalue the cements that will be used in concrete exposed to sulphate attack.

REFERENCES

1. MOTZ, H AND GEISELER, J. Products of steel slags: an opportunity to save natural resources. Waste Management. Vol. 21, 2001. pp 285-293.

2. WILD, S AND TASONG, W A. Influence of ground granulated blastfurnace slag on the sulphate resistance of lime-stabilized kaolinite. Magazine of Concrete Research. Vol. 51, 1999. pp 247-254.

3. HESTER, D, MCNALLY, C AND RICHARDSON M. A study of the influence of slag alkali level on the concrete. Construction and Building Materials. Vol. 19, 2005. pp 661-665.

4. BASHEER, P A M, GILLEECE, P R V, LONG, A E AND MC CARTER, W J. Monitoring electrical resistance of concretes containing alternative cementitious materials to assess their resistance to chloride penetration. Cement and Concrete Composites. Vol. 24, 2002. pp 437-449.

5. DEJA, J. Freezing and de-icing salt resistance of blast furnace slag concretes. Cement and Concrete Composites. Vol. 25, 2003. pp 357-361.

6. LI, G, AND ZHAO, X. Properties of concrete incorporating fly ash and ground granulated blast-furnace slag. Cement and Concrete Composites. Vol. 25, 2003. pp 293-299.

7. KWON, Y J. A study on the alkali-aggregate reaction in high-strength concrete with particular respect to the ground granulated blast-furnace slag effect. Cement Concrete Research. Vol. 35, 2005. pp 1305-1313.

8. SHIH, P H, WU, Z Z, CHIANG, H L. Characteristics of bricks made from waste steel slag. Waste Management. Vol. 24, 2004. pp 1043-1047.

9. SHI, C, QIAN, J. High performance cementing materials from industrial slag- A review. Resour. Conserv. Recycl. Vol. 29, 2000. pp 195–207.

10. SUN, S. Investigations on steel slag cements. In: Collections of Achievements on the Treatment and Applications of Metallurgical Industrial Wastes. Chinese Metallurgical Industry Press, 1983. pp 1-71.

11. SHI, C, DAY, R L. Early strength development and hydration of alkali-activated blast furnace slag/fly ash blends. Advanced Cement Research. Vol. 11, 1999. pp 189-196.

12. TIXIER, R AND MOBASHER, B. Modeling of Damage in Cement-Based Materials Subjectedto External Sulfate Attack. II: Comparison with Experiments Raphae. ASCE Journal of Materials In Civil Engineering. Vol. 15, 2003. pp 314-322.

13. TS-EN 196-1, Methods of Testing Cements - Determination of Strength. Turkish Standard Institute. Ankara. 2002.

14. TS-EN 196-6, Methods of testing cement; Part 6: Determination of fineness Turkish Standard Institute. Ankara. 2002.

15. TORII, K AND KAWAMURA, M. Effects of fly ash and silica fume on the resistance of mortar to sulfuric acid and sulfate attack. Cement and Concrete Research. Vol. 24, 1994. pp 361-370.

16. GOLLOP, R S, TAYLOR H F W. Microstructural and microanalytical studies of sulfate attack: IV. Reactions of a slag cement paste with sodium and magnesium sulfate solutions. Cement and Concrete Research. Vol. 26, 1996. pp 1013- 1028.

17. BIRICIK, H, AKÖZ, F, TÜRKER, F AND BERKTAY, I. Resistance to magnesium sulfate and sodium sulfate attack of mortars containing wheat straw ash. Cement and Concrete Research. Vol. 30, 2000. pp 1189- 1197.

18. WILD, S, KHATIB, J M, AND O'FARRELL, M. Sulphate resistance of mortar, containing ground brick clay calcined at different temperatures. Cement and Concrete Research. Vol. 27, 1997. pp 697-709.

19. AKÖZ, F, KORAL, S AND YÜZER, N. Effects of Magnesium Sulfate Concentration on the Sulfate Resistance of Mortars with and Without Silica Fume. Cement and Concrete Research, Vol. 27, 1997. pp 205-214.

20. NEVILLE, A M. Properties of Concrete, Longman, London, 1995.

21. MEHTA, P K AND MONTEIRO P J M. Concrete microstructure. Properties and Materials, Prentice-Hall New Jersey, 1997.

22. MANGAT, P S AND KHATİB, J M. Influence of fly ash, silica fume, and slag on sulfate resistance of concrete. ACI Materials Journal. Vol. 92, 1995. pp 542-552.

RECYCLING STEEL MILL SCALE AS
FINE AGGREGATE IN CONCRETE MORTARS

S Al-Otaibi

Kuwait Institute for Scientific Research

Kuwait

ABSTRACT: During the processing of steel in steel mills, iron oxides will form on the surface of the metal. These oxides, known as mill scale, occur during continuous casting, reheating, and hot rolling operations. The scale is removed by water sprays and collected then disposed of by dumping. A local steel manufacturing company generates quantities reaching almost 7000 tons/year . This paper presents preliminary findings of a study that investigates the potential for recycling steel mill scale into concrete. The composition of the steel mill scale was determined by XRF. Several mortar mixes were made using the product as a replacement for the fine aggregates. Compressive strength, flexural strength and drying shrinkage were measured for different specimens from the mortar mixes. The results are promising and encourage further study in specific application in concrete, brick, and block manufacturing.

Keywords: Aggregate, Mortar, Steel Mill Scale, Strength.

Dr Saud Al-Otaibi, Associate Research Scientist, Kuwait Institute for Scientific Research, Division of Environmental and Urban Development, Building and Energy Technologies Department.

INTRODUCTION

Waste management is one of the most complex and challenging problems in the world which has a great impact on environment.

A local rolling mill is currently producing reinforcing steel bars in Kuwait. The capacity of this plant is 650,000 tons/year for deformed reinforcing steel bars with diameters from 8 mm to 40 mm. During the processing of steel, iron oxides will form on the surface of the metal. These oxides, known as mill scale, occur during continuous casting, reheating, and hot rolling operations. The scale is removed by water sprays and collected then disposed of by dumping. The plant generates quantities reaching almost 7000 tons/year (Figure 1). The factory approached KISR seeking their assistance in minimizing the environmental impact of dumping this product identifying possible means for recycling.

Steel mill scale is usually recycled in steel plants again in sintering furnaces or hearths (Adams, 1979, and Daiga, et al., 2002). Mill scale that cannot be recycled by steel plants has been used by Portland cement plants as an iron source (Portland Cement Association, 2005, Young et al., 2004). Other researchers (Pradip et al., 1990) carried out on utilizing steel mill scale in the production of alinite cements.

Figure 1   Steel Mill Waste

The steel slag which is somewhat similar to steel scale has been used in civil engineering tenths of years ago (Geiseler, 1999; Neville, 1996; Neville and Brooks 2002; Alizadeh et al., 1996, Asi et al 2005). Portland granulated ground blast furnace slag cement, which is produced from rapidly water-cooled blast furnace slag, has been successfully used in concrete due to the high amount of lime (40-50%), which posses pozzolanic activity (Neville, 2002; Alizadeh et. al., 1996)). Electric arc furnace slag (EAFS) that contains low percentage of amorphous silica and high content of ferric oxides and consequently has low, or no, pozzolanic activities in comparison with blast furnace slag (BFS), is not appropriate to be used in blended cement production (Kamal et. al. 2002). Although many studies have been conducted on the evaluation of steel slag to be used in road construction, there are rare researches regarding the utilization of steel slag in concrete (Kamal et. al. 2002).

Although many studies have been conducted on the evaluation of steel slag to be used in road construction, there are rare researches regarding the utilization of steel slag in concrete (Alizadeh et al). ASTM C33 gives specifications for the use of blast furnace slag as aggregates in concrete, while there is not such a standard for steel slag or steel mill scale.

Alizadeh et al (1996) results of the experiments which are carried out on hardened concrete, indicate that slag aggregate concretes achieved higher values of compressive strength, tensile strength, flexural strength and modulus of elasticity, compared to natural aggregate concretes

Shekarchi et al (2004, 2003) introduced a comprehensive research on the utilization of steel slag as aggregate in concrete. In the study, Shekarchi (2004) concluded that the use of air-cooled steel slag with low amorphous silica content and high amount of ferric oxides it unsuitable to be employed in blended cement. On the other hand, utilization of steel slag as aggregate is advantages when compared to normal aggregate mixes.

Maslehuddin et al (2003) presented a comparative study of steel slag aggregate concrete and crushed limestone concrete. In the study, only part of the coarse aggregate has been replaced by slag aggregate. The study concluded that the compressive strength of steel slag aggregate concrete was marginally better than that of crushed limestone aggregate concrete. Moreover, no significant improvement was noted in the tensile strength.

Manso et al, presented a study in which electric has been used to obtain concrete of better quality. It was concluded that steel slag can be used to enhance concrete properties. However, according to the authors, special attention must be paid to the mixes of concrete to achieve a suitable fine aggregate, which can be obtained by mixing fine slag with filler material. This paper presents preliminary findings of a study that investigates the potential for recycling steel mill scale into concrete.

TEST METHODOLOGY

Materials

Cement: Ordinary Portland cement, conforming to the requirements of ASTM C 150 was used in this investigation.

Aggregates: Local desert sand was used in this investigation.

Steel mill scale obtained form a local steel mill was used (Figure 2)

Water: Potable water was used.

Mortar Mixtures

The mortar mixtures were prepared in cement: fine aggregate ratio of 1:2.5 and a w/c ratio 0.55.

Figure 2   Steel Mill Scale Fine Aggregate

Testing Procedure

Properties of the Fine Aggregates

The specific gravity of fine aggregate (Sand and Steel mill Scale) and water absorption were determined following the ASTM C-127 procedure. The sieve analysis was carried out in accordance with ASTM C-136.

The chemical composition was determined for steel mill scale was determined using X-ray Diffraction (XRF).

Compressive Strength

50 x 50 x 50 mm^3 cubes were prepared to determine the compressive strength according to BS EN 12390-3:2002.

Flexural Strength

Two 75 x 75 x 250 mm mortar prisms were prepared for each of the mixtures. The flexural strength test was carried out in accordance with (BS EN 12390-5:2000), using the two-point loading method.

Drying Shrinkage

Two 25 x 25 x 280 mm mortar bars were prepared for each mixture. The bars were cured for 7 days in water then were left to dry in the controlled temperature and humidity room. The shrinkage was measured using a length comparator in accordance with BS 812: Part 120: 1989.

RESULTS AND DISCUSSION

Characterisation of Fine aggregates

The results of the sieve analysis of both the fine aggregate and the fine steel slag are shown in Table 1, together with ASTM C33 and BS EN 12620 grading limits.

Table 1 Sieve analysis of fine aggregates

SIEVE SIZE (mm)	ASTM DESIGNATION	SAND % Passing	STEEL MILL SCALE % Passing	ASTM LIMITS	BS GRADING REQUIREMENTS		
					Zone C	Zone M	Zone F
9.5	3/8"	100.00	100.0	100	100	100	100
4.75	# 4	99.16	97.7	100	89 - 100	89 - 100	89 - 100
2.360	# 8	96.70	65.8	95 - 100	60 - 100	65 - 100	80 - 100
1.180	# 16	90.32	50.6	80 - 100	30 - 90	45 - 100	70 - 100
0.600	# 30	74.68	36.7	50 - 85	15 - 54	25 - 80	55 - 100
0.300	# 50	30.54	27.0	25 - 60	5 - 40	5 - 48	5 - 70
		9.06			0 – 15*	0 – 15*	0 – 15*
0.150	# 100			10 -30			
			18.2		0 – 20^	0 – 20^	0 – 20^
0.075	# 200	6.26	17.6	NA	NA	NA	NA

* Natural aggregate   ^ Crushed aggregate

It was intended to use the steel mill scale "as received" from the factory without any further screening. Also, it was intended to use the locally available desert sand from near-by sources. This choice would be the most economical for use in concrete mixes when accounting for the costs of transporting and screening.

The Specific gravity and absorption of the aggregates were measured using ASTM C127 and ASTM C128. In each case, three representative samples were taken and tested according to the corresponding ASTM. The apparent specific gravity of the coarse sand and steel mill scale was 2.53 and 2.89 and the absorption was 1.1 and 0.11% respectively. The specific gravity of the steel mill scale is higher than that of normal aggregate. Also, the absorption is smaller (< 1%). These results are acceptable for use in concrete structures, keeping in mind that higher density is expected. Chemical analysis of steel mill scale is shown in Table 2. It is clear that the mill scale has low CaO content indicating that no pozzolanic activity is expected.

Table 1   Chemical analysis of the mill scale used in the study

Oxides	Fe_2O_3	MnO	TiO_2	SiO_2	MgO	CaO	Na_2O	Al_2O_3
%	94.61	1.03	0.014	1.37	0.028	0.111	0.013	0.099

Compressive Strength

The compressive strength development of all the six mixtures is shown in Figure 3. These Results showed the compressive strength of the mortar specimens for all ages increased as the percentage of sand replacement by steel mill scale increased till reaching the 40% level. Then the compressive strength decreases. Overall the steel mill scale introduction gave an increase in strength.

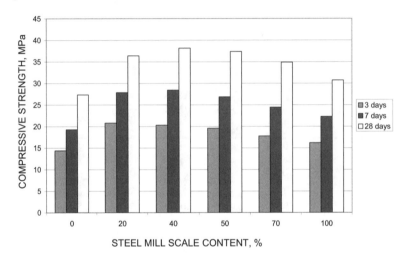

Figure 3   Influence of Mill Scale on Compressive Strength

Flexural Strength

The results in Table 2 indicate that the flexural strength increases in a similar manner to compressive strength when replacing sand by steel mill scale. The optimum value being around the 40% level of replacement.

Table 2   Flexural Strength at 28 days

Mill Scale %	0	20	40	50	70	100
Flexural Strength MPa	6.60	7.05	7.26	6.92	6.90	6.53

Drying Shrinkage

The results in Figure 4 show decrease in drying shrinkage with the increase in the level of replacement of sand by steel mill scale. The control mixture giving the highest value of drying shrinkage and followed by the steel mill mixtures according to their level of replacement. This can be attributed to the denser matrix produced and the less amount on absorbed water.

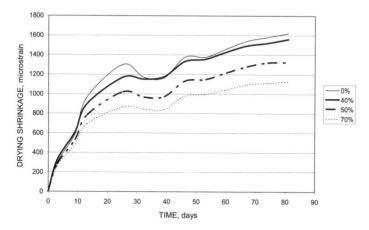

Figure 4   Influence of Mill Scale on Drying Shrinkage

ACKNOWLEDGMENTS

The author wishes to thank the United Steel Company for their support and assistance throughout this research work which is continuing of means for utilising different wastes of the steel industry in construction in Kuwait.

CONCLUSIONS

The main conclusions are:

1. The steel industry in Kuwait has a huge amount of solid waste that need to be dealt with.
2. Replacing 40% of sand with steel mill scale gave the highest increase in compressive strength.
3. Replacing 40% of sand with steel mill scale also increased flexural strength.
4. Drying shrinkage is lower when using steel mill scale.
5. Attention should be given to further research on the influence on concrete durability.
6. Future research work will include other waste forms and applications in construction.

REFERENCES

1.   ADAMS, C.J. (1979) "Recycling Of Steel Plant Waste Oxides - A Review". CANMET Report (Canada Centre for Mineral and Energy Technology) Issue 79-34. p. 11.

2.   ALIZADEH R., CHINI M., GHODS P., HOSEINI M., MONTAZER SH., M. SHEKARCHI M. (1996) "Utilization of electric arc furnace slag as aggregates in Concrete" Environmental Issue, CMI report Tehran. 1996.

3. ASI I., QASRAWI H. AND SHALABI F., 2005 "Use of steel slag in engineering projects", Progress Report No. 1 submitted to United Iron and Steel Manufacturing Company.

4. ASTM C127-07 Standard Test Method for Density, Relative Density (Specific Gravity), and Absorption of Coarse Aggregate

5. ASTM C127-07 Standard Test Method for Density, Relative Density (Specific Gravity), and Absorption of Coarse Aggregate

6. ASTM C128-07a Standard Test Method for Density, Relative Density (Specific Gravity), and Absorption of Fine Aggregate

7. ASTM C136-06 Standard Test Method for Sieve Analysis of Fine and Coarse Aggregates.

8. ASTM C150-07 Standard Specification for Portland Cement

9. ASTM C33-07 Standard Specification for Concrete Aggregates

10. BS 812-120:1989 Testing aggregates. Method for testing and classifying drying shrinkage of aggregates in concrete

11. BS EN 12390-3:2002 Testing hardened concrete. Compressive strength of test specimens

12. BS EN 12390-5:2000 Testing hardened concrete. Flexural strength of test specimens.

13. BS EN 12620:2002 Aggregates for concrete

14. DAIGA, V.R., HORNE, D.A., THORNTON, J.A. (2002) "Steel mill waste processing on a rotary hearth furnace to recover valuable iron units" - Ironmaking Conference, pp. 655-665.

15. GEISELER, J. (1999) "Slag-approved material for better future, Proceeding of International Symposium on the Utilization of Metallurgical Slag (ISUS 99), Beijing, China.

16. KAMAL, M., GAILAN, A. H., HAATAN, A. , HAMEED, H. (2002), "Aggregate made from industrial unprocessed slag" Proceeding of the 6th International Conference on Concrete Technology for Developing Countries, Amman, Jordan.

17. MANSO J., GONZALEZ J AND POLANCO J. "Electric furnace slag in concrete" (2004), Journal of Materials in Civil Engineering, ASCE, November/December, pp639-645.

18. MASLEHUDDIN M., ALFARABI M., SHAMMEM M., IBRAHIM M. AND BARRY M. (2003), "Comparison of properties of steel slag and crushed limestone aggregate concretes", Construction and Building Materials, Vol.17, pp 105-112

19. NEVILLE A. AND BROOKS J. (2002), "Concrete Technology", 2nd edition, Longman, UK.

20. NEVILLE A. M. (1996), "Properties of Concrete", 4th edition, Longman, UK.

21. PORTLAND CEMENT ASSOCIATION (2005), "Iron And Steel Byproducts". Portland Cement Association, Sustainable Manufacturing Fact Sheet.

22. PRADIP, D. VAIDYANATHAN, P. C. KAPUR AND B. N. SINGH, (1990), "Production and properties of alinite cements from steel plant wastes", Cement and Concrete Research, Vol 20, Issue 1, January, Pages 15-24.

23. SHEKARCHI M., SOLTANI M,. ALIZADEH R., CHINI M, GHODS P., HOSEINI M., SH. MONTAZER SH.(2004) "Study of the mechanical properties of heavyweight preplaced aggregate concrete using electric arc furnace slag as aggregate" International Conference on Concrete Engineering and Technology, Malaysia.

24. SHEKARCHI, M., ALIZADEH, R. CHINI, M., GHODS, P. HOSEINI, M. AND MONTAZER, S. (2003), "Study on electric arc furnace slag properties to be used as aggregates in concrete," CANMET/ACI International Conference on Recent Advances in Concrete Technology, Bucharest, Romania.

25. YOUNG; ROM D. AND NORRIS; DAVID, (2004) "Process for using mill scale in cement clinker production", US Patent No. 6709510.

PERFORMANCE ASSESMENT OF INCINERATOR SLAG AS CONTROLLED LOW-STRENGTH MATERIAL

S Naganathan

Sunway University College

H Abdul Razak

University of Malaya

S N Abdul Hamid

Kauliti Alam Sdn Bhd

Malaysia

ABSTRACT. In this paper, the performance of incinerator slag as controlled low-strength material (CLSM) was assessed. Various mix formulations of CLSM were made with the slag and cement. Amount of cement added varied from 0 to 40 percent of weight of slag. Water content was adjusted to satisfy the flow requirements. The CLSM mixes prepared were tested for flowability, setting time, bleeding, and corrosivity in fresh state. Tests like compressive strength, California Bearing Ratio (CBR), Initial Surface Absorption (ISAT), water absorption, and sorptivity were conducted on the hardened CLSM. Compressive strength tests were conducted at 7, 14, and 28 days on 70 mm cubes. Heavy metals analysis for bleed and leachate water was also conducted on the CLSM mixes. The compressive strength of CLSM tested ranged from 0.42 to 9.86 MPa. CBR values ranged from 4 to 23%, and ISAT values from 1.86 to 12 mL/m^2/s. None of the thirteen heavy metals tested showed more than the threshold limits. It was concluded that the performance of incinerator slag as CLSM is good and promising..

Keywords: Controlled low-strength material, Incinerator slag, California bearing ratio, Initial surface absorption, Compressive strength, Leachate, Setting time, Flowability.

S Naganathan is a Lecturer in the American Degree Program of Sunway University College, Malaysia. He is currently doing his PhD in the University of Malaya.

H Abdul Razak is a Professor in the Department of Civil Engineering, University of Malaya. He obtained his PhD from University of Surrey in Structural Engineering and is active in research involving use of indigenous and waste materials for high performance concrete and assessment of deteriorated and damaged structures using modal testing.

S N Abdul Hamid is a manager for Laboratory Management Services in Kualiti Alam Sdn Bhd, Malaysia's sole operator of the integrated waste management facility. She has more than eight years of experience in analysis and characterization of hazardous wastes received from various industries throughout the nation.

INTRODUCTION

One of the ways of using the incineration byproducts is as controlled low-strength material (CLSM). CLSM, in its simplest form is slurry made by mixing sand, cement, fly ash, and water. The amount of water added is adjusted to make the mix flowable. ACI committee 229 [1] defines CLSM as a material having a compressive strength of 8.3 MPa or less. The CLSM is flowable, self compacting, and offers many advantages compared to conventional soil fills.

Some advantages of CLSM over conventional back fills are easy placement with no vibration, less onsite labour requirements, ease of placing in intricate locations, no settlement problems, strong, durable, and flexibility to use any locally available non conventional materials. As CLSM is versatile, and incineration bottom slag is available in plenty, this presents an excellent possibility.

Preliminary tests conducted on CLSM samples showed promising results. In this paper, efforts are made to study the performance of incinerator slag as CLSM. Slag and cement were mixed with water. Weight of cement added ranged from 0 to 40% of the weight of slag. Water content was adjusted to get the required flowability. Various tests were performed on the CLSM in fresh and hardened states. The results were analyzed, and reported.

Corrosivity is designed to identify materials that potentially pose a hazard to human health or the environment due to their ability to mobilize toxic metals if discharged into a landfill environment, to corrode handling, storage, transportation, and management equipment, or to destroy human or animal tissue in the event of inadvertent contact [2]. Corrosion can occur when bleed water or leachate water reacts with metal parts.

A waste exhibits corrosivity if a representative sample of the waste has the property that is aqueous and has a pH less or equal to 2 or greater than or equal to 12.5. The corrosivity of CLSM was studied by measuring pH of bleed water during mixing and leachate water from CLSM cubes at 28 days using Method 9040 in "Test Methods for Evaluating Solid Waste, Physical/Chemical Methods", Environmental Protection Agency, United States, Publication SW-846 (40 CFR Part 261, 1986).

The toxicity of the CLSM mixes was also studied by measuring the concentration of heavy metals in the bleed water and in the leachate. Method 6010 B of EPA [3] using inductively coupled plasma optical emission spectrometer was used for all the heavy metals analysis except mercury, for which method 7473 [3] was used using a mercury analyzer.

METHODOLOGY

Materials used

Cement

Ordinary Portland cement conforming to MS 522 [4] were used in this investigation.

Incinerator slag

Incinerator slag was obtained from Kualiti Alam Sdn Bhd, Malaysia. The slag contained particles ranging from fine particles to porous stones with size up to 60 mm. It comes in a very wet state from the incinerator, and the finer particles make the slag cohesive and difficult to work with. Hence, the slag was first dried in an oven at 105°C until constant mass, and then sieved through a 10 mm size sieve to eliminate unwanted substances and particles larger than 10 mm size. The conditioned slag is as shown in Figure 1. The physical properties of slag are given in Table 1. The slag is found to be a light weight material as the uncompacted bulk density measured in accordance with BS 3797 [5] is less than 1200 kg/m³. The grading of the slag is given in Table 2. As per BS 882 [6], the grading of slag falls in coarse and medium fine aggregate. It also contains 12 percent of materials smaller than 75 microns.

Figure 1   Slag used in the investigation.

Table 1   Physical properties of slag.

PROPERTY	VALUE
Apparent particle density	1840 kg/m³
Compacted Bulk Density	964 kg/m³
Un compacted Bulk Density	911 kg/m³
Grading range	Coarse and medium fine aggregate
Particles <75 μm	12%
Color	Black

Table 2   Sieve analysis of bottom slag.

SIEVE SIZE (mm)	MASS RETAINED(g)	% RETAINED	CUMULATIVE % PASSING	OVERALL LIMITS OF % BY MASS PASSING AS PER BS 822:1992
5	33.6	6.7	93	89-100
2.36	80.1	16.0	77	60-100
1.18	87.8	17.6	59	30-100
0.600	92.7	18.5	40	15-100
0.300	62.6	12.5	27	5-70
0.150	44.7	8.9	18	0-15
0.075	30.1	6.0	12	

Mix formulations, Mixing and casting

Nine mix formulations were used in the investigation. The mix formulations are given in Table 3. The slag and cement were first placed in a concrete mixer, about 50% of the required water was added, and the contents mixed for 2 minutes. The flowability of CLSM was then tested as per ASTM D 6103 [7]. This was done by placing an open ended cylinder 76 mm internal diameter and 150 mm long on a flat surface, pouring CLSM into it up to the top surface, and lifting the cylinder up vertically. The spread diameter of CLSM was then measured. As per ACI 229 [1], the CLSM is considered flowable if the spread diameter is at least 200 mm. More water was then added and the flow measured. This was repeated by adding water until the spread diameter was 200 mm. The contents were then mixed for another 2 minutes. CLSM was then filled in 70 mm cube moulds, 150 mm cube mould for ISAT, CBR mould, and about 800 mL in a 1000 mL plastic measuring jar for bleed water measurement. The balance of the CLSM mix was then sieved in a 5 mm sieve and the mortar that passed through the sieve was filled in small containers for setting time test. Cubes were kept covered with wet burlap in the laboratory environment for one day, and then transferred to the curing environment at 22°C and 95% RH. The curing environment was achieved by filling plastic storage containers with water for a depth of about 50 mm, placing the moulds over the water, closing the containers with airtight cover, and keeping the containers in an air conditioned room at 22°C. The cubes were removed from the moulds on the day of testing.

Table 3   Mix formulations.

MIX ID	BULK PROPORTION (kg/m^3)				PERCENTAGE BY WEIGHT (%)			W/C RATIO
	CEMENT	SLAG	WATER	FRESH DENSITY	CEMENT	SLAG	WATER	
NM 40	348	870	400	1618	22	54	24	1.1
NM 35	296	844	433	1573	19	53	28	1.5
NM 30	249	829	427	1505	17	55	28	1.7
NM 25	199	797	472	1468	14	54	32	2.4
NM 20	177	885	416	1478	12	60	28	2.3
NM 15	132	881	452	1465	9	60	31	3.4
NM 10	88	882	594	1564	6	53	41	6.7
NM5	48	967	592	1607	3	60	37	12.3
NM0	0	1007	600	1607	0	63	37	--

Laboratory Testing

Following tests were performed on the CLSM mix:

1. Compressive strength at 7, 14, and 28 days of curing. Universal testing machine of 100 kN capacity was used for performing the test. The loading rate was kept constant at 0.6 mm/minute. It took 4 to 8 minutes for failure of each cube.

2. Stiffening time test as per BS EN 13294 [8].

3. Initial surface absorption test was done on some mixtures of hardened CLSM. The 150 mm cube was kept in an oven for 21 days and the temperature was maintained at 90°C. The cube was removed from the oven at 27 days, covered with plastic bag, and kept in the air-conditioned room at 22°C. The cube was then tested for ISAT at 28 days as per BS 1881 [9].

4. California Bearing Ration test (CBR). This test was done as per BS 1377 [10] on the hardened CLSM kept in the CBR mould and cured for 28 days.

5. The corrosivity of CLSM by measuring the pH of the bleed and leachate water. A simple electronic pH meter was used for measurement. The bleed water was collected as given in step 3 above. The leachate water was obtained by immersing broken 70 mm cube in distilled water on day 7, and collecting this water on the 28th day.

6. Analysis of bleed and leachate water for the presence of leachable heavy metals. This was done by Kualiti Alam Sdn Bhd, Malaysia. Toxicity Characteristic Leaching Procedure as per Environmental Protection Agency (EPA) (USA), Test number SW-846 2000, 40 CFR Part 261 1986 was used for this analysis.

RESULTS AND DISCUSSION

Density development

The density of fresh and hardened CLSM is given in Table 4. The fresh density shows a decreasing trend with increase in cement content. The fresh density for NM0, NM5, and NM10 are higher than the other mixes. This is because of the addition of more water to satisfy the flow requirements. From the observations, it can be concluded that the fresh density is higher than hardened density. 28 days density varies from 1419 to 1568 kg/m^3 which is much less than density of most of the soils.

Table 4   Density development.

MIX ID	FRESH DENSITY (kg/m^3)	HARDENED DENSITY (kg/m^3)		
		7 Days	14 Days	28 Days
NM 40	1618	1545	1536	1478
NM 35	1573	1508	1499	1484
NM 30	1505	1478	1495	1486
NM 25	1468	1423	1414	1432
NM 20	1478	1448	1485	1419
NM 15	1465	1545	1536	1478
NM 10	1564	1611	1567	1568
NM5	1607	1505	-	1508
NM0	1607	-	-	-

Compressive strength

The results of compressive strength of CLSM are given in Table 5. The 28 days strength of CLSM samples vary from 0.42 MPa to 9.86 MPa. The strength values were calculated from five cubes tested on each day of test. Variations in strength were observed as it is a low strength material. The strength development of various mixes tested is shown in Figure 2. It is found that the compressive strength increases with cement content, and decreases with increase in W/C ratio. The compressive strength tests could not be conducted on NM 0 mix as the cubes were broken while demoulding. This was because there was no cement added in the mix.

Table 5 Compressive strength results.

MIX ID	COMPRESSIVE STRENGTH (MPa)					
	7DAYS	[COV]	14 DAYS	[COV]	28 DAYS	[COV]
NM 40	6.54	[11]	8.10	[5]	9.86	[7]
NM 35	4.26	[6]	5.55	[14]	6.15	[14]
NM 30	3.21	[14]	5.43	[1]	5.88	[14]
NM 25	1.95	[10]	2.71	[12]	3.76	[3]
NM 20	1.41	[6]	2.08	[5]	2.60	[11]
NM 15	0.71	[14]	0.98	[3]	1.40	[7]
NM 10	0.39	[12]	0.44	[7]	0.55	[2]
NM5	0.14	[12]	0.39	[13]	0.42	[9]

[COV] – Coefficient of Variation.

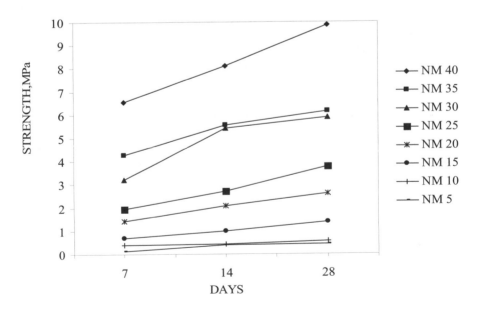

Figure 2 Compressive strength development.

Stiffening time

The stiffening times were measured on the mortars sieved from fresh CLSM and filled in rigid metal containers as per BS EN 13294 [8] The stiffening times as expressed as initial and final setting times for various mix proportions are listed in Table 6. The initial setting time varies from 3 to 8 hours, and the final setting time from 17 to 54 hours. Final set time increases with increase in W/C ratio. Vast variations were observed in measuring the penetration resistance observations. The penetration resistance readings were different on different containers measured at the same time on each mix. Hence, further testing needs to be made before arriving at any conclusion regarding the stiffening time.

Corrosivity

The corrosivity is measured in terms of pH. The pH of bleed water and leachate is shown in Table 7. The pH increases with increase in cement content for both bleed and leachate. The pH of leachate is less than that of bleed for all the mixes. The pH of both bleed and leachate for all mixes are between 11 and 12. The allowable pH for corrosivity as per EPA regulations is between 2 and 12.5. Hence the bleed and leachate samples collected from CLSM were not having corrosive properties.

Table 6   Stiffening time .

MIX ID	SETTING TIME (HOURS)	
	Initial	Final
NM 40	5	17
NM 35	7	19
NM 30	6.5	18
NM 25	7.5	21
NM 20	5	16.5
NM 15	4	23
NM 10	3	51
NM5	6.25	54

Table 7   pH of bleed and leachate water.

MIX ID	pH	
	Bleed water	Leachate at 28 days
NM40	11.91	
NM35	11.78	11.79
NM30	11.91	
NM25	11.81	11.93
NM20	11.78	
NM15	11.66	11.47
NM10	11.54	11.43
NM5	11.44	11.05
NM0	11.34	

Initial surface absorption

The results of initial surface absorption test conducted on some hardened CLSM cubes are reported in Table 8. It is observed that the initial surface absorption decreases with increase in cement content. The 10 minutes surface absorption values fall in the range of 1.82 to 12 mL/m^2/s. A 10 minute absorption value between 0.25 and 0.5 is considered average, and value more than 0.5 is considered high for concrete [11]. This shows that the permeability of CLSM mixes is high when compared with concrete. A more relevant test to assess the permeability of CLSM is to conduct a water permeability test. A new device is being fabricated to conduct permeability testing on CLSM.

Table 8   ISAT, CBR, water absorption, and sorption results.

MIX ID	ISAT (mL/m^2/s)			CBR (%)	WATER ABSORPTION (%)	SORPTIVITY (mm/min$^{0.5}$)
	10 MINUTES	30 MINUTES	1 HOUR			
NM25	1.82	1.13	0.83		20.1	0.47
NM15	4.6	2.93	1.91	23		
NM10	12	10	8.6	8	28	2.4
NM5				6		
NM0				4		

California bearing ratio test

The California bearing ratio test (CBR) is a measure of the supporting value of the subgrade. It is used in pavement design calculations [12]. Results of CBR test showed that the CLSM mixtures exhibited CBR values from 4 to 23% for compressive strength values from <0.42 MPa to 1.4 MPa. A CBR value from 5 to 15% is said to have good subgrade strength, and values from 3 to 5% is said to be normal [13]. CBR value of 4 is for a sandy clay soil, and a value of 20 is for a well graded sandy gravel [12]. Hence the incinerator slag CLSM mix can be used as a good subgrade material.

Water absorption

Test for water absorption was done on 70 mm cubes at 28 days as per BS 1881 [14]. The cubes were taken out from the moulds kept in the curing environment for 21 days and kept in an oven at 105°C. They were taken out from the oven on 27 day and kept covered in plastic bags and kept in air conditioned room. Cubes were tested on 28th day by immersing in water for 30 minutes and measuring the difference in weight. Results showed that water absorption decreases with increase in cement content. The values of water absorption are more than for concrete. This is expected as the surface absorption of CLSM was high.

Sorptivity

Sorptivity test gives an idea about the durability of CLSM. The sorptivity tests conducted on two mixes of hardened CLSM gave values of 0.47 and 2.4 mm/min$^{0.5}$. 70 mm cubes were prepared similar to water absorption test. The cubes were immersed in water and the weight gain was recorded for 4 hours at half an hour intervals.

Sorptivity was the slope of the graph drawn with increase in mass in g/mm^3 in Y axis and (time in minutes)$^{0.5}$ in X axis. Typical values for CLSM mixes range from 0.1 to 1.2 mm/min$^{0.5}$ [15]. Sorptivity decreases with increase in cement content. Further testing needs to be done on sorptivity and water absorption to arrive at more conclusions.

Ultrasonic Pulse Velocity (UPV) Test

The UPV test is used to determine the uniformity of concrete and the location of areas of poor compaction and internal voidage. The ultrasonic pulse velocity test was done on some mixes at 28 days using an ultrasonic pulse equipment which was calibrated for a standard pulse velocity of 26 μs. Results of the test are given in Table 9.

The usual pulse velocities found in concrete is 3.6 to 5 km/s [16]. Hence the CLSM mixes studied were more porous than concrete. This is because of more water addition to maintain the flow requirements. Tests on other mixes could not be done because of the low strength, and cubes getting broken while tested. Due to the fragility of cubes, the UPV test was discontinued from further studies.

Table 9   Ultrasonic pulse velocity test.

MIX ID	UPV AT 28 DAYS (km/s)
NM 40	2.5
NM 30	2.3
NM 20	1.9

Toxicity Analysis

Bleed water during mixing were collected and sent for toxicity analysis. Cubes tested on 7 days for compression strength were immersed in two litres of distilled water, and kept in the curing environment in closed containers. On day 28, the leachate water from the containers were tested for heavy metals.

Results of toxicity analysis on the bleed water for various mix formulations are given in Table 10. Toxicity analysis done on leachate water is given in Table 11.

It is found that all the thirteen heavy metals tested were well within the threshold limits given by EPA [17]. Also the heavy metals content of bleed is more than that of the leachate water. Hence the incinerator slag did not pose any environmental impact out of the thirteen heavy metals studied.

Table 10   Toxicity of Bleed water.

PARA-METER	RESULTS FOR BLEED WATER (mg/l)									THRESHOLD LIMIT (mg/l)
	NM40	NM35	NM30	NM25	NM20	NM15	NM10	NM5	NM0	
Arsenic	0.14	0.12	0.14	0.10	0.18	0.18	0.24	0.13	0.07	5
Barium	7.19	3.62	4.43	19.3	5.61	5.68	5.24	3.76	4.68	100
Boron	0.64	0.12	0.18	0.28	0.09	1.03	1.58	2.56	7.99	400
Cadmium	<0.01	<0.01	<0.01	<0.01	<0.01	<0.01	<0.01	<0.01	<0.01	1
Copper	4.58	3.08	2.08	3.06	3.15	6.00	13.7	6.15	6.49	100
Lead	0.68	0.92	0.58	0.94	0.93	1.68	2.22	1.02	1.20	5
Mercury	0.08	0.06	0.05	0.07	0.03	0.07	0.07	0.10	0.12	0.20
Nickel	0.26	0.35	0.34	0.31	0.50	1.04	1.94	0.70	1.52	100
Selenium	0.02	0.04	0.02	0.02	0.01	0.01	<0.01	0.02	<0.01	1
Silver	0.04	0.06	0.03	0.05	0.11	0.04	<0.01	0.07	0.08	5
Tin	0.57	1.11	0.78	0.76	1.32	2.37	<0.01	0.93	1.45	100
Total chromium	0.82	0.44	0.63	0.46	0.50	0.36	0.62	0.21	0.18	5
Zinc	1.51	1.88	1.37	1.64	2.13	4.06	8.19	3.08	4.67	100

Table 11   Toxicity of Leachate water.

PARAMETER	RESULTS FOR LEACHATE WATER (mg/l)					THRESHOLD LIMIT (mg/l)
	NM35	NM25	NM15	NM10	NM5	
Arsenic	<0.01	0.02	<0.01	<0.01	<0.01	5
Barium	3.45	3.69	1.63	0.93	0.64	100
Boron	<0.01	0.29	0.06	0.35	0.31	400
Cadmium	<0.01	<0.01	<0.01	<0.01	<0.01	1
Copper	0.05	0.16	0.09	0.15	0.16	100
Lead	0.11	0.14	0.09	0.04	0.04	5
Mercury	0.01	0.09	0.01	0.005	0.01	0.20
Nickel	<0.01	<0.01	<0.01	0.02	0.03	100
Selenium	0.02	0.02	0.01	<0.01	<0.01	1
Silver	<0.01	<0.01	<0.01	<0.01	<0.01	5
Tin	<0.01	<0.01	<0.01	<0.01	0.02	100
Total chromium	<0.01	<0.01	<0.01	0.02	0.02	5
Zinc	0.06	0.08	0.04	0.12	0.10	100

CONCLUSIONS

Following conclusions can be drawn from the tests conducted so far:

* Incinerator slag is light weight, and performs well in CLSM. The hardened density of CLSM mixes varied from 1419 kg/m^3 to 1610 kg/m^3.

- It is possible to produce CLSM with compressive strength from 0.3 MPa to 10 MPa by adjusting the cement content.

- The initial surface absorption test values ranged from 1.82 to 12 mL/m^2/s.

- The CBR values ranged from 4 to 23.

- The CLSM produced did not show corrosive characters as evidenced from the pH tests.

- The water absorption and sorptivity tests reveal that the CLSM mixtures were more permeable than that of normal concrete.

- The toxicity tests for thirteen heavy metals showed values well within the threshold limits. Heavy metals concentration on bleed water is more than that of leachate water.

REFERENCES

1.   AMERICAN CONCRETE INSTITUTE. Controlled Low-strength Materials. ACI 229R-99, 1999.

2.   TIKALSKY P J, BAHIA H U, DENG A AND SNYDER T. Excess foundry sand characterization and experimental investigation in controlled low-strength material and hot-mixing asphalt. U.S. department of energy. Final report. Contract No. DE-FC36-01 ID13794. October 2004. 202pp.

3.   www.epa.gov/epaoswer.hazwaste/test/6_series.htm. Accessed in April 2007.

4.   MALAYSIAN STANDARDS. Portland cement (ordinary and rapid-hardening): Part 1 : specification. MS 522: Part 1:2003.

5.   BRITISH STANDARDS INSTITUTION. Specification for Lightweight aggregates for masonry units and structural concrete. BS 3797: 1990, clause 5.2, 3pp.

6.   BRITISH STANDARDS INSTITUTION. Specification for aggregates from natural sources for concrete. BS 882 : 1992. 5pp.

7.   AMERICAN SOCIETY FOR TESTING AND MATERIALS. Standard test method for flow consistency of controlled low-strength material. ASTM D 6103-00. 2003. Annual book of ASTM standards: concrete and aggregates. PA.

8.   BRITISH STANDARDS INSTITUTION. Products and systems for the production and repair of concrete structures - Test methods - Determination of stiffening time. BS EN 13294: 2002.

9.   BRITISH STANDARDS INSTITUTION. Testing concrete. Recommendations for the determination of the initial surface absorption of concrete BS 1881-208:1996.

10. BRITISH STANDARDS INSTITUTION. Methods of test for soils for civil engineering purposes. Compaction-related tests. Clause 7. BS 1377-4:1990.

11. TESTING OF CONCRETE. Surrey University Press. 2nd edition. NY 1989. 153pp.

12. From www.dur.ac.uk/~des0www4/cal/roads/pavdes/pavfound.html downloaded in April 2007.

13. From www.highwaymaintenance.com/cbrtext.htm downloaded in April 2007.

14. BRITISH STANDARDS INSTITUTION. Testing concrete. Method for determination of water absorption. BS 1881-122:1983.

15. RODERICK C K, JONES M ANDAIKATERINI GIANNAKOU. Thermally insulating foundations and ground slabs using highly foamed concrete. Innovations in controlled low-strength materials (flowable fills). American society for testing and materials, STP 1459. Hitch J.L, Howard A.K, Baas W.P. ASTM international, PA, 2004, 105pp.

16. NEIL JACKSON AND RAVINDRA K DHIR. Civil engineering materials. Fifth edition. Macmillan. London. 1996. 267pp.

17. TIKALSKY P J, BAHIA H U, DENG A AND SNYDER T. Excess foundry sand characterization and experimental investigation in controlled low-strength material and hot-mixing asphalt. U.S. department of energy. Final report. Contract No. DE-FC36-01 ID13794. October 2004. Appendix C.

MASONRY WALLS BACKED BY
SELF-COMPACTING CONCRETE

M Resheidat

Jordan University of Science and Technology

M A Al-Sreyheen

Greater Amman Municipality

Jordan

ABSTRACT: This paper presents the experimental and numerical results on the performance of masonry stone wall panels backed by self compacting concrete in comparison with the same panels backed by normal concrete. The purpose of this study is to enhance the bond between the concrete and the building stones under different types of loadings: compression, shear and direct tension. Sixteen half scale models of wall panels were built from stone courses; eight of them backed by self compacting concrete while the other eight were backed by normal concrete. All model wall panels were identical in every respect except for the type of concrete. The panel consists of four stone courses and the dimensions were 500 x 500 mm with 150 mm wall thickness. The panels were loaded by either a uniformly distributed load or a diagonal concentrated load. Shear tests and direct tensile tests on small specimens made of stones and concrete were carried out. The experimental study was supported by a numerical study using a block finite element model using SAP 2000 computer program. The numerical output results in terms of stresses and strains were obtained. Comparisons of results were made between panels using self compacting concrete and those using normal concrete. It is concluded that using self compacting concrete in not only enhances the bond with the stones used but also exhibited high strength and higher load carrying capacity.

Keywords: Masonry walls, Self-compacting concrete, SCC, Bond, Adhesion

Professor Musa Resheidat, Department of Civil Engineering, Jordan University of Science & Technology, Irbid, Jordan. He is ASCE Fellow; Member and Expert of Jordan Engineers Association; IACT Chairman, International Association of Concrete Technology.

Engineer Mohammad Ali Al-Sreyheen, graduated with BSc and MSc degrees in Civil Engineering from Jordan University of Science & Technology, Irbid, Jordan. He is working now in Greater Amman Municipality, Amman, Jordan.

INTRODUCTION

The majority of buildings in Jordan have either masonry stone walls or walls made from concrete or reinforced concrete faced by stones courses. This is a traditional demand by the owners or enforced by the planning and municipal authorities. Accordingly Jordan became unique in this type of construction. It has many advantages in terms of very low or no follow-on cost for maintenance In addition to that this type of construction is highly accepted from the architectural point of view. As building material, stones have been widely used in Jordan since pre-historic times. It has evolved from being a load bearing construction material to one that is merely used for cladding and therefore serves aesthetic purposes. Although each historical cultural period reflects individual building techniques, there are many traditions in the cutting and processing of stones that have been progressed and developed to cope with the construction needs across many decades. Despite of the great demand in using stones, many problems result from using conventional concrete to back the stone walls. These problems are associated with the methods of construction. The contemporary practice is to build 3-4 courses of stones and then back them by concrete This may cause poor contact and hence weak bond between concrete and stones. Upon concrete hardening; this may lead to falling of stone courses On the other hand, if concrete with high water to cement ratio is used, concrete may not comply with specified design strength and hence less bond is attained between concrete and building stones. If concrete as designed for and a specific strength and is to be vibrated to achieve good compaction, this may alter the position of stones and hence may lead to wall collapse. To achieve a dense and well compaction of concrete and to maintain the stability of stone courses became a serious problem. Constructors started to change the method of construction by building the reinforced concrete elements such as beams, columns and slabs, then used the stone courses as cladding facing these elements. The exterior wall panels were sandwich panels; stones courses and hollow blocks filled with concrete in between. This practice is heavily used in multi-storey residential buildings. The main shortcoming lies in a very week and porous concrete as shown in Figure 1.

Figure 1 Sandwich wall panel made of stone, concrete and concrete block layers

Thus it is thought to utilize concrete by which the said problems may be avoided. As a result of this, self compacting concrete (SCC) is considered as a concrete which can be placed and compacted under its self-weight with little or no vibration effort, and which is at the same time, cohesive enough to be handled without segregation or bleeding. It represents one of the most outstanding advances in concrete technology during the last decade. Due to its specific properties, SCC may contribute to a significant improvement of the quality of concrete structures and open up new fields for the application of concrete. Of its advantages, Self-compacting concrete (SCC), fills all recesses, reinforcement spaces and voids, even in highly reinforced concrete members and flows free of segregation nearly to level balance. While flowing in the formwork, method of construction will be eased.

Previous experimental studies were carried out to study the behaviour of masonry stone walls backed by normal concrete [1, 2], but none of these studies used the self compacting concrete. Although self compacting concrete has widely spread in many countries, but objectives did not also focus on its use in masonry stone walls and/or wall panels [3, 4]. Preliminary findings of this study have been published [5]. This study is an application on the use of self compacting concrete that was fully documented in Reference 6.

FINITE ELEMENT ANALYSIS

The wall panel dimensions were 500 x 500 mm and the panel thickness was equal to 150 mm so that the specimen could be housed in the space of the universal testing machine available in the laboratory. Based on that, a numerical analysis of the sandwich panel as composed of two layers; stone and concrete using a mesh of 10 x 10 block finite elements was generated to model both concrete and stone layers. SAP2000 program was employed to estimate the failure load and the output results in terms of stresses and strains for all nodes. Reasonable material properties were assumed to obtain the required approximate failure loads that will be employed in testing the model panels. Finite element analysis was performed for all panels as follows:

- 6 panels backed by normal concrete loaded by an equivalent uniformly distributed load.

- 2 panels backed by normal concrete loaded diagonally by a concentrated load.

- 6 panels backed by self compacting concrete loaded by a uniformly distributed load.

- 2 panels backed by self compacting concrete loaded diagonally by a concentrated load.

Stress trajectories at the interface plane between stones and concrete for one quarter of the panel was shown in Figures 2 and 3. The load was a uniformly distributed load. Figures 4 and 5 show the stress mapping when the specimen was diagonally loaded.

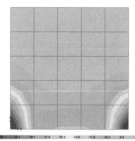

Figure 2   In-plane stress distribution at the interface plane for one quarter of panel.

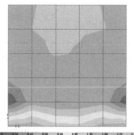

Figure 3   Splitting stress distribution the interface plane for one quarter of panel.

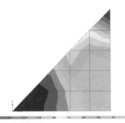

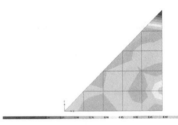

Figure 4   In-plane stress distribution at the interface plane for one quarter of panel diagonally loaded.

Figure 5   Splitting stress distribution at the interface plane for one quarter of panel diagonally loaded.

LABORATORY STUDY

The concrete constituents for both normal concrete(NC) and self compacting concrete(SCC) were basically the same for both concrete types. Due to nature of concrete, the mix proportions vary. For SCC, a dose of viscosity modifying agent (VMA) and superplasticizer (SP) have added to the concrete mix as per the recommendation of manufacturers. The full documentation, description of the materials as well as the mix proportions can be found in Reference 5. In this study, the proportions of concrete mixes for both NC and SCC are given in Table 1.

A total of 16 stone wall panels were constructed. Quality control concrete specimens in terms of cylinders, cubes and prisms were properly sampled, cured and tested. A group of prisms made of concrete and stones have been made to be tested in direct shear and direct tension.

Table1   Basic Concrete Mix Proportioning

CONCRETE TYPE	CEMENT kg	SAND kg	FINE AGGREGATES, kg	COURSE AGGREGATES, kg	WATER Litre
NC	350	500	500	650	160
SCC	350	500	500	650	140

TEST SETUP AND INSTRUMENTATION

Concrete mixes, erection of masonry stone panels, curing, concrete sampling and all tests for compression, tension and shear were carried out in the laboratories of the department of civil engineering at Jordan University of Science & Technology.

Compression and Diagonal Tension Setups

Compression testing machine with capacity of 4000 kN in compression was used to test the stone wall samples. Eight samples with NC and eight samples with SCC have been tested with different cases of loading, ages and compressive strength. Three samples for each type of concrete were tested by loading on stone and concrete, three samples by loading on concrete only, and two samples by diagonal loading. The test setups are shown in Figures 6 and 7.

Figure 6   Loading on stone and concrete

Figure 7   Diagonal loading test

Direct Shear Test

Portable direct shear machine as shown in Figure 8 was used to test the six prisms; each of which has cross section 50 x 50 mm and 100 mm in length. The specimen is made of two parts; concrete and stone. Three specimens made of NC and the other of SCC have been tested with same case of loading, ages and compressive strength. see Specimens are shown in Figure 9.

Figure 8   Stone – Concrete Samples
in Direct Shear Test

Figure 9   Portable Direct Shear Device

Direct Tension Test

A pair of griping devices were designed and then prepared in the engineering workshops as shown in Figure 10. The direct tension setup is shown in Figure 11. Three samples with NC and three samples with SCC have been tested by two methods, loading by gravity and tensile testing machine with same cases of loading, ages and compressive strength.

RESULTS AND DISCUSSION

The maximum normal loads carried by these samples are presented in Table 2. This Table shows that the maximum normal load for the case of loading on concrete and stone together is greater than that for the case of loading on concrete alone, for both normal and self compacted concrete and the maximum normal load for SCC stone walls is greater than that for NC stone walls.

Figure 10   Griping System for Direct   Figure 11   Direct Tension Setup
            Tension Test

Table 2   Maximum Load for Panel Walls.

PANEL NO.	STRENGTH OF CONCRETE MPa	LOADING ON	MAX. LOAD AT SEPARATION OF STONE, kN	MAX FAILURE LOAD, kN
NC 1	17.4	Stone and Concrete	806	806
NC 2	17.4	Concrete	602	602
NC 3	22.0	Stone and Concrete	704	932
NC 4	22.0	Stone and Concrete	652	928
NC 5	22.0	Concrete	650	650
NC 6	22.0	Concrete	503	646
NC 7	22.0	Diagonal	434	434
NC 8	22.0	Diagonal	404	404
SCC 1	17.8	Stone and Concrete	1012	1136
SCC 2	17.8	Concrete	551	1115
SCC 3	22.2	Concrete	942	1192
SCC 4	22.2	Stone and Concrete	969	1208
SCC 5	27.7	Stone and Concrete	1300	1402
SCC 6	27.7	Concrete	1362	1362
SCC 7	22.3	Diagonal	849	849
SCC 8	22.3	Diagonal	737	737

Normal and lateral stresses for all samples is shown in Table 3. This table shows that the finite element results were in agreement with the test results which indicate that the normal and lateral stresses for SCC are greater than those of NC for all cases.

The full documentation of all numerical and test results can be found in Reference 7. Due to space limitation, the load deformation diagrams for a panel loaded by vertical uniformly distributed load as shown in Figure 12 illustrates the load bearing capacity as related to type of backing concrete. Similarly, Figure 13 illustrates the comparison of panel capacity under a diagonal load. All experimental results in terms of compressive stresses and tensile stresses of self compacting concrete are higher than those of normal concrete.

Table 3 Normal and Lateral Stresses for All Panels.

PANEL NO.	STRENGTH OF CONCRETE MPa	NORMAL STRESS, MPa		LATERAL STRESS, MPa	
		Experimental	Numerical	Experimental	Numerical
NC 1	17.4	10.75	11.452	7.164	3.544
NC 2	17.4	10.03	6.782	1.209	2.532
NC 3	22.0	9.38	9.435	4.388	3.623
NC 4	22.0	8.69	6.055	4.005	3.949
NC 5	22.0	10.83	7.353	2.503	2.715
NC 6	22.0	8.39	5.690	2.503	2.101
NC 7	22.0	10.22	7.606	4.956	2.129
NC 8	22.0	9.51	7.073	1.844	1.979
SCC 1	17.8	13.49	14.229	12.159	4.590
SCC 2	17.8	9.18	6.212	4.956	2.314
SCC 3	22.2	15.7	10.654	11.190	3.934
SCC 4	22.2	12.92	12.990	13.987	4.989
SCC 5	27.7	17.33	16.156	11.865	7.929
SCC 6	27.7	18.16	15.479	9.558	5.648
SCC 7	22.3	19.99	14.874	5.310	4.162
SCC 8	22.3	17.36	12.919	2.626	3.615

It could be concluded the bond strength between stones and concrete exhibits high performance in addition to avoiding stability problems during construction. Self compacting concrete gives better bond with building stone than normal concrete when exposed to shear load as in lintels, exposed beams and head walls in constructions. The results tabulated in Figures in Tables 4 and 5 represents the output results of direct shear and direct tensile tests for the interface between stone and concrete. It can be observed that the average value of shear stress for SCC is 63% higher than the average value of NC. Similarly, it can also be observed that the average value of tensile stress for SCC is 35% higher than the average value of NC.

Table 4 Comparison of Shear Stresses at Failure Between NC and SCC of Stone-Concrete Specimens with 50 x 50 mm in cross section.

SAMPLE NO.	CONCRETE TYPE	STRENGTH OF CONCRETE, MPa	SURFACE AREA, mm^2	LATERAL LOAD, kN	SHEAR STRESS, MPa
1	SCC		2500	1.2	0.480
2	SCC	23.54	2250	1.0	0.444
3	SCC		2400	0.95	0.396
4	NC		2352	0.65	0.276
5	NC	23.28	2500	0.90	0.360
6	NC		2254	0.50	0.222

Load Deformation Diagram for Normal panel

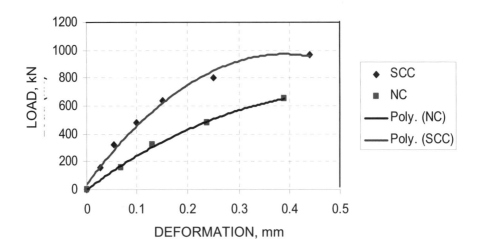

Figure 12   Comparison of Load-Deformation Diagram for Normal Wall Panel

Load Deformation Diagram for Diagonal Panel

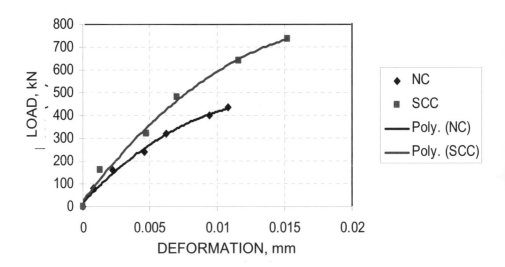

Figure 13   Comparison of Load-Deformation Diagram for Diagonal Wall Panel

Table 5 Comparison of Tensile Stresses at Failure between NC and SCC of Stone Concrete Specimens with 50 x 50 mm in cross section.

SPEC. NO.	CONCRETE TYPE	STRENGTH OF CONCRETE MPa	FAILURE LOAD kN	TENSILE STRESS MPa	AVERAGE TENSILE STRESS MPa
1	SCC		1.72	0.688	
2	SCC	24.4	1.79	0.716	0.703
3	SCC		1.76	0.704	
4	NC		1.37	0.548	
5	NC	25.0	1.20	0.480	0.519
6	NC		1.32	0.528	

CONCLUSIONS

The following conclusions may be drawn from this study:

1. Self compacted concrete has better finishing surfaces. No need for any mechanical vibration of concrete as it is the case in normal concrete.

2. Self compacted concrete has better bond with building stone than normal concrete. The bond between concrete and stones is enhanced by about 40% more than normal concrete.

3. For the same target strength value of concrete, the shear strength of self compacting concrete was significantly increased by 25%.

4. The use of self compacting concrete has increased the panel wall carrying capacity by at least 20% more than the same panel using normal concrete.

5. The finite element analysis is a powerful tool to evaluate the all stresses which was essential for guiding the experimental investigation.

ACKNOWLEDGEMENTS

The authors would like to acknowledge the financial support provided by the Deanship of Scientific Research at Jordan University of Science & Technology. Assistance and help provided by all staff in the Engineering workshops and in the Civil Engineering Laboratories are highly appreciated.

REFERENCES

1. ABDEL-HALIM, M. BASOUL, M., and ABDEL-KARIM, R., "Structural evaluation of concrete-backed stone masonry", ACI Struct. J., V. 86, pp. 608-614, 1989

2. ABDEL-HALIM, M, and BARAKAT S., "Cyclic performance of concrete-backed stone masonry wall", ASCE J. of Structural Engineering, pp. 696-605, May 2003.

3. DEHN, F., "Betontechnologische Voraussetzungen Zur Herstellung Von Selbstverdich-tendem Beton", 10. Leipziger Massivbauseminar, 2000.

4. PETER, J. A. et al, "Studies on Proportioning of SCC and evaluation of its strength and durability characteristics", Proceedings of the 7th International Conference on Concrete Technology in Developing Countries, pp. 116-125, 2004.

5. RESHEIDAT, Musa and AL-SREYHEEN, Mohammad, "Performance of Self Compacting Concrete-Backed Stone Masonry Walls". Proceedings of TCMB 3rd International Symposium, May 21-23, 2007, Istanbul, Turkey, Volume 2, pp.811-819.

6. AL-ZYOUD S.A., "Experimental investigation on self compacting concrete", M.Sc. Thesis, Department of Civil Engineering, Jordan University of Science and Technology, May 2006.

7. AL-SRYHEEN, Mohammad Ali, "Experimental Study On The Performance Of Stone Walls Backed By Self Compacted Concrete", M.Sc. Thesis, Department of Civil Engineering, Jordan University of Science and Technology, May 2007.

THE WORKABILITY OF CONCRETE: IS THERE AN EASY WAY TO PRODUCE SELF-COMPACTING CONCRETE?

A L A Fraaij

M R de Rooij

Delft University of Technology

Netherlands

ABSTRACT. This paper describes a research that has been undertaken to arrive at an easy method to design self-compacting concrete. The basic ideas behind the research goal are highlighted. Test results show that it is possible to use a small (easy) test series to develop self-compacting concrete. However, concrete is not a material that can be handled carelessly, as is also shown in this paper. The paper pleads for development of surface-colloid knowledge in the concrete practice that should be incorporated in the concrete design rules if one designs a concrete mixture upon workability.

Keywords: Self-compacting concrete, Fly ash, Ground limestone, Flow of concrete.

Dr ir Alex L A Fraaij is an associate professor at Delft University of Technology, Department of Civil Engineering, Section Materials Science where he teaches Materials Science. His field of interest is in the application of recycled aggregates in concrete and sustainability aspects of building materials.

Dr ir Mario R de Rooij is as an assist. prof. a member of the Microlab of the Department of Civil Engineering of Delft University of Technology. He is a project leader of research projects on self-healing materials and on the internal structure of hardening concrete.

INTRODUCTION

Self-compacting concrete is made by adjusting the relative volume of fines compared to the volume of the coarser particles. This means that compared to ordinary concrete that is not self-compacting, the volume of the coarser material is smaller than in the normal concrete. Self-compacting concrete cannot be made without the modern types of superplasticizers, but the interaction of the chemicals with the components of the concrete is still not completely understood.

The final estimation of the volumes of the different fractions is a complicated procedure in which, in a sequence of testing steps, some parameters must be fine-tuned. The procedure includes tests on pastes and mortars and involves tests of flowing time through funnels and tests on flow values of mortars (fluidity measurements) and concrete, such as the U-flow test. We know from rheology that the behaviour of a suspension can be described with a yield stress parameter (the minimum shear stress level below which flowing will not occur) and the viscosity parameter (as a coupling factor between the necessary shear stresses and speed of rotation of the suspension); [1, 2]. The "shear stress" - "speed of rotation curves" however are seldom according to the theoretical curves such as the Bingham fluid in Figure 1.

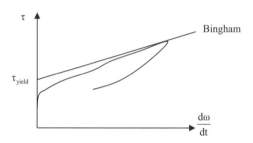

Figure 1   Example of the rheological behaviour of fresh concrete
according to Bingham and the practice.

We have the complication with concrete that the rheological behaviour changes in time after mixing of the concrete. Great effort is done to improve the working time of the super-plasticizers but it is still not possible to control with great confidence the efficiency of the superplasticizers over a period of say 90 minutes after mixing.

We know that superplasticizers lower the yield value of fresh concrete, but do not change the viscosity of the fresh concrete that much. An extra addition of water lowers the yield value and the viscosity, but also diminishes the concrete strength and durability, increases bleeding and causes segregation of the mixture.

It is possible to create self-compacting concrete, but there are some problems:
- The methods that are used to achieve self-compacting concrete in the laboratory are complicated and time consuming.
- The rheological behaviour of fresh concrete is complicated and rather unpredictable.
- We do not fully understand the mechanisms (especially at the micro level) involved.
- The results are highly dependent on the chemicals used to achieve the proper rheological behaviour [3], but we cannot predict this behaviour based on some simple characteristics of the superplasticizers, the cement, the aggregates and fillers and we do not really know what characteristics are of real importance.

The authors try to develop an easy test procedure (while measuring only the flow of concrete) to come to self-compacting concrete. By adjusting the concrete composition, the workability is increased up to the desired level. At the end of the series, a U-flow test can be done to show that the concrete completely levels out (while testing only a couple of those concrete mixtures). The paper describes the present successful findings as well as the failures we were confronted with in our steps in the design of very workable concrete mixtures.

SOME GENERAL IDEAS ABOUT WORKABILITY

A concrete mixture is a complicated suspension with a relative low water volume and a large volume of various particles. The particles range form nanometres to centimetres.

The liquid changes in composition right after mixing with the components of the concrete and high concentration of cations and anions build up due to the dissolution of certain cement components and we are not sure what the other fines and aggregates will do in the water. This change in composition continues for years, but concerning the workability, only the first two hours are of importance for us. In that period, the superplasticizers interact with the liquid and the particle surfaces that also changes in time. In general, the viscosity of the water changes as does the surface energy. Further, the tails of the modern superplasticizers molecules cause a kind of steric hindrance that keeps the fine particles away from each other avoiding (to a certain extent) clogging and clustering, but this mechanism changes in time as well. Up to now, colloid chemical knowledge is not really available in civil engineering concrete practice and this should be overcome in the near future.

In the suspension billions of particles move around, hit each other and stick together or come loose from each other. Those particles show a huge range in size and they do not necessarily have a spherical surface. There is porosity, obscure textures, and electrical surface loadings; there are double layers around surfaces etc.

In a proper self-compacting concrete the movements of the particles must not be accompanied with segregation thus the success and failure of the mixture depends on viscosity behaviour of the mixture as well as the capability of the water-fines paste to glue to bigger particles or to hinder the bigger particles to leave the water-fines suspension.

In a concrete mixture, there are three levels of suspensions: First, the suspension of water with cement and fillers, we call this the paste. The next level is the suspension of paste with sand, we call this the mortar phase. The third level is the suspension of gravel and mortar; this is our concrete. All the three levels show different rheological behaviours. The question is whether we really need to study this behaviour for all three levels in order to make self-compacting concrete.

We know from literature and practice that self-compacting concrete can be achieved by decreasing the volume of the coarser particles and this means that the paste-fines suspension plays an important role in this matter [3 to 10].

Irrespective of what we mean with coarse particles and finer particles we can intuitively suggest that:

- The workability *increases* when the paste layer around the coarser particles becomes thicker.

- The "*stiffness*" of the paste layer influences also the workability: a stiffer water-fines paste will show lower flow values than a more fluid water-fines paste.

- Segregation is found when the stiffness or "*gluing capacity*" of the water-fines paste is not sufficient (too stiff as well as too wet) to hold the gravel.

It looks logical to predict that the second and the third aspect are closely related.

Knowing the densities of all the materials, the particle size distribution of the sand and the gravel, as well as the Blaine values of the filler (such as fly ash) and the cement, the total surface area of all the particles can roughly be estimated with a spreadsheet. As an arbitrary choice, the boundary between fine particles and sand was put on 0.125 mm. See Figure 2 for our definition of paste.

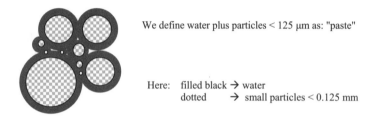

We define water plus particles < 125 μm as: "paste"

Here: filled black → water
        dotted   → small particles < 0.125 mm

Figure 2   Paste: water + all particles < 0,125 mm.

Since in our definition "*paste*" is the suspension of water with fines, the total volume of the paste can be estimated according to:

$$\text{Volume paste} = \text{volume free water} + \sum_{i=1}^{n}\left(\frac{\text{added mass of filler}_i \text{ or cement}_i}{\text{density}_i}\right)$$

As Figure 2 suggests, there may be some air entrapped in the paste, but we disregard this relatively small volume. We suppose that the entrapped air is in between the coarser particles. We also suppose that all the fine particles are surrounded by water and no coagulation or flogging will happen, i.e. that the superplasticizer will do its work well just after mixing. Now the total amount of free water can be estimated from the amount of added water + water in superplasticizers + adhered water on the surface of the aggregates in case of wet aggregates.

The water layer thickness t_{water} around all the particles smaller than 0.125 mm can be calculated according to:

$$t_{water} = \frac{\text{volume of free water}}{\text{total surface of particles} < 125 \mu m} = \frac{\text{volume of free water}}{\sum_{i=1}^{i=n}(\text{Blaine value})_i \times (\text{mass})_i}$$

with Blaine value = specific surface in m^2/kg.

The workability increases with:

1)   the particle roundness and smoothness,

2)   the water layer thickness around the fine particles,

3)   the paste layer thickness around the aggregates,

4)   the efficiency of the superplasticizers in dispersing the particles,

5)   lower temperature.

We assume now that all the particles are spherical, but in reality, this will not be the case. Further, we assume that the bleeding is negligible so that indeed all the sand and gravel particles in the concrete mixture will be surrounded by a suspension of water-fines instead of only plain water or air.

The next step is the estimation of the paste layer thickness t_{paste} around the particles bigger than 0.125 mm with the formula (see also Figure 3):

$$t_{paste} = \frac{\text{volume of paste}}{\text{total surface of particles} > 125\mu m} =$$

$$= \frac{\text{volume of paste}}{\sum\limits_{\text{aggregate 1}}^{\text{aggregate n}} \sum\limits_{\text{sieve fraction j}=1}^{\text{sieve fraction j}=k} (\pi.\overline{d}^2_{\text{sieve fraction j}}) \times (q_j = \text{number of particles per sieve fraction})}$$

$$q_{\text{fraction j}} = \frac{\text{volume of fraction}_j \text{ per 1 m}^3 \text{ concrete}}{\text{volume of 1 sphere in fraction j}} = \frac{\dfrac{\text{mass in fraction j}}{\text{density}}}{\pi.\overline{d}^3_j / 6}$$

Here we assume that all the particles are non-porous, smooth and spherical, which will unfortunately not be the case in reality.

Of course, the workability will increase when the paste layer is thicker and more fluid. However, we cannot make the paste too fluid and too thick because then segregation will occur unavoidably.

Since we don't know how efficient the paste is in wetting the other particles, we assume that 5% of the paste will not contribute to the wetting of the particles.

We define water plus particles < 125 μm as: "paste"

Here: filled black → paste
           dotted   → sand plus gravel particles (> 0.125 mm)

Figure 3   Paste surrounding sand and gravel particles.

THE RESEARCH SET-UP

Based on the ideas of the sections above, a series of concrete mixtures was made and the flow values of the concrete mixture were measured. The mixtures were made to check the relationship between paste layer thickness and flow value. The concrete was made with CEM I 52,5 R ENCI, crushed gravel up to 16 mm, river sand, fly ash and a modified polycarboxylic superplasticizer. The test set-up started with the starting mixture, see Table 1.

Table 1   Starting mixture (kg/m^3 concrete).

Portland cement CEM I 52.5 R	335	Blaine cement m^2/kg	545
River sand (FM = 3.47)	839		
Crushed gravel (FM = 6.55)	946	w/c	0.50
Mixing water	167	w/(c+k.pfa)	0.48
Superplasticizer	0.5% of fines	w/(c+pfa)	0.42
Fly ash (= pfa with k-value = 0.2)	67	Blaine pfa (m^2/kg)	250

The temperature of the mixing water was kept in the range of 20–22 °C. The concrete components were loaded and mixed in a pan-type mixer. Sand and gravel were dry and did not contribute to the free water level.

In the first test series the water-binder ratios were kept constant, thus the increase of paste volume was achieved by adding more superplasticizers, water, cement and fly ash in the same mass ratios. Thus in this series only t_{paste} was modified, not t_{water}.

After the mixing procedure, the slump and flow values have been measured. The concrete density (after compaction in the case of low flow values) and the air content have been measured in order to check if the predicted density and air content were really found in the lab mixture. This appeared to be the case in the trial mixtures.

Figure 4a shows the results of the first five mixtures. Since the idea was to achieve a flow value of roughly 700 mm, these mixtures were not passing the test. So new mixtures were made with higher paste volumes to reach the desired workability level, see Figure 4b. Unfortunately, while those new mixtures were workable enough, we detected segregation. In order to get more data so that we could perform a regression analysis, we made a second test series, series 2. The results are plotted in Figure 4c. The regression line enables us to predict the desired paste layer thickness to achieve a flow value of 700 mm. The last mixture is the big black triangle in Figure 4d and levelled out satisfactory in the U-flow test. To avoid segregation we added starch (for wallpaper paste) in powder form to the mixing water to make the paste more "sticky".

So, the procedure to achieve self-compacting concrete is the following:

1) Make a standard mixture and measure the flow.

2) Increase the volume of the fillers, cements, water and superplasticizers relative to the sand and gravel and calculate t_{paste} at constant water-binder ratio and according to our findings, you may exaggerate a lot in your paste volume.

3) Make this new mixture and measure the flow value. Draw a straight line through the two measured points and calculate the desired t_{paste} at the desired flow level.

4) Make this third mixture and if you expect segregation, add a stabilizer.

5) Test the flow value and check with U-flow test. In the case that the flow is not enough or that the concrete does not level out sufficiently in the U-flow test, you could use somewhat more superplasticizer.

This procedure worked out well for fly ash as filler, but what may happen if you use other fillers, other cements, and other aggregates and if you use another mixer?

For a new series, we asked groups of students (as a task of their second years materials practical) to make mixtures with ground limestone, CEM I 52.5 R ENCI, river sand and smooth river gravel up to 16 mm with a fixed W/C ratio of 0.45 and water/(c+filler) ratios of 0.42 and 0.38 with an ordinary 70 L gravity mixer. The starting point was a concrete with c = 350 kg/m^3.

The Blaine of the ground limestone was unknown, but we estimated the Blaine value to calculate a fictitious paste thickness. We thought that since we would use the same fictitious Blaine value for all the mixtures the basic idea should not be violated. Eight groups of students produced mixtures with a different amount of fines relative to the sand and gravel volume but all with the same w/(c+filler) ratio. The result is presented in Figure 5a. Because even the last mixture (with 380 kg cement and 76 kg limestone filler) did not reach the desired 700 mm flow, the next two groups made mixtures with 1% of superplasticizer instead of 0.5%. The result is presented in Figure 5b, (the open square states for: segregation). So finally, we reached our goal because the last mixture with the coloured square had a flow of 700 mm and it did not segregate. However, it is clear that with the optimistic procedure stated above we did not succeed in three trial mixtures.

It becomes even worse: Four weeks later, seven other groups of students tried to produce the concrete, but now with 0.7% superplasticizer. The result is Figure 5c which is clearly inconsistent with the earlier results.

After 1 month, the students were asked to determine the cube compressive strengths and then they recognized clusters of non-cemented fine powder: the non-well-dispersed limestone powder. Thus, proper mixing, proper wetting of fines is of paramount importance as well as insight in the particle size distribution of the filler and cement.

SOME FINAL REMARKS

It is obvious that some questions should be answered in the future:

• What is the influence of the amount of non-wetted coarse particles and how can we predict this value?
• What is the influence of the amount of improper distributed fines and how can we predict this value?
• How can we predict the efficiency of the superplasticizers right from the start of mixing up to 90 minutes?
• How fast will fine particles coalesce and obscure the workability and why do they do this?
• How can we create concrete that is less sensitive to mixing procedure?
• How can we create concrete that is less sensitive to aggregate properties and filler properties?

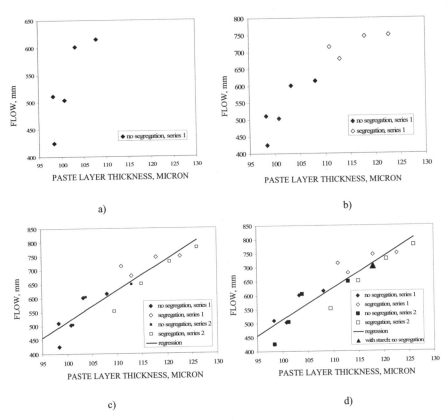

Figure 4   Test results of concrete mixtures with fly ash. The starting (reference) mixture contained 335 kg CEM I 52,5 R, 67 kg fly ash with w/(c+pfa) = 0.42 in all the mixtures. Segregation of mixture ▲ in Figure 4d is avoided by an addition of 60 gram starch/m³.

It is of great importance that we learn to better understand the surface-colloid mechanisms that are happening in the mixture in relation to workability; so that we can we model those surface-effects in the future. Up to now, we are not that far.

CONCLUSIONS

- With the proper mixer (pan type) and with a filler material that is smooth and spherical such as fly ash (figure 4), we can make a concrete mixture with a flow of 700 mm after three trial mixtures and that levels out satisfactory in the U-flow test.
- The mixing procedure is and the type of mixer, are of great importance in order to obtain reliable results.
- The non-consistent results of figure 5c suggest that the working model should be enhanced to surface-colloid mechanisms and to non-spherical particles as well as particles with rough surface texture.

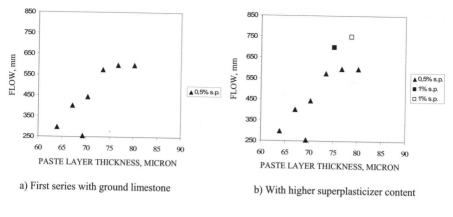

a) First series with ground limestone   b) With higher superplasticizer content

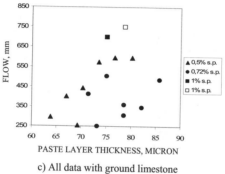

c) All data with ground limestone

Figure 5   Results of mixtures all with CEM I 52.5 R ENCI, ground limestone and fixed W/C = 0.45. Mixtures with 0.5% superplasticizer (s.p.): w/(c+filler) = 0.42. Mixtures with 0.7% and 1% superplasticizer: w/(c+filler) = 0.38.

REFERENCES

1. WALLEVIK, O H. Rheology of Coarse Particle Suspensions, such as Cement Paste, Mortar and Concrete". 2000, TU-Delft.

2. TATTERSAL, G H, BANFILL, P. "The rheology of fresh concrete". Pitman Books Ltd, London 1983.

3. TAKADA, K. Influence of admixtures and mixing efficiency on the properties of self-compacting concrete - Ph.D. Thesis, Delft University of Technology. 2004.

4. OZAWA, K, MAEKAWA, K, UNNISHIMA, M AND OKAMURA, H. High Performance Concrete Based on the Durability Design of Concrete Structures. Proceedings of the 2nd East Asia Pacific Conf. on Structural Engineering and Construction, 1, 1989. pp. 445-450.

5. OKAMURA, H AND OUCHI, M. Self-compacting concrete: Development, Present Use and Future. First Int. Symp. on SCC, Stockholm, Ed. Skarendahl and Petersson, RILEM publications PRO 7, Cochin, 1999. pp. 3-14.

6. OZAWA, K, TAGTERMSIRIKUL, S AND MAEKAWA, K. The filling capacity of fresh concrete. Proceedings of the Fourth CANMET/ACI International Conference on Fly Ash, Silica Fume, Slag and Natural Pozzolans in Concrete, 1992. pp. 122-137.

7. OKAMURA, H AND OZAWA, K. Self Compacting High Performance Concrete. Proceedings of Cairo First Conference on Concrete Structures, Vol. 1. 1996.

8. Oh, S G, Noguchi, T AND Tousawa, F. Towards mix design for rheology of self-compacting concrete. First Int. Symposium on SCC. Stockholm, Ed. by Skarendahl and Peterson. RILEM publications PRO7. Cochan, 1999. pp. 361-372.

9. TOMOSOWA, F, MASUDA, Y AND HAYAKAWA, M. Architectural Institute of Japan's Recommendations for Mix and Construction Practice of high fluid Concrete. International Workshop on Self-Compacting Concrete, 23-26 August, Kochi University of Technology, Kochi, Japan, 1998. pp. 400-437.

10. SU, N, HSU K C AND CHAI, H W. A simple mix design method for self-compacting concrete. Cement and Concrete Research, Vol. 31, 2001. pp. 1799-1807.

COMPARATIVE STUDY OF SELF-COMPACTING CONCRETE WITH FLY ASH AND RICE HUSK ASH

R Malathy

S Saravanan

K Sasi Kumar

Kongu Engineering College

India

ABSTRACT. Self compacting concrete (SCC), is a new kind of high performance concrete (HPC) with excellent deformability and segregation resistance. The major work in the production of SCC is, designing an appropriate mix proportion and evaluating the properties of the concrete obtained. In this study a mix design of SCC for different grades, M_{20} to M_{60} was developed and their flow properties and strength properties were studied. The flow properties such as passing ability, filling ability, viscosity and segregation resistance and compaction factor are checked by conducting various tests. The compressive strengths of concrete are checked after 7, 28, 56 and 90 days curing. Also split tensile strength, flexural strength and impact strength were obtained after 28 days curing.

Key Words: Self compacting concrete, Rice husk ash, Flow ability, Strength

Dr R Malathy, graduated in the year 1992, obtained M.E. (Structural Engineering) in 1998 both from Madras University and PhD from Bharathiyar University in 2005. She has teaching experience of over 20 years and is principal investigator for two R&D projects supported by AICTE and IE (I). She published about 30 papers in journals and conferences. She received Best Teacher Award for 2004 in KEC and Innovative Teacher Award from CLHRD, Mangalore. Presently she is working as Professor in Civil Department, Kongu Engineering College, Perundurai.

S Saravanan, is a final year Bachelor of Civil Engineering student at Kongu Engineering College. He has presented twelve papers at national level symposia and conferences. His areas of interest are e.g. Self Compacting Concrete and post-tensioning in concrete structures.

K Sasi Kumar, is a final year Bachelor of Civil Engineering student at Kongu Engineering College. He has presented four papers at national level symposia. His areas of interest are e.g. Self Compacting Concrete, CADD and Artificial Neural Networks in Civil Engineering.

INTRODUCTION

Self compacting concrete (SCC), is a new kind of high performance concrete (HPC) with excellent deformability and segregation resistance. The major work in the production of SCC is, designing an appropriate mix proportion and evaluating the properties of the concrete obtained. In practice, SCC in its fresh state shows high fluidity, self compacting ability and segregation resistance; all these properties contribute to reduce the risk of honeycombing of concrete. Hence, the SCC produced can greatly improve the reliability and durability of the reinforced concrete structures.

In addition, SCC shows good performance in compressive strength test and can fulfill other construction needs because its production has taken into consideration the requirements in the structural design. In this study, SCC of grades varying from M_{20} to M_{60} are developed and checked for self compactability with Japanese Standard for Civil Engineering specifications. The optimum dosage of superplasticizer is determined for different grades of SCC by conducting Marsh cone test.

LITERATURE REVIEW

Ever since the first report of the development of SCC in Japan in 1988 by Ozawa et al. using superplasticizer and viscosity modifying agent and in 1992, they again identified the factors controlling self compactability namely coarse and fine aggregate content, and developed test methods to check the self compactability and found that the water-powder ratio governs the self compactability [1]. Petersson et al. identified three main criteria for the design of SCC, namely construction criteria, void content and blocking criteria. They also suggested a theoretical model between the blocking volume ratio of aggregate volume and ratio of clear spacing between the reinforcement to aggregate fraction size [2].

Then Mortsell et al. modeled fresh concrete as a two-phase material with matrix of particles < 125 μm and particle phases composed of aggregates. They even developed an expression for the workability parameter was [3]. Sedran et al. reported the mix design procedure for SCC using the rheometer and a mathematical model which optimizes the aggregate packing taking into account the degree of compaction. This model minimizes the water content [4].

Sakata et al. studied the basic properties and the effect of Welan gum on SCC, which was found to be useful in stabilizing the rheological properties and mobility of SCC mix [5]. Nishibayashi et al. investigated the effects of the properties of the constituent materials on the rheological constants of mortar and concrete by using blastfurnace slag and lime powder and four kinds of chemical admixtures [6]. Okamura et al. discussed the procedure for adjusting w/p and SP dosage to achieve desired properties.

He found that for mortars, the ratio of slump flow index to funnel flow index to be almost constant with respect to volume of water to powder ratio for a given value of SP to powder ratio [7]. Saak et al. suggested that interparticle separation is a critical parameter for the design of SCC in addition to the particle packing distance. They proposed a segregation control theory considering static and dynamic (flowing) conditions. They measured yield stress of cement pastes using a shear vane, and the test results showed that concrete had the greatest fluidity at the lowest paste yield stress and viscosity, where segregation is avoided [8].

Malathy and Govindaswamy compared the fresh and hardened properties of SCC containing fly ash, silica fume and metakaolin [9]. Grunewald and Joost Walraven proposed a mix design for SCC reinforced with steel fibers.

They discussed the suitability of test methods and the effect of coarse aggregate content, the content and type of steel fibers on the workability of SCC [10]. Nansu et al. proposed a new mix design method for SCC. According to this method the amount of aggregates required is determined and the paste of binders is then filled in to the voids of aggregates to ensure that the SCC thus obtained has flow ability, self-compacting ability and other desired properties of SCC [11].

Subramanian and Chattopadhyay developed SCC with 50-60 MPa strength using fly ash, SP, naphthalene formaldehyde/acrylic polymer based SP, Welan gum and locally available aggregates in India [12] Gettu et al. proposed a four step methodology for mix proportioning of self compacting concrete with fly ash [13]. As per the EFNARC specifications, while designing the mix, it is most useful to consider the relative proportions of the key components by volume rather than by mass [14].

Jagadish Vengala et al. suggested a sequential method of starting with a high slump, non superplasticized concrete by replacing part of the coarse aggregate with fine fly ash for obtaining SCC. They also concluded that VEA may not be strictly necessary for obtaining SCC and also reported that, when VEA is added the strength increase was not maintained fully, but VEA added SCC showed higher strength to the reference mix [15].

RESEARCH SIGNIFICANCE

The flowing property and self compacting ability of SCC depends on the correct proportioning of ingredients and the dosage of superplasticizer/viscosity modifying agent. Studies are done in different ways for the development of SCC with different types of admixtures and filler materials. Till now there is no proper mix design method was developed for designing SCC of different grades. In this study there was an attempt made to develop the mix design for SCC of different grades varying from M_{20} to M_{60} and tests were conducted for the fresh and hardened properties of SCC. The significance of this research is to develop a chart to obtain the quantity of cement, fly ash, sand and coarse aggregate required for different grades of SCC.

MIX DESIGN PROCEDURE

The mix design for normal concrete is done by both I.S. and A.C.I. methods, for grades varying from M_{20} to M_{40} and by Erntroy and Shacklock's empirical (E.S.) method for grades from M_{45} to M_{60}. For SCC, to keep the powder content in between 160 and 240 litres/m^3, as per EFNARC specifications, fly ash is to be added in addition with the cement content obtained. Extra volume of fly ash added is replaced for coarse aggregate. Some quantity of cement is also replaced by fly ash. The fly ash replacement varies from 5 to 20%. After obtaining the quantity of cement and fly ash for different grades, the optimum dosage of superplasticizer for each grade has been arrived at by conducting the Marsh cone test. The total aggregate content obtained is distributed in the ratio of 50:50 for sand and coarse aggregate.

The coarse aggregate content is reduced by adding fly ash. The water content obtained is taken as it is and the check is to be made that the maximum water content does not exceed 210 liters /m^3. The mix ratios so obtained for normal concrete and SCC are shown in Table 1 giving the quantity of each ingredient for SCC for different grades and Figure 1 gives the quantity of each ingredient required for each grade of SCC.

Figure 2 compares the quantity of cement and water cement ratio (w/c) required for the normal concrete and SCC. From Fig. 2 the cement requirement is reduced by about 5-20%. Even though the w/c ratio is higher for SCC, w/b ratio is less when compared to normal concrete and it is varying from 0.3 to 0.4.

Table 1 Mix Proportions for various grades of normal concrete and SCC

GRADE	NORMAL CONCRETE BY ACI AND ES METHODS	SCC BY ACI AND ES METHODS	NORMAL CONCRETE BY IS AND ES METHODS	SCC BY IS AND ES METHODS
M$_{20}$	0.51:1:2.38:2.65	0.53:1:2.65:2.20	0.49:1:1.48:3.36	0.54:1:2.67:2.19
M$_{25}$	0.49:1:2.26:2.55	0.52:1:2.49:2.18	0.44:1:1.30:2.96	0.51:1:2.51:2.15
M$_{30}$	0.45:1:2.02:2.39	0.49:1:2.36:2.14	0.42:1:1.17:2.65	0.49:1:2.35:2.07
M$_{35}$	0.40:1:1.74:2.12	0.46:1:2.15:1.97	0.40:1:1.09:2.48	0.47:1:2.21:2.03
M$_{40}$	0.36:1:1.46:1.89	0.43:1:1.97: 1.82	0.36:1:1.03:2.32	0.44:1:1.97: 1.85
M$_{45}$	0.35:1:0.90:1.93	0.42:1:1.82: 1.73	0.35:1:0.90:1.93	0.42:1:1.82: 1.73
M$_{50}$	0.34:1:0.79:1.95	0.41:1:1.76: 1.71	0.34:1:0.79:1.95	0.41:1:1.76: 1.71
M$_{55}$	0.32:1:0.69:1.88	0.37:1:1.72: 1.76	0.32:1:0.69:1.88	0.37:1:1.72: 1.76
M$_{60}$	0.30:1:0.60:1.80	0.34:1:1.49: 1.65	0.30:1:0.60:1.80	0.34:1:1.49: 1.65

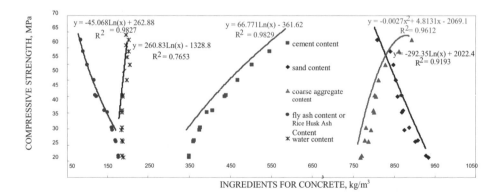

Figure 1 Quantity of various ingredients for obtaining different strength of SCC

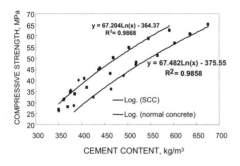
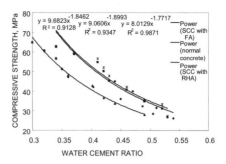

Figure 2   Cement content and w/c ratio for various grades of SCC and normal concrete

EXPERIMENTAL PROGRAMME

Materials

Cement
Ordinary Portland cement of 43 grade conforming to IS: 8112 – 1989 and similar to ASTM type III (C150 – 95) was used for the present experimental investigation. Its specific gravity was 3.15.

Fine aggregate
Natural river sand with fraction passing through the 4.75 mm sieve and retained on 600 μm sieve was used and tested as per IS: 2386. The fineness modulus of sand used was 2.81 with a specific gravity of 2.65.

Coarse aggregate
Crushed granite coarse aggregates of particle shape "average and cubic" was used for the present investigation. The specific gravity was 2.83 and fineness modulus was 6.4.

Water
Potable tap water available in the laboratory with pH value of 7.0±1 and confirming to the requirements of IS: 456 - 2000 was used for mixing concrete and also for curing the specimens.

Fly ash
Fly ash was obtained from Mettur Thermal Power Plant and its properties are given below:

Specific gravity	- 2.00	SiO_2	-60.00%
Surface area	- 2000 m^2/ kg	CaO	-6.00%
Color	- light gray	MgO	-9.00%
Moisture	- 1.76%	Na_2O	-8.50%
Ash	- 98.14%		

Rice husk ash
Rice husk ash was obtained from N.K ENTERPRISES, Orissa and its properties are given over the page:

Appearance	- Very fine powder	SiO_2	-80.00%
Specific gravity	- 2.30	Carbon	- < 4%
Moisture	- 1.76%	CaO	- 1.34%
L.O.I.	- 1.56 %	MgO	- 0.96%

Superplasticizer
In this investigation, superplasticizer Conplast SP 430, based on sulphonated naphthalene polymers, complying with IS 9103- 1999, BS: 5075 part 3 and ASTM C – 494, Type F was used.

Optimum dosage of superplasticizer
The superplasticizer dosage can be optimized for SCC by conducting the Marsh cone test. The test consists of determining the time needed for a certain volume of paste to flow through the Marsh Cone for varying SP/C ratio. An aperture of 8 mm is used and the time taken for 500 ml to flow is measured. The flow time is plotted with respect of SP/C and the optimum SP/C is determined. The optimum is defined as the saturation point beyond which the flow time does not decrease significantly. For all grades of SCC, the optimum dosage was found to be 1.8% by weight of binder content.

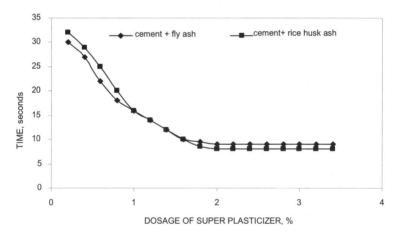

Figure 3   Optimum dosage of Superplasticizer

Test Methods

Flow Tests
Various flow tests were conducted to check the self compactability of the designed SCC of various grades. Slump flow test was conducted to assess the free flow of SCC and to assess the filling ability. V funnel test was conducted to measure the plastic viscosity of SCC. U tube test was conducted to check the passing ability of SCC through reinforcement. L box test was performed to measure the plastic viscosity. To check the homogeneity, a segregation resistance test was conducted.

J Ring test was also done to check flowability as well as passing ability of SCC. Fillability test was done to check the filling capacity. Compaction factor of SCC was also determined found to be similar to that of conventional vibrated concrete factor is 1.

Strength Tests

Compressive Strength Test
As per IS: 516-1959, 150 mm concrete cubes with proportion as per different grades were produced and cured in water in room temperature for 28 days. The 7th and 28th day compressive strengths were tested using an AIMIL compression testing machine of 2000 kN capacity.

Split tensile strength test
The test was carried out by placing a cylindrical specimen horizontally between the loading surfaces of a compression testing machine and the load was applied until failure of the cylinder, along the vertical diameter. Split tensile strength = $2P/\pi LD$, where P is the compressive load on the cylinder, L is the length of cylinder and D is its diameter.

RESULTS AND DISCUSSION

Flow tests

The flow properties were checked for each grade of SCC and it was found that all grades of SCC performed well and conformed to suggested values. Table 2 shows the results obtained for M60 concrete which has inferior results when compared to other grades, but still satisfied the recommendations.

Strength Tests

The compressive strength of normal concrete and SCC were compared for grades between M_{20} and M_{40} by both ACI and IS methods and for M_{45} to M_{60} grades by E.S. method. It was found that there was a slight reduction in strength in SCC with fly ash for all grades and in all methods but there was an increase in strength for SCC with rice husk ash, when compared to normal concrete.

The reduction in strength for SCC designed with fly ash based on ACI method was somewhat less which varied from 2.7 to 6.1% whereas for IS method it was 3.2 to 6.6 %. However when the IS method was followed, the strength of all grades showed higher results than for the ACI method. The reduction in compressive as well as tensile strength is very small, less than 10% for fly ash added SCC.

Regarding tensile strength, the reduction in strength of SCC varies from 2.8 to 8.3%. The tensile strength of normal concrete and SCC at 28 days is shown in Fig. 5. It was observed that the difference in strength is very less and may even be neglected. Fig. 6 shows the relationship between compressive strength and tensile strength of normal concrete and SCC. The relationship for normal concrete is $f_t = 0.378 f_{ck}^{0.662}$ and for SCC this is $f_t = 0.312 f_{ck}^{0.7093}$. As per IS 456-2000, the relationship is $f_t = 0.7 \sqrt{f_{ck}}$ and hence the equivalent relationship for normal concrete is found to be $f_t = 0.693 \sqrt{f_{ck}}$ and for SCC a $f_t = 0.677 \sqrt{f_{ck}}$.

Table 2   Fresh concrete properties of SCC

S.No	TEST	MEASUREMENTS	MINIMUM VALUE OBTAINED		SUGGESTED VALUE
			FLY ASH	RICE HUSK ASH	
1	Slump Flow Test	Diameter of flow	630 mm	600 mm	> 600 mm
2	L Box Test	Flow length	610 mm	600 mm	> 600 mm
3	U Tube Test	Filling Height	300 mm	300 mm	300 mm
4	V funnel Test	Time of flow	9 sec	10 sec	< 10 sec
5	Segregation Test	Segregation Ratio	0.901	0.900	.0.9
6	J Ring Test	Difference in height	9 mm	10 mm	< 10 mm
7	Fillability Test	Filling Capacity	90 %	90 %	80 to 100 %
8	Compaction Factor Test	Compaction factor	1	1	1

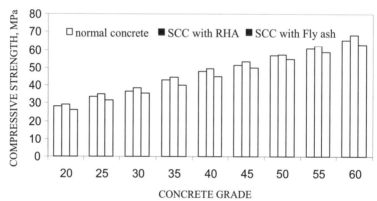

Figure 4   Compressive strength of normal concrete and SCC of all grades at 28 days

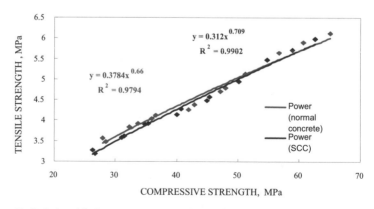

Figure 5   Relationship between compressive strength and tensile strength of SCC and normal concrete

CONCLUSIONS

- The flow properties of the developed SCC for various grades are satisfy the recommended values. The segregation resistance of all SCC grades is also good.

- For the developed mix design, all grades of SCC attained the target mean strength at 28 days.

- Compressive strength of SCC with rice husk ash is more than that of normal concrete and SCC with fly ash.

- The flow properties of SCC with rice husk ash are inferior to that of fly ash with fly ash, but satisfy the recommended values.

- The relationship between compressive and split tensile strength of the designed SCC mixes obeys Power's law, similarly to normal concrete. The equivalent relationship for normal concrete is found to be $f_t = 0.693 \ \sqrt{f_{ck}}$ and for SCC $f_t = 0.677 \ \sqrt{f_{ck}}$ which is very close to the relationship given as per IS 456-2000, that is $f_t = 0.7 \ \sqrt{f_{ck}}$.

- The quantity of materials for SCC for various grades can be directly obtained from the charts.

- The quantity of cement required for SCC is about 5-20% less than that of normal concrete.

- The water content is the same for both normal and SCC and it is varying from 185-205 kg/m^3.

- The water/cement ratio of our SCC may be slightly high for a SCC, but its water binder ratio was much less, it varied from 0.3 to 0.4.

- The optimum dosage of superplasticizer by Marsh cone test was found to be 1.8% by weight of binder i.e. cement and fly ash, but when rice husk was added, the optimum dosage of superplasticizer was 2%.

ACKNOWLEDGEMENT

The authors thank the Management and the Principal of Kongu Engineering College, Perundurai for making the facilities available for the research work. They are also very grateful to AICTE for having sanctioned funds for doing research on Self Compacting Concrete

REFERENCES

1.   OKAMURA, H., AND OZAWA, K. (1994), "Self Compactable HPC in Japan", SP169, ACI, Detroit, pp. 31-44.

2.   PETERSSON, O. BILLBERG, P., AND VAN, B.K., (1996), "A Model for Self-Compacting Concrete", Proceedings of the International RILEM Conference, Paisley, Scotland, June 1996, pp 483-492.

3. MORTSELL, E., MAGGE, M. AND SMEPLAN, S., (1996) "A Particle Matrix Model for Prediction of Workability of Concrete", Proceedings of the International RILEM Conference, Paisley, Scotland, June 1996, pp 428-438.

4. SEDRAN, T., LARRARD, F.D., HURST, F., AND CONTAMINUS, C., (1996), "Mix Design of SCC", proceedings of the International RILEM Conference, Paisley, Scotland, June 1996, pp 439-450.

5. SAKATA, N. MARUYAMA., K., AND MINAMI, M. (1996), "Basic Properties and effects of Welan Gum on SCC", Proceedings of the International RILEM Conference, Paisley, Scotland, June 1996, pp 237-253.

6. NISHIBAYASHI. S, YOSHINO. A, INOUE. S AND KORODA. T, (1996), "Effect of Properties of Mix Constituents of Rheological Constants of Self-Compacting Concrete", Proceedings of the International RILEM Conference, Paisley, Scotland, June 1996, pp 255-261.

7. OKAMURA, H., OZAWA, K AND OUCHI, M. (2000), "SCC Structural Concrete", Vol.1. No. 1, March 2000.

8. SAAK, A.W., JENNINGS, H.M. AND SHAH, S.P., (2001), "New methodology for Designing SCC", ACI Materials Journal, Nov-Dec., pp 429-439.

9. MALATHY, R. AND GOVINDASWAMY, T. (2005) "Properties of Self-compacting concrete with Fly ash, Silica fume and Metakaolin," National conference on New perspectives and paradigm in materials, design and construction of Civil engineering systems, January 10-11, 2005 at C.I.T, Coimbatore. pp 179-188.

10. GRUNEWALD. S AND WALRAVEN. J.C, "Parameter-study on the influence of steel fibres and Coarse aggregate content on the fresh properties of Self-compacting concrete, Cement and Concrete Research, Vol 31, No. 12, pp. 1793-1798.

11. NAN SU, KUNG-CHUNG HSU AND HIS-WEN CHAI (2001), "A simple mix design method for Self-compacting concrete", Cement and Concrete Research 31, pp. 1799-1807.

12. SUBRAMANIAN, S., AND CHATTOPADHYAY, D., (2002), "Experiments for Mix Proportioning of SCC", Indian Concrete Journal, January, pp 13-20.

13. R GETTU et al. (2002), "Development of high strength self compacting concrete with fly ash: a four step experimental methodology", 27th conference on our world in concrete and structure, pp 217 to 224.

14. Specifications and guide lines for SCC, EFNARC, Hampshire, U.K, 2001-2002, pp 4-32.

15. JAGADISH VENGALA, SUDARSHAN M.S. AND RANGANATH R.V. (2003), "Experimental study for obtaining Self-compacting concrete," The Indian Concrete Journal, August, 2003 Vol. 77, No. 8, pp 1261-1266.

OPTIMISING PARTICLE PACKING TO REDUCE THE CEMENT CONTENT IN CONCRETE

S A A M Fennis

J C Walraven

J A den Uijl

Delft University of Technology

The Netherlands

ABSTRACT. Ecological concrete with low cement content can contribute to the decrease of CO_2 emissions. Low cement content concrete with sufficient strength can be produced by optimizing the concrete composition with packing density models and adding by-products like fly ash as a fine filler. In this paper the principles and possibilities of particle packing models are explained and it is shown how they can be used to design ecological concrete. The packing density of aggregate mixtures is modelled and verified by experiments. The modelled values and the experimental results for the packing density show a deviation up to 3%. Eighteen mortars were designed and tested for workability and strength. Workability could not be predicted from the packing density of the sand matrix and the cement paste demand, since particle size and size distribution affected the results. When the packing density of the total particle matrix was related to the workability and the water demand, a good correlation was found. At a fixed water cement ratio, no correlation between strength and particle packing could be found.

Keywords: Cement reduction, Ecological concrete, Mortar, Particle packing models, Rheology, Water demand.

S A A M Fennis, graduated at Delft University of Technology in 2003. Her PhD research is focused on particle packing models applied on the design of ecological concrete.

J C Walraven, PhD civil engineering, is a full professor at Delft University of Technology. His research interests are design of concrete structures and design of innovative concrete mixtures. He is an expert in the field of shear strength of concrete and shear transfer in cracks.

J A den Uijl, is senior researcher and lecturer at Delft University of Technology. His research interests are the bond behaviour of reinforcement bars and pre-tensioned tendons and non-linear numerical modelling of concrete structures.

INTRODUCTION

In concrete production, Portland cement is the component with the highest environmental impact, because of the high amount of energy required for Portland cement production. Energy consumption and CO_2 emissions of concrete can be reduced when by-products from other industries, like fly ash, are applied as cement replacing materials or fillers. By this strategy not only the CO_2 emissions are reduced but residual products from other industries are reused and therefore less material is dumped as landfill and more natural resources are spared. Many by-products from other industries, like silica fume, fly ash or blast furnace slag, have characteristics which can positively influence concrete properties. However, regulations limit the use of large amounts of cement replacing materials, requiring a minimum amount of cement in concrete (e.g. 260 kg/m^3 in the Netherlands) to make sure that concrete properties such as strength and durability are at a sufficient level. Nevertheless, it is believed that by making use of the particle packing optimization techniques, nowadays used in the production of high strength concrete, it is possible to optimize the particle packing in order to lower the cement content in concrete without changing concrete properties in a negative way [1]. This low cement content concrete or so called ecological concrete can be designed by optimizing the concrete composition in such a way that the highest packing density is achieved.

This optimization results in a stiff and strong particle structure, which has a positive influence on the mechanical properties such as shrinkage and creep. Furthermore, the high density of the particle structure leaves less space for voids to be filled with water, which reduces the water demand and increases the strength of concrete. The water demand of the particle structure is very important since a slight increase in water demand in mixtures with low cement content results in high water cement ratios. A high water cement ratio will decrease strength and durability. Therefore, it would be useful to have a design method which can control the water demand of an ecological concrete mixture at a certain desired workability. For this reason a preliminary investigation on mortars was started to determine the effectiveness of 'design methods based on particle packing models' to relate the water or cement paste demand to the workability of mixtures.

PARTICLE PACKING THEORY

The subject of optimizing the concrete composition by selecting the right amounts of various particles has already aroused interest for more than a century. To optimize the particle packing density of concrete, the particles should be selected to fill up the voids between large particles with smaller particles and so on, in order to obtain a dense and stiff particle structure. Most of the early researchers, working on the packing of aggregates, proposed methods to design an ideal particle-size distribution, like Fuller in 1907. Optimizing concrete to a predefined 'ideal' particle-size distribution is still popular today, as it is the method that is most used in practice and applied in most national standards. However, does the ideal particle-size distribution exist? There can be an ideal particle-size distribution for certain types of aggregate, but when various types of aggregate are combined the result can be far from 'ideal'. For instance, when rounded sand is combined with coarse recycled aggregates, the optimal particle-size distribution will differ from the one of a mixture with sand from crushed rock and rounded coarse aggregates. So, how can the ideal particle packing of concrete be found? The answer is: by making use of the geometry of the particles in a particle packing model.

The basic mathematic formulae of almost all particle packing models are the same and purely based on the geometry of the particles. The formulae prescribing the packing density were first introduced by Furnas in 1929 [2]. They are valid for two monosized groups of particles without interaction between the particles. The volume of each of these monosized particle groups can be expressed in its partial volume φ_i, which is the volume occupied by size class i in a unit volume. Furthermore, the relative volume of each size class can be expressed as its volume fraction r_i. By definition:

$$\sum_{i=1}^{n} r_i = 1 \text{ and } r_i = \varphi_i \Big/ \sum_{i=1}^{n} \varphi_i$$

(A list of symbols is presented at the end of this paper.)

When there is only one size class present ($r_1 = 1$), the partial volume of this size class (φ_1) is equal to the total occupied volume or total packing density α_t. When there are two size classes present, two cases can be distinguished:

Case 1: The volume fraction r of the large particles is much larger than the volume fraction of the small particles ($r_1 \gg r_2$).
In this case small particles (diameter d_2) can be added to a container filled with large particles (diameter d_1). By adding the small particles into the voids between the large particles, the voids are filled and thus the total occupied volume and the packing density increase. The total volume occupied by particles in a container [3, 4] is expressed by equation 1.

$$\alpha_t = \varphi_1 + \varphi_2 = \alpha_1 + \varphi_2 \qquad \rightarrow \qquad \alpha_t = \frac{\alpha_1}{1 - r_2} = \frac{\alpha_1}{r_1} \qquad (1)$$

Summarizing: The total packing density is 'the volume of the large particles (which is restricted by the maximum packing density of the large particles)' plus 'the volume of the small particles', in a unit volume.

Case 2: The volume fraction of the small particles is much larger than the volume fraction of the large particles ($r_2 \gg r_1$).
In this case large particles can be added to a container filled with a matrix of small particles. By adding some large particles into a matrix of small particles, the large particles fill up the volume they occupy by 100%. Their contribution to the packing density is therefore equal to their partial volume: φ_1. The small particles can fill up the rest of the unit volume ($1 - \varphi_1$) with their maximum packing density:

$$\alpha_t = \varphi_1 + \varphi_2 = \varphi_1 + \alpha_2(1 - \varphi_1) \qquad \rightarrow \qquad \alpha_t = \frac{1}{r_1 + (r_2/\alpha_2)} \qquad (2)$$

Summarizing: The total packing density is 'the volume of the large particles' plus 'the remaining volume, filled with the maximum amount of small particles (which is restricted by the maximum packing density of the small particles)', in a unit volume.

The packing model proposed by Furnas can be extended to n particle groups by making use of the geometrical and physical relations. It is physically not possible to have a large volume fraction of small particles ($r_2 \gg r_1$) and simultaneously fit all the small particles within the voids between the large particles. Therefore, in this situation automatically case 2 becomes operative and particle group 2 becomes the dominating particle group. Thus, this physical relation comprehends that the packing density is always the minimum value of formula 1 and 2. The same 'minimum'-relation is true for n particle groups. With this additional relation the particle packing model for multi component mixtures, with dominant particle group i and remaining particle groups j, becomes [4]:

$$\alpha_t = \underset{i=1}{\overset{n}{Minimum}}\left\{\alpha_i + (1-\alpha_i)\sum_{j=1}^{i-1}\varphi_j + \sum_{j=i+1}^{n}\varphi_j\right\} \tag{3}$$

Since these formulae only depend on the particle packing and amount of particles of the monosized groups, they are valid for every type of particle, independent of particle characteristics like shape or texture, as long as the particles preserve their shape during packing. Unfortunately, these formulae are only valid to calculate the particle packing of particle groups without any interaction. However, in reality there will always be interaction between the particles:

- Wall effect: The effect of large particles on the packing density of small particles.
  This effect appears close to a wall or close to a large particle, where smaller particles can not be packed as densely as their maximum packing density.

- Loosening effect: The effect of small particles on the packing of large particles.
  This effect appears when the small particles are too large to fit in between the voids between the large particles. The disturbance of the packing of the large particles by the smaller particles is called the loosening effect.

The effectiveness of the particle packing model therefore depends entirely on how these interaction effects are applied in the model. De Larrard [4] implemented these effects quite effectively for various types of rounded and crushed aggregates by introducing interaction parameters a_{ij} and b_{ij}, Equation 4.

$$\alpha_t = \underset{i=1}{\overset{n}{Minimum}}\left\{\frac{\alpha_i}{1-\sum_{j=1}^{i-1}\left[1-\alpha_i + b_{ij}\alpha_i(1-1/\alpha_j)\right]r_j - \sum_{j=i+1}^{n}\left[1-a_{ij}\,\alpha_i/\alpha_j\right]r_j}\right\} \tag{4}$$

In this way particle characteristics like shape and texture are implicitly taken into account in the particle packing models.

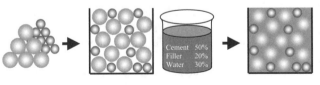

1. Voids between the aggregates are optimised

2. Cement paste composition is optimised

3. Sufficient amount of cement paste is added to fill voids and achieve workability

Figure 1 Concrete design in three steps with existing particle packing models

THE USE OF PARTICLE PACKING MODELS FOR ECOLOGICAL CONCRETE

Particle packing models can be used in two ways. The first method is to predict the maximum packing density of the aggregate structure. This optimal aggregate structure is than combined with a sufficient amount of fluid and stable cement paste to create good concrete, see Figure 1. By creating the highest particle packing density of the aggregates, the lowest amount of cement paste is necessary to fill up the voids in the particle skeleton. In general, the aggregates are stronger than the cement matrix. This means that when the cement paste content is decreased, a stronger concrete is expected as long as all voids between the aggregates are filled [5]. Furthermore, when a strong aggregate structure is achieved, this structure will resist to volume changes, thus reducing shrinkage and creep. In mixtures with a high packing density and a fixed amount of cement, the water and sand content can be reduced. This results in concrete with higher strength [6, 7]. The other way around, this means that a lean cement mixture, like ecological concrete, can still be in a normal strength class [8]. A disadvantage of this design method is the time consuming trial and error design of the cement paste composition.

A second method to use particle packing models to design ecological concrete would be to calculate the particle packing of all particles including cement and fine fillers. In this way only water needs to be added instead of cement paste, which should make it easier to control water demand and thus the material properties of the designed concrete. However, particle packing models are not yet suitable to calculate the packing of cement combined with fine fillers. This is because the existing models only take into account particle characteristics, like shape and texture, implicitly, or they even assume perfectly round particles. The particle characteristics influence the surface forces that work on the small particles, and the surface forces strongly influence the packing density and wettability of the small particles. Therefore, modelling the particle packing and water demand of fine particles is complex. Research in this area is far from finished.

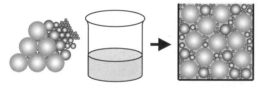

1. *Optimization of the particle structure, consisting of aggregates and fine particles*

2. *Water is added to fill the voids and create workability*

Figure 2   Concept of a new particle packing model to design concrete in two steps

EXPERIMENTAL PROGRAMME

The goal of the experimental program was to examine the influence of the particle packing density and the particle size on the workability of mortars. The reason for this is that a good and workable concrete mixture not only requires that the cement paste fills the voids between the aggregates, but the cement paste should also surround all particles with a thin layer of paste in order to achieve workability. It is believed that the required amount of cement paste and thus the water demand not only depends on the packing density, but also on particle characteristics such as size, shape, texture and surface area. Seventeen mortars with several particle packing densities and varying sizes of sand were designed with a constant water cement ratio. In this way, not only the influence of particle size and particle packing density of the sand on the workability can be determined, but also the effectiveness of design method 1 (as described in the previous paragraph) can be verified.

Mixture Composition

The design of all the mortars was based on design method 1. From a matrix of sand particles the theoretical packing density is determined by particle packing models. Then a cement paste is added. The cement paste consisted of Portland cement CEM I 32,5 R (ENCI Maastricht) and water with a water cement ratio of 0.5. The amount of cement paste that was added to the sand was determined relatively to the packing density of the sand matrix. For the composition of the sand matrix, sand in three different size fractions from the Stevin laboratory of Delft University of Technology was used. The sand fractions 0.25 - 0.5 mm, 0.5 - 1 mm and 1 - 2 mm all have the same origin, rounded river sand, and can be expected to have comparable shape and texture. Density was 2610 kg/m^3 and water absorption 0.3% on average.

The mixture compositions are presented in Table 1. Nine mixtures were designed with only one size fraction of sand. The mixtures vary in the total amount of cement paste which had to be added to fill the voids between the sand and ensure workability. Mixture ten to seventeen were designed at a predefined packing density with either a low amount of fine sand or a high amount of fine sand. Mixture eighteen was added after measuring the packing density of the sand matrix (see results).

Table 1   Theoretical mixture design

				MIXTURE COMPOSITION				
Mix	Sand A 0.25-0.5	Sand B 0.5-1	Sand C 1-2	Packing density sand matrix	Volume sand matrix	Cement paste to fill voids	Extra cement paste for workability	Total cement paste
	dm^3	dm^3	dm^3	-	dm^3	dm^3	dm^3	dm^3
1 A	0.766	-	-	0.581	1.319	0.553	0.109	0.662
2 A	0.766	-	-	0.581	1.319	0.553	0.191	0.744
3 A	0.766	-	-	0.581	1.319	0.553	0.272	0.826
4 B	-	0.766	-	0.603	1.270	0.504	0.109	0.613
5 B	-	0.766	-	0.603	1.270	0.504	0.191	0.694
6 B	-	0.766	-	0.603	1.270	0.504	0.272	0.776
7 C	-	-	0.766	0.597	1.284	0.517	0.109	0.626
8 C	-	-	0.766	0.597	1.284	0.517	0.136	0.653
9 C	-	-	0.766	0.597	1.284	0.517	0.163	0.681
10 AC	0.398	-	0.368	0.672	1.140	0.374	0.109	0.483
11 AC	0.230	-	0.536	0.639	1.199	0.433	0.109	0.542
12 AC	0.529	-	0.238	0.639	1.199	0.433	0.109	0.542
13 AC	0.115	-	0.651	0.618	1.240	0.474	0.109	0.583
14 AC	0.651	-	0.115	0.618	1.240	0.474	0.109	0.583
15 AC	0.398	-	0.368	0.672	1.140	0.374	0.191	0.565
16 AC	0.230	-	0.536	0.639	1.199	0.433	0.191	0.623
17 AC	0.529	-	0.238	0.639	1.240	0.433	0.191	0.623
18 AC	0.529	-	0.238	-	-	0.474	0.109	0.583

Mixing and Testing Procedure

All the mixtures were composed of two kilogram of sand. The cement paste with water cement ratio of 0.5 was premixed. In the mixing process the sand was mixed dry in a three litre Hobart mixer. The cement paste was added during one minute. Then mixing was continued for two minutes after which the mix rested for one and a half minute. If necessary, mortar adhering to the wall was scraped from the bowl during this resting period. Finally mixing was continued for two more minutes. The rheological properties of the mixtures were tested by measuring the slump, the slump flow and the flow value from the flow table test. This was done by a mini cone test (diameter 38-90 mm and height 75 mm) on a 300 mm diameter glass plate flow table (Tonindustrie). The flow value was measured after ten drops with 10 mm free falling height. Furthermore, from some mixtures three 160 × 40 × 40 mm prisms were cast and compacted on the same flow table with ten drops. Three half prisms were used for testing cube compressive strength after 7 days. The other three were used for the 28 day cube compressive strength.

RESULTS AND DISCUSSION

The measured rheological properties of the eighteen mixtures are presented in Table 2. Besides the compressive strength of the mixtures, also the packing density of the sand matrix was measured. Furthermore, because of the premixing of the cement paste, the total added volume of the cement paste was a little less than originally intended. The true values are presented in Table 2 as the total volume of the added cement paste and also as the relative amount of paste present in the final mixture. The particle packing of the sand was measured according to the procedure as described in NEN-EN 1097-3 in a one litre container. Using this procedure, the precision of the packing density is ± 0.002 [-], however, the degree of compaction of the sand matrix can differ depending on the technician performing the experiment. The results of the measured packing densities of the sand matrix of the mixtures ten to seventeen were compared to the theoretical calculated packing densities (according to [9]) which were used for the design of these mixtures. The results are presented in Figure 3.

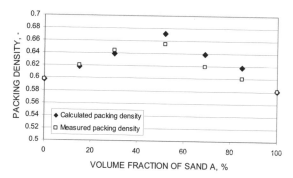

Figure 3   Packing densities of the combined sand matrix of sand A and C

There was a difference of up to 3% between the measured and the calculated values, and this difference was higher with a higher amount of fine particles (sand A 0.25-0.5). This might be explained by the size ratio between the two particle fractions; the mixtures are in fact gap graded. A little segregation occurred in the dry sand mixtures. However, this segregation was not observed during mixing and testing of the mortars. Because of the differences in the measured and calculated values, an extra mixture (18 AC) has been added to the experimental program. In this mixture the total amount of cement paste to be added is calculated from the measured packing density instead of the theoretical one.

Table 2   Experimental results and final cement paste content per mixture

EXPERIMENTAL RESULTS – DESIGN METHOD 1							
Mix	Slump	Slump flow	Flow value	7-day compressive strength	28-day compressive strength	Packing density sand matrix	Volume added cement paste	Relative amount cement paste
	mm	mm	mm	N/mm^2	N/mm^2	-	dm^3	m^3/m^3
1 A	6	95	115	24.2	28.7	0.581	0.651	0.459
2 A	12	95	142	23.5	30.7	0.581	0.729	0.487
3 A	30	98	179	23.2	24.6	0.581	0.809	0.514
4 B	24	95	155	21.3	28.7	0.603	0.598	0.438
5 B	40	105	185	23.3	33.1	0.603	0.685	0.472
6 B	50	140	220	22.0	31.8	0.603	0.768	0.501
7 C	40	120	195	20.6	30.5	0.597	0.616	0.446
8 C	40	120	200	18.5	26.6	0.597	0.638	0.454
9 C	40	125	210	18.8	28.2	0.597	0.666	0.465
10 AC	3	91	111	21.6	31.0	0.655	0.468	0.379
11 AC	22	94	168	28.4	39.9	0.644	0.534	0.411
12 AC	5	90	115	22.4	22.4	0.619	0.529	0.408
13 AC	40	116	200	-	-	0.621	0.574	0.428
14 AC	4	90	124	-	-	0.601	0.573	0.428
15 AC	19	92	159	-	-	0.655	0.555	0.420
16 AC	46	135	215	-	-	0.644	0.613	0.445
17 AC	19	91	160	-	-	0.619	0.617	0.446
18 AC	9	92	129	-	-	0.619	0.572	0.428

In Figure 4, the flow value is related to the relative amount of cement paste. As expected for a certain sand matrix with a fixed packing density, workability increases with an increased amount of cement paste (see Figure 4, for example, sand A).

Furthermore, it seems that the finer the sand particles, the lower the workability. This is not only true for the mixtures with one size fraction of sand (Figure 4 left hand side) but also for the mixtures with two size fractions. Compare, for instance, mixture 13 AC with 15% of fine particles and a flow value of 200 mm with mixture 14 AC with 85% of fine particles and a flow value of 124 mm.

The real (measured) packing density of the sand matrix could have influenced this result, which was the reason to include mixture 18 AC in the experimental program. Mixture 13 AC and 18 AC have almost the same measured packing density and the same amount of cement paste added. However, 13 AC contains 15% of sand A whereas 18 AC contains 69%. This has resulted in a much lower workability for mixture 18 AC.

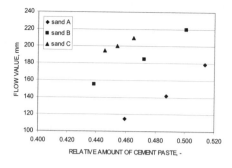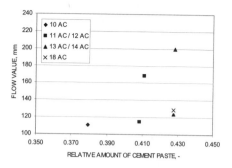

Figure 4   Correlation between flow value and the volume of the added cement paste

The mixtures were composed with a cement paste content to fill the voids between the sand particles and an extra amount of cement paste to ensure workability. In this way, the influence of the packing density of the sand matrix on the workability can be determined.

For the single sand fraction mixtures 1 A to 9 C no correlation between packing density and flow value could be found and also for the mixtures 10 AC to 14 AC (with 0.109 dm^3 extra cement paste) data were scattered. Additionally, no apparent correlation seems to exist between the packing density of the sand matrix and the strength.

It should be mentioned that, if the other rheological results like slump and slump flow would have been plotted against the added volume of cement paste or the relative amount of cement paste, the graphs and correlations would have been similar.

Comparison of Results to Design Method 2

The results of the experiments mentioned before do not substantiate the use of the particle packing concept prescribed in method 1 to design mixtures and predict the workability of those mixtures. The correlation between the relative amount of cement paste and the flow value depends on particle size and not on particle packing density. This means that, for a constant water / cement ratio, design method 1 is not effective to design mortars. To use the particle size distribution for designing concrete mixtures is the most common solution in practice. However, further investigation of the results applied to design method 2 can give important new insights.

The particle packing density of the total particle structure, cement combined with the sand particles, was calculated with a particle packing program [9] and presented in Table 3. The differences between the packing density of mixtures containing sand A and sand C indicate that there is a large interaction between the cement particles and the sand particles.

In both particle matrices about 28% by mass of cement is present. If these cement particles would fit in between the voids of the sand matrix, the theoretical packing density of these mixtures had been the same. Since their particle packing densities differ, this means that the cement particles affect the particle structure of the sand. In other words, the system can not be seen as a sand matrix with cement paste flowing in between the particles as is assumed in method 1.

Table 3   Rheological results and final water content per mixture

EXPERIMENTAL RESULTS – DESIGN METHOD 2

Mix	Slump	Slump flow	Flow value	Theoretical packing density particle matrix	Volume added water	Relative amount of water
	mm	mm	mm	-	dm^3	m^3/m^3
1 A	6	95	115	0.738	0.398	0.281
2 A	12	95	142	0.734	0.446	0.298
3 A	30	98	179	0.729	0.495	0.314
4 B	24	95	155	0.784	0.366	0.268
5 B	40	105	185	0.772	0.419	0.289
6 B	50	140	220	0.759	0.470	0.306
7 C	40	120	195	0.784	0.377	0.272
8 C	40	120	200	0.790	0.390	0.278
9 C	40	125	210	0.787	0.408	0.284
10 AC	3	91	111	0.795	0.286	0.232
11 AC	22	94	168	0.800	0.327	0.251
12 AC	5	90	115	0.773	0.323	0.250
13 AC	40	116	200	0.798	0.351	0.262
14 AC	4	90	124	0.756	0.350	0.262
15 AC	19	92	159	0.786	0.339	0.257
16 AC	46	135	215	0.786	0.375	0.272
17 AC	19	91	160	0.766	0.377	0.273

To further evaluate the results by design method 2, the volume of added water and the relative amount of water in the mixtures was calculated and presented in Table 3. A reasonable correlation was found between the flow value and the packing density for the mixtures with 0.109 dm^3 extra cement paste added: mixtures 1 A, 4 B, 7 C, 10 AC to 14 AC.

However, because these mixtures were designed by method 1, the relative amount of water in the mixtures was not the same. The varying amount of water in the mixtures causes scatter in the results. For this reason it was chosen, following on the relationship between plastic viscosity and normalized solid concentration published by De Larrard [4, 10], to relate the flow value to the packing density and the relative amount of water, described as φ/α_t. In this relation φ is the relative volume of particles in the mortar, φ = (1-relative amount of water) and α_t is the theoretical packing density of the particle matrix as presented in Table 3. The results are presented in Figure 5.

The same relation is presented for method 1 in Figure 5 with $\varphi_{sand}/\alpha_{sand}$ = (1-relative amount of cement paste) / (the theoretical packing density of the sand matrix). With design method 2, the seventeen mixtures show a good correlation between the flow value and φ/α_t. The existence of this correlation means that, in theory, design method 2 can be used effectively to design (the workability of) mortars.

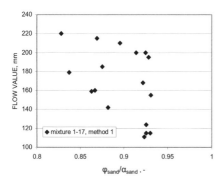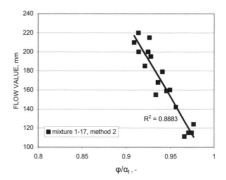

Figure 5   Correlation between the flow value and the fraction of the relative volume of particles in the mortar and the theoretical particle packing density

CONCLUSIONS

In the design of ecological concrete, two different design methods can be distinguished. In the first, one cement paste is added to a particle structure consisting of aggregates. The second method involves the design of the packing structure of all the particles with the addition of water. From the first experimental results obtained from 18 mortar mixtures the following can be concluded:

- Design method 1 is not effective to design a mortar with a prescribed workability. Besides the packing density of the sand matrix, the particle size and the size distribution of the sand seem to have an effect on the workability of mortars. However, this apparent effect can be caused solely by the interaction between the cement and the sand matrix and thus by the packing density of the total particle structure. Since this interaction between the sand matrix and the cement particles can not be modelled with method 1, this method is not suitable to design a mortar with a prescribed workability.

- Design method 2 shows to be effective to design a mortar or concrete at a fixed workability. A relation is found between the packing density of the particle matrix, the relative amount of water and the flow value of the tested mortars. However, the applicability should be checked for different water cement ratios. Furthermore, the particle packing models used for this design method should be extended to make it possible to design mortars and concretes containing more than one fine particle fraction (like cement or fillers).

- At a fixed water cement ratio, no correlation between strength and particle packing could be found so far.

ACKNOWLEDGEMENTS

This research is supported by the Dutch Technology Foundation STW, applied science division of NWO and the Technology Program of the Ministry of Economic Affairs. The experiments were performed by Erik Horeweg, technician at the Stevin Laboratory of Delft University of Technology, which is very much appreciated.

LIST OF SYMBOLS

φ_i Partial volume. This is the volume occupied by size class i in a unit volume.

r_i Volume fraction of particle group i. By definition $r_i = \varphi_i \left/ \sum_{i=1}^{n} \varphi_i \right.$ and $\sum_{i=1}^{n} r_i = 1$.

α_i Packing density of particle group i.

a_{ij} Parameter which describes the loosening effect caused by the particles in class j on the packing of the particles in class i.

b_{ij} Parameter which describes the wall effect caused by the particles in class j on the packing of the particles in class i.

REFERENCES

1. FENNIS S A A M, WALRAVEN J C AND DEN UIJL J A, Optimizing the particle packing for the design of ecological concrete. Proceedings of the 16. Internationale Baustofftagung, Weimar, Bundesrepublik Deutschland, 2006. pp. 1–1313-1–1320.

2. FURNAS C C, Flow of gasses through beds of broken solids. Bureau of Mines Bulletin 307, 1929.

3. JONES, M R, ZHENG L AND NEWLAND, M D. Comparison of particle packing models for proportioning concrete constituents for minimum voids ratio. Materials and Structures. Vol. 35, 2002. pp 301-309.

4. de LARRARD F, Concrete mixture proportioning; a scientific approach, E & FN Spon, London, 1999.

5. RESCHKE T, Der Einfluss der Granulometrie der Feinstoffe auf die Gefügeentwicklung und die Festigkeit von Beton, VBT Verlag Bau+Technik GmbH, Düsseldorf, 2000.

6. DHIR R K, McCARTHY M J AND PAINE K A, Engineering property and structural design relationships for new and developing concretes. Materials and Structures. Vol. 38, No. 275, 2005. pp. 1-9.

7. KRONLÖF A, Filler effect of inert mineral powder in concrete, VTT Technical research centre of Finland, Espoo, Finland, 1997.

8. JOHANSEN V AND ANDERSEN P J, Particle packing and concrete properties. In: Skalny J, Mindess S. Materials Science of Concrete 2, The American Ceramic Society, Westerville, 1996. pp. 111-146.

9. GLAVIND M AND PEDERSEN E J, Packing calculations applied for concrete mix design. Creating with concrete, proceedings of the international congress held at the University of Dundee, Scotland, UK, 1999.

10. FERRARIS C, de LARRARD F, MARTYS N, Fresh Concrete Rheology: Recent Developments. Materials Science of Concrete VI, Mindess S, Skalny J, The American Ceramic Society, Westerville, 2001. pp. 216-241.

FOAMED CONCRETE WITH WASTES OF
PULP AND PAPER INDUSTRY

I A Kozlov

B S Batalin

Perm State Technical University

Russia

ABSTRACT: Large-tonnage wastes products of a pulp and paper industry (WPPI) have recently drawn still more attention of researchers and production workers. With cellulose and kaolin in its composition, these waste products, modified by chemical additives, can be used for manufacturing heat-insulating, finishing and constructive - thermal insulating materials and items. Additional researches carried out in the field of creating building materials on the WPPI basis enabled us to assume the opportunity of using such raw material as a dispersible-reinforcing component in foamed concrete manufacturing. A foaming agent has been developed in the result of our researches. It is based on an animal origin raw material (horns, hoofs and skin of cattle, feather of birds), produce by its boiling in a solution of alkaline component as NaOH or KOH. It is then neutralized by a strong acid and stabilized by salt of a strong acid and transitive metal. Technical characteristics of the received product: foam ratio not less than 20, life-time more than 24 hours. Besides dispersed foamed concrete reinforced with fibres, foamed concrete on the basis of the obtained foaming agent has been produce. It does not give shrinkage and allow reinforcing the structures. Foaming agent in a dry powders condition has been received that make it possible for using in manufacture of dry foam concrete mixes. The received results show the perspectives of using waste products of pulp and paper industry and foaming agent for materials for production.

Keywords: Wastes, Cellulose, Foam, Concrete.

B S Batalin, Doctor of technical sciences (J.D.), professor at the Department of Building Materials of the Perm State Technical University. Born on 10 January 1937.

I A Kozlov, Master of Sciences (M.S.), engineer at the Department of Building Materials of the Perm State Technical University. Born on 25 December 1979.

INTRODUCTION

Large-tonnage wastes of pulp and paper industry (WPPI) have recently drawn increasing attention in both academia and industry. With cellulose and kaolin in their composition, these waste products, modified by chemical additives, can be used for manufacturing heat-insulating and finishing material as well as various other construction industry products. The use the WPPI has long been restrained by its high water content (up to 96%) and therefore high energy demand to turn into building materials. With newly developed methods such as drying with high frequency current, this may be partially solved. It is possible to produce building materials on the basis of WPPI with additives (sawdust, perlite, ashes, antiseptics, fire retardants, bitumen emulsion, etc.), resulting in strength ranging between 1 to 10 MPa, density of 250 to 1200 kg/m^3 and heat conductivity of 0078 W/m·K (for the material with a density of 250 kg/m^3) [1]. Additional research carried out in the field of creating building materials on the WPPI basis enabled us to assume the opportunity of using such raw material as a dispersibly-reinforcing component in foamed concrete manufacturing.

Fibre concrete can be analogous with a variety of cement concrete where fibres are uniformly distributed, such as metal fibres, pieces of steel wire, alkali resistant glass fibre and polymers. The fibre in this case functions as a reinforcing component, and promotes the improvement of the concrete quality, improving its cracks resistance and stress-stain behaviour.

Fibres are commonly used in all types of cement mortar as well as when it is necessary to prevent the formation of deformation cracks resulting from mechanical influence or shrinkage (e.g. floor casting, screeding or pouring into shuttering) [2]. The application of fibres enables to avoid labour-consuming operations on reinforcing.

PROPERTIES OF PAPER INDUSTRY WASTE

Product manufacturing with the use of fibres consists of adding fibres during the mixing process. As a result, the distribution of fibres becomes unorganized, though vibration compacting favours the directed placing of separate fibres. In their parallel to stretching action efforts orientation, the strength of fibre concrete considerably grows in comparison with the chaotic reinforcing when only a smaller part of fibres participates in withstanding the load [2]. The WPPI used by the "Perm paper industry" enterprise, comprises cellulose fibres with impurities of lignin, carbonates of sodium, potassium, magnesium and calcium, and also a small amount of phosphates and nitrates of the same metals. The fibre content was 75-90% of WPPI volume, in grinding thinness: 60-63° [3], the humidity of WPPI before processing into sheets was 19-65%, and pH: 5.9-6.5. The cellulose fibres within the WPPI were up to 150-250 microns in length and a thickness of 1-5 microns. Regarding its chemical composition, WPPI is a relatively harmless material, free from heavy metal contaminants or impurities. Fibres of the waste are either chaotically located or interknit with each other (Figure 1).

It is known from the literature and patents that the WPPI is used as a filler material, but there is no description of its properties of this respect. According to the data in [4], the physical properties of the WPPI make it suitable for manufacturing building materials. In a wet waste-water system and under mechanical influence, the fibres of the WPPI are capable to orientate themselves to a greater or lesser extent depending on the humidity of the system. At humidities lower than 550% by weight, the structure of the waste-water system remains practically unchanged upon external forces due to its high viscosity. At humidities higher than 550 %, moulding causes orientation of the fibres by settling substantially, parallel to each other. Thus, the higher the humidity of the mixture the stronger is the orientation effect resulting in anisotropy of material strength and density. Owing to the ability of fibres to self-orientate, it is possible to achieve high technical parameter products economically.

The mechanical properties of waste-water system define its binding properties, and the changes of these properties to some extent are irreversible. In this respect WPPI obviously plays the role in creating bonds [5], hence, one can suppose the presence of some sort of binding properties attributed to the "waste-water" system.

One of the major characteristics of WPPI as binding agent is its adhesion to aggregates [6]. Due to its fibrous structure, the waste easily interacts with any rough surface and mechanical adhesion is observed. The strength and other properties of composite materials using WPPI (including foam concrete) depend on the adhesion to aggregate. Therefore, it is of great interest to investigate the adhesion of WPPI to those materials which are supposed to be used as aggregate. These may include:
- wood of various kinds (sawdust, shaving, etc.);
- granulated polystyrene;
- crushed gravel and foamed glass sand, granulated slag and others granular materials.

Assessment of adhesion

Adhesion and wetting take place when the structure of many composite building materials is formed. They necessarily affect the hardening of mortars and concrete. A D Zimona's works were devoted to the problems of quantitative estimation of adhesion [7]. Adhesion of a liquid is estimated by the work necessary to separate the liquid from a solid surface, i.e. the restoration of an initial condition of the contacting material. The adhesion work may be assessed by the following methods:
- ultimate wetting angle;
- work required for the separation of the contacting bodies;
- surface tension of the contacting material.

The listed methods cannot measure the adhesion of WPPI as the ultimate wetting angle cannot be defined due to the high viscosity of a concrete mixture. Due to technical reasons, the most feasible is to separate the contacting surfaces and estimate the surface tension. In our case the most accessible method to measure the adhesion is to define the strength when any of contacting materials is shifted to one another. This method is widely used in determining the binding capacity of glues [8].

EXPERIMENTAL

The adhesion between contacting materials was measured on the following samples:
- granulated polystyrene: 100×100 mm;
- wood of fur trees: 150×150 mm;
- wood of pine trees 150×150 mm;
- wood of birch trees 150×150 mm;
- wood of aspen trees 150×150 mm;
- foamed glass: 60×90 mm.

Wood samples were used at natural air humidity. The humidity of WPPI used ranged from 600 to 700% by weight. Such a range of humidity is optimum for good workability [4].

The above materials were covered with a layer of WPPI of 600% humidity at a thickness of 10 mm. The obtained samples were dried to constant weight at room temperature for 5 days. Further, the contact area of waste with a solid material in relation to its shrinkage was defined.

Then, in further experiments, the humidity of the waste was increased to 700%, as it has been established earlier that this also changes its viscosity [4]. In order to measure the adhesion, the samples were tested for shift using a manual press of type MIP-100-2 in the following way. The dried sample, to be more precise its solid part, was fixed on a platform press so that the layer of the waste could move freely under the action of the exerted force. Then a gradually increasing force was applied on the waste layer until separation took place from the solid surface. This defined the strength of adhesion.

RESULTS

The analysis of the results showed that there the adhesion of the waste depended on the character of the solid contact material. Moreover, the adhesion of waste depends not only on the type of filler, but also on the initial humidity of WPPI. The adhesion of the waste with a humidity of 600% changes within the limits of 0.025-0.035 MPa, whereas with 700% humidity, adhesion increases to more than twice, 0.086-0.092 MPa.

The analysis of adhesion of pulp and paper industry waste to foamed glass showed the following results: at 600% humidity the adhesion force was 0.163-0.178 MPa, and at the humidity increased to 700%, the adhesion force ranges from 0.184 to 0.193 MPa. Adhesion of WPPI to wood was very small and could not be determined quantitatively. Dry waste exfoliated without any considerable force. Lignin plays the main role in the adhesion of the waste adhesion to fillers. However, this role needs to be investigated in further research.

Cellulose and lignin are high-molecular substances, or oligomers. Lignin is an amorphous substance which is contained in cellular walls. The amount lignin in coniferous wood reaches 30%. Lignin, as chemical compound, is a reactive polymer. Cellulose is a linear stereo regular homogeneous polymer having a large number of hydroxyl groups (OH) and hence forming hydrogen bonds. This type of chemical connection between hydrogen atom hydroxyl and oxygen atom hydroxyl increases the rigidity of the polymer. Thus WPPI, consisting of cellulose and lignin, shows binding and reinforcing properties.

On the basis of the work carried out it is possible to draw the conclusion that WPPI is a contact hardening binding agent to which amorphous cementing and instable crystal structure are related. It is capable to condense at the moment of contacts between particles, getting closer within the distance of action of surface tension forces.

Maintenance of stronger contacts between the particles of cementation is achieved through the application of external pressure. A small quantity of water is useful for lubrication. The most important issue for this system is the reception of substances in an initially instable crystal or amorphous condition.

Therefore at the first stage of hardening the applied technology has to allow the formation of an irregular structure. System hardening occurs at the moment of a strong bond appearing between particles of the amorphous substance and ordering of the structure on boundaries of contact takes place when metastable conditions transform to steady state. At the second stage of hardening of the matrix substance, in all possible systems to which real cementing substances in micro and macrostructural building conglomerates, processes come to an end with larger or smaller ordering, system energy decreases, the transition of the system becomes rather steady, whenever possible, in a crystalline condition.

The second stage does not come to an end only with the condensation of substances; at this stage the processes of consolidation occur as well – hardening and strengthening of the newly formed structure on micro and macro levels. The process of the second stage hardening is the consequence of the continuous qualitative and quantitative changes of the liquid medium and of the solid phase in the system [2].

At the final stage of hardening, the amount of the liquid medium in the system becomes minimal, and the amount of solid phase reaches maximum, i.e. the size of the relation of the liquid medium with solid phase gradually decreases, coming nearer to some optimum value.

Thus the fibrous structure of WPPI allows assuming its use not only as filling material, but also as an independent binding agent obtained by the stabilization of the fibre - water system with the help of containing dispersed solid phases. The WPPI structure is shown in Figure 1. Due to this structure, with the increase of system humidity, the fibres of the waste easily move away from each other, whilst decreasing humidity make the fibres felt to each other, forming a rigid bond. At high system humidity the fibres occur mainly parallel to each other and this causes the orientation effect. In this state the system is tightened and becomes stronger. The presence of lignin in the system also leads to the formation of rigid connections between fibres which enhance the binding effect.

Being the basic substance of wood, cellulose structurally forms the layered cellular shell (wall) capable to disintegrate into thin cellulose fibrils under mechanical force, and under chemical treatment – to micro fibrils. Fibrils have a crystal structure, typically seen as the regular arrangement of appropriate molecules in a molecular crystal lattice. Micro fibrils basically also contain crystalline areas. In some zones the crystalline phase alternates with a chaotic (amorphous) arrangement of macromolecules, hence well defined orientation in micro fibrils is absent, and the bonds are much shorter. This part is called hemi-cellulose [2].

The degree of polymerization of macromolecules is only 100-200. Short bonds quite often occur among crystal sites of a cellulose fibre and then they are linked to cellulose strongly enough to form cellulosans, but remaining in essence hemi-celluloses. Wood cellulose fibres have spiral structure and contain approximately 55-65% crystalline and 25-35% amorphous (hemi-cellulose) parts. Most researchers tend to assume the existence of a chemical connection between hemi-cellulose and lignin, and also between lignin and cellulose as it is impossible neither to take carbohydrates out from wood without simultaneous partial lignin removal nor to remove lignin in full without extraction of carbohydrates in some degree.

The "waste - water - disperse solid phase" system can hold both fibres and air bubbles when it contains stabilized foam. After mixing, fibre waste becomes uniformly distributed in the obtained system and acts as reinforcement. The stiffened cellular shells form an ultra structure comparable to the structure of reinforced concrete: regarding their properties, cellulose micro fibrils correspond to steel reinforcement, and lignin, with its compressive strength, to concrete [9]. A lightweight thermal insulating material with a density of 50-75 kg/m^3 is obtained after its drying. Compressive strength of the produced foamed concrete reinforced by WPPI ranges between 0.5-1.2 MPa. Shrinkage during hardening is at the level of 10-15%. Time of drying is about 10-12 hours at 80°C. The dried material eventually keeps its size and is not deformed.

Research was carried out to find out the possibility of using additional components of some mineral binders such as gypsum and cement, in order to widen the product nomenclature. The introduction of binding agents promoted the essential reduction of drying terms, shrinkage reduction, and also strength increase. A proposed ratios of components by weight were as follows: waste : gypsum at 1:2; 1:1; 2:1; and 3:1. The same ratios were used for cement, while the WPPI humidity was 300 % by weight. The obtained product was exposed to heat treatment at a temperature of 80°C within 12 hours to constant weight and resulted in the following properties:

At a ratio of components by weight waste : gypsum:

- shrinkage   1-12%,
- compressive strength   1.5 – 8.5 MPa,
- density   300 - 450 kg/m^3.

At a ratio of components by weight waste : cement:

- shrinkage   0.5 - 12%
- compressive strength   2.5 - 10 MPa,
- density   370 – 640 kg/m^3.

It was quite difficult to obtain the above results using existing foaming agents due to a number of disadvantageous properties including: low foam ratio (not more than 15), short life-time (up to 12 hours).

CONCLUSIONS

A foaming agent was developed as a result of our research work. It is based on an animal origin raw material (horns, hoofs and skin of cattle, feather of birds), produced by boiling in alkaline solution containing NaOH or KOH. It is then neutralized by a strong acid and stabilized by the salt of a strong acid and a transition metal. Technical characteristics of the received product are: foam ratio not less than 20, life-time more than 24 hours.

In addition to foamed concrete reinforced with dispersed fibres, reference foamed concrete with only the obtained foaming agent was also produced. Such concrete does not shrink and allows reinforcing the structure. The foaming agent can also be produced as a dry powder, hence it is suitable to make dry foam concrete mixes. The obtained results show the perspectives of using waste products of pulp and paper industry and foam for building material production.

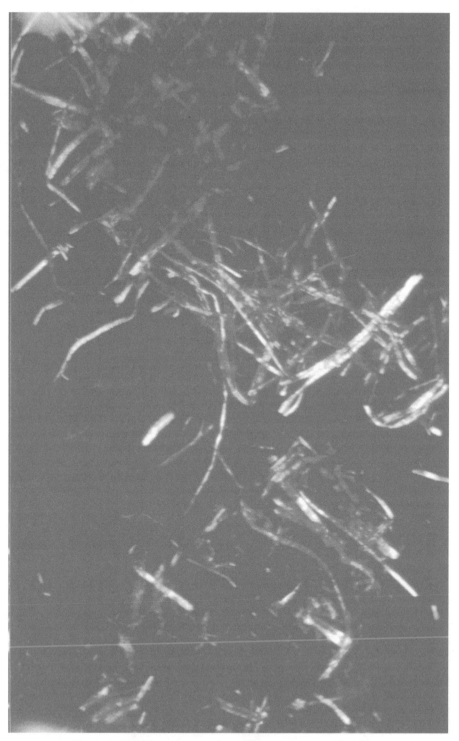

Figure 1   WPPI structure under a microscope, magnified 200×

REFERENCES

1. BATALIN BS AND KOZLOV I A, Materials of building purpose on a scope basis Perm paper industry, Building materials, 2004. No. 1, pp. 42-43.

2. RYBJEV I A, Building materials, High school, 2002, 701 p.

3. GOST 14363.4-89 standard, Method of preparation of tests to physic mechanical tests.

4. BATALIN B S AND KOZIOV I A, Researches of physic mechanical scope properties enterprises Perm paper industry, News of high schools (Construction), 2004, No. 1, pp. 32-34.

5. SYCHYOV M M, Inorganic glues, Chemistry, 1986, 152 p.

6. BATALIN B S AND KOZLOV I A, Researches of adhesive scope properties enterprises Perm paper industry, News of high schools (Construction), 2005, No. 3, pp. 28-30.

7. ZIMON A D, Adhesion of a liquid and wetting, Chemistry, 1974, 412 p.

8. GOST 17580-82 standard, Designs wooden bonded.

9. Internet, www-sbras.nsc.ru/HBC/2004/n03/f09html.

LABORATORY DESIGN OF ROLLER COMPACTED CONCRETE MIXTURE FOR THE CONSTRUCTION OF THE AMATA DAM, SINALONA, MEXICO

A G Gallo

Comision Federal de Electricidad

Mexico

ABSTRACT. The adequate selection of the mixture proportions for a Roller Compacted Concrete (RCC) is a key factor for the construction costs and structural behavior of a dam. In this study case we present the results of the laboratory design of RCC mixture proportions for the Amata Dam, which is a regulator dam for irrigation purposes, located in the North of Mexico, in the state of Sinaloa, with a height of 30 m and a length of 218 m. This paper describes the laboratory procedures for selecting the RCC mixture proportions which was recommended for the construction of the Amata Dam, including the locating, study and selection of natural banks of aggregates, the design of mixtures, the results of the testing of fresh RCC (air content and density), the results of testing the cylindrical samples of RCC (compression, modulus of elasticity and tensile strength) and the analysis of results in order to obtain the optimum water content, the design density for the dam and the minimum density for construction specifications, as well as the strength parameters for designing the dam. The Amata Dam was successfully built in 2005 and the mixture for RCC recommended was used with a few adjustments done in site.

Keywords: Banks, Density, Compaction, Compression Strength, Grading, Roller Compacted Concrete, Water Content.

Amanda Garduno Gallo, is a Civil Engineer, and has worked for the Comision Federal de Electricidad (CFE) for 8 years. She has collaborated on studies regarding Concrete Technology for construction and rehabilitation of infrastructure, basically for production and distribution of power, such as the studies for selecting aggregates banks and the mixture design for the concrete structures of the El Cajon Dam (CFRD) and La Parota Dam (CFRD), the laboratory design for the Roller Compacted Concrete mixtures for Amata Dam (RCCD) and Zapotillo Dam (RCCD). She has participated supervising construction and rehabilitation processes, and has prepared construction specifications for dam projects and repairing procedures for several concrete structures after doing their evaluation.

INTRODUCTION

In Mexico, the Comision Federal de Electricidad (CFE) is the company that generates and distributes power for all the country. The Amata Dam, subject of this study case, was projected and constructed (with the modality of Mixed Contract for Financed Public Works) by CFE. This dam is located in Sinaloa, about 75 km Southeast from Culiacan. This is a regulator dam, part of the hydrologic system of the San Lorenzo River, and its purposes are to optimize the operation of the Comedero Hydroelectric Central and to help on the irrigation of that region.

The project for Amata Dam was proposed as a roller compacted concrete (RCC) gravity dam. During the last period of the construction of the dam it was necessary to change the project, so it was finished with conventional concrete instead of RCC.

The first studies for the RCC, searching for aggregates banks started by 1997, and the construction of the dam started in November 2004 and it was finished by July 2005.

This paper is about the studies for selecting the RCC mixture proportions, including the locating, study and selection of natural banks of aggregates, the design of mixtures, the results of testing fresh and hardened RCC and the analysis of results in order to obtain the optimum water content, the design density for the dam and the minimum density for construction specifications, as well as the strength parameters for designing the dam.

A few aspects of the supervising during the construction of the dam will be described, including the adjustments to the RCC mixture proportions and the changes of RCC to conventional concrete.

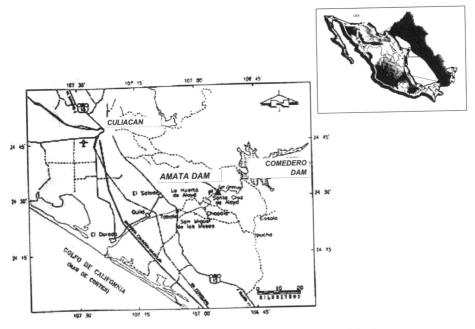

Figure 1   Location of the Amata Dam in western Mexico

ROLLER COMPACTED CONCRETE (RCC) DAMS

For a better understanding of the Amata Dam project, it may be useful to define the RCC. The ACI 116R [1] defines RCC as "concrete compacted by roller compaction; in its unhardened state will support a roller while being compacted". Properties of hardened RCC are similar to those of conventionally placed concrete.

It is important to remember that the RCC is a material with similar structural properties of concrete with the placing characteristics of embankment materials. To achieve the highest measure of cost effectiveness and a high-quality product similar to that expected of conventional concrete structures, the RCC should be placed as quickly as possible, the operations should include as little manpower as possible, the design should avoid multiple mixtures that interfere with production, and the design should minimize complex construction procedures. The design of the structure must be coordinated with the performance requirements for the RCC material and the specification requirements for construction. [2]

RCC construction techniques have made RCC gravity dams an economically competitive alternative to conventional concrete and embankment dams due to the lower costs (from 25 to 50% less than a conventional concrete dam), rapid construction and reduced material quantities, integral spillways, more resistant to internal erosion and overtopping (compared to embankment dam).

In Mexico, the construction of RCC dams started by 1989 with the Trigomil Dam (40 m height) and has been increased since then. The taller one, Corral de Palmas Dam, with 110 m height, was built in 2003. Several new dam projects are in study by CFE, for example the Zapotillo Dam, with 110 m height, and the Arcediano Dam, with 120 m height.

OVERALL LAYOUT OF THE AMATA PROJECT

The layout of Amata scheme is shown in Figure 2. The main components are as follows:

- A test embankment on the left margin, built of RCC, is part of the dam. With 58 m length, 9, 75 m height and 6,5 crest wide.

- The 30 m high RCC gravity dam with a 218.0 m crest length, its crest at el. 140.4 m, requiring about 45,000 m^3 of RCC. Part of it works as the spillway. The maximum design flow rate is 2410 m^3/s with a return period of 1000 years.

- Diversion works, comprising two cofferdams of rockfill with impervious clay core, and the same channel of the intake structure during the construction.

- The intake structure is a concrete lined outlet channel with a length of 46 m and 10 m wide, controlled by 2 tainter gates of 4 x 4 m separated by a central 2 m pier, designed to discharge 160 m^3/sec.

- The spillway is a Creager type weir of free discharging and a deflector bucket, with a total length at crest of 144 m, and its crest at el. 132.7 m. It is designed to discharge 5116 m^3/s corresponding to a 6.45 m height above the crest.

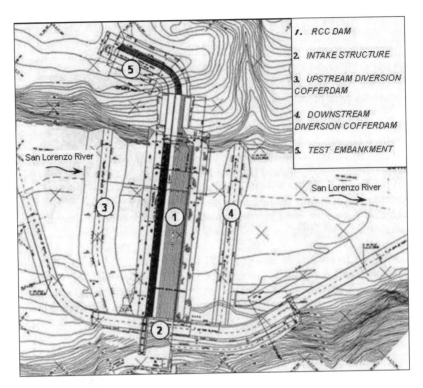

Figure 2   General layout of the scheme

The catchment area is about 8919 km^2. The mean annual runoff is 150,986 m^3 with a mean annual flow of 47.87 m^3/s.

AGGREGATES BANKS

Investigation of banks was necessary in order to assess the availability and suitability of the aggregates needed to manufacture the RCC with qualities meeting the structural and durability requirements.

The RCC typically requires low cementitious contents, therefore larger percentages of fines are used to increase the paste content in the mixture to fill voids and contribute to workability. Additional fines are usually needed, and the study of banks includes the investigation for nonplastic lime banks.

By February 1997 the exploration for aggregates banks was done nearby the site of the project. Searching for aggregates was in the both margins of the San Lorenzo river, and upstream and downstream of the dam axis. There were located 3 gravel-sand banks: Santa Cruz de Alaya, El Tigre and Playon 2, and the bank Humagua for lime. Table 1 shows the volumes of useful material (considering a nominal maximum size of 50.8 mm) and the distance to the dam axis.

Table 1   Banks: location and estimated volumes

BANK	LOCATION	USEFUL MATERIAL, m³			
		Gravel	Sand	Lime	Total
Sta Cruz de Alaya	Right margin River San Lorenzo 6.0 km Downstream	22,820	9427	0	33,247
Playon 2	Right margin River San Lorenzo 1.5 km Upstream	15,680	6105	0	21,785
El Tigre	Right margin River San Lorenzo, 2.5 km Downstream	14,887	5840	0	20,691
Humagua	3.5 km Dowstream	0	0	Enough (> 2500)	

AGGREGATES PROPERTIES

As with conventional concrete, aggregates for RCC were evaluated for quality and grading. The gravel-sand banks were sampled and the physical and chemical properties were identified in the laboratory. The tests included: grading, unit weight, absorption, organic impurities, lost of fines by washing, soundness, abrasion, potential alkali reactivity. Initially, it was intended to use a nominal maximum size of 76.2 mm for the RCC mixtures, but the grading of the banks did not fulfil the recommended limits and the RCC mixtures presented segregation. Then it was needed to remove the particles retained on sieve 50.8 mm. The laboratory results showed that the three aggregate banks had very similar materials and that they were adequate for RCC, so it was possible to mix the samples of the three of them in order to have one integrated sample. Table 2 shows the properties of this integrated sample. The samples from the lime bank were also studied at the laboratory, verifying its low plasticity, adequate for RCC mixtures. In order to fulfil the recommended grading limits [3] for the RCC mixtures, it was needed to separate gravel (passes sieve 50.8 mm and is retained in sieve 4.75 mm) and sand (passes sieve 4.75 mm and is retained in sieve No. 200), and mix them, with the lime, in this proportion, by weight: 57% gravel, 37% sand and 6% lime. Figure 3 shows the grading after this adjustment.

Table 2   Properties of the integrated sample.
(Proportion: 34% Santa Cruz de Alaya, 33% El Tigre y 33% Playon 2)

PROPERTIES	STANDARD FOR TESTING	INTEGRATED SAMPLE	
		Gravel	Sand
Density, g/cm³	ASTM-C-127, 128 - 01	2.62	2.61
Absorption, %	ASTM-C-127, 128 - 01	1.56	1,.52
Fines by washing, %	ASTM - C -117 - 95	----------	6.2
Organic impurities	ASTM- C - 40 - 99	----------	Acceptable
Soundness, %	ASTM - C - 88 - 99	0,.35	0.30
Abrasion,%	ASTM - C - 131- 03	12.67	----------
Potential AR	ASTM - C - 289 - 02	Innocuous	Innocuous

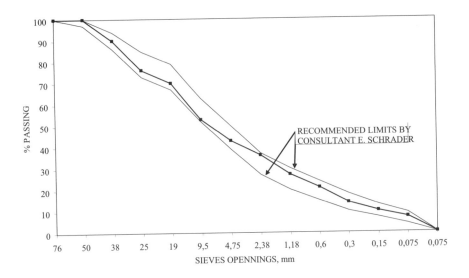

Figure 3   Grading of the integrated material: 57% gravel, 37% sand and 6% lime

DESIGN OF MIXTURE PROPORTION FOR RCC

The proper selection of RCC mixture proportions is very important for obtaining an economical and durable concrete.

The RCC mixture proportioning procedures are similar to those of conventional concrete. The differences are due to the low water content and no-slump consistency of RCC. During construction, the RCC mixture must be stable enough to support the weight of a vibratory roller, yet workable enough to allow some aggregate reorientation; voids between aggregates particles will be filled with paste during the compaction operations. In the laboratory the compaction for molding the cylinders was performed using a pneumatic pole tamper and plastic molds with steel sleeves, as shown on Figure 4.

Figure 4   Views of the laboratory equipment, compaction and preparation of RCC cylinders

In order to determine the mixture proportion for the construction of the Amata dam, a two phases program for mixtures at laboratory was established. It is important to mention that in all the mixtures a Portland-pozzolan cement, type IP according to ASTM C-595 [4], was used, so it was not necessary to add natural pozzolans, as usual for RCC mixtures.

On the first phase, the optimum water content was determined: 5 mixture proportions were done with the same cement content (120 kg/m^3) and with different water contents (2.8%, 3.2%, 3.4%, 3.6% and 4.0%). The main properties to fresh RCC were determined, density (ASTM C-138 [5]) and air content (ASTM C-231 [6]); and the mechanical properties to hardened RCC cylindrical samples at different ages (3, 7, 28, 90 and 180 days), compression strength (ASTM C-39 [7]), tensile strength (ASTM C-496 [8]), secant modulus of elasticity to 40% of the ultimate compression strength and modulus of elasticity to ultimate strength (ASTM C-469 [9]).

Table 3 shows the results of density and air content of fresh RCC. Considering those densities and the workability of the mixtures, the optimum water content was established as 3.2%.

On the second phase 4 mixture proportions were prepared. This time, the lime content (6%) and the water content (3.2%) were fixed, and the cement content changed on each proportion (70, 100, 150 and 180 kg/m^3). The same properties for fresh and hardened RCC than in the previous phase were determined. Table 3 shows the results of the fresh RCC testing.

Table 3   Results of density and air content of fresh RCC

| MIXTURES (DEFINED AS: CEMENT (kg/m^3) – POZZOLAN (kg/m^3) – WATER (%) | | | | | |
| PHASE 1 | | | PHASE 2 | | |
Mixture	Density kg/m^3	Air content %	Mixture	Density kg/m^3	Air content %
120-0-2,8	2402	1.10	70-0-3,2	2429	1.10
120-0-3,2	2436	1.10	100-0-3,2	2432	1.10
120-0-3,4	2424	0.70	120-0-3,2	2436	1.10
120-0-3,6	2432	1.10	150-0-3,2	2422	0.95
120-0-4,0	2414	1,10	180-0-3,2	2440	1.20
			Average	2432	1.09
The results are the average of two batches.			σ	7	0.089

Specified Density for RCC

Density is defined as mass per unit volume; it was determined according to ASTM-C-138 [5]. Density of RCC depends primarily on aggregate density and the degree of compaction. The lower water content of RCC mixtures result in a higher density than conventional concrete.

For construction purposes, it was necessary to establish the specification for the density of the RCC Amata dam, and it was done with the results shown on Table 3, as follows:

With the average density, calculated as the average minus the standard deviation, and the same assumption for the average air content, the optimum density for RCC was 2450 kg/m^3.

By experience from other dams, it was considered a 97.5% of compaction during construction, and knowing that the air content can have normal variation within a 100%, the average density during construction was calculated, 2332 kg/m³. The minimum density during construction, 2278 kg/m³, was calculated as the 93% of the optimum density. For construction specification purposes, the average density during construction was fit as 2330 kg/m³, and the minimum density during construction was specified as 2280 kg/m³.

Strength parameters

The strength parameters for the RCC, compression strength, secant modulus of elasticity to 40% of the ultimate strength and to ultimate strength, and tensile strength, needed for the designer of the dam, were calculated based on the test results. With these results (Table 4), it was possible to obtain graphics of each parameter depending of the age and with different cement contents.

Table 4   Results of compression strength, tensile strength, and secant modulus of elasticity of the RCC mixtures.

| MIX | COMPRESSION STRENGTH, MPa | | | | | TENSILE STRENGTH, MPa | | | | |
| | Age, days | | | | | Age, days | | | | |
	3	7	28	90	180	3	7	28	90	180
70-0-3,2	3.26	4.37	5.07	6.53	7.78	0.39	0.59	0.78	0.98	1.47
100-0-3,2	4.72	7.20	8.87	9.95	12.04	0.74	0.96	1.28	1.57	1.96
120-0-3,2	7.69	8.38	10.70	9.98	13.04	1.09	1.18	1.67	1.86	2.35
150-0-3,2	10.29	12.26	14.54	14.89	16.71	1.53	1.55	1.77	2.35	2.45
180-0-3,2	11.58	11.76	16.54	14.97	17.88	1.73	1.97	2.75	1.96	2.65

The results are the average of two samples. The results are from one sample.

MIX	Secant Modulus of Elasticity, MPa									
	40% Ultimate Strength					Ultimate Strength				
	Age, days					Age, days				
	3	7	28	90	180	3	7	28	90	180
70-0-3,2	2635	5692	5874	5229	7389	189	503	1147	1159	2883
100-0-3,2	2881	7496	9045	16084	14277	440	906	2589	3857	6915
120-0-3,2	6699	11444	13438	17847	8292	1230	1673	6381	3150	3293
150-0-3,2	4157	12940	18572	19178	19288	2669	2883	6735	7473	9377
180-0-3,2	6321	8490	19431	16978	10575	2614	1969	6521	6781	5993

The results are the average of two samples.

The final graphics for the compression strength were obtained as follows:

From the strength results, averages shown on Table 4, the graphic compression strength – age, for the different cement contents of the mixtures tested on laboratory was established. Those curves showed tendencies, but some results were eliminated in order to reduce the dispersion. With this, the curves were fitted and their representative equations were obtained. From those equations, the ratio compression strength/cement content (with no measuring units) was calculated for the different ages, which is an efficiency cement ratio used in Mexico. With a conservative criterion, the average efficiency cement ratio less the standard deviation, the curves compression strength-age-cement content shown on Figure 5, were established for cement contents from 70 to 180 kg/m³. These are the curves used to select the RCC mixture proportion for the design and construction of the Amata Dam. Following a similar procedure from the results of the testing showed on Table 4, the graphics that represent the variation of the following parameters were established:

- Tensile strength-age-cement content, shown on Figure 5,
- Secant Modulus of Elasticity (to 40%) -age-cement content, shown on Figure 5,
- Secant Modulus of Elasticity (to ultimate strength)-age-cement content, on Figure 5.

Those curves provided the properties of the RCC, once the proportion mix was selected, for the design of the dam, including the spillway.

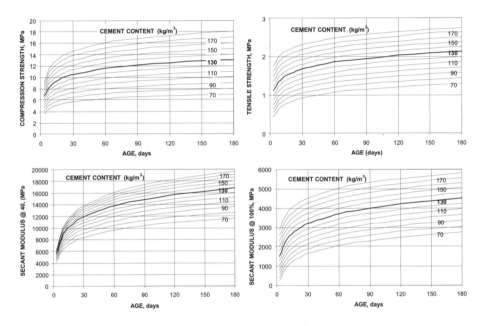

Figure 5   Representative curves of the development of the compression strength, tensile strength, and modulus of elasticity depending on the age and cement contents

Mixture proportion for RCC

The selection of the RCC mixture proportion was required by the designing needs, a compression strength of 11.8 MPa was needed at 90 days. From the curves shown on Figure 5 it can be seen that the curve for a cement content of 130 kg/m³ is the one that fulfils the strength requirement. With this, the RCC mixture proportion for the construction of the Amata dam was recommended as shown on Table 7. The mechanical properties for the specified age of 90 days, and for a cement content of 130 kg/m³, was obtained from the graphs shown on Figure 5, and are mentioned on Table 5. Those were the parameters used by the designer. In similar way, the mixture proportions for the RCC that was needed for the spillway was determined (200-0-3,2).

Table 5   Selected RCC mixture proportion and its properties

MIXTURE PROPORTIONS: 130-0-3,2			
Material	Content, kg/m³	Theoretical Properties at 90 days used for the	
Cement IP	130	design of the dam	
Water	78	Density, kg/m³	2330
Gravel	1270	Compression strength, MPa	12.2
Sand	824	E_{40}, MPa	14,883
Lime	134	E_{100}, MPa	3990
Total	2436	Tensile strength, MPa	2.0

Field Adjustment of Mixture Proportions

The Contractor (Canoras) and the supplier of concrete (Cemex) adjusted the mixture proportions by August 2004, before the construction of the dam. The first consideration was to adjust the grading with the banks that were in exploitation (El Playon 2 and El Tigre); then the mixture proportion was established as 23% gravel (passing sieve 50 mm and being retained in sieve 9.5 mm), 69% gravel-sand (passing sieve 9.5 mm and being retained in sieve No. 200), 8% of lime, 120 kg/m³ of cement type IP, water content 3.4% and 1.3% of a plasticizing and retarding admixture designed for RCC (Plastiment RCC-M).

CONSTRUCTION OF THE AMATA DAM

Aggregates production

The exploitation of the banks El Playon 2 and El Tigre was done with two sieve systems, one of them fixed at the project site and a mobile one, close to the banks. The construction specifications required 50% of the aggregates already stockpiled (approximately 30,000 m³), before starting the dam construction, but it started with 30% of the volume so the aggregate production had an accelerated rhythm.

RCC production plant

The RCC production plant was a continuous mixing plant (pugmill) brand Vince Hagan, model HTCTB8300. This plant had two boxes for the aggregates with conveyors, and two continuous weight scales, one for aggregates and the other one for the RCC. The production rate capacity was 150 m³/h. The cement used for construction was Portland pozzolan, type IP.

RCC test embankment

After the RCC production plant was installed by Cemex on the right margin, the RCC test embankment, on the left margin was started (October 2004). The total volume of fresh RCC was 2300 m³, and compacted was 1700m³.

The construction of the RCC test embankment, helped to verify the behavior of the mixture proportion, to calibrate the nuclear density gauge, to evaluate the operation of the production plant, to train personnel on the transportation, placing, spreading, compaction and curing of the RCC, to verify the equipment for each activity, to establish the bedding mix proportion, to determine the preparations for joints, and to verify the results of density and compression strength of the RCC.

Transporting, placing and compaction of RCC

A conveyor system was used to bring near the RCC to the dam site, and then it was moved with trucks of 7 and 14 m^3 volume capacity. The placing and spreading was done with a Positrack (bulldozers were discarded because they caused excessive segregation), the lift thickness was about 0.35 m before compaction. The compaction was done with a single drum 10 ton roller (CS 553), with 8 passes; in tight areas, such as adjacent to the downstream face concrete panels, it was achieved with a small roller and tamper jumping jack compactors.

Change in project

The Amata dam was designed as a gravity RCC dam but during its construction some problems between the Contractor and CFE forced a change in the project: from RCC to conventional concrete. Finally about 80% was RCC (32,930 m^3) and 20% conventional concrete (8900 m^3). The conventional concrete mixture was designed in order to have similar mechanical properties than the RCC, with f'c = 11.8 MPa.

Figure 6   Views of the Amata dam during and after construction

CONCLUSIONS

The investigation of banks showed the availability of aggregates and lime close to the site of the dam project, and with the quality required for producing RCC. For the laboratory studies the grading of the aggregates had to be fitted in order to fulfil the limits recommended. It was needed to be separated and then mixed by weight as follows: 57% gravel, 37% sand and 6% lime. For this project the maximum size had to be 50.8 mm because larger material caused segregation on the mixtures. The use of cement Portland IP, available in the construction area, prevented the addition of pozzolans. The selection of the RCC mixture proportion was done in two phases, the first one, to establish the optimum water content. On the second one, the mechanical properties were determined and with the designing requirements the adequate mixture proportion was established.

The field adjustments were needed before construction, due to the banks in exploitation, the production and transporting processes, the weather conditions, etc. An adequate RCC study on laboratory is necessary to establish the construction specifications for the density. During the Amata dam construction the quality control of the density was held according to the values provided by the laboratory. A double-prober nuclear gauge was used, for each 200 m^2 of RCC, five tests were done, and the average had to be higher than 2330 kg/m^3 and none of the single readings could be less than 2280 kg/m^3. If that specification was not fulfilled, the recompaction of the area was ordered, and if this was not enough, the material of that lift had to be removed. A complete RCC mixture proportions laboratory study needs to be reliable in order to give the information needed for an adequate dam design, safe, durable and economical, and also will help to establish the optimum construction procedures.

REFERENCES

1. AMERICAN CONCRETE INSTITUTE. ACI 116.R-00. Cement and Concrete Terminology.

2. US ARMY CORPS OF ENGINEERS. EM-1110-2-2006. Engineering and Design Roller-Compacted Concrete.

3. AMERICAN CONCRETE INSTITUTE. ACI 207.5R-88. Roller-Compacted Mass Concrete.

4. AMERICAN SOCIETY FOR TESTING AND MATERIALS. ASTM C 595-00. Standard Specification for Blended Hydraulic Cement.

5. AMERICAN SOCIETY FOR TESTING AND MATERIALS. ASTM C 138-01. Standard Test Method for Density (Unit Weight), Yield and Air Content (Gravimetric) of Concrete.

6. AMERICAN SOCIETY FOR TESTING AND MATERIALS. ASTM C 231-97. Standard Test Method Air Content of Freshly Mixed Concrete by the Pressure Method.

7. AMERICAN SOCIETY FOR TESTING AND MATERIALS. ASTM C 39-01. Standard Test Method for Compressive Strength of Cylindrical Concrete Specimens

8. AMERICAN SOCIETY FOR TESTING AND MATERIALS. ASTM C 496-00. Standard Method for Splitting Tensile Strength of Cylindrical Concrete Specimens.

9. AMERICAN SOCIETY FOR TESTING AND MATERIALS. ASTM C 469-02. Standard Test Method for Static Modulus of Elasticity and Poisson's Ratio of Concrete in Compression.

10. AMERICAN SOCIETY FOR TESTING AND MATERIALS. ASTM C 33-99. Standard Specification for Concrete Aggregate.

11. AMERICAN SOCIETY FOR TESTING AND MATERIALS. ASTM C 1017-98. Standard Specification for Chemical Admixtures for Use in Producing Flow Concrete.

SELF COMPACTING CONCRETE TECHNOLOGY IN JORDAN

M Resheidat

S A Alzyoud

Jordan University of Science and Technology

Jordan

ABSTRACT This paper presents experimental investigation on introducing the technology of self compacting concrete-SCC in Jordan using local materials. Development of SCC in Jordan is highly needed to overcome many problems in the construction sector. Trial concrete mixed have been carried aiming at studying fresh and hardened concrete properties. Local crushed limestone and Portland cement were used in addition to fly ash and limestone powder. Additives in the form of carboxylated ether based super plasticizers and viscosity modifying agents were employed. Properties of fresh concrete were evaluated using slump cone, L-box , U-box, V-funnel tests and sieve stability analysis. Performance of hardened concrete is determined by the classical strength tests. It is expected that results of this study will introduce a guidelines and recommendations to serve the concrete industry sector in general and to ready mix plants in particular.

Keywords: Self compacting concrete, Concrete technology, SCC.

Professor Musa Resheidat works in the Department of Civil Engineering, Jordan University of Science & Technology, Irbid, Jordan. He is ASCE Fellow; Member and Expert of Jordan Engineers Association; IACT Chairman, International Association of Concrete Technology.

Engineer Sukina A Alzyoud has graduated with BSc and MSc degrees in Civil Engineering from Jordan University of Science & Technology, Irbid, Jordan. She is working now in London, UK.

INTRODUCTION

Concrete is the most widely consumed material in the world, after water. Placing the fresh concrete requires skilled operatives using slow, heavy, noisy, expensive, energy-consuming and often dangerous mechanical vibration to ensure adequate compaction to obtain the full strength and durability of the hardened concrete. Self compacting concrete (SCC) represents one of the most outstanding advances in concrete technology during the last two decades and represent a solution to all placing problems. SCC is a concrete which flows to a virtually uniform level under the influence of gravity without segregation, during which it deaerates and completely fills the formwork and the spaces between the reinforcement even in highly reinforced concrete members and flows free of segregation nearly to level balance.

SCC was developed by Okamura [1] in 1986 for the first time in the world, then prototype was developed in 1989 by Ozawa [2] and has been spread all over the world. Due to its specific properties, SCC may contribute to a significant improvement of the quality of concrete structures and open up new fields for the application of concrete. The use of SCC offers many benefits to the construction practice: eliminating compaction work, reducing placement cost and construction time, improving the productivity, reducing noise during casting and providing better working conditions in inner city areas. Other advantages of SCC are improving homogeneity of concrete production and achieving high quality of finished surface. SCC is a generic term for mixes designs that differ from traditional concretes at the molecular interface between the cement compounds and the admixture polymers. The fluidity of SCC ensures a high level of workability and durability whilst the rapid rate of placement provides an enhanced surface finish. SCC's high strengths overnight strengths typically reach 30-40N/mm^2 and 2-day strengths can break the 100 N/mm^2 barrier enable easier and more reliable remolding. SCC is certainly the way forward for both in situ and precast concrete construction. The health and safety benefits and the improved construction and performance results make it a very attractive solution. In early 1990s there was only limited knowledge of SCC because large corporations kept it as secret for commercial benefits and it was known under commercial names of each company. Simultaneously research studies and developments continued on mix design where new admixtures were used to produce SCC mixes with performance matching Japanese SCC mixes. Then the technology spread out to Europe and all over the world [5-14]. Documentation of historical background could be found in Reference 15.

Today self compacting concrete is being studied worldwide, and there are a lot of papers presented to any related concrete conferences. Until now, there is no universally adopted standardized test method for evaluating compactability of this concrete. Studies are continuing for evaluating its properties to spread this technology to the world although it is being rapidly adopted by large number of countries. The fundamental objective of this work is to provide information on the fresh and hardened properties of self-compacting concrete produced using available local raw materials in Jordan, and to facilitate the introduction of SCC technology by general construction practice all over the country. The engineering properties of SCC in sampled specimens were to be compared with those of normal concrete. The specimens made from local aggregate, pozzolanic cement. High range water reducing agent and viscosity modifying agent were manufactured in Jordan. The concrete was produced in construction materials laboratory. The following fresh properties such as filling ability, passing ability, segregation and bleeding resistance were evaluated by the special equipment manufactured for this task. The mechanical properties of sampled specimens were tested such as compressive and flexural strength, stress-strain curves and elastic modulus.

COMPOSITION OF SELF COMPACTING CONCRETE

The basic components of SCC are same as those used in the conventional concrete. However, to obtain the requested properties of fresh concrete, in SCC a higher proportion of ultra fine materials and the incorporation of chemical admixtures, in particularly polycarboxylated polymer based superplasticizer, and viscosity modifier are necessary. Ordinary and approved filler materials like fly ash, limestone powder, blast furnace slag, and silica fume and quartzite powder.

There are wide variations of materials and proportions that can be used to produce SCC, but these proportions must fall within the limits listed in Table 1. For achieving economical mixes powder content (cement and fillers) should be kept to minimum cause of there expensive prices. The upper limit of water can be used if viscosity enhancer is used.

Table 1   Range of mix proportions per 1 m^3 of concrete.

CONSTITUENTS	BY VOLUME	BY WEIGHT
Coarse aggregate	30-34% of concrete volume	750-920 kg
Fine aggregate	40-50% of mortar volume	710-900 kg
Powder	-	450-600 kg
Water	150-200 l/m^3 of concrete	150-200 kg
Paste	34-40% of concrete volume	-

Cement

The SCC mixes investigated in this study were prepared with pozzolanic Portland cement produced by Jordan cement factory according to Jordan standard specifications-219-1995.

Aggregates

A continuously crushed limestone aggregates and silica sand brought from AlHusun quarries in Irbid district. Preliminary tests were performed on the available aggregates and sand according to ASTM standard methods.

Fly Ash

Fly Ash is a fine residue of burned coal, it is not available in Jordan markets but it was brought from India for Abdoun Bridge construction project in Amman. The experiments use fly ash of class C whish complied with the requirements of ASTM C 168. It is used to avoid segregation and to provide flow ability to concrete mix.

Limestone Powder

Limestone powder was also brought from the same quarries and it was chosen because it is available with very high amounts and low prices in Jordan and there is a previous experience of using it. It was used to enhance viscosity and segregation resistance and it can inhibit the temperature rise of concrete if it is used in high proportions.

Chemical Admixtures

Two types of chemical admixtures were used: super plasticizers from a new generation of copolymer-based superplasticisers designed for production of SCC. With a density of 1.3 and it was used within a range of 0.5-3.5 L/100 kg of cementing materials including cement, limestone powder and fly ash. The other type is viscosity modifying admixture (VMA) which were used to increase water viscosity to increase segregation resistance and to reduce the changes of concrete properties caused by variation of material properties and environmental conditions, VMA has a specific gravity of 1.4 and it was used at dosage range of 1-5 L/m^3 of concrete.

EXPERIMENTAL WORK

Self Compacting Concrete Mix Design

The major challenge of developing SCC mix is to achieve the balance between fluidity and high segregation resistance which is related to mix viscosity. Three ways to increase viscosity of the mix are reducing water to powder ratio, using large quantities of powdery materials, and addition of viscosity modifiers. Strength of concrete mainly corresponds to water to cement ratio for given cement and other components of certain mix. Maximum permissible water to cement ratio for durability should be considered, so this ratio should be achieved based on these two parameters. Fillers can be incorporated in higher amounts for durability consideration than achieving high strength. VMA must be incorporated to achieve higher viscosity and reduce the changes of concrete properties caused by variation of material properties and environmental conditions.SP dosage is determined using slump cone test.

In this study we use the available mix proportions from available mixes in literature. 15 mixes were carried out to study the effect of W/C ratio, type and content of filler, and VMA content. Where the first five mixes investigate the effect of the content of both FA and LSP on SCC properties where other parameters are fixed (W/P = 0.39, VMA = 1.5 L/m^3), the next five mix proportions for SSC mixes were tabulated in Table 2.

FRESH AND HARDENED PROPERTIES OF SCC

Two variables (yield value and viscosity) are therefore required for a complete description of the rheological properties of SCC. The place-ability and stability were examined as follows:

Slump flow test (total spread and T50 time)

L-Box Test

The L-Box test allows measurement of the filling ability, passing ability, and resistance to segregation of SCC mixes. The vertical part of the box is filled with fresh concrete and left at rest for 60 seconds to allow any internal segregation to occur. The gate is opened, and the concrete flows out into the horizontal part of the box.

Table 2 Mix proportions of self compacting concrete mixes.

ID	C kg	FLY ASH, kg	LSP kg	SAND kg	CA kg	W kg	W/P	VMA (L)	SP (L)
1	610	0	0	760	700	240	0.39	1.5	2
2	490	93	31	760	700	240	0.39	1.5	2
3	397	130	98	760	700	245	0.39	1.5	2
4	537	0	94	760	700	240	0.39	1.5	2
5	537	94.8	0	760	700	240	0.39	1.5	2
6	427	97.5	97.5	760	720	240	0.39	1.5	2
7	816	96	48	720	700	240	0.25	0.5	3.6
8	680	80	40	720	700	240	0.3	1	2.6
9	583	68.6	34.3	740	700	240	0.35	1.2	2.2
10	510	60	30	740	700	240	0.4	1.5	2
11	408	48	24	760	700	240	0.5	3	0.8
12	582.9	68.6	34.3	740	700	240	0.35	0.5	2.2
13	582.9	68.6	34.3	740	700	240	0.35	1	2.2
14	582.9	68.6	34.3	740	700	240	0.35	2	2.2
15	582.9	68.6	34.3	740	700	240	0.35	3	2.2

U-Box Test

The U-flow test measures the filling ability, and blocking ability of SCC. It is considered as the most appropriate for determining the self consolidating abilities of a concrete mix. The U-box apparatus consists of two chambers separated by a gate and row of vertical reinforcing bars The maximum height percentage is the ratio of the filling height to the final height of concrete in the first chamber (H_2/H_1).

J-Ring Test

The J-Ring is used in conjunction with the slump flow test, the V-funnel test or any other apparatus that provides a discharge of concrete. These combinations should test the flowing ability and, with the contribution of the J-Ring, the passing ability of the concrete.

V-Funnel Test

The V-funnel is used to evaluate the filling capacity of a concrete, and the capacity to pass through narrow spaces without blocking the flow.

Sieve Stability Test

It is used to quantify the resistance to segregation of a mix.

Visual Stability Index

The resistance to segregation of SCC can be visually evaluated in a lesser or greater degree in almost every test mentioned above. Visual Stability Index (VSI) test helps to quantify the stability of SCC mixes.

The VSI test is recommended to be implemented with the slump flow test; although, the parameters evaluated in the VSI test can be found as well in the L-flow test, U-flow test and in general every test that allows the observation of a significant volume of SCC. The range of the values for the VSI is from 0 through 3, with zero being a highly stable mix, and 3 designates a highly unstable mix. The apparatuses for L, U, J and V shapes were designed, built and assembled for this study and shown in Figure1.

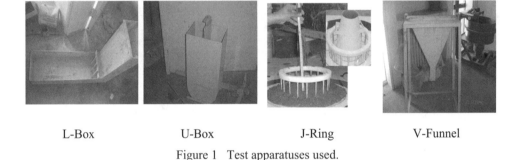

L-Box   U-Box   J-Ring   V-Funnel

Figure 1   Test apparatuses used.

The properties of hardened SCC were evaluated as follows:

a) Compressive Strength Test for Cubes.
b) Compressive Strength Test for Cylinders.
c) Splitting Tensile Test.
d) Modulus of Rupture.

Experimental set-ups are shown in Figure 2.

Cube Test   Cylinder Test   Splitting Test   Rupture Test

Figure 2   Test set-ups of hardened concrete.

RESULTS AND DISSCUSSION

Results of Fresh Properties

Three key properties of SCC which are the filling ability, passing ability and segregation resistance have been evaluated using different tests, namely slump flow diameter (S), T50, the measured V-Funnel time (t), the J-ring, Boxes test, and sieving analysis. Table 3 presents the results of these tests on all investigated mixes where the mix proportions were given in Table 2.

Table 3   Results of the fresh properties of SCC.

Mix	Slump Flow mm	T50 (sec)	J-Ring mm	V-Funnel (sec)	L-Box (h_2/h_1)	U-Box (h_2-h_1) mm	*Sieving Analysis %
1	700	02:30	7.5	07:30	0.93	22	7.5
2	705	02:00	6.0	06:30	0.95	26	8.0
3	720	03:30	6.5	07:00	0.95	19	8.2
4	690	02:30	7.0	07:30	0.96	25	7.8
5	710	03:00	6.5	09:00	0.99	19	8.0
6	710	03:00	6.5	08:30	0.98	18	8.2
7	760	04:00	2.0	09:30	0.99	9	16.0
8	745	03:30	3.0	9:30	0.99	12	14.0
9	735	03:30	8.0	08:30	0.98	16	11.0
10	700	03:00	12.0	06:30	0.96	20	10.0
11	650	02:30	13.5	06:00	0.92	25	7.7
12	690	03:30	9.5	06:30	0.95	26	9.1
13	695	03:30	7.5	07:00	0.96	19	8.7
14	710	03:30	6.5	07:30	0.97	15	8.2
15	720	04:00	4.0	08:00	0.98	11	7.5

*Segregation portion

Tests of Filling Ability

Slump flow test

It measures the workability and consistency of self compacting concrete; it is easy to use in laboratory and construction site.

The results from Table 3 show that Slump flow values are within acceptable ranges (650-760 mm), and the value is affected by several parameters such as: water to powder ratio (W/P), type and amount of filler and viscosity modifying agent (VMA). Different values of W/P ratio were used (0.25, 0.30, 0.35, 0.40, 0.50). As the W/P ratio decreases the paste volume increases causing higher plasticity, flow ability and cohesion of mix giving higher slump and T50 values.

When the mixtures with mineral additives are examined (first five mixes where other parameters were constant) it can be seen that mixtures containing FA showed better flowability and deformability when compared to LSP. Moreover, the increased amounts of FA improved the workability of the mixtures, while the increased amounts of LSP slightly limited the self compacting characteristics. When the particle sizes of the mineral additives are compared, it can be observed that LSP is coarser than FA. However, both of these mineral additives increased the workability of SCC mixes. A partial replacement of cement by FA results in higher volume of paste due to its lower density and this increase in the paste volume reduces the friction at the fine aggregate-paste interface and improves the plasticity and cohesiveness, and thus leads to increased workability. Moreover, the spherical shape of FA particles is also reported to improve the workability.

The spherical shapes reduce the friction at the aggregate-paste interface producing a "ball-bearing effect" at the point of contact. Therefore, it can be concluded that fineness is not the only parameter of a mineral additive to improve the workability of a SCC mixes, as seen in Figure 3.

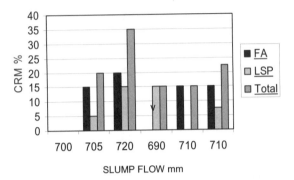

Figure 3   Effect of filler type and quantity on slump flow.

The results show that all SCC mixes with VMA meet the required slump flow value as well as the T50 flow time . Figure 4 shows that the increase of VMA is associated with an increase of cohesion and plasticity and hence slump values. the content of VMA was used at different contents varied from (0.5-3) l/m^3.

V-Funnel Flow Time

The V-funnel flow time T serves as a measured variable for describing the viscosity of SCC. The viscosity of a suspension is dependent mainly on the water/solids ratio and the overall grading curve. where high W/P ratio leads to faster flow and reduced flow time. Viscosity is increased by incorporating VMA leading to increase funnel flow time. Correlations between the slump flow values and each of the T50, and V-funnel time. None of the relationships were closely correlated, confirming a degree of independence between the two properties

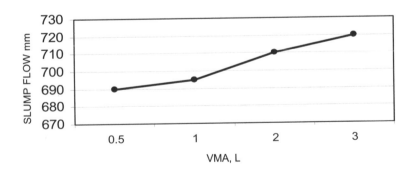

Figure 4   Effect of VMA on slump flow.

Tests of Passing Ability

J-Ring Test

The J-Ring test is used to determine the flow behavior of SCC in the face of obstructions. It is carried out in the same way as the slump flow test, but the J-Ring is also required to simulate placement conditions in which the concrete has to overcome obstacles to flow. The extent of the tendency to blocking is determined as a rule by measuring a difference in level. Some mixes have a difficulty of achieving the self compacting degree of passing ability through obstacles of reinforcement due to effect of the characteristics of used coarse aggregates, there shape and size. For the mixes where the W/P ratio was the only variable, passing ability of concrete through obstacles increased as powder content was increased and enhancing coating of aggregates and decreasing the friction between aggregates.

Boxes Tests (U-Box, L-Box)

The box tests are tests for assessing the place-ability of SCC. In both methods a closed vertical chamber is filled with the concrete to be tested so that a hydrostatic pressure head is produced. The difference in levels after opening gate determines the tendency to blocking. Passing ability through the two boxes tests was achieved because the spacing between obstacles was higher than maximum size of aggregate and the paste was coating each aggregate particle preventing friction between aggregate and bars. As the W/P ratio and VMA were increasing the blocking value from the two tests was decreasing.

Hardened Properties of Self Compacting Concrete

Splitting tensile test

The results of testing for the standard cylindrical specimens (6" x 12") under line load were listed in Table 4. It can be seen from the results of the 15 mixes that all parameters which influence the characteristics of the microstructure of the cement matrix and of the interfacial transition zone are of decisive importance in respect of the tensile load bearing behavior.

Table 4   Splitting tensile strength values.

MIX ID	SPLITTING STRENGTH MPa	MIX ID	SPLITTING STRENGTH MPa
1	4.4	9	5.33
2	3.75	10	4.74
3	3.6	11	3.33
4	4.07	12	5.12
5	4.82	13	5.53
6	3.7	14	5.53
7	6.63	15	5.76
8	6.13		

Compressive strength test

The results of standard compressive strength at 28 days are presented in Table 5.

Modulus of rapture

The results of flexural strength are presented in Table 6.

Table 5   Compressive strength values for cubes and cylinders.

MIX ID	CUBE STRENGTH, MPa	CYLINDER STRENGTH MPa	$f_{cube}/f_{cylinder}$
1	47.98	38.73	1.24
2	39.49	32.18	1.23
3	37.63	31.07	1.21
4	41.79	34.69	1.20
5	46.08	38.11	1.20
6	38.28	31.08	1.23
7	68.36	57.42	1.19
8	61.14	51.15	1.19
9	54.59	45.17	1.20
10	46.89	38.55	1.21
11	36.97	30.2	1.22
12	51.33	42.22	1.21
13	51.93	43.11	1.20
14	52.51	43.39	1.21
15	53.97	45.02	1.21
Average			1.21

Table 6   Flexural strength results.

MIX ID	f_r (MPa)	MIX ID	f_r (MPa)	MIX ID	f_r (MPa)	MIX ID	f_r (MPa)
1	5.09	5	5.44	9	5.78	13	6.15
2	4.12	6	4.29	10	5.32	14	6.17
3	3.88	7	7.54	11	3.71	15	6.34
4	4.43	8	6.84	12	5.95		

SCC modulus of elasticity

The test set up for the recording the deformations and the axial compressive applied load on a cylindrical specimen is shown in Figure 5. Having calculated the compressive stresses and the corresponding strains and plotting the stress strain curve, the estimated modulus of elasticity is then obtained. Stress strain diagrams were drawn for three different mixes of different W/P ratio. Results are seen in Figure 6. Even though SCC mix has comparatively low coarse aggregate content its modulus of elasticity was almost similar to values of CVC from literature this may attribute to the nature of SCC and the presence of large content of mineral admixtures which make the microstructure of SCC mix more dense and homogenous, and transition zone near aggregate is almost absent.

Figure 5   Strain gauge set-up.

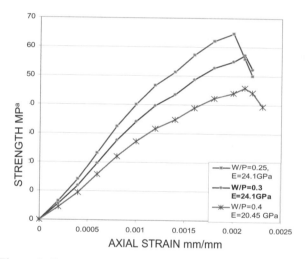

Figure 6   Stress-strain curves of SCC at different W/P ratios.

CONCLUSIONS AND RECOMMENDATIONS

Conclusions

The following conclusions may be drawn from this study:

1. The usefulness of the tests usually is dependent on the nature of the mix itself. Similar to the case for conventional concrete, it might be advisable to select one set of tests for high slump flow range and another one for low slump flow range for SCC mixes.

2. A 'good' SCC was defined and used to determine the acceptance range of test methods. Three criteria other than filling ability were established to make this definition:

3. Passing ability: a H_2/H_1 ratio (standard L-box) higher than or equal to 0.8 was required or $H_2-H_1 < 30$ mm (standard U-box). This is normally considered to identify a satisfactory level of passing ability. Passing ability/segregation resistance: absence of clustering of aggregates behind the bars of the L-box or U-Box.

4. Segregation resistance: the homogeneity of SCC was assessed by visual observations from the slump flow test and the appearance of SCC in the mixer. and depending on slump value a test for segregation is needed.

5. None of the test methods for filling ability was able to adequately cover all key characteristics of SCC as a single test. Generally, concrete with a slump flow smaller than 720 mm and a blocking ratio higher than 0.8 (difference in heights < 30) is likely self-compacting. Surface conditions and material of the base plate were found to affect, principally, filling ability, and passing ability to a lesser extent.

6. The basic mechanism of blocking was found to depend principally on the paste volume of the concrete, and on the characteristics of coarse aggregate (size, shape and quantity) relative to the gaps through which the concrete has to flow.

7. L-box test is preferred than U-box cause it is more correlated to filling ability and other passing ability tests. The L-box could not assess all three functional requirements simultaneously although both filling ability and segregation resistance could influence the measured H_2/H_1 ratio in the L-box test.

8. SCC has better finishing surfaces with less air voids on surfaces.

9. At similar w/c ratio, strengths of the SCC mixes, using the limestone powder as filler, were significantly lower than the mixes contain fly ash because it is coarser and its purity is low.

Recommendations

The following recommendations have been outlined to form some remarks to be followed and points of further research:

1. Slump flow and T50 can be used to determine workability ranges for construction projects, each project use a value that suites its requirements.

2. For densely reinforced structures, e.g. civil engineering structures, the L, U-box with three bars is recommended; for less densely reinforced structures the box with two bars; in both cases even spacing of all gaps is recommended. H_2/H_1 ratios of at least 0.8 or H_2-H_1 of less than 30 mm are recommended with both bar spacing.

3. Whenever the slump flow exceeds 720 mm a check must be done on segregation resistance by a suitable segregation stability tests.

4. To spread this technology in Jordan a study on packed stone walls is recommended to show practical SCC advantages in Jordan.

REFERENCES

1. OKAMURA H, OUCHI M. Self compacting concrete. J Advanced Concrete Technology, 2003; 11(1): pp. 5–15.

2. K OZAWA, K MACKAWA AND H OKAMURA, "High performance concrete based on the durability design of concrete structures", Proceedings of the 2nd East Asia Pacific Conference on Structural Engineering and Constructor, Chiang-Mai, 1989, pp. 445–450.

3. ATTIOGBE E; SEE H T AND DAZKCO J; "Engineering Properties of Self-Consolidating Concrete," www.masterbuilders.com/eprise/main/MBT/Content/Products/Rheodynamic/ACBMpapers/EngineeringProperties.pdf

4. PERSSON B; "A Comparison between mechanical properties of Self-Compacting concrete and the corresponding properties of normal concrete", Cement and Concrete Research, vol. 31, No.2, February 2001, pp. 193-198.

5. EFNARC, "Specification and Guidelines for Self-Compacting Concrete," EFNARC. Hampshire, United Kingdom, Feb. 2002, 32 pp., www.efnarc.org

6. GRUBE, GRUBE. "Self compacting concrete - another stage in the development of the 5-component system of concrete" ACI Materials Journal, v.96, No. 3, May-June 1999, pp. 346-353.

7. K OZAWA, K MACKAWA AND H OKAMURA, "High performance concrete based on the durability design of concrete structures", Proceedings of the 2nd East Asia Pacific Conference on Structural Engineering and Constructor, Chiang-Mai, 1989, pp. 445– 450.

8. K H KHAYAT AND A YAHIA, "Effect of Welan gum-high range water reducer combinations on rheology of cement grout", ACI Mater. J. 94 (5) (1997) 367–372.

9. KHAYAT K H; "Workability, Testing and Performance of Self-Consolidating Concrete," ACI Materials Journal, v.96, No. 3, May-June 1999, pp. 346-353.

10. KHAYAT K H AND GUIZANI Z, "Use of viscosity-modifying admixture to enhance stability of fluid concrete", ACI Mat. J., pp. 332, 1997.

11. KHAYAT K H; HU C AND MONTY H; "Stability of Self-Consolidating Concrete, Advantages, and Potential Applications," Proceedings of the First International RILEM Symposium; Cachan Cedex, France; Ed. RILEM Publications, 1999, pp.143-152.

12. KÖNIG G; HOLSCHEMACHER K; DEHN F AND WEIßE D, "Self-Compacting Concrete - Time Development of Material Properties and Bond Behaviour." Proceedings of the Second International Symposium on Self-Compacting Concrete, Tokyo, (2001), pp. 507-516.

13. KURDUS WOLFGANG, "Assessment of the fresh concrete properties of self compacting concrete" Library of JSCE No. 31, pp. 107-120, 2001

14. KURDUS WOLFGANG, Controlling the workability properties of self compacting concrete used Ferraris, C.; Browner, L.; Ozyildirim, C.; and Daczko, J.; "Workability of Self-Compacting Concrete," Proceedings of PCI/FHWA/FIB International Symposium on High-Performance Concrete: The Economical Solution for Durable Bridges and Transportation Structures; Orlando, FL; Sept. 25-27, 2000; pp. 398-407.

15. SUKINA A AL-ZYOUD, "Experimental Investigation On Self Compacting Concrete", MSc. Thesis, Faculty of Graduate Studies, Jordan University of Science and Technology, Irbid, Jordan, May 2006.

INFLUENCE OF MINERAL ADDITIONS ON DURABILITY OF SELF-COMPACTING CONCRETE

R Cioffi

F Colangelo

University of Naples

M Marroccoli

University of Basilicata

Italy

ABSTRACT. In this study the durability of self-compacting concretes (SCCs) made with different mineral additions has been evaluated. Five concrete mixtures have been designed, four of these as self-compacting and one as conventional concrete. Water to cement ratio equal to 0.5 has been kept constant in all the concrete mixtures and the same aggregate size distribution has been employed. All the prepared SCCs have been studied in terms of chloride penetration depth, sulphate attack resistance and carbonation depth. Chloride penetration has been determined employing a 0.27 M $CaCl_2$ solution in contact with concrete cylinders for six months. The damage due to sulphate attack has been determined on cubic specimens immersed into a 5% $MgSO_4$ solution for 12 months. Carbonation depth has been evaluated after 3, 6 and 12 months of air exposure.

Keywords: Durability, Self-compacting concrete, Mineral additions, Chloride penetration, Sulphate attack, Carbonation

Raffaele Cioffi is full professor at the University of Naples "Parthenope" Italy. His main research activities concern the stabilization/solidification of industrial solid wastes and sediments and the preparation concrete made with coal fly ashes, MSWI ashes, etc.

Francesco Colangelo is a research civil engineer at the Department of Technologies at the University of Naples "Parthenope", Italy. He is working in the area of recycling of industrial waste in concrete and in geo-environmental applications.

Milena Marroccoli is Associate Professor of Materials Science and Technology at the Engineering Faculty of the University of Basilicata, Italy. Research interests are in the broad field of cementitious materials with particular emphasis in environmental applications.

INTRODUCTION

The evolution of performance of concrete in presence of aggressive environment is of concern in order to estimate the life cycle cost of a building [1-3]. The life cycle cost is a holistic design approach to durability that can provide to rational way to obtain a good quality concrete.

The following information is typically required to perform life cycle cost analysis:
1. nature of environment in contact with concrete;
2. effect of aggressive media on concrete performance;
3. cost of protection systems and repairs.

At this scope, the best means of maximizing the probability that concrete will be durable is to produce concrete that will provide the desired service for the desired service life in the environment in which it will be placed and used. Every concrete mixture should be proportioned in accordance with exposure conditions (nature of chemical environment), construction considerations and structural criteria. The value of the concrete will be defined in terms of maturity, permeability, air-void structure quantification, resistance to aggressive media (sulphate resistance, chloride penetration etc.), strength and in situ performance [4-6]. In addition, generally, to improve the durability of reinforced concrete a low water to cement ratio and high cover over the reinforcing steel bars must be used. Moreover, fundamental factor in casting durable concrete is the use of pozzolans (coal fly ash, silica fume) [7, 8] and ground granulated blast-furnace slag [9] and chemical admixtures in combination with portland cement, and the proper selection of aggregate (proportion, hardness, grading, shape, size and phase composition).

In the field of high performance and durable concrete Self-Compacting Concrete [10] represent a special concrete that should not only flow under its own weight but should also fill the entire form and achieve uniform consolidation without segregation [11]. SCC is used in structures with closely spaced reinforcing bars and should be able to flow through and completely fill the form without vibration.

It should be stressed that durability of SCC is a factor of some concern as the type of cement, the grain size distribution, quantity of cement and amount of chemical admixtures and mineral addition must be borne in mind [12]. In this study, similar quantities of three different mineral additions were used, namely powdered limestone, silica fume and blast furnace slag. The aim was to investigate the influence of mineral additions on resistance of concrete to aggressive environmental through the determination of depth of carbonation, chlorine penetration and loss in strength in presence of sulphate solution.

EXPERIMENTAL

Materials

The cement used in the experiments was CEM II/A-L 42.5 R (according to European Standards EN-197/1). The coarse aggregates were crushed limestone, with specific gravities of 2520 and 2600 kg/m^3, and water absorptions of 2.05% and 0.63%, respectively. They were separated into different size fractions that were than recombined to a specific grading curve. The blast furnace slag was a granulated product ground to a Blaine fineness of about 4500 cm^2/g and a particle size range of 1-10 μm and its basicity coefficient [Kb = $(CaO+MgSO_4)/(SiO_2+Al_2O_3)$] was equal to 1.08. The silica fume used was commercial material produced by ELKEM – Norway. The chemical composition and specific surface values of all the above materials are detailed in Table 1. Powdered limestone, 95% passing the 125 μm sieve, was also used.

Table 1   Chemical composition and specific surface of the limestone cement, slag, silica fume and limestone powder.

OXIDES (%)	CEM II	GBFS	SF	L
SiO_2	20.56	35.16	93.00	0.30
Fe_2O_3	3.21	1.40	0.70	-
Al_2O_3	4.88	10.76	0.80	0.10
CaO	60.93	41.91	0.20	55.10
MgO	1.62	7.68	0.60	0.50
Na_2O	-	0.11	0.60	-
K_2O	-	0.14	1.00	-
SO_3	2.78	1.92	-	-
TiO_2	-	0.32	-	-
L.O.I.	5.76	1.78	1.50	43.00
Specific surface area, cm^2/g	3970	4500	200000	4250

The superplasticizer used was an acrylic one with 40% solid content and a specific gravity of 1.2 kg/dm^3. It was used in all SCC mixtures and its amount was proportioned to ensure a constant workability. The water content of the superplasticizer (SP) was considered during the mix-design phase. An inorganic Viscosity Modifying Agent (VMA) was also used in all SCC mixtures to enhance stability.

Methods

Mixtures preparation

Four SCC mixtures were designed in addition to a conventional concrete characterized by 240 mm of slump. In the case of conventional concrete, external vibration was used. Table 2 shows mixture proportions for all the concrete specimens. In this table the first part of the mixture code gives information on the mineral additions contained in the concrete (L, SF and BS indicate limestone, silica fume and blast furnace slag, respectively), while the second part indicates the kind of concrete (SC for self-compacting and CC for conventional concrete). The mixtures used presented similar mixture proportions in terms of the aggregate to cement ratio.

The W/C ratio was kept constant at 0.5 for all the concrete mixtures; moreover the water/(cement + powder) is similar for all SCCs. The amount of total binder (including cement, ground limestone, blast furnace slag and silica fume) was about 400 kg/m^3.

Table 2   Concrete mixture proportions

Mix code	Aggr. 0-3 mm	Aggr. 3-6 mm	Aggr. 6-8 mm	Aggr. 8-20 mm	CEM II	BS	SF	L	VMA	SP
				COMPOSITION, kg/m^3						
CC	663	338	341	478	350					2.11
SC	793	203	290	431	394				1.58	1.98
L-SC	800	210	290	442	365			36.5	1.61	2.65
SF-SC	791	203	290	430	378		30.2		1.63	2.86
BS-SC	800	212	292	440	365	36.5			1.61	2.65

Fresh concrete properties

In order to prepare all the concrete mixtures, a 0.09 m^3 capacity concrete mixer was used. After mixing, fresh concrete was employed for the determination of slump, slump flow, L-box, air content and unit weight. Italian standards were followed to determine slump, air content and unit weight (UNI 9418, UNI 6395-72). The slump flow test measures the mean diameter of concrete spread after the removal of the slump cone. A spread of at least 600 mm is required for SCC. All the mixtures prepared did not show any segregation phenomena.

The L-box test is carried out to test for the blocking of concrete in very reinforced sections. The apparatus used is 700 mm in length and 200 mm in width. It has three vertical bars with spacing of 4.25 cm. The vertical box is 600 mm in height while the horizontal box is 150 mm. When the concrete reaches the border of the horizontal part of the box, its height at the furthest point (H_1) and the height in the vertical box (H_2) are measured. The H_1/H_2 ratio is required to range between 0.8 and 1. The slump flow test gives information on filling ability while the H_1/H_2 ratio provides information on passing ability in restricted spaces.

Compressive strength

To evaluate the strength development, compressive tests were carried out after 28, 90, 180 and 365 days of water curing. The specimens were kept covered with a wet cloth and a plastic sheet during the first 24 hours; after this period, they were cured in water until the test age. The specimens were dried in air for two hours before testing. Three specimens (100-mm cube) were cast for each data point. A 3000 kN capacity compressive testing machine was used.

Dynamic modulus of elasticity

Dynamic modulus of elasticity was determined using 28-day water cured cylinder specimens of 150 mm diameter x 300 mm height. The dynamic modulus of elasticity was recorded by ultrasonic pulse velocity measurements. This method links the velocity with which an ultrasonic pulse passes through a material, with its elastic modulus. This test was carried out according to RILEM NDT 1 Standard. The specimens were tested at age of 28, 90, 180 and 365 days.

Chloride penetration measurement

UNI 7928 was used as a basis for the concrete chloride penetration test. Concrete cylinders were cast and, one day later, they were put in contact with a 0.27 M $CaCl_2$ solution. The specimens were covered with paraffin in their lower part, while in the higher one was placed a PVC pipe, tied to the specimens by a metallic clip. It is shown in Figure 1.

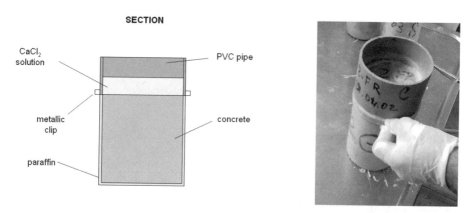

Figure 1   Details of specimens used for chloride penetration test

The solution in containers was replaced every month. There were made three specimens for each concrete mixture. Cylinders were broken after three and six months, through splitting test. It was measured chloride penetration depth, using silver nitrate solution as a colorimetric indicator, as shown in Figure 2.

Figure 2   Details of chloride penetration measurement

Sulphate attack resistance

Concrete cubes (150 x 150 x 150 mm³) were cast and immersed, one day later, into a solution containing 5% of $MgSO_4$. Solutions were replaced every month for the first 4 months, and later at 7, 10 months of exposure. The mass variation was measured every month. The compressive strength was measured after 1, 3, 6 and 12 months. A total of three cubes were tested for each data point.

The companion specimens, cured in water, were also tested to evaluate the compressive strength. In this way it is possible to make a simulation of concrete attitude cast-in situ, where it came into contact with sulphate environment, before hydration is completed.

The damage, expressed as percentage loss in strength, was calculated as follows:

Damage (D) = (1-Rs/Rw) · 100

Where:

Rs = the average (in MPa) of compressive strength of three specimens cured in sulphate solutions;
Rw = the average (in MPa) of compressive strength of three specimens cured in water.

Carbonation depth measurement

The phenolphthalein test was used to determinate the carbonation depth. Phenolphthalein is a colourless acid/base indicator which turns purple when the pH is above a value in the range of 8.4-9.8, that is, when the concrete is alkaline. Phenolphthalein is prepared as a 1% solution in 70% ethyl alcohol. The solution is sprayed onto a freshly broken surface which has been cleaned of dust and loose particles. The measurement is carried out immediately after the broken surface has been exposed, alkaline areas of concrete turning a vivid purple colour. If no coloration occurs, carbonation has taken place and thus the depth of the carbonated surface layer can be measured.

RESULTS AND DISCUSSION

Fresh Concrete Properties

The properties of fresh mixtures such as the values of slump-flow, L-box, air content, and mass weight are given in table 3.

Table 3   Properties of fresh concrete

MIX CODE	UNIT WEIGHT (kg/m^3)	SLUMP (mm)	SLUMP FLOW (mm)	L-BOX $(H_1/H_2$ ratio)	AIR CONTENT (%)
CC	2332	240	-	-	2.0
SC	2324	-	620	0.81	2.6
L-SC	2284	-	600	0.81	2.6
SF-SC	2314	-	600	0.83	2.1
BS-SC	2307	-	640	0.82	2.4

It can be seen that the H_1/H_2 ratio in the L-box test was greater than 0.8 for all SCC mixtures, even if the bar spacing to maximum aggregate size ratio was low. The quantity of air content ranged from 2% to 2.6%. The lowest air content value was equal to 2% for CC, this was followed by SF-SC (2.1%), BF-SC (2.4%), L-SC (2.6%) and SC (2.6%). The air content of the SF-SC mixture was very similar to the CC one, even if a conventional concrete mixture was put in moulds in two steps and consolidated on a vibrating table to remove entrapped air. In the case of other mixtures the percentage of air was not too high and this confirms that all SCCs had high self-compacting properties. The amount of superplasticizer in all SCC mixtures was proportioned to give a slump-flow value of approximately 600-640 mm. The bleeding of all mixtures was absent.

Mineral additions, together with chemical admixtures, gave the mixtures a good workability and flowability. In this regard, compared to the SCC mixture, the other self-compacting mixtures needed a larger amount of superplasticizer for a fixed slump flow value, as shown in Table 3. This can be attributed to the greater fineness of the mineral additions, whose specific surface area ranged from 4250 to 200,000 cm^2/g. For instance the amount of superplacizer added to the SF-SC mixture was 1.35 times as much as the one added to SC (see Table 2). In particular the SF-SC mixture required more superplasticizer than the others but showed the lowest slump flow value; which is consistent with data published by other researchers.

Compressive Strength

The compressive strength development up to one year's curing for all the mixtures is shown in Figure 3. The 365-day strengths obtained ranged from 62.5 to 79.0 MPa.

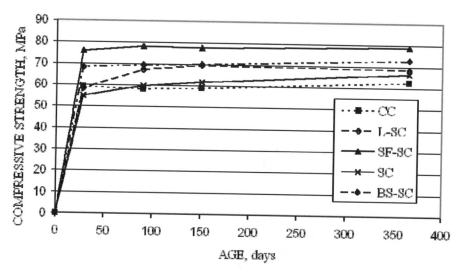

Figure 3   Compressive Strength development with age

The maximum strength value was recorded for the SF-SC mixture, while the lowest was for CC. The addition of silica fume gave a higher compressive strength, at both medium (28 days) and later ages, compared to other mineral additions. For each mixture there was no remarkably different strength values between medium and later ages, probably due to the high fineness of the mineral additions employed. However SCCs displayed a higher compressive strength then the conventional one, regardless of the addition used, as stated in literature.

Dynamic Modulus of Elasticity

The dynamic modulus values are shown in Figure 4. After 365-day curing, SF-SC showed the highest dynamic modulus value (52.7 GPa) in comparison with the other self-compacting mixtures and it was quite similar to that of the CC (53.9 GPa). This is of great importance because the SF-SC showed compressive strength values that were 20 MPa higher and for this reason, as is well known, for a given strain, its cracking phenomena decrease.

Also in the case of BF-SC the dynamic modulus values were higher than SC and equal to 51.9 GPa.

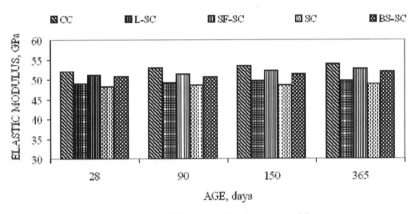

Figure 4   Modulus of elasticity development with age

Chloride Penetration

The maximum chloride–ion penetration depth was measured in SC and L-SC mixture and it is respectively 27.8 and 26.7 mm. The SF-SC specimens showed the highest resistance to chloride ions penetration (11.0 mm). Figure 5 presents results of chlorine-ion penetration.

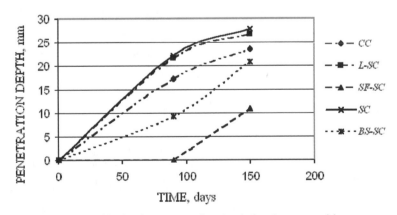

Figure 5   Chlorine-ion penetration depth development with age

Sulphate Attack

The maximum damage (D), due to sulphate attack, was noted in SF-SC specimens (D=24.7) while it was the lowest in BS-SC specimens (D=12.7). SC concrete showed a better behaviour (D=13.6) than CC concrete (D=20.1). Silica fume is one of the major admixtures for HSSCC (High Strength Self-Compacting Concrete), but this can not be used in magnesium sulphate environmental [5,6]. The results of sulphate attack are shown in Figure 6.

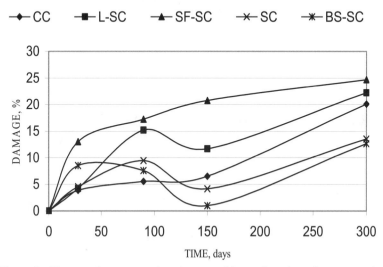

Figure 6 – Damage (percentage loss in strength) as a function of exposure time.

Carbonation

A summary of the results for carbonation depth (Cd) at the three measurement periods (3, 6, 12 months) is shown as follows. SF-SC specimens did not show slight signs of carbonation (after 12 months Cd_{SF-SC}= 0.8 mm). SC, L-SC and CC concrete carbonated more than the BS-SC concrete (Cd_{BS-SC} = 1.2 mm) [14]. The carbonation depths of the mixtures SC, L-SC and CC are not very different and can be considered nearly equal (after 12 months Cd_{SC}= 1.9, Cd_{L-SC}=2.0, Cd_{CC}=1.9).

CONCLUSIONS

The work carried out in this paper has allowed the following conclusions to be drawn:

1. The chloride penetration depth is maximum for the systems SC and L-SC with values of 27.8 and 26.7 mm, respectively.

2. The maximum damage due to the sulphate attach has been measured in the case of SF-SC, while the minimum has been noted for BF-SC. Furthermore, SC showed a better behaviour than CC. The above results have shown that silica fume cannot be employed in presence of magnesium sulphate solutions, even if silica fume is one of the major admixtures for high strength self-compacting concrete (HSSCC).

3. As regarding carbonation depth, the lowest values have been found for the systems SF-SC and BF-SC, while, the highest values have been noted for the systems SC, CC and L-SC.

REFERENCES

1. J S KONG AND D M FRANGOPOL, Journal of Structural Engineering, Vol.129, 6, 818, 2003.

2. M A EHLEN, Journal of Infrastructure System, Vol. 3, 4, 129, 1997.

3. N S Berke, M C Hicks, J Malone AND K A Rieder, Concrete International, Vol. 27, 8, 63, 2005

4. U SCHNEIDER AND S W CHEN, Cement and Concrete Research, Vol. 35, 1705, 2005.

5. M COLLEPARDI, Concrete International, Vol. 21, 1, 69, 1999.

6. M COLLEPARDI, Proceedings of Second International Symposium on Concrete Technology for Sustainable, Hyderabad, India, 55, 2005.

7. C S POON, S AZHAR, M ANSON AND Y. L. WONG, Cement and Concrete Research, Vol. 31, 1291, 2001.

8. L JIANG, Z LIU AND Y YE, Cement and Concrete Research Vol. 34, 1467, 2004.

9. G J OSBORNE, Cement and Concrete Composites, Vol. 21, 11, 1999.

10. B PERSSON, Cement and Concrete Research, Vol. 33, 1933, 2003.

11. W ZHU AND P J M BARTOS, Cement and Concrete Research, Vol. 33, 921, 2003.

12. B.PERSSON, Cement and Concrete Research, Vol. 33, 373, 2003.

13. M LACHEMI, K M A HOSSAIN, V LAMBROS, P C NKINAMUBANZI AND N BOUZOUBAA, Cement and Concrete Research, Vol. 34, 917, 2004.

14. B ANDENAERT AND G DE SCHUTTER, in: HPC Congress, G Ko"nig, F Dehn, T Faust (Eds.), Leipzig, 853, 2002.

CRUSHED DUST IN SELF-COMPACTING CONCRETE

P Kumar

Rajasthan Technical University

S K Kaushik

Indian Institute of Technology

India

ABSTRACT. Availability of river sand is becoming scarce in many parts of the world. Economical considerations warrant investigations on devising an alternative to its use. On the other hand, the crushed dust, produced while breaking big stone boulders in stone crushers for producing coarse aggregates is available at a cheaper rate. Usually, most of this dust produced is not used in concrete due to the high amount of fine fraction and associated higher water demand. The additional water demand of the dust may be compensated through the use of fly ash as the same is known to result in reduction of water demand in concrete. The additional finer fraction, usually below 150-micron size, may contribute as a supplementary powder material in the case of self-compacting concretes (SCC), mix proportioning of which requires high amount of powder fraction. An experimental investigation on development of SCC mixes with crusher dust as the fine aggregate together with low calcium fly ash by weight equal to about 40% of the weight of the total powder is reported in the paper. Compressive strengths of 40-70 MPa were obtained at the age of 28 days with the ratio of volumes of water to powder in the range of 0.96 to 0.70. The paper describes the properties of the raw materials; the mix proportions employed and the test results obtained.

Keywords: Crushed dust, Self compacting concrete, Fly ash, Micro-silica.

Dr Praveen Kumar is Associate Professor at the Department of Civil Engineering, University College of Engineering, Rajasthan Technical University, Kota, Rajasthan, India. His main research interest is in investigating innovative uses of fly ash in concrete and performance in terms of mechanical properties and durability of concrete containing low calcium fly ash. His experimental work includes mix proportioning of self-compacting concrete, use of statistics in its modelling and study of microstructure of SCC through scanning electron microscopy. He is also involved in characterization and investigation of use of waste materials from stone processing industries catering for concrete, road and flooring construction.

Dr S K Kaushik is former Professor and Head of Department of Civil Engineering at the Indian Institute of Technology, Roorkee, India. He has published more than 200 technical papers. His main work area has been cement based composites.

INTRODUCTION

River sand has been extensively used as fine aggregate in the production of concrete. However, unabated mining of the sand from the river beds contributed to several environmental concerns including greater erosion in the associated catchments areas. This has forced local Governments to put regulatory measures on mining of sand at several places. The contractors and the persons involved in the sand collection and related business, in turn, sometimes resort to illegal mining. On the other hand, construction activities involving the use of concrete have increased particularly in the third world countries including India in the last few years. The entire scenario has led to a sharp increase in the demand and the price of river sand. Economical considerations warrant investigations on devising an alternative to its use.

Crusher dust is invariably produced when crushing the stones for making it suitable for use in road and concrete construction. The proportion of this dust and chips (with average size of 3 mm) may be as high as 22% of the total aggregate crushed [1]. Usually, the dust produced at stone crushers is not used in concrete due to the high amount of fine fraction and associated higher water demand. Hence, this dust is available in abundance at a cheap rate. In many areas of India, its price is presently about one third of that of the sand.

Inclusion of low calcium fly ash of suitable physical and chemical properties in sufficient quantity in concrete containing the dust can partially offset the increased water demand arising due to the presence of the dust. Simultaneous use of high range water reducing admixtures may be made to offset the remaining water demand. The additional cost incurred on employing super-plasticiser can be accommodated in the saving accrued through replacement of river sand by the dust.

In mix proportioning of the self compacting concrete (SCC), a higher fraction of powder material (material below 125 micrometre size) is necessitated to keep the mix cohesive along with a high flow-ability. The finer fraction of the dust may contribute as a supplementary powder material in the case of SCC.

An experimental study presented herein investigated the development of SCC mixes on these lines, with an aim to achieve medium to high strengths at early age as well as high ultimate strengths and help in mitigating the ecological problem of disposal of crusher dust and fly ash.

EXPERIMENTAL PROGRAMME

Raw Material Characteristics

Ordinary Portland cement, Grade 53 conforming to IS 12269-1987 and a low calcium fly ash conforming to ASTM C618-1994, EN 450, class I of JIS A 6201 and Grade I of IS 3812-2003 [2-4] were used in the present investigations. The un-densified micro-silica used conformed to ASTM C1240-2001, EN 13263 and IS 15388-2003 [5, 6]. The characteristic properties of the three cementitious materials are provided in Table 1 to Table 3.

Crushed aggregates and dust obtained from local stone crushers were used. The particle size distributions of the different size fractions of the aggregates used are shown in Table 4. It shows that the dust contained only 13% material finer than 150 μm. Thus, the crusher dust was found to be suitable as fine aggregate in concrete as per the relevant Standards, IS 383 and BS 882 [7, 8]. When compared with the particle size specifications for the aggregates as per ASTM C 33-92a, the fraction passing 150 μm is marginally higher than the maximum limit of 10% [9]. The other particle size fractions of the crushed dust do conform to the ASTM specifications as shown in Table 4 [9].

Table 1   Physical characteristics of the cementitious materials used

PROPERTY → MATERIAL ↓	BLAINE FINENESS, cm²/g	SPECIFIC GRAVITY, g/cm³	% COARSER THAN 45 μm	AVERAGE PARTICLE SIZE, μm
Cement	2950	3.14	22.5	27.89
Fly ash	3500	2.24	5.6	10.10
Micro-Silica	220,000*	2.20	0.7	0.25*

* as provided by the manufacturer

Table 2   Chemical characteristics of fly ash and micro-silica

PROPERTY	FLY ASH	MICRO-SILICA
Silicon Dioxide (SiO_2), % by mass	57.5	95.1
$SiO_2+Al_2O_3+Fe_2O_3$, % by mass	91.0	95.1
Loss on Ignition, % by mass	0.57	2.79

As evident from Table 4, two commercially available size ranges, i.e. 2-6 mm and 6-12 mm contributed to the coarse aggregate size range. The 2-6 mm aggregate, commonly termed as the zero-size grit, is another marginal material (in addition to the dust) produced at the stone crushers, which has a price almost similar to that of the dust and significantly lower than the 6-12 mm aggregate fraction. However, the zero-size grit contained a higher proportion of flaky and elongated particles than the 6-12 mm aggregate. The percentages of such particles (taken together) were about 30% and 10%, respectively, in the two size ranges.

Poly-carboxylate (PCE) based super-plasticiser and a polysaccharide based viscosity modifying agent (VMA) were also used.

Table 3   Other important characteristics of cement

S. N.	CHARACTERISTICS	TEST RESULTS
1.	Setting time –	
	Initial, minutes	150
	Final, minutes	210
2.	Compressive strength, MPa	
	3 days	38.0
	7 days	48.0
	28 days	57.0

Table 4   Test results of sieve analysis of aggregates

SIEVE SIZE, mm	PARTICLE SIZE, mm			
	0-4.75 (CRUSHER DUST)		2-6	6-12
	PERCENTAGE PASSING			
	ACTUAL USED	ASTM C33-92a		
12	100	100	100	95.0
10	100	100	100	34.7
4.75	99	95-100	34	1.7
2.36	91	80-100	3	—
1.18	60	50-85	–	—
0.600	43.5	25-60	–	—
0.300	25	10-30	–	—
0.150	13	2-10	–	—

Mix Proportioning

SCC is usually employed in situations where the cross section thickness is small or reinforcement is highly congested. In some such situations, SCC with low maximum size of aggregates (MSA), say, limited to the 2-6 mm size aggregates, may be useful. The present investigations aimed at development of two types of SCC mixes, one with MSA as 6 mm and another with MSA as 12 mm. In the former case, the coarse aggregate fraction (>4.75 mm) is available in the concrete from the zero-size grit only. The grit comprises sizeable proportion of flaky and elongated particles, this increases the surface area of the aggregates to be wetted. Hence the coarse aggregate fraction was kept very low in the SCC with 6 mm MSA.

The empirical guidelines proposed by Okamura and Ozawa forms the basis of mix proportioning of SCC by most of the investigators [10-13]. It suggests fixing the water-powder ratio by volume (Vw/Vp) between 0.90 and 1.10. The initial mix M1 in the present study employed Vw/Vp equal to 0.96, including the water present in the super-plasticiser and the liquid VMA. The ratio of the water to the cementitious materials by weight (w/p) was adopted as 0.36 in the mix M1. As the coarse aggregate content was increased in the subsequent mixes, the water content was reduced to possibly the minimum limit of w/p, i.e. 0.256; that could be attained with the set of materials. Table 5 shows the details of the four optimum mixes obtained.

The powder composition was selected with three major considerations:

1. Flow-ability
2. Resistance to segregation
3. Economy

A mortar flow test was carried out to ascertain the effect of variation in the powder composition on the flow characteristics. A powder composition of cement : fly ash : micro-silica of 1.0 : 0.7: 0.05 was selected on the basis of the test results obtained [14].

The segregation tendency of concrete increases with the presence of relatively high proportion of angular shaped, flaky and elongated particles contributed by the crusher dust and zero-size grit as compared to the concrete made with river sand and corresponding size fraction of the 12 mm nominal size aggregates. This was checked through the use of 5% micro-silica addition by weight of cement in the mix. The synergetic effect of high amount of fly ash and little amount of micro-silica also helped in maintaining cohesion in the mix at the high level of super-plasticiser, particularly in the mixes M3 and M4.

The cost per unit volume of fly ash was about one fifth of the cost of the cement. Hence, by adopting fly ash content equal to about 40% of the powder weight, economy was achieved even with significant quantity of the super-plasticiser in the mixes M3 and M4.

Many of the ordinary Portland cements (including the one used in this study) commercially available presently in India, exhibit bleeding when PCE based super-plasticisers are used even in nominal dosages. However, the bleeding does not take place in the ternary mix of cement - fly ash - micro-silica, even at sufficiently high dosages of the super-plasticiser employed. The large powder volume selected in this study also contributed to the prevention of bleeding.

Table 5   Mix proportions

MIX	QUANTITIES, kg/m³									RATIOS		
	Powder							Aggregates				
	Cement	Fly ash	MS*	Water	SP	VMA	Dust	2-6 mm	6-12 mm	w/p	V_w/V_p	V_f/V_m
M1	357	253	18	220	6.2	1.3	893	476	0	0.359	0.961	0.426
M2	361	257	18	210	5.6	0	925	502	0	0.337	0.901	0.445
M3	378	262	19	177	11.4	2.0	690	281	481	0.286	0.767	0.373
M4	385	262	19	160	11.6	2.0	702	286	490	0.256	0.686	0.384

* Micro-Silica

Coarse aggregates play an important role on the passing ability of SCC in congested areas; their volume needs to be limited. When the volume of coarse aggregate in concrete exceeds a certain limit, the opportunity for collision or contact between the coarse aggregate particles increases rapidly and there is an increased risk of blockage when the concrete passes through the spaces between steel bars. Okamura suggested limiting the coarse aggregate volume to 50% of the solid volume in general [10-13]. This limit varies with characteristics of the aggregates. With well shaped aggregates, a relatively high proportion of coarse aggregates can be accommodated [10].

The pioneering investigations on SCC in India by Subramanian and Chattopadhyay suggested the same to be limited to 46 percent instead of 50% [15]. In the present study, the mixes with MSA of 12 mm were proportioned with 46% volume content, whereas, the M1 and M2 mixes (with MSA 6 mm), contained coarse aggregates equal to only half of this percentage. The low percentage was employed because of the presence of a large quantity of elongated and flaky particles in the 2-6 mm size grit.

The volume ratio of fine aggregate to mortar (Vfa/Vm) was suggested as 40% by Okamura and Ozawa [10]. The linear optimization method proposed by Domone et al. extends the guidelines of Okamura and Ozawa. It specifies Vfa/Vm to be in the range 0.40-0.47 if w/p lies within 0.30-0.40. For SCC with w/p less than 0.30, the Vfa/Vm is suggested as 0.40 [16]. In the present study, the values adopted are shown in Table 5. In the determination of the volume of fine aggregate (Vfa), the particles smaller than 90 micron present in the dust are excluded, as these are assumed to be a part of the "powder" in line with Okamura and Ozawa [10]. Table 4 shows that 13% of the crusher dust was finer than 150 micron (by weight).

About 75% of this part was observed to be finer than 90 micron and is excluded from the weight of the dust considered for calculation of the Vfa for use in the ratio Vfa/Vm.

The mixes were prepared in a laboratory tilting drum mixer with a total mixing time of about three minutes. For mix M2 with reduced water content, a higher mixing time of about five minutes was employed to obtain similar flow than in the mix M1.

Behaviour of Mixes in Fresh and Hardened States

Table 6 provides the details of the slump flow, height difference observed in the U box test and the GTM screen stability indices obtained for the SCC mixes in accordance with EFNARC guidelines [17]. The compressive strength values obtained at various ages are also shown in Table 6.

Table 6   Properties of different mixes in fresh and hardened states

MIX	SLUMP FLOW, mm	SEG. INDEX, %	U BOX H2-H1, mm	COMPRESSIVE STRENGTH, MPa				
				7 d	28 d	90 d	135 d	215 d
M1	630	7.6	0	21.4	45.7	53.1	60.6	66.6
M2	610	10.6	10	24.7	44.3	63.1	68.7	75.7
M3	700	9.3	0	37.8	59.1	70.1	74.6	82.6
M4	700	18.3	0	48.1	68.3	76.6	82.7	88.0

DISCUSSION OF TEST RESULTS

Rheological Behaviour

The following inferences can be drawn in respect of the rheological properties based upon the mix proportions shown in Table 5 and the test results shown in Table 6:

1. All the four mixes satisfy the qualifying specifications for a concrete mix to be termed as SCC [17]. The mix M4 can be grouped under SR2 class of SCC as specified in the European guidelines, as the segregation index is greater than 18%, but less than 23% [17].

2. When the mix constituents and rheological performance of mixes M1 and M2 are compared, it is observed that the segregation resistance decreased in M2, as exhibited by a higher value of the segregation index. It may be noted that a longer mixing time was tried in mix M2 with reduced water content, as well as the use of VMA was withdrawn. This observation indicates that VMA in small quantity is beneficial in maintaining the segregation resistance of the concrete mixes. The mixing time was increased with an aim to get a possibly higher flow-ability, but visual observations showed no substantial increase in the same. The increase in the segregation index may also be partially attributed to the increase in the mixing time.

Compressive Strength

Figure 1 shows the variation of compressive strength data with Vw/Vp. An attempt has been made to fit a straight line of the type "y = Ax + B" to the experimental values, with "y" representing the strength in MPa and "x" representing the Vw/Vp ratio. The coefficient of correlation "R²" value for the straight line fit is quite high (0.941-0.986). This leads to the conclusion that the compressive strength may be assumed to be linearly varying with Vw/Vp, if the mix satisfies the rheological specifications of SCC. The values of the slope "A" and the constant "B" obtained for the five different sets of data corresponding to the different ages are shown in Table 7. The trend of the slope, represented by the values of "A" indicates that the effect of the increase/decrease in Vw/Vp decreases with an increase in the age. The effect of decrease in the Vw/Vp ratio is to mainly increase the early age strength.

Table 7   Curve fitting data for strength with Vw/Vp at various ages

Age, days	"A"	"B"	R^2
7	-97.8	114.0	0.986
28	-88.12	127.5	0.941
90	-78.50	10.8	0.952
135	-73.20	132.3	0.962
215	-72.05	137.9	0.954

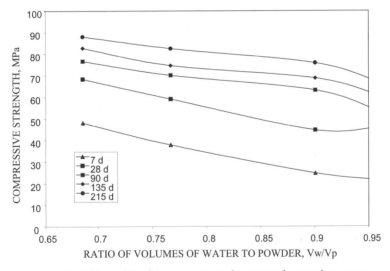

Figure 1   Effect of Vw/Vp on compressive strength at various ages

The percentage increases in the strength beyond the corresponding strength values at the age of 7 days till the age of 215 days are found to be 211%, 206%, 118% and 83% for the mixes M1, M2, M3 and M4, respectively. It may be noted that the mixes M1 and M2 contained very low volume of coarse aggregate (half of those in M3 and M4). This indicates that the coarse aggregate content in the self compacting concrete is not an important factor for the ultimate strength.

CONCLUSIONS

1. SCC can be produced employing crusher dust as the fine aggregate, in lieu of river sand in conjunction with low calcium fly ash and 5 to 7% micro-silica.

2. A wide range of compressive strengths at seven and twenty eight days age can be obtained with a variation in the coarse aggregate content and the ratio of the volume of water to the powder. However, the ultimate strength appears to be affected less with variation in these two parameters in such SCC.

ACKNOWLEDGEMENTS

The authors acknowledge the material support received by Elkem India Pvt. Ltd. and MC-Bauchemie (I) Pvt Ltd for the experimental work reported in this paper.

REFERENCES

1. MANNAN M A, GANAPATHY C, ACHYUTHA H, KURIAN V J, ASRAH H, ZAKARIA I AND BOLONG N, A study on high performance concrete using sandstone dust as fine aggregate. International Conference on Fibre Composites, High Performance Concretes and Smart Materials, ICFRC, Chennai, January, 8-10, 2004. pp. 601-612.

2. IS 12269. Specification for 53 grade ordinary Portland cement, Bureau of Indian Standards 1987, New Delhi.

3. ASTM C618. Standard specification for coal fly ash and raw or calcined natural pozzolan for use as a mineral admixture in Portland cement concrete, 1994, Philadelphia, USA.

4. IS 3812. Indian standard specifications for pulverised fuel ash Part 1 as pozzolana and admixture in cement mortar and concrete, Bureau of Indian Standards, 2000, New Delhi.

5. ASTM C1240. Standard specifications for use of silica-fume as a mineral admixture in hydraulic cement concrete, mortar and grout, ASTM International, 2001, Philadelphia, USA.

6. IS 15388. Indian standard silica fume specification, Bureau of Indian Standards, 2003, New Delhi.

7. BS: 882 British standard specification for aggregates from natural sources for concrete, British Standards Institution, 1992, London.

8. IS: 383. Indian standard specification for coarse and fine aggregates from natural sources for concrete. Bureau of Indian Standards, 1970, New Delhi.

9. ASTM C 33-92a. Standard specification for concrete aggregates, ASTM International, Philadelphia, 1992, USA.

10. OKAMURA H AND OZAWA K, Mix design for self compacting concrete, Concrete Library of JSCE, No. 25, June, 1995, pp. 107-120.

11. OKAMURA H, OZAWA K AND OUCHI M, Self compacting concrete, Structural Concrete, No.1, March, 2000, pp. 3-17.

12. OKAMURA H AND OUCHI M, Self compacting concrete, Journal of Advanced Concrete Technology, Vol. 1, No. 1, 2003, pp. 5-15.

13. DOMONE P L, JIN J AND CHAI H W, Optimum mix proportioning of self compacting concrete, Innovation in Concrete Structures: Design and Construction, Proceedings of Creating with Concrete, University of Dundee, Dundee, September, 1999, pp. 277-285.

14. KUMAR P, Development and structural properties of self compacting concrete with ternary mixes, Ph. D. thesis, 2005, Indian Institute of Technology, Roorkee, India.

15. SUBRAMANIAN S AND CHATTOPADHYAY D, Experiments for mix proportioning of self compacting concrete, Indian Concrete Journal, Vol. 76, No. 1, 2002, pp. 13-20.

16. GAIMSTER R AND DIXON N, Self compacting concrete, Advanced Concrete Technology, Elsevier, Vol. 3, Processes, pp. 9/1-9/23.

17. The European Guidelines for Self Compacting Concrete, Specification, Production and Use, May, 2005, Website: www.efnarc.org.

BAROTECHNOLOGY: A NEW BREATH FOR FOAMED/AERATED CONCRETE TECHNOLOGY

Yu G Kovalchuk

National Research & Development Concern

G Yu Kovalchuk

V D Glukhovsky State Scientific Research Institute for Binders and Materials

Ukraine

ABSTRACT: The "Barotechnology" of making foamed concrete is based on mixing in a hermetic mixer at increased air pressure. By analogy with the caisson disease, a pressure drop during discharging the mixer provides for additional air volume entrained within the mix. Using this principle side by side with a mechanical activation by turbulent high-speed mixing makes for increased stability of the foamed mix, decreased foaming agent content and a considerable growth of strength. The whole cycle of making the stable foamed mix takes place in a single mixer within 3 minutes, after that the same mixer begins functioning as a concrete pump transporting the foamed mix directly into the distant (up to 40 m) mould or the formwork. When being applied for the technology of aerated concrete, "Barotechnology" makes for an increased stability of a foamed mix and a 30-40% economy of an expensive aluminium powder. Based upon these principles, industrial-scale mixers were created for both aerated concrete plants (with the volume of the mixer up to 11.5 m^3) and for making a foamed concrete "in-situ" using mobile "mini" mixers (0.25-0.5 m^3). Case studies of making foamed concrete stronger than traditional one in both cellular concrete factories and in "in-situ" industrial-scale projects are also pointed out.

Keywords: Aerated concrete, Cellular concrete, Foamed concrete, Equipment, Mixers.

Professor DSc Yu G Kovalchuk is a President of the National Research & Development Concern "Institute for Porous Concretes", also known as Concern "SILICATE" (Academy of Civil Engineering of Ukraine). For more than 30 years he specializes in the field of designing of equipment for cellular concrete plants. He is one of the authors of "Barotechnology".

Dr G Yu Kovalchuk is a senior research worker in the V. D. Glukhovsky State Scientific Research Institute for Binders and Materials, Ukraine. His scientific interests are special cements and concretes, in particular, cellular concretes and heat or/and fire resistant materials.

INTRODUCTION

Nowadays, cellular concrete technology is an intensively developing area of building materials. This is caused by a permanent growth of energy costs (i.e. oil, natural gas) resulting in rising standards for heat insulation of buildings in European countries. Correspondingly, it is expected that the demand for heat-insulating materials will constantly increase. It regards both main directions of cellular concrete technologies: lime-based autoclaved aerated concretes produced in highly productive plants and OPC-based non-autoclaved foamed concretes produced in small-scale plants or "in-situ" as monolithic lightweight constructions (Figure 1). Despite the aerated concretes are the most widely developed cellular material around the world, foamed concrete use grew fast in the last few years [1] for important large-scale projects associated with mine stabilizations, road bases, etc., due to a progress in building chemistry and novel equipment.

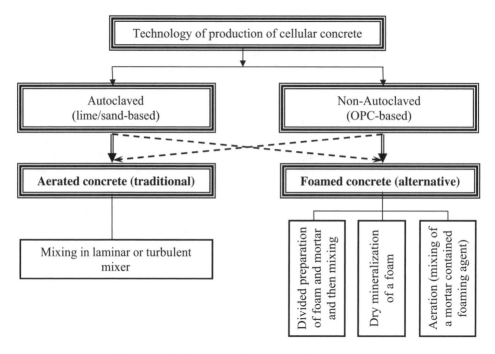

Figure 1   Technologies for making cellular concretes

However, some kind of technological problems restrain these materials. In particular, regarding foamed concrete, destruction of the foamed mix during transportation and casting occurs; hence there is a strong need to use a complicated 3-step technology (making foam/ making mortar, and mixing) and increased foaming agent contents etc. to be mentioned as important problems.

The progress in the technology of cellular concretes is traditionally expected to be connected with a new generation of superplasticizers and improved cutting equipment. At the same time, an improved technology of making aerated mix (mixing) is a powerful reserve for development of the technology, especially in view of some problems facing the producers of mixing equipment for cellular concretes. In particular, some part of the aluminium powder (up to 30%) concentrates in the upper part of the aerated concrete mixers and thus does not take part in the process. From the other side, existing technologies for preparation of a foamed mix have serious disadvantages, thus an ordinary technology for separated preparation of foam and mortar leads to degradation of the initial foam. The technology of dry mineralization of a foam is a very difficult task for real-scale equipment, and the aeration technology may result in seriously retarding the hydration process due to adsorption of foaming agent molecules on the surface of cement particles [3]. Thus, all of existing technologies require an excess quantity of air entraining or foaming agents to be used, which particularly affects the porous structure destructed during transportation and casting, lowered strength values and other problems.

THE ESSENCE AND ADVANTAGES OF "BAROTECHNOLOGY"

"Barotechnology" is a patented novel technology for preparation of porous (cellular) concrete mixes, independently on their nature (i.e. this technology is universal and can be easily applied for both aerated and foamed mixes). The technology was invented in the former USSR in 1980, and up to now is being developed by two research groups in Kiev (Ukraine) and Moscow (Russia) [4]. The principle of "Barotechnology" is based on mixing in a hermetic mixer at increased air pressure. First of all, it provides for additional air volume entrained into the mix, but additionally, it results in improved stability of a porous mix: a pressure drop during discharging from the mixer provides for some increased mix volume (similarly to the effect called "caisson disease") compensating for the possible destruction of the porous structure during transportation and moulding. Using this principle side by side with a mechanical activation by turbulent high-speed mixing makes for increased stability of the foamed mix, decreased foaming agent demand, and considerable growth of strength. Turbulent mixing leads to a cavitation effect, making additional air being incorporated and mechanical activation of the cellular concrete mix.

The whole cycle of making the stable foamed mix takes place in a single mixer within approximately 3 minutes, after that the same mixer begins functioning as a concrete pump transporting the foamed mix directly into a distant mould or formwork. A pressure up to 1 bar is sufficient for the "Barotechnology", so the mixers are not classified as highly hazardous equipment.

The main advantages of "Barotechnology" are listed below:

- increased compressive strength;
- decreased size and improved uniformity of pore sizes which improves some special properties of cellular concrete, such as frost resistance etc.;
- universality: the same mixers may be applied for making both aerated and foamed concrete;
- improved air entrained volume: "Barotechnology" makes for decreased aerating or foaming agent contents, it also improves conditions for the hardening of cement;
- compensation of destruction of the porous mix during discharging and moulding;
- improved homogenization results in enhanced service properties of cellular concrete;
- the technology allows the porous mix to be transported over a long distance (at least 40 m horizontally or 10 m vertically) directly to the mould or formwork;
- simplification of construction of mixers: these consist of only one mixing tank (without any additional mixers for mortar or dry foam).

When being applied to aerated concrete, the "Barotechnology" resulted in 15 to 25% higher compressive strength, 15 to 20% lower water absorption and 10 to 15% lower humidity sorption [2]. Similar results were obtained for "Barotechnology"-based foamed concretes: in some cases, the compressive strength was twice as high as that of traditional foamed concrete (Table 1). The increased compressive strength is probably related to additional mechanical activation during turbulent mixing, improved homogeneity of the mix and lowered aerating or foaming agent content.

Table 1 The effect of "Barotechnology" on the mechanical properties of foamed concrete
(data obtained in industrial-scale projects)

DENSITY OF A FOAMED CONCRETE, kg/m^3	COMPRESSIVE STRENGTH, MPa	
	Using traditional technology*	Using "Barotechnology"
400	1.0	1.2
500	1.5	2.5
900	3.5	7.5

* Standard value from Ukrainian national standard DSTU B V.2.7-45-96.

The effect of "Barotechnology" was found to be dependent on various constructional factors such as pressure, construction of blades, peculiarities of air feed points, mixing rotation speed and regime, sequence of batching of raw materials, etc [2].

For example, increasing the pressure in a hermetic mixer from 0.1 to 0.4 MPa provides for 1.7 times smaller pore sizes, 10 times higher number of pores and 2.4 times higher concentration of the gas phase in the mix. Hence, the higher the pressure within the mixer, the smaller the pore size and the higher the volume of the air entrained to the mix.

Complex investigation in both laboratory and real industrial conditions made for optimization of various technological parameters. It resulted in a wide family of cellular concrete mixers projected and applied in industry [6-8].

EXPERIENCE OF INDUSTRIAL APPLICATION OF "BAROTECHNOLOGY"-BASED EQUIPMENT

Laboratory-scale equipment

Laboratory-scale mixers based on "Barotechnology" were created in order to study details of the technology and elaborate mix proportions of cellular concretes. "Barotechnology" requires specific mix design due to differences in the efficiency of air entraining agent and improved compressive strength (it allows using increased filler content). Once established by using laboratory equipment, the mix design is ready to be applied for industrial-scale equipment and technology.

Figure 2   Laboratory-scale mixers based on "Barotechnology": BS-1 (12 l) and UPS-L (10 l)

Mobile equipment for making foamed concretes "in-situ"

Mobil mini-mixers [4, 5] based on "Barotechnology", first offered by Prof. I. Udachkin (Russia), were specially designed for making constructions from monolithic foamed concrete "in-situ" (Figure 3). This technology provides for important economy in transport (there is no need for transportation of porous blocks comprising 50-80% air) and workability, but additionally, these structures are more effective since they do not contain junctions made with traditional mortar. Moreover, there is an important opportunity to transport a porous mix directly to a distant formwork (for example, to another storey) by a hose.

a)

b)

c)

Figure 3   Mobile equipment for making foamed concrete based on "Barotechnology":
UPS-1M (Ukraine, 0.3 m^3), UPS-2 (Ukraine, 0.25 m^3), SiC (Russia, 0.16 m^3)

Mobile mixers require only 2 workers, they have a mass of up to 350 kg and their size allows transporting in a trailer of a car. Productivity is over 2 m^3 per hour. Mobile mixers are applied more frequently for creating heat-insulation constructions of floors, roofs and external walls (between two layers of a construction material such as bricks). They have become also popular in Ukraine for creation of small foamed concrete plants (with a productivity of up to 10 m^3 per day). There are over one hundred mobile mixers used in different plants and constructions in the Ukraine and Russia to date (Figure 4). The term of recoupment of these mixers is approximately 3 month (in Ukrainian conditions).

a) b)

Figure 4 Producing of foamed concrete using a "Barotechnology"-based mobile equipment: moulding in a small foamed concrete plant (a), concreting "in-situ" (b)

Industrial mixers for small plants (up to 25 000 m^3/year)

For the bigger plants, with a productivity of over 10 m^3 per hour, semi-automatic mixers with a capacity of over 0.5 m^3 were elaborated (Figure 5) [4, 5]. These plants have become popular in towns that lay far from big fabrics of aerated concrete.

Figure 5 Small industrial mixers based on "Barotechnology":
PBS-0.5 (Ukraine, 0.5 m^3), UPB (Russia, 1.0 m^3)

Large-scale industrial mixers (up to 13 m^3) for making aerated concrete in large plants

"Barotechnology" has opened a very important opportunity related to large factories for making aerated concrete: very large capacity mixers can be designed (Figure 6). For example, there is now experience for applying "Barotechnology"-based mixers with a capacity of up to 11.5 m^3, whereas the largest "normal" mixers in the former USSR had a capacity of 6.2 m^3.

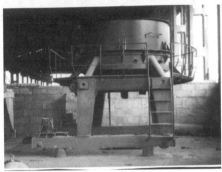

Figure 6   Industrial mixers based on "Barotechnology":
BS-3.2 (3.2 m³), BGS-1 (6.5 m³), BGS-2 (9 m³), BGS-3 (11.5 m³)

Using such large size mixers does not cause any problem to the properties of the aerated concrete produced, moreover, these lead to additional positive effects achieved (for example, for densities over 700 kg/m³, up to 100% of aluminium powder may be saved) and provide for maximal automation and industrialization of the plant.

Therefore, only one mixer of 11.5 m³ may meet the requirements of a very large factory with an annual output of up to 160 000 m³. It is also to be noted that the mixer BGS-2 based on "Barotechnology" has 50% higher capacity than the SMS-40B, the largest serial mixer in the former USSR (9.0 m³ compared to 6.2 m³), but at the same time, the mixer BGS-2 has 33% lower electric capacity than the traditional one (37 kW compared to 55 kW).

More than 150 000 m³ of aerated concrete were produced in the Plant for Porous Articles of Nikolaev (USSR/Ukraine) using a BGS-2 mixer. For such a large mixer, special moulds and cutting machine were specially designed and manufactured (Figure 7).

The service properties of the materials were found to be better than those of traditionally produced ones in the same plant. Mixer BGS-3, the largest in the world, in turn, was projected to be installed in the new technological line CONREX-240 for novel plants with an annual output of 240 000 m³ of aerated concrete.

Figure 7   Massifs of aerated concrete, 9 m³ of volume, 600 mm in height, made using the BGS-2 mixer

CONCLUSIONS

"Barotechnology" is a novel technology for preparation of porous (cellular) concrete mixes, independently on their nature (i.e. this technology is universal and can be easily applied for both aerated and foamed mixes), and is based on mixing in a hermetic mixer at increased air pressure. Cellular (porous) concretes made using "Barotechnology" represent significant economy for the aerating/foaming agent (at least 30% savings) and notable increases in mechanical strength (up to 50%), increased entrained air volume at decreased expenditure on gas-producing agent, compensation for the destruction of a porous mix during transportation and discharging.

Additionally, when being applied in a plant for making aerated concrete, "Barothechnology" provides for simplification of the technological process and the introduction of various additives as well as enables of making mixers of very high capacity (up to 11.5 m³) which at the same time run using 33% lower electrical energy than traditional ones.

When the technology is applied to make foamed concrete constructions "in-situ", "Barotechnology" also reduces transport costs significantly, accelerates construction, improves workability and enables transportation of a porous mix directly to a distant formwork (up to 40 m horizontally and up to 12 m vertically). There is successful experience with "Barotechnology" for making cellular concretes both in plants and "in-situ".

ACKNOWLEDGEMENTS

The paper is dedicated to Prof. Igor Udachkin (1942-2006), one of the authors and the ideologist of "Barotechnology". It was a great pleasure to collaborate with him all these years. The authors believe that his ideas regarding monolithic foamed concrete made by "Barotechnology" have a brilliant future.

This research is particularly based on the INTAS Collaborative Call Nr 04-82-7055 "Energy and Natural Resources Saving Heat Insulating Materials Made Using Local Raw Materials and Industrial By-Products and Wastes".

REFERENCES

1. DHIR R K, NEWLANDS M D AND McCARTHY A, (eds.), Use of Foamed Concrete in Construction. Proceedings of the International Congress "Global Construction: Ultimate Concrete Opportunities", Scotland, UK, 2005, 160 pp.

2. KOVALCHUK YU G, Homogenization and Air Entraining into the Cellular Concrete Mix by Supplying a Redundant Pressure, Ph.D. thesis, Kiev, 1990.

3. MARTYNENKO V A, Cellular and Porous Lightweight Concretes, Porogi, Dnipropetrovsk, 2002, 172 p.

4. KOVALCHUK YU G, Barotechnology: Development and Perspectives, Proceedings of the 2nd International Seminar "Theory and practice of production and application of a cellular concrete in construction industry", Dnipropetrovsk, 2005. pp. 39-52.

5. UDACHKIN I B, Small-Scale Lines for Making Articles Based on Cellular Concrete: Building Materials, Vol. 2, 1993. pp. 30.

6. Pat. 1263329 USSR, The Method for Homogenization of Wet and Dry Mixes, 1985, (the patent has analogies in the Ukraine and Russia).

7. Pat. 1583301 USSR, The Mixer, 1987, (the patent has analogies in the Ukraine and Russia).

8. Pat. 1599351 USSR, The Method for Preparing a Cellular Concrete Mix, 1988, (the patent has analogies in the Ukraine and Russia).

OPTIMISATION OF COMPOSITION OF LIGHTWEIGHT CONCRETE WITH ORGANIC FILLER IN THE FORM OF *PHRAGMITES AUSTRALIS* REED

D Malaszkiewicz

M Bołtryk

Białystok Technical University

Poland

ABSTRACT. The aim of the study was testing the possibility of application of fast growing species of common reed grass *Phragmites australis,* which is a very common plant in Poland growing on waterlogged lands, as organic filler in lightweight concrete. The first step was mineralization of the plant material in order to neutralize organic compounds, which may retard or obstruct cement hydration. The experimental results of the optimization of reed blade length, amount of superplasticizer and method of reed-concrete mixture consolidation are presented. Conifer sawdust or natural quartz sand was applied as fine aggregate. Lightweight concrete with a compressive strength exceeding 4 MPa was obtained.

Keywords: Organic aggregate, Lightweight concrete, Common reed, Consolidation of concrete mixture

Professor M Bołtryk, is a civil engineer, professor in the Faculty of Civil and Environmental Engineering at Bialystok Technical University. His research interest focuses mainly on concrete technology, concrete durability and consolidation techniques.

D Malaszkiewicz, is a doctor of civil engineering and assistant professor in the Faculty of Civil and Environmental Engineering at Bialystok Technical University. Her research interest focuses mainly on lightweight concrete with organic fillers and on recycled aggregate concrete.

INTRODUCTION

The idea of sustainable development is currently the main philosophy of economic development in the European Union. According to this concept, economic development should be equal in the different regions of our planet and should secure the increase of social wealth. At the same time it should not lead to devastation of the natural environment and to depletion of natural resources. The best available low energy-consuming technologies should be used. The reduction of CO_2 emission, energy saving and preservation of natural resources e.g. fossil fuels, belong to the most serious problems environmental protection faces. Of a special concern is the reduction of energy consumption by individual households and consumers, e.g. improvement of thermal insulation of houses. Research in this area has naturally grown considerably over the past 20 years. So far the problems lay in finding and testing suitable locally available plant material, as well as reasonable bonding agents for the production of building materials with good insulating characteristics. Attempts to test the suitability of sawdust and straw as fillers for lightweight concrete production were undertaken in Poland in the early fifties [1, 2].

Different plants such as conifers or *Miscantus x giganteus* can be considered as raw material for the production of plant-based building materials. Lightweight concrete is made of biomass mixed with special cements [3, 4]. Former studies proved that the plant material *Miscanthus* has fiber constituents which are considerably stronger than straw [5] and because of its chemical composition it makes a good raw material for building industry [3, 6, 7].

The reasonable alternative for plant materials characterized above may be fast growing species of common reed grass e.g. *Phragmites australis* [8]. It is a very common plant in Poland growing on waterlogged lands. It is also cultivated on sludge drying bed to abate chemical pollution. The young plant is used as a feed. There are many evidences testifying that peatland biomass with commercial potential can be harvested for different applications. That enables an economic exploitation combined with the conservation of natural environment [9]. Reed may be used as a raw material for energy production, roofing material, or insulation material. Reed grass contains approximately 10.8 % of mineral compounds, mainly silica. It also contains wax and saponin. It has high fire resistance, frost resistance, high pH and salinity. All these properties encouraged of its use for many years in the construction industry in the form of insulating boards, under-plaster mats, roofing material or even as reed-concrete structural elements.

MINERALIZATION OF REED GRASS

The research on mineralizing reed grass was conducted in Poland in the sixties of the last century. It was considered that it would be possible to find the proper materials and technology to mineralize reed/grass. This should increase its durability, decrease its hygroscopic feature and stabilize volume changes of concrete. However in that research the influence of such components as cellulose, hemicellulose, pectins and many other organic substances upon cement hydration was omitted. Some organic bindings present in grass in acid or alkali environment may decompose into monosaccharides – hexose and pentose. These monosaccharides dissolve in water easily and as a result cause formation of hydrophilic adsorption layers upon cement grains. These layers retard hydration of cement, especially in the initial period. Another harmful result of the adsorption layers is loss of ability of cement grains to coagulate.

The research conducted in Białystok Technical University [10], concerning the use of wood shavings as filler for lightweight concrete, showed that tri-calcium silicate ($3CaO \cdot SiO_2$) is the most susceptible cement compound to harmful influence of saccharides. This is the reason why saccharides are purposefully eliminated from grass (especially from its surface layer) and to bind them into compounds neutral to cement hydration.

The authors selected four types of mineralizing agents for initial investigations: $CaCl_2$, $Al_2(SO_4)_3$ with $Ca(OH)_2$, $MgSO_4$ with CaO and bituminous paste with a high-boiling organic solvent. All these mixtures were selected on the analogy of wood shavings mineralization [3]. Mineralization allows obtaining the organic filler resistant to decay, with higher durability and lower water absorption, better adhesion to cement paste and lower susceptibility to volume changes. The best results were obtained for the solution of $Al_2(SO_4)_3$ with $Ca(OH)_2$ and these results are presented in this study.

Materials

Phragmites australis straw for the experiment was harvested from a protected mire area of the National Park Narew, NE Poland. In March 2004 plant material was collected along transect extending from the valley margin across the floodplain up to the river. The average ash content in the reed was 4.9% with the silica content being 4.6% on average. After drying, the straw was cut into 0.5 to 2.0 cm long pieces. Bulk density of dry straw (dried at a temperature of 60°C) was 53.5 kg/m^3. Water absorption after 10 minutes of saturation was 142% and maximal saturation after 48 hours was 194%. The overall sugar content was 338.0 g/kg and the reducing sugar content was 2.43 g/kg in the dry material.

In order to neutralize the harmful organic compounds contained in the reed and to stabilize its properties, the reed was mineralized with a mixture of 4% solution of $Al_2(SO_4)_3$ and hydrated lime $Ca(OH)_2$ mixed in the proportion 1:2. The proportions of the composition in relation to reed mass are given in Table 1.

Table 1   Proportions of mineralizing agents in relation to reed mass

MIX NUMBER	PROPORTION OF AGENTS OF REED MASS, %	
	$Al_2(SO_4)_3$	$Ca(OH)_2$
1	3	6
2	6	12
3	9	18

Portland cement CEM I 32.5 was used as binder. The chemical composition and the parameters of this cement are given in Table 2.

The composition of reed-concrete was proportioned on the basis of the triangle for proportioning wood shavings concrete [3] assuming the constant consistency of the concrete mixtures for all compositions. The final concrete mixture composition was: cement – 382 kg/m^3, reed – 130.5 kg/m^3, and water – 356 dm^3/m^3.

Table 2   Chemical composition and parameters of cement

COMPOSITION, %	AVERAGE VALUE	PARAMETER	AVERAGE VALUE
CaO	63.89	Compressive strength, MPa	
SiO_2	21.70	after 2 days	23.8
Fe_2O_3	2.35	after 28 days	45.3
Al_2O_3	4.91	Setting time, minutes	
MgO	2.50	initial	133
SO_3	3.14	final	195
$Na_2O_{eq.}$	0.89	Specific surface, cm^2/g	3264
Insoluble residue, %	1.00		
Loss on ignition, %	1.38		

When concrete mixtures were prepared, the sequence of operations was strictly observed. The reed was mixed with a solution of $Al_2(SO_4)_3$ and part of mixing water, then hydrated lime mixed with the rest of mixing water was added and finally cement was added. This sequence was necessary because the reed first had to be saturated with the mineralizing agent, then the mixture had to be neutralized (because of acid reaction of $Al_2(SO_4)_3$ solution) and at the same time plasticized with lime. The duration of mixing the reed with $Al_2(SO_4)_3$ solution did not exceed 5 minutes.

Specimens in 100 mm cubic steel moulds were consolidated in three layers using a hand rammer. The concrete was kept in the moulds for 48 hours and then in air-dry condition (temperature 20±2°C and R.H. 50-60%). After 28 days the following tests were conducted: compressive strength, bulk density and water absorption. Also reference specimens without mineralizing agent were prepared and tested.

EXPERIMENTAL RESULTS

The experimental results of compressive stress of reed-concrete mineralized with $Al_2(SO_4)_3$ solution and hydrated lime are presented in Figure 1. The series without mineralizing agent is denoted in Figure 1 as "0". Series I.1, I.2 and I.3 contained $Al_2(SO_4)_3$ and hydrated lime in the following proportions: 3% and 6%, 6% and 12%, 9% and 18% of reed mass, respectively.

Because the damage course of lightweight concrete with organic filler differs from the damage of ordinary concrete, compressive stress at 10 mm deformation is considered as the compressive strength of such lightweight concrete.

As it can be seen from Figure 1, the mineralization caused an 8-time increase in the compressive strength of reed-concrete in relation to specimens not subjected to mineralization. The most efficient proportion of mineralizing agent is: 9% of $Al_2(SO_4)_3$ solution and 18% of $Ca(OH)_2$ in relation to reed mass. The values of bulk density and water absorption of reed-concrete are given in Table 3.

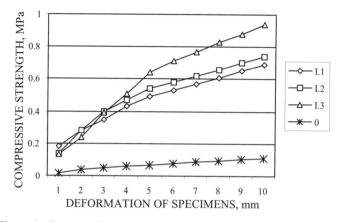

Figure 1   Compressive strength of reed-concrete mineralized with a
solution of $Al_2(SO_4)_3$ and hydrated lime

Table 3   Bulk density and water absorption of reed-concrete

MIX NUMBER	BULK DENSITY ρ_O, kg/m^3	WATER ABSORPTION n_W, %
1	763	31.96
2	744	31.44
3	735	33.09

For further research the following proportion of mineralizing agent was taken: 9% of $Al_2(SO_4)_3$ solution and 18% of $Ca(OH)_2$ in relation to organic filler mass.

OPTIMIZATION OF LENGTH OF REED BLADE

Optimization of the length of reed blade was performed on the following concrete composition: cement: 382 kg/m^3, organic aggregate (reed and conifer sawdust): 130.5 kg/m^3, water: 356 kg/m^3. Reed was screened and divided into two fractions: 2/10 mm and 10/20 mm. No mineral aggregate was added. Proportions of the organic aggregate are given in Table 4.

Table 4   Proportions of organic filler in concrete mixture

SERIES NUMBER	AVERAGE GRAIN SIZE, mm	FRACTION CONTENT, %		
		sawdust 0/2 mm	reed 2/10mm	reed 10/20 mm
1	10.5	0	50	50
2	9.69	9	45	46
3	8.79	18	41	41
4	7.89	27	37	36
5	7.41	33	33	34
6	6.49	37	37	26
7	5.57	41	41	18
8	4.65	45	45	10
9	3.50	50	50	0

After 28 days of hardening the following concrete parameters were determined: bulk density, water absorption and compressive strength. Experimental results are presented in Table 5.

Table 5   Experimental results of selected performance parameters of reed-concrete depending on grain size distribution

SERIES NUMBER	AVERAGE GRAIN SIZE mm	BULK DENSITY, ρ_p, kg/dm^3	WATER ABSORPTION, n_w, %	COMPRESSIVE STRENGTH, MPa
1	10.5	0.733	39.95	1.52
2	9.69	0.740	40.66	1.58
3	8.79	**0.727**	41.72	1.60
4	7.89	0.786	38.96	1.91
5	7.41	0.797	37.82	2.40
6	6.49	0.810	37.31	2.60
7	5.57	0.816	38.72	2.90
8	4.65	0.807	**36.91**	**3.11**
9	3.50	0.806	38.40	2.65

The highest compressive strength and the lowest water absorption were obtained for the composition number 8 (45% sawdust, 45% reed 2/10 mm and 10% reed 10/20 mm). This proportion is considered to be optimal for reed-concrete based only on the organic filler though bulk density of this composition (0.807 kg/dm^3) is about 11% higher than the lowest one (0.727 kg/dm^3) and it may affect heat-insulating performance of this concrete.

OPTIMIZATION OF CONSOLIDATION METHOD

In order to improve strength features of reed-concrete quartz sand and superplasticizer were introduced. The content of superplasticizer varied from 0 to 4% in relation to cement mass. The concrete mixture composition was as follows: cement: 400 kg, water: 439.6 dm^3, quartz sand: 300 kg, sawdust: 84.6 kg, reed 2/10 mm: 44.0 kg and reed 10/20 mm: 45.4 kg.

Three consolidation methods were applied: ramming in three layers using a hand rammer, consolidation using brake press and vibration with brake press. Five series were prepared for each consolidation method depending on the superplasticizer content (0, 1, 2, 3 and 4% in relation to cement mass).

As the standard methods of testing concrete mixture consistency turned out to be inadequate for reed-concrete mixture, the cone method for mortars was applied (according to the Polish standard PN-85/B-04500). A vessel in the form of cone was filled with reed-concrete mixture, placed in the apparatus and then a metal cone was lowered until it touched the surface of the mixture. Then the metal cone was released so that it could penetrate the mixture. The depth of immersion is the measure of the consistency. The test was repeated five times and the results are given in Table 6.

Table 6   Consistency of reed-concrete mixture

SUPERPLASTICIZER CONTENT, %	CONE IMMERSION, cm					
	1	2	3	4	5	AVERAGE
0	2.2	2.6	2.9	2.4	2.5	2.52
1	3.2	2.8	3.3	3.9	3.7	3.38
2	4.2	4.8	4.4	4.6	4.8	4.56
3	6.1	6.5	6.3	5.9	6.3	6.22
4	7.2	7.1	6.6	7.4	7.8	7.22

As it was expected, the consistency increased with the increase of superplasticizer content. The difference between mixtures without and with superplasticizer (3%) is shown in Figure 2.

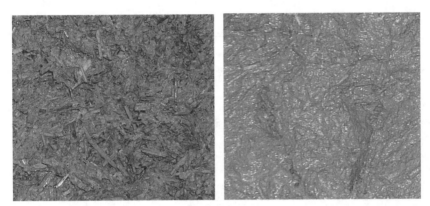

Figure 2   Reed-concrete mixture with 0% (left) and 3% (right) of superplasticizer

Test results of bulk density, water absorption and compressive strength of reed-concrete depending on the consolidation method are given in Tables 7, 8 and 9, respectively, for hand consolidation, brake press and vibration with brake press.

Table 7   Test results of reed-concrete consolidated using a hand rammer

SUPERPLASTICIZER CONTENT, %	BULK DENSITY ρ_p, kg/dm^3	COMPRESSIVE STRENGTH, MPa	WATER ABSORPTION, n_w, %
0	0.925	**4.22**	46.1
1	0.886	4.10	44.3
2	0.908	4.09	**41.1**
3	**0.874**	2.59	51.8
4	0.877	3.09	44.9

Table 8   Test results of reed-concrete consolidated using a brake press

SUPERPLASTICIZER CONTENT, %	BULK DENSITY, ρ_p, kg/dm^3	COMPRESSIVE STRENGTH, MPa	WATER ABSORPTION, n_w, %
0	0.881	3.27	44.7
1	0.894	**3.91**	**42.0**
2	0.821	2.91	47.3
3	0.835	2.86	47.9
4	**0.815**	2.21	51.2

Table 9   Test results of reed-concrete consolidated by vibration with a brake press

SUPERPLASTICIZER CONTENT, %	BULK DENSITY, ρ_p, kg/dm^3	COMPRESSIVE STRENGTH, MPa	WATER ABSORPTION, n_w, %
0	0.896	**3.90**	**44.3**
1	0.842	2.87	46.3
2	0.797	2.39	52.9
3	0.789	1.74	55.3
4	**0.779**	0.78	63.7

Addition of superplasticizer affected not only the consistency but also the bulk density. With an increase of superplasticizer quantity the density lowered. This might have been caused by aeration of the concrete mixture. The decrease in compressive strength may also testify to this phenomenon. The amount of superplasticizer addition should not exceed 2%. Any further increase is not only detrimental to strength, but also causes difficulties with consolidation of the mixture and segregation of constituents. A high superplasticizer content caused leakage of the cement paste and adhesion of the organic filler to the moulds when mechanical methods of consolidation were applied. The example of specimens with different amounts of superplasticizer is presented in Figures 3 and 4. The specimens consolidated in a brake press had even and smooth surface and they most resembled ordinary concrete (Figure 4).

Figure 3   Specimens with 1% (left) and 4% (right) superplasticizer
consolidated using a brake press with vibration.

Figure 4   Specimens after 28 days of hardening consolidated using a hand rammer, brake press and vibration with brake press, respectively (at 4% superplasticizer content).

CONCLUSIONS

- There is a possibility of using waste materials (conifer sawdust) and plant material renewable within a human life period (*Phragmites*) for the production of insulating and insulating-structural lightweight concrete.

- Application of $Al_2(SO_4)_3$ with hydrated lime as mineralizing agent for the organic filler in the form of *Phragmites* increased the compressive strength of reed-concrete by 8 times compared to the non-mineralized material.

- The compressive strength of reed-concrete varies depending on the grain size distribution of the plant aggregate (length of reed blade) and may differ more than twice (from 1.52 to 3.11 MPa).

- The application of superplasticizer improves the workability of reed-concrete mixture, but an excessive amount causes segregation of mixture constituents and adhesion of plant filler to steel moulds.

- Static methods of reed-concrete mixture consolidation give higher compressive strength, lower water absorption and smoother surface to concrete.

ACKNOWLEDGEMENT

The study was carried out under the research project number W/IIB/1/05.

REFERENCES

1. PROCHASKA W, Reed-concrete Experimental Building In Oliwa (Trzcinobetonowa budowa doświadczalna w Oliwie), Scientific Publications of Building Research Institute, No. 39, PWT, Warsaw, 1951, in Polish.

2. ROZMEJ Z, Research on Possibility of Reed Mineralization (Badania nad możliwością mineralizacji trzciny), Scientific Publications of Building Research Institute, No. 52, PWT, Warsaw, 1952.

3. BECK A, SCHAB H, DRACH V, HÖHN H AND FRICKE J, Wärmeleitfähigkeit von Leichtbetonbauteilen aus Miscanthus, in: 3. International Miscanthus-Tagung, Bonn, März, 2004.

4. HÖHN H, Mineralisch gebundene Baustoffe – Miscanthus, in: Anbau und Ververtung von Miscanthus in Europa, Pude R. (red.), Beiträge zu Agrarwissenschaften, Bd. 26, Verlag Wehle, Bad Neuenahr, 2002, pp. 30-32.

5. HESCH R, Bau- und Dämmplaten sowie Schüttdämmung aus Miscanthus x giganteus, In: Pude R. (ed.) Miscanthus – in: Anbau bis zur Verwertung, Beiträge zu Agrarwissenschaften, Bd. 19, Verlag Wehle, Bad Neuenahr, 2000, pp. 63-67.

6. HUTH H-V, Ermittlung bauphysikalischer und baustatischer Eigenschaften von Miscanthus – Leichtbeton, in: Anbau und Ververtung von Miscanthus in Europa, Pude R. (red.), Beiträge zu Agrarwissenschaften, Bd. 26, Verlag Wehle, Bad Neuenahr, 2002, pp. 33-38.

7. PUDE R, TRESELER C-H AND NOGA G, Morphological, Chemical and Technical Parameters of Miscanthus Genotypes, Journal of Applied Botany and Food Quality, 78, 2004, pp. 58-63.

8. PUDE R, BANASZUK P, TRETTIN R AND NOGA G, Suitability of Phragmites for lightweight concrete, Journal of Applied Botany and Food Quality, 79, 2005, pp. 141-146.

9. JOOSTEN H AND CLARKE D, Wise use of mires and peatlands, Background and principles including a framework for decision-making, International Mire Conservation Group and International Peat Society, 2002.

10. BOŁTRYK M, ABDULLA T AND LELUSZ M, Optymalizacja składu wiórobetonu w aspekcie zmniejszenia zużycia cement, Zeszyty Naukowe Politechniki Białostockiej, Budownictwo, Zeszyt 10, 1991.

11. DAWDO C, Wood Shavings Concrete in Building Engineering: Properties, Design, Production, Application (Wiórobeton w budownictwie. Właściwości, projektowanie, produkcja, zastosowanie), Białystok Technical University, 1994, in Polish.

USE OF MARBLE WASTE IN THE COMPOSITION OF HYDRAULIC CONCRETES

M Belachia   S Bensebti   H Hebhoub

University of Skikda

H Aoun

University of Annaba

Algeria

ABSTRACT. The discharge of waste materials presents many problems (the place occupied by storage sites, importance of the costs, environmental impact). Thus, the industry of construction does not pose a problem only at the end of the life cycle of these products but also at the beginning. Therefore, it is necessary to find a means for the valorisation and the re-use of this waste and consequently find another source of aggregates. The main purpose of this study is to demonstrate technically the possibility of using marble waste aggregates as substitute in hydraulic concrete. The article presents the study methodology and the characterization of marble waste and various formulations of concretes. The study consists of analysis of the results obtained and their comparison with a pilot test concrete (100% natural aggregates).

Keywords: Concrete, Hydraulic, Waste, Marble, Substitute, Pollution.

Dr M Belachia, Department of Civil Engineering, University of Skikda, Algeria.

Dr H Aoun, Department of Hydraulic, University of Annaba, Algeria.

Dr S Bensebti, Department of Civil Engineering, University of Skikda, Algeria.

Dr H Hebhoub, Department of Civil Engineering, University of Skikda, Algeria.

INTRODUCTION

The building materials industry generates secondary products or waste, which have a direct effect on the environment. The storage of such waste in dumps pollutes the air and contaminates water sources and agricultural fields. Therefore, it is necessary to proceed at the elimination of these by-products by finding a means to value and to re-use them. Every year, millions of tons of wastes are collected, e.g. concrete from the demolition of old constructions, and those approaching the end of their design life, as well as a large quantity of red bricks not conforming to the standard specifications. The ecological considerations require taking into account the environment by either avoiding further use of natural materials, or eliminating by-products and wastes in dumps which can pollute the natural environment. Either way, it requires a large investment. In addition, the ecological and economic demands require the valorisation and waste recycling [1].

The aim of this study is to contribute to the re-use of the marble waste in the manufacture of hydraulic concrete which will eventually allow:

- Elimination of marble waste by recycling and re-use, therefore, protecting the environment,
- Solution to the problems related to aggregates shortages.

RECYCLED AGGREGATES USED AS A SUBSTITUTE

In Algeria, the aggregates demand has considerably increased. This situation is certainly linked to the country's development. The south regions of the country experience a shortage in the supply of aggregates. The annual demand for sand in the building sector is about 15 million tons. In the future, we can expect approximately 0.25% of buildings demolished annually which will result in over 1 million cubic meters of concrete waste.

The degraded substance, the effluent or by-products which must be discharged because these are contaminated or damaged, are considered as waste [1]. In 1996, 80% of some 1 million tons of construction sites wastes, have been treated by preparation and sorting installations, and were able to be recycled in Switzerland. Several types of residues, by-products and other wastes can be used according to their cost effectiveness and their properties as aggregates [2]. Among various studied materials, we find slag, recovered concrete, waste from power stations and waste coming from the exploitation of mines or quarries [3, 4]:

a) Recovered concrete is material containing a large proportion of hydrated cement particles
b) Flying ashes are produced during the combustion of coal in thermal power stations
c) The blast furnaces slag is a by-products of the steel industry
d) Crushed bricks are waste produced by brickyards, then crashed,
e) Marble is a metamorphic rock resulting from the metamorphism of limestone, composed mostly of calcite (a crystalline form of calcium carbonate, $CaCO_3$). It is extensively used for sculpture, as a building material, and in many other applications.

In our present study, we are interested in the marble waste of Fil-Fila quarry near Skikda. We have two types of waste; the first is discharged from the marble block quarry as falls and rubble, the rate of waste being 56% (about 70,000 m^3) of the production in 2003; the second, discharged from factory transformation, arising as block size falls and falls of floor stone and paving stones, the rate of waste being about 22% (about 20,000 m^3) in 2003. Currently, the recycling of this waste is processed by a private company. We present the physical and chemical characteristics of the marble from FIL-FILA quarry.

Table 1   Physical and chemical characteristics of the marble

CHARACTERISTICS	MARBLE TYPE		
	White	Grey	Green Mignonette
Physico-mechanical characteristics			
True density, g / cm^3	2.736	2.746	2.752
Bulk density, g / cm^3	2.684	2.738	2.717
Compactness, %	98.03	99.70	98.70
Porosity, %	1.96	0.30	1.30
Absorption rate, %	0.39	0.11	0.16
Saturation rate, %	0.87	0.50	0.77
Compressive strength (dry state), MPa	94.3	135.2	93.1
Compression strength (after cooling and reheating), MPa	94.8	100.7	102.7
Wear resistance, g / cm^2	1.82	0.719	0.996
Impact resistance, cm·kg/cm^3	40	56	83.50
Chemical characteristics			
CaCO$_3$	99.05	97.73	97.22
MgO	1.03	0.99	3.05
CaO	54.86	54.00	51.05
Fe$_2$O$_3$	0.04	0.22	0.04
Al$_2$O$_3$	0.08	0.56	0.08
SiO$_2$	0.15	0.90	0.15
P.C	44.26	43.65	44.26

CHARACTERISATION OF AGGREGATES MIXTURES

The used gravel is a 5/15 and 15/25. It is calcareous and crushed. The sand is a 0/5 and rounded (sea sand).

Analyses in the laboratory for natural aggregates and recycled aggregates (0% of substitute and 100% marble waste materials respectively) show the following results (Table 2).

Table 2   Characterization of basic aggregates

MATERIALS	BULK DENSITY, g/cm^3		TRUE DENSITY, g/cm^3		FLAKINESS INDEX, %		ES, %		CLEANLINESS TEST, %	
	0%	100%	0%	100%	0%	100%	0%	100%	0%	100%
Sand 0/5	1.734	1.615	2.597	2.666			88	96		
Gravel 5/15	1.554	1.461	2.666	2.672	11	12			0.5	1.41
Gravel 15/25	1.421	1.566	2.666	2.666	16	12			1.15	1.11

Table 3 represents the characteristics of a mixture between natural and recycled aggregates with the following proportions 25%, 50%, and 75%.

Table 3   Characterization of mixture-aggregates

MATERIALS	TRUE DENSITY, g/cm³			FLAKINESS INDEX, %			ES, %		
Proportion, %	25	50	75	25	50	75	25	50	75
Sand 0/5	2.631	2.649	2.666				92	94	95
Gravel 5/15	2.666	2.666	2.666	14	17	15			
Gravel 15/25	2.666	2.666	2.666	11	13	13			

Figure 1 shows that the fineness modulus increases in line with the substitution rate. This variation ranges from 1.91 (0% of recycled aggregates) to 4.33 (100% of recycled aggregates). The variation influences conversely the workability of concrete.

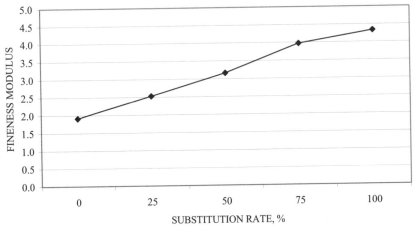

Figure 1   Influence of the substitution rate on the fineness modulus

According to the characterization results, we conclude that from the grain size distribution analysis that the obtained sand from the grinded marble waste is coarse, therefore it has no fines. On the other hand, the sand equivalent test (E.S) confirms that the recycled aggregates are very clean. The optimum fineness modulus is at 25% of the substitution rate.

EXPERIMENTAL STUDY

The objectives are as follows:

- Total or partial substitution rate of natural aggregates by the recycled aggregates,
- The influence of water/cement ratio E / C (0.45; 0.55; 0.65),
- The compressive and tensile strengths.

The experimental program is divided into two parts:
1st part: Constant concrete workability (cement content 350 kg/m³), Table 4.
2nd part: constant W/C ratio (cement content 350 kg/m³), Table 5.

Table 4   Concrete mixtures

NOTATION	CONCRETE TYPE
B1.1(0%)	Pilot test concrete of crushed aggregates - AIN SMARA quarry - and rounded sand - CHATT quarry - with A_C=8
B1.2(25%)	Crushed aggregates concrete - AIN SMARA quarry - and rounded sand - CHATT quarry - with replacement of 25% of crushed aggregates by marble waste - FILFILA quarry - with A_C = 8
B1.3(50%)	Crushed aggregates concrete - AIN SMARA quarry - and rounded sand - CHATT quarry - with replacement of 50% of crushed aggregates by marble waste FILFILA quarry - with A_C = 8
B1.4(75%)	Crushed aggregates concrete - AIN SMARA quarry - and rounded sand - CHATT quarry - with replacement of 75% of crushed aggregates by marble waste - FILFILA quarry- with A_C = 8
B1.5(100%)	Crushed aggregates concrete: 100% of crushed aggregates by marble waste - FILFILA quarry - with A_C = 8

Table 5   Concrete mixtures

NOTATION	CONCRETE TYPE
B2.1.1(0%).W/C=0.45 B2.1.2(0%).W/C=0.55 B2.1.3(0%).W/C=0.65	Pilot test concrete of crushed aggregates - AIN SMARA quarry - and rounded sand - CHATT quarry
B2.2.1(25%).W/C=0.45 B2.2.2(25%).W/C=0.55 B2.2.3(25%).W/C=0.65	Crushed aggregates concrete - AIN SMARA quarry - and rounded sand - CHATT quarry - with replacement of 25% of crushed aggregates by marble waste - FILFILA quarry
B2.3.1(50%).W/C=0.45 B2.3.2(50%).W/C=0.55 B2.3.3(50%).W/C=0.65	Crushed aggregates concrete - AIN SMARA quarry - and rounded sand - CHATT quarry - with replacement of 50% of crushed aggregates by marble waste FILFILA quarry
B2.4.1(75%).W/C=0.45 B2.4.2(75%).W/C=0.55 B2.4.3(75%).W/C=0.65	Crushed aggregates concrete - AIN SMARA quarry - and rounded sand - CHATT quarry - with replacement of 75% of crushed aggregates by marble waste - FILFILA quarry
B2.5.1(100%).W/C=0.45 B2.5.2(100%).W/C=0.55 B2.5.3(100%).W/C=0.65	Crushed aggregates concrete: 100% of crushed aggregates by marble waste - FILFILA quarry

The concrete mixtures were formulated by the Dreux-Gorisse method. The cement content was 350 kg/m^3 for all mixes.

RESULTS AND DISCUSSION

The first part of experiments presents the variation of the substitution rate of concrete type (B1.i) with constant concrete workability. Therefore, different water quantities were used every time. The second part presents 3 concrete types (B2.i.1, B2.i.2 and B2.i.3) with different substitution rates and a constant W/C ratio.

Our main research concerned both concrete states, the fresh state is defined by slump test and density measurement, the hardened state is characterised by compressive and tensile strengths.

The curves (Figure 2) show the density variation according to the substitution rate is ineffective for the concrete B1.i,, while for the three others (B2.i.1, B2.i.2, B2.i.3), there is a relatively considerable variation. The highest densities were noticed at 50% substitution rate, this is explained by a regular granulometry at this rate. Then, the density decreases when the substitution rate reaches 100%.

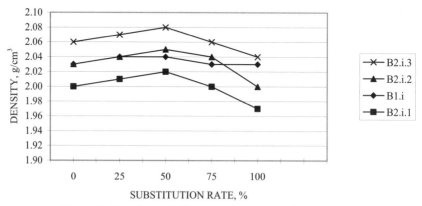

Figure 2   Density variation according to substitution rate

The substitution rate influences conversely the concrete workability (Figure 3), except the concrete B1.i, where the slump test shows a constant level. The other concretes, we notice a decrease of the workability after a light increase for the 25% rate. The explanation is that from 50% of substitution the concrete matrix is almost constituted by cement paste only.

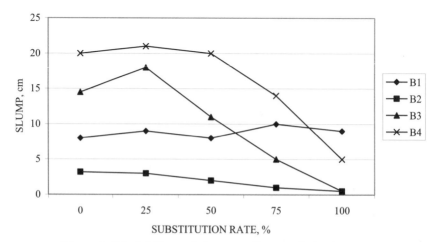

Figure 3   Workability variation according to substitution rate

Figures 4 and 5 represent the concrete compressive strength variation. The concrete of W/C = 0.55 ratio shows a constant compressive strength for all substitution rate. The highest compressive strengths were registered at 25% of substitution rate for the other concretes.

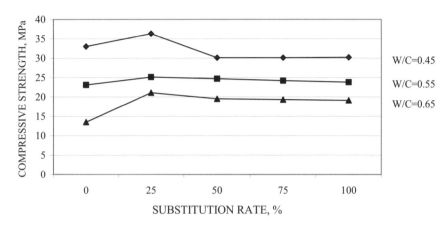

Figure 4   Compressive strength according to W/C ratio

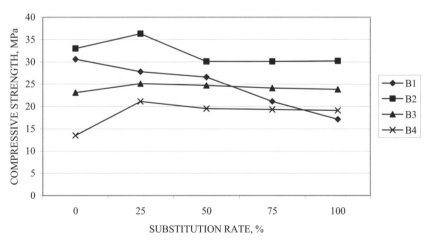

Figure 5   Compressive strength according to substitution rate

The minimal tensile strength (Figure 6) was registered for the concrete B2.i.3 with a substitution rate of 100%.

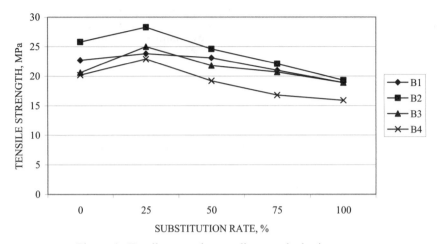

Figure 6   Tensile strength according to substitution rate

CONCLUSIONS

Our study concerned the possibility of the reuse of marble waste in concrete. The intervention in this study is an experimental approach to substituting natural aggregates by the recycled aggregates. The study has determined the performances of various concrete mixtures which help to understand the behaviour of recycled aggregates used in concrete.

Four parameters were needed to predict concrete performances such as density, workability and compressive and tensile strengths. The high concrete strengths were obtained by the 25% substitution rate. The maximum density was obtained for concretes of 50% substitution rate. We can say that satisfactory workability and acceptable strengths were obtained in this study. The results enable us to conclude that recycled aggregates, coming from marble waste can be used as alternative aggregates for their economic aspects. Therefore, technico-economic evaluations and recommendations are required.

REFERENCES

1.   RAMACHANDRAN V S, Utilisation des déchets et sous produits comme granulats du béton, Conseil national de recherche Canada, CBD-215-F, 1981.

2.   DREUX G, Nouveau guide du béton, Edition Eyrolles, 1985.

3.   DUPAIN R, LANCHON R AND SANTATTOMAN J C, Granulats, sols, ciments et bétons: caractéristiques des matériaux de génie civil par les essais de laboratoire, Edition Casteilla, Paris, 2000.

4.   BARON J AND SAUTEREY R, Les bétons: bases et données pour leur formulation. Edition Eyrolles, 1996.

EFFECTS OF WET PROCESSED CRUSHED CONCRETE FINES AS SECONDARY AGGREGATES IN BUILDING MATERIALS

K Weimann

Federal Institute for Materials Research and Testing

A Müller

Bauhaus-University Weimar

Germany

ABSTRACT: In many countries construction and demolition waste consists largely of concrete rubble. Many studies, therefore, have investigated the possibilities for reusing concrete. Regarding a close-loop recycling of concrete rubble, there have been different attempts to recycle concrete demolition waste in concrete production. A demonstration project, funded by the European Union, investigated the possibilities of improving the quality of crushed concrete fines by a wet treatment in a pilot plant. The procedure included the abrasion of the binder matrix in a stirring unit, a wet classifying to remove abraded cement and a density separation in a jig. Various adjustments of the pilot plant were tested to optimize the processing. Thus the original crushed concrete fines were sorted in fine, light and heavy fractions. The heavy fraction was supposed to be of higher material quality. Laboratory tests showed an improvement of material properties such as loss by washing and binder matrix content as well as water absorption for the investigated materials. The tests for building materials on mortar and concrete specimens also showed that wet treated crushed concrete fines can be used as secondary aggregates – up to a certain proportion. A life cycle analysis confirmed the environmental compatibility of wet processing.

Keywords: Recycled aggregates, Crushed concrete fines, Wet treatment, Jig, Building material properties.

Karin Weimann works as a scientist for the Federal Institute for Materials Research and Testing (BAM) in Germany, Division IV.3, Waste Treatment and Remedial Engineering.

Prof Dr Anette Müller, Bauhaus-University Weimar, Germany, Chair of Mineral Processing of Building Materials and Reuse.

INTRODUCTION

In Europe, about 50% of domestic and industrial waste comes from the construction industry. On average about 500 kg of construction and demolition (C & D) waste is generated per capita annually [1]. Taking into account that concrete will continue to be the dominant construction material in Europe, concrete demolition waste will remain a major part of the total waste.

In Germany, more than 60 million tons of demolition waste is generated annually. Approximately 70% of the construction and demolition waste is currently recycled [2]. The largest percentage of the recycled C & D waste is reused in road construction, for example in roadbed substructures. Considering the amount of concrete produced in Germany since 1945, it is clear that the amount of concrete in C & D waste will continue to increase in the coming decades (Figure 1) [3].

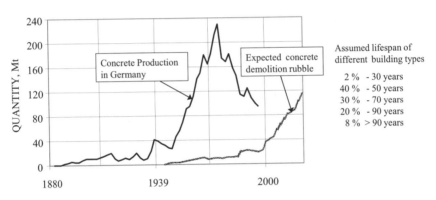

Figure 1   Approximation of the future increase of the amount of concrete demolition rubble in Germany, based on the estimated lifespan of buildings [3]

In terms of a close-loop recycling of concrete rubble, several methods were implemented to use the crushed concrete as a source of aggregate for the production of new concrete. The main advantages of reusing concrete rubble as secondary aggregates for concrete production are:

- reduction of waste accumulation and land filling
- reduction of the consumption of primary aggregates – saving natural resources
- in several cases also a reduction of energy consumption for gaining and transporting aggregates as raw material [4, 5].

Depending on the crushing process, about a third of the broken concrete is currently generated as crushed concrete sand. The material properties of coarse recycled concrete aggregates often enables the substitution of a certain percentage of natural aggregates in concrete production. In contrast, recycled crushed concrete fines can usually not be used. Besides a generally higher content of undesired substances, there is also a higher content of adherent old cement paste which is the main cause for inferior building material properties of crushed concrete fines [6, 7]. The comparatively high surface area of fine grains causes this accumulation.

A high content of adherent binder matrix causes a loss of quality in the material properties of the secondary aggregates, for example lower density, higher water absorption and porosity [8, 9]. In the majority of cases this results in inferior properties of building materials, which employ those secondary aggregates [10, 11].

A demonstration project, RECDEMO – funded by the LIFE-Programme of the European Union – investigated the possibilities of improving the quality of crushed concrete fines by a wet treatment in a pilot plant. The tests were carried out to optimize recycling techniques for gaining a concrete sand fraction which can be used as secondary aggregates to replace natural sand.

WET TREATMENT OF CONCRETE FINES AND LABORATORY INVESTIGATIONS

The concept of the wet treatment included different techniques to improve the building material properties of crushed concrete fines:

- a wet treatment
  → to reduce the content of loss by washing, soluble chlorides and sulphates and to elute remaining contaminants
- a digestion in a stirring unit (Eirich Mixer)
  → to abrase adherent binder matrix
- a classification step in a hydrocyclone
  → to remove abraded binder matrix and other particles below 100 μm

and most importantly

- a density separation in a jig
  → to separate grains with a comparatively high apparent density, preferably close to the density of natural grains, from lighter material. The heavy fraction was supposed to be of superior building material properties.

A simplified diagram of the wet processing is shown in Figure 2. Applying such a wet treatment, the crushed concrete fines are separated into three fractions: finest, light and heavy material.

For the study, different sand fractions (0-4 mm) were treated in a pilot plant for the wet treatment of mineral wastes. The sand fractions were all gained by selective dismantling processes to ensure low concentrations of unwanted or harmful substances. The concrete rubble was crushed in impact or jaw crushers. Laboratory tests demonstrated the environmental compatibility of all test materials.

The pilot plant was operated continuously. The mass flow rate of the test runs mostly ranged between 0.55–0.65 t per hour output material (dry substance). Before entering the stirring unit, the crushed concrete fines were mixed with water. Different stirrer settings were tested to vary the application of energy used for the abrasion of the binder matrix. Rotations of the stirring between 100 and 1000 rotations per minute (rpm) were tested. In the next step a hydrocyclone was used to remove grains smaller then 100 μm. The finest fraction 0–0.1 mm consisted mainly of loss by washing and abraded binder matrix.

The remaining sand fraction of 0.1-4 mm passed through the air-pulsed jig. In the jig the sand fraction was separated by density and partly by classification. The sorting in a jig (jigging) is mainly based on the different densities of the treated materials. Natural grains like quartz or dolomite have higher densities (quartz: 2.65 t/m³, dolomite: 2.85-2.95 t/m³) than concrete (2.2-2.5 t/m³) and hardened cement paste (1.5-1.8 t/m³). In water, natural grains therefore sink faster than particles of the same size of concrete or of binder matrix.

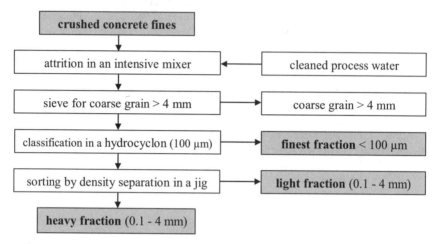

Figure 2   Flow scheme of the wet treatment process for crushed concrete fines

In a jig, pulsating water causes the particles to float in the water. Particles of different density but of the same grain size stratify within the material flow: heavy grains sink and lighter particles rise to an upper zone. The resulting layers of heavy and light materials can be discharged separately. Other influencing factors are grain shape and grain size. Flat particles, that are undesired in concrete or mortar production, sink slower, hence spherical grains accumulate in the heavy fraction. The sinking of grains is also affected by the grain size, smaller grains sink more slowly. Depending on the regulation of the jigging process, the density separation is accompanied in most cases by a measurable classification.

Different test runs were carried out with various jig and stirring unit settings to optimize the wet treatment. A mass balance for every test run was calculated. To evaluate the wet treatment for a life cycle analysis, the energy consumption of every unit (at different settings) was measured.

Samples of input material (dry treated concrete fines) and output fractions from every test run were taken to evaluate the processing. Material properties of the fractions were investigated by chemical and physical analyses. Microscopic investigations were carried out on samples of different materials and fractions. For all fractions, except the finest fraction, tests on workability, e.g. consistency and water absorption, and tests on strengths, elasticity and shrinkage were carried out. Most of the tests were conducted on mortar test specimens of the same mortar mixture, containing only crushed concrete fines as aggregates. Three test series were carried out on concrete test specimens of the same concrete mixture and also of the same aggregate grading curve containing crushed concrete fines (input material and heavy fraction) in a proportion from 0 to 50% of the total amount of aggregates. Details of mortar mixtures and concrete mixtures are shown in Table 1.

Table 1   Mortar and concrete mixtures for wet treated crushed concrete fines, heavy fraction, material A

	MORTAR	CONCRETE
Cement type	CEM I 32.5	CEM I 32.5
Amount of Cement (kg/m³)	338 - 413	310
w/c ratio	0.6	0.6
Amount of crushed concrete fines (%)	100	10 - 50

RESULTS

Laboratory Investigations

The chemical and the microscopic investigations showed that the material properties of the gained heavy fractions improved against the material properties of natural grains. The content of loss by washing, of binder matrix, measured as acid-solubles, and the water absorption were reduced. Correspondingly, the content of natural grains, measured as acid-insolubles, and the apparent density increased. An overview of results from selected test runs (sufficient recovery of heavy fraction) of two different crushed concrete fines from different origins is shown in Tables 2 and 3. Material A was gained from concrete demolition waste from the selective dismantling of a thermal power plant, Material B emanated from the deconstruction of a motorway.

Table 2   Material properties of wet treated and untreated crushed concrete fines, material A in comparison to natural sand

	INPUT MATERIAL	HEAVY FRACTION	LIGHT FRACTION	FINEST FRACTION	NATURAL SAND
loss by washing < 100 μm, % [1]	5.53	0.53	5.76	100.00	0.55
acid-insolubles, % [1]	78.45	84.04	70.85	57.41	99.70
acid-solubles, % [1]	21.55	15.96	29.15	42.59	n.m.
water absorption, % [2]	7.59	5.53	11.38	n.m.	n.m.
apparent density, g/cm³ [3]	1.85	1.93	1.64	n.m.	2.49

Table 3   Material properties of wet treated and untreated crushed concrete fines, material B

	INPUT MATERIAL	HEAVY FRACTION	LIGHT FRACTION	FINEST FRACTION
loss by washing < 100 μm, % [1]	3.34	0.49	1.47	100.00
acid-insolubles, % [1]	82.41	84.90	74.02	62.22
acid-solubles, % [1]	17.59	15.10	25.93	37.78
water absorption, % [2]	5.59	4.16	8.51	n.m.
apparent density, g/cm³ [3]	1.89	1.86	1.68	n.m.

[1] percentage of dry matter   [2] percentage after 10 minutes
[3] (10 × 10 × 10) cm³ mortar-cube after 28 days   n.m.: not measured

The central aim, to get a heavy fraction with a reduced binder matrix content and higher density, was achieved in every test run. As the results of the acid leaching tests show, an accumulation of particles with a comparatively high binder matrix content could be observed in the light fractions. In the heavy fractions also a certain amount of adherent old cement paste still remained, while in the finest fractions of all test materials parts of quartz and other rock components were found as well. Figure 3 shows the changes in the cement paste content of the different wet treated fractions with regard to the input materials A and B.

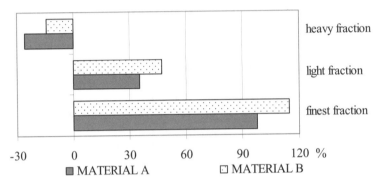

Figure 3 Changes in the content of acid-solubles in the output fractions of material A and B, referenced to the input materials

As expected, the content of old binder matrix corresponded with the water absorption and the porosity of the investigated materials. In Figure 4, mean values of the water absorption of fractions from different test runs with material A und B are shown in comparison to the water absorption of one test run with material C. Material C, gained by the selective dismantling of a concrete building, was treated in a jigging process twice. Although the reduction of the water absorption from the input material C to the heavy fraction is almost not measurable, still a certain amount of particles with a higher content of binder matrix can be separated from the heavy fraction.

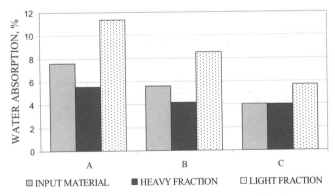

Figure 4 Water absorption of crushed concrete fines from different sources

Additional laboratory analyses showed, that the wet treatment also changed the granulometric composition of the materials. The mechanical application of energy in the stirring unit caused the desired abrasion of binder matrix and, consequentially, an increase of finer grains in the total mass of the treated output material. As expected, throughout the jigging process a classification of the concrete fines also occurred. Finer particles accumulated in the light fractions, bigger grains in the heavy fractions.

Wet treated and untreated crushed concrete fines were also investigated with an environmental scanning electron microscope (ESEM). The examinations showed an accumulation of grains with less binder matrix and more particles with favourable grain shapes in the heavy fractions. In Figure 5, a thin-ground section photo of the input material, crushed concrete fines A, shows a natural grain, which is nearly completely covered with binder matrix. For comparison, Figure 6 shows a thin-film section photo of particles from a heavy fraction of material A. While some particles are still embedded in old cement paste, natural grains without or almost without binder matrix could only be found in the heavy fractions.

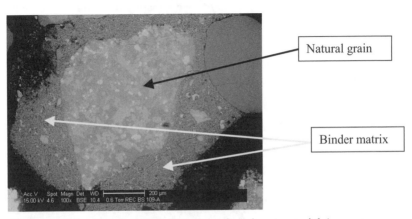

Figure 5   Thin-ground section of particles from input material A,
natural grain completely embedded in old cement paste

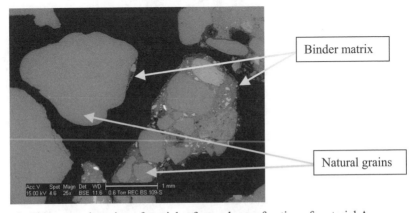

Figure 5   Thin-ground section of particles from a heavy fraction of material A,
natural grains with different contents of adherent binder matrix

Tests on mortar and concrete

The properties of mortar made from wet-treated heavy and light fraction from the crushed concrete fines of type A and B showed no significant improvement in compressive and flexural strength compared to the input material, although measured values of compressive strength could be up to 32% and for flexural strength up to 60% higher than for the input material. The results of the strength tests of mortar made of both materials A and B were similar.

On the other hand, the measurements of the dynamic modulus of elasticity, which is especially influenced by the content of old cement paste [6, 10, 12], showed a clear enhancement of the heavy fractions and lower values for mortars made of light fractions for material A. The results of the dynamic modulus of material B showed an improvement of most of the heavy fractions, too. Figure 7 shows the results of the measurements of dynamic modulus of elasticity of different fractions of material A and B related to the proportion of old cement paste on the total solid addition to the mortar. The correlation between the results of the elasticity tests and the content of binder matrix of the recycled concrete fines can be seen in the stability index R^2.

Additional material tests on concrete specimen made with different contents of wet-treated heavy fraction and input material of crushed concrete type A were carried out. The results of measurements of compressive strength, of dynamic modulus of elasticity and of shrinkage showed an improvement of the material properties of the wet treated crushed concrete fines. A correlation between the old cement paste content in the recycled aggregates and the values of the building material properties could be observed. The quality of the replaced natural sand was not met.

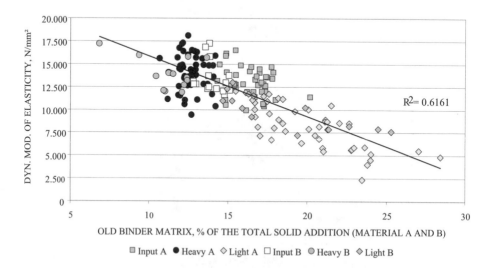

Figure 7   Dynamic modulus of elasticity after 28 days, in relation to
the proportion of old cement paste in the total solid addition

Measurements of the shrinkage of concrete made with a content of 20% and 50% wet-treated (heavy fraction A) and untreated (input material A) crushed concrete fines on the total amount of aggregates are shown in Figure 8. The measurement results are shown below in comparison to the values of concrete made of only natural aggregates. The measuring period was 270 days.

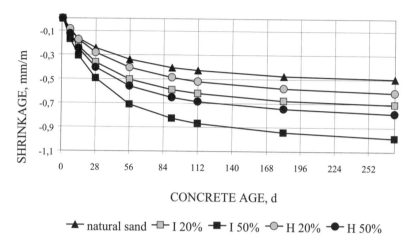

CONCRETE AGE, d

── natural sand ─□─ I 20% ─■─ I 50% ─○─ H 20% ─●─ H 50%

Figure 8   Shrinkage of concrete samples, containing different proportions of untreated (I)
and wet treated (H) concrete fines (Material A)
in comparison to concrete without concrete fines

In contrast to the tests on mortar specimen, all building material tests on concrete showed a clear improvement of all tested building material properties by the wet treatment. This is probably due to the fact that the grading curves were identical for all concrete mixtures (grading curve B16). Negative effects of the classification of the crushed concrete fines during the wet processing, which influenced the results of the tests on mortar specimens, were excluded.

Environmental evaluation

Besides technical feasibility, for a recycling process and its product the environmental compatibility plays a significant role. Therefore the wet processing for the production of a heavy fraction of crushed concrete fines was environmentally evaluated in comparison to the production of natural sand.

Basis for the environmental evaluation was the mass balance of the wet processing and the energy consumption of the pilot plant and its single aggregates. The aim of the appraisal was the comparison of the ecological impacts of substituting natural sand with material from a heavy fraction (replacing 20% of the total aggregates) in a concrete C20/25 of the same mixture. The parameters for the evaluation were greenhouse gases (CO_2 and CO_2-equivalents), acidification (SO_2 –equivalents), ozone precursor substances, dust and land consumption.

The calculation of material flows and energy consumption of the pilot plant was carried out with a special software for ecobalance and lifecycle analysis [13, 14]. Data for the mining of natural sand were taken from literature and databases [4, 15].

The results of the environmental evaluation showed that there was no significant difference in energy consumption and emissions in the production of concrete made of natural aggregates in comparison to the production of concrete with a content of 20% wet treated crushed concrete fines as secondary aggregates. The main environmental factors in concrete production are energy consumption and emissions from the production of cement. If crushed natural sands are substituted by crushed concrete fines, slightly less energy is consumed. In any case, the land consumption is reduced.

CONCLUSIONS

In this study, a wet treatment of crushed concrete fines processing – including the steps of mechanical removal of the binder matrix and a density separation by a jig – was tested. The wet treated crushed concrete fines were investigated in comparison with only dry treated input material with regard to their applicability as an aggregate for building materials. The main results and conclusion can be summarized as follows:

- The wet treatment can improve the material properties of the gained heavy fraction: a reduction of the binder matrix content (measured as acid-solubles) as well as of water absorption and of loss by washing. Accordingly, the natural grain content and the density increased.

- The tests for building materials on mortar specimen made with the heavy fractions of different crushed concrete sands varied. The values of the dynamic modulus of elasticity, which is most influenced by the old cement paste content, showed the best improvements after the wet treatment. The measured strength properties (compressive strength and flexural strength) of the mortars were surely influenced by the uncorrected grading curves of the heavy fractions after the wet treatment.

- The tests on concrete specimen made with crushed concrete fines showed an improvement in all tested building material properties after the wet treatment: compressive strength, dynamic modulus of elasticity and shrinkage. There was a clear correlation between the content of old cement paste and the building material properties.

- Generally the building material quality of natural sand could not be reached by the wet treated crushed concrete fines. This can be explained with binder matrix which remained in the heavy fraction and also with the observed unfavourable grain shapes after the crushing process. Nevertheless, the wet treatment can improve the building material properties of crushed concrete fines so that natural sand could be replaced by wet-treated crushed concrete fines up to a certain level.

- An environmental evaluation of the wet processing showed no significant difference between energy consumption and emissions for the production of concrete containing 20% wet treated crushed concrete fines as secondary aggregates and concrete solely made of natural aggregates.

ACKNOWLEDGEMENT

The work on the RECDEMO project was funded by the European Union. It is part of the LIFE Environment program (Demonstration project LIFE00 ENV/D/000319).

REFERENCES

1. SYMONDS GROUP LTD, Construction and demolition waste management practices and their economic impacts, European Commission, Final Report, 1999, 104 p.

2. KWTB 4, Monitoring-Bericht Bauabfälle (Erhebung 2002), Arbeitsgemeinschaft Kreislaufwirtschaftsträger Bau, Berlin, Düsseldorf, Duisburg, 2005, 83 p.

3. GÖRG H, Bauen für den Umweltschutz: Es gibt viel zu tun - nur wissen wir nicht wann! Teil 2, Altlasten Spektrum, Vol. 10, 2001, pp. 152-153.

4. EYERER P AND REINHARDT H-W, Ökologische Bilanzierung von Baustoffen und Gebäuden, Baupraxis, Birkhäuser Verlag, Basel, 2000, 229 p.

5. HEINZ D, Vom Baustoff zum Bauwerk - Chancen für eine nachhaltige Entwicklung, 3, Münchner Baustoffseminar, München, 2003, pp. 21.

6. MÜLLER A, Baustoffkreisläufe - Stand und Entwicklung, ibausil, Weimar, F.A. Finger-Institut für Baustoffkunde (Publisher), Conference Proceedings, Vol. 15, 2003, pp. 1289-1308.

7. WEIMANN K AND MÜLLER A, Baustoffeigenschaften von nass aufbereiteten Betonbrechsanden, ibausil, Weimar, 2006, pp. 12.

8. MAULTZSCH M, MELLMANN G AND MEINHOLD U, Eigenschaften hochwertiger Betone aus aufbereitetem Bauschutt, ibausil, Weimar, F.A. Finger-Institut für Baustoffkunde (Publisher), Conference Proceedings Vol. 15, 2003, pp. 33-47.

9. KATZ A. Properties of concrete made with recycled aggregate from partially hydrated old concrete, Cement and Concrete Research, Vol. 33, 2003, pp. 703-711.

10. HANSEN T C, Recycled aggregates and recycled aggregate concrete second state-of-the-art report developments 1945-1985, Materials and Structures / Materiaux et Constructions, Vol. 18, 1986, pp. 201-246.

11. BEHLER K AND MEYER A, Untersuchungen zum Einsatz von Betonbrechsanden in sandreichen Betonen, Materialprüfungsanstalt Bremen, AiF-Forschungsbericht Nr. 12349, Bremen, 2002, 68 p.

12. FRIEDL L, Experimentelle Untersuchungen zum Transport von Wasser und Chlorid in rezykliertem Beton und zu der daraus ableitbaren Gefahr der chloridinduzierten Stahlkorrosion, Fakultät für Bauingenieur- und Vermessungswesen, TU München, Dissertation, 2004.

13. IFU, Umberto, ifu Hamburg GmbH, Internet, 2004.

14. WEIL M, JESKE U AND SCHEBEK L, Beton mit und ohne rezyklierten Zuschlag im mineralischen Baustoffstrom, Wasser- und Geotechnologie, Vol. 1, 2002, pp. 93-105.

15. UMWELTBUNDESAMT, Probas - Prozessorientierte Basisdaten für Umwelt-management-Instrumente, Internet, 2005.

PROPORTIONING CONCRETE MIXTURES WITH CRUSHED CONCRETE PAVER BLOCKS AS RECYCLED AGGREGATES

M C Nataraja   S S Ur-Rahman   L Das   R Sandeep
S J College of Engineering
T S Nagaraj
R V College of Engineering
India

ABSTRACT. In India huge quantities of construction and demolition (C&D) waste are produced in urban areas and most of these wastes possess the potential for recycling with concrete as repository. These recycled aggregates are weaker, more porous with higher values of water absorption. Hence appropriate technology is needed to know their typical aggregate crushing strength in concrete and proportion mortar strengths to higher concrete strength or to limit the strength of concrete to that of the aggregate strength for optimal use of cementing materials. Since failure of concrete with these marginal aggregates is influenced by aggregate crushing, conventional mix design methods and country's standard code would not be adequate as the concrete strength development is not wholly dependent on water- cement ratio. This investigation demonstrates the potential of systematic approach to proportion concrete mixtures where synergy among different ingredients can be accounted for appropriately.

Keywords: C&D waste, Recycled aggregate, Crushed concrete paver blocks, Re-proportioning concrete mixtures, Compressive strength and workability.

M C Nataraja is presently Assistant Professor, S.J. College of Engineering, Mysore. He obtained his Ph.D from the Indian Institute of Technology, Kharagpur, India. His main areas of research interest are fiber composites, concrete mix proportioning, high performance concrete and use of marginal materials in concrete. He has published over 80 technical papers in journals and conferences. He is the recipient of best paper award from Indian concrete Institute and International council for steel fiber reinforced concrete, India.

T S Nagaraj is presently Distinguished Professor, RV College of Engineering, Bangalore and Adjunct Professor, University of Massachusetts, Amherst, USA. He is Professor Emeritus of the Indian Institute of Science, Bangalore. His main areas of research interest are geo- and concrete materials. He has over 290 publications and three books to his credit. He is an elected Fellow of the Indian National Academy of Engineering.

S S Ur-Rahman is presently the Vice Principal S. J. College of Engineering, Mysore, India. He has many publications to his credit. He is working in the area of concrete materials and structural concrete.

L Das and **R Sandeep** were former postgraduate students.

INTRODUCTION

In India, large quantity of construction and demolition (C&D) waste is released every year. This includes substantial amount of used concrete paving blocks in urban areas. The disposal of waste has become a severe social and environmental problem in the country. The possibility of recycling of waste by the construction industry is thus of increasing importance. In addition to the environmental benefits in reducing the demand on land for disposing the waste, the recycling of C&D wastes can also help to conserve natural materials and to reduce the cost of waste treatment prior to disposal. C&D wastes are normally composed of concrete rubble, bricks, blocks and tiles, sand and dust, timber, plastics, cardboard and paper, and metals. Concrete rubble usually constitutes the largest proportion of C&D waste. It has been shown that crushed concrete rubble, after separation from other C&D waste and sieved, can be used as a substitute for natural coarse aggregates in concrete or as a sub-base or a base layer in pavements. This type of recycled material is called recycled aggregate. Successful application of recycled aggregate in construction projects has been reported in some European and American countries, as reviewed by Poon et al. [1]. While this type of material has been used in large amounts in non-structural concretes or used as road bases, its use in structural concrete is limited. Only a few cases have been reported on the use of recycled aggregates in structural concrete, and the amount of recycled aggregate used has generally been limited to a low level of replacement of the total weight of coarse aggregate. Use of recycled aggregate in India is very limited and is picking up slowly [2]. The limited use of recycled aggregate in structural concrete is due to certain inherent deficiencies of this type of material. In comparison with natural normal weight aggregates, recycled aggregates are weaker, more porous and have higher values of water absorption [3-4].

This paper presents a recent study at SJCE, Mysore, India which aims to develop a technique for using recycled aggregates obtained from broken paving blocks in non-structural concrete. The blocks were obtained from a local concrete paving industry. These blocks were crushed to 20 mm or less for use in concrete.

EXPERIMENTAL INVESTIGATION

Materials Used

In the present experimental study, the following materials were used: cement, fine aggregate, crushed granite as coarse aggregate, crushed paver block as coarse aggregate, superplasticiser and water

Cement

53 Grade OPC conforming to IS: 12269-1987[5] was used. The results obtained from the tests are shown in Table 1.

Crushed granite

Crushed granite obtained from a nearby quarry is used as coarse aggregate. The aggregate passing through 20 mm and retained on 4.75 mm are used with 70:30 proportion which satisfies the requirements as per IS: 383-1970 [7]. The specific gravity, bulk density and water absorption are 2.70, 1570 kg/m^3 and 0.45% respectively.

Table 1   Properties of OPC

Sl No.	PROPERTIES	TEST RESULTS	IS: 12269-1987 REQUIREMENTS
1	Standard consistency, %	31	No standard value
2	Setting time, minutes a. Initial setting time b. Final setting time	 92 220	 Not less than 30 Not more than 600
3	Specific gravity	3.15	
4	Compressive strength, MPa 3 days 7 days 28 days	 32.0 46.0 65.0	 27 37 53
5	Fineness by Blaine's Air Permeability method, m^2/kg	265	Not less than 225

Crushed paver blocks

Paver blocks that are disposed as wastes from the local manufacturing factory are brought and hand crushed. These crushed paver blocks are sieved to get the required fraction aggregate passing through 20 mm and retained on 10 mm sieves, and passing through 10 mm and retained on 4.75 mm sieves. These fractions are used in 70:30 proportion. The specific gravity, bulk density and water absorption are 2.44, 1250 kg/m^3 and 8.8% respectively.

Water

Tap water is used for mixing and curing of concrete and mortar cubes.

Superplasticiser

In the present investigation Conplast-SP 430 [8] superplasticising admixture is used, which complies with IS: 9103: 1979. Conplast SP 430 is based on sulphonated naphthalene polymers and is supplied as a brown liquid instantly dispersible in water. Conplast–SP 430 has been specially formulated to give high water reduction up to 25% without loss of workability. Its specific gravity is 1.22 (at 30° C) and chloride content is Nil. Air entrainment is approximately 1%.

MIX PROPORTION

In the present investigation, for the trial mix proportioning, the method given in draft IS: 10262-2004 [9], which is in wide circulation by the Bureau of Indian Standards (BIS) is used.

The following steps are used to get the trial mix proportions.

- Fix the water cement ratio at 0.5 to obtain the compressive strength of the trial mix for use as reference state for subsequent re-proportioning of mixes.
- Fix the water content in unit volume of concrete based on the aggregate size and required workability.

- Knowing the water-cement ratio and water content, calculate the cement content. Check for exposure conditions from IS:456-2000 [10].
- Find the volume of coarse aggregate corresponding to the size of the coarse aggregate and fine aggregate zone. Volume fractions of coarse aggregate and fine aggregate content are selected from the IS draft code.

Arrive at the volume of all the constituents

    a) Volume of concrete = 1 m^3
    b) Volume of cement = weight of cement/specific gravity of cement × 10^3
    c) Volume of water = weight of water/specific gravity of water × 10^3
    d) Volume of admixture = weight of admixture/ specific gravity of admixture
    e) Volume of aggregate = a - (b + c + d)
    f) Weight of coarse aggregate = e × specific gravity × volume of fraction × 10^3
    g) Weight of fine aggregate = e × specific gravity × volume fraction × 10^3

For the trial mix, determine the workability and compressive strength at 7 days and 28 days.

RE-PROPORTIONING OF CONCRETE MIXTURES

In concrete technology, the concrete mix proportioning mainly depends on the Abrams' law according to which it has been categorically stated that, *for a given set of materials*, the strength development is solely dependent on free water–cement ratio. In other words as cement or combinations of cementitious materials and/or aggregate characteristics such as size, shape and surface characteristics change, even if the water – cement ratio is the same the strength development is not the same. Owing to this, a trial mix is arrived at based on empirical considerations and tested for its strength. The strength obtained for this trial mix might not meet the practical requirements. Hence an adjustment to water–cement ratio has to be made until it is possible to arrive at the water–cement ratio required to arrive at the final mixture proportions so as to meet the practical strength requirements envisaged.

In concrete, mortar is regarded as the matrix and coarse aggregate is the distributed phase. It is obvious that the paver waste is not as strong as conventional natural aggregate. According to the law of mixture of the composite material, the behavior of concrete in terms of the properties of the individual phases and their proportions can be analyzed. Assuming a unit cell model composed of mortar and coarse aggregate by law of mixtures appropriate properties of the constituents can be computed. For the limiting case of no bond, the assumption of identical stresses in the matrix and coarse aggregates is reasonable, if the particles are more rigid than the matrix. If the particles are less rigid than the matrix, as in the case of weak aggregate and high mortar matrix strength, the bond is of less significance than composite behavior. The coarse aggregate experiences the deformation of the matrix up to the limit of its compressive strength. For the assumption of a perfect bonding between the coarse aggregate and matrix without any slippage at the interface, the strains experienced between the coarse aggregate, matrix, and concrete are the same. For a unit cell model, the relation involving the stress acting on each of the two phases (matrix, σ_m and coarse aggregate σ_a) loading and their volume fractions, (matrix, v_m and that of coarse aggregate, v_a) is:

$$\sigma_c = \sigma_m v_m + \sigma_a v_a \quad\quad (1)$$
$$\text{for } \varepsilon_c = \varepsilon_a = \varepsilon_m \text{ and } v_m + v_a = 1$$

Where σ_m and σ_a are the strength of matrix and aggregate respectively; v_m and v_a are the volume fraction of matrix and aggregate respectively; ε_c, ε_a, and ε_m are the strains in concrete, aggregate and matrix respectively.

To advance the generalized approach to proportion concrete mixes taking into account of coarse aggregate in concrete the possibility of using the above relations merits examination.

1. From the strength data of concrete, where aggregate fracture has been observed, along with the compressive strength of constituent mortar matrix, typical strength of paver aggregate in concrete is calculated.
2. Using the same law of mixtures with the typical strength of paver aggregate known, the required compressive strength of mortar matrix is calculated for the specific concrete strength of concrete.
3. The water cement ratio required to get this mortar strength is calculated by using the Generalized Abrams' Law [11].

This exercise is designated as 'Re-proportioning Method' [12, 13].

CONCRETE AND MORTAR SPECIMEN CASTING, CURING AND TESTING

Test on Fresh Concrete

All mixes are tested for workability in terms of slump and compacting factor (CF) as per the Indian Standard IS: 1199-1959 [14]. The main purpose of these tests is to check the consistency and the uniformity of concrete from batch to batch. The workability of all mixes is quite satisfactory.

Test on Hardened Concrete and Mortar

The concrete cubes are prepared and tested for compressive strength as per IS: 516-1959 [15]. 7 day and 28 day strengths presented are the average of 3 cubes' strength, respectively. Non-destructive tests namely Rebound hammer and ultrasonic pulse velocity tests have been conducted as per IS: 13311(PT1)-1992 [16] and 13311(PT2) -1992 [17]. These testes are also repeated for equivalent mortar cubes.

EXPERIMENTAL RESULTS

Paving aggregates

Based on the properties of materials used, trial mix proportioning was arrived at. From the results of the trial mix, re-proportions were obtained for different strengths of concrete as explained in the following sections. The constituent mortar and concrete cubes were cast for all mixes and tested at 7 and 28 days.

1. Trial mix details for concrete containing paving aggregates and corresponding mortar cubes for a water cement ratio of 0.5 (reference) and water content as 190 kg/m³ (reference) are presented in Tables 2 and 3. The workability and the compressive strength of concrete at 7 days and 28 days are shown in Table 4. The non-destructive test results are presented in Table 5.

Table 2   Trial mix design details for granite and recycled paving aggregates

DETAILS	GRANITE AGGREGATE	PAVING AGGREGATE
Water content, kg/m^3	190	190
Superplasticiser, %	1	1
Water / cement ratio	0.50	0.50
Fine aggregate	Natural river sand	Natural river sand
Coarse aggregate	Crushed granite	Crushed granite
Specific gravity of cement	3.15	3.15
Specific gravity of sand	2.64	2.64
Specific gravity of coarse aggregate	2.70	2.44
Specific gravity of Superplasticiser	1.145	1.145
Volume of coarse aggregate	0.64 (Table-3, IS -10262 Draft code)	0.64 (Table-3, IS -10262 Draft code)
Volume of fine aggregate	0.36	0.36

Table 3   Trial mix details per cubic meter concrete

CONCRETE	CRUSHED GRANITE AS COARSE AGGREGATE	CRUSHED PAVER BLOCKS AS COARSE AGGREGATE
Water / Cement ratio (W/C)	0.5	0.5
Water content, kg/m^3	190	190
Cement content, kg/m^3	380	380
Fine Aggregate, kg/m^3	656	656
Coarse Aggregate, kg/m^3	1192	1077
Aggregate / Cement ratio (A/C)	4.86	4.55
Superplasticiser, %	1.0	1.0
Mortar mix		
Water / Cement ratio (W/C)	0.5	0.5
Fine Agg. / Cement ratio (FA/C)	1.73	1.73

Table 4   Compressive strength and workability of trial mix concrete and mortar

TYPE OF MIX	COMPRESSIVE STRENGTH, MPa				WORKABILITY	
	7 days	SD*	28 days	SD	Slump, mm	CF
Crushed Granite as coarse aggregate (SSD)	36.70	0.44	58.03	0.38	180	0.92
Crushed Paver blocks as coarse aggregate (SSD)	35.22	0.34	45.46	0.69	Zero	0.82
Equivalent Mortar **	31.58	1.65	54.10	1.56	Collapse	1

*Standard deviations, * *Mortar proportion is the same for both aggregates (1:1.73)

Table 5   N D T results of trial mix concrete and mortar

TYPE OF CONCRETE	REBOUND HAMMER READING				PULSE VELOCITY, m/s			
	7 days	SD	28 days	SD	7 days	SD	28 days	SD
Crushed granite as CA	43.15	3.88	49.02	6.10	4441	40.45	4574	35.60
Crushed Paver blocks CA (SSD)	39.28	3.28	46.89	7.03	4346	29.76	4364	74.34
Mortar	33.07	3.48	50.22	5.98	3951	15.71	4216	8.55

2. Re-proportioning of mixes is done for M25 for concrete containing paving aggregates. The results obtained are shown in Table 6.

3. Trial mix details for concrete containing paving aggregates for water cement ratios of 0.4, 0.5 and 0.6 with water content as 190 kg/m^3 along with 7 days and 28 days compressive strength and workability results are shown in Tables 7-9.

Granite aggregates

From the comparison point of view, all the above series are repeated using 20 mm and downsize granite aggregates. The results are presented in Tables 2-10.

DETERMINATION OF AGGREGATE CHARACTERISTIC STRENGTH

The aggregate characteristic strength has been determined from composite mechanics consideration, using the equation,
$$\sigma_c = \sigma_m V_m + \sigma_a V_a$$
Where,

σ_c = Strength of concrete

σ_a = Characteristic strength of aggregate

V_a, V_m = Volume fraction of aggregate and mortar

σ_m = Strength of mortar

$V_m = 1 - V_a$

V_a = Weight / Specific gravity

Thus,

V_a = Weight/ Specific gravity
     = 1077 / (2.44 * 1000)
     = 0.441

V_m = 1- 0.441
     = 0.559

$\sigma_a = [\sigma_c + \sigma_m V_m] / V_a$
     =[45.46-54.10*0.559]/ 0.441 = 34.50 MPa

With this characteristic strength, concrete mixes are re-proportioned for different levels of compressive strength as illustrated below. Initially, concrete for strength of 25 MPa is designed using paving aggregate. The w/c ratio is found using the method of re-proportioning and in this case the designed concrete strength is less than the aggregate strength.

$$S/S_{0.5} = -0.2 + 0.6 *(C/W) \quad \text{for } S_{0.5} > 30 \text{ MPa}$$
$$25 / 45.46 = -0.2 + 0.6 *(C/W)$$

Solving, $W/C = 0.8$. Keeping the water content as 190 kg/m^3, the cement content is 238 kg/m^3. Fine and coarse aggregates are 698 kg/m^3 and 1145 kg/m^3, respectively. The test results are presented in Table 6.

Table 6   Compressive strength and workability of re-proportioned mix

TYPE OF MIX	COMPRESSIVE STRENGTH, MPa				WORKABILITY	
	7 days	SD*	28 days	SD	Slump, mm	CF
Re-proportioned for 0.80 w/c ratio	16.9	0.21	22.2	0.75	Zero	0.82

In addition, trial mixes are designed for two different w/c ratios of 0.4 and 0.5 using crushed granite and paver as coarse aggregate and the details of the mixes are presented in the Tables 7 and 8. Test results are shown in Tables 9 and 10.

Table 7   Trial mix details per cubic meter of concrete for granite as coarse aggregate

CONCRETE	CRUSHED GRANITE AS COARSE AGGREGATE		
Water Cement ratio (W/C)	0.40	0.5	0.60
Water content, kg/m^3	190	190	190
Cement content, kg/m^3	475	380	317
Fine Aggregate, kg/m^3	627	656	675
Coarse Aggregate, kg/m^3	1141	1192	1227
Aggregate-cement ratio (A/C)	3.72	4.86	6.00
Superplasticiser, %	1.0	1.0	NIL

Table 8   Trial mix details per cubic meter of concrete for crushed paver blocks as coarse aggregate

TRIAL MIX CONCRETE DETAILS	CRUSHED PAVER BLOCKS AS COARSE AGGREGATE		
Water Cement ratio (W/C)	0.40	0.5	0.60
Water content, kg/m^3	190	190	190
Cement content, kg/m^3	475	380	317
Fine Aggregate, kg/m^3	627	656	675
Coarse Aggregate, kg/m^3	1031	1077	1109
Aggregate-cement ratio (A/C)	3.49	4.55	5.63
Superplasticiser, %	1.0	1.0	NIL
Proportion	1:1.32:2.17	1:1.73:2.83	1:2.13:3.50

Table 9 Compressive strength and workability of trial mix concrete (granite)

TYPE OF MIX	COMPRESSIVE STRENGTH, MPA				WORKABILITY	
	7 DAYS	SD*	28 DAYS	SD	SLUMP, MM	CF
0.40	50.6	0.88	63.7	1.12	45	0.86
0.50	37.3	0.23	52.8	0.38	180	0.92
0.60	29.4	0.26	36.1	0.94	15	0.86

Table 10 Compressive strength and workability of trial mix concrete (paver)

TYPE OF MIX	COMPRESSIVE STRENGTH, MPA				WORKABILITY	
	7 DAYS	SD*	28 DAYS	SD	SLUMP, MM	CF
0.40	35.8	0.60	50.3	0.91	Zero	0.77
0.50	28.4	0.85	41.3	0.82	Zero	0.82
0.60	20.5	0.30	32.9	0.45	Zero	0.84

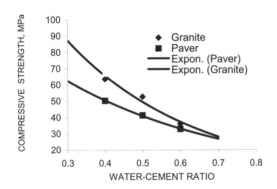

Figure 1   Variation of 28 days compressive strength with respect to different water cement ratios for granite and paver aggregate concrete

ANALYSIS OF RESULTS

Compressive Strength and Workability

An experimental investigation has been made to study the workability and compressive strength of concrete containing paver block as recycled coarse aggregate. The same is compared with that of normal concrete, which contains granite as coarse aggregate. Trial mix proportions are done for crushed paver aggregates ranging from 20 mm to 4.75 mm for water cement ratio of 0.5 and water content of 190 kg/m^3. The details of the proportions are presented in Tables 2-3.

The test results on compressive strength, rebound hammer and ultrasonic pulse velocity of concrete and equivalent mortar cubes are presented in Tables 4 and 5. From Table 4, it is observed that the strength of concrete with paver aggregate at 7 days is almost the same as that of the normal concrete whereas at 28 days the strength of the recycled aggregate concrete is substantially less that that of normal concrete.

This is possibly due to the low strength of paver aggregates which controls the strength of concrete at later ages. Based on the results of the trial mix concrete and equivalent mortar, the aggregate characteristic strength is determined by composite mechanics approach. Using these results, re-proportioning is done for concrete with paver aggregates for strength of 25 MPa.

To analyze the behavior of paver aggregate as coarse aggregate at different w/c ratios and to compare the same with that of natural aggregate, concrete with different w/c ratios (0.4, 0.5 and 0.6) are proportioned. The compressive strength and workability results are presented and using theses strength results w/c ratio curves are developed for both concretes as shown in Figure 1. It can be seen from Table 10 that the workability of paver aggregate is very low compared to that of normal aggregate. This is due to the movement of water from the matrix to the pores on the surface of the aggregate. The irregularity and the minor pores on the surface of the paver aggregates increase the total surface area of the aggregates. This increased surface area reduces the water available for mixing of concrete. This is the main reason for the reduction in workability. Tables 7 and 8 shows the difference in the mix proportion results between paver aggregate and normal aggregate concrete. The difference is mainly due to the reduced bulk density and specific gravity of the paver aggregate. This results in reduced weight per m^3 of concrete. The characteristic strength of paver aggregate as obtained by the use of linear law of mixture is 34 MPa. With this characteristic strength, the water-cement ratio required to obtain 25 MPa concrete is 0.80. The 7 and 28-day results are presented in Table 6. The 28-day strength is slightly less than (or very close to) 25 MPa. This suggests that the re-proportioning method can be applied to concrete which contains paver aggregates.

Non-destructive Tests

Non destructive tests based on rebound hammer number and ultrasonic pulse velocities are presented in Table 5. There is some variation in the rebound hammer results and is mainly due to the surface characteristic of the specimen. The results are affected due to different surface and internal moisture conditions of the concrete. The pulse velocity readings are more or less consistent and are more than 3500 m/s showing that the concrete cubes cast were of good quality.

DISCUSSION AND CONCLUSIONS

Mix Design as per IS: 10262 Draft Code

An attempt has been made to use the Indian draft code for mix design. This code permits the use of many admixtures (chemical and mineral). A systematic study on mix design is done and many mixes using granite and paving aggregates (which are relatively light) are designed and tested for wide range of strength and workability. Importance of the mix design based on the draft code is realised.

Concrete with Granite and Recycled Paving Aggregate

Aggregates obtained from broken paving blocks are relatively lighter and porous compared to conventional granite aggregates. Their surface is rough and their characteristic compressive strength is rather less. In spite of this, relatively good concrete can be produced from these paving aggregates.

The compressive strength of concrete with paving aggregate is relatively less compared to concrete with granite concrete. In addition, the workability of the concrete decreased drastically when paving aggregates were used which is mainly due to its surface characteristics. To compensate this, superplasticiser needs to be added. Though the workability of concrete is very low compared to granite concrete, the mix could be compacted well by vibration. In case higher workability is required, the mix should be redesigned with increased cement content in addition to the use of superplasticiser. However concrete of medium strength in the range of 20 MPa to 40 MPa can be easily produced using paving aggregates as seen from Table 10.

In case the required strength is not achieved, the mix can be redesigned using re-proportioning method. This has been demonstrated fully in this work. For this, one concrete mix with paving aggregates is re-proportioned using the results of the trial mix and this re-proportioned mix exhibited the desired strength. In this way concrete with a wide range of compressive strengths and desired with workability can be designed based the results of one trial mix. Water cement ratio curves are generated for paving aggregate and these curves can be used as a reference for further mix design for different strength. However, adjustments in the workability should be done depending on the situation.

Strength of aggregate plays an important role for the production of high strength concrete. For low strength concrete ($w/c > 0.60$, structural or non structural), the strength of aggregate is not very significant. However, for high strength concrete, the use of low strength aggregates decreases the compressive strength significantly, as seen from water cement ratio curve. However, redesigning the mix using the re-proportioning method can compensate this loss in strength. This investigation stresses the need to determine the constituent mortar strength along with that of concrete strength when unconventional coarse aggregates are used. Trial concrete and mortar mixes are proportioned at a water-cement ratio of 0.5 to obtain synergy of all concrete ingredients including the interfacial and characteristic strength of coarse aggregate. When the strength of mortar is greater than that of concrete, it is likely that failure would be predominantly by aggregate crushing. This mode of failure of concrete specimens is to be confirmed by examination of failed specimens. If it corroborates, one has to calculate the characteristic strength σ_a, by use of linear law of mixtures with known values of volume fractions of the mortar, v_m and coarse aggregate, v_a, and compressive strength of concrete, σ_c and mortar, σ_m from equation (1).

For the specified strength of concrete, which is higher than that of the characteristic strength of aggregate, the same linear law of mixtures calculates the compatible mortar strength. To obtain that mortar strength, the water-cement ratio is calculated by use of the Generalized Abrams' law using the strength value of mortar at the water-cement ratio of 0.5. With the calculated water-cement ratio, concrete and constituent mortar mixes are proportioned. The compressive strength of concrete and mortar can be checked to ascertain that the specified requirements are met. However in the present work, re-proportioning of concrete for strength less than the characteristic strength of paving aggregate is done and the application of composite mechanics is demonstrated. The paving block wastes generated in the factory or elsewhere can be recycled to produce reasonably good aggregates, which can be used to produce medium strength concrete. This has been demonstrated successfully in this work.

REFERENCES

1. POON, C S, KOU, S C AND LAM, L. Use of recycled aggregates in molded concrete bricks and blocks. Construction and Building Materials. Vol. 16, 2002. pp 281-289.

2. NAGARAJ, T S AND ZAHIDA BANU, A F. Proportioning of Concrete Mixes with Recycled Aggregate. International Conference on concrete technology for Developing Countries, 1998. Turkey.

3. KUMAR MEHTA, P. Concrete- Structure, Properties and Materials, Prentice Hall- Inc., Englewood Cliffs, New Jersey. 1985.

4. NEVILLE, A M. Properties of Concrete, ELBS and Longman Singapore, Third edition. 1989.

5. IS: 12269-1987. Specification for 53-grade Ordinary Portland Cement. Bureau of Indian Standards, New Delhi. 2001.

6. IS: 2386-1963. Method of Tests for Aggregate for Concrete. Bureau of Indian Standards, New Delhi. 1982.

7. IS: 383-1970. Specification for Coarse and Fine Aggregate from Natural Sources for Concrete. Bureau of Indian Standard, New Delhi. 1993.

8. FOSROC Product Brochure, Conplast SP 430.

9. IS: 10262, Draft "Indian Standard Recommended Guidelines for Concrete Mix Design" Bureau of Indian Standard, New Delhi, India. 2004.

10. IS: 456-2000. Indian Standard-Plain and Reinforced Concrete Code of Practice [Fourth Revision]. 2002.

11. NAGARAJ, T.S. AND ZAHIDA BANU, A. F. Generalization of Abram's law. Cement and Concrete Research. Vol. 26, No.6, 1996. pp. 933- 942.

12. NAGARAJ, T S, SASHI PRAKASH, S G AND RAGHUPRASAD, B K. Re-proportioning Concrete Mixes. ACI Materials Journal. Vol. 90, No. 1, 1993. pp 50-58.

13. NATARAJA, M C, NAGARAJ, T S AND ASHOK REDDY. Proportioning Concrete Mixes with Quarry Waste. International Journal of Cement, Concrete and Aggregates. ASTM. Vol. 23, No. 2, 2001. pp 1-7.

14. IS 1199: 1959. Methods of sampling and analysis of concrete. Bureau of Indian Standard, New Delhi. 1991.

15. IS: 516-1959. Method of Tests for Strength of Concrete [Eleventh reprint]. Bureau of Indian Standard, New Delhi. 1991.

16. IS: 13311[PT1]-1992. Methods of nondestructive testing of concrete: part 1 ultrasonic pulse velocity. Bureau of Indian Standards, New Delhi. 1999.

17. IS: 13311[PT2]-1992. Methods of nondestructive testing of concrete: part 1 Rebound hammer. Bureau of Indian Standards, New Delhi. 1999.

EXPERIMENTAL RESEARCH ON THE MECHANICAL PROPERTIES OF STRUCTURAL CONCRETE MADE WITH RECYCLED AGGREGATES

A L Materazzi

S Monotti

A D'Alessandro

University of Perugia

Italy

ABSTRACT. The results of an experimental and theoretical research on the structural use of concretes manufactured with recycled aggregate are introduced. Particularly the statistic properties of the compression strength of these concretes have been studied using a meaningful number of concrete cubes made with different percentages of recycled aggregate. On the basis of these experimental results a methodology for the calibration of the partial safety coefficient to be used with these concretes has been developed, taking as reference the results of a Level 2 reliability analysis. The first results obtained by means of a parametric experimentation allowed to underline that the use of the recycled concretes in structural elements requires a particular design procedure and cannot be limited only to quantitative prescriptions on the acceptable maximum percentage of recycled aggregate.

Keywords: Mechanical properties, Aggregate, Sustainability

Annibale Luigi Materazzi: University of Perugia, Italy – Department of Civil and Environmental Engineering (Full Professor).

Simone Monotti: Polytechnical University of Marche, Italy – Department of Materials and Environment Engineering and Physics / University of Perugia, Italy – Department of Civil and Environmental Engineering (PhD).

Antonella D'Alessandro: University of Perugia, Italy – Department of Civil and Environmental Engineering (Researcher).

INTRODUCTION

As it is known, it is possible to use aggregate derived by grinding of concrete for the partial substitution of natural aggregate in concrete production. This type of technology allows recycling prefabrication industry wastes and rubble produced during the demolition of structural concrete, with evident advantages for the environment and preservation of natural resources. The use of concrete so produced is expected to occur via a very limited number of normative national and international techniques [1-4] even if some countries, for example the United States of America, where privilege is given for instance to construction works realized by technologies respectful to the environment. It is however necessary to note that while numerous bibliographical references exist on the matter [5-10], there are only a few procedures available to calculate how to achieve the same safety level of use as with normal concrete. In any case it is evident that the mechanical behaviour of the concretes manufactured with recycled aggregate depends on the quality and quantity of the recycled aggregate used and as a consequence the safety analysis procedures used for the calculation of normal concrete structures must be adapted accordingly.

The Faculty of Engineering of Perugia has been active in a relevant research programme for a long time investigating into both theoretical and experimental aspects. The experimental part involved preparation of a statistically designed array of standard specimens in a meaningful number, containing recycled concrete, to measure the strength achieved. The analysis of the obtained results allowed determining the statistic characteristics of specimen strength relative to the percentage of recycled material, individualizing particularly the value of the mean strength of its standard deviation and of its coefficient of variation. The comparison with the results of analogous tests conducted on conventional concretes has confirmed that the mechanical strength of the material made with recycled aggregate is more scattered. In order to test the influence of this scatter on the structural safety, a reliability analysis has been performed using the β safety distance technique.

With this aim, the general programme of the research considered sections submitted to a combined compressive and bending stress. These sections were made of reinforced concrete, with typology and recurrent dimensions. In the present communication the first results of this part of the study are shown with reference to a transversal only section and at a range of eccentricity of the applied load. The probabilistic analysis was carried out in the double hypothesis to produce the above mentioned structural elements with normal concrete and with recycled concrete, all having the same characteristic strength. As reference, we used the probability of failure value applicable to the partial safety coefficient of normal concrete as indicated by the Italian standard. In this way, we could determine the partial coefficient which allows designing sections with the same reliability and, vice-versa, the required increase in characteristic strength of the concrete with recycled material content to obtain sections of the same performance as with normal concrete.

INTERNATIONAL SPECIFICATIONS

European Union

As early as the 1970s, with the directive 75/442 CEE, the sensibility towards the problem of construction and demolition wastes (C&D) began to gain certain importance in the European Community [11]. Since then work groups such as the "Construction and Demolition Waste

Project Group" or the "Demolition and Reuse of Concrete" of RILEM (International Union of Laboratories and Experts in Construction Materials, Systems and Structures) are preparing the editing guidelines in order to encourage the states to adopt tools and initiatives to implement the correct politics for the management of C&D wastes.

Some European countries, particularly Denmark, Holland, Belgium and Germany, have been working for some years on studies how to utilise recycled materials of demolition. In these countries there are laws and prescriptions in force which impose recycling of materials. Germany is the greatest producer of C&D waste in the European Union. In this country, there has been approved a standard in 1995 which defines the requirements for the use of construction wastes and recycled demolition materials. Another standard has been elaborated in 1998 dealing with the use of recycled aggregate in concrete. In Holland, the scarce availability of aggregate sources and landfill spaces provide incentives for the recovery and recycling of building wastes much more than anywhere else in Europe, hence the partial substitution of natural aggregates (gravel) in concrete production exceeds 20% in the country [12]). In Spain a group of study has been created within the Spanish Standard Technical Committee AEN/CTN-146 "Aridos" regarding the recycling of demolition materials as aggregates for new concrete. Simultaneously there has been a large-scale project concerning the planning, management and incentive of activities of recycling. In addition, several technical recommendations on the use of the recycled aggregate have been developed and emanated by the CUR (Commissie voor Uitvoering van Research) in Holland in 1986, by the NIIzbh (Research Institute Concrete for and Reinforced Concrete) in Russia in 1984 and by the DCA (Danish Concrete Association) in Denmark in 1989. The Dutch standards (CUR No. 4 and 5) and the Danish one (DCA No. 34) seem to allow direct use of recycled aggregate also in structural concrete.

The United States of America and Japan

In the United States of America the ASTM (American Society for Testing Materials) defined some requirements in 1979 (ASTM C125) and in 1982 (ASTM C33) for the acceptance of the material coming from structural demolition, as aggregate to make new concrete. Subsequently the United States Army Corps of Engineers has modified its own specifications and recommendations in order to promote the use of recycled concretes. The consequence of this is that there exists no particular limit in the United States for the use of recycled aggregate in concrete.

The use of recycled aggregate has been the object of a series of recommendations and proposals of law also in Japan. Recently, in 1998, the Japan Concrete Institute created a joint technical committee to establish a new plan of law for recycled materials in concrete (JIS/TR). Research activity in Japan on C&D materials has been developed during the past 25 years largely through an extensive national project "Research for the Future" with the aim to set practical methods for the production of structural concrete starting from recycled aggregate of large and small sizes and a system of recycling in order to reduce the related costs with the life cycle of construction materials.

Italy

In comparison with other European countries, there is a deep gap to be noticed in the use of recycled materials in Italy: only 9% of wastes are recovered and over 90% are used without any revaluation. In other countries, the unavailability of raw materials and the excessive production of demolition rubble to be used have led to the development of recycling processes. Recently, nevertheless, greater attention has been given to the problems of sustainable development and to environmental impact also in Italy, with the demand to raise awareness of renewable raw materials as aggregate sources. The first experiences of application of recycling of materials from C&D origin were at the end of the '80s. The Italian standard has followed the evolution of the European legislation with the law 22/77 (Ronchi Decree) by introducing the priorities of prevention, recovery and discharged in the management of the wastes. In the last three years the Italian legislation has taken important steps about the recycling of rubble and demolition waste as aggregate for new concrete. In 2002 the Technical Committee CEN/TC has compiled the standard EN 12620 (UNI EN 12620:2002 in Italy) "Aggregates for Concrete" that, unlike the preceding UNI 8520, also considers aggregates or fillers of artificial origin or from recycling, not only from natural sources. An impulse to the use of recovered materials, including aggregate, was given by the Decree May 8th 2003, No. 203 containing: "Norms so that the public offices and the societies in predominantly public capital cover the annual requirement of manufactured articles and goods with a ratio of products obtained from material recycled in the non inferior measure to the 30% of the same requirement" and from the circular of the Ministry of the environment and the defence of the Territory No. 5205/2005. The D.M. 203/2003 has also founded the "Repertoire of the Recycling", with a series of technical prescriptions, with regard to the use and the characteristics of the admissible materials. More recently, the Technical Standards for the Constructions 2005 [13] have introduced the possibility to use recycled aggregates for manufacturing concrete, also for structural use, provided that the mixtures are documented and proven through fit for purpose tests in the laboratory. The standard defines, for orientation, the percentages of recycled aggregate to use for given final strength values of the obtained concrete, and it prescribes the control of the processes of recycling. However, there is no guidance given regarding the specific calculation to apply when concretes with recycled aggregate are made and used.

NOTES ON THE METHODS OF EVALUATION OF SAFETY

Methods of Level 2

Unlike complete probabilistic methods, Level 2 methods use, as safety assessment parameter, the so-called "safety index" β, tied to the probability of failure from the relationship:

$$P_f = \Phi(-\beta) = \int_{-\infty}^{\beta} N_{0,1}(x)\,dx \qquad (1)$$

in which Φ it is the function of division of the variable aleatory normal standard [14].

The constituents of the specific problem of eccentric pressure in sections of reinforced concrete are, in this sense, modelled as n independent aleatory variables, for convenience in an opportune vector X. The possible statistic dependence among some of them can be removed by using appropriate procedures, as well as density distributions different from that of normal probability can be considered [15, 16].

The limit state considered is therefore obtained as expressed through an appropriate function of limit state G(X) = 0. The space R^n in which the aleatory variable of base are defined is, therefore, divided by the function of limit state in two dominions:
- the dominion of failure U in which it occurs G(X) < 0;
- the safety dominion S in which it occurs G(X) > 0.

It is comfortable to study the problem, instead of the original aleatory variable X, by using the opportune variable aleatory normal standards auxiliary Y. To such purpose the aleatory variables X are projected in the space of the variable aleatory normal standards through the relationship:

$$Y_i = \frac{X_i - \bar{X}_i}{\sigma_{X_i}}$$
(2)

in which $\bar{X}_i$ is the middle value of the i aleatory variable and σ_{X_i} is the corresponding standard deviation. Similarly the function of limit state is projected in the same space obtaining:

$$Z(\mathbf{Y}) = 0$$
(3)

The new condition of limit state can be developed in Taylor series which has the initial point in the origin, stopping, in first approximation, to the terms of the first order:

$$Z(\mathbf{Y}) \cong Z(\mathbf{Y})_0 + \boldsymbol{\alpha}^T \mathbf{Y} = 0$$
(4)

in which

$$\alpha_i = \left[\frac{\partial Z}{\partial Y_i} \right]_0$$
(5)

we observe, therefore, that

$$Z(\mathbf{Y})_0 + \sum_n \alpha_i Y_i = 0$$
(6)

is the equation of an hyperplane, tangent to the surface Z(Y) in the point of least distance from the origin (point of projection). Such distance is really the value of β, which assumes the notable geometric meaning of ray of the tangent sphere to the surface of limit state:

$$\beta = \min\left\{ \sqrt{\sum_n Z(Y_i)^2} \right\}$$
(7)

At this point the value of β can be determined by using any technique of optimization. Examples of applications of this method for structural elements made in reinforced concrete can be found, for example, in [17-19].

Methods of Level 1

The methods of Level 1 represent the version more simplified of the procedures of evaluation of the safety and they are those incorporated in the normative techniques. They are semi-probabilistic methods, because they exclusively use the characteristic values of the aleatory variable in question and opportune partial safety coefficients.

The problem of the measure of safety reduces therefore to the control of the inequality:

$$\gamma_S \cdot S_{k(0.95)} \leq \frac{R_{k(0.05)}}{\gamma_R}$$
(8)

in which $S_{k(0.95)}$ and $R_{k(0.05)}$ are the characteristic values of the effect of the actions and the strengths, respectively, and γ_S and γ_R are the partial safety coefficients on the actions and on the materials.

The role of the partial safety coefficients is that to decrease the implicit value of the probability of failure in the use of the only characteristic values. For such purpose these are required to be opportunely "calibrated" by using superior level analysis methods.

In the present job, a methodology is pointed out to for the calibration of the partial safety coefficients to use in the verification of sections on a combined compressive and bending stress which are obtained with recycled concretes.

EXPERIMENTAL WORK

Programme of Work

Present research has been articulated in various phases, through experimental tests and theoretical investigations. In the experimental part, the aggregate to be recycled was first crushed. The material used comprised reject precast reinforced concrete elements, stripped from steel and having a characteristic compressive strength at start, equal to $R_{ck} = 55$ N/mm^2.

The crushed aggregate was sifted from fines and was then further separated to different size fractions. The material used for making the recycled conglomerate fell within the inclusive dimensions between 0 and 30 mm. Subsequently, the particle size distribution curve and water content were determined, both for recycled and natural aggregates through standard procedures of measurement, washing, desiccation and weighing. The recycled aggregate had 4.6% moisture content.

Consequently, there were four different mix designs characterized by different percentages of recycled aggregate: with 0% recycled aggregate, with 30% large size recycled aggregate, with 50% and 100% recycled aggregates. Standard 150 mm cubic specimens were made. Additional specimens prepared were: 20 specimens with 100% natural aggregate, 20 with 50% recycled aggregate of dimensions 0 to 30 mm, 10 specimens with 30% large size recycled aggregate (>5 mm) and finally 8 specimens with 100% of recycled aggregate. In the mix design with 30% recycled aggregate the largest size used was only 5 mm. The aggregate proportions were selected to reflect an opportune granulometric assortment in the mix design: for the small size part, sand between 0 and 4 mm was used, while for large aggregate, sizes between 5 and 12.5 mm were used.

The compositions of the different mix designs are reported in Table 1. The four mix designs contained the same quantity of cement with the aim to get the same workability. To such purpose in the mix design with recycled aggregate the dosage of superplasticizer has been increased slightly from 0.8% to 1% of the weight of cement. Nevertheless, the mixes with recycled aggregates needed more water in comparison to normal concrete, especially in the case of 100% recycled aggregate. The specimens were compacted using a vibrating table, removed from the moulds after one day and cured in 20°C water until test age.

Prior to testing, the specimens were capped according to the standard UNI EN 12390 [20]. Before the compressive strength test, geometric and physical characteristics of the specimens were determined as required by the standard UNI 6132-72 [21], including linear dimensions, section, mass, preventive operations and possible defects. The tests were carried out with standard instrumentation and showed classical failure for all specimens, characteristic of a good concrete (bipyramidal failure surfaces).

Table 1   Mix design of various mixtures of concrete, kg/m^3

RECYCLED AGGREGATES	CEMENT kg	WATER kg	W/C	RECYCLATE kg	SAND kg	SIZE 1 kg	SUPER-PLASTICIZER, kg
0%	400	160	0.40	0	830	830	3.2
30%	400	170	0.42	498	830	332	4
50%	400	172	0.43	830	747	83	4
100%	400	180	0.45	1660	0	0	4

Table 2   Statistics propriety of compressive strength

RECYCLED AGGREGATES	MEDIUM, N/mm^2	STANDARD DEVIATION, N/mm^2	V	R_{min}, N/mm^2	R_{max}, N/mm^2
0%	60.6	46	0.076	50.0	68.5
30%	49.3	41	0.084	41.0	53.0
50%	41.5	39	0.095	34.5	49.5
100%	38.1	41	0.107	32.0	44.0

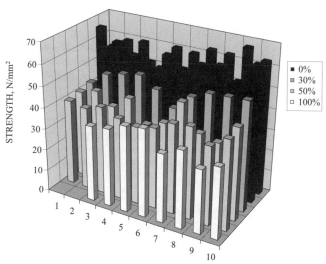

Figure 1   Experimental results of compressive strength tests

Statistic Analyses of the Results

The analysis of the obtained results has allowed, relatively to the percentages of the recycled material used and to the characteristics of the initial concrete, to determine the statistic characteristics of mechanical strength, by individualizing, particularly, the value of the middle strength, of the standard deviation and the coefficient of variation.

The results underlined a diminution of the middle strength of concretes, when recycled aggregate was present in the mixes, from 60.6 N/mm^2 in the case of 100% natural aggregate, to 38.1 N/mm^2 for 100% recycled aggregate. In the same way, the coefficient of variation (V) increased from 0.076 for the natural aggregate, to 0.107 for the recycled aggregate (Table 2). Such results are in line, as regards the coefficients of variation of strength, with those reported in [5] in which the authors analyzed the statistic proprieties of compressive strength of normal concretes and recycled concretes, projected to equal characteristic strength.

EMPLOYMENT OF THE EXPERIMENTAL RESULTS IN THE PLANNING OF REINFORCED CONCRETE SECTIONS

General Considerations

The obtained experimental results were used for the planning of sections on combined compressive and bending stresses realized with recycled concrete. Two alternative approaches were followed:

- determine, naturally in the limits allowed by the available experimental data, the partial safety coefficient of the concrete, γ_C, to use in order to plan a section that guarantees the same safety level of one realized with normal concrete of the same characteristic strength R_{ck};
- determine the necessary increase of the characteristic strength so that a recycled concrete guarantees the same safety level of a normal concrete of assigned characteristic strength.

The first approach corresponds, evidently, to a typical problem of calibration of a partial safety coefficient, in this case of that of the concrete, whose coefficient is opportunely increased for neutralizing the negative effects of the greatest deviation of strength that is observed in recycled concretes.

The second approach, in turn, can be taken as a doublet of the first one and it consists of the determination of a table of correlations between the characteristic strength of normal concrete and that of recycled concrete, built in such a way that two corresponding concretes, one normal and the other recycled, guarantee the same performance i.e. the same ultimate strength (by calculation) and the same safety.

This requires that an increased value of the characteristic strength of concrete R^*_{ck} to be used simultaneously with an additional safety coefficient γ^*_C. All that in order to have a ratio R^*_{ck}/γ^*_c equal to the characteristic strength value of the corresponding normal concrete, which can be regarded as the maximum compressive strength of concrete:

$$\sigma_{c,ult} = 0.85 \frac{0.83 R_{ck}}{1.6} \tag{9}$$

The above equation is applicable both to normal concrete and, with a higher characteristic strength γ^*_C, also to recycled aggregate concrete.

Considered Cases and Hypothesis of Base

The experiments were carried out on a rectangular section of reinforced concrete with the dimensions of b = 300 mm and H = 500 mm, reinforced with 4 + 4 Ø20 (As' = As = 1257 mm^2) (Figure 2). The concrete cover was 40 mm.

Three different conditions of solicitation were considered:

   1) compressive stress: e = 0;
   2) combined compressive and bending stress: e = H/4;
   3) combined compressive and bending stress: e = H/2.

Conditions of solicitation with greater eccentricity were not taken into consideration since preliminary investigations have shown that in such cases the load baring capacity of structural elements is only weakly influenced by the coefficient of variation of the concrete. The deprived concrete of recycled aggregate considered was of C25/30 (R_{ck} = 30 N/mm^2) class, while the steel of reinforcement was FeB44k (fyk = 430 N/mm^2). The partial safety coefficient used in the calculations for normal concrete was γ_C = 1.6 and for steel the partial coefficient was γ_S = 1.15.

In the course of the reliability analyses three aleatory variables were considered: the compression strength of concrete, the tensile strength of steel and the normal stress or the bending moment of projection. The statistic proprieties of the strength R_c of the deprived normal concrete of recycled aggregate (R_{ck} = 30 N/mm^2) and of those of the steel f_y are given in Table 3. For the external solicitations a normal distribution has been assumed with a variation coefficient equal to 0.30.

Compressive stress: e = 0

The load bearing capacity, calculated according to Level 1, was N_{ult} = 2942.2 kN. From such value that corresponds to N_d, the middle value of normal load of projection $N_{d,m}$ = 1306.2 kN and the corresponding standard deviation, equal to 391.9 kN were obtained. The equation of limit state assumes the form:

$$0.85 \cdot 0.83 \cdot R_c \cdot b \cdot H + f_y \left(A_s + A_s{'} \right) - N_d = 0 \tag{10}$$

Using the aleatory variables, reported in Table 3, in the above equation, the limit state was obtained as index β equal to 7.230.

Combined compressive and bending stress: e = H/4

In correspondence of the eccentricity e = H/4 = 125 mm, the point that individualizes the collapse to Level 1 on the dominion of interaction N-M of the section has coordinates N_{ult} = 1737.8 kN and M_{ult} = 217.2 kNm. Since eccentricity is constant, the bending moment is considered as the unique aleatory solicitation. Its middle values, therefore, are equal to $M_{d,m}$ = 97.0 kNm, which corresponds the middle value of the normal load $N_{d,m}$ = 776.3 kN.

The collapse of the section happens in this case with the procedures of the field 4 (compressed concrete with maximum deformation 3.5‰ and tense steel not yielded). The distance of the neutral axis from the compressed area and the equation of limit state is expressed from the following system, with the index of reliability obtained to be 7.476.

$$0.5715 \cdot R_c by + A_s{'} f_y - 0.0035 \cdot \frac{(d-y)}{y} \cdot EA_s - \frac{M_d}{e} = 0 \tag{11}$$

$$0.5715 \cdot R_c by \cdot \left(\frac{H}{2} - 0.416y \right) + A_s{'} f_y \cdot \left(\frac{H}{2} - d{'} \right) +$$
$$+ 0.0035 \cdot \frac{(d-y)}{y} \cdot EA_s \cdot \left(\frac{H}{2} - d{'} \right) - M_d = 0 \tag{12}$$

Combined compressive and bending stress: e = H/2

In relation to the eccentricity and H/2 = 250 mms, the following characteristic values of solicitation of the projection were identified N_d = 1163.3 kNs and M_d = 290.8 kNms, to which a middle value of solicitation equal to $M_{d,m}$ = 129.9 kNms and a middle value of normal $N_{d,m}$ = 519.6 kNs correspond. Since the collapse of the section takes place according to Level 1 also in this case, with the maximum deformation of compressed concrete being 3.5‰ and the steel in tension not yielding, the limit state condition can be expressed from Equations 11 and 12. The coefficient is 7.339.

Calibration of the Partial Safety Coefficient of the Concrete

Methodology used

On the basis of the obtained experimental results, the partial safety coefficient was determined whose utilization leads to the planning of a recycled concrete section that has the same probability of failure as the corresponding section made with normal concrete of the same characteristic strength R_{ck}.

Table 3   Characteristics of the aleatory variables common to all cases

VARIABLE	DISTRIBUTION	MEDIA	STANDARD DEVIATION	V
f_y, N/mm^2	Lognormal	468.45	23.42	0.05
$R_{c,0}$, %	Lognormal	34.28	2.615	0.076

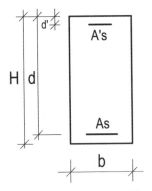

Figure 2   Definition of the geometry of the section

Table 4   Partial safety coefficients for recycled concretes

RECYCLED AGGREGATES	E = 0		e = H/4		e = H/2	
	$A_s = A_s'$ mm^2	γ_C	$A_s = A_s'$ mm^2	γ_C	$A_s = A_s'$ mm^2	γ_C
0%	1257.0	1.6000	1257.0	1.6000	1257.0	1.6000
30%	1358.1	1.6634	1287.5	1.6228	1274.6	1.6172
50%	1520.6	1.7765	1333.2	1.6580	1302.5	1.6450
100%	1729.0	1.9462	1387.2	1.7012	1336.7	1.6798

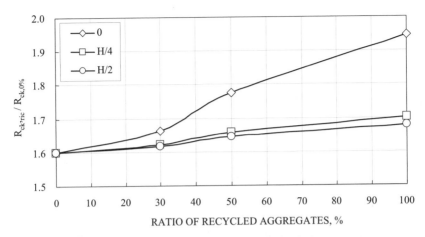

Figure 3   Partial safety coefficients for recycled concretes

With reference to the section represented in Figure 2, the adopted methodology of calibration was based on a reliability analysis to Level 2, carried out by calculating index β by the procedure proposed by Rackwitz and Fiessler [16], which also allows taking account of not normal distributions. Solution of the problem for minimum was obtained by using known algorithms from the literature [22]. This methodology comprises the following phases:

1) a section of reinforced concrete is assigned through first of all its geometry and its reinforcement as well as the constituent materials, resulting in a normal characteristic strength R_{ck} and steel strength f_{yk};

2) preset the value of the eccentricity and the solicitations of calculation, the value of the maximum moment of projection M_d and the normal load of projection N_d compatible with the strength of the section are determined;

3) modelling the applied loads, the strength of normal concrete and that of the steel as aleatory variable determine the value of safety index β which corresponds to the section in examination. Such prefixed value of β is assumed as reference for the operation: $\overline{\beta} = \beta$;

4) arriving therefore to the use of concrete compositions with a preset percentage of recycled aggregate but having the same characteristic strength as normal concrete;

5) the partial safety coefficient value of the concrete is increased: $\gamma_c^* = \gamma_c + \Delta\gamma_c$;

6) by using the usual algorithms of level 1, the metallic reinforcement section is updated with the aim to keep the load bearing capacity of the section constant ($M_d = N_d$);

7) with reference to the section so modified and keeping in mind the statistic properties of recycled concrete used, the safety index value β* is determined;

8) checking whether the safety index of the section with recycled concrete reaches the same safety index as with normal concrete. If so, the procedure ends and γ*c is the sought after value of the partial safety coefficient: $\gamma_c = \gamma_c^*$;

9) in opposite case, an iterative procedure starts, which involves the repetition of the procedure beginning from point 5.

Numerical results

The application of the procedure described before has led to the determination of the partial safety coefficients given in Table 4 and in Figure 3. For every level of recycled aggregate content, the experimentally determined strength variation coefficients were used.

Correlation between Strengths of Normal Concrete and Recycled Concrete

Parallel to the analyses of calibration of the partial safety coefficient γ_C to be applied to the compressive strength of recycled concrete, a correlation table was set up comparing normal concrete and recycled aggregate concrete so that the latter can replace natural aggregate concrete without any alteration to the reliability of the structural element.

Methodology used

The procedure used to create the correlation table was similar to the one explained in section "Calibration of the Partial Safety Coefficient of the Concrete", with the differences that in point 4 the concrete composition is adjusted to contain the same quantity of reinforcement; whereas in point 5 it is the characteristic strength value of concrete which is increased: $R_{ck}^* = R_{ck} + \Delta R_{ck}$. Similarly, if the safety index of the section with recycled concrete reaches the same safety index of the original normal concrete, the procedure ends and $R_{ck} = R_{ck}^*$ is the sought after value as the required characteristic strength for recycled aggregate concrete. In opposite case, again, an iterative procedure starts, which involves the repetition of the procedure beginning from point 5.

Numerical results

The application of the procedure described before has led to establish a correlation between the strengths of normal concrete and that of concrete manufactured with recycled aggregate. The results are given in Table 5 and Figure 4. For every level of recycled aggregate content, the experimentally determined strength variation coefficients were used.

CRITICAL ANALYSIS OF THE OBTAINED RESULTS

The examination of the numerical results shows that the partial safety coefficient for the concrete is always greater than the value prescribed by the Italian standard for reinforced concrete and, depending on the section considered, can also exceed the value 2.00. Moreover, it increases especially with the percentage of recycled aggregate.

We can observe that the partial coefficient is a function of the eccentricity of the external load: it is greater in the case of centred compression and tends to decrease progressively as soon as the eccentricity increases.

This behaviour evidently depends on the role developed by the strength of concrete in comparison to the steel armour regarding the load bearing capacity of the section. Naturally, since it is complicated to use varying partial coefficients with the eccentricity of external loads, the highest value shall be taken as reference.

Table 5   Correspondence between the strength of recycled concretes

RECYCLED AGGREGATES	$E = 0$		$e = H/4$		$e = H/2$	
	R_{ck}	$\dfrac{R_{ck,0\%}}{R_{ck,ric}}$	R_{ck}	$\dfrac{R_{ck,0\%}}{R_{ck,ric}}$	R_{ck}	$\dfrac{R_{ck,0\%}}{R_{ck,ric}}$
0%	30.000	1.0000	30.000	1.0000	30.000	1.0000
30%	31.187	1.0396	30.377	1.0126	30.290	1.0097
50%	33.588	1.1196	30.968	1.0323	30.766	1.0255
100%	38.020	1.2673	31.721	1.0574	31.373	1.0458

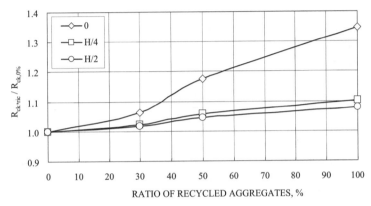

Figure 4   Correspondence between the strength of recycled concretes

An entirely analogous running is observed as in Figure 3. This confirms the idea that the structural use of recycled aggregate requires appropriate adjustment of the formalities of calculation that considers the specific mechanical properties of recycled concrete.

CONCLUSIVE OBSERVATIONS

In present communication the first results of an on-going theoretical-experimental research work were presented, as carried out at the University of Perugia on the use of recycled aggregate in structural concrete production. The experimental investigation included the determination of the statistic properties of compressive strength in a meaningful number of standard specimens made with recycled concrete in various percentages. This underlined sensitive differences in comparison to concrete manufactured with natural aggregate. Particularly, the mechanical strength results were lower and affected by greater scatter. Particular attention was paid to study the correction factors in order to guarantee the same reliability as for normal aggregate concrete. For this purpose, a methodology to calibrate the partial safety coefficient is proposed, based on the reliability analysis conducted to Level 2.

Notwithstanding the results obtained so far are limited to a number of load conditions and reduced typologies, they allowed to stress that structural use of these concretes requires opportune changes in the formalities of calculation and it cannot limit itself to quantitative prescriptions on the maximum recycled aggregate content to use as a function of the required concrete strength.

REFERENCES

1. DIN 4226-100:2002 Aggregates for concrete and mortar - Part 100: recycled aggregates, Berlin.

2. JIS A5021:2005 Recycled aggregate for concrete-class H, Tokyo.

3. JIS A5023:2006 Recycled concrete using recycled aggregate class L, Tokyo.

4. JIS TS A0006:2004 Concrete using recycled aggregate, Tokyo.

5. XIAO L, LI J AND ZHANG C, On statistical characteristics of the compressive strength of recycled aggregate concrete, Structural Concrete, Vol. 6, No. 4, 2005, pp. 149-153.

6. DIOTALLEVI P P, SANDROLINI F AND COSTANTINO A B, Calcestruzzo strutturale con aggregati naturali e riciclati: indagini teoriche e sperimentali per la sostenibilità, Atti Giornate AICAP 2004, Patron, Bologna, 2004, pp. 181-185.

7. OIKONOMOU N D, Recycled concrete aggregates, Cement & Concrete Composites, Vol. 27, 2005, pp. 315-318.

8. AJDUKIEWICZ A AND KLISZCZEWICZ A, Influence of recycled aggregates on mechanical properties of HS/HPC, Cement & Concrete Composites, Vol. 24, 2002, pp. 269-279.

9. XIAO J, SUN Y AND FALKNER H, Seismic performance of frame structures with recycled aggregate concrete, Engineering Structures, Vol. 28, 2006, pp. 1-8.

10. DOS SANTOS J R, BRANCO F AND de BRITO J, Mechanical properties of concrete with coarse recycled aggregates, Structural Engineering International, IABSE, Vol. 3, 2004, pp. 213-215.

11. BRESSI G, Recupero di risorse dai rifiuti da costruzione e demolizione: problematiche e prerogative, Atti Convegno "Il recupero di materiale ed energia dai rifiuti", Gubbio 14-15 febbraio, 2002.

12. GIMÉNEZ C AND URSELLA P, Esperienze di laboratorio nel possibile riutilizzo dei materiali provenienti da demolizioni, Recycling, Vol. 1, 1998

13. D.M. 14 Settembre 2005 - Norme Tecniche per le Costruzioni.

14. HASOFER A M AND LIND N, An exact and invariant first-order reliability format, Journal of Engineering Mechanics, 100 (EM1), 1974, pp. 111-121.

15. ROSENBLATT M, Remarks on a Multivariate transformation, Annals of Mathematical Statistics, Vol. 23, No. 3, 1952, pp. 470-472.

16. RACKWITZ R AND FIESSLER B, Structural reliability under combined random load sequences, Computer and Structures, Vol. 9, 1978, pp. 489-494.

17. GIUFFRÈ A AND GIANNINI R, Il metodo della distanza di sicurezza per il progetto probabilistico a livello 2, Giornale del Genio Civile, 1978, pp. 159-180.

18. KASZYNSKA M, LAUMET P AND NOWAK A S, Material quality and reliability of beams in flexure and shear, Millpress, Rotterdam, Proc. ICOSSAR, 2005, pp. 687-694.

19. SZERSEN M M, NOWAK A S AND SZWED A, Reliability-based sensitivity analysis of RC columns resistance, Millpress, Rotterdam, Proc. ICOSSAR, 2005, pp. 2525-2530.

20. UNI EN 12390:2002 Prova sul Calcestruzzo indurito.

21. UNI 6132-72:1972 Prova distruttive sui Calcestruzzi. Prova di Compressione.

22. NOWAK A S AND COLLINS K R, Reliability of Structures, McGraw-Hill, 2000.

OIL SHALE AS AN AGGREGATE IN MAKING CONCRETE MIX

B Z Mahasneh

Mu'tah University

Jordan

ABSTRACT: This paper presents the benefits gained from using oil shale in concrete technology as part of the mixing aggregate. The effect of oil shale on both compressive strength and tensile strength has been investigated. Results from several cylinder specimens as well as cube specimens have been presented in this study. The results presented herein indicated that the strength of specimens having oil shale aggregate is controlled by the composite action of both oil shale and natural aggregate that were used to make the concrete. The presence of shale in aggregate enhances the composite material properties. A comparison between the usage of natural aggregate and oil shale aggregate was investigated in this study, and the results have shown satisfactory improvement.

Keywords: Oil shale, Oil shale aggregate, Plain Concrete, Splitting, Compressive strength, Composite material.

Bassam Mahasneh, Ph.D. Associate Professor of Geotechnical Engineering. Ph.D. in Pile foundation, Zhe Jiang University, PRC 1998. Master of engineering in pile foundation, Tongji University, PRC 1994. Bachelor of Science in Civil engineering, Jordan University 1990.

INTRODUCTION

Jordan has a very large resource of oil shale, especially in the southern part; Al-Lajjon district, Al-Kerak area. Various demonstration projects have been carried out using oil shale as a raw material to provide crude oil and fly ash after being burned.

Fly ash comes from the retort residue of oil shale used to utilise its pozzolanic activity [1]. Tests on compressive strength of cement and lime mortars blended together indicated that up to 20% of cement could be replaced with ash used as an additive.

In literature, some researches investigated the potential use of Jordanian oil shale ash as raw material or additive to Ordinary Portland Cement mortar and concrete [2]. Results showed that the replacement of cement, sand, or both by an about 10% by weight would yield the optimum compressive strength, and its replacement of cement by up to 30% would reduce its compressive strength. It was found that Jordanian oil shale ash on its own possesses a limited cementitious value contributing to the strength of mortar and concrete.

The composition and the physical properties of oil shale ash pastes were discussed and compared with the corresponding properties of Ordinary Portland Cement paste [3, 4]. In the ash paste, the porosity remained unchanged with time, and the mechanical properties remained satisfactory. Usually, the inorganic portion of the oil shale is high in silica; therefore, the by-product possesses pozzolanic properties.

The effect of addition of fly ash and burnt clay pozzolana on the engineering properties of cement composites was discussed [5]. It was noticed that Portland cement based mortars though harden rapidly and attain high strength but possess relatively poor early age properties. Composite mortars incorporate lime along with Portland cement whose presence improves early age properties. Laboratory investigations revealed the possession of good early age properties and better strength at later ages.

Cementing properties of oil shale; moist curing of cast and compacted samples were discussed [6,7]. The investigations made with bed and cyclone oil shale ash samples prepared in a pilot plant fluidized bed at 650°C and 750°C. Relatively high compressive strength was obtained in compacted samples by controlling the water/cement ratio.

The development of pozzolana and the development of a carrier fluidifying agent to control the loss of slump for high strength concrete with shale ash were discussed [8]. An increase in the compressive strength from 5% to 10% for 10% replacement of cement with oil shale fly ash, and an increase in the compressive strength of 10% to 15% at 56 days was noted.

The variability of fractional, chemical, and mineralogical composition of oil shale fly ash was investigated [9]. The investigations showed that cement could be substituted by oil shale fly ash, arising as the waste of burning oil shale dust at power plants.

Excavations left huge amounts of unusable oil shale piles. The use of these piles in industry is not commercial. The aim of this study is to investigate the feasibility of using oil shale as aggregate in concrete. The parent (oil shale rock) brought from the Al-Lajjon Quarry and crushed into different grain sizes using a jaw crusher and the required aggregate were classified based on ASTM classification.

The crushed aggregates were classified based on coarse aggregate passing sieve size 18.75 mm and retained on sieve size 12.5 mm, medium aggregate passing sieve size 12.5 mm and retained on sieve size 9.375 mm, and fine aggregate passing sieve No.8.

TEST SETUP

In this work, the effect of oil shale as an aggregate on both compressive strength and tensile strength is presented. Laboratory results of several cylindrical test specimens of different sizes and concrete cubes are given. The test was divided into two main sets. The first set used oil shale aggregate as filling material while the second one used plain concrete as filling material. Furthermore, different grain size portions (oil shale aggregate) were used to investigate the effects on the characteristics of the concrete mix made by oil shale aggregate. The composite concrete mix was divided into three main design mixes. In the first mix, coarse lime aggregate was replaced with coarse oil shale aggregate of the same size. For the second mix; medium lime aggregate was replaced with medium oil shale aggregate. Fine lime aggregate was replaced with fine oil shale aggregate in the third mix. Cylindrical specimens of a diameter of 100 mm and height of 150 mm were studied. Two specimen lengths were used; 200 and 300 mm as well as normal cubes of 150 mm. A lean concrete mix was used in the test. In the test setup, plain and composite concretes were subjected to the same environment to reveal the effect of oil shale aggregate on composite concrete strength at different grain sizes used in the mix.

Strength of both plain and composite concrete was recorded for various grain sizes of oil shale aggregate: coarse, medium and fine. Each time, the specified oil shale grain size was selected and added to the concrete mix to replace the same size of that for plain concrete mix. A fixed rate of loading of 1.33×10^{-3} MN/s was applied for all specimens in this test.

OIL SHALE SPECIFICATIONS

The oil shale parent rock was brought from the Al-Lajjon district, Al-Kerak area and crushed into different grain sizes using a jaw crusher. The specified grain sizes of the crushed material were as follows: coarse aggregates passing sieve size 18.75 mm and retained on sieve size 12.5 mm; medium aggregate passing sieve size 12.5 mm and retained on sieve size 9.375 mm; and fine size aggregate passing sieve No. 8.

The specific gravity of oil shale was equal to 1.953 g/cm^3, the impact value was equal to 28.3%, and the absorption value was equal to 4.36%, whereas the specific gravity, the impact value, and the absorption of lime aggregate were: 2.64, 21.4%, and 1.13%, respectively.

MIX DESIGN

The following mix proportions were used to produce the plain and the composite concrete: coarse aggregates weighing 28.5 kg passing sieve size 18.75 mm and retained on sieve size 12.5 mm; medium aggregate of 28.5 kg passing sieve size 12.5 mm and retained on sieve size 9.375 mm; and fine size aggregate of 38 kg passing sieve No. 8; cement of 14.0 kg; and free water of 11.75 litre. Calculation were carried out to find out the unit weight of each mix and arranged as follows: The plain concrete unit weight was 2385 kg/m^3. Replacing the coarse aggregate with oil shale decreases the unit weight to 2133 kg/m³. But replacing the medium aggregate with oil shale reduces the unit weight to 2088 kg/m³, and a distinguished reduction in the unit weight was very clear using fine oil shale for lime stone to a value equal to 18899 kg/m³. Cement weight for each batch, even when replacing the aggregate type, remained constant to figure out the effect of replacing limestone aggregate with oil shale on concrete strength.

COMPRESSION TEST

Strength of 150 mm cubes was tested in compression according to ASTM, for both plain and composite concretes in which coarse lime aggregate was replaced with oil shale aggregate. Figure 1 (a, b, and c) presents the effect of replacing the coarse, medium and fine lime aggregate with coarse, medium and fine oil shale aggregate at test ages of one, two, three and four weeks. Test results were compared at each test age.

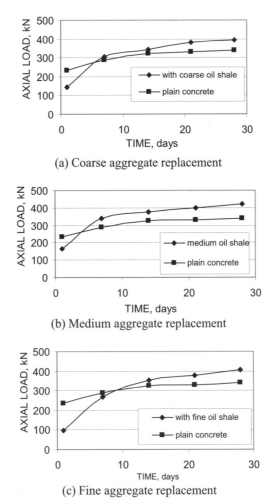

(a) Coarse aggregate replacement

(b) Medium aggregate replacement

(c) Fine aggregate replacement

Figure 1   Failure loads of 150 mm cube concretes with and without oil shale aggregate

SPLITTING TEST

Splitting test was carried out according to the ASTM method with different cylinders, Ø150 mm × 300 mm and Ø100 mm × 200 mm in dimension. The concrete was made with and without replacing the lime aggregate with oil shale aggregate. Figure 2 (a, b, and c) presents the effect of replacing lime aggregate with oil shale aggregate on the tensile strength of concrete at 1, 2, 3, and 4 weeks.

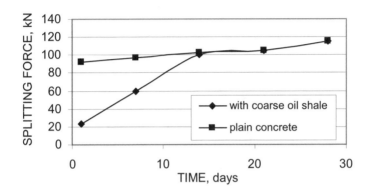

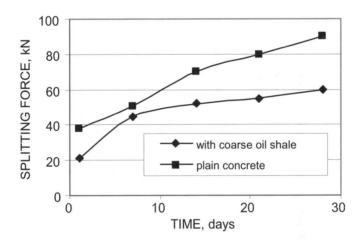

(a) Coarse aggregate replacement
Ø150 mm × 300 mm (top) and Ø100 mm × 200 mm (bottom)

Figure 2   Splitting force at failure for concrete with and without oil shale aggregate, cont'd

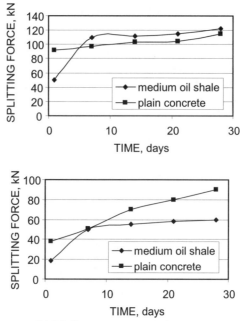

(b) Medium aggregate replacement
Ø150 mm × 300 mm (top) and Ø100 mm × 200 mm (bottom)

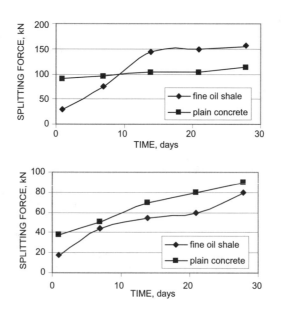

(c) Fine aggregate replacement
Ø150 mm × 300 mm (top) and Ø100 mm × 200 mm (bottom)

Figure 2   Splitting force at failure for concrete with and without oil shale aggregate

RESULTS AND DISCUSSION

The results show that as oil shale aggregate is included in the composite matrix instead of lime aggregate, the compressive strength enhances and concrete performance improves. This is related to the good bond between the oil shale aggregate and the mortar paste that gives a strong combination. Figure 1 (a, b, and c) presents the benefit gained from replacing coarse, medium and fine size lime aggregate mix with oil shale aggregate on concrete cube (150 mm) compressive strength. Figure 1 shows that: replacing the coarse size lime aggregate with oil shale increases the compressive strength by 14% at 28 days, and replacing the medium size lime aggregate with oil shale increases the compressive strength by 19%, and replacing the fine size lime aggregate with oil shale increases the compressive strength by 16%.

Figure 2 (a, b, and c) presents the relation between the splitting force versus time for concrete cylinders Ø150 mm × 300 mm and Ø100 mm × 200 mm made with lime aggregate and oil shale aggregate. The results show that the size of the cylinder affects the tensile strength of concrete made with oil shale (coarse and medium) replacing lime with oil shale aggregate. Concrete made with oil shale and cast as Ø150 mm × 300 mm cylinders shows better performance compared to Ø100 mm × 200 mm specimens. This illustrates that the behaviour of oil shale becomes apparent only on large specimens regarding tensile strength. In fine grain size aggregate, oil shale shows moderately low tensile strength in comparison with that of lime aggregate in making concrete.

Tests on raw materials of lime aggregate and oil shale aggregate show that the specific gravity of lime aggregate is 2.65, while that of oil shale is 1.95. This reflects the difference in weight between concretes made using oil shale and using lime aggregate.

CONCLUSIONS

Based on the above results and discussion, the following points could be included as a conclusion for this work.

1. The use of oil-shale aggregate on concrete design mix is feasible.

2. The compressive strength enhanced and better concrete performance encountered using coarse oil shale aggregate in compare with concrete made of lime aggregate. This related to the good bonds between the oil shale aggregate and the mortar paste that give a strong combination.

3. Tensile strength for big size of concrete cylinder enhanced using oil shale aggregate. The size of the cylinder affect the tensile strength of the concrete made with oil shale (coarse and medium) replacing lime with that for oil shale aggregate.

4. The weight of the concrete made using oil shale aggregate is lighter than that made using lime aggregate; because the specific gravity of oil shale is less than that for lime aggregate.

5. More work should be done to evaluate the durability of concrete made by oil-shale aggregate.

ACKNOWLEDGEMENTS

The author would like to acknowledge Dr. Emhaidy Gharaibeh for reviewing this manuscript and the technicians of the Properties of Concrete Laboratory Eng Wafa Sihaimat, Hussein Al-Saraireh, and Ahmad Al-Habashneh for their help in the experimental part. Finally I would like to thank my students for their help and support.

REFERENCES

1. KHEDAYWI T, YEGINOBALI A, SMADI M AND CABRERA J, Pozzolanic Activity of Jordanian Oil Shale Ash, Cement and Concrete Research, 1990, Vol. 20, pp. 843-852.

2. SMADI M AND HADDAD R, The Use of Oil Ash in Portland cement concrete, Cement and Concrete Composites, Elsevier 2003, Vol. 25, pp. 43-50.

3. BAUM H, BENTUR A AND SOROKA I, Properties and structure of Oil Shale ash pastes; I: Composition and Physical features, Cement and Concrete Research, 1985, Vol. 15, No. 3, pp. 391-400.

4. BAUM H, BENTUR A AND SOROKA I, Properties and structure of Oil Shale ash pastes; II: Composition and Physical features, Cement and Concrete Research, 1985, Vol. 15, No. 2, pp. 303-314.

5. MALHOTRA S K AND DAVE N G, Investigation into the effect of addition of fly ash and burnt clay Pozzolana on certain engineering properties of cement composites, Cement and Concrete Research, 1999, Vol. 21, pp. 285-291.

6. BENTUR A, ISH-SHALOM M, BON-BASSAT M AND GRINBERG T, Properties and application of oil ash, Spec. ACI Publication, 1985; Vol. 37, pp. 779–802.

7. BENTUR A, ISH-SHALOM M, BON-BASSAT M AND GRINBERG T, Cementing properties of Oil ash; II: Moist curing of cast and compacted samples, Cement and Concrete Research, 1981, Vol. 10, pp. 799-807.

8. FENG N Q, CHEN Y N, HE Z S AND TSANG K C, Shale Ash Concrete, Cement and Concrete Research, 1997, Vol. 27, No. 2, pp. 279-291.

9. COSTANTIN F, Influence of Variability of Oil shale fly ash on Compressive Strength of cementless building compounds, Construction and Building Materials, 2004, Elsevier, Article in press.

COMPRESSIVE STRENGTH AND DEFORMATION MODELLING OF RECYCLED AGGREGATE CONCRETE

A E B Cabral
CEFET/CE
D C C Dal Molin
J L D Ribeiro
UFRGS
Brazil

ABSTRACT: Improper disposal of Construction and Demolition (C&D) waste is a problem that afflicts the authorities and damages the population of several countries of the world. On the other hand, the demand for raw materials for the construction industry is growing. The recycling of C&D waste appears as a solution for these problems. With the intention of contributing to this knowledge area, two of the main properties (the compressive strength and the modulus of elasticity) of recycled aggregates concrete (RAC) was mathematically modelled. A factorial experimental study was done using three types of recycled aggregates (concrete, mortar and red ceramic-brick) originated from Porto Alegre/RS's Metropolitan Area, Brazil. Besides the aggregate type and content, the water/cement ratio varied from 0.40 to 0.80. The mathematical model presented for the compressive strength shows that recycled coarse aggregates have a more detrimental effect than recycled fine aggregate. However, the use of red ceramic recycled fine aggregates increased the strength. The substitution of the natural aggregates by recycled aggregates resulted in a reduction of the modulus of elasticity. The red ceramic recycled coarse aggregate and the concrete recycled fine aggregate exercised the largest and the smallest influence, respectively.

Keywords: C&D waste, Modelling, Compressive strength, Modulus of elasticity.

Dr A E B Cabral, Doctor Environment and Science (EESC/USP), M. Sc. Civil Construction (UFRGS), B. Sc. Engineering (UFC/CE), Lecturer CEFET/CE

Dr D C C Dal Molin, Doctor Civil Construction (EPUSP), M. Sc. Civil Construction (UFRGS), B. Sc. Engineering (UFRGS), Head Researcher UFRGS.

Dr J L D Ribeiro, Doctor Civil Construction (EPUSP), M. Sc. Civil Construction (UFRGS), B. Sc. Engineering (UFRGS), Researcher UFRGS

INTRODUCTION

The improper disposal of construction and demolition (C&D) wastes is a problem that afflicts the authorities and reaches the population not only in Brazil, but also in other countries of the world. Such deposition affects directly the environment, being co-responsible with other factors for inundations, for damages to the landscape, obstruction of roads, proliferation of diseases, among other damages to health and human life [1].

In addition, the super-exploration of the mineral beds, from where the aggregates that are used in the construction sites are extracted (natural and non-renewed resources), gradually causes damage to the environment. The construction industry consumes between 14% and 50% of the extracted natural resources in the planet. In Brazil, it is esteemed that there is an annual consumption of 210 million tons of aggregates only for the production of mortar and concrete, and this does not include paving and losses [1]. In the United Kingdom, in the year of 1992, about 240 million tons of aggregates were consumed, the great majority being obtained through dredging or extraction in quarries [2].

A solution to these problems could be the recycling of C&D waste and its reutilization in the construction industry as an alternative material. There is a critical shortage of natural aggregates for concrete production in many urban areas. At the same time, an increasing amount of C&D waste is generated in the same areas [3, 4].

There is also lack of dumping sites for this waste, making the distance between the demolition sites and the disposition areas longer, and thus increasing the disposal costs. Therefore, with recycling, another source of material is obtained, besides it propitiates a reduction in the final volume of the residues to be disposed [4, 5].

The construction industry is a productive sector contributing considerably to the Brazilian economy. Between 1980 and 1996, this sector was responsible for 65% of all investment in the country. In 1999, this sector already reached the mark of 70%. In 2001, this sector was responsible for 15.6% of GDP, and the construction of residential buildings represented a value between 6% and 9% of the national GDP [6].

To promote such magnificence, the construction industry is now the largest consumer of natural resources of the society, absorbing from 20 to 50% of those resources explored in the world [1]. This industry consumes approximately 66% of the whole natural wood extracted of the nature [7].

The construction activities demand a significant amount of natural materials, such as sand and gravel. The extraction of those materials modifies the profile of the rivers and their balance, introducing environmental problems. The extraction of rocks from mountains is also a dangerous activity to the environment, once it also alters the landscape and provokes stability problems [8].

As in every industrial process, the use of the inputs of the construction industry generates wastes, and in great scale, these residues need to be managed. The construction macro-complex is responsible for 40% of the wastes generated in the economy [1]. In Taiwan it is estimated that an annual generation of 640,000 tons of concrete wastes on its own [9] while in Iran, in 1994, about 350 thousand tons of red ceramic wastes were produced as bricks and tiles [10].

In Australia the estimated annual production of construction and demolition waste is more than 3 million tons, and in the great Australian cities, this waste corresponds to 11 to 15% of the total waste produced [11]. In the United Kingdom, this type of waste corresponds to 30% [12] of the whole solid waste produced, while in Brazil this number approaches 50% [13, 14].

To have an idea of the magnitude of this waste generation, in the United Kingdom this sector produces around 109 million tons per year corresponding to 60% of the total wastes produced in the country, and this amount correspond to 66% of the 165 million tons of natural aggregates consumed annually by the construction industry. However, recycled aggregates contribute to less than 1% of the total of aggregates used in concrete production [15]. It is estimated that only approximately 123 million tons of concrete waste are reused in the European Community, United States and Japan [4].

C&D wastes vary in their composition in relation to location and time. However, concrete, mortar and red ceramic wastes appear as the main components of C&D wastes, reaching above 70% by weight [13, 7, 16, 17]. That heterogeneity influences the characteristics of recycled aggregates. Therefore, it is necessary to know the concrete properties produced with those recycled aggregates when the tenor of these mean constituents varies.

To introduce the sustainability concept in traditional sectors, like the construction industry, it is requested to break certain development paradigms. Some environmental subjects already contemplated in the first steps of the planning of their consumption goods. This is a challenge that the researches should take for themselves, looking for alternatives to satisfy the producers' demands to maintain the quality of their products and continue satisfying their customers, as the society wants a solution to the continuous degradation of natural resources.

OBJECTIVES

The performance of concrete when natural aggregates are changed for recycled aggregates is modified, therefore it is necessary to understand the behaviour of those concretes regarding some mechanical and durability properties. However, this work is limited to the compressive strength and the modulus of elasticity of the concretes containing recycled coarse and fine aggregate, having various water/cement ratios.

MATERIALS AND METHODS

Experimental program

Seven independent variables (factors) were identified, namely: fine and coarse recycled aggregates of red ceramic, fine and coarse recycled aggregate of mortar, fine and coarse recycled aggregate of concrete and the water/cement ratio. The complete experimental project to study the united effect of all those 7 factors on the dependent variables is the experiments' factorial project 2^k [18]. The execution of that project consists of accomplishing 27 concrete mixtures, in other words, 128 mixtures. In function of limitations of time and cost, the solution found to make possible the execution of the experimental phase, with high degree of reliability of the results, was the use of the project composed of second order.

The base of the project composed of second order is a factorial project 2^k, fractional or complete, where it is added to this last one, all the 2^k vertices of a star and the central points of the star and of the factorial project [18]. For this experimental program, a fractional factorial project and the central points were adopted.

The fractional factorial experiments are very useful when there is a great number of factors to be investigated, there is a great number of independent variables and an optimization of time and costs is wanted to obtain the results, because the division consists of splitting the complete project to two or more blocks and execute only one of those blocks, with random choice [19]. This type of study using statistical tools was already executed previously by other several national and international researches [9, 16, 17]. For this complete project, it was divided to 4 blocks, being executed only a block, in other words, 32 mixtures that consists the mixtures 1 to 32 in Table 1.

This experiment has 7 factors, so the star has 14 vertices that correspond to the mixtures 33 to 46. The mixtures 47 and 48 correspond to the central points. They are the same because the central point of the fractional factorial project is the same as the central point star. The mixtures 49 and 50 were included in the experimental project, once they represent the moments where all the aggregates (recycled and natural) are present in the concrete, attributing as values for the water/cement ratio, the two averages of the inferior and superior thirds, in other words, 0.46 and 0.74. Table 1 summarizes all of the concrete mixtures used in the study.

Table 1   Concrete mixtures as defined accomplishing the division of the experiment

MIXTURES	W/C	COARSE AGGREGATE, %				FINE AGGREGATE, %			
		NAT.	CONC.	RED CER.	MORTAR	NAT.	CONC.	RED CER.	MORTAR
01	0.46	100	0	0	0	100	0	0	0
02	0.74	100	0	0	0	0	0	100	0
03	0.74	100	0	0	0	0	100	0	0
04	0.46	100	0	0	0	0	50	50	0
05	0.74	0	0	0	100	0	0	0	100
06	0.46	0	0	0	100	0	0	50	50
07	0.46	0	0	0	100	0	50	0	50
08	0.74	0	0	0	100	0	33	33	33
09	0.46	0	0	100	0	0	0	0	100
10	0.74	0	0	100	0	0	0	50	50
11	0.74	0	0	100	0	0	50	0	50
12	0.46	0	0	100	0	0	33	33	33
13	0.74	0	0	50	50	100	0	0	0
14	0.46	0	0	50	50	0	0	100	0
15	0.46	0	0	50	50	0	100	0	0
16	0.74	0	0	50	50	0	50	50	0
17	0.46	0	100	0	0	0	0	0	100
18	0.74	0	100	0	0	0	0	50	50
19	0.74	0	100	0	0	0	50	0	50

Table 1 Concrete mixtures defined accomplishing the division of the experiment. – cont'd

MIXTURES	W/C	COARSE AGGREGATE, %				FINE AGGREGATE, %			
		NAT.	CONC.	RED CER.	MORTAR	NAT.	CONC.	RED CER.	MORTAR
20	0.46	0	100	0	0	0	33	33	33
21	0.74	0	50	0	50	100	0	0	0
22	0.46	0	50	0	50	0	0	100	0
23	0.46	0	50	0	50	0	100	0	0
24	0.74	0	50	0	50	0	50	50	0
25	0.46	0	50	50	0	100	0	0	0
26	0.74	0	50	50	0	0	0	100	0
27	0.74	0	50	50	0	0	100	0	0
28	0.46	0	50	50	0	0	50	50	0
29	0.74	0	33	33	33	0	0	0	100
30	0.46	0	33	33	33	0	0	50	50
31	0.46	0	33	33	33	0	50	0	50
32	0.74	0	33	33	33	0	33	33	33
33	0.60	0	50	25	25	0	33	33	33
34	0.60	0	0	50	50	0	33	33	33
35	0.60	0	25	50	25	0	33	33	33
36	0.60	0	50	0	50	0	33	33	33
37	0.60	0	25	25	50	0	33	33	33
38	0.60	0	50	50	0	0	33	33	33
39	0.60	0	33	33	33	0	50	25	25
40	0.60	0	33	33	33	0	0	50	50
41	0.60	0	33	33	33	0	25	50	25
42	0.60	0	33	33	33	0	50	0	50
43	0.60	0	33	33	33	0	25	25	50
44	0.60	0	33	33	33	0	50	50	0
45	0.80	0	33	33	33	0	33	33	33
46	0.40	0	33	33	33	0	33	33	33
47	0.60	0	33	33	33	0	33	33	33
48	0.60	0	33	33	33	0	33	33	33
49	0.46	25	25	25	25	25	25	25	25
50	0.74	25	25	25	25	25	25	25	25

The compressive strength and the modulus of elasticity of the produced concretes are the answer variables, in other words, dependent variables. They were made according to the Brazilian Standards ABNT NBR 5739/94 and NBR 8522/03, respectively. Some other variables were fixed, such as the specimens' age (28 days) and the recycled aggregates pre-humidification water, once these could absorb the mixing water, thus modifying the water/cement ratio (w/c).

The water absorption of the fine recycled and fine natural aggregates were measured through the method proposed by ABNT NBR NM 30/00 and the water absorptions of the coarse recycled and coarse natural aggregates were measured through the method proposed by NM 53/02. For each aggregate, the absorption was determined twice, through two samples. The averages of the results for the fine and the coarse aggregates are showed in the Tables 2 and 3, respectively.

The specific gravity of the fine and coarse aggregates was also determined through the method proposed by the Brazilian standards ABNT NBR 9776/87 and ABNT NM 53/02, respectively. For each aggregate, the specific gravity was determined twice, through two samples. The averages of the results are also shown in Tables 2 and 3 for fine and coarse aggregates, respectively.

The bulk density of all the aggregates was determined through the method proposed by the Brazilian Standard ABNT NM 45/00. For each aggregate, the bulk density was determined twice, through two samples. The objective of the determination of the bulk density would be to discover which aggregate generates a better packing, in other words, which distribute better in the concrete mass, leaving the minimum of voids in a certain volume.

But as the recycled aggregates are coming from different raw materials, with different specific gravities, the results obtained cannot be compared, since the specific gravity's influence is not isolate. So, the best way to compare them is to make a parametric comparison between the natural aggregate and the recycled one. This approach was made for both groups and the bulk density already corrected is shown in Tables 2 and 3 for fine and coarse aggregates, respectively.

Mixing Process

The proportion of natural aggregate was selected through the method of IPT/EPUSP [20], fixing the slump value at 120±20 mm and determining the dosage diagram. With this last one, the execution of the 50 experimental project mixtures began. However, some adjustments were necessary to be done in the mixtures with substitution of recycled aggregate for natural aggregate.

Table 2   Characteristics of fine aggregates

AGGREGATE	METHOD		
	NM 30/00 Absorption, %	NBR 9776/87 Specific gravity, g/cm^3	NM 45/02 Bulk density, kg/m^3
Natural	0.42	2.64	1440
Recycled concrete	7.55	2.56	1540
Recycled mortar	4.13	2.60	1440
Recycled red ceramic	10.69	2.35	1460

Table 3   Characteristics of coarse aggregates

AGGREGATE	METHOD		
	NM 53/02 Absorption, %	NM 53/02 Specific gravity, g/cm^3	NM 45/02 Bulk density, kg/m^3
Natural	1.22	2.87	1560
Recycled concrete	5.65	2.27	1430
Recycled mortar	9.52	2.01	1390
Recycled red ceramic	15.62	1.86	1260

Firstly, a volume compensation was made on the recycled aggregates to be used in the pre-determined mixtures [3, 16, 17, 21, 22], because the simple substitution of recycled aggregate mass for natural aggregate mass would result in mixtures with larger volumes of recycled aggregate, since the specific gravity of recycled aggregates is smaller than that of the natural aggregates, thus demanding more water and cement to produce equivalent mixtures to the mixture with natural aggregates. The volume compensation of recycled aggregates in the experimental project mixtures was done according to the Equation 1:

$$M_{AR} = M_{AN} \cdot \frac{\gamma_{AR}}{\gamma_{AN}} \tag{1}$$

where: M_{RA}= recycled aggregate mass (kg)
γ_{RA}= specific gravity of the recycled aggregate
M_{NA}= natural aggregate mass (kg)
γ_{NA}= specific gravity of the natural aggregate

The mass of water that would be added to the recycled aggregates before mixing was also determined. So these aggregates were put already moist in the pan-mixer, thus avoiding absorption of the mixing water, which would disturb the hydration process. Some researchers [16, 17] used tenors around 40% to 50% of the total of absorbed water in 24 hours. Others [23-25], in turn, used larger tenors of pre humidification water, reaching saturation.

In this research, the recycled aggregates were moistened 10 minutes before mixing, with 80% of the water that would be absorbed in 24 hours by the recycled aggregate mass. This value was arrived at by preliminary testing, which showed that recycled aggregates absorb 80% of their total water absorption, on average, in the first 120 minutes after mixing.

After the presaturation, the coarse aggregate was placed in a vertical axis pan-mixer with part of the mixing water. Soon afterwards, cement was added and the rest of the mixing water. Finally the fine aggregate was added. In some mixes it was also necessary to add some superplasticizer, until it reached the pre-determined workability of 120 mm slump.

Once reached the intended workability, four cylindrical specimens of 100 mm of diameter and 200 mm of height were moulded for each produced mix, according to the procedures of the Brazilian Standard NBR 5738/03.

After casting, the cylinders were exposed to environmental temperature and humidity for 24 hours, then they were demoulded and placed in a controlled environmental chamber (20°C, 100% R.H.), until one day before the test date.

RESULTS

The results of the treatments were in agreement with those of the factorial project, which allowed testing linear and quadratic terms. The accomplished tests also allowed testing linear and no linear models, for the dependent variables. For a better understanding of the models, abbreviations of the names of the independent and dependent variables were introduced, as shown in Table 4.

Table 4   Independent and dependent variables symbols

SYMBOL	VARIABLE	
	Name	Type
Rmc	percentage of natural coarse aggregate replaced by recycled mortar coarse aggregate	independent
Rmf	percentage of natural fine aggregate replaced by recycled mortar fine aggregate	independent
Rcc	percentage of natural coarse aggregate replaced by recycled concrete coarse aggregate	independent
Rcf	percentage of natural fine aggregate replaced by recycled concrete fine aggregate	independent
$Rrcc$	percentage of natural coarse aggregate replaced by recycled red ceramic coarse aggregate	independent
$rrcf$	percentage of natural fine aggregate replaced by recycled red ceramic fine aggregate	independent
w/c	water/cement ratio	independent
f_c	compressive strength	dependent
E_c	modulus of elasticity	dependent

The collected data allowed establishing models relating the dependent variables with the independent variables, setting up simple models, such as multiple linear regression, or complex models, such as non-linear regression.

An analysis of the standardized residues was made with the construction of the models. The collected data that generated standardized residues larger than 3 in module were eliminated of the analysis and these values were obtained only for mixture 3.

The analysis presented was developed leaning in a linear regression routine, which presented good results despite being relatively simple. Some complex models were also tried, but the statistical growth was small, so the use of the simplest model was opted for.

Compressive strength – analysis and discussion

The model that was rather adapted to represent the compressive strength of the concretes with recycled aggregates is shown in Equation 2.

$$fc = \left(\frac{102.43}{5.38^{w/c}} \right).[1 - (\%replaced)] \tag{2}$$

The first term between parentheses refers to the concrete strength without the substitution of natural aggregates by recycled ones and it is a function of the water/cement ratio. This term was previously defined, starting from an analysis of the obtained values when the water/cement ratio was 0.46, 0.60 and 0.74, being defined from ways to generate the best possible adjustment, in other words, to minimize the forecast mistakes.

The second term, among brackets, defines a percentage to be applied on the original strength, usually reducing it in line with the natural aggregate that was replaced by the recycled ones. That strength loss is a function of the percentage of the natural aggregate being replaced by the recycled aggregate, as shown below:

$$Strength\ loss = f(\%\ replaced)$$

Multiple regression was then made and it was identified which were the dependent variables that exercised a significant effect on the strength loss of the recycled concretes. All the dependent variables were considered significant, except the water/cement ratio (w/c).

The model of the strength loss found had a quite satisfactory correlation coefficient (98.04%), being described in Equation 3.

$$Strength\ loss = 0.338.rmc + 0.153.rmf + 0.275.rcc + 0.067.rcf + \\ 0.371.rrcc - 0.138.rrcf \tag{3}$$

The attempt of inserting square terms (as $rmc.rmc$, for example) or interactions (as $rmc.rmf$, for example) didn't improve the adjustment, so these more complex terms were left out.

Then, the final model that estimates the strength as a function of the percentage of substitution of natural aggregates by those recycled and of the water/cement ratio is shown in Equation 4. In that model, the percentage of substitution of the fine or coarse aggregates by those recycled should be informed in the scale of 0 (0%) to 1 (100%), while the water/cement ratio is expressed in the usual scale, varying from 0.40 to 0.80. The maximum sum of the percentage of substitution of natural aggregates by recycled ones should be 1 (100%) for each aggregate type (coarse and fine).

$$fc = \left(\frac{102.43}{5.38^{w/c}}\right).[1 - (0.338.rmc + 0.153.rmf + 0.275.rcc + 0.067.rcf + 0.0371.rrcc - 0.138.rrcf)] \tag{4}$$

According to the presented model, the substitution of natural aggregates by recycled ones results in a strength reduction, except in the case of the fine recycled aggregate of red ceramic that provides an increase in strength.

It is also observed that coarse aggregate substitution produces an effect larger than fine aggregate substitution, as a function of the coefficients whose magnitude is the same.

With the above model, some graphs were generated to illustrate the influence of each type of recycled aggregate, using 0%, 50% and 100% as tenors of substitution and intermediate water/cement ratios, in other words, 0.46, 0.60 and 0.74. Also a table was set up using the losses and earnings for the strength in each case. The graphs and the table are shown in Figures 1, 2 and 3 and in Table 5.

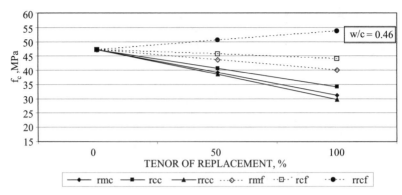

Figure 1   Compressive strength behaviour when varying the tenor of replacement and the recycled aggregate type, for a water/cement ratio equal to 0.46

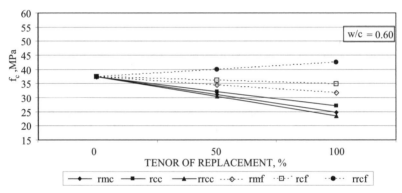

Figure 2   Compressive strength behaviour when varying the tenor of replacement and the recycled aggregate type, for a water/cement ratio equal to 0.60

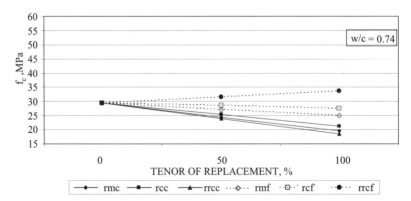

Figure 3   Compressive strength behaviour when varying the tenor of replacement and the recycled aggregate type, for a water/cement ratio equal to 0.74

Table 5   Compressive strength performance of recycled aggregate concretes

REPLACEMENT TENOR, %	TYPE OF RECYCLED AGGREGATE					
	rmc	rcc	rrcc	rmf	rcf	rrcf
0	1.00	1.00	1.00	1.00	1.00	1.00
50	0.83	0.86	0.81	0.92	0.97	1.07
100	0.66	0.72	0.63	0.85	0.93	1.14

Figures 1 to 3 show the influence of the water/cement ratio on the compressive strength of the concretes. It is also observed that the influence of the fine recycled aggregates on compressive strength is the lowest. All the recycled aggregates exercise a negative influence in the strength, except for the fine recycled aggregate of red ceramic, which produced an improvement in the strength, arriving to 7% for 50% of replacement and 14% for 100%.

This improvement in the strength could be due to pozzolanic reactions, improving the interfacial transition zone between the paste and the aggregates and consequently improving the mechanical properties of the concretes and mortars produced with this type of fine recycled aggregate [16]. The strength increase is partly due to the roughness of particles in the recycled ceramic aggregates that imparts a better bond between the cement paste and the aggregates. Another possibility is that the water absorbed by the recycled aggregate becomes available for continued hydration of cement [2].

However, for the natural coarse aggregate replacement by the recycled coarse aggregate of red ceramic, the worst result was obtained. According to the model, this replacement induces a 19% loss of strength for 50% substitution and 38% for 100% replacement. Such reduction values are quite coherent with reported research [25].

This behaviour is probably due to the angular aggregate shape, even if having the same grading as others. This shape does not provide an efficient grain package and thus it produces concretes with more voids [2]. Table 3 shows that the bulk density of the red ceramic coarse aggregate is lower than that of the natural coarse aggregate.

The coarse aggregate that obtained a better performance in the strength behaviour was the recycled concrete aggregate, although it still showed a reduction in the strength around 14% for 50% of substitution and 28% for 100%.

According to the model, as much the fine recycled aggregates of mortar as of concrete have only a little influence on the performance of concretes made with them, showing a reduction of only 8 and 3%, respectively, for 50% replacement and 15 and 7%, respectively, for 100%.

Modulus of elasticity – analysis and discussion

The model that was rather adapted to represent the modulus of elasticity of concrete with recycled aggregate is shown in Equation 5.

$$E_c = \left(\frac{21.69}{w/c^{0,5}} \right).[1 - (\%replaced)]$$ (5)

The procedure to obtain the loss of the modulus of elasticity model was the same used to obtain the strength loss model using multiple regression tools. So the modulus of elasticity loss model is described below in Equation 6. The correlation coefficient of this model was 99.64%.

$$\textit{Modulus of elasticity loss} = 0.352.rmc + 0.158.rmf + 0.231.rcc + 0.11.rcf + 0.44.rrcc + 0.113.rrcf \tag{6}$$

Again, the attempt of inserting square terms (as *rmc.rmc*, for example) or interactions (as *rmc.rmf*, for example) did not improve the adjustment, so these more complex terms were left out.

Then, the final model that estimates the modulus of elasticity as a function of the percentage of substitution of natural aggregates by recycled ones and of the water/cement ratio is shown in Equation 7. In this model, the percentage of substitution of fine or coarse aggregates by recycled ones should be informed in the scale of 0 (0%) to 1 (100%), while the water/cement ratio is expressed in the usual scale, varying from 0.40 to 0.80. The maximum sum of the percentage of substitution of natural aggregates by recycled ones should be 1 (100%) for each aggregate type (coarse and fine).

$$E_c = \left(\frac{21.69}{w/c^{0.5}} \right).\left[1 - \left(0.352.rmc + 0.158.rmf + 0.231.rcc + 0.11.rcf + 0.44.rrcc + 0.113.rrcf\right)\right] \tag{7}$$

According to the presented model, the natural aggregate substitution by the recycled ones results in a reduction of the modulus of elasticity for all aggregates types, which is coherent with findings of other researches [3, 25, 27, 28].

It is also observed that the coarse aggregate substitution produces a larger effect in the modulus loss than the fine aggregate substitution, with the magnitude of the coefficients being the same. This behaviour is coherent, since the modulus of elasticity of concrete is intrinsically linked to the volumetric fraction, to the specific gravity, to the moduli of elasticity of aggregates and cement matrix, and to the characteristics of the interfacial transition zone [26].

The modulus of elasticity of the aggregates is mainly linked to their porosity and, at a little lower influence, to the maximum diameter of the aggregate, to their forms, texture, grading and mineralogical composition. The rigidity of the aggregate controls the capacity of the cement matrix to restrict deformation and this rigidity is determined by the porosity of the aggregate.

The specific gravity of the recycled fine aggregate is smaller than that of the recycled coarse aggregate, so it is coherent with the modulus of elasticity of concretes made with the former being smaller than the modulus of elasticity of concretes made with the latter.

With the model described in Equation 7, some graphs were generated to illustrate the influence of each type of recycled aggregate, using 0%, 50% and 100% as tenors of substitution and intermediate water/cement ratios, in other words, 0.46, 0.6 and 0.74. Also a table was set up using the losses and earnings for the modulus of elasticity in each case. The graphs and the table are shown in Figures 4, 5 and 6 and in Table 6.

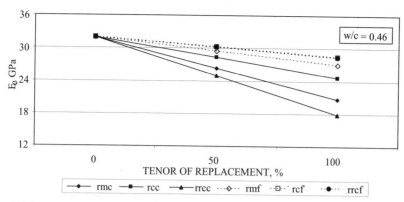

Figure 4   Elastic modulus behaviour when varying the tenor of replacement and the recycled aggregate type, for a water/cement ratio equal to 0.46

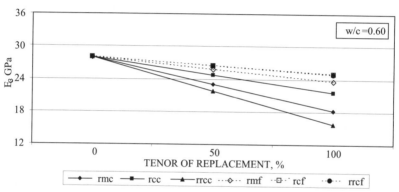

Figure 5   Elastic modulus behaviour when varying the tenor of replacement and the recycled aggregate type, for a water/cement ratio equal to 0.60

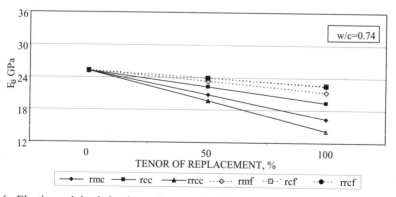

Figure 6   Elastic modulus behaviour when varying the tenor of replacement and the recycled aggregate type, for a water/cement ratio equal to 0.74

Table 6　Elastic modulus performance of recycled aggregate concretes

REPLACEMENT TENOR, %	TYPE OF RECYCLED AGGREGATE					
	rmc	rcc	rrcc	Rmf	rcf	rrcf
0	1.00	1.00	1.00	1.00	1.00	1.00
50	0.82	0.88	0.78	0.92	0.95	0.94
100	0.65	0.77	0.56	0.84	0.90	0.88

Through the Figures 4 to 6, the influence of the water/cement ratio can be observed on the performance of the modulus of elasticity of concretes produced with recycled aggregates. The smallest influence of the fine recycled aggregates on the modulus of elasticity is also noted.

All the recycled aggregates exercise a negative influence on the modulus of elasticity of concretes produced with them and the coarse recycled aggregate of red ceramic exercises the largest influence, reaching a loss of 22% in the value of the modulus for 50% substitution and 44% for 100%. Such behaviour could be explained considering the characteristics of the recycled aggregates, since the coarse recycled aggregate of red ceramic has the smallest specific gravity and the largest water absorption of all the used aggregates, seeming to be the most porous of all.

Among the coarse recycled aggregates, recycled concrete exhibits the smallest influence on the value of the concrete made with it, nevertheless showing a considerable loss of 12% for 50% substitution and 23% for 100% substitution. This biggest deformation presented by concretes with recycled concrete aggregates is a function of the high tenor of mortar (around 40% of the volume) being present in the same [21]. These results are also coherent with those reported by other researchers [3] who also substituted natural coarse aggregates by recycled coarse aggregate of concrete, finding a modulus of elasticity around 19% smaller than in concrete with natural aggregates.

Although the fine recycled aggregate of concrete exhibits the smallest influence, with a reduction of only 5% in the modulus of elasticity, for 50% substitution and 10% for 100%, it is known that this type of aggregate has a high tenor of natural aggregate in its composition as a result of the crushing of concrete with natural aggregates, having a high specific gravity and the smallest water absorption among the recycled aggregates used.

In agreement with these results, other researchers [2, 29] noted that the modulus of elasticity of concretes with recycled red ceramic aggregates reached only 50 to 66% of the modulus of elasticity of natural aggregate concretes, for a same level of resistance. Other researchers [3, 25] presented the moduli of elasticity of concretes made with recycled concrete aggregates being 15 to 45% smaller than the modulus of conventional concretes.

CONCLUSIONS

In general, the concretes produced with recycled aggregates had lower compressive strengths and moduli of elasticity compared to the control concrete with natural aggregates. According to the previously determined model, recycled coarse aggregates have more significant detrimental influence on compressive strength. For recycled fine aggregates, strength decreased is also observed when replacing natural aggregates in concrete. However, it was observed that the recycled fine aggregate of red ceramic improved the compressive strength of concrete. Concretes produced with recycled aggregates showed reduced modulus of elasticity values when compared with that of conventional concrete. According to the previously determined model, recycled coarse aggregates exhibit a larger influence on the modulus of elasticity than fine recycled aggregates. Among all the tested aggregates, the red ceramic recycled coarse aggregate exhibited the largest influence on the modulus of elasticity and the recycled fine concrete aggregate did the smallest.

ACKNOWLEDGEMENT

The authors thank to the Oriented Nucleus for the Construction Innovation (NORIE) of the Federal University of Rio Grande do Sul (UFRGS), in Brazil, for supporting the experiments accomplished and to the Brazilian Research Supporting Agency (CAPES) for the financial support through the PQI 106/08-03 agreement (CEFET/CE-EESC/USP).

REFERENCES

1. JOHN V M, Utilization of solid waste like construction materials, In: Recycling of construction and demolition waste to produce construction materials / Editors: Alex Pires Carneiro, Irineu Antônio Schadach de Brum and José Clodoaldo da Silva Cassa. Salvador: EDUFBA; 312 p., 2001, pp. 27-45 (in Portuguese).

2. KHALAF F M AND DeVENNY A, Recycling of demolished masonry rubble as coarse aggregate in concrete: review. Journal of Materials in Civil Engineering, Vol. 16, No. 4, 2004, pp. 331-340.

3. HANSEN T C AND BØEGH E, Elasticity and drying shrinkage of recycled-aggregate concrete, ACI Journal, Vol. 82, No. 5, 1985, pp. 648-652.

4. LAMOND J F, CAMPBELL R L, CAMPBELL T R, CAZARES J A, GIRALDI A, HALCZAK W, HALE H C, JENKINS N J, MILLER R AND SEABROOK P T, Removal and reuse of hardened concrete, ACI Materials Journal, May-June, 2002, pp. 300-325.

5. SANI D, MORICONI G, FAVA G AND CORINALDESI V, Leaching and mechanical behaviour of concrete manufactured with recycled aggregates, Waste Management, Vol. 25, 2005, pp. 177-182.

6. NETO J C M, Management of construction and demolition wastes in Brazil, São Carlos: Rima, 2005. 162 p. (in Portuguese).

7. ZORDAN S E, Utilization of C&D waste as aggregate to make concrete. Campinas-SP, 1997 140 p., Master Theses, Civil Engineering College of the State University of Campinas (in Portuguese).

8. BIANCHINI G, MARROCCHINO E, TASSINARI R AND VACCARO C, Recycling of construction and demolition waste materials: a chemical-mineralogical appraisal, Waste Management, Vol. 25, 2005, pp 149-159.

9. LIU Y, TYAN Y, CHANG T AND CHANG C, An assessment of optimal mixture for concrete made with recycled concrete aggregates, Cement and Concrete Research, Vol. 34, 2004, pp. 1373-1380.

10. KAHLOO A R, Properties of concrete using crushed clinker brick as coarse aggregate, ACI Materials Journal, July-August, 1994, pp. 401-407.

11. SHAYAN A AND XU A, Performance and properties of structural concrete made with recycled concrete aggregate, ACI Materials Journal, Vol. 100, No. 5, 2003, pp. 371-380.

12. ADAMS K T, PHILLIPS P S AND MORRIS J R, A radical new development for sustainable waste management in the UK: the introduction of local authority Best Value legislation, Resources, Conservation and Recycling, Vol. 30, 2000, pp. 221-244.

13. PINTO T P, Differ management methodology for construction and demolition wastes, São Paulo-SP, 1999. 203 p., Doctorate Thesis, Polytechnic College of the University of São Paulo (in Portuguese).

14. FREITAS C S, CORREIA R F, FRANÇA K P, SANTANA F G AND LEITE M B, Clandestine discards diagnostics of the construction and demolition wastes in Feira de Santana/BA: prior research, In: VI Sustainable development seminar and recycling in civil construction, Proceedings, São Paulo, IBRACON, CT-206, 2003 (in Portuguese).

15. DHIR R K, PAINE K A AND DYER T D, Recycling construction and demolition wastes in concrete, Concrete, March, 2004, pp. 25-28.

16. LEITE M A, Evaluation of mechanical properties of concretes produced with recycled aggregates of construction and demolition wastes, Porto Alegre-RS, 2001, 270 p. Doctorate Thesis, Civil Engineering Pos Graduate Program of the Federal University of Rio Grande do Sul (in Portuguese).

17. VIEIRA G L, Research of the corrosion process under the action of chloride ions in recycled aggregate concretes, Porto Alegre-RS, 2003, 151 p., Master Thesis, Civil Engineering Pos Graduate Program of the Federal University of Rio Grande do Sul (in Portuguese).

18. RIBEIRO J L D AND CATEN C S T, Experiment projects, Porto Alegre: FEEng/UFRGS, 2001, 128 p. (in Portuguese).

19. WERKEMA M C C AND AGUIAR S, Planning and analysis of experiments: how to identify and to evaluate the main influential variables in a process, Belo Horizonte: Fundação Christiano Ottoni, 1996, 294 p. (in Portuguese).

20. HELENE P R L AND TERZIAN P, Manual of dosage and control of concretes, São Paulo: PINI, Brasília: SENAI, 1992, 349 p. (in Portuguese).

21. HANSEN T C AND NARUD H, Strength of recycled concrete made from crushed concrete coarse aggregate, Concrete International, Vol. 5, No. 1, 1983.

22. MASCE N O, MIYAZATO S AND YODSUDJAI W, Influence of recycled aggregate on interfacial transition zone, strength, chloride penetration and carbonation of concrete, Journal of Materials in Civil Engineering, Vol. 15, No. 5, 2003, pp. 443-451.

23. ZAHARIEVA R, BUYLE-BODIN F, SKOCZYLAS F AND WIRQUIN E, Assessment of the surface permeation properties of recycled aggregate concrete, Cement and Concrete Composites, Vol. 25, 2003, pp. 223-232.

24. de BRITO J, PEREIRA A S AND CORREIA J R, Mechanical behaviour of non-structural concrete made with recycled aggregates, Cement and Concrete Composites, Vol. 27, No. 4, 2005, pp. 429-433.

25. XIAO J, LI J AND ZHANG C, Mechanical properties of recycled aggregate concrete under uniaxial loading, Cement and Concrete Research, Vol. 35, 2005, pp. 1187-1194.

26. MEHTA P K AND MONTEIRO P J M, Concrete: structure, properties and materials, São Paulo, Ed. PINI, 1994. (in Portuguese).

27. AJDUKIEWICZ A AND KLISZCZEWICZ A, Influence of recycled aggregates on mechanical properties of HS/HPC, Cement and Concrete Composites, Vol. 24, 2002, pp. 269-279.

28. GÓMEZ-SOBERÓN J M V, Porosity of recycled concrete with substitution of recycled concrete aggregate: an experimental study, Cement and Concrete Research, Vol. 32, 2002, pp. 1301-1311.

29. RILEM RECOMMENDATION, Specifications for concrete with recycled aggregates, 121 - DRG guidance for demolition and reuse of concrete and masonry, Materials and Structures, Vol. 27, 1994, pp. 557-559.

FLEXURAL BEHAVIOUR OF HYBRID REINFORCED CONCRETE BEAMS

G F Kheder
J M Al Khafaji
University of Mustansiriya
Iraq

ABSTRACT. In this research work a new type of reinforced concrete beams were investigated, the beams were hybrid made from two types of concrete with different compressive strengths, namely 20 and 70 MPa. Three categories of beams were investigated. The first was cast with 20 MPa concrete, the second with 70 MPa concrete, while the third set (hybrid) was cast with two layers of different concrete mixes, the lower 185 mm was cast with concrete with 20 MPa concrete, while the upper 90 mm. was cast with 70 MPa concrete. The beams were reinforced with steel ratios between 0.95 and 3.55%. The beams were tested simply supported with third point loading. Their load carrying capacity, deflection and flexural crack widths were measured. The hybrid beams showed an increased load capacity compared to normal strength beams, this increase was up to 70.2% more than the ordinary strength concrete beams, while corresponding high strength beams showed an improvement up to 88.6% only. Deflection of hybrid beams was very close to that of high strength concrete and showed lower deflection compared to normal strength beams. On average, the deflection of hybrid beams was 82.7%, and only about 5% higher than high strength beams. Flexural cracks widths in hybrid beams were narrower than both normal strength and high strength beams at all loading stages. At yielding stage cracks width of hybrid beams were 64% (on average) of corresponding normal strength beams and 82% (on average) of corresponding high strength beams. Finally, a cost-moment capacity comparison showed that hybrid strength beams proved to be more economical compared to the other two types of beams especially for high steel ratios.

Keywords: Beams, Behaviour, Capacity, Concrete, Cracking, Flexural, Hybrid.

Ghazi F Kheder, Professor of Civil Engineering, University of Mustansiriya, Iraq. Published several papers on flexural and volume change cracking, self compacting concrete and nondestructive testing of concrete structures.

Jasim M Al Khafaji, Assistant Professor of Civil Engineering, University of Mustansiriya, Iraq. Consultant of several engineering projects, published several papers on repair of concrete members, beam – column strength and flexural behaviour of concrete beams.

INTRODUCTION

Concrete is an excellent construction material, with some superior properties as strength in compression, mouldability, and durability. But one of the few drawbacks of concrete is its low strength in tension and its brittleness. If concrete members are subjected to flexural loading, part of the concrete cross section will be subjected to compressive stress while the other part of the section will be subjected to tensile stresses which the concrete cannot resist and thus will crack and fail in tension.

In order to solve this problem, steel reinforcement is provided in positions where the concrete cross section is subjected to tensile stresses. The steel reinforcement will carry the tensile stresses in which the concrete fails to resist. As a result the concrete cross section will be cracked and these flexural cracks will sometimes extend to more than two thirds of the concrete cross section depth. The cracked depth of the section will not contribute to the load carrying capacity of the member and also will only has a minor role in determining the deflection as the cracked moment of inertia of the concrete cross section will play a major role in determining the equivalent concrete cross section moment of inertia.

Thus the main role for the cracked concrete in flexure is to fix and keep the steel reinforcement in position and to make the concrete in compression and the steel reinforcement in tension acting together to support the externally applied moment.

As a result of this drawback, concrete members subjected to flexural loading will not benefit much from using high strength concrete ($f'_c > 40$ MPa), and usually these members are cast using normal strength concrete, because about 60 to 80% of the concrete cross section subjected to tensile stresses will have a negligible role in enhancing the load carrying capacity of the concrete beam.

In this research work, it is aimed at introducing a new concept in reinforced concrete beams subjected to flexural loading. This concept is to use the high strength concrete in the compression zone only of the beam, while the rest of the beam is cast with normal strength concrete. Thus it makes benefit of the high strength concrete in the compression zone and reduces the cost of the beam by using normal strength concrete in the tension zone, where its effect is negligible.

EXPERIMENTAL WORK

In this research work twelve reinforced concrete beams were cast. These beams were divided into three categories, normal strength (N) high strength (H) and hybrid strength (HY).

The beam's cross section was 175×275 mm with total length of 3000 mm. These beams were reinforced with 2 or 3 deformed reinforcing bars with diameters of 16 or 25 mm. the yield strength of the reinforcement was 460 MPa and ultimate strength (at failure strength 635 MPa). Also 6 mm plain bars were used as stirrups, these stirrups were provided at spacing so that the beams will not fail in shear.

Two types of concrete mixes were used; the first with nominal normal compressive strength of 20 MPa normal strength mix, while the second with nominal high strength of 70 MPa. The concrete mix by weight proportions used for the 20 MPa mix was 0.63:1:2.03:3.26 and for the 70 MPa mix was 0.27:1:1.13:1.94 (water: cement: sand: gravel). A super-plasticizer was added to the latter mix, the dosage of addition was 1.5% by the weight of cement. Siliceous sand was used; the sand had a fineness modulus of 2.94. Also crushed siliceous gravel with maximum aggregate size of 14 mm was used as coarse aggregate, both types of aggregates conformed to the ASTM grading requirement [1].

The category N beams were cast with the 20 MPa mixes, while the category H beams were cast with the 70 MPa mix. On the other hand beams of category HY were cast using both mixes, the lower 185 mm. part of the beams were cast using the normal strength concrete, while the upper 90 mm was cast using the high strength concrete. The upper part of the HY beams was cast immediately after casting the normal strength lower parts, and the two mixes were consolidated together using a vibrating table, so that to prevent the formation of weak layer between the two concrete casts.

With each beam, 5 cylinders ($\varnothing 150 \times 300$ mm) and two prisms ($150 \times 150 \times 750$ mm) were cast to measure the concrete compressive strength; modulus of elasticity and modulus of rupture were tested according to ASTM standards [2 - 4].

The beams were cured in water for 28 days in a controlled temperature curing tank. Then the beams were taken out of the curing tank and left to dry in air for two days. Demec points were fixed on the lower edge of the beam at spacing of 50 mm. These demec points were used to measure crack widths during loading stages using portable extension-meter with accuracy of 0.002 mm / division. Then the beams were tested in simply supported conditions with third point loading over a net span of 2800 mm. Tests were carried out till complete failure of the beams. During the tests the deflection at the centre of the beams and the cracks width and spacing were recorded to investigate the structural behaviour of the hybrid beams as compared to the normal strength beams and high strength beams.

Figure 1 and Table 1 show the details of the beams investigated as well as the measured compressive strength, modulus of rupture and modulus of elasticity of the concrete cast in each beam.

EXPERIMENTAL RESULTS

Flexural Strength of Concrete Beams

The flexural strength of the beams was measured to evaluate the benefit of using hybrid concrete beams, as compared to normal strength or high strength concrete beams. Table 2 shows the load carrying capacities of the three categories of beams investigated, a comparison was carried out at three levels of loading, these levels are at first cracking stage, at yielding of steel reinforcement stage and finally at ultimate load stage (failure).

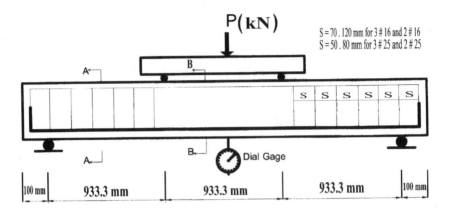

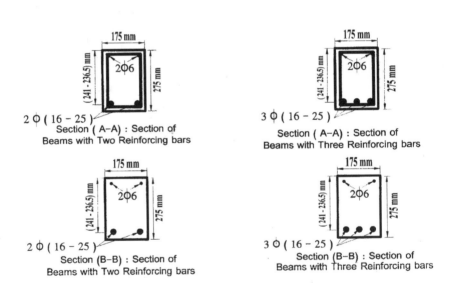

Figure 1   Concrete beam details

Table 1   Beams details and measured concrete properties

BEAM NOTATION	PROPERTY							
	A_s mm^2	H mm	B mm	D mm	ρ %	f'$_c$ MPa	f$_r$ MPa	E$_c$ GPa
N2016	402	275	175	241	0.95	23.6	3.6	24.9
N3016	603	275	175	241	1.43	23.0	3.7	24.8
N2025	981.7	275	175	236.5	2.37	21.0	3.5	23.2
N3025	1472.6	275	175	236.5	3.56	22.0	3.6	23.6
H2016	402	275	175	241	0.95	65.4	7.4	37.9
H3016	603	275	175	241	1.43	71.7	8.0	40.5
H2025	981.7	275	175	236.5	2.37	74.9	8.1	42.2
H3025	1472.6	275	175	236.5	3.56	68.8	7.8	38.8
HY2016	402	275	175	241	0.95	22.0/71.4*	3.9/8.2	24.0/41.9
HY3016	603	275	175	241	1.43	22.0/71.1	3.6/8.0	23.7/41.5
HY2025	981.7	275	175	236.5	2.37	21.2/68.3	2.9/7.8	24.5/39.9
HY3025	1472.6	275	175	236.5	3.56	22.3/73.7	3.6/8.4	24.2/42.5

*Properties of normal and high strength concrete used in the hybrid beam.

From Table 2, it can be seen that the benefit of using hybrid concrete is manifested at high steel ratios. At low steel ratio beams with steel ratio of 0.95% (N2016, H2016 and HY2016) there was no significant increase in load carrying capacity of the beams by using hybrid or even high strength concrete, the yielding moment of these three beams were 46.7, 52.3, and 51.3 kN·m, which means that the increase was only 12% and 10% compared to the normal strength beam. The ultimate moment capacities showed similar results, the ultimate moment capacities of these beams were 53.2, 56.0, and 54.1 kN·m, this corresponds to an increase of only 5.3%, and 1.7% compared to normal strength beams. On the other hand, the high steel ratio beams investigated with steel ratios of 3.56% (N3025, H3025 and HY3025) gave yielding moments of 105.5, 168.0 and 158.7 kN·m, thus resulting in an improvement in load capacity in comparison with the normal strength beam of 59.2% and 50.4% respectively. The ultimate moment capacities of these beams were 106.4, 200.7 and 181.1, thus the increase in ultimate moment capacity were 88.6% and 70.2% compared to the reference concrete beam N3025.

In order to verify the flexural behaviour of the hybrid concrete beams concept. The cracking moment of each beam was calculated using ACI 318-02 [5] concept:

$$M_{cr} = f_r I_g / y_b \qquad (1)$$

The calculated values for the cracking moments showed good agreement with those obtained experimentally. In order to be able to calculate the cracking moment of the hybrid beams with the two different types of concrete, the high strength at the upper part of the beam was transformed to normal strength concrete by multiplying the section by the modular ratio of the two types of concrete. From the obtained results, it can be concluded that the transformed section concept proved to result in good predicted values, and can be applied for hybrid concrete beams.

Table 2   Flexural strength of concrete beams

BEAM NOTATION	MEASURED			CALCULATED	
	CRACKING MOMENT kN·m	YIELDING MOMENT kN·m	ULTIMATE MOMENT kN·m	CRACKING MOMENT kN·m	ULTIMATE MOMENT kN·m
N2016	9.0	46.7	53.2	9.3	51.0
N3016	10.6	69.1	70.9	10.2	68.9
N2025	10.3	--	99.9	10.5	78.2*
N3025	12.0	--	106.4	12.1	87.1*
H2016	18.2	52.3	56.0	17.8	56.9
H3016	20.0	77.0	98.0	20.1	82.9
H2025	21.8	122.3	134.9	21.5	126.9
H3025	21.7	168.0	200.7	22.4	173.5
HY2016	12.7	51.3	54.1	12.4	57.2
HY3016	12.0	74.7	93.3	12.3	83.4
HY2025	13.1	121.3	124.1	13.3	125.0
HY3025	14.3	158.7	181.1	13.9	176.3

* Beams failed by compression failure mode.

The ultimate moment capacity was also calculated and checked with the values obtained experimentally. Two types of stress distributions were tried; the first was the parabolic stress distribution [5] for N, H, and HY beams and the triangular stress distribution [6] for H and HY beams only.

In the parabolic stress distribution the design stress intensity coefficient α_1 is taken as 0.85 and the neutral axis depth is multiplied by the factor β_1 which is taken as 0.85 and 0.65 for normal and high strength concrete respectively. While for triangular stress distribution for high strength concrete α_1 is taken as 0.75 and β_1 as 0.65. From the calculated ultimate moments obtained using both stress distributions mentioned above, a comparison was carried out with experimental results. From the obtained results it was found that the parabolic stress distribution gave better predicted values as compared to triangular stress distribution. Table 2 shows predicted values using parabolic stress distribution. The measured ultimate values were slightly higher that the calculated ones. The reason for this was because the calculations of the ultimate capacity of the beams were based on the yielding strength of the steel (460 MPa), while the ultimate strength of this steel was 635 MPa.

Deflection and Ductility of Concrete Beams

The mid span deflections of the three types of concrete beams were monitored from the on setting of loading till the complete failure of the beams. Figure 1 shows the deflection of the investigated beams. From this figure it can be seen clearly that the deflection of the hybrid beams and the high strength beams were almost identical, and lower than that of the normal strength concrete beams at all loading levels. It can also be seen that using high strength concrete beams and hybrid concrete beams resulted in increasing the ductility of the concrete beams. Brittle compression failure was observed in beams N2025 and N3025 (because the steel ratio in these beams exceeded the balanced steel ratio for the 20MPa concrete), this type of failure changed to tension failure in both the corresponding high strength concrete beam (H2025, H3025) and the hybrid concrete beam (HY2025, HY3025). From Figure 2 and Table 3 it can be seen clearly that the deflection of the high strength beams and the hybrid strength beams at yielding were very similar to each other, beyond the yielding of steel reinforcement the high strength beams exhibited the highest ultimate deflection accompanied with the highest load carrying capacity compared with the other two types of beams.

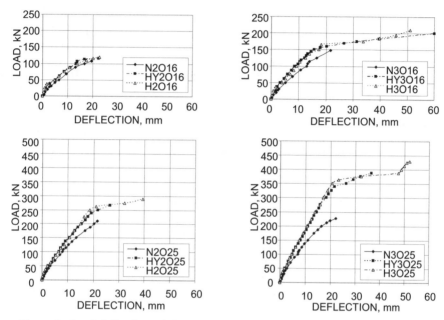

Figure 2   Load deflection relationship for normal, high and hybrid strength beams

Table 3   Ductility index of concrete beams

BEAM NOTATION	DEFLECTION AT SERVICE MOMENT, mm	DEFLECTION AT YIELDING MOMENT, mm	DEFLECTION AT ULTIMATE MOMENT, mm	DUCTILITY INDEX
N2O16	9.40	16.7	19.8	1.19
N3O16	7.62	21.6	23.0	1.06
N2O25	8.21	--	21.6	1.00
N3O25	10.70	--	22.2	1.00
H2O16	11.40	19.0	24.7	1.30
H3O16	12.54	18.1	51.0	2.82
H2O25	10.36	21.1	39.5	1.87
H3O25	11.90	23.2	52.0	2.24
HY2O16	12.30	13.6	22.4	1.65
HY3O16	9.62	20.2	60.0	2.97
HY2O25	13.20	21.7	28.5	1.31
HY3O25	14.47	21.7	36.4	1.68

In addition, Table 3 shows the ductility index of the all the concrete beams investigated. The ACI committee 363 [6] defines the ductility index as the ratio of the deflection at ultimate load to the deflection at the yielding. The ductility indexes of both the high strength concrete and hybrid concrete beams were higher than that of the corresponding normal strength concrete beams. This proves that using high strength concrete in the compression zone of concrete beams is beneficial and increases the ductility of the beams, thus permitting the use of higher steel ratios as compared to normal strength concrete. The ductility indexes for the normal strength beams ranged between 1.00 and 1.19 while for the high strength beams ranged between 1.30 and 2.82 and for the hybrid beams between 1.31 and 2.97. In order to evaluate the load capacity versus mid span deflection relationship of the three types of beams, a new concept for comparison was implied. This concept was simply the deflection to applied moment ratio. This ratio was calculated at three loading levels, namely, service, yield and ultimate loads (service load is taken 60% of the ultimate load capacity). The result of this comparison is given in Table 4. From this table it can be seen that for all the three types of the investigated beams, the deflection to moment ratio increased with the increase of the applied moment on the beam, that is the rate of increase in the deflection of the beam was greater than the increase in the applied load. On the other hand this ratio decreased with the increase in the amount of steel reinforcement, for all the three types of beams investigated.

Flexural Crack Widths of Concrete Beams

Table 5 shows the crack width of the three categories of beams investigated (N, H and HY), this table gives the crack widths at three different loading stages, that are, at service load, yielding load and ultimate load stages. Table 5, also shows the number of cracks and average crack spacing forming within the beam span (within a length of 2800 mm.). From these results, it can be seen that the hybrid strength beams exhibited the narrowest cracks compared to the other two types of beams; this behaviour took place in all three levels of loading, i.e. service load stage, yielding load stage and ultimate load stage. This behaviour can be attributed to the following two reasons:

The hybrid beams benefited from the high strength compression zone to increase its rigidity and decrease its curvature during loading stages as compared to normal strength concrete beams. The lower deflection values recorded in these beams are simply due to the higher rigidity of the hybrid beams compared to normal strength beams. On the other hand, there was no great difference in the rigidity of the hybrid beams and the high strength concrete beams, this is because the compression zone of the beam as well as the steel reinforcement are the two main parts of the beam cross section that determines its equivalent cracked moment of inertia, and the concrete in the tension zone only have a small effect on the moment of effective moment of inertia of the cracked beam.

In high strength concrete beams as compared to normal strength beams, it can be said that with the increase in concrete compressive strength, both the flexural strength of the concrete as well as the bond strength between the steel reinforcement and the concrete will be increased, but the flexural strength of concrete was found to increase at a greater rate as compared with the bond strength. The combined effects of the two properties of concrete (flexural strength and bond strength) will result in smaller number of cracks, wider crack spacing and wider crack widths in high strength concrete as compared to normal strength concrete. In the hybrid concrete beams, the lower part of the beam (tension zone) is cast using normal strength concrete, thus as cracking will take place, the beam will benefit from the superior behaviour of the normal strength concrete as compared to the high strength concrete, as the latter exhibited wider crack spacing [9, 10].

Table 4   Deflection to load ratio for normal, high and hybrid strength beams

BEAM NOTATION	DEFLECTION TO SERVICE MOMENT RATIO	DEFLECTION TO YIELDING MOMENT RATIO	DEFLECTION TO ULTIMATE MOMENT RATIO
N2016	0.294	0.358	0.503
N3016	0.179	0.312	0.312
N2025	0.137	--	0.220
N3025	0.151	--	0.209
H2016	0.247	0.363	0.405
H3016	0.213	0.235	0.520
H2025	0.128	0.173	0.298
H3025	0.133	0.137	0.259
HY2016	0.235	0.265	0.414
HY3016	0.172	0.270	0.643
HY2025	0.177	0.187	0.201
HY3025	0.122	0.137	0.229

Table 5   Crack widths in concrete beams at different loading levels

BEAM NOTATION	CRACK WIDTH AT SERVICE LOAD, mm	CRACK WIDTH AT YIELDING LOAD, mm	CRACK WIDTH AT ULTIMATE LOAD, mm	TOTAL NUMBER OF CRACKS	AVERAGE CRACK SPACING, mm
N2016	0.69	1.23	1.44	26	107.7
N3016	0.49	1.15	1.28	31	90.3
N2025	0.37	--	0.62	32	87.5
N3025	0.34	--	0.56	40	70.0
H2016	0.58	1.14	1.35	23	121.7
H3016	0.46	1.00	1.05	26	80.0
H2025	0.39	0.59	0.61	27	103.7
H3025	0.29	0.44	0.53	32	87.5
HY2016	0.29	0.91	1.22	29	96.6
HY3016	0.27	0.69	0.97	39	71.8
HY2025	0.25	0.57	0.59	27	103.7
HY3025	0.23	0.39	0.46	36	77.8

In order to evaluate the cracking behaviour of the types of beams in reference to their cracking characteristics and load carrying capacity, the ratio of crack width to applied load was found, and a comparison is given in Table 6 below.

From this table it can be seen that the hybrid beams always gave the lowest (crack width to load ratio) as compared to the other two types of beams; normal and high strength beams. Therefore it can be said that the hybrid beams exhibited better cracking behaviour as compared to the other two types of beams investigated.

Table 6   Ratio of crack width to applied moment at different loading levels

BEAM NOTATION	CRACK WIDTH TO SERVICE MOMENT RATIO, mm/kN·m	CRACK WIDTH TO YIELDING MOMENT RATIO, mm/kN·m	CRACK WIDTH TO ULTIMATE MOMENT RATIO, mm/kN·m
N2016	0.022	0.026	0.027
N3016	0.012	0.016	0.018
N2025	0.006	--	0.006
N3025	0.005	--	0.005
H2016	0.017	0.022	0.024
H3016	0.008	0.013	0.011
H2025	0.005	0.005	0.005
H3025	0.002	0.003	0.003
HY2016	0.009	0.018	0.023
HY3016	0.005	0.009	0.010
HY2025	0.003	0.005	0.005
HY3025	0.002	0.002	0.003

COST EVALUATION OF HYBRID CONCRETE BEAMS

In order to evaluate the reduction in cost by using hybrid reinforced concrete beams as compared to normal strength or high strength beams, the cost of each material was assumed according to present local market prices. A relative comparison was carried out, putting the prices of each material in random unit price and is shown in Table 7.

Table 7   Price of material

MATERIAL	COST, units/kg
Cement	0.20
Aggregates (fine and coarse)	0.01
Steel reinforcement	0.60
Superplasticizer	4.00

From the mix proportions used, each mix contained the quantities of materials given in Table 8. This table also gives the cost of the two concrete mixes used in this research.

Table 8   Concrete mix costs

MATERIAL	Mix 20 (i.e. 20 MPa strength)	Mix 70 (i.e. 70 MPa strength)
Cement	333 kg/m^3	437 kg/m^3
Aggregate	1761 kg/m^3	1917 kg/m^3
Superplasticizer	--	6.5 kg/m^3
Total cost per 1 m^3 concrete	84.2 units	132.5 units

According to the above information, the cost of the concrete only in the normal, high and hybrid strength beams is 12.16 units, 19.14 units and 14.44 units respectively. This cost did not include steel reinforcement. By including the cost of steel reinforcement provided in each of the investigated beams, the total cost will be as shown in Table 9. Also Table 9 shows a comparison made between the three types of concrete beams by implying the "Cost / Ultimate moment capacity" ratio, in order to give a clear picture of the economical use of hybrid concrete beams. From this comparison the following conclusions can be withdrawn:

1- At low steel ratios, the high strength beams are more expensive than normal strength beams, while the hybrid beams have very close cost to normal beams.
2- At steel ratios of 1.43% and above, hybrid strength beams are cheaper than the normal strength beams, while the high strength beams remains more expensive that normal strength beams even when load capacity of the beam is considered.
3- Further reduction in cost of the hybrid strength beams can be obtained when the high strength concrete part in the hybrid beams is designed exactly with a depth equals to the compression zone of the beam.

Table 9   Cost – Moment capacity analysis of normal, high and hybrid strength beams

BEAM NOTATION	STEEL RATIO, %	TOTAL BEAM COST, units	COST TO MOMENT CAPACITY RATIO, units/kN·m
N2016	0.95	17.84	0.460
N3016	1.43	20.68	0.384
N2025	2.37	26.03	0.333
N3025	3.56	32.96	0.378
H2016	0.95	24.81	0.578
H3016	1.43	27.65	0.438
H2025	2.37	33.00	0.340
H3025	3.56	39.93	0.289
HY2016	0.95	20.12	0.469
HY3016	1.43	22.96	0.364
HY2025	2.37	28.31	0.292
HY3025	3.56	35.24	0.255

CONCLUSIONS

1. The use of hybrid concrete beams resulted in an increase in the ultimate load capacity of the beams; the effect of using high strength concrete in the compression zone was beneficial in highly reinforced concrete beams. Improvement in load carrying capacity of 70% was obtained in hybrid beams while the improvement in similar high strength beams was 88.6% as compared to normal strength beams. The use of hybrid beams permitted increasing the maximum ratio of steel reinforcement that can be used in reinforced concrete beams.

2. The deflection up to yielding of steel reinforcement of both the hybrid and high strength concrete beams were very similar to each other, and less than that of the corresponding normal strength beams, this behaviour was because using normal strength concrete in the

tension zone of the hybrid beams did not have an adverse effect on its flexural rigidity (cracked moment of inertia). On average the deflection of hybrid beams was 82.7% of that of normal strength beams while that of the high strength beams was 78% at similar loading levels (only about 5%) lower than hybrid beams).

3. The ultimate deflection of the hybrid beams was between those of the normal strength and high strength beams. The ductility index of the hybrid beams and the high strength were higher than that of the corresponding normal strength beams, in all the investigated beams. The ductility indexes obtained were 1.00-1.19, 1.30-2.82 and 1.31-2.97 for normal, high and hybrid strength beams respectively.

4. Crack widths in hybrid concrete beams were narrower than those of normal and high strength concrete beams but very close to high strength beams, and at all loading stages. This behaviour was due to the higher flexural rigidity of the hybrid beam as compared to normal strength beams, and to the closer crack spacing in normal strength concrete as compared to high strength concrete. At yield, the maximum crack width in hybrid beams were only about 64% and 82% (on average) of corresponding normal and high strength beams respectively. As a result the crack widths to applied moment ratio at all loading stages were lower for hybrid concrete beams as compared to both the normal and high strength concrete beams.

5. Using hybrid concrete beams proved to be more economical compared to normal strength and high strength beams, the higher the flexural steel ratio level the more economical the beam will be.

REFERENCES

1. ASTM C33-93a, "Specification for Concrete Aggregates", ASTM Annual Book, Vol. 04.02.

2. ASTM C39-93a, "Test Method for Compressive Strength of Cylindrical Concrete Specimens", ASTM Annual Book, Vol. 04.02.

3. ASTM C469-94, "Test Method for Static Modulus of Elasticity and Poisson's Ratio of Concrete in Compression", ASTM Annual Book, Vol. 04.02.

4. ASTM C78-94, "Test Method for Flexural Strength of Concrete Using Simple Beams With Third Point Loading", ASTM Annual Book, Vol. 04.02.

5. ACI COMMITTEE 318M-02, "Building Code Requirements for Reinforced concrete and Commentary ", (ACI – 318M – 02/ ACI – 318 RM – 02), American Concrete Institute, 2002, 443 p.

6. ACI COMMITTEE 363-92, "State of the Art Report on High Strength Concrete", ACI Manual of Concrete Practice Part 1, Detroit Michigan, 55 p.

7. "High Performance Concrete Structural Designers Guide", U.S. Department of Transportation, Federal Highway Administration, First Edition, March 2005, 128 p.

8. TEPFERS, R., "A Theory of Bond Applied to Overlapped Tensile Reinforcement Splice for Deformed Bars", Publication 73.2, Division of Concrete Structures, Chalmers University of Technology, Goteborg, Sweden, 1973, 328 p.

9. TENG AND YE, Z., "Bond Behaviour of Deformed Bars in High Strength Concrete", Bond in Concrete: From Research to Practice. Proceedings of the CEB international Conference, Riga Technical University, Riga Latvia, Oct. 1992, Vol. 2, Topic 3-7, pp. 4-18.

DURABILITY OF CONCRETES CONTAINING TWO NATURAL POZZOLANS AS SUPPLEMENTARY CEMENTING MATERIALS

A A Ramezanianpour

A Rahmani

Amirkabir University of Technology

Iran

ABSTRACT: Concrete deterioration and corrosion of reinforced concrete structures is a major problem in marine environments. Therefore, there is a substantial need for the construction of different concrete structures in such severe environments. Survey of newly built and old concrete structures in such regions show that many of these structures are not able to satisfy their minimum service life requirements. The use of special cements and pozzolans in corrosive regions has shown desirable performance of reinforced concrete and enhanced the durability of concrete. Results of accelerated and long term tests in simulated environments show that supplementary cementing materials can enhance the durability of concrete. In this study, concrete specimens containing two natural pozzolans, namely Trass and Pumice have been thoroughly investigated. Tests conducted include compressive strength, permeability, chloride diffusion, corrosion of reinforcing bars and sulfate attack all at different ages. Supplementary cementing materials are considered as variables. Results of one year tests are presented in this paper. The performance of concrete mixtures depends upon the type of pozzolan. Concrete mixtures containing natural pozzolans showed better performance in terms of sulfate attack, chloride permeability and corrosion when compared with plain cement control concrete mixtures.

Keywords: Natural pozzolan, Corrosion, Permeability, Sulfate attack, Durability

Professor A A Ramezanianpour was born in Tehran, Iran in 1951. He obtained his M.Sc from Tehran University and his Ph.D degree from Leeds University in Civil Engineering. He has been involved in teaching for the last 30 years in the field of concrete technology. He has served as an advisor in many national projects. He is a member of several national and international organizations. He has been awarded as distinguished professor and researcher in Iran. He is very active in the field of concrete technology and concrete durability and has published 42 books and 212 papers in International journals and conferences.

Mr A Rahmani is a researcher within Amirkabir University of Technology, Iran.

INTRODUCTION

Concrete is one of the most widely used construction materials, because of its good durability to cost ratio. However, when subjected to severe environments its durability can significantly decline due to corrosion of embedded reinforcement and/or degradation of the concrete. As the demand for construction in harsh environments increases, so does the concern for long service lives of these structures. Typically, concrete structures are designed to perform, even in aggressive environments, for 50 to 100 years with minimal maintenance. Polymer modified mortars are known as chemically resistance concrete materials have not been widely used due to economical problems and environmental restrictions. Instead, mineral admixtures or supplementary cementitious materials are commonly used in concrete because they may improve durability. Apart from an obvious economic benefit, the use of pozzolans improves the ITZ of aggregates and therefore creates a more durable concrete. It is recognized that pozzolanic activity is influenced by the chemical and mineralogical composition of both the pozzolan and the cement, their relative finenesses, environment conditions, curing time and the effect of admixtures.

As for pozzolans in general, substitution of Portland cement reduces the strength at early ages, but the difference reduces at later ages. The heat of hydration of concretes containing natural pozzolans is less than that of plain Portland cement concretes. The permeability of concretes containing natural pozzolans is also less than for plain Portland cement concretes. Therefore, concretes containing natural pozzolans are more durable based on several reported results especially in sulfate attack, alkali aggregate reaction and corrosion of bars [1, 2, 3, 4]. Gollop, Taylor and Al-Amoudi et al. reported that the better resistance of the blended cements in sulfate environments was due to their lower calcium hydroxide content [5, 6]. Malhotra investigated the effect of cement type on corrosion rate. In this research, cements containing natural pozzolans showed better performance when compared with other cements [7]. Corrosion of concretes containing plain and various pozzolanic materials have been investigated by many researchers [8, 9, 10, 11]. In many cases the corrosion of reinforced concretes due to chloride attack was less in concretes containing pozzolanic cements.

In the present study, usage of two natural pozzolans, namely Trass and Pumice, as supplementary cement materials have been thoroughly investigated. Tests conducted include compressive strength in water and sulfate solutions, chloride diffusion, corrosion of reinforcing bars and carbonation all at different ages. The water-cementitious material ratio is considered constant for all specimens. Supplementary cementing materials are considered as variables. Cement replacement percentage were varied from 10-30%. Results of one-year tests are presented in this paper. The performance of concrete mixtures containing natural pozzolans was better than for the plain cement control concrete mixtures in terms of sulfate attack, chloride permeability, carbonation depth and corrosion. Also concretes containing natural pozzolans with 22.5% replacement level showed the best performance in all durability tests when compared with other concrete mixtures.

EXPERIMENTAL PROGRAM

Materials

The materials used in this investigation were locally sourced and they satisfied the requirements of respective Iranian Standards.

Cement – ASTM Type II Portland cement was used in this investigation.

Pozzolan – Trass and Pumice as two natural pozzolans were used in the mixtures. Table 1 presents the results of typical chemical composition of the Type 2 Portland cement and pozzolans.

The coarse and fine aggregates used in this investigation were 5-20 mm and 0-5 mm siliceous crushed river gravel and silica river sand respectively. The properties of aggregates are shown in Table 2.

Table 1   Chemical composition of cement and cement replacement materials

	SiO_2	Al_2O_3	Fe_2O_3	MgO	CaO	SO_3	$Na_2O(eq.)$
Cement	21.24	4.32	4.51	3.35	60.08	1.9	1.5
Trass	67.8	12.2	2.22	1.58	3.67	0.067	3.28
Pumice	64.6	17.3	3.86	1.34	4.6	0.35	5.56

Table 2   Aggregate properties

	Specific gravity	(%) Absorption
Sand	2.51	1.90
Gravel	2.57	2.95

Superplasticizer – The superplasticizer was a conventional melamine-based admixture with a solids content of 40% and a PH of 8. The superplasticizer used in the concrete mixes to achieve a slump between 50-80 mm.

Steel Reinforcing Bars – The steel reinforcing bars met the requirements of Grade 60 of ASTM A615/A 615 M.

Concrete Mixtures

Three concrete mixtures namely SC (control mix), CT (Trass replacement) and CP (pumice replacement) were designed with varying cement replacement percentage for investigation of their durability. Water-cementitious materials ratios (w/cm) were constant for all mixtures and equal to 0.45. Cement replacement levels were considered equal to 10, 22.5 and 30%. Mixture proportions of concrete are summarized in Table 3. The slump of the fresh concretes was kept between 5 to 8 cm. The $10 \times 10 \times 10$ cm cubes were used for compressive strength in water and 5% sodium sulfate environments and carbonation tests. The 10×5 cm and 10×25 cm cylindrical specimens with 10 cm diameters were used for rapid chloride penetration tests and corrosion potential, respectively. In order to measure the corrosion potential, ϕ 12 steel bars were embedded at center of cylindrical specimens. After casting, the specimens were cured in water, at laboratory environment, 5% sulfate environment and 11% NaCl tidal environment related to the test methods.

Table 3   Mixture Proportions

Mixture	CRM type	w/cm	Cement (kg)	CRM (kg)	Sand (kg)	Gravel (kg)
SC	-	0.45	420	-	815	995
CT1	10% Trass	0.45	378	42	815	995
CT2	22.5% Trass	0.45	325.5	94.5	815	995
CT3	30% Trass	0.45	294	126	815	995
CP1	10% Pumice	0.45	378	42	815	995
CP2	22.5% Pumice	0.45	325.5	94.5	815	995
CP3	30% Pumice	0.45	294	126	815	995

Test Methods

Chloride diffusion in concrete mixtures was investigated with Rapid Chloride Penetration Test (RCPT). In this test the sum of the transmitted electron charge was measured and compared with standard values for evaluating the quality of concretes. A portable corrosion measurement device measured half-Cell potential. Polarization resistance and concrete resistance were measured by DC impedance technique (Galvanostatic Pulse Technique). The applied currents were normally in the range of 10 to 100 μA and typical pulse durations were between 5 to 30 seconds [12].

The Micro-Cell corrosion current was estimated based on the measured polarization resistance using the following expression:

$$I_{corr}=B/R_p$$

Where, I_{corr} = the micro-cell corrosion current density in μA/cm^2, B = an empirical constant assumed to be 25 mV for actively corroding steel and 50 mV for passive steel and R_p= the polarization resistance.

Carbonation depth of the specimens was measured by spraying a 1% phenolphthalein solution on freshly cut surfaces.

The compressive strength in the water and sulfate environments was measured for all mixtures.

RESULTS AND DISCUSSION

Corrosion Potential, Electrical Resistance and Micro-Cell Corrosion

The corrosion potential, Electrical Resistance and Micro-Cell corrosion current density were measured at 90, 180, 270 and 360 days. Test results are illustrated in Tables 4-5 and Figures 1-3.

Table 4   Current density in the tidal region ($\mu A/cm^2$)

code	90 days	180 days	270 days	360 days
SC	0.81	0.88	0.89	1.35
CT1	0.59	0.69	0.63	0.93
CT2	0.66	0.71	0.66	0.69
CT3	0.63	0.62	0.70	0.69
CP1	0.71	0.70	0.62	0.97
CP2	0.51	0.55	0.51	0.59
CP3	0.60	0.56	0.68	0.66

Table 5   Concrete resistance in the tidal region (kohm)

code	90 days	180 days	270 days	360 days
SC	1.2	1.15	1.05	1
CT1	1.2	1.2	1.1	1.05
CT2	1.3	1.25	1.2	1.2
CT3	1.4	1.3	1.2	1.15
CP1	1.25	1.15	1.05	1.05
CP2	1.4	1.35	1.25	1.25
CP3	1.4	1.3	1.25	1.2

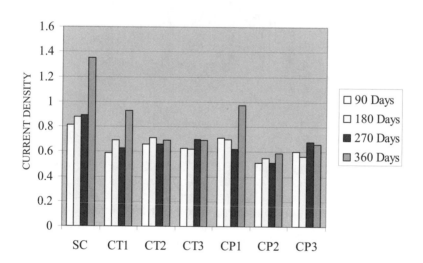

Figure 1   Current density in the tidal region.

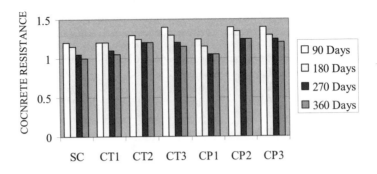

Figure 2   Concrete resistance in the tidal region.

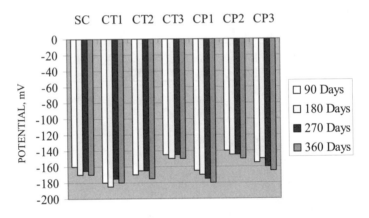

Figure 3   Half-cell potential in the tidal region.

For the tidal region, concrete made with 22.5% and 30% natural pozzolan showed more negative potential. For concretes with 10% natural pozzolan negative potential was also observed. The control mixture was the worst in terms of negative potential. Higher current density and lower electrical resistance were observed for the specimens made plain cement when compared with other specimens. For all concrete mixtures, current density increased and electrical resistance decreased with time. Concretes containing 22.5% Pumice and 22.5% Trass showed lower current density and higher electrical resistance, respectively.

Carbonation Depth and Chloride-Ion Diffusion

Test results of chloride-ion diffusion at different ages are included in Table 6 and and plotted in Figure 4. Obviously, the charge passed is substantially reduced with incorporation of pozzolans as compared to control concretes at the different ages. The incorporation of 10% and 22.5% of Trass and Pumice reduces the charge passed. These reductions are negligible when replacement level is considered equal to 30%. The long-term depth of carbonation of the mixtures is shown in Figure 5. Concretes containing 22.5% and 30% Trass and Pumice showed low carbonation depth too. In these tests, control concretes showed the worst performance.

Table 6   RCPT test results (coulomb)

Code	90 days	180 days	270 days	360 days
SC	3311	3008	2754	2405
CT1	2401	2159	2002	1894
CT2	2311	2042	1896	1693
CT3	2154	1853	1745	1596
CP1	2673	2211	1915	1887
CP2	2400	2010	1705	1552
CP3	2443	2018	1805	1608

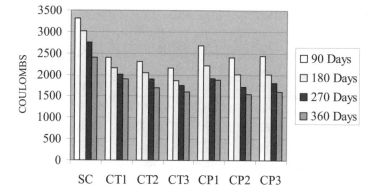

Figure 4   RCPT test results.

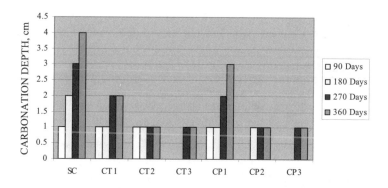

Figure 5   Carbonation depth (cm).

Compressive Strength in Water and Sulfate Solution

Compressive strength test results for concrete mixtures under standard curing condition in the water and 5% sodium sulfate solution are summarized in Tables 7 and 8. The results are also shown in Figures 6-7. The strength development of control concrete is rather good. The 7, 28 and 90-day strengths are 27, 40.8 and 50.2 MPa, respectively. At 10% replacement, the strengths of concretes containing Trass and Pumice are also high between 103-105% and 81-97% of those of control concrete at the same age respectively. For 22.5% replacement level, reductions in strength at 7 days are apparent for the mixtures. Their 7-day strengths are 82% of that of control concretes at the same age. At the age of 28 days, their strengths are 90–102% of that of control concretes. The low early and later age strengths development is the common feature of pozzolanic materials.

It can be seen that the sulfate solution has affected the control concrete mixtures by reducing its compressive strength after one year. No reduction in the compressive strength of concrete mixtures containing pozzolans was observed after one year. This clearly reveals the fact that concretes containing natural pozzolans show better performance in sulfate environments. Also the 22.5% replacement level of Trass and Pumice has better performance than the other mixtures at other replacement levels.

Table 7   Compressive strength in the water (MPa)

Code	7 days	28 days	90 days	180 days	270 days	360 days
SC	27.0	40.8	50.2	50.4	54.7	55.5
CT1	27.8	42.5	52.8	54.0	59.8	58.8
CT2	22.2	36.8	42.5	43.2	47.2	46.0
CT3	19.8	34.0	38.5	40.2	45.8	43.2
CP1	21.8	37.0	48.5	46.2	46.0	53.2
CP2	22.2	41.5	51.2	52.5	54.5	58.8
CP3	18.8	31.0	37.5	36.8	43.0	46.2

Table 8   Compressive strength in the sodium sulfate solution (MPa)

Code	7 days	28 days	90 days	180 days	270 days	360 days
SC	26.0	39.3	44.7	35.5	41.0	42.0
CT1	27.8	44.0	57.0	59.0	57.0	58.0
CT2	23.5	37.5	40.8	45.0	48.2	49.0
CT3	21.0	37.0	36.8	40.2	44.2	45.0
CP1	22.5	36.8	47.0	45.0	53.8	51.8
CP2	21.2	40.2	50.2	51.5	54.8	57.0
CP3	19.2	30.5	40.5	39.2	46.5	45.0

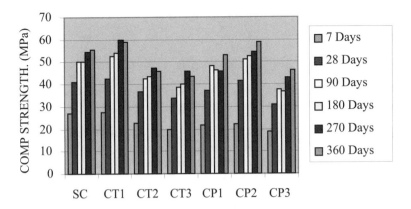

Figure 6   Compressive strength in the water .

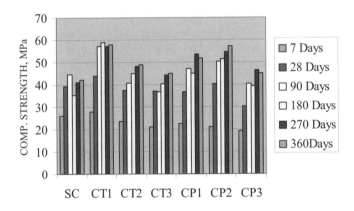

Figure 7   Compressive strength in the sodium sulfate solution .

CONCLUSIONS

From the results obtained in this investigation, the following conclusions can be drawn:

1. Compressive strength of concretes containing pozzolan was lower at early ages when compared with control concrete mixtures. However the difference is less at longer ages.

2. The use of Trass and Pumice as supplementary cement material reduces the permeability of concretes and consequently increases its durability.

3. Concretes containing natural pozzolans with 22.5% replacement level showed the best performance in all durability tests when compared with other concrete mixtures.

4. Concretes containing natural pozzolans with 10% replacements level showed better performance of than the control mix in all durability tests.

5. The worst performance was observed for the control concrete mixtures in sulfate and chloride solutions.

REFERENCES

1. R.N.SWAMY. Cement Replacement Materials, Surrey University Press Vol. 3, 1986.

2. V.M. MALHOTRA, Supplementary Cementing Materials for Concrete, Minister of Supply and Services and Services Canada, Ottawa, Canada, 1987.

3. RAMAZNIANPOUR, A.A, HILLEMEIER, B, POURKHORSHIDI, A.R. PARHIZKAR, T, RAIES GHASEMI, A.M. Performance of Pozzolanic Cement Concretes in hot and Aggressive Environments, 7th International Conference Concrete in Hot and Aggressive Environments, Bahrain, October 2003

4. RAMAZNIANPOUR, A.A, RADFAR, MOSLEHI, MAGHSOODI, Performance of a different Pozzolanic Cement Concretes under Cyclic Wetting and Drying, 6th CANMET/ACI International Conference on Fly Ash, Silica Fume, Slag and Natural Pozzolans in Concrete, ACI SP 178-39, 1998, pp. 759-777.

5. R.S.GOLLOP, H.F.W.TALOR. Micro Structural and Micro Analytical Studies of Sulfate Attack: IV. Reaction of a Slag Cement Paste with Sodium and Magnesium Solutions, Cement and Concrete Research 26 (7), 1996, pp. 1013-1028.

6. O.S.B.AL-AMOUDI, M. MASLEHUDDIN, M.M. SAADI, Effect of Magnesium and Sodium Sulfate on the Durability Performance of Plain and Blended Cements, ACI Mater J 92 (1), 1995, pp.7-12

7. MALHOTRA, V.M.CARETE, G.G & BREMNER, T.W. Current Status of CANMET'S Studies on the Durability of Concrete Containing Supplementary Cementing Materials in Marine Environment", ACI Special Publication, SP109, 1988, pp. 31.

8. THOMAS, M.D.A., JONES, M.R. A Critical Review of Service Life Modeling of Concrete Exposed to Chlorides, Concrete in the Service of Mankind: Radical Concrete Technology, (eds, R.K. Dhir and P.C. Hewlett), E. and F.N. Spon, London, 1996, pp. 723-736

9. BERKE, N. S., HICKS, M.C. Estimating the life Cycle of Reinforced Concrete Deck and Marine Piles Using Laboratory Diffusion and Corrosion Data", Corrosion Forms and Control for Infrastructure, ASTM STP 1137, V. Chaker, ed., American Society Testing and Materials, Philadelphia, 1992.

10. MOHAMMED, T.U, YAMAJI, T, TOSHIYUKI, A. HAMADA, H. Corrosion of Steel Bars in Cracked Concrete Made with Ordinary Portland, Slag and Fly Ash Cement, Proceeding of the 7th CANMET/ACI International Conference on Fly Ash, Silica Fume, and Natural Pozzolans in Concrete, Madras, India, 2001.

11. MOHAMMED, T.U, OTSUKI, N, HISADA, M, HAMMADA, H. Marine Durability of 23-Year-Old Reinforced Concrete Beams, Fifth CANMET/ACI International Conference on Durability of Concrete, Barcelona, Spain, ACI SP 192-65, 2000, pp. 1071-1088

12. ASTM C876, Standard Test Method of Half-Cell Potentials of Uncoated Reinforced Steel in concrete, ASTM Publication 1995.

MODIFIED SAND CONCRETE

R M Kettab   A Bali

Polytechnic National School

Algeria

A Alliche

University of Paris

France

ABSTRACT. Sand concrete is considered as one of the newly used materials in civil engineering permitting to valorise natural resources such as sand. Present study aimed to find a formula for sand concrete to be use in construction. The proposed methodology to obtain the optimal formula was based on the criteria of strength, workability and compactness. Performance of concrete modified using rubber powder was also studied. Ground rubber was added to the granular skeleton. The optimal proportion of modified sand concrete which gives the best performance was identified and its main characteristics determined.

Keywords: Sand concrete, Radioactivity, Dune sand, Formulation, Polymers, Rubber, Valorisation, Characteristics.

Dr R Mitiche Kettab is lecturer in Civil Engineering and researcher of the Construction and Environment Laboratory in the Polytechnic National School (E.N.P.) in Algiers. His research interests focus on mechanical behaviour of concrete, performance of road and building materials, dune sand concrete, and the use of local materials and reuse of materials.

Dr A Bali is professor in Civil Engineering and head of the Construction and Environment Laboratory in Algiers. He supervised many research students on various topics related to the development of concrete constituent materials. His research interests include building materials and their impacts on the environment, natural fibre reinforced concrete, and valorisation of demolition and industrial wastes in concrete, the use of local materials and cementitious additions, the effect of high temperature on durability of concrete and reuse of materials.

Dr A Alliche, Lecturer (HDR) in the University of Paris, France. He directed several work on concrete behaviour. His research interests focus on mechanical behaviour of concrete, modelling, durability, performance of building materials, dune sand concrete, the use of industrial by-products in construction, fatigue and fracture of concrete, recycled materials.

INTRODUCTION

Our objective was to determine the influence of the addition of the rubber powder at low percentages on the sand concrete behaviour in order to use it as a structural material. To achieve this goal, we adopted a methodology based on the criteria of strength, workability and compactness (consistency).

SAND CONCRETE

The definition of "sand concrete" is different from that of mortar owing to the fact that from the cement consumption point of view, this material contains an identical amount of binder as traditional concrete and indeed in certain cases replaces it in structural concrete [1]. Sand concrete is distinguished from traditional concrete by its large sand content, the absence or low content of fine gravel and the incorporation of additives. Sand concrete is thus a fine concrete, made up of a mixture of sand, cement, additives and water. With regard to this basic composition and to meet the needs for certain uses, other specific additives (fillers, fibres, dyes, etc.) can be used [1]. Sand concretes have then the same cement content as traditional concretes, 250 to 400 kg/m^3 with compactness being reached by complementary fines addition, generally limestone, which was replaced in this study by rubber powder. In certain uses, special concrete is required, with such characteristics where traditional concrete does not perform particularly well and sand concrete can better satisfy demands, notably regarding the following characteristics:

- Consistence: sand concrete being workable, its capacity to fill the moulds with a low degree of compaction, constitutes a significant asset.
- Cohesion and absence of segregation: the variation between dimensions of the particles is not significant; hence this material presents a rather good cohesion, low bleeding and especially a total absence of segregation problems.
- Aggregate grading and small particle size: the material would be useful in injection works, for sections with congested reinforcement and for structure repair.
- Transportation: sand concrete from the central plant to the site of use is transported by the same means as for traditional concrete. Sand concretes are generally more plastic with no risk of segregation during transport and can thus be transported for long distances.
- Surface quality, finish: the continuous grading of sand concrete results in very aesthetic, architectonic effects either on the shape of the elements or on the surface aspect.

The use of rubber powder in concrete is an alternative suggested by many authors. The rubber of tyres is suggested when the mechanical resistance is not the principal characteristic, but shock resistance is.

RUBBER

Rubber is a natural or synthetic compound characterized by its:
- Elasticity,
- Impermeability and electric resistance; the resistance of rubber to water and the majority of chemical liquids make it in fact a product adapted for coatings,
- Abrasion resistance: the coefficient of friction of rubber is high on dry surfaces and low on wet surfaces
- Flexibility

Vulcanized rubber compared to pure rubber has increased solidity and elasticity, as well as a greater resistance to temperature changes.

EXPERIMENTAL STUDY

Introduction

The study comprised three phases [2, 3]:

- Identification of the basic materials used for making sand concrete, by studying their physicochemical characteristics. In fact, the knowledge of certain parameters is necessary and constitutes a base to establish the formulation of the mixture.

- Determination of mechanical characteristics.

The first series of tests consists of elaborating various formulae of concrete of fine sand corrected using two materials (crushed sand and fine gravel) at various percentages. Tensile and compressive strength tests at different ages, as well as measurements of compactness and density were carried out [2]. In the second series of tests, rubber powder was introduced within the granular skeleton at different proportions (1 to 4%) based on formulations having given the best performance [3].

Identification of Materials Used

The determination of the physico-chemical properties of materials allows reaching a good formulation of concrete, and thus facilitates the interpretation of the obtained results. The tests were carried out in accordance with the existing standards.

Materials

Mechanical characteristics and the chemical analysis are given in Tables 1, 2 and 3. Dune sand is of siliceous nature (Figure 1), crushed sand originated from a calcareous tender rock; whereas fine gravel was sources from a hard calcareous rock.

Cement

A single type of cement, CPJ-CEM II A 32.5, was used in the various formulations. Its absolute density is equal to 3144.33 kg/m^3.

Table1   Mechanical characteristics and chemical analysis of fine gravel

		CHARACTERISTICS		
Flatness coefficient	ρ, g/cm^3	Micro Deval, %	Friability, %	Compactness, %
3.24	2.64	1.00	30.74	48.3
		Chemical Analysis, %		
Insolubles	$Fe_2O_3 +$ Al_2O_3	$CaCO_3$	CO_2 Composition water	Sulphates	Chlorides
1.08	1.52	91.58	39.98 2.14	trace	trace

Table 2   Mechanical characteristics and chemical analysis of dune sand used

m_f	ρ, g/cm^3	Sand equivalent, %	VBS blue methylene	Friability, %	Compactness, %
		CHARACTERISTICS			
0.88	2.5 1.5	88.0	0.17	13.4	59.9

Insolubles	$Fe_2O_3 + Al_2O_3$	$CaCO_3$	CO_2	Composition water	Loss on ignition	Sulphates	NaCl
			Chemical Analysis, %				
94.60	0.78	5.13	2.34	0.50	2.38	trace	0.23

Table 3   Mechanical characteristics and chemical analysis of crushed sand used

m_f	ρ, g/cm^3	Sand equivalent, %	VBS blue methylene	Friability, %	Compactness, %
		CHARACTERISTICS			
2.75	2.6 1.5	91.1	0.06	58.0	59.9

Insolubles	$Fe_2O_3 + Al_2O_3$	$CaCO_3$	CO_2	Composition water	Loss on ignition	Sulphates	Chlorides
			Chemical Analysis, %				
1.50	1.33	93.0	40.6	4.40	4.05	trace	trace

Table 4   Chemical analysis of cement

ELEMENTS	SiO_2	CaO	MgO	Fe_2O_3	Al_2O_3	SO_3	P.F	Free CaO	R.lns	Cl$^-$	Na_2O	K_2O
Content, %	21.01	63.89	1.10	3.07	5.40	2.42	1.46	0.23	1.15	0.013	0.25	0.77
Standard NA 2185 equivalent to NF P15 301	*	*	< 5.0	*	*	<4.0	*	*	*	<0.05	*	*

*Non-normalized.

Table 5   Physical-mechanical analysis of cement

ELEMENTS	DENSITY, g/l	SSB, cm^2/g	CN, %	SETTING TIME		EXP, mm	STRENGTH, MPa					
							FLEXURAL			COMPRESSIVE		
				Initial	Final		2d	7d	28	2d	7d	28d
Results	913	3043	28	1h49	3h37	0.69	5.1	7.3	8.2	22.4	43.6	58.1
Standard N.A 442/2000 equivalent to NF P15 301	*	*	*	>1h30	<10h	<10mm	*	*	*	>10	*	CPJ-CEMII/A32.5

*Non-normalized

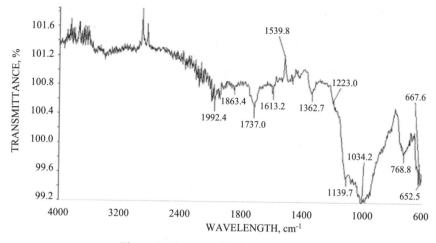

Figure 1   Dune sand Infra Red Analysis

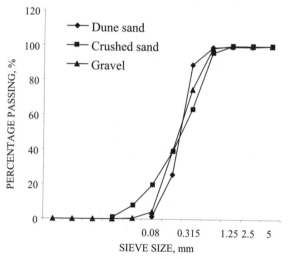

Figure 2   Grading curves

Water

Tap water was used in all tests, containing very low quantities of salts, sulphates and acids which in turn could be harmful to concrete and particularly to its durability.

Rubber powder

Rubber crumb was used as additive, which originates from the falls of the rubber used in the manufacture of car carpet. These falls are processed to granules of max. 2 mm size. Another source of crushed rubber waste had a maximum size of 1.25 mm and 31% passing 0.08 mm. The purity of this waste was about 45%.

The additive used in this study was a vinyl acetate copolymer (EVA) having a general formula $(CH_2)_n$ (Figure 3). EVA is compatible with the aggregate. This polymer is an elastic material at room temperature and has low viscosity at coating and placing temperatures of concrete. The powder shows a very fine grading (0.1 to 1 mm), its fusion temperature is between 200 and 220°C and its density is 0.8 g/cm^3 at ambient temperatures.

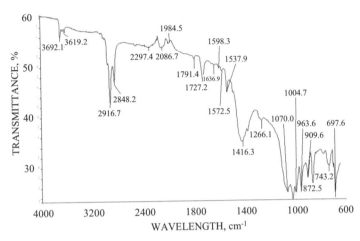

Figure 3   Rubber Infra Red Analysis

The admixtures

One type of water-reducing superplasticizer, suitable for low W/C and high performance concrete, was used. Its density was 1.18 and its normal range of use varied from 0.3 to 1% by cement weight.

Remarks on sands used

When analysing the obtained results, the following remarks can be made.

According to Table 6, neither dune sand or crushed sand can be used on its own for making concrete; because of certain characteristics which are out of the specifications of standards. Their mixture can, however, lead to good results following a mutual correction to their characteristics.

Table 6   Comparison between dune sand characteristics and crushed sand characteristics

CHARACTERISTICS	DUNE SAND	CRUSHED SAND
Grading	Gap-graded	Continuous
Fineness modulus	Lower than the standards	In the standards
Cleanliness	Very good	Out of standards
Hardness	Good	Weak
Chemical analysis	Good	-

Sand Concrete Design

Design of concrete consists of finding the best proportions of the various components once their compatibility is ascertained. Then the evaluation stage involves using predictive tests to establish the behaviour of sand concrete. The choice of these compositions is made starting from the two principal criteria of structural concrete:

- Workability
- Compactness.

Plastic sand concrete had a time of flow ranging between 7 and 10 s [5].

Test specimens used

Two types of test specimens were used:

Compressive strength values were obtained by crushing 40 mm cubes obtained by cutting the $40 \times 40 \times 16$ mm prisms to the required size with a saw.

Tensile strength, by flexure, was determined on prismatic specimens. Cubic specimens of dimensions $100 \times 100 \times 100$ mm were used for certain compositions which were of particular interests (optimal compositions) in order to determinate some rheological characteristics (Young's modulus and Poisson's ratio).

Results and their Interpretation

The basic formula has been retained, and then compressive strength tests were carried out at 7 and 28 days, whereas for the tensile strength, the tests were carried out at 28 days only. The test results are given in Table 7.

Table 7   Composition and characteristics of the basic formulation

COMPOSITION, kg/m³				WORKABILITY LCPC, s	ρ_{th}, kg/m³	ρ_a, kg/m³	R_c, MPa		R_t, MPa
Cement	Additive	Water	Sand				7d	28d	
375	1.5	232.5	1333	9	2222	2084.2	8.03	16.7	1.24

where:

ρ_{th}	theoretical density.
ρ_a	apparent bulk density.
R_c	compressive strength.
R_t	tensile strength

It was noticed that the properties of the basic concrete design were not satisfactory. To overcome this deficiency and to achieve the defined objectives of use in construction, it was necessary to seek a more continuous grading sand sort (Figures 2 and 3). In the second series of tests: rubber powder has been introduced within dune sand concrete with crushed sand addition (50% of crushed sand BSD by mass), and in sand concrete with crushed sand and gravel additions (30% gravel, 20% crushed sand BSDC). These two mixtures gave the best performance.

Remark: The W/C ratio and additive content was identical for both series.

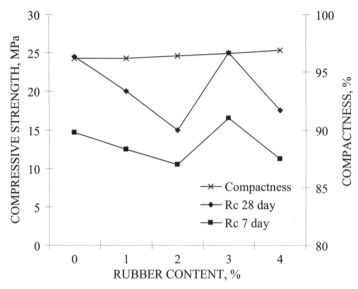

Figure 4   Performances of modified dune sand concrete with crushed sand additions

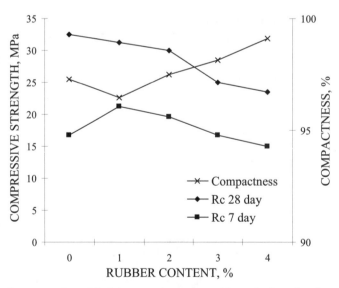

Figure 5   Performance of modified dune sand concrete with crushed sand and gravel additions

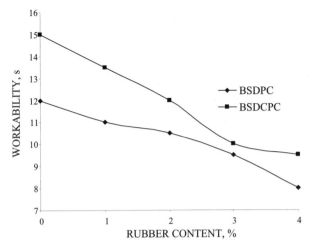

Figure 6   Workability of modified sand concrete

Determination of rheological characteristics

The main rheological parameters were measured, determining the longitudinal and transversal stress – strain behaviour on the same test specimen (cylindrical Ø160 × 320 mm) equipped with electrical gauges in the two directions. The transversal to longitudinal strain ratio, under the same compressive stress, gives the value of the Poisson's ratio (Table 8).
The elastic modulus was determined as the slope of the stress – strain curves.

Table 8   Rheological characteristics of the concretes

COMPOSITION	MODULUS OF ELASTICITY, MPa	POISSON'S RATIO
BSD0	14792.9	0.110
BSD PC3%	15822.8	0.220
BSDC0	26385.2	0.205
BSDC PC2%	20491.8	0.160
BSDC PC3%	16722.4	0.365

Compactness increases in an almost uniform way with the increase of the rubber powder content, reaching the value of 97% with more than 2% addition. This makes it possible to say that mixtures (*BSDC* PC 2%, *BSDC* PC 3%, *BSDC* PC 4%) are optimized in their solid skeleton, the particles of the two sands find a great facility to be rearranged thanks to the voids created between the grains of fine gravel. The rubber fillers intervene to fill these voids and to form a compact mixture.

In the same way, for mixture *BSD* PC, workability is improved with the increase of rubber powder content, thanks to the very low water absorption by rubber particles.

It was noticed that strength became almost constant between 0% and 2%. Beyond 2% a fall of strength is observed, this is due to the lack of intergranular cohesion caused by the rubber particles. The decrease in strength for 4% powder content is about 22%.

CONCLUSIONS

The experimental study on the tested compositions made it possible to show that the addition of rubber powder, even at low percentages, within sand concrete is disadvised when the mechanical strength is the principal target. The estimation of the decrease in strength for a 4% of rubber powder is about 22%.

However, the addition of rubber powder may provide sand concrete a better resistance towards shocks and aggressive environments. In other words it can lead to a durable sand concrete.

The modulus of elasticity of dune sand concrete corrected only with the crushed sand is lower than that of charged sand concrete, which in turn is comparable to that of ordinary concrete, because the modulus is a function of the coarser aggregate diameter.

The modulus of the witness charged concrete is more significant than that of charged concretes modified by the rubber powder.

For sand concrete the modulus of elasticity as well as Poisson's ratio increase with the addition of rubber powder.

In terms of confinement and durability, the modified rubber powder sand concrete presents the following two essential characteristics:

- a very high resistance to external aggressions over a period of 300 years;
- a high confinement capacity.

REFERENCES

1. SABLOCRETE. Béton de sable, Presse de l'école nationale des ponts et chaussées, Paris, France 1994.

2. GUENOUN R AND KETTAB R, Etude et formulation d'un béton de sable de dune, Ecole Nationale Polytechnique Alger, Algérie 2003.

3. HAMMADACHE K AND KETTAB R, Etude et formulation d'un béton de sable de dune modifié à la poudrette de caoutchouc, Ecole Nationale Polytechnique Alger, Algérie 2005.

4. NEVILLE A M, Propriétés des bétons, édition EYROLLES, Paris, France, 2000.

5. DURIEZ M AND ARRAMBIDE J, Nouveau traité de matériaux de construction, tome 1, Edition DUNOD, Paris, France, 1961.

6. CHAUVIN J J, Le béton de sable, LRPC, Bordeaux, France, 1990.

EFFECT OF SHAPE OF AGGREGATE ON WORKABILITY, STRENGTH AND DURABILITY OF CONCRETE

P Agrawal

Hindustan Construction Company Ltd

Y P Gupta

BCEOM-LASA JV

India

ABSTRACT. The shape of aggregates plays an important role in achieving the desired properties of concrete. Aggregates include angular / rounded, flaky (FI), elongated (EI) or some odd shape pieces depending upon the type of the stone and crusher plant. Out of all, flakiness and elongation affect the most the workability and permeability of the concrete mix, its cement and water demand as well as its compressive and flexural strengths. Improper shape of aggregates reduces the compactibility and cohesion characteristics of pumpable concrete. In the present investigation, the effect of shape of aggregate was investigated. Aggregate was sorted in three categories i.e. Flaky, Elongated and Normal Angular shapes. These were mixed in different proportions and their effect was observed in tests such as Aggregate Impact Value and Los Angeles Abrasion Value. Various proportions of such aggregate were mixed in preparing M 30 grade concrete mixes. The effect of different percentages of Flaky and Elongated particles on concrete workability, cube strength, flexural strength and permeability was studied. The results indicate that with the increase in flakiness, workability gets affected. However, the effect of the portion of elongated particles is less than that of flaky particles. The details of findings and their effect on compressive and flexural strength and permeability, influencing durability are reported in this paper.

Keywords: Concrete, Aggregate, Durability, Flakiness Index, Elongation Index.

Mr Prashant Agrawal, QC MANAGER, M/s Hindustan Construction Company Ltd. Allahabad Bypass Project, Allahabad, India

Dr Y P Gupta, Material Consultant, BCEOM-LASA JV, Allahabad Bypass Project, Allahabad, India

INTRODUCTION

It is well known that aggregate plays an important role in achieving the desired properties of concrete. Although aggregates constitute 70 to 80% of the total volume of concrete, very little attention is given to controlling the shape and surface texture to optimize the properties of concrete. The shape of aggregates includes angular / rounded, flaky (FI), elongated (EI) or some odd shapes depending upon the type of stone and stone crusher. Improper shape of aggregate influences the water demand for a given mix and affects workability, compactibility, and cohesion characteristics of pumpable concrete. It also influences compressive strength, flexural strength and other properties such as permeability and durability of concrete. The effect of various shapes of aggregate is given in this paper.

Shape Parameters

There are five major shape parameters which are the measures of roundness of aggregate as given below.

i.	Flakiness Ratio (FR)
ii.	Elongation Ratio (ER)
iii.	Sphericity (SPH)
iv.	Shape Factor (SF)
v.	Convexity Ratio (CR)

where,

Flakiness Ratio: The thickness to breadth ratio is defined as flakiness ratio. It is also called flatness ratio or Flakiness Index. As per definition, a particle is flaky if its thickness (least dimension) is less than 0.6 times the mean sieve size of the size fraction to which the particle belongs. i.e.

FI = Thickness / Breadth of aggregate

Elongation Ratio: The breadth to length ratio of an aggregate particle is defined as elongation ratio. It is also called Elongation Index. As per definition, a particle whose length (largest dimension) is more than 1.8 times the mean sieve size of the size fraction is said to be elongated. i.e.

EI = Breadth / Length of aggregate

Out of all, flakiness and elongation affect the most the workability and permeability of a concrete mix as well as its cement and water demand. Therefore, these two parameters of aggregate were investigated here.

Two sizes of aggregate particles: i.e. 20 and 10 mm size, which are generally used in a standard concrete mix, were chosen for the investigations. They were all dolomite limestone in the crushed form.

REVIEW OF PROVISIONS IN DIFFERENT SPECIFICATIONS

Indian Specifications

In India, there are different specifications governing the concrete mix design and its quality. These are from the Bureau of Indian Standard, Indian Road Congress etc. Some of the provisions are given below:

IS 383: This standard deals with the specification for coarse and fine aggregates from natural sources for concrete. There is no prescribed limit of FI and EI in the code.

SP 23: The Hand Book on Concrete Mixes recommends that the limit for flakiness should not be greater than 25%, when the aggregate is to be used for concrete, but there is no defined limit for elongation.

Ministry of Highways and Bridges: The Specification for Road & Bridge Works defines that the maximum value for flakiness index for coarse aggregates should not exceed 35%. However, no limit is given for elongation.

IRC 21: The Standard Specification for Road & Bridge Works only mentions that coarse aggregates shall not contain pieces of disintegrated stones, soft, flaky, elongated particles etc. However, no limits are mentioned.

US Scenario: The US specifications lay emphasis on the particle index test and flat or elongated particles test. The 'New Super seed Guidelines' allow very low value for aggregates to be flat or elongated with an aspect ratio (length to thickness ratio) greater than 5:1.

British Scenario before 2004: Prior to January 2004, the procedure for conducting the flakiness index test and the elongation index test were given in BS 812. The specification limited the flakiness index of the coarse aggregate to 50 for natural gravel and 40 for crushed or partially crushed coarse aggregate; however, for paving surfaces, lower flakiness index was recommended. In the British Standards, no limit for the elongation index was mentioned.

Present Status of British Standards: From January 2004, the British Standard for aggregates used in the construction sector has been replaced with a new series of European standards for aggregates (BS EN 13242). As per the new code, a maximum flakiness index of 35% has been specified for aggregates used for concrete in British Specifications.

EXPERIMENTAL INVESTIGATION

The coarse aggregate taken was dolomite limestone in crushed form, having a nominal size of 20 mm. It has been sorted into three categories i.e. Flaky, Elongated and normal Angular shapes.

These were mixed in different proportions and their effect was observed in terms of the Aggregate Impact Value and the Los Angeles Abrasion Value. Both elongation and flakiness ratios were varied from 0 to 50% whilst keeping the other parameter constant. The combined value of FI + EI has also been limited to 50%.

Various proportions of such aggregate were mixed in a laboratory mixer of 0.1 m^3 capacity to prepare M 30 grade concrete. Cubes (150 × 150 × 150 mm), cylinders (Ø150 mm × 160 mm) and beams (150 × 150 × 700 mm) were cast for the determination of compressive strength, flexural strength and permeability.

The Selected Concrete Mix

Concrete Grade	:	M30 (characteristic strength: 30 MPa)
Water/Cement Ratio	:	0.45
Cement	:	OPC 53 grade (350 kg)
Aggregate to Cement Ratio	:	5.52
Admixture	:	Super plasticizer (about 1% of cement)
Fine Aggregate	:	River sand (777 kg)
Coarse Aggregate	:	Crushed stone (1155 kg)

Material Properties

The properties of materials used are given in Tables 1 to 3.

Table1   Properties of cement

FINENESS, m^2/kg	SOUNDNESS	SETTING TIME, min.		SPECIFIC GRAVITY, g/m^3	COMPRESSIVE STRENGTH, MPa
		Initial	Final		
342	1.5	135	205	3.15	60.4

Table 2   Properties of course and fine aggregate

MATERIAL	WATER ABSORPTION,%	SPECIFIC GRAVITY, g/m^3	SOUNDNESS		DELETERIOUS CONTENT,%
			$Na_2 SO_4$	$Mg SO_4$	
Coarse aggregate	0.263	2.800	2.210	3.030.	0.400
Sand	1.129	2.602	4.000	-	2.500

NB as far as reactivity is concerned, both materials were innocuous.

Table3   Combined gradation of concrete mix

IS SIEVE, mm	AGGREGATE			COMBINED GRADING	GRADING LIMITS AS PER NEVILLE	
	20 mm	10 mm	river sand		lower limit	upper limit
40	100	100	100	100.00	100	100
20	92.26	100	100	97.21	95	100
10	4.99	91.09	100	63.66	55	75
4.75	0.73	9.73	96.19	41.07	36	48
2.36	0.00	1.25	83.10	33.54	30	42
1.18	0.00	0	65.90	26.36	22	34
0.60	0.00	0	51.63	20.65	16	27
0.30	0.00	0	15.85	6.34	5	12
0.15	0.00	0	3.53	1.41	0	3
0.75	0.00	0	0.60	0.24	0	2
F. M.	7.05	6.03	2.82	-	-	-

OBSERVATIONS AND DISCUSSION OF RESULTS

Table 2 gives all the observations recorded during the experimental investigations. The effect of variations in FI and EI on workability, strength, and permeability etc. is discussed below.

Workability of the Concrete Mix

The workability of the concrete mix was measured with the help of a 300 mm standard size slump cone. A small quantity of admixture (about 1%) was added to the concrete mix. Each concrete mix was examined for slump, segregation and bleeding etc. It was observed that the slump had a tendency of shearing type behaviour with the increasing percentage of flaky particles beyond 30%. No segregation was observed.

Table 2   Summary of results for the concrete mix

CHARACTERISTIC STRENGTH: 30 MPa	TARGET STRENGTH: 42 MPa
CEMENT (OPC 53 GRADE): 350 kg/m^3	AGGREGATE / CEMENT RATIO: 5.52
W/C: 0.45	FINE AGGREGATE / RIVER SAND: 0.42

Sl. No.	EI,%	FI,%	FI + EI,%	AIV,%	LAAV,%	Concrete density, g/cm^3	28 day strength, MPa compressive	28 day strength, MPa flexural	Permeability coefficient, $\times 10^{-4}$
1	0	0	0	10.82	8.60	2.557	54.44	5.65	4.20
2	0	10	10	10.86	9.30	2.553	49.35	4.92	4.40
3	0	20	20	13.14	10.60	2.547	45.89	4.46	4.50
4	0	30	30	13.71	11.32	2.522	43.87	4.19	5.00
5	0	40	40	15.43	11.60	2.528	40.85	3.42	5.10
6	0	50	50	15.71	11.82	2.515	39.75	3.03	5.30
7	10	0	10	10.83	9.20	2.551	53.78	5.45	4.00
8	10	10	20	11.21	9.36	2.545	49.10	4.87	4.10
9	10	20	30	13.24	10.72	2.544	45.72	4.40	4.20
10	10	30	40	14.00	11.40	2.541	43.56	4.11	4.90
11	10	40	50	15.55	11.72	2.539	40.62	3.31	5.00
12	20	0	20	10.88	9.44	2.547	53.22	5.38	2.80
13	20	10	30	11.71	9.70	2.545	48.87	4.72	4.00
14	20	20	40	13.51	10.72	2.544	45.59	4.35	4.20
15	20	30	50	14.50	11.50	2.540	43.36	3.60	4.60
16	30	0	30	11.70	9.60	2.544	52.89	5.32	2.50
17	30	10	40	11.81	9.72	2.543	48.77	4.68	3.40
18	30	20	50	13.70	10.74	2.542	45.42	4.29	4.10
19	40	0	40	12.76	9.80	2.531	52.22	5.26	2.60
20	40	10	50	12.86	9.80	2.525	48.61	4.50	3.30
21	50	0	50	13.71	9.82	2.503	52.00	5.18	3.00

The Effect of FI and EI on the Aggregate Impact Value and Los Angles Abrasion Value

In the case when EI = 50% and FI = 0, the following values were measured: AIV = 13.71% and LAAV = 9.82%.

Figures 1 to 5 show that the flakiness index had a larger impact on the aggregate impact value and Los Angeles abrasion value than the elongation index. This effect was increasing as the number of flaky aggregate pieces increased from 0% to 50% by a factor of 9% on an average for each 10% increase in flakiness for a fixed value of elongation. Further increasing flakiness, more than 40%, had very little effect on AIV. Figures 1 to 5 also show that the flakiness index had a larger impact on LAAV than the elongation index. It was increasing by a factor of an average 7.48% for each 10% increase in the flakiness for a fixed value of elongation. On the other hand, the elongation index had a considerably smaller effect on AIV and LAAV. It changed by about 5.3% and 2.83% for AIV and LAAV, respectively for each 10% increase in EI and fixed value of flakiness.

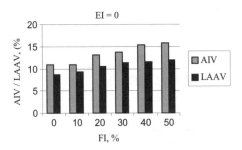

Figure 1   Value of AIV and LAAV for varying FI and keeping EI = 0%

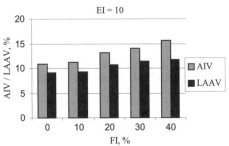

Figure 2   Value of AIV & LAAV for varying FI and keeping EI = 10%

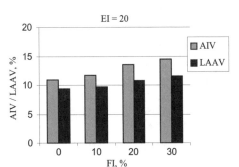

Figure 3   Value of AIV and LAAV for varying FI and keeping EI = 20%

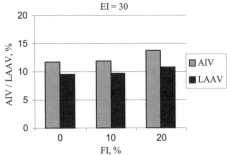

Figure 4   Value of AIV and LAAV for varying FI and keeping EI = 30%

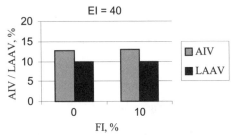

Figure 5   Value of AIV and LAAV for varying FI and keeping EI = 40%

The Effect of FI and EI on the Density of Concrete

Each cube was weighed using an electronic balance. The density of concrete was thus calculated. The variation of density is given in Figures 6 to 10 for different cases.

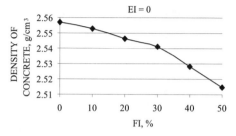

Figure 6   Density of concrete for varying FI and keeping EI at 0%

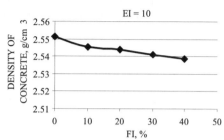

Figure 7   Density of concrete for varying FI and keeping EI at 10%

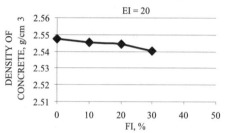

Figure 8   Density of concrete for varying FI and keeping EI at 20%

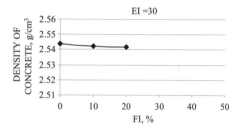

Figure 9   Density of concrete for varying FI and keeping EI at 30%

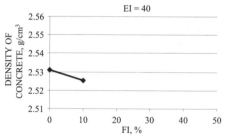

Figure 10   Density of concrete for varying FI and keeping EI at 40%

When EI = 50%, the density of concrete was 2.503 g/cm^3 keeping FI = 0, which was the minimum value in all observations. While this was only 2.515 g/cm^3 for FI = 50% and keeping EI = 0. From Figures 5 to 10, it can be seen that the elongation index had a larger impact on the density of concrete than the flakiness index. As the flakiness index increased from 0 to 50%, the density decreased by about 0.3% for every increment of 10% in the flakiness index. The elongation index also had a similar impact on density.

The Effect of FI and EI on Compressive Strength

From Figures 11 and 15, it can be seen that the flakiness index had a larger impact on the 28 day compressive strength. The decrease in strength was about 9.2% for each 10% increase in flakiness index and it decreased by 27% when FI changed from 0 to 50% for EI = 0. However, for an increase in elongation, the decrease was only about 0.5 to 2% for each increase of 10% in elongation index. The compressive strength for EI at 50 was 52.0 MPa when FI was zero.

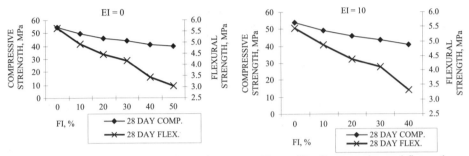

Figure 11 Compressive and flexural strengths for varying FI and keeping EI at 0%

Figure 12 Compressive and flexural strengths for varying FI and keeping EI at 10%

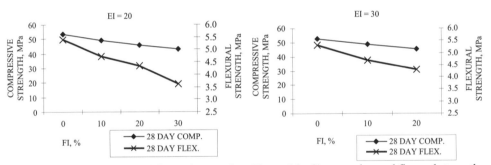

Figure 13 Compressive and flexural strengths for varying FI and keeping EI at 20%

Figure 14 Compressive and flexural strengths for varying FI and keeping EI at 30%

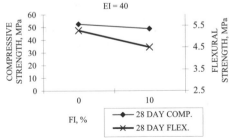

Figure 15 Compressive and flexural strengths for varying FI and keeping EI at 40%

The Effect of FI and EI on Flexural Strength

Flexural strength was calculated from the 28 days testing of the beam sized 150 mm × 150 mm × 700 mm by using the following formula.

Flexural strength $= P \times 1000 \times L / (b \times d \times d)$, for a > 200 mm
$= P \times (3000 \times a) / (b \times d \times d)$, for a > 170 mm, but less than 200 mm
$=$ Result discarded when a < 170 mm

where,

b = width of sample beam, 150 mm.
d = depth of sample at the point of failure, 150 mm.
a = distance between the line of fracture and the nearest support.
P = failure load.
L = total support length of specimen, 600 mm

The variation of flexural strength with respect to different parameters is also given in Figures 11 to 15. It can be seen that the flakiness index had a larger impact on the 28 days flexural strength than the elongation index. As the flakiness index increased from 0 to 50%, the 28 day flexural strength decreased by an average 11% for each increase of 10% of flakiness. The elongation index had a considerably smaller effect (2.0% decrease) on the flexural strength when EI increased from 0 to 50% taking FI as zero. The flexural strength for EI at 50% was 5.18 MPa when FI = 0, while it was 3.03 MPa when FI = 50% and EI = 0.

The Effect of FI and EI on the Permeability of Concrete

The permeability of concrete was determined using cylindrical specimens having 150 mm diameter and 160 mm height. These were exposed to a water pressure equivalent to 7 kg/cm² for 96 hours in the permeability apparatus shown in Figure 16.

After 96 hours, the cylinders were split under line load. The depth of penetration of water in the cylinder was measured as well as the volume of water lost was recorded. The results were interpreted as follows:

1. Average depth of water penetration in cylinder.
2. Coefficient of permeability was calculated as volume of water lost divided by volume of concrete penetrated with water i.e. Permeability coefficient = volume of water lost / Average volume of concrete showing the effect of water penetration.

Figure 16 Permeability apparatus

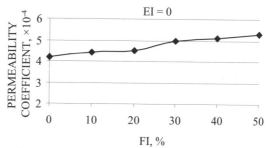

Figure 17 Permeability coefficient of concrete for varying FI and keeping EI at 0%

The permeability coefficient is plotted in Figures 17 to 21. It can be seen that the flakiness index had a larger impact on permeability than the elongation index. As the flakiness index increased, the permeability coefficient also increased considerably. This increase in the permeability coefficient was 6.5% for every 10% increment of flakiness index. The elongation index also had the same impact on the permeability coefficient.

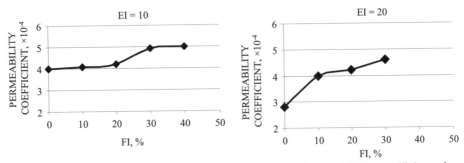

Figure 18   Permeability coefficient of concrete for varying FI and keeping EI at 10%   Figure 19   Permeability coefficient of concrete for varying FI and keeping EI at 20%

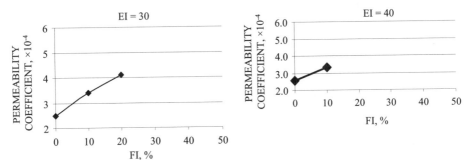

Figure 20   Permeability coefficient of concrete for varying FI and keeping EI at 30%   Figure 21   Permeability coefficient of concrete for varying FI and keeping EI at 40%

The Effect of Combined FI and EI as 50% on Various Properties of Concrete

From Figure 22, it is evident that even by increasing the EI from 0 to 50% and decreasing FI from 50 to 0%, the AIV and LAAV for the aggregate decreased. This decrease was 2.5% on the average for AIV and 3.5% for LAAV for every 10% decrease in FI. Figure 23 shows that the density of concrete increased by about 1% and then decreased with the decrease in FI. It was optimum at about FI = 25%.

Figure 24 shows an increasing trend of compressive strength and flexural strength with the decrease in FI and increase in EI. The permeability coefficient decreased with the decrease in FI and increased in EI as shown in Figure 25. This decrease was about 22%. From Figures 22, 23 and 25, it can be observed that an optimum value of FI was about 25% and EI 20%. These values in the range of 20-30% gave the best results.

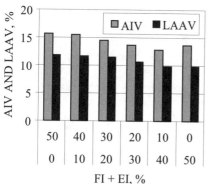

Figure 22   Value of AIV & LAAV for varying
EI from 0 to 50% and FI from 50 to 0%

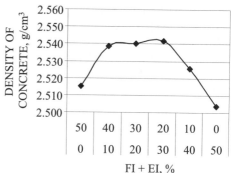

Figure 23   Density of concrete for varying
EI from 0 to 50% and FI from 50 to 0%

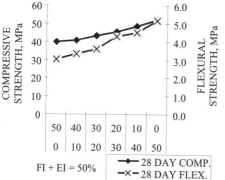

Figure 24   Compressive and flexural strength
of concrete for varying EI from 0 to 50%
and FI from 50 to 0%

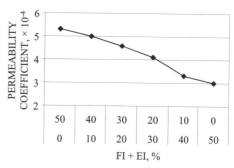

Figure 25   Permeability coefficient
of concrete for varying EI from 0 to 50%
and FI from 50 to 0%

Failure pattern of beams and cubes

1. It is generally seen that failure occurs at the interface of aggregate and mortar.
2. In flaky aggregate, some voids are observed at the interface of such aggregate and mortar. This will affect permeability and durability.
3. Elongated aggregate pieces are broken.
4. Mortar matrix is generally crushed.
5. As there is no aggregate which has zero percent flaky or elongated particles, a better combination of FI as 25% or less and EI as similar value will give the optimum values of density, permeability and strength.

CONCLUSIONS

The results indicate that with the increase in flakiness index, workability changes considerably. Increased flakiness also affects the cement demand as shown by the compressive strength results. The details of findings and their effect on strength, permeability, and durability are reported in this paper. Some of the observations are given below:

- The flakiness index has a larger impact on the aggregate impact value than the elongation index. as the flakiness index increases, the aggregate impact value increases considerably.
- The flakiness index also has a larger impact on the Los Angles abrasion value than the elongation index. As the flakiness index increases, the Los Angles abrasion value increases slightly. An optimum value of AIV and LAAV was obtained when FI was 20% and EI was 30%.
- The flakiness index has a larger impact on the 28 day compressive strength than the elongation index. As the flakiness index increased, the 28 day compressive strength decreased by about 27% when the FI increased from 0 to 50%. However, the increase in the elongation index has a very little effect on strength, which reduces by only about 2% for each 10% increase in EI from 0 to 50%.
- The flakiness index has a larger impact on the 28 day flexural strength than the elongation index. As the flakiness index increases, 28 day flexural strength decreases by about 11% when the FI is increased by 10% from 0 to 50%.
- The flakiness index has a similar impact on the permeability of concrete as the elongation index. As the flakiness index increased, the permeability coefficient changed by about 6.5% for every 10% increase in FI from 0% to 50%.
- The flakiness index also affects the density of concrete. It decreased by about 0.3% as the FI increased in 10% steps from 0 to 50%. This was mainly because the voids in concrete increased due to the large flaky particles. Maximum density was obtained when FI was about 25% and EI was kept at a similar value.
- An optimum value of strength was obtained when FI was in the range of 20 to 25% and EI was similar, if a total of 50% of FI and EI was taken as seen from Figures 17 and 18. The maximum value of compressive and flexural strength was obtained when the flakiness index was zero for a reasonable value of the elongation index.

ACKNOWLEDGEMENT

The work has been carried out in M/S HCC Ltd. Site Laboratory. The authors are thankful to them and QC staff of M/S BCEOM and HCC for their help. Special thanks to Mr. Suryakanta Bal, QC Engineer in M/S HCC Ltd. Allahabad.

REFERENCES

1. NEVILLE A M, Properties of concrete, IV edition, Pearson Education Pvt. Ltd., 2005.

2. MEHTA P K AND MONTEIRO P J M, Concrete microstructure, properties and materials, ICI, 1999.

3. KAPLAN M F, Flexural and compressive strength of concrete as affected by the properties of coarse aggregate, Journal of ACI, No. 55, 1959.

4. IS: 383-1970, Specifications for coarse and fine aggregate from natural sources for concrete, BIS, New Delhi.

5. SP: 23-2001, Handbook on concrete mixes based on Indian standards, BIS, New Delhi.

6. IRC 2001, Specifications for Road and Bridge Works, Indian Road Congress, New Delhi.

7. BS 882: 1992, Aggregates from Natural Sources for Concrete, British Standard.

8. BS: EN 13242, Aggregate for unbound and hydraulically bound material for use in Civil Engineering Works and Road Construction, European Standard for Aggregates.

9. AASHTO M-283, Coarse aggregates for highway and airport construction.

MODELLING OF THE DYNAMIC BEHAVIOUR OF FERROCEMENT COMPOSITE PLATES

Y B I Shaheen

A A Hamada

M A El-Sissy

Menoufia University

Egypt

ABSTRACT. The main advantages of composite structures are lightness and anisotropic properties, which can be utilized to optimize dynamic behavior. This paper highlights interactions between modelling and experimentation for the dynamic behavior of ferrocement composite structure. Mathematical modeling based on response surface methodology (RSM) has been established. The validation of these models and the dynamic characterization of materials have been examined. The effect of boundary conditions and ferrocement composite properties such as elastic modulus, volume fraction and open mesh area on the frequency and damping factor values has been studied. The established mathematical models have been found to be quite unique, powerful and flexible. These models reflect the complex, interactive and higher-order effects of the various input parameters. Through these models it can expect any experimental results of frequency and damping factor value with any combination of input parameters. The boundary condition and open mesh percentage play an important role for controlling the dynamic behavior of ferrocement structure.

Keywords: Dynamic, Frequency, Damping, Ferrocement, Modeling, Response surface methodology (RSM)

Yousry Bayoumy Ibrahim Shaheen is Professor of Strength and Testing of Materials, Department of Civil engineering, Faculty of Engineering at Menoufiya University, Shebin El-Kom, Egypt.

Dr A A Hamada is Associate Professor, Department of Production Engineering and Machine Design, Faculty of Engineering at Menoufiya University, Shebin El-Kom, Egypt.

M A El-Sissy is Graduate student, Department of Civil engineering, Faculty of Engineering at Menoufiya University, Shebin El-Kom, Egypt.

INTRODUCTION

The main advantages of composite structures are lightness and anisotropic properties, which can be utilized to optimize dynamic behavior. But the increase in stiffness and decrease in mass emphasize vibrating problems, thus the prediction of composite structure dynamic behavior is a very topical subject. Varied fields such as transportation, civil engineering and sports are particularly concerned. Natural frequencies and mode shapes, damping capacities, as well as responses to harmonics, shocks and random excitations are studied [1-3].

Numerical simulation of the behavior of cement-based materials has been a subject of intense research. The past two decades have seen a plethora of diverse mathematical models using finite element FE analyses to simulate the behavior of plain concrete. The theory of plasticity has been extensively employed for these simulations [4-11]. It has been recognized that the theory is sufficiently general to incorporate the complexities of concrete. This study used anisotropic elasto-plasticity with material parameters derived from simple experiments for finite-element simulation of the mechanical behavior of ferrocement. Since ferrocement is a composite material with distinct directional properties, it is imperative that an anisotropic material model be employed to describe the material. While a number of anisotropic plasticity models have been proposed in theory, only a few have actually been applied to composite materials. Detailed anisotropic modelling requires the determination of a large number of material parameters, through a series of sophisticated experimental tests.

Arif et al [12] conducted a large number of experiments to explore the behavior of ferrocement in tension, compression, and flexure. The experiments included tension and compression tests on mortar used in ferrocement specimens and tension tests on the mesh used as reinforcement. This study used the above experimental data for numerical simulation. The elastic parameters for ferrocement were derived from in-plane tests on constituent materials and the inelastic properties from similar tests on ferrocement specimens. This set of anisotropic material properties was used to simulate the tensile and flexural behavior of ferrocement. The experimental program comprised tests with woven and welded meshes at different orientations and with a varying number of mesh layers. Ferrocement can be approximated as a laminated composite in which the nonlinear response of layers with meshes and the behavior of mortar in compression and tension need an appropriate representation across the thickness. Here, this is effectively accomplished using a layered approach. This approach permits the gradual development of plastic zones through the plate thickness [13-15]. From the above literature review it is clear that a little research on the modeling of dynamic behavior for ferrocement plates has been published.

The main objective of this paper is to highlight interactions between modelling and experimentation for the dynamic ferrocement structure. Based on the response surface methodology, the validation of predictive models and the dynamic characterization of materials have been examined. The experimental modal analysis focuses on the frequencies and damping factor. The effects of boundary condition and ferrocement composite properties such as elastic modulus, volume fraction and open mesh area on the frequency value and damping factor have been studied.

EXPERIMENTAL PROCEDURE

Material Preparation

The experimental program was designed to investigate the effect of reinforced steel mesh configurations and boundary conditions on the dynamic behavior of ferrocement plates. Five different patterns of mesh reinforced were used. Different materials were used to produce the plates including: mortar, steel meshes, silica fume, super-plasticizer, fly ash and polypropylene fibers. The mix proportions are shown in Table 1. Figure 1. shows the configurations of steel meshes used in the present work, each of which has a fixed dimension of 150 x 150 mm.

Table 1   Mix proportions for mortar

MATERIALS	TYPE	PROPERTIES PER 750 cm^3
Sand	Fine sand passing sieve	984
Cement	Ordinary Cement type (I)	295
Water	Potable water	175
Mineral admixtures	Silica fume SF	49.2
Superplasticiser	Euco-Eypet CASTM C 494	9.8
Olypropylene fibers	Fibrillated fiber 18 μm length, 50 μm diameter	147.6

Table 2   Expanded steel metal specification

STYLE	Wt Kg	DIAMOND SIZE mm	OVERALL THICKNESS mm	PERCENT OPEN AREA %
838	1.5	11	2	80
1038	1.88	11	2.3	82
1550	2.4	22	3.8	72
2050	3.42	22	4.8	70
2038	2.4	16	3.7	68

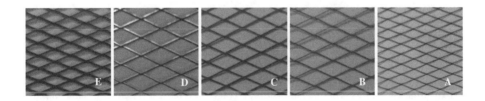

Figure 1   Configurations of steel meshes

Table 2 declares the specifications of the used steel meshes according to B.S 405, 1997. The prepared mesh of each specimen was placed into wooden molds (150 x 150 x 10 mm) and cast in vertical position, consolidation of the mortar was carried out using external and internal vibration. To minimize the effect of bleeding at an earlier stage, about 5mm more mortar was placed on the top in excess of specimen dimensions. This excess mortar was removed when finishing of the top surface was done. The mold was covered with damp burlap to prevent moisture loss. The specimens were stripped off the molds seven days after casting and were air cured in the laboratory before testing.

Test Conditions and Measurements

The experimental apparatus is shown in Figure 2. The mounting of the specimens was either clamped or simply supported edges in the test rig. The excitation and measuring procedure including calibration, transducers using the dual channel signal analyzer (B&K 2034) in conjunction with the fast Fourier transform (FFT), gives the mathematical connection between time and frequency. The frequency and damping factor measured for the fundamental frequency and associated damping factor was carried out for each specimen. The experimental results are taken as an average of five measurements of each. The damping factor of a particular resonance can be calculated from the width of the resonance peak in the magnitude of the (FRF) [16] where;

$$D = \frac{1}{2Q}, \quad Q = \frac{w}{\omega_d} \tag{1}$$

The resonant frequency (ω_d), the quality factor (Q) and the bandwidth (w) can be found from the magnitude diagrams using the reference cursor.

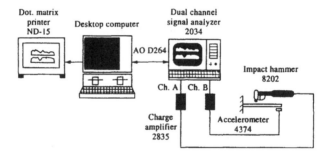

Figure 2   Diagrammatic sketch of instrumentation set-up

Experimental Design

The accuracy and effectiveness of an experimental programme depends on careful planning and execution of the experimental procedures [17,18]. With a view to achieving the above-mentioned aim, in this study experiments were carried out according to a central composite second order rotatable design based on response surface methodology (RSM).

Design of experiment (DOE) features was utilized to obtain the central composite second order rotatable design and also to determine the coefficients of mathematical modelling based on the response surface regression model. These developed models are very helpful for the users to predict frequency and damping factor for the proposed values of input variables, to select an optimum combination of input variables for the minimum frequency and maximum damping factor.

The testing parameters used in the present work are boundary condition (BC), elastic modules (E), Volume fraction, (V$_f$) and Open mesh, (OM), which are arranged in five levels (Table 3). Table 4 shows the various combinations of experimental testing parameters and parameter levels in the L31 RSM orthogonal array. The selection of the orthogonal array is subject to the condition that the degree of freedom for the orthogonal array should be greater than or at least equal to those for the input parameter.

Table 3   Experimental parameters and levels for dynamic testing

INPUT PARAMETERS	LEVEL					OUTPUT PARAMETERS
	-2	-1	0	+1	+2	
Boundary condition, (BC)	CFFF	HSFF	CSFF	CCFF	CCCC	
Elastic modulus, (E), GPa	24	26	28	30	32	- Frequency (Hz)
Volume fraction, (V$_f$), %	2	2.15	2.3	2.45	2.6	- Damping factor
Open mesh, (OM), %	68	72	76	80	84	

C: Clamped, F: Free, H: Hinged, S: Simply supported

MATHEMATICAL MODELLING

Response surface modelling was used to establish the mathematical relationship between the response (Y_u) and the various input parameters [16,17]. The general second order polynomial response surface mathematical model, which analyses the parametric influences on the various response criteria, can be described as follows:

$$Y_u = b_0 + \sum_{i=1}^{k} b_i X_i + \sum_{i=1}^{k} b_{ii} X_i^2 + \sum_{j>i}^{k} b_{ij} X_i X_j \qquad (2)$$

where; Y_u is the response and the X_i (1, 2, k) are coded level of k quantitative variables. The coefficient b_o is the free term, the coefficients b_i are the linear terms, the coefficients b_{ii} are the quadratic terms, and the coefficients b_{ij} are the interaction terms. Applying the least square technique, the values of these coefficients can be estimated by using the observations collected (Y_1, Y_2, Y_n) through the design points (n). The relationship between the coded and actual parametric values for the different x-variables can be expressed as follows:

$$Y_u = b_0 + b_1 X_1 + b_2 X_2 + b_3 X_3 + b_4 X_4 + b_{11} X_1^2 + b_{22} X_2^2 + b_{33} X_3^2 + b_{44} X_4^2 + b_{12} X_1 X_2 + b_{13} X_1 X_3 + b_{14} X_1 X_4 + b_{23} X_2 X_3 + b_{24} X_2 X_4 + b_{34} X_3 X_4 \qquad (3)$$

The actual and coded parametric values for each parameter are listed in Table 6. In our analysis, the boundary conditions (X_1), elastic modulus (X_2), volume fraction (X_3) and open mesh (X_4) are taken as controlling variables. Frequency (F) and damping factor (D) are obtained through a series of experiments according to the experimental plan based on central composite rotatable second order design, as shown in Table 3, to develop the equations of the response surfaces.

Based on Eq. 3, the effects of the above-mentioned variables on the magnitudes of the frequency have been evaluated by computing the values of various constants using the relevant experimental data from Table 4. The mathematical relationship for correlating the frequency and the considered various input parameters can be expressed as follows:

$$F=55+36.7X_1 +5.29X_2 +0.875X_3 - 2.04X_4 +10.05X^2_1 - 0.8X^2_2 - 0.4479X^2_3 -1.448 X^2_4+3.437X_1X_2 +0.4375X_1X_3 -0.688X_1X_4-1.063X_2X_3 - 0.688X_2X_4 +0.313 X_3X_4$$

(4)

Similarly, the mathematical relationship between the damping factor and the considered input parameters can be expressed as follows:

$$D = 0.056 - 0.029 X_1 - 0.026 X_2 - 0.01 X_3 - 0.001 X_4+ 0.0087 X^2_1 + 0.015 X^2_4 + 0.003 X_1X_2 - 0.001 X_1X_3 + 0.01 X_2X_3 - 0.0047 X_2X_4$$

(5)

It should be noted from the final equations that there are some coefficients omitted. These coefficients are non-significant according to Student's t-test that determined the significant and non-significant parameters. Also, from Tables 5 and 6, it is evident that the final models tested by analysis of variance (ANOVA) indicated that the adequacy of models was established.

RESULTS AND DISCUSSION

Effect of Parameters on Frequency

Figure 3 shows the frequency value-elastic modulus relationship at different boundary conditions (BC) based on mathematical modeling using surface response methodology under constant value of volume fraction and open mesh area (2.3% and 76% respectively). From Fig.3 it is clear that the frequency values dramatically increases with the increase of elastic modulus for all boundary conditions. This result can be attributed to the increased potential energy with the higher elastic modulus. The high value of frequency under clamped fixation from all sides of plate (CCCC) state of boundary condition may be attributed to the high level of potential energy at this condition. It must be mentioned that at the highest value of elastic modulus, frequency value for CCCC is about 35% higher than that for clamped-free-free-free (CFFF) state of boundary condition. The minimum value of frequency achieved under CFFF condition is due to the low level of potential energy at this condition. Furthermore, from Fig. 3 it is clear that the boundary condition has little influence on the frequency value especially at low values of elastic modulus. Hence, elastic modulus is the most significant parameter on the frequency values especially in the high range of E-values.

Figure 4 shows the effect of volume fraction on frequency value at different boundary conditions under constant value of elastic modulus and open mesh area at central values 28 GPa and 76% respectively. From Fig. 4 it is clear that the frequency value significantly increases with the increase of volume fraction. Furthermore, under all boundary conditions, frequency values remain nearly unchanged simply because volume fraction masked the effect of boundary condition on frequency.

Table 4   Experimental design matrix and results

Exp. No.	BC Actual	Coded	E Actual, GPa	Coded	V_f Actual, %	Coded	OM Actual, %	Coded	FREQUENCY, Hz	DAMPING FACTOR
1	HSFF	-1	26	-1	2.15	-1	72	-1	25	0.160
2	CCFF	1	26	-1	2.15	-1	72	-1	87	0.110
3	HSFF	-1	30	1	2.15	-1	72	-1	32	0.080
4	HSFF	-1	30	1	2.15	-1	72	-1	110	0.030
5	HSFF	-1	26	-1	2.45	1	72	-1	27	0.140
6	CCFF	1	26	-1	2.45	1	72	-1	91	0.070
7	HSFF	-1	30	1	2.45	1	72	-1	32	0.080
8	CCFF	1	30	1	2.45	1	72	-1	110	0.030
9	HSFF	-1	26	-1	2.15	-1	80	1	24	0.170
10	CCFF	1	26	-1	2.15	-1	80	1	83	0.120
11	HSFF	-1	30	1	2.15	-1	80	1	30	0.070
12	CCFF	1	30	1	2.15	-1	80	1	104	0.020
13	HSFF	-1	26	-1	2.45	1	80	1	27	0.140
14	CCFF	1	26	-1	2.45	1	80	1	91	0.070
15	HSFF	-1	30	1	2.45	1	80	1	27	0.070
16	CCFF	1	30	1	2.45	1	80	1	104	0.020
17	CFFF	-2	28	0	2.30	0	76	0	24	0.150
18	CCCC	2	28	0	2.30	0	76	0	168	0.020
19	CSFF	0	24	-2	2.30	0	76	0	43	0.118
20	CSFF	0	32	2	2.30	0	76	0	60	0.004
21	CSFF	0	28	0	2.00	-2	76	0	51	0.080
22	CSFF	0	28	0	2.60	2	76	0	53	0.030
23	CSFF	0	28	0	2.30	0	68	-2	55	0.102
24	CSFF	0	28	0	2.30	0	84	2	44	0.140
25	CSFF	0	28	0	2.30	0	76	0	55	0.060
26	CSFF	0	28	0	2.30	0	76	0	55	0.060
27	CSFF	0	28	0	2.30	0	76	0	56	0.050
28	CSFF	0	28	0	2.30	0	76	0	54	0.060
29	CSFF	0	28	0	2.30	0	76	0	56	0.050
30	CSFF	0	28	0	2.30	0	76	0	54	0.060

Table 5   Analysis of variance for frequency

Source	DF	Seq SS	Adj SS	Adj MS	F	P
Regression	14	34079.2	34079.15	2434.225	878.52	0.000
Linear	4	30465.0	88.57	22.143	7.99	0.001
Square	4	3356.2	3356.15	839.038	302.81	0.000
Interaction	6	258.0	258.00	43.000	15.52	0.000
Residual Error	16	44.3	44.33	2.771		
Lack-of-Fit	10	40.3	40.33	4.033	6.05	0.019
Pure Error	6	4.0	4.00	0.667		
Total	30	34123.5				

Table 6   Analysis of variance for damping factor

SOURCE	DF	Seq SS	Adj SS	Adj MS	F	P
Regression	14	0.061646	0.061646	0.004403	43.55	0.000
Linear	4	0.050150	0.004419	0.001105	10.93	0.000
Square	4	0.009821	0.009821	0.002455	24.28	0.000
Interaction	6	0.001675	0.001675	0.000279	2.76	0.049
Residual Error	16	0.001618	0.001618	0.000101		
Lack-of-Fit	10	0.001446	0.001446	0.000145	5.06	0.030
Pure Error	6	0.000171	0.000171	0.000029		
Total	30	0.063263				

The influences of the open mesh on frequency value at different boundary conditions are determined based on a non-linear mathematical model, at constant volume fraction (2.3%) and elastic modulus (28 GPa) and is shown in Fig. 5. A quite similar behavior was observed for the effect of open mesh area on frequency under different boundary conditions (Fig. 5). So one can say that O.M and V_f have a dominant influence on frequency as compared with boundary condition. So, to control the frequency of a composite structure both volume fraction and open mesh area could be considered a useful tool in this regard.

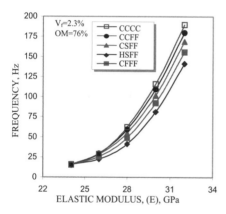

Figure 3   Frequency-elastic
modulus relation

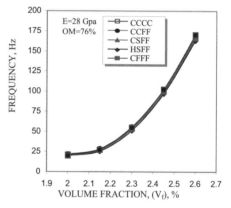

Figure 4   Frequency-volume
fraction relation

Figure 6 demonstrates the frequency value-volume fraction relationship at different elastic modulus under constant value of open mesh and boundary condition (72% and CSFF respectively). From this figure it is clear that the frequency values increases with the increase of volume fraction with all elastic modulus. However, the rate of increase depends on the elastic modulus values. At low E (24 GPa), frequency value drastically increases with V_f about 1.97 times. Meanwhile, when E is high (32 GPa), frequency increases with V_f about 1.45 times. This is because the contribution of E on stiffness is remarkable at low level of V_f.

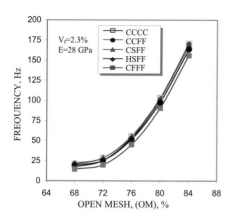

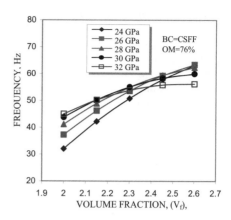

Figure 5   Frequency-open mesh area relation at different boundary conditions

Figure 6   Frequency-volume fraction relation at different elastic modulus

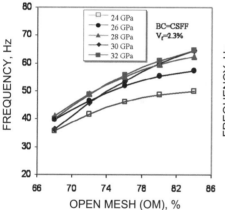

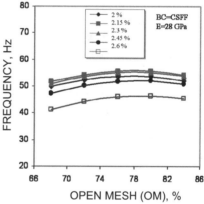

Figure 7   Frequency-open mesh area relation at different elastic modulus

Figure 8   Frequency-open mesh area relation at different volume fractions

Figure 7 displays the effects of the open mesh area and different elastic modulus on the frequency value at constant boundary condition (CSFF) and volume fraction (2.3%). As mentioned previously, this figure clearly shows the great influence of OM on frequency value. At all values of E, the frequency value increases with OM up to a certain limit (76%) beyond which smoothly increases. No remarkable effect of E on the frequency value at low value of OM. On the other hand, this effect was noticeable at high value of OM. It is may be due to a strong adherent between steel mesh and matrix with the resultant increase in stiffness and so frequency. This is quite true at high values of OM (82 %) where the open areas are large enough to be filled completely with the mortar the adherence is strong enough.

Figure 8 shows the estimated frequency values in relation to the control parameters open mesh at different volume fractions under constant value of elastic modulus and boundary condition (28 GPa and CSFF respectively). From Fig. 8, frequency value slightly increases with the increase of OM at all levels of V_f up to certain limit (76 %). Simply because of the strengthening effect of fibers, which enhances the adherences between steel mesh and matrix, which reflects on stiffness value. However, at high V_f-value (2.6%), one can presume that the high fiber volume fraction is associated with cracks and debonding interfaces that will certainly enhance damping as will be seen later.

Effect of Parameters on Damping Factor

First of all, the dependence of damping on the investigated parameters is quite clear as compared with frequency. So a well-defined conclusion could be arrived. Figure 9 declares the effect of elastic modulus (E) and boundary condition (BC) on damping factor under constant value of volume fraction and open mesh area (2% and 76% respectively). It is clear that the damping factor values significantly varied with the elastic modulus under all the investigated states of boundary condition but in a reverse trend as compared with frequency (Fig. 3). Fig. 9 reflects that the damping factor quasi linearly decreases with the increase of elastic modulus due to the increased stiffness. The low value of damping factor occurs under CCCC condition. Which may be attributed to the minimum dissipated energy at this condition. On other hand, the maximum value of damping factor observed under CFFF state is due to the maximum dissipated energy at this condition.

Figure 10 illustrates the effect of volume fraction at different boundary conditions on damping factor under constant value of elastic modulus and open mesh area at central values 28 GPa and 76% respectively. The damping factor significantly changed with the volume fraction. The higher the volume fraction, the lower the damping factor since the fibers represents passage energy transmission with the resultant decrease in damping factor. Meanwhile, at low level of V_f the effect of BC is limited since BC overwhelms the damping factor.

The influences of the open mesh at different boundary conditions on damping factor are determined based on a non-linear mathematical model, at constant volume fraction (2.3%) and elastic modulus (28 GPa) as shown in Fig. 11. From Fig. 11 it can be noted that, the open mesh has a dominant influence compared with BC. The damping factor decreases with the OM up to certain limit (80%) and then unchanged. This can be attributed to the varied strengthening effect of fiber according to open mesh type. Decreasing open mesh area results in an increase in mortar content at a decrease in fiber ratio both of which yield lower damping.

Figure 12 demonstrates the effect of volume fraction on damping factor at different elastic modulus under constant value of open mesh and boundary condition (76% and CSFF respectively). From this figure it is clear that the damping factor linearly varied with V_f up to certain limit (2.45%). At low V_f (2%) damping factor drastically decreases with E about 2.57 times for low modulus composite plate. Meanwhile, when V_f is high (2.45%), damping factor remains unvaried with E. This is because the effect of E on stiffness is remarkable at low level of V_f.

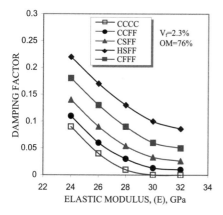

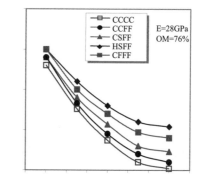

Figure 9   Damping factor-elastic modulus
relation at different boundary conditions

Figure 10   Damping factor-volume
fraction relation at different boundary

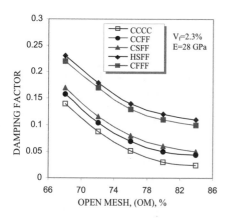

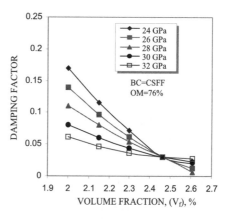

Figure 11   Damping factor-open mesh area
relation at different boundary conditions

Figure 12   Damping factor-volume fraction
relation at different elastic modulus

Figure 13 displays the effects of the open mesh area at different elastic modulus on the damping factor at constant boundary condition (CSFF) and volume fraction (2.3%). As mentioned previously, this figure clearly shows the great influence of E on damping factor specially, at higher values of elastic modulus. At a fixed value of E the damping factor slightly decreases with OM and an explanation was given before.

Figure 14 shows the effect of open mesh area at different volume fractions on damping factor under constant value of elastic modulus and boundary condition (28 GPa and CSFF, respectively). From Fig. 14, damping factor linearly decreases with the increase of open mesh area at both low and high V_f values due to the strengthening effect of fibers.

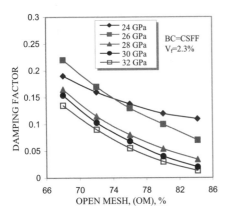
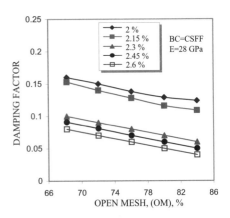

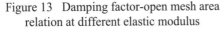

Figure 13   Damping factor-open mesh area
relation at different elastic modulus

Figure 14   Damping factor -open mesh
area relation at different volume fractions

CONCLUSIONS

1. The established mathematical models have been found to be quite unique, powerful and flexible. These models reflect the complex, interactive and higher-order effects of the various input parameters. Through these models it can expect any experimental results of frequency and damping factor value with any combination of input parameters.

2. The open mesh percentage plays an important role for controlling the dynamic behavior of ferrocement structure.

3. The maximum value of frequency is achieved under clamped fixation from all sides of plate (CCCC) and is attributed to the high level of stiffness at this condition.

4. The developed models are very helpful for the users to predict the dynamic behavior of ferrocement structure and to select an optimum combination of input variables for the minimum frequency and maximum damping.

REFERENCES

1. CHAO, W.C., REDDY, J.N., "Analysis of laminated composite shells using a degenerated 3D element", Int J Num Meth Eng; 20, pp. 1991-2007, 1984.

2. SARAVANOS, D.A., "Integrated damping mechanisms for thick composite laminated and plates", J Appl Mech, 61, pp. 375-83, 1994.

3. KOO, R.N., LEE, I., "Vibration and damping analysis of composite laminated using shear deformable finite element" ALU J; 31(4), pp. 728-35, 1993.

4. BICANIC, N., AND PANKAJ., "Some computational aspects of tensile strain localization modelling in concrete", Eng. Fract. Mech., 35(4/5), pp. 697–707, 1990.

5. PANKAJ., "Finite element analysis in strain softening and localization problems", PhD thesis, Univ. College of Swansea, Swansea, U.K., 1990.

6. WILLAM, K.J., ET AL., Topic 7, "Computational aspects of finite element analysis of reinforced concrete structures", Technical Rep. CU/SR-9, 1993.

7. ETSE, G., AND WILLAM, K., "Integration algorithms for concrete plasticity", Eng. Comput., (13/8), pp. 38–65, 1996.

8. FEENSTRA, P.H. AND DE BORST, R., "Composite plasticity model for Concrete, Int. J. Solids Struct", 33/51, pp. 707–730, 1996.

9. HOFSTETTER, G., AND MANG, H.A., "Computational plasticity of reinforced and pre-stressed concrete structures", Computational Mech., Berlin, 17(4), pp. 242–254, 1996.

10. MALVAR, L.J., CRAWFORD, J.E., WESEVICH, J.W., AND SIMONS, D., "Plasticity concrete model for DYNA3D", Int. J. Impact Eng., 19(9/ 10), pp. 847–873, 1997.

11. PANKAJ, M. ARIF, AND SURENDRA K., "Kaushik, M.ASCE. Mechanical Behavior of Ferrocement Composites: Numerical Simulation", Journal of Materials in Civil Engineering, (14:2), pp. 156-163, 2002.

12. ARIF, M., KAUSHIK, S.K., AND PANKAJ., "Estimation of mortar modulus: a comparison of strategies", J. Ferrocement, 29(3), pp. 79–187,1999.

13. ARIF, M., PANKAJ, AND KAUSHIK, S.K., "Mechanical behavior of ferrocement composites: an experimental investigation", Cem. Concr. Compos., 21(4), pp. 301–312, 1999.

14. SARAVANOS, D.A., "Integrated damping mechanisms for thick composite laminated and plates", J Appl Mech, 61, pp. 375-83, 1994.

15. HUANG, H.C., "Static and dynamic analysis of plates and shells", Springer, Berlin, 1989.

16. RANDALL, R.B., "Application of B&K equipment to frequency analysis", Bruel and Kjear, 1977.

17. MONTGOMERY, D.C., "Design and analysis of experiments", Wiley, New York, 2001.

18. DAS, M.N. AND GIRI, N.G., "Design and Analysis of Experiments", Wiley and Sons, New York, Second Edition, 1986.

SELF-COMPACTING CONCRETE IN SOUTH AFRICA

J P Jooste

University of Johannesburg

Y Ballim

University of Witwatersrand

South Africa

ABSTRACT. Self-compacting concrete (SCC) has found recent application in South Africa. The first notable project is the use of SCC in the construction of the pylons for the Nelson Mandela Bridge in Johannesburg. The successful use of SCC in this first project has caused this technology to be increasingly used in various specialized projects where the characteristics of SCC are considered appropriate. As a contribution to the development of SCC technology in South Africa, this paper present details of a research programme undertaken to assess the opportunities and potential for using local concrete-making materials to produce SCC. Three different crushed aggregate types (granite, dolomite and andesite) were used in order to consider their properties and suitability when used in SCC. Two mixture design models were assessed for use in designing SCC mixtures using local materials. The Tattersall two point test apparatus was used to determine the basic rheological parameters of the concrete mixtures investigated. These results were compared with results obtained from the slump flow test. The results indicate that SCC can be made with local crushed aggregates, but careful mixture design is required. This investigation provided a basis for further research work on SCC in South Africa.

Keywords: Self-compacting concrete, Rheology, Workability.

Mr J P Jooste, is Head of Department at the Department of Civil Engineering Technology, University of Johannesburg.

Prof Y Ballim, is the Deputy Vice Chancellor (Academic) at the University of Witwatersrand.

INTRODUCTION

Even though SCC has been used since 1988 in most countries, it is still a relatively new material in South Africa. SCC has been used in a few specialized projects, and local research in this concrete type is in the infancy stages. In many cases, research has been focused on the rheology and workability of SCC only.

This paper presents an overview of projects where SCC was used as a construction material. The projects are the Nelson Mandela bridge, Bridge 2235, a spiral staircase and a culvert repair. The paper also presents results from a test programme aimed at investigating the use of local concrete material in SCC. Two mixture design methods were used to determine a suitable method for the design of SCC mixtures in South Africa.

CASE STUDIES

The Nelson Mandela Bridge Project

The first project where SCC was used in South Africa was the Nelson Mandela Bridge, which was constructed in 2002. The placement method used in this project, which entailed pumping from the bottom up, was a first for South Africa.

The Nelson Mandela bridge (Figure 1) is a cable-stayed bridge allowing access to the cultural precinct of Johannesburg from the northern part of the city. SCC was used in the steel pylons to introduce the stiffness required. Because of the height of these pylons (31.1 m and 43.9 m) access constraints (due to operating railway lines), stressing chambers at the top of the pylons and free fall limits, the concrete could not be placed conventionally from the top [1].

Figure 1   The Nelson Mandela Bridge

In addition, mechanical vibration was not possible due to limited access and external vibration was inappropriate because of the large amount of energy needed to overcome the pylon inertia. To overcome the placing problems it was decided to pump SCC into the pylons from the bottom. The concrete was pumped through a special pipe and valve arrangement at the bottom of each pylon.

Bridge 2235

Bridge 2235 (Figure 2) forms part of an off ramp from the Bakwena highway. The Bakwena highway, which extends from Pretoria to Botswana, is part of the east-west link across the southern part of the continent.

Figure 2   Bridge 2235

The bridge deck is a post-tensioned two-cell box girder type structure (Figure 3), unlike the conventional metal drum void formers used in similar bridges. In the case of the latter design, the bottom slab and webs are cast first and have to harden before the top slab can be cast. To save time and labour costs, it was decided to cast the deck of Bridge 2235 in one operation. Since compaction and placing was a problem in the reinforcing congested bottom slab [2]. SCC was used successfully to complete the bridge deck

Spiral Staircase

In 2003, a spiral staircase (Figure 4) at an office complex in Pretoria was constructed using SCC. The position and geometry of this staircase made vibration difficult. The staircase also had to be cast in a single operation since no joints were allowed. At first, the formwork was not strong enough to withstand the concrete pressure and adjustments to the formwork were required. With the formwork problems solved, the construction of the staircase was successful and the appearance acceptable [3].

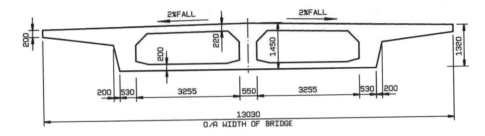

Figure 3   Deck cross section [2]

Figure 4   Spiral staircase

Culvert repair

SCC was also used on a project close to Cape Town for the repair of a culvert where the soffit had deteriorated to the extent that the steel reinforcement became exposed. To repair this, timber shuttering was placed below the soffit leaving enough room for extra reinforcing steel and concrete. SCC was placed through openings drilled from the top. Inspection openings were also provided at the other end of the slab to check if the space was filled completely.

The operation was completed quickly and successfully. The only problem that was encountered was that rain affected the mix on one of the days and the superplasticizer dosage had to be amended. An alternative to using SCC in this case was to build a detour and rebuild the culvert. With the use of SCC the problem was solved in a shorter time and more cost effectively [4].

Figure 5   Repaired culvert in Cape Town

MIXTURE DESIGN APPROACHES

First mixture design model

The first mixture design method used was adapted from a mixture design used for conventional concrete. The aim was to design an adaptable mixture with a specific water to binder ratio and target strength. In the development of these mixtures, three different aggregate types from the Gauteng area were used to determine the most suitable type as well as the best blend of these aggregates (Table 1 and 2).

To achieve the required flowability with no segregation, superplasticizers and viscosity modifiers were used at an appropriate dosage in relation to the mixture design. The dosage was determined using the manufacturer's guidelines as a starting point and then by assessing the mixture visually. After mixing was stopped, the flowability and resistance to segregation was assessed using a steel float. This mixture design method is largely based on experience and difficult to quantify. A mixture was prepared, mixed and tested using the slump flow test. The slump flow result and behaviour of the mixture was then assessed visually to determine if the admixture dosage was acceptable.

To determine the flowability of the mixtures, the slump flow test was used. Each mixture was tested in the Tattersall Two Point Tester to determine the rheological properties. The same mixtures were used for the slump flow test and the results were compared with the rheology test results.

The proportions of aggregates used were based on experience and adjustments of trial mixtures. Many trial mixtures were tested and the final mixture proportions used in this part of the investigation are given in Table 3.

Table 1   Fine Aggregate Gradings

PARTICLE SIZE	FILLER SAND % Passing	ANDESITE % Passing	GRANITE % Passing	DOLOMITE % Passing
9.5	100	100	100	100
6.7	100	99.8	100	100
4.75	99.8	98.5	98.8	99.7
2.36	95.2	71.5	75.7	62.7
1.18	91.8	47.7	54.8	38.2
0.6	86.8	32.8	37.8	27
0.425	78.8	27.4	30.7	23.6
0.3	62.3	22.8	24.5	20.9
0.15	16.8	15.9	15.1	17.1
0.075	7.7	12.6	10.6	14.9

Table 2   Coarse Aggregate Gradings

PARTICLE SIZE	ANDESITE % Passing	GRANITE % Passing	DOLOMITE % Passing
19	100	100	100
13	91.6	92.6	83.1
9.5	42.6	34.1	30.2
6.7	12.1	2.8	3.3
4.75	3.2	0.4	0.4
3.35	1.6	0.2	0.3
2.36	1.3	0.2	0.3

Table 3   Concrete mixture proportions used for the first method

MATERIAL	MIXTURE A1	MIXTURE G1	MIXTURE D1
Cem I 42.5	196 kg/m³	204 kg/m³	183 kg/m³
Fly ash	138 kg/m³	143 kg/m³	130 kg/m³
Filler sand	400 kg/m³	372 kg/m³	354 kg/m³
Fine aggregate	804 kg/m³ (Andesite)	698 kg/m³ (Granite)	823 kg/m³ (Dolomite)
Coarse aggregate	810 kg/m³ (Andesite)	810 kg/m³ (Granite)	835 kg/m³ (Dolomite)
Water	183 kg/m³	190 kg/m³	171 kg/m³
Super plasticizer	6.3 kg/m³	6.2 kg/m³	3.8 kg/m³
Viscosity modifier	0.03 kg/m³	0.03 kg/m³	0.02 kg/m³
W:C	0.55	0.55	0.55

Second mixture design model

This model is based on a method developed in Taiwan by Su, et al[5]. The main objective of this method is to determine the amount of paste required to fill the openings between loosely piled aggregate. The reason for using this particular model is its simplicity and adaptability. Because of limited design procedures and testing this method is more acceptable to SCC producers who do not have special facilities and testing equipment. This model could also be used for different aggregate types and any required designed strength. Table 4 shows the mixture proportions of the mixtures used in this investigation. These mixtures were designed for the three different aggregate types using crusher sand as fine aggregate to assess the suitability of the design model for these three different aggregate types.

Table 4   Concrete mixture proportions used for the second method.

MATERIAL	MIXTURE A2	MIXTURE G2	MIXTURE D2
Cem I 42.5	218 kg/m³	218 kg/m³	218 kg/m³
Fly ash	182 kg/m³	179 kg/m³	205 kg/m³
GGBS	46 kg/m³	45 kg/m³	51 kg/m³
Fine aggregate	1129 kg/m³ (Andesite)	1022 kg/m³ (Granite)	917 kg/m³ (Dolomite)
Coarse aggregate	667 kg/m³ (Andesite)	659 kg/m³ (Granite)	814 kg/m³ (Dolomite)
Water	197 kg/m³	196 kg/m³	208 kg/m³
Super plasticizer	3.1 kg/m³	3.1 kg/m³	2 kg/m³
W:C	0.55	0.55	0.55

RESULTS

The design strength for both methods was taken as 30 MPa, but the compressive strength results given in Tables 5 and 6 indicate the discrepancies in these two methods. For the first method the water:binder ratio was used to determine the strength and for the second method the water:cement ratio. The effect of the extenders on the compressive strength was not taken into account when designing the mixture using the second method. This is clearly an unreasonable approach and an inefficient use of the cement extenders.

This investigation has emphasized the necessity to use rheology in evaluating the flow properties of SCC. Rheology describes and verifies the important qualities needed for SCC to be self-compacting, low yield stress and sufficiently high plastic viscosity [6]. The rheological data is therefore needed in the mixture design of SCC. The two values, g (yield stress) and h (plastic viscosity) characterise the total physical effort required to place and compact fresh concrete. The yield value (τ_0) quantifies the effort to start movement and plastic viscosity (μ) the extra effort to sustain the movement at a reasonable speed [7]. The Tattersall Two Point Tester with an interrupted helical impeller was used to investigate and improve SCC mixture designs. To convert g to τ_0 and h to μ the following factors were used:

$$\tau_0 = 72.86 * g$$
$$\mu = 121.95 * h$$

If the yield stress is too high it indicates that more superplasticizer is required and the mixture needs to be adjusted. The plastic viscosity results will depend on the application that the mixture is intended for. SCC designed for vertical members, like columns, requires a cohesive mixture and therefore a lower plastic viscosity than for horizontal members (floors) where a very flowable mixture with a high plastic viscosity is suitable. An increase in the fines content (including cement and extenders) or the addition of a viscosity modifier increases the plastic viscosity which ensures sufficient stability [8]. The results from the Tattersall Two Point test and the slump flow test for both methods are given in Table 7.

Table 5　Compressive strength results for method 1,

Mixture No.	Date	COMPRESSIVE STRENGTH RESULTS					
		7 Days			28 Days		
		Density kg/m^3	Individual MPa	Average MPa	Density kg/m^3	Individual MPa	Average MPa
A1	17 Jun 2004	2514	14.5	14.5	2516	33.6	33.1
					2514	33.1	
					2511	32.6	
G1	17 Jun 2004	2330	17.6	17.6	2280	36.7	37.8
					2320	39	
					2330	37.8	
D1	10 Jun 2004	2454	12	12	2448	26.6	27.8
					2484	28.4	
					2480	28.4	

Table 6　Compressive strength results for method 2,

Mixture No.	Date	COMPRESSIVE STRENGTH RESULTS					
		7 Days			28 Days		
		Density kg/m^3	Individual MPa	Average MPa	Density kg/m^3	Individual MPa	Average MPa
A2	27 Jun 2005	2434	28.4	27.7	2461	56.9	55.1
		2446	26.6		2478	53.1	
		2460	28		2458	55.2	
G2	27 Jun 2005	2239	27	26.9	2259	48.5	48.4
		2246	27.6		2243	46.8	
		2248	26		2255	50	
D2	26 Jul 2005	2465	22.6	21.7	2471	51.8	50.6
		2452	21.6		2476	49	
		2471	20.9		2495	50.9	

Table 7   Workability results

MIXTURE NUMBER	TATTERSALL RESULTS		SLUMP FLOW RESULTS	
	τ_o Pa	μ Pas	T50 sec	Final mm
A1	11	39	5	620
G1	1.5	59	3	715
D1	8	37	4	610
A2	14.8	15	5	615
G2	2.3	21	5	600
D2	7.9	20	4	670

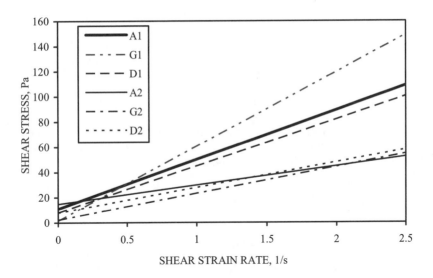

Figure 6   Bingham lines based on the results from the Tattersall Two Point tester.

Figure 6 presents a summary of the results from the Tattersall Two Point test, showing the straight line relationship between shear stress (τ) and shear strain rate ($\dot{\gamma}$). The lines in Figure 6 were created from the yield stress (τ_o) and plastic viscosity (μ) readings substituted into the Bingham equation ($\tau = \tau_o + \mu\,\dot{\gamma}$)[7] to produce a linear relationship. The results show that all the mixtures have a low yield stress, but the viscosities vary between mixtures. This indicates that all the mixtures can be classified as SCC (low yield stress and sufficiently high plastic viscosity [6]), but may not be equally applicable to different applications for SCC.

CONCLUSIONS

The case studies included in this paper demonstrate that although SCC is a relatively new technology in South Africa, its limited use thus far, in specialised applications, has been very successful. From the workability results it is clear that both mixture design methods can be used to design SCC. The second method would take preference, since this method can be quantified and is more structured. When using the first method the designing process relies on experience and the use of trail and error to achieve an appropriate mixture. The result of the case studies and research projects confirms that SCC can be made using locally available materials. However, it is important to note that the only research undertaken to date on SCC in South Africa was specifically focused on workability and compressive strength. However, the strength relationships should be adjusted to account for the strength contribution of all cement extenders in the mixture. Furthermore, the additional benefits in using rheology to describe the workability of concrete are that the results are based on fundamental properties, have numerically similar values and are reproducible. Unlike the commonly used terms, like "wet" or "plastic", rheology can be used to describe concrete flow more accurately.

ACKNOWLEDGEMENTS

The authors would like to thank Lafarge (SA), Holcim (SA) and Grinaker-LTA, for allowing them to publish the information given and for the materials donated for the testing.

REFERENCES

1. PARROCK, A AND JERLING, W. The Nelson Mandela Bridge: The use of concrete on a predominantly steel structure. Developing concrete to serve practical needs Conference. 14 Oct. 2004. Midrand, South Africa, pp 244 – 245.

2. THE CONCRETE SOCIETY OF SOUTHERN AFRICA. Significant project: Bridge 2235 – N4 Platinum Toll Highway, Concrete Beton Nr.106, May 2004, Midrand, South Africa, 2004, pp 5.

3. KLEINHANS, E. Private communication, Johannesburg: Lafarge, March 2004.

4. GAZENDAM, M. Private communication, Cape Town: University of Stellenbosch, June 2004.

5. SU, N., HSU, K-C. AND CHAI, H-W. A simple mix design method for self-compacting concrete. Cement and Concrete Research, vol.31, 2001, pp 1799-1807.

6. BILLBERG, P. Self-compacting concrete for civil engineering structures – The Swedish experience. CBI Report 2:99, Swedish Cement and Concrete Research Institute, SE-100 44 Stockholm, Sweden, 1999, p 34.

7. TATTERSALL, G.H. Workability and quality control of concrete, 1st ed. London, E & FN Spon., 1991, pp 54-77

8. MICHIGAN:AMERICAN CONCRETE INSTITUTE. ACI 211.1-91(1997), Standard practice for selecting proportions for normal, heavyweight, and mass concrete, Farmington Hills, 1997.

USING MARGINAL MATERIAL FOR ROLLER COMPACTED CONCRETE BASES IN INDIA

C Hazaree

Iowa State University

United States of America

ABSTRACT. With enormous ongoing construction activity, good quality aggregates are rapidly depleting in India. Compacted dry density and resulting inconsistencies in strength of the Roller compacted concrete bases are areas of concern in assuring the uniformity of support to the wearing course. This paper analyses the results of effects of using marginal aggregates (shape of coarse aggregates and higher fines content of fine aggregates) both naturally formed and crushed stone aggregates on the compacted density and strength. Cement was substituted up to 45% using lagoon ash. Proctor curves, fresh properties such as Vebe time, density, air content, yield, etc and mechanical strengths (cube compressive, flexural and split tensile) up to 180 days were measured. The paper reports typical results. It was found that the grading and the shape of the aggregates make a great impact on its workability, compaction characteristics and strengths. The changes in the water demand could be up to 30% with respect to the control concrete with a proportionate variation in density and strength. The overall analyses shows that marginal aggregates could cautiously be used for production of roller compacted concrete for pavement bases.

Keywords: Shape index, Microfines, Lagoon ash, Specific strength, Embrittlement ratio

Chetan Hazaree, obtained his BS degree from the University of Pune in 2002, where he was awarded the Prof. P.K. Mehta fellowship for carrying out research on concrete incorporating high volumes of fly ash and lagoon ash. Subsequently he worked for three years as Materials Engineer on various Highway and infrastructure construction projects. These include BOT and World Bank funded projects. He has also earned the 'Advanced concrete technology' certificate from the City and Guilds of London Institute. He has published several papers on mechanical aspects, abrasion resistance of roller compacted concrete, and concrete with marginal materials, and gap graded concrete and high volume coal ash concrete. Currently he is pursuing his graduate studies in Geotechnical and Materials engineering at the Iowa State University, USA. He is working on the freeze/thaw, permeability and sorptivity aspects of concrete as a part of research under a Dwight David Eisenhower Research Fellowship awarded by the National Highways Institute, USA. He is also serving on two Transportation Research Board committees.

INTRODUCTION

Use of roller-compacted concrete (RCC) sub-base has been increasing in popularity in recent years in India due to its obvious advantages. The National Highways Authority of India (NHAI) golden quadrilateral projects have made an extensive use of this technology. A massive undertaking of such infrastructure projects is leading to escalating shortages manifested in the form of soaring prices of virgin and energy intensive materials like river aggregates, quarried and crushed stones and manufactured cements and steel. Additionally, pressure from the environmental community and several other factors including sustainability and the fact that India is a developing country are forcing the civil engineering community to search for cost-effective, energy efficient and sustainable construction materials.

Moreover, there exists a need for using the existing marginal materials, which the specifications do not allow. Hence there is a need to research such materials and make practical changes in the construction specifications and drive them towards sustainable specifications. RCC sub-base provides one such avenue due to its low strength requirements and substantial volume of applications. This research presents various studies carried out to accommodate such marginal materials into RCC sub-base.

LITERATURE REVIEW

RCC is a sort of super-dry concrete that presents its engineering credentials with high density, improved strength and superior performance. Moreover, as a construction technique, a large amount of concrete and hence pavements could be quickly placed with minimum amount of labour, equipment, and no reinforcing steel, thus making it attractive to the construction economics and could lead to savings in the range of 15-40% over conventional concrete.[1-3].

In addition to this, it is hypothesized that RCC pavements have credentials that surpass those of conventional pavements for public transport and industrial applications in relatively harsher environments. [4, 5]. At the same time, it would be pertinent to enlist a few limitations of RCC which are its relatively poor aesthetics, surface texture, high permeability and uncontrolled cracking [6, 7], that need further research and improvement in the technology.

Incorporating class F fly ash (FA) into RCC to make RCC with fly ash (FRCC) can further reduce costs with a concurrent improvement in the performance in terms of its workability, compactibility, heat evolution and prolonged strength gain [8], and permeability [9]. Moreover, the use of fly ash increases the fines content of the mixture, which is further augmented by replacing partly the fine aggregate in concrete, ultimately leading to handling ease, improved strength and surface texture [3].

In addition to this, recent research on particle packing of filler materials has shown that for lower cement contents, the voids ratio decreases rapidly with filler addition depending upon the fineness of the filler (fly ash in current study). [10]. Lastly, it has been proved that the fly ash's contribution to strength increases as the specific area by Blaine increases. [11]. Researchers have found that lagoon or pond ash can be used for dry lean concrete (DLC), used as pavement base in India with 55% cement replacement, with the total binders' content of 277 kg/m^3 in the mix [12].

Studies on the compaction characteristics of RCC mixes using compaction energy as a variable have found that there exists an optimum sand/aggregate ratio at which the compaction effort is minimum; the effect of increasing the water content is reduction of the energy required to attain maximum compaction; the optimum amount of water for compaction may be obtained by selecting a water content beyond which the compaction energy reduction is minimal [13, 14]. Further to this it has been found that the compressive and splitting tensile strengths are the functions of density and the splitting tensile strength is strongly correlated to the compressive strength [15].

RESEARCH SIGNIFICANCE

Sustainability is becoming increasingly important in today's world. Judicious usage of available natural resources with sufficient recycling and reuse are becoming the key drivers of a balanced growth. This becomes more important in the developing countries like India, where the population explosion and rapid urbanization are calling for a substantial infrastructural demand. With mammoth infrastructure development projects undertaken all over India, good quality, virgin resources are becoming scarce driving civil engineers towards the usage of marginal materials. With timely research and technical development, the problem could be acutely analysed and resolved. This research was undertaken with the motivation of utilizing very fine fine aggregates for RCC. Moreover, the author experienced fluctuations in the compacted density due to the variations in the shape indices of coarse aggregates. Utilizing the aggregates once produced is mandatory. Hence variations in shape indices are primarily overcome by using a slightly greater quantity of cement so that the mix proportion becomes resilient to such variations. Further to this limited research on lagoon ash has been conducted until date and its applications is primarily in brick manufacturing. Thus, this research was undertaken to solve this multidimensional problem.

SOME IMPORTANT DEFINITIONS

Embrittlement Ratio: is defined as the ratio of unconfined uniaxial compressive strength (UUCS) to either flexural (E_b) or splitting tensile strength (E_t).

Specific strength of cement in concrete (SP_{con}): is defined as the contribution of 1% cement to the concrete strength and is given by strength at a particular age divided by the percent of cement of the total binders.

Specific strength of cement in concrete with fly ash or pond ash (SP_{fcon}): is the specific strength of cement in fly ash or pond ash concrete and is given by strength at a particular age divided by the percent cement when compared with respect to the control trial.

Specific effect of pozzolanic effect (SP_{po}): is defined as the specific contribution of fly ash or pond ash to the strength of FRCC and is calculated for a particular age as the difference between SP_{fcon} and SP_{con}.

Contribution rate of pozzolanic effect to strength (R_{po}): is defined as the percent of the strength portion supplied by fly ash or pond ash with its pozzolanic effect and is calculated by dividing SP_{po} by SP_{fcon}.

MATERIALS

Ordinary Portland cement (OPC) meeting Indian Standard, IS 12269:1987 (Equivalent to BS 12: 1991) was used in the investigations. Class F Fly ash (FA) directly collected from electrostatic precipitator (ESP) (used in notation as E) from one of the local Thermal Power Station meeting IS 3812:2003(BS 3892: 1997) was used along with OPC in the control cement-replaced mixtures. Typical lagoon ash (used in notation as L) with average LOI of the contemporary lagoon ash piles was used in the remaining cement-replaced mixes The physical characteristics and chemical composition of OPC, Fly ash and lagoon ash are summarized in Table 1.

Table 1   Physical and chemical properties of binders

TESTS	CEMENT	COAL ASHES		ASTM C618-94	IS 3812-1981
		E	L		
Physical tests					
Specific gravity, g/cm^3	3.15	2.26	2.11		
Passing 45 μm sieve, %	NA	83.6	53.4	>66	NA
Blaine fineness, m^2/kg	299	385	231		Min. 250
Initial setting time, min.	160	NA	NA		
Final setting time, min.	225	NA	NA		
Autoclave expansion, %	0.07	0.02	0.04		
Activity with cement, %	NA	84.2	68.1	>75	
Compressive strength, MPa					
3d	31.4	NA	NA		
7d	42.2	NA	NA		
28d	60.6	NA	NA		
Chemical composition, % by mass					
Silicon dioxide	20.7	60.72	55		>35
Aluminium oxide	5.08	27.5	25.2		
Ferric oxide	3.86	5.32	1.26		
$SiO_2+Al_2O_3+Fe_2O_3$	29.64	93.54	81.46	≥70	≥70
Calcium oxide	64.2	1.42	1.17		
Magnesium oxide	1.02	0.48	0.94		
Sulphur as SO_3	2.1	0.18	0.22	≤5	≤2.75
Alkalis as Na_2O	0.42	1.71	1.75	<1.5	<1.5
LOI	1.76	0.8	13.74	<6	<12
C	NA	0.28	8.1		
Moisture	0	0.5	4.5		

The coarse aggregates were crushed Deccan trap Basalt aggregates with nominal maximum size of 26.5 mm. Three different levels of shape index (combined flakiness and elongation index) were used. These were 30, 40 and 50%. The combined index was carefully chosen so that the flakiness and elongation indices had approximately 1:1 ratio. These were chosen to cover the usual extent of shape indices obtained when the screens had excessive wear and tear for a typical highway project. The internal combination of the coarser and finer fractions of coarse aggregates was kept constant for all the mixes. The effects on the water demand, compacted dry density and strengths were observed.

Two types of fine aggregates were used in the trials. Natural sand was uncrushed-clean, siliceous and natural from the Krishna River; while crushed stone sand was crushed Basalt and was obtained from the same crusher from which the coarse aggregates were obtained. No Mica was found in the sand. It was ensured that the particle size distribution of both the fine aggregates were almost identical. All these aggregates complied with the requirements of IS 383-1970 (BS 882:1992). The average grading used in the local area was used for the control mix. Three different levels viz. 0, 10 and 20% of 75 μ were used to access the effects of higher fines content on the mixture properties. These were directly replaced for equivalent weight of sand. The specific gravity and water absorption values of coarse aggregate (CA), natural sand (NS) and crushed sand (CS) were 2.95, 1.1%; 2.77, 1.5% and 2.88, 1.9%, respectively. The particle size distribution of the aggregates along with the target-combined aggregates' grading is shown in Figure 1.

Well-water confirming to IS456: 2000 and used in Ready Mix concrete (RMC) plant supplying concrete for pavements was used for the study. No admixture was used.

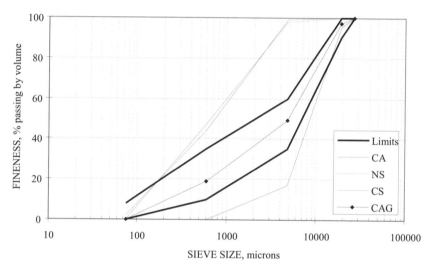

Figure 1   Particle size distribution of aggregates

MIXTURE PROPORTIONING

After examining the bulk characteristics, the particle size distribution and shape indices the maximum density curve (MDC) was plotted using the following relation (9):

$$CPP = (d/D)^{0.45} \times 100$$

where   d: sieve size opening under consideration;
D: nominal maximum size of the aggregate (NMSA) grading.

Table 2   Mix proportions for control mixes

MIX	CEMENT, kg/m³	ASH, kg/m³	CA, kg/m³	FA, kg/m³	WATER, kg/m³	VBT, s	NOMC, %	MDD, kg/m³
Control	160	0	1275	845	152	30	6.67	2550
CFA35	105	82	1250	835	155	29	6.82	2520
CFA45	88	108	1240	820	157	27.5	6.96	2490

The aggregate grading for the control mixes were kept almost parallel to MDC. Proctor curves were obtained at fixed aggregate/binder ratios to obtain the nominal optimum moisture content (NOMC) and maximum dry density (MDD). A uniform cement super-substitution factor of 1.5 was used while replacing the cement.

The fine aggregate content was chosen after optimizing the fine aggregate/aggregate ratio for minimum compaction energy. Mixes were proportioned for a constant Vebe consistency of 30±5s. Table 2 shows the control mixture proportions and the fresh properties.

A total of 3 control (including one OPC and two cement replaced with ESP fly ash); 2 with cement replaced with lagoon ash combinations; three combinations with variations in the coarse aggregates' shape index and three variations in fines passing 75 μm each with the natural sand and crushed sand as well. These combinations made a total of 36 mix proportions. Figure 2 shows the tree diagram for these investigations.

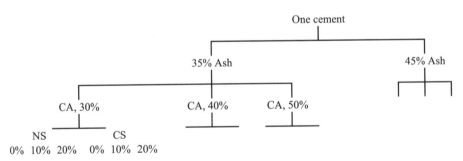

Figure 2   Tree diagram showing the trials

SPECIMEN FORMULATION AND TESTING SCHEDULE

The specimens were cast using the standard compacting hammer described in Ministry of road transport and highways — Guidelines for roads and Bridges. (16) The cube specimens were compacted in three layers with each layer compacted for 60±5 seconds. The beam specimens were also compacted in three layers for a time period sufficient enough to achieve equivalent density as cube density ±25 kg/m³ (avg.). The cylindrical specimens were cast using a circular compacting foot instead of the rectangular based on same principle. The consistency was measured in terms of Vebe time.

Table 3 gives the specimen description and testing schedule adopted for the testing program. The Table also provides information about the testing standards followed for various tests.

Table 3   Testing schedule

TEST	SPECIMEN TYPE	SPECIMEN SIZE, mm	AGE, days	STANDARD USED
UUCS	Cube	150	7, 28, 90, 180	IS 516-1959
MOR	Beam	700 × 150 × 150	7, 28, 90, 180	IS 516-1959
CST	Cylinder	∅ 300 × 150	28, 90	IS 5816-1999

Note: MOR: Modulus of rupture; CST: Cylinder splitting tensile strength

RESULTS AND DISCUSSIONS

Water demand, NOMC-MDD, compaction characteristics

As can be seen from Figure 3, the water demand for constant VB consistency increased with the increase in the fines content and the shape index of the coarse aggregates. It also increased with the increase in the lagoon ash content, probably due to higher LOI. Similar trends were observed for the crushed sand mixes. The range of increase in the water demand in NS mixes was 1 to 24% while the same increase ranged from 3.5 to 30% for CS mixes. In addition to this the NOMC increased as the shape index of coarse aggregates increased. On the other hand the relative density (calculated as the ratio of dry density of a mix under consideration to the dry density of control mix) decreased with the increase in the shape index of the coarse aggregates. This was observed both for the NS and CS mixes. This could be attributed to the increased shape index, which in turn reduced the compactibility of the aggregate grading with a constant binders' content.

Relatively longer compaction effort per unit volume of concrete was required for beam specimens. On the contrary, the compaction effort was least for cylindrical specimens but the densities were having a wider scatter than the other two specimen shapes. In addition to this, the densities for CS mixes were invariably greater than the densities of NS mixes. This could be partially attributed to the difference in the specific gravity and partially due to better packing.

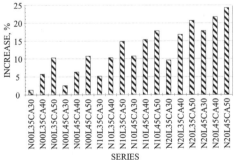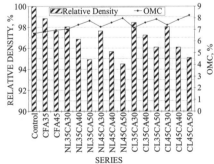

Figure 3   Increase in water demand (left) and relative density and NOMC change (right)

Nature of Strength Development and Correlations

Mechanical strength

UUCS results for NS and CS mixes are shown in Figures 4 and 5. As can be observed that the strength reduced with the increase in the lagoon ash content, and the increase in the shape index of coarse aggregate and fines content of the fine aggregate. The strength reduction at 7 days varied from 32 to 77% and 47 to 76% for NS and CS mixes respectively.

Although the strength reductions are comparable, the strengths obtained for CS mixes were invariably higher than the corresponding NS mixes. This was observed even though the water/binder ratio for CS mixes was invariably higher than the corresponding NS mixes.

This could be attributed partially to the mineralogy and partially to a better aggregate-paste interlock. There was a gradual strength gain in the mixes due to the pozzolanic action of lagoon ash at subsequent ages. This is discussed in detail in the subsequent sections.

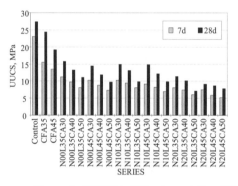 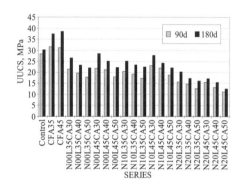

Figure 4   UUCS in NS mixes at 7 and 28 days (left) and at 90 and 180 days (right)

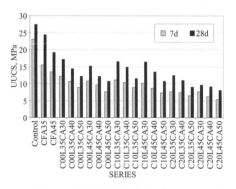 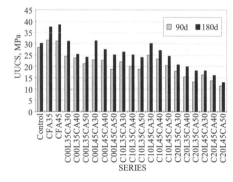

Figure 5   UUCS in CS mixes at 7 and 28 days (left) and at 90 and 180 days (right)

Embrittlement Ratio (ER)

ER is very important in pavement design. It primarily decides the relationship between the compressive and tensile strength of concrete. In these investigations, the ER's were calculated to analyse the comparative strength gain of tensile strength over the compressive strength. Figure 6 shows the embrittlement ratios in flexure and in splitting tension for various mixes. The ER's for all the mixes were initially found to be higher, but with age, a gradual reduction was observed, thus showing better tensile performance with age. The ER's in flexure varied from 9.33 to 7.79 and from 7.43 to 6.00 at 7 and 180 days respectively for NS mixes. While CS showed a variation from 9.76 to 7.71 and 8.43 to 7.27 at 7 and 180 days, respectively. Similar figures were obtained for ER's in splitting tension. The ER's decreased with the increase in the shape index of coarse aggregate. This nature of reduction of ER could be attributed to better aggregate-paste interlock development with age due to the pozzolanic effect of lagoon ash.

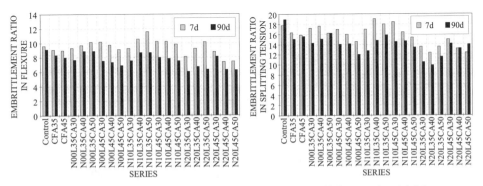

Figure 6   Embrittlement ratio in Flexure (left) and Splitting tension (right)

Pozzolanic activity

Specific strength of cement in Ash Concrete (SP_{fcon})
From Figure 7, it can be seen that the strength contribution of OPC in the control mix is normal and increases slowly, while its strength contribution in developing the long term strength in lagoon ash concrete is greater; depends on ash quantity, its pozzolanic activity and improves with age. Thus cement performance improves in the presence of pozzolanic action. This analysis can be utilized in selecting the right combination of cement and coal ash without appreciating the complete mechanics of hydration.

Specific strength of pozzolanic effect (SP_{po})
SP_{po} (expressed as the difference between SP_{fcon} and $SP_{con)}$ is a way of quantifying the relative contribution of ash in the strength development of ash concrete with respect to the control mixture. Referring to Figure 7, it can be observed that there exists an increasing tendency of strength contribution by ash with age, depending on the pozzolanic activity of the ash. The initial SP_{fa} is negative in almost all the lagoon ash mixes and might be attributed to the dilution effect produced by lagoon ash in terms of the cementitious materials. While at later ages the trend is reversed and the pozzolanic effect becomes predominant. This analysis might be applied in optimizing cement content and deciding the age at which the characteristic strength should be defined and specified. Furthermore, this analysis can be used in selecting appropriate coal ash and as a complementary data to the pozzolanic activity of coal ash.

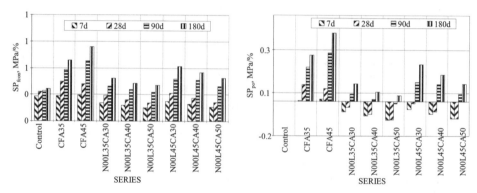

Figure 7   Specific contribution to strength by cement (left) and pozzolanic effect (right)

Contribution rate of pozzolanic effect to strength (R_{fa})
R_{pa} (expressed in %) is a way of normalizing the pozzolanic effect of ash over SP_{fcon}. Referring to Figure 8, it can be seen that the initial contribution rate of the pozzolanic effect to the strength is negative due to the dilution effect but is dominated by the pozzolanic effect at latter ages. R_{po} increases gradually with the contribution rate of pozzolanic action to UUCS increasing up to 40% for NS mixes and 47% for CS mixes.

This effect is again a function of the pozzolanic effect and the type of ash used in concrete. This analysis can be utilized in further limiting the cement and fly ash contents and making mixes leaner (thus saving natural resources and reducing environmental pollution), wherever possible. Additionally with an established application history this analysis could further be used in strength prediction at later ages.

Irrespective of the binders content of the mixes, at later ages, all the mixtures showed the following relationship in terms of the contribution rate of pozzolanic activity compressive strength < flexural strength < splitting tensile strength. This might be attributed to the fact that compressive strength is relatively less sensitive to the internal structure of concrete than the tensile strength. The nature of rupturing of the aggregate-paste bond is different in these modes of failure of concrete.

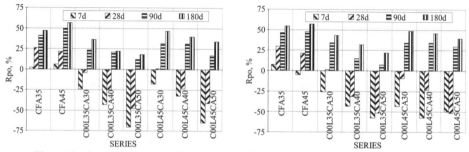

Figure 8   Contribution rate of pozzolanic effect to UUCS (left) and MOR (right)

CONCLUSIONS AND RECOMMENDATIONS

The following conclusions and recommendations could be made based on the presented study. For a given consistency:

1. Using the method of super-substitution of fly ash, the water demand increases with the increase in the cement replacement level, aggregate shape index and the proportion of the fraction passing 75 micron.

2. The embrittlement ratio (as defined in this paper) is dependent on the quantity of ash. But it was found that the embrittlement ratio invariably decreases with the age of concrete, substantiating the facts that there is a better tensile strength development with respect to the compressive strength and that the pozzolanic effect has a better influence on the tensile strength than the compressive strength.

3. The pozzolanic effect can be analyzed using the notions of specific strengths. It was observed that the early age strength of concrete was less but picked up at later ages, when the pozzolanic effect started manifesting itself in the process of strength gain. The analysis shows that the pozzolanic effect is negative initially due to the cement dilution effect but becomes positive at later ages, when the pozzolanic effect dominates and this depends on the ash type and content.

4. The contribution rate of pozzolanic effect is more pronounced in the tensile strength than in the compressive strength; depends on the type of ash, its fineness and water-binder ratio; but is found to remain unaffected by the relative amount of cement content in the mix.

5. Lagoon ash could be used in RCC for pavement bases with sufficient care exercised. The quantity to be used in a unit volume of mixture should be decided by lab trials.

6. The fines content also needs careful evaluation. In this study, the natural sand was clay-free resulting in good strength development. Crushed sand with high fines can be used with prior evaluation.

REFERENCES

1. AHMED A E AND EL-KOURD A A, Properties of Concrete Incorporating Natural and Crushed Stone very fine sand, ACI Materials Journal, Vol. 86, No.2, 1989, pp. 417-424.

2. CELIK T AND MARAR K, Effects of crushed stone dust on some properties of concrete, Cement and Concrete Research, Vol. 26, No.7, 1996, pp. 1121-1130.

3. ACI 325.10R95 Report on Roller Compacted Concrete Pavements. Re-approved 2001, pp. 31-51.

4. PITTMAN D W AND RAGAN S A, Drying Shrinkage of RCC for Pavement Applications, ACI Materials Journal, Vol. 95, No. 1, 1998, pp. 19-26.

5. NAIK T R, Strength and Durability of Roller Compacted High Volume Fly Ash Concrete Pavements, Practice Periodical on structural design and construction, Nov. 2001, pp. 155-165.

6. DELVA K L, Rennick Yard RCC Pavement Design and Construction, Proceedings of Conference on RCC II, Edited by Hansen and Guice, ASCE, San Diego, California, USA, Feb. 29 – Mar. 2, 1988, pp. 410-418.

7. QASRAWI H Y, ASI I M AND WAHHAB H I, Proportioning RCCP Mixes under Hot Weather Conditions for A Specified Tensile Strength, Cement and Concrete Research, Vol. 35, No. 2, 2005, pp. 267-276.

8. GHAFOORI N AND CAI Y, Laboratory-Made RCC Containing Dry Bottom Ash: Part I-Mechanical Properties, ACI Materials Journal, Vol. 95, No. 2, 1998, pp. 121-130.

9. LI Q, ZHANG F, ZHANG W AND YANG L, Fracture and Tension Properties of Roller Compacted Concrete Cores in Uniaxial Tension, Journal of Materials in Civil Engineering, September-October 2002, pp. 366-373.

10. CHENG C, SUN W AND QIN H, The Analysis on Strength and Fly Ash Effect of Roller-Compacted Concrete with High Volume Fly Ash, Cement and Concrete Research, Vol. 30, No. 1, Jan 2000, pp. 71-75.

11. MALHOTRA V M AND PIGEON M, Frost Resistance of Roller Compacted High Volume Fly Ash, Journal of materials in Civil Engineering, November 1995, pp. 208-211.

12. JONES M R, ZHENG J L AND NEWLANDS M D, Estimation of Filler Content Required to Minimize Voids Ratio in Concrete, Magazine of Concrete Research, Vol. 55, No. 2, 2003, pp. 193-202.

13. HWANG K, NAGUCHI T AND TOMOSAKWA F, Prediction Model Of Compressive Strength of Fly-Ash Concrete, Cement and Concrete Research, Vol. 34, No. 12, 2004, pp. 2269-2276.

14. BAPAT J D, SABNIS S S, HAZAREE C V AND DESHCHOUGULE A D, Eco-Friendly Concrete with High Volume of Lagoon Ash, Journal of Materials in Civil Engineering, Vol. 18, No. 3, 2006, pp. 453-461.

15. KOKUBU K, KONDOH T AND UENO A, Studies on compactibility and test methods of extremely dry concrete used for roller compacted concrete pavements, Fifth international conference on concrete pavement design and rehabilitation, 1993, pp. 271-280.

16. KOKUBU K, CABRERA J G AND UENO A, compaction properties of roller compacted concrete, Cement and Concrete Composites, Vol. 19, 1996, pp. 109-117.

17. DELATTE N, AMER N AND STOREY C, Improved management of RCC pavement technology, University Transportation Center for Alabama, UTCA Report No. 01231, 2003.

18. Guidelines for Roads and Bridges, Ministry of Road Transport and Highways, 3e, 2003.

SOME EXPERIENCE IN THE PRODUCTION OF SCC IN SCG

Z Romakov   **D Bojovic**
K Jankovic   **Z Kacarevic**
L Loncar
IMS Institute
Serbia

ABSTRACT: In the last few years, new technologies, techniques and materials began to regain their original status in the region of the former Yugoslavia. The need for fast economic progress and foreign donations through various projects enabled the use of new technologies and materials in new building constructions. New bridges over the Bosna river in Doboj and Modriča, from the appreciated donation of the Japanese government, provided the first practical steps in accepting self compacting concrete as a new material.

Keywords: Aggregate, Cement, Admixture, Concrete, Self compacting concrete, Bridge, Main girder

Zoran Romakov, BSc CE is working in the Laboratory for Materials Testing at the IMS Institute, Belgrade, Serbia. His 30 years work experience includes both on-site (supervision, concrete quality control) and laboratory testing. He also worked in Russia, Iraq and Libya.

Dragan Bojovic, BSc CE is working in the Laboratory for Materials Testing at the IMS Institute, Belgrade, Serbia. He is a postgraduate student at the Faculty of Civil Engineering at Belgrade University. His research interests focus on SCC.

Dr Ksenija Jankovic, is a senior research fellow in the Laboratory for Materials Testing at the IMS Institute, Belgrade, Serbia. Her research interests include polymer modified concrete, recycled aggregate concrete, lightweight concrete, concrete quality control, properties of precast concrete elements, fibre reinforced concrete. She has published about 80 papers.

Zoran Kacarevic, BSc CE is working in the Materials Department at the IMS Institute, Belgrade, Serbia. He worked according to BS, ASTM, DIN, GOST, and JUS standards in the country and abroad (Philippines, Iraq). He also has experience in investigating special cements, lightweight concrete and fibre reinforced concrete.

Ljiljana Loncar, BSc CE is working in the Materials Department at the IMS Institute, Belgrade, Serbia. She is especially interested in concrete quality control, control of concrete factory and production of precast elements. She is a part of the team for implementation of new EN standards in that field.

INTRODUCTION

In the last few years, self compacting concrete began to regain its original status in the region of the former Yugoslavia. With the appreciated donation of the Japanese government, two concrete bridges over the Bosna river in Doboj and Modriča (Republic of Srpska) are under construction (expected completion of construction is October 2006.).

Both bridges are designed by Nippon Koei Co. Ltd in consortium with the Central Consultant Inc., Tokyo, Japan. The main contractor is Obayashi Co. Japan, and subcontractors are "Integral Inženjering" Laktaši and "ŽGP" Sarajevo.

The Institute for testing materials (IMS) from Belgrade, Serbia, took a significant part in the construction of the mentioned bridges, including preliminary investigation of materials, preliminary trials of concrete mixes for all concrete classes, production control and final evidence of concrete properties, both in fresh and hardened state.

DESCRIPTION OF BRIDGES

The bridges over the Bosna river in Doboj and Modriča are classically designed and constructed bridges. The Doboj bridge is 200 m long (five spans of 40 m each), perpendicular on the main flow of the river; whereas the Modriča bridge is 240 m long (six spans of 40 m each), at an angle of 60° between the bridge and the main flow of the river.

The bridge substructures are composed of massive reinforced footings, abutments, piers and pier heads, made of class C 25/30 concrete. The superstructures are composed of four main girders (prestressed single beams) in each 40 m span, supported on abutments and pier heads by means of neoprene bearings. Each single beam is 2.5 m high, classical double T shape in cross section, with a volume of concrete being 52 m^3, approximately 130 tons in mass. All single beams in one span are connected above bearings and in the quarters of the span with prestressed concrete cross beams.

Despite three heavy floods in the summer of 2005, all substructure works were successfully finished in the autumn of 2005. Main girders production started on 01.09.2005 on both sites, and finished on 28.04.2006 in Doboj and 21.05.2006 in Modriča, interrupted during the winter season 2005/2006.

First intention of the designer was to produce the main girders in 10 m long sections in the precast concrete plant 18 km from the Doboj site, transport to the site, connect sections with epoxy resins in their final position and then apply prestressing force. After carefully examining all relevant factors, this idea was abandoned and main girders were produced in situ over heavy scaffolding.

After structural design, the following classes of concrete were required:
- Footings, piers and pier heads: C 25/30
- Concrete piles: C 30/37
- Main girders: C 50/60
- Cross beams and bridge deck slab: C 30/37
- Compressive strengths on preliminary trials were required to be 20% greater than the required strengths.

All concretes were required to have entrained air of 4.5±1.5% in fresh state for durability reasons (this requirement was later adjusted to be 3.5±1.5% for main girders concrete because of strength requirements).

This work represents all activities, problems, solutions and final results of main girders production and quality control.

CHOICE OF CONCRETE COMPOSITION

First investigations of materials for the construction of the bridges in Doboj and Modriča started in the late autumn of 2004. After careful examination of all available materials and some pilot trials on concrete, the following materials were chosen for preliminary trials:

- Cement: CEM I 52.5N from the factory Našicement, Croatia,
- Aggregates: 0-4 mm natural from Bosna river (for all concretes)
                4-8 and 8-16 mm natural from Bosna river (for concrete classes up to C 30/37)
                4-8 and 8-16 mm crushed from Vrbas river for concrete class C 50/60.
- Admixtures: Glenium 51 (hyperplasticiser, produced by MBT Italia),
                Meta Air (air entraining agent, produced by MBT Italia).
- Water: from city supply

Table 1   Typical properties of cement CEM I 52.5N Našice

CHEMICAL COMPOSITION %		PHYSICAL PROPERTIES	
Hydraulic constituents		Fineness	
SiO_2	19.88	Grains < 0.09 mm, %	1.16
Al_2O_3	6.26	Specific surface, cm^2/g	5211
Fe_2O_3	2.45	Density, g/cm^3	3.09
CaO	56.22	Setting time	
Insoluble residue	0.44	Standard water, %	29.30
Moisture at 105°C	0.30	Initial	2h 20min
Loss on Ignition	0.79	Final	3h 40min
CO_2 in $CaCO_3$	0.61	Volume stability	
Free CaO	2.34	Cakes	stable
CaO in $CaCO_3$	0.78	Le Chatelier, mm	0.25
CaO in $CaSO_4$	2.66	Linear deformations, mm/m	
SO_3 in $CaSO_4$	3.80	after 4 days	-0.06
CaS	0.00	after 7 days	-0.17
MgO	2.17	after 14 days	-0.25
Na_2O	0.27	after 21 days	-0.39
K_2O	0.70	after 28 days	-0.43
Alkalis as Na_2O	0.73	Mechanical properties	
MnO	0.00	Compressive strength, MPa	
FeO	0.00	after 1 day	24.2
P_2O_5	0.00	after 3 days	39.7
Degree of saturation	86.80	after 28 days	60.5

All concrete constituents were under permanent control of a certified testing house. Basic tests on all materials for making preliminary trials were also carried out.

Fractions 0-4 mm (both natural and crushed) from all suppliers in a wide region around Doboj and Modriča showed lack of fine particles, so the cement content used in the preliminary trials was somewhat higher than usual, in order to replace lacking particles. This question was more a matter of economics than technique.

PRELIMINARY TRIALS

Preliminary trials for all classes of concrete were carried out during the winter 2004/2005 in the laboratory for concrete of the Institute IG in Laktaši, Republic of Srpska, in presence of representatives of Nippon Koei Co. and Obayashi Co. On the basis of test results at preliminary trials and work trials at selected concrete plants, mix proportions were adopted.

The air entraining level was prescribed for durability reasons, bearing in mind the high strength requirement for main girders and that the entrained air decreases compressive strength. A reference concrete mix of class C 50/60 without entrained air was also made and all necessary tests were carried out on this hardened concrete. Compared to the air entrained mix, the compressive strength increased from 75.3 MPa to 92.1 MPa, and the unit weight of concrete increased from 2377 to 2437 kg/m^3. Further tests had shown that loss of compressive strength after 250 cycles of freezing and thawing was less than 2% (the standard JUS M.M1.016 allows a maximum permissible strength loss of 25%) and the concrete satisfied the requirements of the standard JUS U.M1.055 regarding resistance against freezing and de-icing salts. A proposal for an alternative concrete mix without entrained air (for main girders only) with additional tests was submitted to the Contractor but this was not accepted.

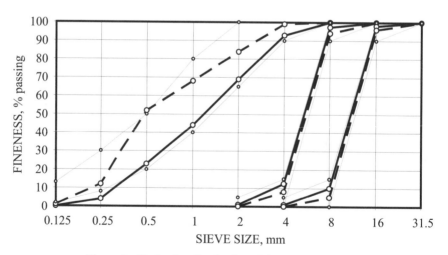

Figure 1   Grain size distribution of the aggregate used

Note: Full line – aggregate used for preliminary trials PP-08 and PP-11
Dashed line – aggregate used for the preliminary trial 08D and work trials, later used in actual production

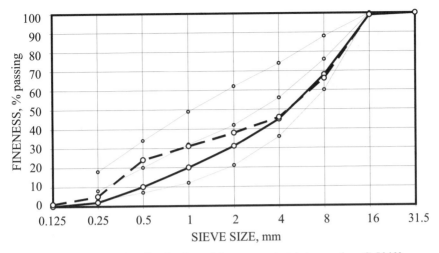

Figure 2   Grain size distribution of the aggregate mixtures, class C 50/60

Note: Full line – aggregate used for preliminary trials PP-08 and PP-11
       Dashed line – aggregate used for the preliminary trial 08D and work trials, later used
       in actual production

Table 2   Summary on preliminary and work trials

TRIAL MARK	PRELIMINARY TRIAL			WORK TRIAL	
	PP-08	PP-11	08D	MODRIČA	DOBOJ
Date of execution	24.12.2004.	27.12.2004.	27.07.2005.	09.08.2005.	10.08.2005.
Cement CEM I 52.5N, kg	578	575	567	567	567
Aggregate total, kg	1622	1682	1615	1615	1615
Fraction 0–4 mm, kg	730 *	757 *	727	727	727
Fraction 4–8 mm, kg	324	336	323	323	323
Fraction 8-16 mm, kg	568	589	565	565	565
Hyperplasticizer, kg	2.53 **	3.17 **	4.56	4.56	4.56
Air entrainer, kg	0.29	0.00	0.06	0.06	0.06
Water, kg	191	190	198	198	198
Slump, cm	22.0	23.0	25.0	27.0	25.5
Entrained air, %	3.5	0.7	2.7	3.0	2.8
Temperature, °C	17	16	31	26	26
Unit weight, kg/m^3	2393	2450	2385	2385	2385
7 days strength, MPa	57.9	75.4	71.7	64.2	65.6
28 days strength, MPa	75.3	92.1		77.4	76.6

*   from Bosna river (trials PP-08 and PP-11); from Sava river (trial 08D and work trials in
     Modriča and Doboj)

** Glenium 51 (trials PP-08 and PP-11); Dynamon SX (trial 08D and work trials in Modriča
     and Doboj)

CONCRETE PLANTS

The choice of concrete plants for the construction of bridges was very important because of the required properties for both fresh and hardened concrete and the total amount of concrete. Carefully examining the available concrete plants in the region of Doboj and Modriča, the following concrete plants have been chosen:

- Doboj bridge:
  - "Širbegović" concrete plant, 18 km from Doboj (abandoned after few weeks for problems in transport and retention of fresh concrete properties),
  - "ŽGP" Sarajevo, plant Doboj (25 m³/h, 1.5 km away from the site) as main plant,
  - "Autokomerc" Doboj (15 m³/h, 100 m away from the site) as spare plant, sometimes used simultaneously with main plant.
- Modriča bridge:
  - "ŽGP" Sarajevo, plant Modriča (25 m³/h, situated on the site) as main plant,
  - "ŽGP" Sarajevo, plant Doboj (25 m³/h, 47 km away from the site) as spare plant.

Those setups proved to be satisfactory, both for capacity and fresh concrete uniformity. None of concrete plants have automatic devices for batching of admixtures, even so, due to the very good control, fresh concrete was uniform regarding the required properties.

PREPARATION AND START OF CONCRETING

Concreting of both bridges started in the spring of 2005. During the first months of concreting some problems concerning placeability and workability of fresh concrete have been noticed. This regarded mainly the relatively rapid loss of workability (slump loss, fresh concrete became sticky, supplementary addition of admixtures on-site for improving fresh concrete properties was not permitted). As the first blocks of concrete (footings, pier lifts) were relative massive, there was no great problem in placing concrete (by concrete pumps and buckets) and in terms of the final appearance and quality. Although cement and admixtures (hyperplasticizer, polycarboxylate ester based, applied within the recommended limits) used, were of high quality, it seemed that something was wrong in their combination. Mutual opinion of Consultant, Engineer and Contractors was to not waste time to establish what was wrong in the applied combination of materials, but to find a better combination of materials which would suit all requirements. Trial concreting of one 10 meter long section of a main girder had shown that the firstly adopted mix proportion for concrete class C 50/60 was not suitable for main girders production:

- lot of vibrations (both with external and internal vibrators) had to be applied.
- lower flange and partial girder web (25 cm thick), being congested with reinforcing bars and tendon ducts, could not be satisfactory filled with concrete, large honeycombs occurred.

As a response to such problems, Concrete Engineers from IMS and IG Institutes proposed the following changes in fresh concrete composition to the Consultant, which would not affect adversely the required concrete properties:

- water to cement ratio to keep as low as 0.35,
- retain the same cement quantity (on the demand of the Consultant, although the cement content could be decreased),
- increase the workability of fresh concrete (increase slump up to 25±3 cm) by adequately adjusting the amount of plasticizer,

- replace existing aggregate fraction 0-4 mm with the already identified, much better grain size distribution fraction from the Sava river, produced by MGM Tolisa, Orašje,
- identify a better combination of cement / hyperplasticizer.

Several preliminary trials were carried out in the period May - July 2005, followed by another trial concreting of a main girder section in the beginning of August 2005 with a mix proportion adopted by the Consultant. The replacement hyperplasticizer was Dynamon SX (modified acryl polymer based, produced by MAPEI, Italia). Compressive strengths, both of preliminary trials and work trials were of the same level as in preliminary trials with precedent combination of materials (about 75 MPa after 28 days). The fresh concrete used for work trials was of flowing consistency: Abrams cone slump 25-27 cm, Abrams cone spread 60-65 cm. Despite relatively high ambient temperatures (about 30°C), the concrete retained the same consistency for more than one hour. During work trial concreting, settlement of the steel mould bottom (made of wood) occurred and about 4 m^3 of concrete leaked through the created gap to the distance of about 6 meters to the left and right side. During leaking of concrete neither segregation nor lamination were observed. After repairing the mould, concreting was successfully finished, with the concrete surface being quite uniform after removing the mould.

Comparative compressive strength tests were carried out on cubes cast with slight vibration on vibrating table and without any vibration (concrete was simply poured into moulds and levelled with a steel ruler). No significant difference in unit weight (less than 10 kg/m^3) and compressive strength (about 1-2 MPa) were observed.

Although the concrete used for making main girders on both bridges could be deemed and used as "self compacting concrete" for plain and light reinforced structures, some light vibrations (much less than usual) were applied, on demand of the Consultant, because of the great amount of reinforcing bars (about 200 kg/m^3) and the presence of a total of six prestressing tendon ducts, 8.5 cm in diameter, so for the greatest part of the main girder there was only 2.5-3 cm space left for concrete to pass into the lower part of the girders. Slight vibrations were applied just to be sure that each reinforcing bar and each tendon duct were totally wrapped and covered with concrete. So, this concrete could be designated as "quasi self compacting concrete" and experience and knowledge from these structures was a good start point for future use of self compacting concrete in various types of structures.

CONCRETING OF MAIN GIRDERS

Concreting of main girders started, as mentioned, on 01.09.2005, finished on 28.04.2006 in Doboj and 21.05.2006 in Modriča using materials and mix proportion as in work trials. Only one mould was used per each site, if more than one mould had been available, scheduled work on main girders could have been significantly shorter. The duration of each main girder concreting varied from 4 to 7 hours, depending on the number of concrete plants engaged (more time was necessary in the beginning of works, less after some experience has been obtained). Concrete was transported by means of truck mixers (two per each concreting plus one as spare) and concrete pumps (one active pump as main and one pump or tower crane with 1.0 m^3 bucket as spare). All main girder concreting was carried out without any interruption; no honeycombs, leakage or any imperfections were observed on the outer surfaces after removal of moulds. Curing of concrete had been performed by covering with blankets and spraying with water for at least seven days.

QUALITY CONTROL OF CONCRETE WORKS

A very important part of the Doboj and Modriča bridge constructions was the well-organized and strictly performed quality control at all stages of construction, under permanent supervision of the Consultant and Designer representatives. Site laboratories for concrete for basic tests of fresh and hardened concrete were established at both sites prior to start of concrete works.

The quality control consisted of:

- Control of all materials delivered on the site (visual inspection, sampling and testing in laboratories of the IMS and IG Institutes, providing of certificates issued by certified Testing Houses),
- Control of materials on concrete plants prior to concreting (quantities of each material, visual inspection, moisture content in aggregate),
- Permanent control during batching and mixing of concrete by experienced technicians of the IMS Institute (quantity of constituents, consistency, entrained air),
- Control of fresh concrete from each truck mixer delivered to the site (consistency, entrained air, temperature, chloride content),
- Making of concrete specimens (15 cm cubes and 15/30 cm cylinders for compressive strength test), curing specimens to the moment of testing,
- Compressive strength tests at 7, 28 and 91 days.

All control tests of concrete constituents were carried out at the Laboratory for concrete of the Institute IG in Laktaši.

Compressive strength tests were carried out at the site laboratory in Doboj (15 cm cubes), the laboratory for concrete of IG Laktaši (15 cm cubes) and the laboratory for concrete at IMK Banja Luka (15/30 cm cylinders).

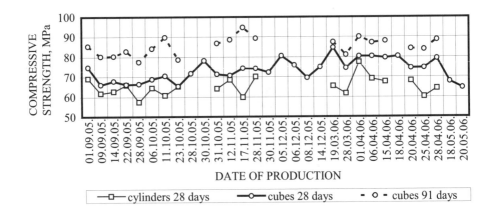

Figure 3   Doboj bridge – strength test chart, concrete class C 50/60

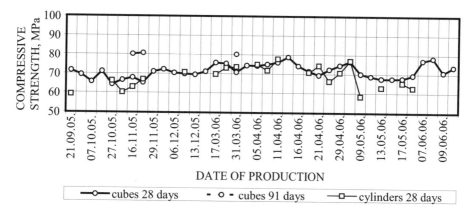

Figure 4   Modriča bridge – strength test chart, concrete class C 50/60

Table 3   Compressive strengths – statistical parameters

	15 cm cubes			15/30 cm cylinders	f_{91}/f_{28}
DOBOJ BRIDGE					
Age, days	7	28	91	28	
Number of results	30	30	20	20	
Minimum strength, MPa	58.4	64.6	77.5	57.4	
Mean strength, f_{cm}, MPa	66.5	73.3	85.4	65.2	1.18
Maximum strength, MPa	74.1	84.6	94.7	77.5	
Standard deviation, MPa	4.6	5.4	4.4	4.6	
Characteristic strength, MPa	59.8	65.3	78.9	58.4	
MODRIČA BRIDGE					
Age, days	7	28	91	28	
Number of results	38	38	3	21	
Minimum strength, MPa	57.3	64.3	80.1	58.3	
Mean strength, f_{cm}, MPa	66.3	71.4	80.3	68.3	1.18 ·
Maximum strength, MPa	77.5	78.7	80.6	77.7	
Standard deviation, MPa	5.9	3.7		5.8	
Characteristic strength, MPa	57.6	66.0		59.7	

Notes: Strength test results are represented as mean strength of three or more consecutively taken specimens,

Characteristic strength: $f_{ck} = f_{cm} - 1.48\ \sigma$ (after DIN EN 206-1: 2000)

Due to the very good control in concrete production, the fresh concrete had very uniform properties when delivered to the site:

- Variations in slump were low (24 to 27 cm),
- Entrained air content varied mainly from 2.5 to 3.5%.

The hardened concrete satisfied all requirements regarding compressive strength and appearance. It is worth mentioning that the cement used for concrete works on both bridges had a significant strength gain from 28 to 91 days (18%), this can be useful for structural analysis on future projects.

Shrinkage and creep of concrete has been determined at the IMS Institute in Belgrade, according to the standards JUS U.M1.027 and JUS U.M1.029 using specimens cast on the site of the Doboj bridge. Mean values of deformation after 91 days were:

- shrinkage: 0.27 mm/m.
- creep: 0.85 mm/m.

AKNOWLEDGMENTS

This paper is the result of the project TR – 6503 B supported by the Ministry of Science and Environment Protection.

OPTIMISATION OF A LIGNOCELLULOSIC COMPOSITE FORMULATED WITH A LOCAL RESOURCE : THE DISS (*AMPELODESMA MAURITANICA*)

M Merzoud

Centre Universitaire de Souk-Ahras

M F Habita

Université de Annaba

Algeria

R M Dheilly   **A Goullieux**   **M Queneudec**

Université de Picardie Jules Verne

France

ABSTRACT. Diss (*Ampelodesmos mauritanicus*, family of Poaceae) is a very luxuriant plant growing in wild state around the Mediterranean North Africa and dry areas of Greece and Spain. It grows in France, mainly, in pockets of the Alpes-Maritimes, the Var, the South of Corsica and Herault. In the past, it was used as building material because of its mechanical and hydrous qualities. The use of such a fibrous plant in a cementitious matrix leads to lightweight materials with very attractive tensile behaviour that can be used as advantageous filling materials for structures subjected to seismic effects. This paper is focused on the optimisation of this kind of material on the basis of mechanical and hydrous properties. The basic vegetable material, very fibrous, presents indeed an absorption of about 90% that would be corrected. Moreover, we noted a considerable retardation of setting and very low resistances during the composite tests with natural crushed diss, despite the fact that the fibres have considerable tensile strength. To improve the contribution of fibres in cementitious composites, we have carried out a treatment by boiling the fibres of diss to extract the substances responsible for the adverse reaction between fibres and cement paste. We have also carried out a treatment with linseed oil and bitumen to attenuate the absorption rate of fibres. The results obtained are encouraging and enable to foresee a later development of this material.

Keywords: Lignocellulosic diss, Ampelodesma mauritanica, Fibres, Absorption, Treatment.

Mr M Merzoud, Lecturer, Master degree, Centre Universitaire de Souk-Ahras, 41000 Algeria

Dr R M Dheilly, Senior Lecturer, Laboratoire des Tehnologies innovantes, UJPV, Département de Génie Civil, France

Dr A Goullieux, Senior Lecturer, Laboratoire des Tehnologies innovantes, UJPV, Département de Génie Civil, France

Prof M Queneudec, Laboratory Director, Laboratoire des Tehnologies innovantes, UJPV, Département de Génie Civil, France

Prof M F Habita, Senior Lecturer, Laboratoire de Génie Civil, Université de Annaba, Algeria

INTRODUCTION

Concrete containing lignocellulosic materials has been, in the last decades, subjected to numerous interesting research works. Indeed, the vegetable matter represents a source of renewable products.

Savastano Jr et al. [1] used residues of sisal, banana tree and eucalyptus as reinforcement in cementing composites. The composites thus obtained present acceptable mechanical performance. The works of Kriker et al. [2], based on the use of four types of date palm fibres in a cementitious matrix, showed that the increase of length and percentage of fibres improves the flexural strength and hardness of the composite, but decreases the compressive strength. The use of banana tree, sugar cane and coconut fibres, as epoxy polymer concrete reinforcement [3], shows that the coconut fibre, unlike the two other types of fibres, makes it possible to obtain a slight increase of the composite flexure. Ledhem et al. [4] showed in their work on cementitious composites containing wood shavings that the thermal treatment of wood could increase the mechanical resistance, thermal conductivity, and reduce extreme dimensional variations of the composites. Aamar Daya [5] used dust from the stripping of flax fibres as aggregate in a composite with cementing matrix. His works showed that the treatment of flax dust with boiled water considerably improves the mechanical resistance of the composites.

However, the presence of a plant within a cementitious matrix gives rise to concerns of increased sensitivity to water, in addition to other mechanical disorders, and thermal disperformance (Piementa et al. [6]). The strongly alkaline environment developed by the hydration of cement causes hydrolysis reactions and solubilises some compounds, as sugars, hemicelluloses and pectins (Simatupang [7]). Garci Juenger et al. [8] and Bilba et al. [9] have studied the influence of sugars on the setting of the cementitious composites and showed that sugar retards cement hydration.

The existing literature concerning diss fibres seems to indicate that there is a lack of technological valorisation of this plant, in particular in the field of cementitious composites. Species of this plant exist in wild state and in large quantity around the Mediterranean countries, and their fibrous nature seems to offer as much qualities to the cementitious composites as traditional fibres. However many studies reported a great oil accumulation in the vegetable matter, which is likely to interact with the cementing paste [10]. This is why further treatments must be carried out.

In order to reduce the water absorption of this plant, which is about 90%, and the retardation of setting observed during the mechanical tests on composites containing crushed natural diss, we have carried out various treatments of fibres: by boiling (extraction of the soluble substances), coated with bitumen or linseed oil (reduction in the rate of absorption).

MATERIALS AND EXPERIMENTAL METHODS

Materials

The diss material as aggregate in our composites has been crushed with a 10 mm mesh Retsch type cutting mill. The different stages of the study of crushed diss fibres are as follows:

- Tr1: fibres of natural diss, dried in the oven at 50°C and 100°C
- Tr2: fibres boiled in water, then dried in the oven at 50°C
- Tr3: fibres boiled in water and coated with linseed oil
- Tr4: fibres boiled in water and coated with bitumen.

During the treatment with boiling water, we have preserved the boiled water residue to study its influence on the setting of cement. The cement used was a CPA type CEM I 52.5 (standard EN 196-1).

Experimental methods

The morphology of various fibres was studied by scanning electron micrographs (SEM). The images were taken with the following devices:

- video microscope (Controlab®) VH-Z25 provided with a zoom 25× to 175×,
- an annular light with cold lighting appliance, non-diffuse lighting, semi-shaving, positioned on the video microscope, allowing the description of the relief of the samples,
- a high resolution screen 507×688 pixels,
- a PC,
- a system of vision VIDEOMET (Controlab®) allowing digitalisation and visualisation of images.

The test specimens made with various fibres, treated or not, were preserved for 28 days in a storage room (R.H. = 95%, T = 20°C), and then dried at 50°C until a constant mass before testing.

Mechanical tests were carried out, according to the European standard EN 196-1, on prismatic specimens of 4 × 4 × 16 cm. Tensile strengths was measured using a three point flexural test bench, equipped with a system of acquisition. Compressive strength tests were carried out on the halves retained from flexural testing, with a standard Perrier 68.7 machine.

The dynamic elastic modulus was determined by a sonic method, according to the standard E0641 Ultrasonic Tester, on prismatic specimens of 4 cm × 4 × cm × 16 cm. The principle was based on the determination of the propagation velocity of the ultrasonic waves (celerity) in the composite. The dynamic elastic modulus is given by the relation:

$$Ed = \rho C_L^2 \tag{1}$$

Where Ed = dynamic elastic modulus, MPa; ρ = Bulk density of the specimen, kg/m^3; C_L = celerity of wave, m/s

Water absorption tests of fibres were carried out by immersing a previously dried and weighed quantity of diss M (0). The fibres were wrapped over a fine canvas to avoid any loss of material. Weighing was carried out at definite times of immersion M(t), after drying the sample to remove any adsorbed water. The test was carried out until stabilisation of the mass. The rate of water absorption according to the root of time was determined by the relation:

$$W\ (\%) = [M\ (t) - M\ (0)]* 100/M\ (0) \tag{2}$$

Formulations

The systematic study of mechanical strength according to the Water/Cement ratio (W/C), allowed choosing, for each type of fibres, the W/C ratio corresponding to the optimal compressive and flexural strengths of the composites. During this work, the W/C ratio was varied between 0.5 and 0.9.

The diss fibre / cement ratio was set to 4:1 (by volume) for all formulations. The obtained experimental results for boiled diss composites are shown in Figure 1.

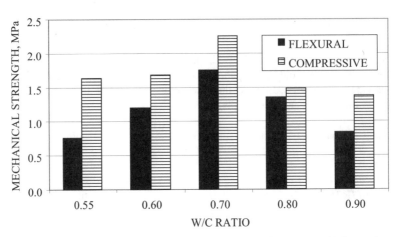

Figure 1 Mechanical strength (flexural and compressive) versus W/C for various formulations of boiled diss

Table 1 Water absoption percentage and the optimum W/C ratio for various treatments applied to diss fibres

VEGETABLE FIBRES	% OF WATER ABSORPTION	OPTIMAL W/C
Tr1	92. 38	0.7
Tr2	90. 00	0.7
Tr3	negligible	0.4
Tr4	negligible	0.5

RESULTS AND DISCUSSION

Dynamic elastic modulus

The average values obtained during this work for the dynamic elastic modulus and bulk density are represented in Figure 2 for various formulations.

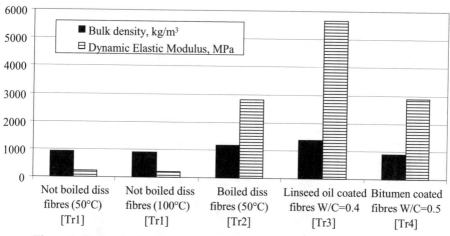

Figure 2   Dynamic elastic modulus and bulk densities for various treatments

It can be observed that for the same bulk, the dynamic elastic modulus values are very low for the untreated diss fibre composites. This can be attributed to the absence of adherence of diss fibres to the cement paste because of the bad cement hydration. They are medium for boiled diss and boiled diss coated with bitumen composites. The most important values were obtained for boiled diss coated with linseed oil, because of the strong adhesion of fibres to the cementing matrix.

Mechanical strengths

The average mechanical strength in compression and flexion, and bulk densities of various composites are represented in Figure 3 for various formulations.

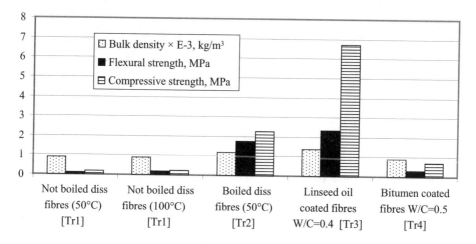

Figure 3   Mechanical strength and bulk densities for various treatments

Case of untreated vegetable fibres

The mechanical strength results of composites containing unboiled diss dried in the oven at 50 or 100°C remain very low. This phenomenon is certainly due to the changes occurring at the matrix-fibres interface and the hydrolysis and solubilisation reactions of some compounds such as sugars, hemicelluloses and pectins caused by the highly alkaline environment developed by cement hydration. The presence of these soluble fractions was also confirmed by adsorption tests of diss fibres in vapour phase. During these tests, some mould appeared on unboiled diss fibres at the end of the eleventh day, whereas no development was noted in boiled diss fibres during all test periods. It can be noted that water-soluble fractions are the cause of weak adhesion of materials.

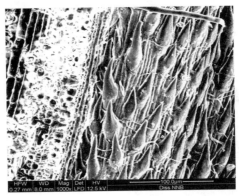

Figure 4   Scanning electron micrograph of unboiled diss fibres, magnification × 1000

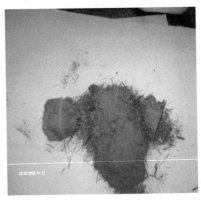

Figure 5   Unboiled diss composite specimen after flexural test

Figure 6   Unboiled diss composite specimen after compression test

Influence of hydrothermal treatment: Boiling treatment

In accordance with the results obtained by Ledhem et al. [4] and Aamar Daya [5], the boiling treatment of diss fibres in the treatment water allows to eliminate soluble matters, which are the cause of retardation of setting and lack of cohesion between aggregate and matrix. Indeed the use of water treatment in the cement paste has showed an important initial retardation of setting. The results are summarized in Table 2.

Table 2 Tests of initial setting time with various mixing waters

SAMPLE	INITIAL SETTING TIME, hours
Cement + Tap water	4.50
Cement + Water residue of the boiled diss	8.00

Strong strengths are of course due to the elimination of the detrimental substances, but it is also necessary to point out to the influence of the diss fibres skin. Indeed the skin of diss fibres is composed of tiny spines which will allow a better adhesion to the cement paste as shown in Figure 7. This type of fibre / matrix cohesion offers better tensile strength in flexural tests and better lateral tension in compressive tests.

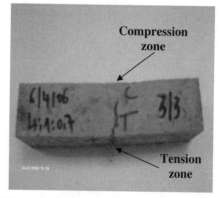

Figure 7   Scanning electron micrographs of boiled diss fibres, magnification × 1000

Figure 8   Boiled diss composite, W/C=0.7 ratio, after flexural test

The high strength of these composites is also due to the fact that the fibres are placed longitudinally, which enables them to adhere well to the cement paste, and behave as reinforcement (Figure 8). Moreover, at the flexural crack level at the tensional section, we noted that the diss fibres were well coated in packages by the cement paste, which enabled them to resist well to the tensile stress (Figures 9 and 10).

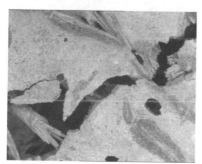

Figure 9   Video microscopic image of the crack, magnification × 50

Figure 10: Video microscopic image of fibres at tensional crack of specimen, magnification × 175

Composite containing diss boiled and coated with linseed oil

To attenuate high water absorptions of fibres, they were coated with linseed oil. This treatment allows coating all fibres without covering spines (Figure 11).

During flexural tests, no cracks were observed under the first maximum loading. It took six full cycles of loading to reach the tensile failure of the specimen. The fibres of diss coated with linseed oil acted as reinforcement in the composite (Figure 12).

The strong resistance obtained for the diss coated with linseed oil was mainly due to the thorny structure of fibres even after coating, and to the good link of the cement paste with fibres, as shown in the video microscopic images of Figures 13 and 14.

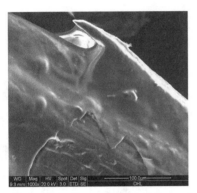

Figure 11   Scanning electron micrograph of boiled diss fibres, coated with linseed oil, magnification ×1000

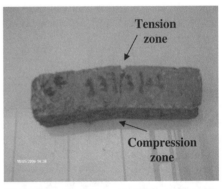

Figure 12   Specimen containing boiled diss coated with linseed oil, after flexural test

Figure 13   Digital image of fibres of the crack surface, magnification × 50

Figure 14: Fibre detail of the crack, magnification × 100

Composite containing diss boiled and coated with bitumen

To decrease the water absorption rate of diss fibres, we also carried out a coating with bitumen. In this case the scanning electron micrographs showed that this treatment modified the surface of the fibres, they became smooth and the spines disappeared (Figure 15).

The obtained strengths in tensile tests became very low because of the reduction of adherence. This was due to the morphology of fibres coated with the bitumen, but also because probably to the random disposition of fibres coated in the cementing matrix, which did not leave fibres resist the traction efforts to which they were submitted. The compressive strength of the composites was also low probably because of the low lateral tension strength of fibres.

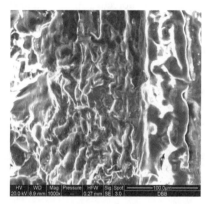

Figure 15   Scanning electron micrograph of boiled diss fibres, coated with bitumen, magnification × 1000

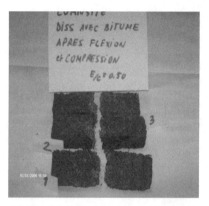

Figure 16: Rupture of the specimens containing diss coated with bitumen

CONCLUSIONS

This work related to the optimization of a cementing composite containing diss with the application of various treatments.

It is clearly seen that boiling water treatment improved the performance of the composite due to the elimination of the water-soluble compounds, but coating the previously boiled fibres with linseed oil was by far the most powerful way among those which were tested.

For similar composite bulk densities, the compressive and tensile strengths of diss fibres coated with linseed oil composites (Tr3) were higher than those obtained with boiled diss fibres (Tr2). This was explained by the fact that the thorny structure was still present after the treatment (Tr3). Moreover, the fibres were placed longitudinally and in parallel, which tends to increase their role as reinforcement.

Diss fibres coated with bitumen (Tr4) gave lower compressive and traction strengths, because of the disappearance of the spines and the random disposition of fibres in the composite. These results clearly show the interest to use such material as resistant filling and even as structural element for structures situated in seismic zones, because of their lightness, resistance and high ductility.

REFERENCES

1. SAVASTANO Jr H, WARDEN P G AND COUTTS R S P, Brazilian waste fibres as reinforcement for cement-based composites, Cement and Concrete Composites, Vol. 22, 2000, pp. 379-384.

2. KRIKER A, DEBICKI G, BALI A, KHENFER M M AND CHABANNET M, Mechanical properties of date palm fibres and concrete reinforced with date palm fibres in hot-dry climate, Cement and Concrete Composites, Vol. 27, 2005, pp. 554-564.

3. REIS J M L, Fracture and flexural characterization of natural fiber-reinforced polymer concrete, Construction and Building Materials, 2005, article in press.

4. LEDHEM A, DHEILLY R M, BENMALEK M L AND QUÉNEUDEC M, Properties of Wood-based composites formulated with aggregate industry waste, Construction and Building Materials, Vol. 14, 2000, pp. 341-350.

5. AAMR DAYA E H, Contribution à la valorisation de co-produits du lin, poussières obtenues par aspiration lors du teillage, dans une matrice cimentaire, Thèse de doctorat de l'Université de Picardie Jules Verne, 2004.

6. PIEMENTA P, CHANDELLIER J, RUBAUD M, DUTRUEL F AND NICOLE H, Etude de la faisabilité des procédés à base de bétons de bois, Cahier du CSTB 2703, Janvier-Février 1994.

7. SIMATUPANG, Abbaureaktionen von Glucose, Cellobiose und Holz unter dem Einfluss von Portlandzementmörtel, Holzforschung, Vol. 40, 1986, pp. 149-155.

8. GARCI JUENGER M C AND JENNINGS H M, New insights into the effects of sugar one the hydration and microstructure of cement pastes, Cement and Concrete Research, Vol. 32, 2002, pp. 393 –399.

9. BILBA K, ARSENE, M A, OUENSANGA A, Sugar cane bagasse fibre reinforced cement composites, Part I. Influence of the botanical components of bagasse on the setting of bagasse/cement composites, Cement and Concrete Composites, Vol. 25, 2003, pp. 91-96.

10. VILÀ M, LLORET F, OGHERI E AND TERRADAS J, Positive fire-grass feedback in Mediterranean Basin woodlands, Forest Ecology and Management, Vol. 147, 2001, pp. 3-14.

11. DREUX G AND FESTA J, Nouveau guide du béton et ses constituants, Ed Eyrolles, 8ème Edition, 2ème tirage 2002.

USE OF SILICA RESIDUE FROM CHEMICAL INDUSTRY AS A MINERAL ADDITIVE IN CONCRETE

K Rübner

U Meinhold

BAM division Building Materials

Germany

ABSTRACT. Silica residue is a filtration product of waste water treatment of chemical industry. Its chemical composition and physical characteristics could enable the use of the filtrate as a mineral additive for concrete and thus the reduction of the amount of residues to be landfilled. To evaluate the utilisation potential of the silica residue as mineral additive, mortars and concretes containing different proportions of the residue as a replacement for cement were produced. Their engineering properties, such as workability, compressive strength, dynamic modulus of elasticity, shrinkage, and some durability characteristics were studied. Considering the results, up to 10% of cement can be replaced by silica residue in a mortar or concrete, whilst the engineering and durability properties remain almost unchanged. Furthermore, the findings of pore structure analyses, calorimetric experiments and thermogravimetric measurements show that the addition of the filtration residue influences the hydration process of the cement.

Keywords: Silica, Waste Material, Concrete, Mineral Additive, Pozzolan, Engineering Properties.

Dr K Rübner is head of the working group Saving Resources by Material Recycling of the BAM division Building Materials. As a chemist, her particular interests are in the fields of physical and chemical analysis of building materials, durability aspects and use of residues in cementitious materials.

Mr U Meinhold received his practical experience by working in the ready-mixed concrete industry before joining the BAM division Building Materials. Before his retirement this year he was responsible for the concrete laboratory. His further interests are cements and additives.

INTRODUCTION

Today, modern reprocessing techniques lead to materials (ashes, slags, slurries), which bear the potential to be used as aggregates or additives in mineral building materials. Because of a highly sophisticated reprocessing technique residues with relatively stable composition over a certain range and relatively defined properties can be produced. At the same time, new German and European regulations demand a waste management to decrease waste volume stored at landfills, to save natural resources and to enhance sustainable development by recycling of diverse residues.

Fly ashes and silica fumes, which are residues from flue gas cleaning of coal-fired power stations or electro-melting furnaces, have been used as components of blended cements or as additions to concrete for years. The engineering and durability properties of mortars and concretes can be optimised by means of these pozzolanic materials [1-4]. In Germany almost the total quantity of fly ash of 4.1 million tonnes per year is applied in concrete and cement production as well as in road constructions [5]. About 28 thousand tonnes per year of silica fume are used by the building industry [6]. However, there are other waste materials and residues available, which can be reused in concretes due to their amorphous siliceous or silica components. Such materials are sediments from process gases of silicone and silane production [7], residues from aluminium sulphate and silica production [8, 9], used-up zeolite based catalysts [10-14] as well as silica-rich residues from heat exchangers of geothermal power plants [15].

The study reported here focused on the assessment of the use of a silica residue (SR) from waste water treatment of chemical industry as mineral additive for concrete. This was explored by means of comprehensive characterisation of the silica residue and testing several engineering and durability properties of mortars and concretes made with SR. Furthermore, pore structure analyses, calorimetric experiments and thermogravimetric measurements were used to indicate the pozzolanic properties of SR.

EXPERIMENTAL PROGRAMME

Materials

Residue

The silica residue (SR) was obtained from a chemical plant in Thuringia, Germany. It was a filtration product of waste water treatment from silica and zeolite production. The residue was available as a dry coarse grained material with a maximum grain size of 2 mm.

Mix design of mortars and concretes

To test engineering and durability properties, mortars and concretes with addition of silica residue were produced in accordance with the German guidelines for inorganic concrete additives [3]. Mortars were made using Portland cement CEM I 32.5 R and CEN reference sand [16]. Concretes were made with Portland cement CEM I 32.5 R and siliceous aggregates according to the grading curve of B8 [17]. Different proportions of silica residue were added as a replacement for cement so that the total binder content was kept constant. To achieve an almost constant workability when keeping the water/binder ratio constant, a Na-naphthalene/melamine sulphonate based superplasticizer Woerment FM 21 was added. Details of the mixes are shown in Table 1.

Table 1   Mortar and concrete mix proportions and flowability spread of mixtures.

Specimen	CONTENT							
	Cement (kg/m³)	SR[1] (%)	Total binder (kg/m³)	Aggregate (kg/m³)	Max. grain size (mm)	w/b	Super-plasti-cizer[1] (%)	Flow-ability spread (cm)
M-0% SR	450	0.0	450	1350	2	0.50	0.0	15.3
M-5% SR	428	5.0	450	1350	2	0.50	1.3	14.0
M-10% SR	405	10.0	450	1350	2	0.50	2.9	10.9
C-0% SR	360	0.0	360	1700	8	0.57	0.0	12.0
C-5% SR	342	5.0	360	1700	8	0.57	1.3	11.6
C-10% SR	324	10.0	360	1700	8	0.57	2.9	10.7

M - mortar; C - concrete
[1] in terms of binder content

Methods

Characterisation of silica residue

The chemical composition of the silica residue was analyzed according to German guidelines for inorganic concrete additives [3]. The mineral phases were analysed by X-ray powder diffraction (XRD) using CuK_α radiation.

To determine the physical characteristics of the residual material, the density were determined with a gas pycnometrical method using helium according to DIN 66137-2 [18]. The pore structure was studied with nitrogen sorption measurements according to DIN ISO 9277 [19] and DIN 66134 [20]. The specific surface area was calculated with the BET method [19]. The pore size distribution was determined with the BJH method [20]. Furthermore, images were taken with an environmental scanning electron microscope (ESEM).

The water demand of cement/silica residue mixtures with different contents of SR was determined with the Vicat method according to DIN EN 196-3 [21].

Tests on mortars and concretes

The workability of fresh mortars and concretes was tested measuring the flowability spread with a flow table as a function of time according to DIN EN 1015-3 [22] and DIN EN 12350-5 [23], respectively.

Comprehensive tests were performed on hardened mortars and concretes. The strength development was determined using mortar prisms (4 x 4 x 16 cm) according to DIN EN 196-1 [16] and 15 cm concrete cubes according to DIN EN 12390-3 [24]. All specimens were stored under water until testing. The dynamic elastic modulus (E-modulus) was determined on 4 x 4 x 16 cm prisms cured under water according to [25]. Shrinkage was tested on prisms of the same size, but cured one day under water and then at a climate of 20°C and 65% r.h. according to [25].

With regard to durability aspects, carbonation was tested with the phenolphthalein test on prisms stored 7 days under water and then at 20°C and 65% r.h. according to [3]. Beside the depth, the velocity of carbonation was calculated by linear regression of the plotting of carbonation depth versus square root of time.

To estimate the frost resistance of the concretes, 15 cm cubes were examined using the CIF test according to [26]. The specimens were stored in demineralised water during the test. The mass loss and the loss in dynamic E-modulus were determined for 56 freeze-thaw cycles.

Studies on pore structure and pozzolanic activity

The mortar and concrete remnants from compressive strength testing were crushed and sieved to 3-8 mm granulated samples for pore structure measurements. The samples were dried by isothermal smooth drying at 22°C and 4 kPa above a cold trap according to [27, 28].

The pore structure was measured by mercury intrusion porosimetry (MIP) applying a maximum pressure of 200 MPa according to ISO 15901-1:2005 [29]. By assuming a contact angle of 140° and a mercury surface tension of 0.48 N/m a range of pores and pore entrances, respectively, from 60 μm to 4 nm pore radius was assessable.

The specific surface area of the samples was determined by nitrogen adsorption measurements according to the BET method [19].

The total porosity of the samples was calculated from the ratio of bulk density and density according to DIN 66137-1 [30]. Therefore, the bulk density was measured on granulated samples with a analyser working on the principle of the displacement technique using a quasi-fluid composite of small, rigid spheres having a high degree of flowability [31]. The density analysis was performed on ground samples according to [18]. All samples were dried at 105°C before the density measurements.

The quantitative analysis of calcium hydroxide $Ca(OH)_2$ was determined for mortar samples with thermogravimetric measurements (TG). The samples were ground and dried at 105°C. The mass loss of the samples was determined during constant heating from 20 to 1050°C with a heating rate of 5 K/min in nitrogen atmosphere. According to [32] the mass loss of the temperature range between 480 and 550°C can be related to degradation of $Ca(OH)_2$.

Effects on cement hydration were studied on cementitious pastes without and with SR addition using an isothermal heat conduction calorimeter. The water/binder ratio was kept constant at 0.50. After mixing, the pastes were cast into glass vials, which where inserted into the calorimeter. The measurements of heat of hydration started about 10 minutes after water addition. The tests were performed at 20°C.

RESULTS AND DISCUSSION

Characteristics of Silica Residue

The results of chemical analysis and physical characterisation of the silica residue are shown in Table 2. With the exception of too high contents of sulphate and chloride, the chemical composition of the source material (second column) meets the requirements of the German guideline for inorganic concrete additives [3]. The silica, which ranges from 70 to 80 M.-%, is a amorphous compound according to the XRD analysis. The sulphate and chloride compounds could be almost completely removed be wet processing. After the processing, the silica residue was dried and ground to a particle size smaller than 0.2 mm. The analysis of processed SR is shown in the third column of Table 2.

Table 2 Chemical and physical characteristics of silica residue (SR).

CHARACTERISTICS		SILICA RESIDUE			LIMITS[1]
		source material	wet processed	multi sampling[2]	
Particle size	(mm)	≤ 2 mm	≤ 0.2 mm	--	≤ 0.2 mm
Density	(g/cm³)	2.35	2.26	2.17-2.22	--
Spec. surface area	(m²/g)	196	289	279-295	$> 18 / < 25^{3)}$
Pore volume[4]	(cm³/g)	0.77	0.90	0.76-0.91	--
Av. Pore radius	(nm)	16	15	16-18	--
	SiO_2	72.46	80.20	73.38-75.75	--
	MgO_{total}	4.68	4.53	4.22-4.71	7.0
Chemical	CaO_{total}	1.45	1.78	1.39-1.77	20.0
composition	SO_3	5.60	0.17	3.48-5.84	1.5-3.5
(M.-%)	$K_2O/Na_2O^{5)}$	1.66	0.35	1.05-2.00	4.5
	Cl^-	0.44	0.03	0.11-0.32	0.1-0.3
	Loss on ignition	4.52	3.83	3.09-5.73	5.0-8.0

[1] According to German guideline for inorganic concrete additives [3].
[2] 9 different samplings within a month.
[3] Valid for current knowledge about silica fume.
[4] Mesopores with pore width of 2-50 nm.
[5] Na_2O equivalent

To assess the variation of composition in dependence on time, SR samples, which came from nine different samplings within a month, were examined. The results of the analyses are summarized in the fourth column of Table 2. In terms of a residue, the variations in chemical composition seem to be acceptable.

The density of the silica residue is about 2.3 cm³/g. The specific surface area of the material reaches in dependence on the processing degree from 200 to 300 m²/g. These huge values are caused by porous agglomerates, which consist of colloidal and micronised silica particles. Figure 1 shows an ESEM image of such silica agglomerates, which look fluffy with irregular rims. The high mesopore volume (pore width 2-50 nm) of about 0.8 cm³/g also confirms this assumption.

The high specific surface area of the residue causes a high water demand. Tests on cementitious pastes with the Vicat method [21] show that the water demand of the paste is up to 0.5 times higher if 10% of cement is replaced by SR.

Figure 1 ESEM image of silica residue wet processed, dried and ground to particles < 0.2 mm.

Properties of Mortars and Concretes

Figure 2 shows the development of workability of fresh mixtures of mortars and concretes in terms of flowability spread as a function of time. The mortars and concretes containing silica residue obtain faster stiffening than the control mixes without SR. However, the compaction on the flow table is quite good even for the mortars and concrete mixes containing 10% SR. But the addition of a superplasticizer is required in order to achieve sufficient workability. The SR mixtures show a slightly thixotropic behaviour.

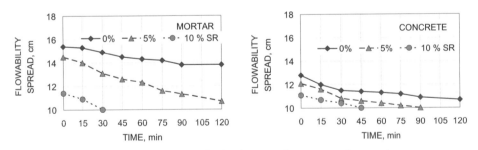

Figure 2 Workability of fresh mixtures of mortars and concretes with and without addition of silica residue.

The development of compressive strength of hardened mortars and concretes containing SR follows that of the control mixes. The compressive strengths are plotted versus the testing age in Figure 3. It shows that 5% as well as 10% of cement can be replaced by silica residue in a mortar or concrete whilst the compressive strength remains almost unchanged.

In accordance with this finding, the development of dynamic E-modulus does not principally differ if the cement is partially substituted by SR. The E-modulus after 28 days ranges between 35 and 40 GPa for both mortars and concretes. Furthermore, the shrinkage behaviour shown in Figure 4 is almost similar to that of the control specimens. The final shrinkage values of 0.8-0.9 mm/m for mortars and 0.5-0.6 mm/m for concretes at the age of 90 days are in the accepted range.

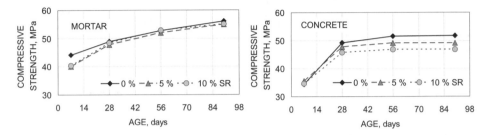

Figure 3   Development of compressive strength of mortars and concretes with and without addition of silica residue.

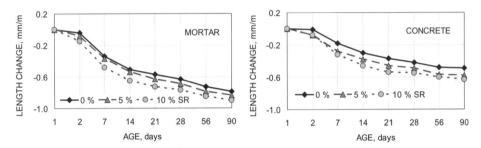

Figure 4   Shrinkage of mortars and concretes with and without addition of silica residue.

The engineering properties of the hardened mortars and concretes containing silica residue meet the requirements of the German guideline for inorganic concrete additives [3]. The criterion is a compressive strength, which reaches 70 to 100% of that of the control material without additive. Additionally, the E-modulus and the shrinkage should not be significantly different from that of the control specimens. Furthermore, the findings suggest that the silica residue has a certain pozzolanic activity.

With regard to durability aspects, the carbonation and frost resistance behaviour of mortars and concretes is not significantly influenced by the addition of silica residue. Table 3 summarises the results of carbonation and freeze-thaw testing. The depth and the velocity of carbonation of SR mortars and concretes are slightly lower than that of the control specimens. The addition of silica residue has no effect on the frost resistance behaviour of the concretes. All concretes have minor mass losses below the failure criteria of 1500 g/m² after 56 freeze-thaw cycles.

However, the loss in E-modulus due to partial destruction of inner structures exceeds the limits of a relative loss of 40% after 56 cycles. This indicates a weak frost resistance for both, the control concrete without residue and the SR concrete.

Table 3 Results of tests of carbonation and frost resistance.

Specimen	CARBONATION		FREEZE-THAW TEST (CIF)	
	Depth[1] (mm)	Velocity (mm/d$^{-0.5}$)	Mass loss[2] (g/m²)	Rel. dyn. E-modulus[2] (%)
M-0% SR	2.7	0.31	--	--
M-5% SR	2.7	0.24	--	--
M-10% SR	2.5	0.26	--	--
C-0% SR	3.4	0.45	292	65
C-5% SR	3.4	0.38	379	63
C-10% SR	2.8	0.24	601	62

M - mortar; C - concrete
[1] after 90 days of storing
[2] after 56 freeze-thaw cycles

Pore Structure and Pozzolanic Properties

The results of porosity and pore structure measurements on mortar and concrete samples are summarised in Table 4. As supposed by the results of compressive strength tests, porosity, total pore volume and specific surface area of SR mortars and SR concretes do not significantly differ from those of control samples without residue. However, the addition of SR shifts the average pore radius to smaller radii. This finding indicates a slight refinement of pore structure due to SR addition, which is well-known for the effect of pozzolanic additives [32, 33].

Table 4 Results of pore structure analyses and thermogravimetric measurements (TG) of mortars and concretes containing different proportions of silica residue at the age of 90 days.

Sample	PORE STRUCTURE ANALYSES				TG	
	Total porosity (%)	Total pore volume (MIP) (mm³/g)	Average pore radius (MIP) (nm)	Specific surface area (BET) (m²/g)	$Ca(OH)_2$ content (M.-%)	
					calculated[1]	measured
M-0% SR	15.1	63.0	59.6	12.4	5.76	5.76
M-5% SR	15.6	62.6	49.2	10.5	5.47	4.48
M-10% SR	16.1	62.4	49.0	11.7	5.18	3.58
C-0% SR	13.8	54.2	71.4	4.7	--	--
C-5% SR	13.5	55.2	48.7	5.8	--	--
C-10% SR	14.6	54.3	40.5	5.9	--	--

M - mortar; C - concrete
[1] in terms of cement content

The calorimetric measurements on cementitious pastes without and with SR give further evidence for a pozzolanic reaction. Figure 5, left hand side, shows the influence of silica residue on the rate of heat evolution during the first 36 hours of hydration. While the second hydration step starts slightly later, the addition of silica residue accelerated the hydration. The SR particles with irregular surfaces presumably act as nucleation sites for hydration products. The values of heat release shown in Figure 5, right hand side, confirm this finding. The additional released heat decreases fast and from 24 until 48 hours the values of heat of hydration are almost alike for the pastes with and without SR. But afterwards, the paste containing SR released a higher heat again. This could be an indication of a pozzolanic reaction of the silica residue, which starts 24 to 36 hours after addition of water.

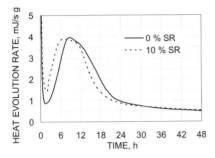 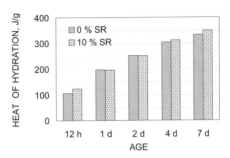

Figure 5   Influence of silica residue on the heat evaluation rate in terms of total solid content (left) and on the total heat of hydration in terms of cement content (right) of cementitious pastes (w/b = 0.50)

Furthermore, the pozzolanic activity of the silica residue is confirmed by quantitative analysis of $Ca(OH)_2$ in mortars containing different SR amounts by thermogravimetric measurements. The results are written in the right column in Table 4 in comparison to calculated values, which were estimated in terms of lesser cement content of mortars containing SR. They show that an increasing SR content in the mortars causes a reduction in $Ca(OH)_2$, which is disproportionate to the decrease due to cement removal.

CONCLUSIONS

Based on the results of the characterisation of the silica residue and the tests on mortars and concretes containing SR, the following conclusions can be drawn:

The high amorphous silica content ranging from 70 to 80 M.-% and the specific surface area of about 200 m²/g could enable the use of the silica residue as a mineral additive for concrete.

The chemical composition of residues coming from different samplings shows an acceptable variation.

Sulphates and chlorides inherent in the filtrate, which are said to cause damage in concrete, can be removed by wet processing.

Up to 10% of cement can be replaced by silica residue in a mortar or concrete, whereas the engineering and durability properties remain almost unchanged. However, the addition of superplasticizer is required in order to achieve sufficient workability.

The addition of SR slightly accelerates the cement hydration process. It causes a slight refinement of the pore structure of mortars and concretes and a reduction of their $Ca(OH)_2$ content.

These findings suggest that the silica residue may act both as a filler and as a pozzolan.

With regard to further improvement of the properties of mortars and concretes containing SR, the fineness and particle size distribution of the residue should be optimized. Furthermore, the mix proportion of concretes, the type of cement and/or superplasticizer and the way of mixing needs to be studied more fully.

ACKNOWLEGDEMENTS

The authors would like to appreciate the support of Chemiewerk Bad Köstritz GmbH, Thuringia. The work of the laboratory staff of the division Building Materials of BAM is gratefully acknowledged. Especially, thanks are given to Mathias Lindemann, Claudia Hagemeister and Jürgen Götze for the numerous chemical and physical analyses of the materials and to Frank Haamkens for his assistance in the production and testing of mortars and concretes. Many thanks are also due to Simone Hempel, Institute for Construction Materials of Technische Universität Dresden, who carried out the ESEM analysis.

REFERENCES

1.   DIN EN 450. Fly ash for concrete, May 2005.

2.   DIN EN 13263. Silica fume for concrete, October 2005.

3.   DEUTSCHES INSTITUT FÜR BAUTECHNIK (DIBt). Zulassungs- und Überwachungsgrundsätze für Anorganische Betonzusatzstoffe (Guidelines for inorganic concrete additives). Schriften des Deutschen Instituts für Bautechnik, Reihe B, Heft 17, Dezember 2004, DIBt, Berlin, 2005, 28pp.

4.   AGARWAL, S K. Pozzolanic activity of siliceous materials. Cem. Con. Res. Vol. 36, 2006. 5pp, in press.

5.   RÖMPP. Flugasche (Fly ash). In: RÖMPPOnline, Version 2.10, Georg Thieme Verlag, Stuttgart, March 2006.

6.   MÜLLER, C. Beton als kreislaufgerechter Baustoff (Concrete as closed loop compatible building material). Deutscher Ausschuss für Stahlbeton (DAfStb), Heft 513, Beuth-Verlag, Berlin, 2001, pp 157.

7. CURBACH C, HEMPEL R, SPECK K. Ein Abprodukt der chemischen Industrie als Betonzusatzmittel (A residue of chemical industry as additive for concrete). In: Curbach M, Große W, Haim H D, Opitz H, Schorn H, Stritzke J (Eds.). Jahresmitteilungen 2000, Schriftenreihe des Instituts für Tragwerke und Baustoffe, Heft 12, Technische Universität Dresden, Dresden, 2000, pp 49-65.

8. FU X, WANG S, HUANG S, HOU X, HOU W. The influences of siliceous waste on blended cement properties. Cem. Con. Res. Vol. 33, 2003. pp 851-856.

9. ANDERSON D, ROY A, SEALS R K, CARTLEDGE F K, AKHTER H, JONES S C. A preliminary assessment of the use of an amorphous silica residual as a supplementary cementing material. Cem. Con. Res. Vol. 30, 2000. pp 437-445.

10. PACEWSKA B, WILINSKA I, BUKOWSKA M, NOCUN-WCZELIK W. Effect of waste aluminosilicate material on cement hydration and properties of cement mortars. Cem. Con. Res. Vol. 32, 2002. pp 1823-1830.

11. HSU K-C, TSENG Y-S, KU F-F, SU N. Oil cracking waste catalysts as an active pozzolanic material for superplasticized mortars. Cem. Con. Res. Vol. 31, 2001. pp 1815-1820.

12. PACEWSKA B, BUKOWSKA M, WILINSKA I, SWAT M. Modification of the properties of concrete by a new pozzolan. - A waste catalyst from the catalytic process in a fluidized bed. Cem. Con. Res. Vol. 32, 2002. pp 145-152.

13. PAYA J, MONZÓ J, BORRACHERO M V, VELÁZQUEZ S. Evaluation of the pozzolanic activity of fluid catalytic cracking catalyst residue (FC3R). Thermogravimetric analysis studies on FC3R-Portland cement pastes. Cem. Con. Res. Vol. 33, 2003. pp 603-609.

14. SU N, FANG H-Y, CHEN Z-H, LIU F-S. Reuse of waste catalysts from petrochemical industries for cement substitution. Cem. Con. Res. Vol. 30, 2000. pp 1772-1783.

15. ESCALANTE J I, MENDOZA G, MANCHA H, LÓPEZ J, VARGAS G. Pozzolanic properties of a geothermal silica waste material. Cem. Con. Res. Vol. 29, 1999. pp 623-625.

16. DIN EN 196-1. Methods of testing cement - Part 1: Determination of strength, May 2005.

17. DIN 1045-2. Concrete, reinforced and prestressed concrete structures - Part 2: Concrete; Specification, properties, production and conformity; Application, July 2001.

18. DIN 66137-2. Determination of solid state density - Part 2: Gaspycnometry, December 2004.

19. DIN ISO 9277. Determination of the specific surface area of solids by gas adsorption using BET method, May 2003.

20. DIN 66134. Determination of the pore size distribution and the specific surface area of mesoporous solids by means of nitrogen sorption - Method of Barrett, Joyner and Halenda (BJH), February 1998.

21. DIN EN 196-3. Methods of testing cement - Part 3: Determination of setting time and soundness, May 2005.

22. DIN EN 1015-3. Methods of test for mortar for masonry - Part 3: Determination of consistence of fresh mortar (by flow table), April 1999.

23. DIN EN 12350-5. Testing fresh concrete - Part 5: Flow table test, June 2000.

24. DIN EN 12390-3. Testing hardened concrete - Part 3: Compressive strength of test specimens, April 2002.

25. BUNKE N. Prüfung von Beton. Empfehlungen und Hinweise als Ergänzung zu DIN 1048 (Testing concrete. Recommendations and information as addition to DIN 1048). Deutscher Ausschuss für Stahlbeton (DAfStb), Heft 422, Beuth-Verlag, Berlin, 1991, 53pp.

26. SETZER M J. RILEM Draft Recommendation, RILEM TC 176 IDC (2001) CIF-Test - Capillary suction, Internal damage and Freeze Thaw test - Reference method and alternative methods A and B. Materials and Structures. Vol. 34, 2001. pp 515-525.

27. ADOLPHS J, SETZER M J, HEINE P. Changes in pore structure and mercury contact angle of hardened cement paste depending on relative humidity. Materials and Structures. Vol. 35, 2002. pp 447-486.

28. RÜBNER K, HOFFMANN. D. Characterization of Mineral Building Materials by Mercury Intrusion Porosimetry. Particle and Particle Systems Characterization. Vol. 23, 2006. pp 20-28.

29. ISO 15901-1:2005. Pore size distribution and porosimetry of solid materials by mercury porosimetry and gas adsorption. Part 1: Mercury porosimetry. ISO International Organization for Standardization, 2005.

30. DIN 66137-1. Determination of solid state density - Part 1: Principles, November 2003.

31. WEBB P A, ORR C. Analytical methods in fine particle technology. Micromeritics Instrument Corp, Norcross, GA, 1997, pp 205-213.

32. MARSH B K. Relationship between engineering properties and microstructural characteristics of hardened cement pastes containing pulverized fuel ash as a partial cement replacement. PhD Thesis, The Hatfield Polytechnic, Cement and Concrete Association, 1984.

33. RÜBNER K, SCHMIDT C. The microstructure of high-strength mortars and concretes. Fresenius' Journal of Analytical Chemistry. Vol. 349, 1994. pp 243-245.

THEME THREE:

ARCHITECTURE AND ENGINEERING: APPROPRIATE DESIGN

THE CO_2 EMISSION AND UPTAKE IN THE LIFE CYCLE OF CONCRETE

C Pade

Danish Technological Institute

M Guimaraes

Aalborg Portland Cement

Denmark

ABSTRACT. More than 50% of the CO_2 emitted during cement production is originated from the calcination of limestone. This CO_2 is reabsorbed during the life cycle of cement based products such as concrete and mortar in a process called carbonation. The methodology and the impact that concrete carbonation has in the assessment of CO_2 emissions from cement production has not been fully documented. Specifically, there is a lack of knowledge about the carbonation of demolished and crushed concrete. The existing models for calculating carbonation do not take into account the secondary life of concrete, i.e. what takes place after the concrete has been demolished. Consequently, the contribution of the cement and concrete industry to net CO_2 emissions may be strongly overestimated. This study encompasses theoretical work, laboratory studies, surveys and calculations. Based on work carried out as part of a Nordic Innovation Centre project it is argued that a significant proportion of the CO_2 emitted by calcination in a 100-year perspective may be taken up again by concrete through carbonation [1-4]. This is based on the assumption that the concrete structure has a service life of 70 years and that the concrete is crushed for recycling after demolition of the structure. Also, the impact of the CO_2 uptake on life cycle inventory of concrete products is illustrated using a concrete roof tile and a highway bridge edge beam as examples.

Keywords: CO_2 uptake, Carbonation, Life cycle inventory.

Claus Pade, Danish Technological Institute, Concrete Centre, Taastrup, Denmark.

Maria Guimaraes, Aalborg Portland, Research and Development Centre, Denmark.

INTRODUCTION

More than 50% of the CO_2 emitted during cement production is originated from the calcination of limestone. This CO_2 is reabsorbed during the life cycle of cement based products such as concrete and mortar in a process called carbonation.

The methodology and the impact that concrete carbonation has in the assessment of CO_2 emissions from cement production has not been fully documented. Specifically, there is a lack of knowledge about the carbonation of demolished and crushed concrete. The existing models for calculating carbonation do not take into account the secondary life of concrete, i.e. what takes place after the concrete has been demolished. Consequently, the contribution of the cement and concrete industry to net CO_2 emissions may be overestimated. The emission and uptake of CO_2 during the life cycle of concrete is compared to that of bio fuel in Figure 1. A fundamental difference between bio fuel and concrete is the time when the CO_2 uptake takes place in the life cycle. In the bio fuel cycle the uptake takes place early, whereas in the concrete cycle the uptake takes place late.

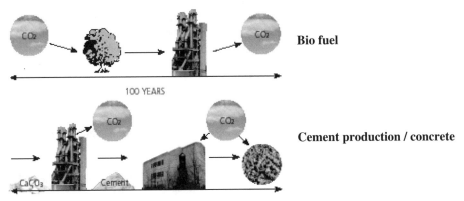

Figure 1   Bio fuel CO_2 life cycle and concrete CO_2 life cycle.
Bio fuel is considered CO_2 neutral – the calcination/carbonation process in concrete is not.

METHODOLOGY OF ESTIMATING THE CO_2 UPTAKE IN CONCRETE

The general methodology for estimating the percentage of a given concrete that carbonates during its service life is illustrated in Figure 2. It is necessary to know the concrete composition, the exposure conditions as well as the dimensions of the concrete (the exposed surface area). Based on the concrete composition and the exposure conditions, a carbonation rate constant can be assigned to the concrete, and depending on the exposed surface area and the service life, the volume of concrete carbonated during its service life can be estimated by multiplying the exposed area of concrete and the carbonation depth. The depth of carbonation (d_c) can be approximated as:

$$d_c = k \cdot t^{0.5} \tag{1}$$

where "k" is the carbonation rate factor and "t" is time.

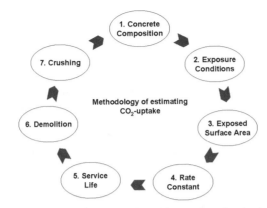

Figure 2   Methodology for estimating CO_2 uptake of a single concrete

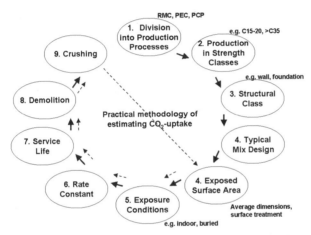

Figure 3   Practical methodology used for estimating the CO_2 uptake in the Nordic countries

If following its service life the concrete is demolished, crushed and reused e.g. as road sub-base, another round in the circle is needed to include the CO_2 uptake during the secondary life. Here, obviously, crushing of the concrete will have a major impact on the exposed surface area, and hence substantial amounts of concrete will carbonate in a relatively short period of time.

The total amount of CO_2 absorbed by concrete can be estimated according to:

$$CO_2 \text{ uptake } (kg\ CO_2 / m^3 \text{ carbonated concrete}) \approx 0.75 \times C \times CaO \times \frac{M_{CO_2}}{M_{CaO}} (kg / m^3) \qquad (2)$$

where "C" is the mass of Portland cement clinker per m³ concrete; "CaO" is the mass fraction of CaO in the cement clinker; and "M" is the molar mass of CO_2 and CaO, respectively. Different CO_2 uptakes can be calculated depending on which fraction of CaO is considered as being able to carbonate. The 0.75 factor is based on the assumption that all CaO in calcium hydroxide, AFt and AFm is carbonated and that 50% of the CaO in C-S-H is carbonated. Other studies [5, 6] assume that only the CaO present in calcium hydroxide carbonates which results in a factor of 0.3 instead of 0.75 in Equation 2.

Based on the estimated volume of carbonated concrete, the CO_2 uptake in the concrete can be estimated by multiplying the volume of carbonated concrete by the CO_2 uptake per cubic meter as estimated by Equation 2. The methodology used to determine an inventory of concrete production is illustrated in Figure 3. In the Nordic countries the available statistical information makes it most convenient to divide concrete into ready-mixed concrete, precast concrete element (hollow core slabs, columns and beams, wall elements, etc.), and precast concrete product (paving stone, roof tile, etc.). Within each "production class" the concrete production is divided into strength classes and classes of standard concrete products (e.g. wall, foundation, and paving stone). Each strength class (e.g. C15-C20 or >C35) is assigned to a typical mix design and each class of standard concrete product is assigned to typical dimensions. Standard exposure conditions are subsequently assigned to the individual standard concrete products. Based on the strength class and exposure condition a carbonation rate constant is assigned according to Table 1. The values in Table 1 are based on data reported in the literature for CEM I concretes. A correction factor was applied to cement types containing fly ash, microsilica, slag or limestone. All these additives have been found to increase the rate of carbonation. Also, a correction factor accounting for the surface treatment of part of the exposed surface was applied in order to most accurately estimate the CO_2 uptake.

Table 1 Carbonation rate constants in mm/(year)$^{0.5}$ for various concrete strengths (CEM I) and concrete exposure conditions as used in the Nordic investigation of CO_2 uptake [1, 4]

EXPOSURE CONDITION	COMPRESSIVE STRENGTH, MPa			
	< 15 MPa	15-20 MPa	25-35 MPa	> 35 MPa
Exposed	5	2.5	1.5	1
Sheltered	10	6	4	2.5
Indoors	15	9	6	3.5
Wet	2	1.0	0.75	0.5
Buried	3	1.5	1.0	0.75

The CO_2 uptake from the annual production of concrete after X years of service can be estimated by Equation 3. Also the effect of demolition and crushing in the second life of concrete can be estimated using the Equation 3.

$$CO_2 \, uptake \, (kg \, CO_2) =$$
$$\sum \left(Eq. \, 2 \times A_{RMC, \, slabs} \times d_c\right) + \left(Eq. \, 2 \times A_{RMC, \, walls} \times d_c\right) + \left(Eq. \, 2 \times A_{RMC, \, foundations} \times d_c\right) + ... \quad (3)$$

Where Eq. 2 is the CO_2 uptake per m^3 of the particular type concrete, $A_{x,y}$ is the exposed area of this type of concrete, and d_c is the depth of carbonation at the end of the service life of the particular type of concrete.

CO_2 UPTAKE IN THE NORDIC COUNTRIES

Concrete carbonates during the service life of the structure and also after being demolished and recycled for secondary use. In the Nordic study [4] the service life of concrete was estimated to be 70 years, after which the concrete would be demolished. Carbonation of the demolished concrete was calculated for a 30 year life accounting for the crushing of the concrete. The percentage of demolished concrete being crushed varies substantially among the Nordic countries from 0% in Iceland to over 90% in Denmark, being crushed to sub 32 mm.

The results of the calculations indicate that the volume of concrete that is carbonated during its service life ranges between 28% and 37%. The percentage of concrete that carbonates after demolition increases to 37% to 86% because of the reduction in particle sizes. This difference reflects the various recycling rates of crushed concrete among Nordic countries.

In a North American study by the Portland Cement Association [5, 6] the percentages of carbonated concrete in the U.S. were estimated based on a service life of 100 years. The annual concrete production (year 2000) was divided into three categories based on 28-day compressive strength (21, 28 and 35 MPa) each with a typical mix design containing 86% Portland cement and 14% fly ash. The corresponding carbonation rate constants were estimated at 8.5, 6.7, and 4.9 mm/(year)$^{0.5}$. Based on these assumptions and accounting for surface coating, the percentage of concrete carbonated during its service life was the percentage of concrete carbonated in 100 years, estimated to be 29-38% depending on binder composition, a range rather similar to the 28-37% estimated in the Nordic study.

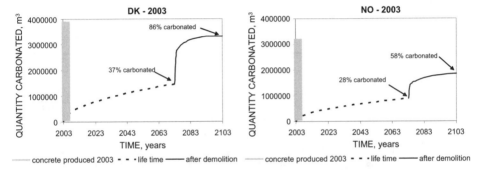

Figure 4   Carbonation of concrete produced in Denmark and Norway in the year 2003. Notice the effect of demolition and crushing after the 70 years of service life [4]

Applying Equation 3 to the production of concrete in year 2003 in the Nordic countries indicates that between 33% and 57% of the CO_2 emitted due to calcination in the cement production is taken up again by the concrete manufactured in a service life period of 70 years and a secondary life period of 30 years (Table 2).

Table 2   Comparison of the CO_2 emission from calcination in cement production and the CO_2 uptake from concrete carbonation during 100 years in four Nordic countries [4]

2003	DENMARK	NORWAY	SWEDEN	ICELAND
CO_2 emission from calcination, million metric tons	0.600	0.675	0.730	0.050
CO_2 uptake in 100 years, million metric tons	0.34	0.22	0.24	0.02
CO_2 uptake relative to CO_2, emission from calcination, %	57	33	33	42

CO$_2$ UPTAKE IN LCI

The CO$_2$ balances of 1 m of concrete highway edge beam produced in Denmark and of 1 m^2 of concrete roof tile produced in Norway are shown in Figures 5 and 6, respectively. The CO$_2$ balance of the edge beam is estimated based on 70 years of service life and 30 years of secondary life following demolition and crushing, whereas the balance of the roof tile is calculated for 50 years of service life and 50 years of secondary life following demolition and crushing. The compositions of the edge beam and of the roof tile is shown in Table 3.

Table 3   Concrete composition of an edge beam and of a roof tile

MATERIAL	EDGE BEAM	ROOF TILE
Cement, kg/m^3	238	480
Fly ash, kg/m^3	135	
Water, kg/m^3	133	166
Fine aggregate, kg/m^3	579	1760
Coarse aggregate, kg/m^3	1160	
Other, kg/m^3	27	2

By taking CO$_2$ uptake from carbonation into account, the CO$_2$ balance over 100 years is reduced by 18% and 28% for edge beam and roof tile, respectively. As both the edge beam and roof tile are made from mix compositions corresponding to strengths higher than 35 MPa, the carbonation rate constants are low. Thus the vast majority of the carbonation takes place after demolition and crushing, i.e. the particle size reduction from crushing is essential for obtaining a high CO$_2$ uptake for these two particular concrete products.

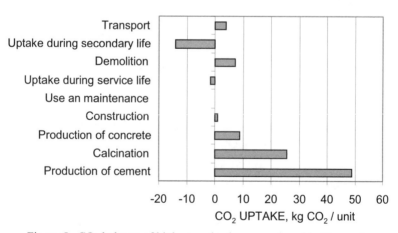

Figure 5   CO$_2$ balance of highway edge beam produced in Denmark.
The mass of one unit is 502 kg corresponding to 1 m of edge beam.

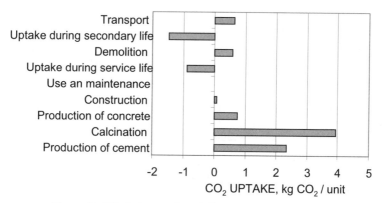

Figure 6 CO_2 balance of roof tile produced in Norway.
The mass of one unit is 42 kg corresponding to 1 m² of roof tile.

CONCLUSIONS

A Nordic and a North American study using somewhat different approaches both arrived at similar percentages of concrete volume to be carbonated during its service life, i.e. 28-37% and 28-39% respectively.

In the Nordic study taking into account the secondary life involving crushing of demolished concrete, the percentage of carbonated concrete in 100 years increased to up to 86% for Denmark.

The CO_2 uptake corresponding to the volume percentage of carbonated concrete depends on the clinker content and on the mineral phases that are considered to carbonate. If only calcium hydroxide is considered to carbonate around 30% of the CaO in the clinker carbonates, whereas if all the CaO in the AFm and AFt phases and 50% of the CaO in the C-S-H phase are considered to carbonate, then around 75% of the CaO in the clinker will carbonate.

Using the latter set of assumptions, the Nordic study estimated that between 33% and 57% of the CO_2 emitted from calcination in cement production will have been taken up during 100 years.

In a life cycle inventory perspective, accounting for the CO_2 uptake by concrete results in significantly reduced CO_2 balances for a highway bridge edge beam (18%) and for a roof tile (28%), and similar values must be expected for all concrete products.

ACKNOWLEDGEMENTS

This paper is primarily based on the project: "CO₂ uptake during the concrete life cycle" partially financed by the Nordic Innovation Centre.

REFERENCES

1. LAGERBLAD B, Carbon Dioxide Uptake During Concrete Life Cycle, state of the art, Swedish Cement and Concrete Research Institute - CBI, Stockholm, www.cbi.se, ISBN 91-976070-0-2, 2005.

2. JONSSON G AND WALLEVIK O H, Information on the use of concrete in Denmark, Sweden, Norway and Iceland, Icelandic Building Research Institute, Reykjavik, www.ibri.is, ISBN 9979-9174-7-4, 2005

3. ENGELSEN C J, MEHUS J AND PADE C, Carbon dioxide uptake in demolished and crushed concrete, Norwegian Building Research Institute, Oslo, www.byggforsk.no, ISBN 82-536-0900-0, 2005.

4. KJELLSEN K, GUIMARAES M AND NILSSON Å, The CO_2 balance of concrete in a life cycle perspective, Danish Technological Institute - DTI, Copenhagen, www.teknologisk.dk, ISBN 87-7756-758-7, 2005.

5. GAJDA J AND MILLER FM, Concrete as a sink for atmospheric carbon dioxide: A literature review and estimation of CO_2 absorption by Portland cement concrete, PCA, Chicago, R&D serial No. 2255, 2000.

6. GAJDA J, Absorption of atmospheric carbon dioxide by Portland cement, PCA, R&D, Chicago, serial No. 2255a, 2001.

MASONCRETE: THE NEW METHOD OF CONSTRUCTION OF MASSIVE ENGINEERING STRUCTURES

G V Gopalarao

Consultant

India

ABSTRACT. The author evolved a new method of construction from his extensive experience in the construction of Nagarjunasagar Dam, which is the largest masonry dam in the world. This new method of construction is named Masoncrete and, as can be seen from the statements enclosed, it saves considerably both on cost as well as quantity of cement (which is a valuable commodity) as compared to conventional concrete.

Keywords: Masoncrete, New method, Construction.

G V Gopalarao obtained his BA degree (Mathematics) from Andhra University, India, BE (First class) degree in Civil Engineering from Madras University, India, and post graduate diploma DIIT in Concrete Technology and Design from the Indian Institute of Technology, Kharagpur, India. He is a member of the Institution of Engineers, India and also a professional engineer of the same institution. The author has a working experience of 33 years in the construction of various river valley projects in the Irrigation department of Andhra Pradesh, India out of which 6 years in the quality control job in the construction of Nagarjunasagar Dam, which is the world's largest masonry dam.

INTRODUCTION

This paper discusses about a new method of construction called Masoncrete which has got distinct advantages over conventional concrete in the construction of massive engineering structures like dams, abutments etc.

MASONCRETE

Masoncrete is the method of setting rubble stones in cement concrete, instead of cement mortar, and vibrating the concrete with a needle vibrator in the interstices of the rubble stones. The actual method of construction is suggested as follows. While starting the work, a layer of 6" to 9" of cement concrete of suitable proportion and size of aggregate is laid and vibrated. Over that layer, rubble stones of suitable sizes are closely arranged similar to that in random rubble masonry (RRM). In the interstices of the rubble stone, concrete is poured and vibrated with a needle vibrator, so that concrete flows into the sides and fills up all voids thereby making it void proof. Here there is no possibility of human failure to fill up the voids since vibration will fully pack the concrete. Thus, Masoncrete obviates the drawbacks of RRM and has the positive advantages.

Background

Random rubble masonry (RRM) and cement concrete are the two chief construction materials for the construction of dams. The following is a brief note about the two systems of construction.

Random rubble masonry

Random rubble masonry was the construction materials for most of the dams in olden time so also some in modern times. Some examples of dams in olden times which were constructed with RRM are:

1. Masonry dam across Cauvery river in India (2nd century AD).
2. Curved dam built by Romans at Kassorine, 225 km south west of Tunis (2nd century AD).
3. Band e Saraj dam in Afghanistan, 27 km from Ghazni (7th century).
4. Masonry dam at New Brunswick in USA in 1681.
5. Crosbois Dam in France in 1837.
6. Masonry Arched Dam called Meer Allum dam near Hyderabad, India, in 1800.
7. Masonry anicuts were constructed by the eminent British engineer Sir Arthur Cotton across the rivers Godavary and Krishna in India, around 1850.
8. Roosevelt dam in USA in 1911.
9. Nagarjunasagar Dam about 90 km from Hyderabad, India which is the world's largest masonry dam, 1956-68.

Cement concrete was adopted in the construction of dams from 1900. Some examples are:

1. Hoover Dam in the USA in 1916.
2. Bhakra Dam in India in 1948.

In modern times cement concrete is acclaimed to be a universal construction material.

Drawbacks of Concrete Construction

Concrete, if examined closely, may not be an economical construction material, especially for massive engineering structures. For procuring the coarse aggregate, we have to break down massive stones from the quarry into a size, say, 75 mm or 40 mm etc. and again we are using cement and sand to build them up into a massive structure. Evidently, the energy to break up the large stones into smaller ones is wasteful and can be avoided. The more massive the stones we use for construction, the more economical it will be. If we examine ancient structures, most of them are constructed with massive stones or cut stone slabs.

Nevertheless, in the present circumstances, the size of stones that can be used in a structure has got its own limitation both from the point of view of feasibility of transport and also the size of the structure. For example, we cannot use stones weighing a tonne since it is very difficult at every stage to handle it. We cannot even use a stone of a size more than the dimensions of the structure. To satisfy these conditions, random rubble masonry in cement mortar using random rubble stones of 2 to 3 eft. volume is adopted successfully in the construction of various gigantic structures like the Nagarjunasagar dam, India.

Drawbacks of Masonry Construction

The cement mortar used in RR masonry forms the weakest link and is the controlling factor. The main drawback in the construction of RR masonry is that there is large variation of strength between the Random Rubble stone and that of the cementing material. The strength of mortar is several times less than that of the rubble stone thus the strength of mortar is the deciding factor in the assessment of the strength of the masonry the mortar. Even though we use very strong rubble stone such as granite, the strength of the structure is controlled only by the strength of that of the cement mortar used therein.

The soundness of construction of RR masonry mainly depends on the packing of spalls and chips with the mortar in the interstices of big rubble stones. This is dependent on the workmanship of the masons and packers, which at present is not quite good and is not dependable. It requires very close supervision stone to stone to ensure that packing is properly done between the rubble stones. In spite of every care taken, post construction grouting is required since many voids will be left unnoticed either by the mason or by the supervisory staff.

To obviate the above drawbacks, a method of construction, **via media**, or we can say a **hybrid** between RR masonry and cement concrete is suggested, which is named as Masoncrete in which concrete is the binding material instead of mortar, thereby increasing the strength of the structure since concrete is 3 to 4 times stronger than the mortar with the same cement / aggregate ratio.

Cost Analysis

The rates of (i) Masoncrete, with cement concrete 1:5:10 as binding material, (ii) Conventional random rubble masonry with cement mortar 1:5 and (iii) Cement concrete 1:5:10 are worked out as per schedule of rates of 1999-2000 and standard data of irrigation department of Andhra Pradesh, India and are appended herewith (Table 1). From the rate structure the following inferences can be drawn.

1. Masoncrete is cheaper than concrete by about 20.7%.
2. Masoncrete consumes LESS CEMENT i.e., only 52.8 kg (1.06 bags) of cement per cubic meter against 132 kg (2.65 bags) of cement for concrete and 98 kg (1.96 bags) of cement for random rubble masonry.
3. Masoncrete gives nearly three times stronger structure than conventional random rubble masonry for a mere 22% more cost. Since the binding material for Masoncrete is cement concrete 1:5:10, being nearly three times stronger than the binding material of random rubble masonry, which is cement mortar 1:5.

With the above distinct advantages, Masoncrete is definitely more advantageous than the conventional types of concrete construction for massive engineering structures.

Table 1   Data cost analysis

QUANTITY	DESCRIPTION OF WORK	RATE	UNIT	AMOUNT, Rs.
1. Masoncrete using cement concrete 1:5:10				
1.00 m³	RR granite stone	87.5	1 m³	87.50
0.40	Cement concrete 1:5:10 with 20 mm granite material	1062.0	1 m³	424.80
1.80 Nos.	Mason	102.7	each	184.86
1.40 Nos.	Men mazdoor (labourer)	65.3	each	91.42
1.40 Nos.	Women mazdoor (labourer)	65.3	each	91.42
				880.00
Requirement of cement per 1 m³ = 0.40 × 0.132 = 52.8 kg or 1.06 bags				
2. RR masonry in cement mortar 1:5 – 1 m³				
1.10 m³	RR granite stone	87.5	1 m³	96.25
0.34 m³	Cement mortar 1:5	760.0	1 m³	258.40
1.80 Nos.	Mason	102.7	each	184.86
1.40 Nos.	Men mazdoor (labourer)	65.3	each	91.42
1.40 Nos.	Women mazdoor (labourer)	65.3	each	91.42
				722 .35
Requirement of cement:0.34 × 288 kg = 98 kg or 1.96 bags				
3. Cement concrete 1:5:10 — 1 m³				
0.92 m³	20 mm granite material	465.0	1 m³	427.80
0.46 m³	Cement mortar 1:5	760.0	1 m³	349.60
0.20 Nos.	Mason	102.7	each	20.54
1.80 Nos.	Men mazdoor (labourer)	65.3	each	117.54
1.40 Nos.	Women mazdoor (labourer)	65.3	each	9142
1.00 m³	Mixing and vibrating concrete (27.5 + 27.5)	55.0	1 m³	55.00
				1061.90 or 1062
Requirement of cement per 1 m³ of cement concrete 1:5:10 = 0.46 m³ × 288 = 132 kg (2.65 bags)				
4. Cement mortar 1:5 — 1 m³				
288 kg	Cement	2400.0	1 tonne	691.20
1 m³	Sand	41.3	1 m³	41.30
1 m³	Mixing charges	27.5	1 m³	27.50
				760.00
Requirement of cement per 1 m³ of cement mortar 1:5 = 288 kg = 5.76 bags				

Table 2 Comparative statement

		COST, USD per m³ Rs.	CEMENT DEMAND, kg / m³ concrete	STRENGTH COMPARATIVE NUMBER
1.	Masoncrete using cement concrete 1:5:10	880 / 20	52.8	3
2.	Cement concrete 1:5:10	1062/24	132.0	3
3.	Random rubble masonry in cement mortar 1:5	722/16.40	98.0	1

The strength of cement concrete 1:5:10 is approximately 3 times that of cement mortar 1:5
Monetary savings by adopting Masoncrete over plain concrete = 20.7%
Savings in cement = 60%

SUMMARY

Masoncrete which is evolved as a hybrid between masonry and concrete has got the distinct advantages of both masonry and concrete and eliminates their disadvantages. Masoncrete saves on overall cost and also saves on cement which is a scarce commodity. With the same cement / aggregate ratio, Masoncrete is several times stronger than RR masonry since the binding material is concrete, which is much stronger than mortar.

BIBLIOGRAPHY

GOPALARAO M, Nagarjunasagar, Bharatiya Vidya Bhavan, India

RAO K L, Calculation design and testing of reinforced concrete, Charotor Bookstall, India

Concrete Manual of Bureau of Reclamation, Denver, Colorado, USA

IS456: Code of practice for plain and reinforced concrete, Indian standard

REQUIRED STEEL CONFINEMENT FOR A PERFORMANCE BASED DESIGN OF RC COLUMNS

N Djebbar
N Chikh

Constantine University

Algeria

ABSTRACT. The deformation and resistance capacity of reinforced columns is a determinant factor in the overall performance of RC moment resisting frames. Deformations resulting from an interstorey drift or concordant plastic rotations at the column ends are widely used by the evaluation procedures of the seismic vulnerability of buildings. Since the transverse volumetric ratio is recognised as the factor governing the deformation capacity of these structural elements, simple formulations of this ratio are proposed function of a moderate damage level guarantying both human life and the structural integrity with a non costly eventual repair. These expressions introduce an adjustable transverse reinforcement ratio ρ_{sh} for different columns subjected to a constant interstorey drift, leading thereby to a coherent global behaviour. This is because columns of the same level will exhibit a similar deformation capacity level. In this respect, we can refer to storey deformation capacity like storey deformation demand.

Keywords: Transverse reinforcement, Damage, Ductility, Performance, Column, Interstorey drift.

Dr N Djebbar is an Associate professor at the Constantine University, Algeria where he obtained his Doctorat d'Etat. He obtained his MSc from Cardiff University, Great Britain. His primary research is in the PBD of moment resisting RC frames and rehabilitation of RC structures.

Pr N Chikh is a Professor at the Constantine University, Algeria. He received his PhD from the University of Leeds, England. Field of research: early age concrete properties, strengthening of RC members with FRP and earthquake resistant design of RC structures.

INTRODUCTION

Performance based design (PBD) has been experiencing a rapid development in recent years towards practical design implementation. A multi level performance based design consists on a selection of design criteria for a defined structural system, are expressed in a sated level of seismic hazard with a defined reliability level, this system will not be damaged beyond limit states or other considered limits. For design applications, it is more desirable to establish performance limits in terms of curvature and drift. Although buildings designed to current codes performed well during recent earthquakes from life safety perspective (strength and ductility), the damage level of structures, economic loss due to loss of use and cost of repair were unexpectedly high [1, 2]. For this reason, the damage control must be more explicit within seismic provision of building codes in order to avoid the resulting prejudice especially in urban sites. In general the damage state of a column is determined through a damage index, which is usually related to displacement ductility defined as the maximum lateral displacement to the yield displacement, dissipated energy, stiffness degradation and deformation of columns.

The new seismic code generation based on the performance design such as ATC-40 [3] recommends for reinforced concrete columns acceptance criteria based on plastic rotation limits assigned for three levels of performance: IO (immediate occupancy), LS (life safety) and SS (structural stability); corresponding in general to damage states namely: light $(d(\delta) < 0.1)$, moderate $(0.1 < d(\delta) < 0.5)$ and heavy $(d(\delta) > 0.5)$. These rotation limits may be converted in terms of corresponding interstorey drift. The RPA99 [4] code attributes for the interstorey drift a single value $(\delta \leq 1\%)$ to be complied with, which could be related to performance level LS guarantying human life and preserving structural integrity of the building.

This performance level agrees with a moderate damage state $(0.1 < d(\delta) < 0.5)$; however it is worth mentioning that for $(d(\delta) > 0.3)$ the reparation will be costly. As the current procedures for the performance assessment of buildings are based on the acceptance criteria for the structural elements, some expressions have been developed [5] on the basis of Erduran et al. damage curves [6] to verify the acceptability of reinforced concrete columns in structures frames type. These expressions introduce an adjustable transverse reinforcement ratio ρ_{sh} for different columns subjected to a constant interstorey drift while taking into account the resulting ductility demand level and column's flexibility, leading thereby to coherent global behaviour. This is because columns of the same level will exhibit a similar deformation capacity level. In this respect, we can refer to storey deformation capacity like storey deformation demand.

PERFORMANCE LEVELS

The basic principle established within seismic design philosophy consists of considering that it is not economically justified that, in a seismic area all the structures should be conceived to survive the strongest possible ground motion without any damage. It is more reasonable to take the point of view that the structures must exceed a moderate earthquake without damage and to tolerate a certain level of damage for a severe seism as long as structural collapse is prevented; it is the objective sought by the seismic regulations. However there is no general agreement on the acceptable damage level, but only criteria of acceptance to characterize these performances:

a   Life safety: is the primary requirement;
b   Reparable damage: a distinction is made between structural damage which can be repaired and damage which cannot be repaired;
c   Collapse prevention: the structure can undergo serious damage but must stand after the ground motion in order to avoid loss of life, injuries and damage of the contents of buildings.

For a structural performance defined in terms of a state of damage, strain and deformation are better indicators of damage than stresses [7]. In term of displacement the structural response can be related to strain-based limit state, which in turn is assumed to be related to a certain damage level. Table 1 gathers some recommended performance levels expressed in states of damage and corresponding relative displacements [1].

Table 1   Performance levels, corresponding states of damage and drift limits

PERFORMANCE LEVEL	DAMAGE STATE	INTERSTOREY DRIFT, %
*Fully operational, immediate occupancy	No damage	<0.2
*Operational, damage control, Moderate	Repairable	<0.5
*Life safe – Damage state	Irreparable	<1.5
*Near collapse, limited safety	Severe	<2.5
*Collapse		>2.5

The states of damage can be also defined by the terms: -negligible, light, moderate and severe. The three first states are expressed quantitatively through their crack widths, except for the columns under a strong axial loading where the observed cracks can be contained. The severe damage is given by using the load-deformation $(P-\delta)$ curves of the columns tested under cyclic loading rather than the curves of capacity, being able to over-estimate ultimate ductility because the analysis pushover does not take account of the resistance degradation induced by the action of the cyclic loading, and the level of damage corresponding to the ultimate index of ductility is selected as a superior limit of the severe damage recording the value of 90%.

DAMAGE CURVES

Exploiting numerical investigations Yakut et al. [6] developed damage curves for concrete columns based on the drift ratio and taking into account:
- the modified Kent and Park model [8] was used for confined concrete.
- the only failure mode observed is flexure
- flexural failure of the columns occurred before shear failure
- longitudinal reinforcement was properly done so that no bond and lap splice problems occur

The parametric study carried out revealed that the most parameters that affect the deformation limits of the reinforced concrete columns are:
- f_y : the yield strength of the longitudinal reinforcement
- (L/i) : slenderness of the column
- (N/N_0) : the axial load level
- ρ_{sh} : the volumetric ratio of the transverse reinforcement

Of these four parameters, the first two namely f_y and (L/i) affect the yield drift ratio δ_y significantly [9], however the last two have an effect on the ultimate ductility, and the deformation capacity of the columns. The ratio of the amount of transverse reinforcement to the axial load level $\rho_{sh}/(N/N_0)$ was introduced to characterize ductility level of columns, and the damage curves were established for three ductility classes. The damage equation is proposed in its following final form for three ductility classes:

$$d(\delta) = f(\delta).g(\delta)$$

$$f(\delta) = 1 - e^{-\left(\frac{\delta}{a(c_s).(c_{fy})}\right)^b} \tag{1}$$

$$g(\delta) = 0.5\left[1 - \cos\left(\frac{\pi\delta}{c(c_s).(c_{fy})}\right)\right] \text{ if } \frac{\delta}{(c_s).(c_{fy})} \leq c, \text{ and } g(\delta) = 1.0 \text{ if } \frac{\delta}{(c_s).(c_{fy})} \succ c \tag{2}$$

a, b, c: equation parameters varying according to ductility levels. They were determined by least squares curve fitting using mean and extreme values for each damage state.

$g(\delta)$: correcting function introduced in order to take into account very small deformations

δ : represents the interstorey drift

TRANSVERSE REINFORCEMENT

Regulation confrontation

The yielding of dissipative zones of energy is reached as a consequence of plastic diffusion through a potential length called length of the plastic hinge. The deformation state is intimately related to behaviour laws of the constitutive materials namely; unconfined concrete, confined concrete and steel. However confined concrete is found as the governing parameter. An additive transverse reinforcement, well disposed laterally, has a double function in resisting shear forces as well as providing confinement for the concrete core; enhancing then both section resistance and its deformation capacity, conferring hence a ductile behaviour.

Although the RPA fixes a transverse steel ratio ρ_{sh} to respect, unfortunately no allusion is made to confinement; factor governing the ductile aptitude of concrete sections [10];

$$\lambda_g \geq 5 \qquad\qquad \rho_{sh\,min} = 0.3\% \qquad\qquad \rho_{sh} = A_{sh}/s.b_1 \tag{3}$$

$3 < \lambda_g < 5$ $\qquad\qquad$ linear interpolation

$\lambda_g \leq 3$ $\qquad\qquad\qquad$ $\rho_{sh\,min} = 0.8\%$

$s \leq Min(150mm,10d_{bl})$ zone 1 and 2 and $s \leq 10cm$ in zone 3

The geometrical slenderness λ_g, introduced as a parameter conditioning the selection of a suitable transverse steel ratio without considering ductility level; is far from guarantying the required safety level.

It is noticeable, that the European regulations; EC-8 [11] include various geometrical and mechanical parameters and materials used affecting the local behaviour, prescribes hence values of the volumetric ratio, according to the ductility class level:

$$\rho_{sh} \geq max\left[\frac{k_0}{\alpha}(0.9v_d + 0.10)\left(0.35\frac{A_g}{A_c} + 0.15\right), \rho_{sh,min}\right] \tag{4}$$

k_0 and $v_{d,max}$ are function of the ductility class $v_d = \dfrac{P_e}{A_g.f_{cd}} \leq v_{d,max}$

$$\rho_{sh} = \frac{\text{confinement steel volume}}{\text{confined concrete volume}} \cdot \frac{f_{yd}}{f_{cd}} \tag{5}$$

$$s \leq Min\left(\frac{b_c}{3}, 150mm, 7d_{bl}\right) \text{ (class M)}$$

The confined steel required by the NZS 3101 code [12] is given in terms of the transverse steel area in one direction function of the material characteristics, transverse steel spacing and axial load intensity:

$$A_{sh} = 0.3sh^c\left(\frac{A_g}{A_{nh}} - 1\right)\frac{f'_{co}}{f_{yh}}\left(0.5 + 1.25\frac{P_e}{\phi f'_{co}A_g}\right) \tag{6}$$

Case studies

A practical illustration [13] was carried out on 2 regular frame buildings having 5×3 spans with 5 and 4 levels respectively, without infill participation. The formwork of the principal and secondary beams is taken respectively equal to 30×40 cm^2 and 30×35 cm^2, for the two considered cases. However that of the columns is taken equal to 40×40 cm^2 if the building is R+4 and 35×35 cm^2 if this last is R+3. In order to evaluate the capacity of resistance and deformation μ_φ^c of the column base sections, a computer program was established by using the Modified Kent and Park's confined concrete model. A typical section of transverse steel configuration commonly used in Algeria (square + lozenge) is adopted.

The obtained results show that the column sections at the base of the first level of the considered structures exhibit a medium ductile behaviour with a capacity of deformation higher than the demand except for those where the transverse reinforcement ratio is introduced by observing the RPA code (zone 2); especially central and end supporting columns. Furthermore [13] the deformation of the central row of the R+4 resisting frame shifted in the low ductility class **L**. The building will perish in this case by bursting of the transverse reinforcement since the central row remained in the incapacity to satisfy the necessary deformation demand, while the reserve of resistance reserve is rather consequent. If the NZS and the EC8 codes are considered, both the resistance and the deformation reserves remain satisfactory. However the corner column contains a much more pronounced deformation capacity and seems to have a highly ductile behaviour (**H**). This reserve would be reduced if higher modes of vibration induced by torsion effects are taken into account, reducing then the confinement steel deformability in other words the section ductility.

It was found [14] that the confinement reinforcement recommended by the EC8 corresponds rather to the mean ductility factor; i.e. that on average 50% of the columns provides the required ductility. This could be satisfactory for highly redundant structures, since a significant number of columns will participate to dissipate the totality requested energy whatever is the quantity dissipated by each one of them. Nevertheless this flexibility will not be granted to structures slightly redundant. This example will be reconsidered in order to estimate the degree of structural damage while supposing the interstorey drift recommended by the RPA99 code ($\delta = 1\%$) as being reached. Initially, a categorization of the ductility level is reflected through the $\rho_N = \rho_{sh}/(N/N_0)$ ratio [6]:

- $\rho_{sh}/(N/N_0) \prec 5\%$ low ductility (L)
- $5\% \prec \rho_{sh}/(N/N_0) \prec 10\%$ moderate ductility (M)
- $\rho_{sh}/(N/N_0) \succ 10\%$ high ductility (H)

and then the degree of damage is evaluated by considering its average value.

Table 2 gathering different values of the column damage of the considered structures (R+3 and R+4) function of the seismic zoning of the RPA99 code; allows establishing the following remarks:

- in zone 2, the recorded damage for the central and end supporting columns of the R+3 structure is moderate but requiring an expensive repair (20/24 is 5/6 of the whole). For the R+4 structure the end supporting and corner columns record a moderated damage necessitating an expensive repair (16/24 or 2/3 of the whole), however central columns have an excessive damage leading to structural instability (8/24 or 1/3 of the whole).

- in zone 3, the recorded damage is moderate requiring an expensive repair concerning central columns of the R+3 structure (8/24 or the 1/3 of the whole) and central and end supporting columns for the R+4 structure (20/24 or 5/6 of the whole).

Table 2 Evaluation of the column damage by considering the RPA99 code

| Column 35 × 35 | | STRUCTURE R+3 | | | | | | | |
| | | Zone 2 | | | | Zone 3 | | | |
	v	ρ_{sh} %	ρ_N %	C.D	$d(\delta)$	ρ_{sh} %	ρ_N %	C.D	$d(\delta)$
Corner	0.092		8.80	M	0.39		12.28	H	0.31
End sup.	0.151	0.81	5.36	M	0.39	1.13	7.48	M	0.39
Central	0.200		4.05	L	0.54		5.65	L	0.39

| Column 40 × 40 | | STRUCTURE R+4 | | | | | | | |
| | | Zone 2 | | | | Zone 3 | | | |
	v	ρ_{sh} %	ρ_N %	C.D	$d(\delta)$	ρ_{sh} %	ρ_N %	C.D	$d(\delta)$
Corner	0.086		8.14	M	0.31		11.28	H	0.23
End sup.	0.146	0.7	4.79	L	0.47	0.97	6.64	M	0.31
Central	0.205		3.4	L	0.47		4.73	L	0.47

We can notice from table 3, when considering the NZS and EC-8 codes that the recorded damage is moderate and easily reparable for the whole of the columns of the R+3 structure, however it will require an expensive repair for central and end supporting columns of the R+4 structure (5/6 of the whole); and this for the 2 considered codes.

Table 3   Evaluation of the column damage by considering the NZS and EC-8 codes

Column		STRUCTURE R+3							
35 × 35		NZS				EC-8			
	v	$\rho_{sh}\%$	$\rho_N\%$	C.D	$d(\delta)$	$\rho_{sh}\%$	$\rho_N\%$	C.D	$d(\delta)$
Corner	0.092	0.97	10.54	H	0.23	0.89	9.67	H	0.23
End sup.	0.151	1.08	7.15	M	0.31	1.08	7.15	M	0.31
Central	0.200	1.22	6.10	M	0.31	1.08	5.40	M	0.31
Column		STRUCTURE R+4							
40 × 40		NZS				EC-8			
	v	$\rho_{sh}\%$	$\rho_N\%$	C.D	$d(\delta)$	$\rho_{sh}\%$	$\rho_N\%$	C.D	$d(\delta)$
Corner	0.086	1.03	11.98	H	0.31	1.03	11.98	H	0.31
End sup.	0.146	1.13	7.74	M	0.39	1.13	7.74	M	0.39
Central	0.205	1.13	5.51	M	0.39	1.13	5.51	M	0.39

Influence of the slenderness ratio

While varying the column's slenderness ratio L/i and the transverse reinforcement ratio - reduced axial load level, curves of Figure 1 are obtained which illustrate the interaction of the flexibility of the element and the ductility class associated to the damage level.

We can notice that the RPA99's limit recommended code to avoid short columns, namely $\frac{L}{h} \geq 4$ which corresponds to $\frac{L}{i} \geq 13.9 \approx 14$ has to be reconsidered because it agrees only for a highly ductile section case recording a moderate damage requiring an expensive repair and this for a an interstorey drift of 0.75%.

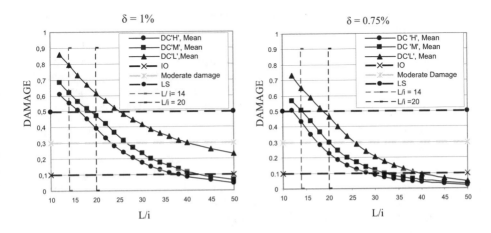

Figure 1   Variation of the damage level according to the slenderness ratio and the ductility level

However the transverse steel ratio as recommended by RPA code confers for central columns only a low ductile behaviour which combined with an interstorey drift of 1% will inevitably lead to the ruin of the elements having a slenderness ratio lower than 20, which is corresponding to $\frac{L}{h} = 5.8 \approx 6$. As an indication, table 4 gathers some average damage values for usual column section for a current storey height of 3.06 m and a 30×40 cm^2 beam section.

Table 4 Mean damage for usual section columns

DUCTILIT Y CLASS	L = 2.66 m				
	30×30	35×35	40×40	45×45	50×50
	Slenderness ratio L/i; (L/h)				
	30.715	26.327	23.036	20.477	18.429
	(8.87)	(7.6)	(6.65)	(5.91)	(5.32)
H	0.165	0.234	0.306	0.377	0.439
M	0.223	0.309	0.391	0.462	**0.510**
L	0.378	0.472	**0.542**	**0.603**	**0.658**

These results illustrate clearly the influence of the level of ductility and that of the column's dimensions on the damage level. Since the volumetric ratio of the transverse reinforcement is recognized as being the factor governing the capacity of deformation of these elements, a simple formulation of this ratio is proposed by considering a moderate damage level guaranteeing the human safety as well as the structural integrity with a non expensive repair. Table 5 summarizes the various expressions of ρ_{sh} jointly articulated on N/N_0 and L/i factors, defining by the way the applicability of the recommended limit relating to the interstorey drift according to the flexibility of columns, function of a ductility class and the desired resulting damage level.

The results obtained by using the proposed expressions for an interstorey drift of 1% for the 2 structure cases (R+3 and R+4) are gathered within table 6. It shows clearly that the level of ductility is the same one for the columns constituting each structure type, leading then to a better coherence of the global behaviour. To introduce a transverse steel ratio variable or more exactly adjustable for different column types by taking into account their corresponding axial load intensity as well as a constant global displacement agrees with the idea of overall deformation capacity conducting to the desired damage level. For this, the transverse volumetric ratio such as previously proposed $\rho_{sh} = f\left[\left(v, \frac{L}{i}\right), \delta\right]$ adapts well with the performance based seismic design approach. However, it seems more suitable to reduce the interstorey displacement to 0.75% in order to avoid an eventual loss of structural stability since the damage criterion is not taken into account by the RPA99 code, furthermore the control of this damage level for an interstorey displacement of 1% requires generally highly ductile sections (for elements having $17 \leq L/i \leq 23$).

Table 5 Proposed expressions; resulting damage function of the flexibility of columns

DUCTILITY LEVEL	INTERST. DRIFT δ, %	EXPRESSIONS	DAMAGE	
			$d(\delta)\leq 0.3$	$0.3 \prec d(\delta) \prec 0.5$
H	1.0	$\rho_{sh}/(N/N_0)=13-\sqrt{(L/i-17)0.5}$	$L/i \succ 23$	$17 \leq L/i \leq 23$
	0.75	$\rho_{sh}/(N/N_0)=13-\sqrt{(L/i-14)0.5}$	$L/i \succ 18$	$14 \leq L/i \leq 18$
M	1.0	$\rho_{sh}/(N/N_0)=10-\sqrt{(L/i-20)}$	$L/i \succ 26$	$20 \leq L/i \leq 26$
	0.75	$\rho_{sh}/(N/N_0)=10-\sqrt{(L/i-16)0.5}$	$L/i \succ 20$	$16 \leq L/i \leq 20$

Table 6 Evaluation of column ductility by using the proposed equations

| Column 35 × 35 L/i=26.327 | ν | STRUCTURE R+3 | | | | | | | |
| | | Ductility level H | | | | Ductility level M | | | |
		$\rho_{sh}\%$	μ_φ^c	$\mu_\varphi^c G$	$\mu_\varphi^d G$	$\rho_{sh}\%$	μ_φ^c	$\mu_\varphi^c G$	$\mu_\varphi^d G$
Corner	0.092	1.00	14.58			0.69	10.97		
End sup.	0.151	1.64	16.29	16.33	9.10	1.13	11.51	11.45	9.10
Central	0.200	2.17	17.28			1.50	11.60		

| Column 40 × 40 L/i=23.036 | ν | STRUCTURE R+4 | | | | | | | |
| | | Ductility level H | | | | Ductility level M | | | |
		$\rho_{sh}\%$	μ_φ^c	$\mu_\varphi^c G$	$\mu_\varphi^d G$	$\rho_{sh}\%$	μ_φ^c	$\mu_\varphi^c G$	$\mu_\varphi^d G$
Corner	0.086	0.97	14.50			0.71	11.39		
End sup.	0.146	1.64	16.80	16.75	8.96	1.20	12.50	12.51	8.96
Central	0.205	2.30	17.79			1.69	13.10		

CONCLUSIONS

The transverse volumetric ratio recommended by the Algerian code is far away of satisfying the requested local deformation demand of RC columns. Taking into account the interstorey drift, a formulation of this ratio is required function of a moderate damage level guaranteeing as well as life safety and structural integrity with an eventual easy repair.

The suggested formulation of ρ_{sh} jointly articulated on (N/N_0 and L/i) is integrated within this framework; and allowed to emphasize the limits of the column's flexibility i.e. of its dimensions. These expressions introduce an adjustable transverse reinforcement ratio ρ_{sh} for different columns subjected to a constant interstorey drift, taking into account the resulting ductility demand and column's flexibility, leading thereby to coherent global behaviour.

This is because columns of the same level will exhibit a similar deformation capacity level. In this respect, we can refer to storey deformation capacity like storey deformation demand.

REFERENCES

1. GHOBARAH A, Performance based design (PBD) in earthquake engineering: State of development, Engineering Structures, Vol. 23, 2001, pp. 878-884.

2. GIONCU V AND MAZZOLANI F M, Ductility of Seismic Resistant Steel Structures, Spon Press, London, U.K., 2002

3. APPLIED TECHNOLOGY COUNCIL, ATC-40, Seismic Evaluation and Retrofit of Concrete Buildings, California, 1996.

4. REGLES PARASISMIQUES ALGERIENNES, RPA 99, CGS, Algiers, 2000.

5. DJEBBAR N, Contribution to the Study of Seismic Performance of Concrete Linear Elements, (in French), PhD Thesis, University of Constantine, Algeria, April 2006.

6. ERDURAN E AND YAKUT A, Drift Based Damage Functions for Reinforced Concrete Columns, Computers and Structures, Vol. 82, 2004, pp. 121-130.

7. MOEHLE J P, Displacement-based Seismic Design Criteria. In Proceeding of 11th World Conference on Earthquake Engineering, Acapulco, Mexico, paper N°2125, Oxford, Pergamon, 1996.

8. PARK R, PRIESTLEY M J N AND GILL W D, Ductility of Square Confined Concrete Columns, Journal of the Structural Engineering, Vol. 108, No. 4, April 1982.

9. PAULEY T, An estimation of displacement limits for ductile systems, Earthquake Eng. Struct. Dyn. Vol. 31, 2002, pp. 583-599.

10. DJEBBAR N, BOUSALEM B AND CHIKH N, Performance Parasismique des Portiques en Béton, Aspect Théorique et Constat Réglementaire, 3ème Partie : *Exigences sur les Aciers';* Revue Algérie Equipement, No. 36, Décembre 2002.

11. EUROCODE 8: Design of Structures for Earthquake Resistance - Part 1: General Rules, Seismic actions and Rules for Buildings, December 2003.

12. NZS 4203 General Structural Design and Design Loadings for Buildings, Wellington, Standards Association of New Zealand, 1992.

13. DJEBBAR N, BOUSALEM B AND CHIKH N, Notion de Comportement global - Comportement local dans la Performance Parasismique des portiques en Béton, *Etude Comparative Réglementaire'*. Revue Sciences et Technologie B – No. 20, Décembre 2003, pp. 63-69. Université Mentouri, Constantine

14. TREZOS C G, Reliability Considerations on the Confinement of RC Columns for Ductility, Soil Dynamics and Earthquake Engineering, Vol. 16, 1997, pp. 1-8.

COMPARISION OF EXPERIMENTAL AND THEORETICAL TORSIONAL STIFFNESS OF REINFORCED CONCRETE BEAMS

A Barbosa Ferraz Cavalcanti

M T Gomes Barbosa

Federal University of Juiz de Fora

E de Souze Sanchez Filho

P Barbosa

Fluminese Federal University

Brazil

ABSTRACT. This paper presents a study of the torsional behaviour of reinforced concrete beams with normal and high strength concrete subjected to pure torsion. Two theoretical approaches are studied for adjustment of the curves torsion *vs.* torsion angle per unit of length, and the secant torsional stiffness is obtained in the cracked stage of the beams. One approach was found in the literature, and another is proposed by the authors. These two approaches are analyzed and compared with the experimental results found in the literature. This analysis shows that the theoretical adjustment for these author's approaches curves gives good results and more consistent values to secant torsional stiffness.

Keywords: Torsion; Secant torsional stiffness; Reinforced Concrete.

Adriana Barbosa Ferraz Cavalcanti, M.Sc. Her fields of interest include structural concrete and structural strengthening with composite materials.

Emil de Souza Sánchez Filho, D.Sc. Professor of Fluminense Federal University, Brazil, member of ACI, PCI, FIP, IABSE, IBRACON. His fields of interest include structural concrete and structural strengthening with composite materials.

Maria Teresa Gomes Barbosa, D.Sc. Professor of Federal University of Juiz de Fora, Brazil, member of ACI and IBRACON. Her fields of interest include concrete structures and new materials.

Plácido Barbosa, M.Sc. Professor of Fluminense Federal University, Brazil. His fields of interest include structural concrete and structural strengthening with composite materials.

INTRODUCTION

The Spatial Truss Analogy to reinforced concrete torsion design is a consistent theoretical model, and is corroborated by experimental tests. The code [1] adopted this model in their provisions [2-4]. Several models for torsional rigidity of reinforced concrete beams found in the literature use this truss approach.

This article analyzes the torsional rigidity approach proposed in [5], and proposes an approach to obtain the formula to secant torsional stiffness with basis in the tri-linear diagram to the curve torsion vs. angle of torsion per unit length. It uses three torque values for the analysis, the crack torque, the yield torque, and the ultimate torque.

The results of the proposed approach are confronted with experimental results of reinforced concrete beams with normal strength and high strength concrete.

SPATIAL TRUSS ANALOGY

The Space Truss Analogy is adopted and accepted internationally by concrete codes and standards, where the torsion design is done in the Ultimate Limit State, applicable to concrete with normal strength and high strength. This theoretical model contemplates only the uniform torsion: equilibrium torsion and compatibility torsion.

Figure 1 shows the aims parameters of the Space Truss Analogy to pure torsion, where the beam section is admitted similar to a thin-walled tube, with the shear flow resisted by the longitudinal and transverse reinforcement (closed stirrups). The cracked concrete membrane has a strut angle α.

The strut angle α and ultimate torsion T are given by:

$$tg\,\alpha = \sqrt{\frac{\left(A_{st}\,f_{yt}\right)}{\left(A_{s\ell}\,f_{y\ell}\right)}\cdot\frac{u}{s}} \tag{1}$$

$$T = 2\,A_0\sqrt{\frac{\left(A_{st}\,f_{yt}\right)\left(A_{s\ell}\,f_{y\ell}\right)}{u\,s}} \tag{2}$$

where

A_u, A_ℓ – transverse and longitudinal reinforcement, respectively;

$f_{ty}, f_{\ell y}$ – yield strength of transverse and longitudinal reinforcement, respectively;

s – closed stirrups spacing;

A_0, u – area and perimeter enclosed within the centre line of the thin-walled (Bredt's theory), respectively.

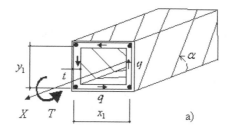

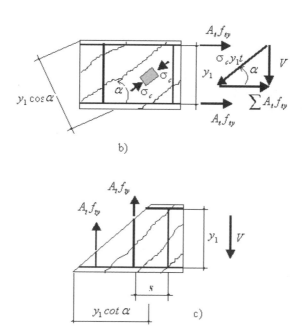

Figure 1   Space Truss Analogy

EXPERIMENTAL DATA

The study [6] aim to examine experimentally the values of the cracking strength, ultimate strength, ductility, torsional stiffness, and the strains of concrete of normal and high-strength concrete beams subjected to pure torsion. Sixteen reinforced concrete beam specimens with a cross section of 350 × 500 mm and the clear concrete cover of 20 mm were fabricated. The overall length of the beams was 3100 m, and the middle of them was approximately 1600 mm. The transverse and longitudinal reinforcement were arranged according design provisions of [7]. The total steel ratios of the longitudinal and transverse reinforcements varied from 1.2% to 4.0%. The normal and high-strength concrete beams were designed for the 28 day cylinder compressive strength of 35 MPa and 70 MPa, respectively.

The beams were simply supported at two bearings under which the roller supports were installed to release the restraint of longitudinal elongation of specimens during the test. The vertical load was applied using a 5000 kN universal testing machine through a spreader beam on torsional arms to produce the torque.

The load was initially applied in greater increments of approximately 10 kN/m torques. When 80% of the predicted ultimate strength was reached, the load increment was then reduced to produce an equivalent torque of approximately 5 kN/m until rupture.

These authors showed that from design provisions of [7] the cracking torque given by Theory of Elasticity is underestimated, and the value given by Skew Bending Theory is overestimated.

Table 1 gives the experimental torques of [6].

Table 1   Experimental torques [6]

BEAM	T_{CR}, kNm	T_u, kNm	T_u, kNm
H-06-06	70.6	79.7	92.0
H-06-12	75.0	83.5	115.1
H-12-12	77.1	116.8	155.3
H-12-16	79.3	157.0	196.0
H-20-20	76.0	0	239.0
H-07-10	70.5	91.1	126.7
H-14-10	61.8	100.0	135.2
H-07-16	65.3	90.3	144.5
N-06-06	43.2	71.5	79.7
N-06-12	51.8	80.9	95.2
N-12-12	49.3	113.0	116.8
N-12-16	57.1	125.0	138.0
N-20-20	55.0	0	158.0
N-07-10	41.6	93.8	111.7
N-14-10	41.8	108.0	125.0
N-07-16	40.0	94.9	117.3

H – high strength concrete. N – normal concrete.

TORSIONAL RIGIDITY

This paper aims to study two different theoretical approaches to evaluate the torsional rigidity of reinforced concrete beams.

The theoretical approach of [4] gives the following expressions:

– torque:

$$T = T_n - (T_n - T_{cr}) \left(\frac{\theta_u - \theta_a}{\theta_u - \theta_{cr}} \right)^\phi \tag{3}$$

– torsional rigidity:

$$(GC)_e = \frac{T_a}{\theta_u - (\theta_u - \theta_{cr}) \left(\dfrac{T_n - T_a}{T_n - T_{cr}} \right)^{1/\phi}} \tag{4}$$

or

$$(GC)_e = \frac{T_a}{\left(\dfrac{T_n - T_a}{T_n - T_{cr}} \right)^{1/\phi} \theta_{cr} + \left[1 - \left(\dfrac{T_n - T_a}{T_n - T_{cr}} \right)^{1/\phi} \right] \theta_u} \tag{5}$$

where

T_a, T_{cr}, T_n – applied, cracking, and ultimate torque, respectively;

θ_{cr}, θ_u – angle of torsion per unit of length, at cracking and at ultimate strength, respectively;

$(GC)_e$ – ultimate effective torsional rigidity;

ϕ – exponent calibrated by experimental results.

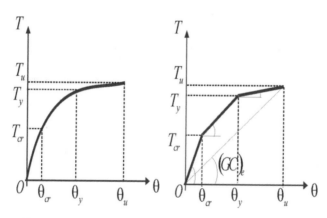

Figure 2 Curve $T \times \theta$ and theoretical tri-linear approach

The proposed approach gives:

$$T = T_n - (T_n - T_{cr}) \psi \left(\frac{\theta_u - \theta_a}{\theta_u - \theta_{cr}} \right)^{1.02} \tag{6}$$

were coefficient ψ is calibrated by experimental results, and adopts the expressions 4 and 5 with $\phi = 1.02$. The experimental values of the beams of [6] were used in the above equations. Table 3 shows the results with the exponent ϕ to [5] approach and to the proposed approach.

Table 3 Theoretical results for beams of [6]

BEAM	TORQUE, kNm		TORSIONAL RIGIDITY, kNm^2	
	$\phi = 2$	$\phi = 1.02$	$\phi = 2$	$\phi = 1.02$
H-06-06	74143	71497	7305	6853
H-06-12	86549	80900	8121	7362
H-12-12	115038	113004	7492	5967
H-12-16	132864	124992	8024	5940
H-07-10	96,558	93797	7893	7006
H-14-10	114018	107997	5675	4927
H-07-16	103469	94898	7805	7119
N-06-06	84743	79697	7305	6853
N-06-12	103955	94228	8121	7362
N-12-12	131111	116786	7492	5967
N-12-16	168503	157028	8024	5940
N-07-10	104188	91102	7893	7006
N-14-10	119178	100006	5675	4927
N-07-16	111608	90314	7805	7119

H – high strength concrete. N – normal concrete.

The tangent rigidity is gives by:

– $\phi = 2$ for [5] approach:

$$GC = 2 \cdot (T_u - T_{cr}) \cdot \frac{(\theta_u - \theta)}{(\theta_u - \theta_{cr})^2} \qquad (7)$$

– $\phi = 1.02$ for proposed approach:

$$GC = 1.02 \cdot (T_u - T_{cr}) \cdot \psi \cdot \frac{\left[\frac{(\theta_u - \theta)}{(\theta_u - \theta_{cr})} \right]^{(2 \cdot 10^{-2})}}{(\theta_u - \theta_{cr})} \qquad (8)$$

Table 4 shows the values of the tangent rigidity for beams of [6].

Table 4 Tangent rigidity for beams of [6]

BEAM	$\phi = 2$, kNm²	$\phi = 1.02$, kNm²
H-06-06	1001	865
H-06-12	1938	1693
H-12-12	3076	1945
H-12-16	4549	2190
H-07-10	2431	1818
H-14-10	2904	2118
H-07-16	3166	2388
N-06-06	1078	395
N-06-12	1698	862
N-12-12	2304	541
N-12-16	3188	1413
N-07-10	2177	757
N-14-10	2443	856
N-07-16	2444	1067

H – high strength concrete. N – normal concrete.

The [8] tested in the Laboratory of Structures and Materials of PUC–Rio one beam subject to pure torsion (Figure 3), and obtained the results: $T_{cr} = 14.70\,kNm$ and $\theta_{cr} = 0.00244\,rad/m$; $T_y = 19.07\,kNm$ and $\theta_y = 0.02653\,rad/m$; $T_u = 21.57\,kNm$ and $\theta_u = 0.04\,rad/m$.

The adjustments of the experimental curve $T \times \theta$ of this beam gives:

–approach of [5]:

$$T = 20.687\,kNm\,;\ (GC)_e = 779.63\,kNm^2\,;\ \frac{T_{y,theor}}{T_{y,exp.}} = 8.48\%$$

– proposed approach:

$$\phi = 1.02\,;\ \psi = 1.036\,;\ T = 19.07\,kNm\,;\ (GC)_e = 731.82\,kNm^2$$

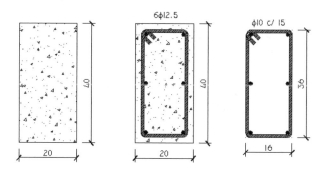

Figure 3 Beam VR 20 cm × 40 cm [8]

CONCLUDING REMARKS

The beams with high-strength concrete present values of $T_{y,teor}$ greater than $T_{y,exp.}$, and the beams with normal concrete present values equal to $T_{y,teor}$, and a small curvature in the graph $T \times \theta$. In the post-cracking stage approach of [5] gives an equation of 2 degrees, with best agreement to experimental results for beams with normal strength concrete. The values obtained with the proposed approach applied to beams with high-strength concrete give good agreement with values of the experimental curves $T \times \theta$, because it always contains the point $T_{y,exp.}$. These experimental curves are nearly linear in the interval $T_{cr} - T_u$, and the proposed approach represents more correctly the results. The results to beams with normal strength concrete obtained with [5] approach present best agreement with the experimental values of T_y, when compared with the results obtained to beams with high-strength concrete. The calculated values to tangent rigidities for beams with high-strength concrete are not good for two approaches. The beam tested by [8] had the theoretical value of T_y equal the experimental value, and agreement between theoretical and experimental values prove the efficiency of the proposed approach.

REFERENCES

1. EUROCODE 2, Design of Concrete Structures. Comité Euro-International du Béton, 1992.

2. CEB-FIB, Practical Design of Structural Concrete. MC 90, Comité Euro-International du Béton, 1999, pp. 52-58.

3. CEB-FIB, Model Code for Concrete Structures, MC 90, Comité Euro-International du Béton, Thomas Telford Services Ltd., UK, 1993.

4. REGAN P, Basic Design for Moment, Shear and Torsion, FIB Bulletin 2: Structural Concrete – Textbook on Behaviour, Design and Performance, Vol. 2. 1999, pp. 156-169.

5. TAVIO AND TENG S, Effective Torsional Rigidity of Reinforced Concrete Members, ACI Structural Journal, Proceedings Vol. 101, No. 2, 2004, pp. 252-260.

6. FANG I K AND E SHIAU J K, Torsional Behavior of Normal and High-Strength Concrete Beams, ACI Structural Journal, Proceedings Vol. 101, No. 3. 2004, pp. 304-313.

7. AMERICAN CONCRETE INSTITUTE, ACI 318 – Building Code Requirements for Structural Concrete, 2005.

8. SILVA FILHO J J H, Theoretical and Experimental Study of RC Beams Strengthened to Torsion with Carbon Fibre Composites (manuscript of the PhD thesis), PUC-Rio, Rio de Janeiro, 2005.

STRENGTH OF SHEAR SECTION PRE-TENSIONED CONCRETE STRUCTURES SUBJECTED TO FLEXURAL MOMENTS

J Valivonis

G Marčiukaitis

B Jonaitis

A Šneideris

Vilnius Gediminas Technical University

Lithuania

ABSTRACT. Steel strands as reinforcement for concrete structures are used in many cases without additional anchorage. In such cases the length of anchorage of the strands is quite long and in the support part of the beams there is a zone in which the strands are not completely introduced in interaction with the concrete. In strength calculations of the support zone reduction in strength and prestressing force provided by the longitudinal reinforcement has to be taken in to account. Flexural resistance of reinforced concrete flexural member at this zone is less than that in the other parts. Therefore, in spite of the quite small value of the acting bending moment, such support zone can fail due to the action of this particular bending moment. The influence of the bending moment and anchorage length of prestressed reinforcement on the strength of shear section of concrete structures is analyzed in this paper. Theoretical investigations in the strength of shear section of prestressed concrete flexural members taking into account the anchorage length, prestressing force and transverse reinforcement were performed. It was determined by investigations that in design of such structures the decrease in effect of longitudinal reinforcement must be compensated by an adequate amount of shear reinforcement. Results of theoretical investigations indicated that the strength of concrete structures under the action of sufficiently large loads is governed by the resistance of the shear section to the action of the bending moment. Hollow core prestressed concrete slabs were tested. A comparison of the results obtained by experiments and by theoretical calculations of prestressed concrete stabs was made.

Keywords: Pre-tensioned concrete, Shear section, Shear reinforcement, Bending moment.

Dr Juozas Valivonis, Professor, Head of Department of Reinforced Concrete and Masonry Structures of Vilnius Gediminas Technical University, Lithuania. Scientific interest – composite, reinforced and masonry structures.

Hab Dr Gediminas Marčiukaitis, Professor in the Department of Reinforced Concrete and Masonry Structures of Vilnius Gediminas Technical University, Lithuania. Scientific interest – composite, reinforced concrete and masonry structures.

Dr Bronius Jonaitis, Assoc. Prof in the Department of Reinforced Concrete and Masonry Structures of Vilnius Gediminas Technical University, Lithuania. Scientific interest – analysis of structures, state assessment, renovation.

Dr Arnoldas Šneideris, Assoc. Prof in the Department of Reinforced Concrete and Masonry Structures of Vilnius Gediminas Technical University, Lithuania. Scientific interest – reinforced concrete and masonry structures, renovation.

INTRODUCTION

Prestressed reinforcement is often used in the design of concrete structures. For the manufacture of concrete members, prestressed reinforcing bars, wires or strands can be used. The behaviour of prestressed members is highly influenced by the anchorage of the reinforcement. When in separate cases ribbed bar reinforcement is used, anchorage generally is provided by the ribs of the bars. For wire or strand reinforcement anchorage can be provided by the friction. Strands or wires can be used with special external or internal anchors. In such cases the reinforcement is introduced into common action with the concrete instantly. However when a large amount of reinforcement is provided, the arrangement of anchors in the support zone becomes quite complicated. In such cases the anchors have to be installed in different sections or one has to rely on self-anchoring of the strands in the concrete. The anchorage length of the strands without special anchors is quite large. At the zone of anchorage of strands the reinforcement is only partially introduced into the common action with the concrete. Therefore a risk arises for the flexural member to fail at the support zone. Failure in the support zone may occur in the cross or shear [1-3] sections due to the action of the shear force (Figure 1a) or in the shear section due to the action of the bending moment (Figure 1b).

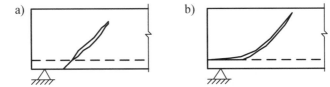

Figure 1   The mode of failure of the shear section of a reinforced concrete member due to the shear force action (a), and due to the bending moment action (b)

It is important for the structures to be manufactured by the long-line process [4-7]. The failure depends on the weaker section [8-10]. For reinforced concrete members without shear reinforcement the greater risk is to fail in the shear section due to the action of the bending moment. Members with shear reinforcement can fail in the cross section since the shear section may be strengthened sufficiently by the shear reinforcement.

The strength of the support zone depends significantly on the anchorage of the longitudinal reinforcement. It determines the location of the section at which the reinforcement is completely introduced into common action with the concrete.

In spite of the fact that reinforced concrete slabs during their test quite often fail in the shear section due to the action of the bending moment, in design of such structures the resistance of the shear section to the action of the bending moment in most cases is not verified [11, 12].

GENERAL PRINCIPLES FOR CALCULATION

It is suggested in design of reinforced concrete structures to verify strength of the shear section in relation to the bending moment:

$$M_{Ed} < M_{Rd} \tag{1}$$

where: M_{Ed} – bending moment in the section 1–1; M_{Rd} – flexural strength of the member.

It is especially important for the prestressed reinforcement within a prestressed concrete structure without additional anchors. There is a risk that the reinforcement may slip in the anchorage zone if there are no additional anchors. Therefore in the course of the analysis for strength of the anchorage zone of the member, it is not possible to take into account the full effect of the reinforcement. The diagram for strength analysis of the shear section of a reinforced concrete member without the shear reinforcement under the action of the bending moment is shown in Figure 2, and that with the shear reinforcement in Figure 3. The strength of the members without shear reinforcement can be determined by:

$$M_{Rd} = \gamma_x \cdot f_{yd} \cdot A_{p1} \cdot z_s . \tag{2}$$

The lever arm of internal forces z_s is determined by

$$z_s = d - \frac{\gamma_x \cdot f_{yd} \cdot A_{p1}}{2 \cdot f_{cd} \cdot b} . \tag{3}$$

Where γ_x – coefficient taking into account the degree of introduction of the longitudinal reinforcement into the common action with the concrete

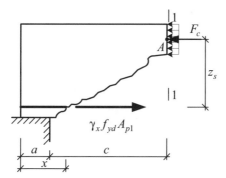

Figure 2   Diagram for strength analysis of the shear section of a reinforced concrete member without the shear reinforcement under the action of the bending moment

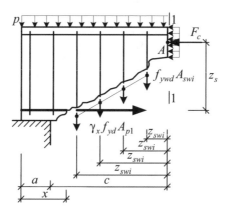

Figure 3   Diagram for strength analysis of the shear section of a reinforced concrete member with the shear reinforcement under the action of the bending moment

$$\gamma_x = \frac{x}{l_{pt2}} \tag{4}$$

Where x – length of the support zone; l_{pt2} – the length of stress transfer.

The strength of the members with the shear reinforcement can be determined by

$$M_{Rd} = \gamma_x \cdot f_{yd} \cdot A_{p1} \cdot z_s + \sum_{i=1}^{n} f_{ywd} \cdot A_{swi} \cdot z_{swi} \tag{5}$$

The distance, c, from the edge of the support to the section 1–1 for simply supported reinforced concrete structures may be expressed by

$$c = \frac{V_{Ed,\max}}{v_{sw} + p} \tag{6}$$

Where $V_{Ed,\max}$ – the shear force at the support; p – uniformly distributed load on the member.

The intensity of the shear reinforcement is:

$$v_{sw} = \frac{f_{ywd} \cdot A_{sw}}{s_w} \tag{7}$$

The investigations showed that the strength of the shear section to the action of the bending moment in reinforced concrete flexural structures without the shear reinforcement depends on the length of stress transfer by the longitudinal reinforcement, the length of the support zone, the amount of the longitudinal reinforcement in the member as well as on the magnitude of the acting load. For the members with shear reinforcement in addition to the said factors, the amount of the shear reinforcement in the support zone is of great influence.

THEORETICAL ANALYSIS

Reinforced concrete hollow core slabs, manufactured by the long line process, are without shear reinforcement. In this case the strength of the shear section is provided by the concrete and by the pre-compression force due to the prestressing of the reinforcement. In strength verification of the shear section of a slab for the action of the bending moment, the amount of longitudinal prestressed reinforcement and the length of stress transfer are very important. Arrangement of special additional anchorages in the slabs of such type is impossible. The reinforcement is anchored by the friction and irregularities in the strands. It results in quite large length of the zone of prestress transfer.

Theoretical calculations made using formulae (1) to (5) revealed that the strength of the shear section for the action of the bending moment increases proportionally with the amount of longitudinal prestressed reinforcement (Figure 4). It was observed in the investigations that an increase in the percentage of the longitudinal reinforcement by 2% resulted in 1.8 times increase in the value of the moment of resistance M_{Rd}. This indicates that the strength of the shear section depends quite greatly on the longitudinal reinforcement of a reinforced concrete member, in the same way as it does for the strength of the cross section.

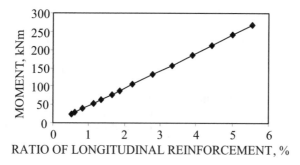

Figure 4   Relationship between the strength of the shear section for the action of the bending moment and the percentage of the longitudinal reinforcement

The theoretical analysis revealed that when the amount of prestressed reinforcement is kept constant, but with increased prestressing, the length of the zone of the stress transfer increases resulting in the decrease in the strength of the shear section for the action of the bending moment (Figure 5). When the prestressing of the reinforcement is sufficiently small (600-900 N/mm^2) its influence on this strength is greater than that at larger prestressing (900-1300 N/mm^2). If the increase in prestressing from 600 N/mm^2 to 800 N/mm^2 results in 25% decrease in the design strength, then the increase in prestressing from 900 N/mm^2 to 1100 N/mm^2 results in just 17% decrease in the design strength (Figure 5). Since the value of prestressing effects the strength of the shear section for the action of the bending moment via the length of stress transfer, the performed theoretical calculations revealed that the relationships "strength vs. the length of the stress transfer" (Figure 6) and "strength vs. the prestressing value" (Figure 5) are similar in their character.

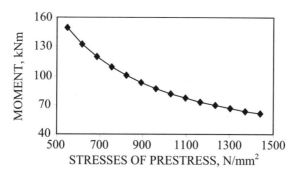

Figure 5   Relationship between the strength of the shear section for the action of the bending moment and the value of prestress of the longitudinal reinforcement

By considering the analysis of the formulae (2) to (4) one can state that the strength of the shear section of reinforced concrete members for the action of the bending moment depends on the length of their support zone. Inspection of buildings under construction showed that the actual lengths of supports of reinforced concrete members can be very different.

As the length of the support zone is small, the section that is used to verify the strength of the member is located quite close to the end of the structure. It means that in such a section, introduction of reinforcement into the common action with the concrete is quite poor. Thus the strength of the shear section for the action of the bending moment varies with the length of the support zone. The theoretical analysis revealed that an increase by 10 mm in the support zone length for a 400 mm high prestressed concrete slab resulted in about 12-13% increase in strength of the shear section for the action of the bending moment (Figure 7). A theoretical analysis of strength of the shear section of reinforced concrete beams for the action of the bending moment was carried out. In reinforced concrete beams reinforced with longitudinal and shear reinforcement, the strength of the shear section for the action of the bending moment is provided by both longitudinal and shear reinforcements. The strength of the member increases proportionally with the ratio of the longitudinal reinforcement (Figure 8), whereas the strength of the shear section of reinforced concrete members increases with the amount of shear reinforcement (Figures 8 and 9).

Theoretical analysis however pointed out that using the calculation method presented above, the strength of the shear section for the action of the bending moment does not increase proportionally with increased shear reinforcement. When the amount (intensity v_{sw}) of the shear reinforcement does not exceed 0.2 the theoretical strength of the member increases with the amount of the shear reinforcement. Once this amount of shear reinforcement is exceeded, the theoretical strength starts to decrease with the amount of the shear reinforcement (Figure 9).

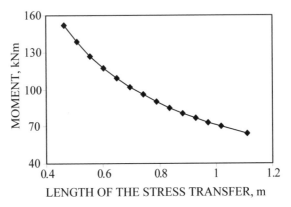

Figure 6 Relationship between the strength of the shear section for the action of the bending moment and the length of the stress transfer of the longitudinal reinforcement

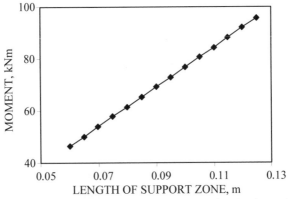

Figure 7 Relationship between the strength of the shear section for the action of the bending moment and the length of support zone of a slab

It is affected by the decrease in the theoretical length of projection c of the critical shear section into the horizontal axis. In this case the value of the following expression decreases:

$$\sum_{i=1}^{n} f_{ywd} \cdot A_{swi} \cdot z_{swi}$$

It leads to the reduction in the value of design strength of the member M_{Rd} determined by the formula (5). The relationships between the strength of the inclined section of reinforced concrete beams to the action of the bending moment and the value of prestessing, the length of the zone stress transfer and the length of the support zone of the beams, respectively, are very similar to the appropriate relationships for reinforced concrete slabs without the shear reinforcement. Theoretical analysis revealed that 50 mm increase in the length of the support zone of a reinforced concrete beam resulted in about 14-15% increase of its strength. The prestressing in the reinforcement is small, hence its increase from 600 N/mm² to 800 N/mm² results in 15% design strength decrease, but an increase in the prestressing from 900 N/mm² to 1100 N/mm² results in 10% design strength decrease (Figure 9).

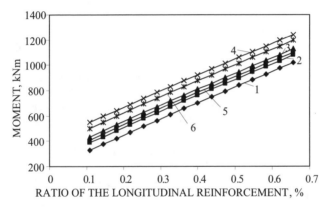

Figure 8   Relationship between the strength of the shear section of reinforced concrete members reinforced with the shear reinforcement to the action of the bending moment and the ratio of the longitudinal and intensity of the shear reinforcement (v_{sw}):
1 – 90 kN/m, 2 – 120 kN/m, 3 – 140 kN/m, 4 – 360 kN/m, 5 – 450 kN/m, 6 – 710 kN/m

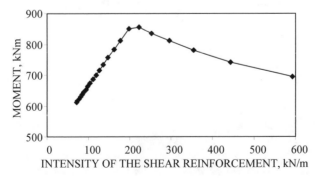

Figure 9   Relationship between the strength of the shear section of reinforced concrete members reinforced with shear reinforcement to the action of the bending moment and amount of the shear reinforcement

EXPERIMENTAL INVESTIGATIONS

Experimental investigations in hollow core reinforced concrete slabs were performed. The slabs were reinforced with steel strands and there was no shear or transverse reinforcement. The slabs varied in the depth and the shape of the cross-section. All slabs were 1200 mm in their nominal width. In the tests, the structures were loaded with concentrated loads in the support zone. The span length (*l*) between the supports of the slabs was different. The distance between the force application point and the nearer support was $a = 2.5h$ (Table 1). The diagram for the tests of the slabs is shown in Figure 10.

Figure 10   Diagram for test of reinforced concrete slabs

The slabs were tested with static load. During the test the behaviour of the strands in the concrete (slip), the occurrence of cracks and their development were examined and the load of failure was recorded. The tests of slabs revealed that the character of failure of the slabs significantly depends on the self-anchorage of the reinforcement. It was determined that the reinforcement of the slabs presented in the Table 1 slipped in relation to the concrete during failure. It permits to conclude that failure took place in the inclined crack due to the action of the bending moment. The character of failure of the experimental slabs is shown in Figure 11.

Using formulae (2) to (4), the theoretical strength of the experimental slabs was calculated. Comparison of the theoretical and experimental strengths of the slabs gave 4% to 17% difference in these values. The theoretical and experimental values of strengths of experimental slabs are presented in the Table 1.

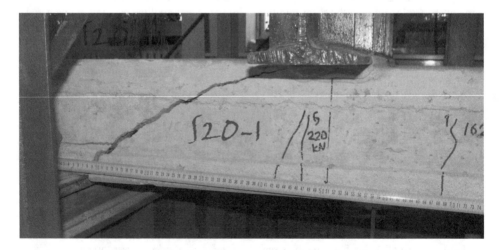

Figure 11   The character of failure of a slab in the shear section due to the action of the bending moment

Table 1   Characteristics and strengths of the experimental slabs

PROPERTY	SLABS				
	S20-1	S30-1	S30-2	S40-1	S27-1
Cross-sectional depth of the slab, h, mm	205	300	298	396	263
Design length of the slab, l, mm	4400	4440	3890	6175	4000
Distance to the force F, a mm	500	750	750	1000	662
Area of pre-stressed reinforcement, A_{p1}, mm^2	813	1470	1250	1470	774
Average compressive strength of the concrete, f_{cm}, n/mm^2	45.2	45.4	44.5	44.1	50
Experimental strength, $m_{e,obs}$, kNm	89.4	137	186	269	170
Theoretical strength, $m_{r,cal}$, kNm	75	132	155	281	142
$M_{R,cal}/M_{E,obs}$	0.84	0.96	0.84	1.05	0.83

CONCLUSIONS

Performance of prestressed concrete members is highly influenced by the anchorage of reinforcement. The anchorage length of the strands without special anchors is quite large. In the zone of the anchorage of the strands the reinforcement is not fully introduced into action with the concrete and because of this reason there is a risk of failure of inclined sections in the support zone due to the action of the bending moment.

The theoretical calculation carried out using formulae (1) to (5) revealed that the strength of the shear section of flexural members to the action of the bending moment increases proportionally with the amount of longitudinal prestressed reinforcement, or with the length of their support. However using a constant amount of prestressed reinforcement and an increasing value of prestressing, there is going to be an increase in the length of the stress transfer zone and consequently there will be a reduction in the strength of the shear section to the action of the bending moment.

The strength of the shear section of reinforced concrete members increases with the amount of shear reinforcement, but the theoretical analysis revealed that using the method of calculation described above, the strength of the shear section to the action of the bending moment does not increase proportionally. When the amount (intensity v_{sw}) does not exceed 0.2, the theoretical strength of the member increases, but when this amount of the shear reinforcement is exceeded, the theoretical strength begins to reduce.

Experimental investigations pointed out that the members with prestressed reinforcement can fail in shear section due to the action of the bending moment and therefore in design of such members it is necessary to verify the strength of the shear section for the action of the bending moment. Comparison of strengths of reinforced concrete slabs obtained from tests with those calculated by the formulae presented in this paper demonstrated sufficiently good agreement of the values of these strengths.

ACNOWLEDGEMENT

The work was sponsored by the Lithuanian State Science and Studies Foundation (G-181 and G-09/06).

REFERENCES

1. BECKER R J AND BUETTNER D R, Shear Tests of Extruded Hollow Core Slabs, PCI Journal, Vol. 30, No. 2, March–April 1985.

2. PAJARI M AND KOUKKARI H, Shear Resistance of PHC Slabs Supported on Beams, I: Tests. Journal of Structural Engineering, ASCE, Vol. 124, No. 9, Sept. 1998, pp. 1050–1061.

3. HAWKINS N M AND GHOSH S K, Shear Strength of Hollow-Core Slabs, PCI Journal. January–February 2006. pp. 110–114.

4. PAJARI M. Resistance of prestressed hollow core slabs against web shear failure, VTT Building and Transport, ISSN 1455.0865 (URL: www.vtt.fi/inf/pdf/), VTT 2005. 68 p.

5. HOOGENBOOM P C J, Analysis of hollow-core slab floors, HERON, Vol. 50, No. 3, 2005. pp. 173–185.

6. ANDERSEN N E AND LAURIDSEN D H, Hollow Core Concrete Slabs, Technical report X 52650, Part 2. April 1999, 29 p.

7. LUTZ C, Prestressed Hollow-Core Concrete Slabs – Problems and Possibilities in Fastening Techniques, Otto-Graf-Journal, Vol. 13, 2002, pp. 53–64.

8. ORETA A AND WINSTON C, Simulating size effect on shear strength of RC beams without stirrups using neural networks, Engineering Structures, Vol. 26, No. 5, 2004, pp. 543–691.

9. MANSOUR M, DICLELI Y M, LEE J Y AND ZHANG J, Predicting the shear strength of reinforced concrete beams using artificial neural networks, Engineering Structures, Vol. 26, No. 6, 2004, pp. 693–856.

10. CLADERA A AND MARÍ A R, Shear design procedure for reinforced cross and high-strength concrete beams using artificial neural networks, Part I: beams without stirrups, Engineering Structures, Vol. 26, No. 7, 2004, pp. 857–1012.

11. Eurocode 2: Design of concrete structures – Part 1: General rules and rules for buildings, Central Secretariat: rue de Stassart, 36 B-1050 Brussels, 2005, 230 p.

12. JOKŪBAITIS V AND KAMAITIS Z, Shear cracks modelling for assessing the stresses in concrete shear reinforcement, Journal of Civil Engineering and Management, Vol. VIII, Suppl. 1, 2002, pp. 24–28.

CONFINED AND HIGH STRENGTH SELF-COMPACTING CONCRETE WITH BLENDED CEMENTS

T Kibriya

National University of Sciences and Technology

Pakistan

ABSTRACT. This experimental study aimed at carrying out wide ranging testing for evaluating various properties of self-compacting concrete (SCC) with blended cements comprising of agricultural waste. The study has been carried out in two parts. Part 1 of this study comprises of blended cement SCC in confined conditions while Part 2 comprises of SCC with blended cement in the high strength range. Due to restricted length of paper, only essential data has been included. In part 1, blended cement SCC confined in steel tubes has been studied for use as columns for new construction as well as for rehabilitation/ replacement of short columns of old bridges, buildings as well as deep foundation applications. Blended cement containing 75% ordinary Portland cement and 25% rice husk ash was used for this study. Almost 350 to 450% increase in strength was observed by filling steel tubes with SCC (containing blended cement) in square and circular tubes respectively. Failure strains reduced by 40 to 45%. Smaller steel sections along with use of blended cement containing rice husk ash will reduce the costs of construction and repairs/renovation works along with reducing the disposal problems of this massively produced agricultural waste in the rice growing regions. In part 2, tests on specimen caste from high strength SCC with blended cements using slightly higher content of rice husk ash i.e. 50% OPC blended with 50% rice husk ash, natural aggregates and sand have been carried out for various mixes for compressive strengths of 60 to 100 N/mm^2 which have proved to be satisfactory with regards to strength whilst improved durability and acid resistance have been observed. Higher compressive strengths with higher elastic moduli and reduced permeability along with better sulphate and acid resistance have been observed.

Keywords: Blended cement, Rice husk ash, High strength, Self-compacting concrete, Steel tubes, Repair/rehabilitation works, Deep foundations.

Dr Tahir Kibriya completed his doctorate in civil engineering from The City University, London, UK in 1991 and has research interests in concrete and concrete materials. He worked as Chief Consulting Engineer/Principal Designer at Directorate of Design and Consultancy of Engineer in Chief in Pakistan for many years after which he joined National University of Sciences and Technology as Head of Civil Engineering Department/Chief Instructor where he taught subjects of Design of Concrete and Steel Structures to M.Sc. and B.E civil engineering courses along with conducting research. He has published a number of research papers on concrete and concrete materials.

INTRODUCTION

Massive quantities of rice husk are generated as agricultural waste from rice processing units worldwide. This waste material has no usage and creates disposal problems. Its high silica content makes it suitable for use with cement. Limited investigations on rice husk ash mixed with OPC have suggested improved strength/durability however performance of SCC with blended cements in confined state have not been studied yet nor SCC with blended cements in the high strength has been widely investigated to-date [1-6]. This experimental study aimed at evaluating the above mentioned properties of SCC made from blended cements using rice husk ash, Portland cement, natural aggregates. Rice husk ash used for blending was obtained by burning rice husk in an industrial furnace with temperatures maintained around 500 to 600°C. Ashes were then cooled to 20°C and subsequently ground in a laboratory ball mill for about 120 minutes and thereafter sieved through a #325 sieve. Ashes passing 100% through the sieve were then used for blending with ordinary Portland cement.

PART 1: SCC WITH BLENDED CEMENT IN CONFINED CONDITIONS

Consolidation of concrete is difficult in deep and smaller concrete sections with limited accessibility. Over-consolidated concrete segregates whilst under-consolidated concrete can have voids and cavities affecting the integrity and uniformity of the concrete cross-section thereby adversely affecting the strength. Self-consolidating concrete (SCC) consolidates under its own weight with no vibration and without exhibiting segregation or bleeding assuring the structural integrity and uniformity of the cross-sectional area of deep sections. Recent research on use of blended cement containing agricultural waste i.e. rice husk ash has proved to increase compressive strengths of normal/high strength concretes by about 30% along with reducing the permeability, shrinkage and improving durability of such concrete to acid, sulphate and chloride attack. Selection of 25% replacement of OPC with rice husk ash has been on the basis of excellent performance of such blended cements and concrete in the recent studies. The advantages of the concrete filled steel tubes as columns over other composite systems are that the steel tube provides reinforcement and formwork for concrete, prevents concrete spalling, environmental damage and from offensive agencies whilst concrete in the steel tube supports thin walls and prolongs/prevents local buckling of the steel tubing. The composite column adds significant stiffness to the structure as compared to more traditional steel frame.

TEST SAMPLING

Four groups comprising two groups of circular shapes and two of square shapes were tested with a total of 20 series with 60 specimens in all. 3 specimens in each group were hollow, 3 filled with normal concrete, 3 filled with SCC with blended cement, 3 braced and filled with normal concrete and 3 braced and filled with SCC containing blended cement. Braced tubes were hollow steel tubes cross braced internally with #3 deformed bars welded in transverse direction, alternately at a distance of 150 mm centre to centre and filled with concrete. The height of all specimens was 750 mm. A superplasticizing admixture along with a viscosity modifying agent was used in preparing SCC with blended cement.

The dimensions of the square columns selected aimed at maintaining similar cross sectional area as of corresponding circular columns. Other details are given in Table 1. Steel pipes used for this experimental investigation were made of 250 MPa steel.

Ordinary Portland cement was used with crushed granite coarse aggregates and medium grading sand for normal concrete. 75% ordinary Portland cement blended with 25% rice husk ash was used as cementitious material for SCC. Concrete strengths of 30 N/mm^2 were used as shown in Table 2. The test strengths of different columns were also compared with strengths given by various codes and also the design equation proposed by Georgios Giakoumelis and Dennis Lam.

Table 1   Properties of steel tubes used

GROUP	COLUMN TYPE	OUTER DIAMETER, mm	INNER DIAMETER, mm	THICKNESS mm	D/T OR B/T	L/D OR L/B	λ
A	Circular	160	155	2.5	64	4.7	19
B	Circular	111	106	2.5	44	6.7	27
C	Square	125	120	2.5	50	5.9	21
D	Square	87	82	2.5	35	8.6	30

λ – slenderness ratio, D – diameter, t – thickness, b – width, L – length

EXPERIMENTAL TESTING

The columns were tested in a 2000 kN capacity Universal compression testing machine. The specimens were centred in the testing machine in order to avoid eccentricity. Loading rate was maintained at 4.5 kN/s. vertical displacement was measured by displacement transducers. The top and bottom faces of specimen were grinded and made smooth and levelled to remove surface imperfections and maintain uniformity of loading on the surface.

Table 2   Properties of concrete used

No.	SAMPLE	SLUMP, mm	f_c' 28 DAYS N/mm^2	f_c' 90 DAYS N/mm^2
1	Normal Concrete	60	31	37
2	SCC with bended cement	560	37	46

DISCUSSION OF TEST RESULTS

The test results are listed at Table 5. Most results observed for normal concrete filled steel tubes are compatible with the recent research carried out [7-10].

Table 3   Proportions of concrete used

TYPE	W/C RATIO	CEMENT kg/m^3	WATER l/m^3	SAND kg/m^3	GRAVEL kg/m^3	VMA %	SP l/m^3
Normal concrete	0.45	360	162	710	1040		
SCC with blended cement	0.45	380	171	860	860	0.02	3

Columns - Hollow steel

All columns behaved in almost similar way with yielding strain observed to be between 0.004 and 0.006 and stress in steel around 250 MPa at failure. Deformation behaviour was also similar. Square columns started yielding at lower loads as compared to circular ones.

Table 4   Limiting values of b/t

TYPE	LRFD	EUROCODE	ACI CODE
Square	40	50.6	49.16
Circular	40	85	80

Table 5   Results of experimental testing and comparison with codes

No.	COLUMN TYPE	ACTUAL CAPACITY kN	Pu LRFD kN	Pu ACI kN	Pu EUROCODE kN	GEORGIOS EQUATION kN
1	A circular hollow	290	341	341	309	309
2	A-do- filled NC	1296	698	784	909	986
3	A-do- filled & braced NC	1025	698	784	909	986
4	A-do- filled SCC	1640	698	784	909	986
5	A-do- filled & braced SCC	1370	698	784	909	986
6	B circular hollow	237	235	235	213	213
7	B-do- filled NC	591	380	443	454	531
8	B-do- filled & braced NC	474	380	443	454	531
9	B-do- filled SCC	722	380	443	454	531
10	B-do- filled & braced SCC	647	380	443	454	531
11	C Square hollow	183	339	339	307	307
12	C-do- filled NC	670	551	681	788	830
13	C-do- filled & braced NC	595	551	681	788	830
14	C-do- filled SCC	840	551	681	788	830
15	C-do- filled & braced SCC	765	551	681	788	830
16	D Square hollow	241	234	234	212	212
17	D-do- filled NC	505	293	393	428	455
18	D-do- filled & braced NC	497	293	393	428	455
19	D-do- filled SCC	594	293	393	428	455
20	D-do- filled & braced SCC	565	293	393	428	455

NC – Normal concrete, SCC – Self compacting concrete with blended cement.

Columns - Concrete filled only

About 450% increased strengths were observed for 160 mm diameter circular columns filled with SCC with blended cement. Similar columns filled with normal concrete showed increased strengths by about 300% as compared to the load carrying capacity of 160 mm diameter hollow steel columns. Failure strains were observed to be in the range of 0.006 to 0.007. The stress strain curve indicates ductile behaviour. About 200% increased strengths were observed for 112 mm diameter circular columns filled with SCC with blended cement whilst similar tubes filled with normal concrete were observed to have increased strengths by about 150%. Failure strains in both cases were observed to be in the range of 0.009 to 0.01.

The strains indicate an extremely ductile behaviour. 360% higher strengths were observed for 125 mm square columns filled with SCC with blended cement whilst similar tubes filled with normal concrete were observed to have increased strengths by about 260%. Failure strains were observed to be in the range of 0.007 to 0.008.

146% higher strengths were observed for 87 mm square columns filled with SCC with blended cement whilst similar tubes filled with normal concrete were observed to have increased strengths by about 100%. Failure strains were observed to be in the range of 0.01.

Figure 1   Specimen at failure

Columns - Concrete filled & braced

About 360% increased strengths were observed for 160 mm diameter circular braced columns filled with SCC with blended cement as compared to 250% increased strength for similar braced columns filled with normal concrete. Failure strains in both cases were observed to be in the range of 0.006 to 0.007 with ductile behaviour.

About 170% increased strengths were observed for 111 mm diameter circular columns braced filled with SCC with blended cement whilst similar tubes filled with normal concrete were observed to have increased strengths by about 95%. Failure strains in both cases were observed to be in the range of 0.009 to 0.01 with extremely ductile behaviour.

310% higher strengths were observed for 125 mm square columns filled with SCC with blended cement whilst similar tubes filled with normal concrete were observed to have increased strengths by about 225%. Failure strains were observed to be in the range of 0.007 to 0.008. 134% higher strengths were observed for 87 mm square columns filled with SCC with blended cement whilst similar tubes filled with normal concrete were observed to have increased strengths by about 100% as compared to the load carrying capacity of hollow steel columns. Failure strains were observed to be in the range of 0.01.

COMPARING STRENGTHS BY DESIGN CODES

The values calculated by method given in various codes are listed at Table 6. It can be observed that all codes give conservative values whereas capacities calculated by using proposed equation by Georgios Giakoumelis and Dennis Lam [4] are more realistic for normal concrete filled circular columns.

For square tubes, this equation gives reasonable values for smaller sections but give excessively large values for larger sections. The proposed equation is: $N_u = 1.3 A_c f_c' + A_s f_s$. SCC with blended cement filled in tubes gives 60 to 65% higher strengths.

CONCRETE AREA AND CONFINEMENT - EFFECTS

For concrete filled circular sections, the confinement effect of concrete increases the concrete strength. Due to concrete confinement the stress bearing capacity of concrete increased 3.5 to 4.5 times in circular columns. It implies that when diameter is kept constant and steel thickness is increased, the confinement factor increases thereby increasing the compressive strength of concrete. It has also been observed that for circular sections, increase in concrete area increased the strength of the column.

SHAPE/MODE OF ULTIMATE FAILURE

Almost all columns failed due to local buckling of steel followed by crushing of concrete. The failure was a ductile one. The failure mode of almost all columns was due to steel plate buckling along with typical crushing failure of concrete where the steel wall was pushed out by the concrete core, which was confined by the steel as shown in Figure 1. When the steel was removed from the specimen after failure, the concrete was found to have taken the shape of the deformed steel tube, which illustrates the composite action of the section.

CONCLUSION

SCC with blended cement increased the load carrying capacity of hollow steel columns by 360 to 450% for square and circular columns respectively. The failure strains observed for SCC with blended cement filled columns were similar to the failure strains observed for tubes with normal concrete thereby indicating the reduced strains for SCC with blended cement filled specimen due to higher loads. Almost 100 to 150% increased load carrying capacity was observed for SCC with blended cement filled steel tubes as compared to similar steel tubes filled with normal concrete. Circular sections with larger concrete area showed higher increase in strength and reduced strains as compared to smaller section with lesser concrete area and square sections. Such higher strengths of SCC with blended cement filled steel tubes imply the use of smaller sections in repair/renovation works. Use of efficient smaller sections along with cheaper rice husk ash blended in cement for self compacting concrete reduces the cost resulting into cheaper and economical cost of repairs and rehabilitation works.

PART 2: SCC WITH BLENDED CEMENT IN HIGH STRENGTH RANGE

Wide ranging investigations covering most aspects of mechanical behaviour and permeability were carried out for various mixes for compressive strengths of 60, 80 and 100 N/mm². Compressive strengths of high strength SCC specimen with blended cements were observed to be higher by about 4 to 10% than the control specimen, for concrete with 50% Portland cement blended with 50% rice husk ash. Higher elastic moduli and reduced permeabilities were observed along with better sulphate and acid resistance. Better strengths and improved durability of such high strength SCC makes it a more acceptable material for major construction projects.

CONCRETE MIX DESIGN

Three high strength concrete mixes for characteristic strengths of 60, 80 and 100 N/mm^2 were designed using ordinary Portland cement blended with 50% rice husk ash, crushed natural calcareous limestone aggregates (maximum 20 mm diameter) and medium grade sand. Control mix contained 100% Portland cement [11]. Table 6 gives the details of mixes.

TEST SAMPLING

Three specimens each from three different batches were used in all tests. Specimen used for different tests were as follows:

Compressive strength/density	150 mm cubes, 150 mm diameter, 300 mm long cylinders
Flexural strength	150 mm × 150 mm × 750 mm beams
Stress/strain behaviour	150 mm diameter, 300 mm long cylinders
Static modulus of elasticity	150 mm diameter, 300 mm long cylinders
Dynamic modulus of elasticity	150 mm × 150 mm × 750 mm beams
Ultrasonic pulse velocity	150 mm cubes
Initial surface absorption	150 mm cubes
Sulphate and Chloride resistance	150 mm cubes (immersed in 5% H_2SO_4 and 5% HCl solutions for 90 days and measuring weight loss)

All specimens were cured in water at 20°C for 42 days before testing.

Table 6   Design of high strength SCC mixes

STRENGTH N/mm^2	W/C RATIO	CEMENT kg	SAND kg	WATER kg	AGGREGATE kg	SP l/m^3	VMA %
60	0.36	465	515	168	1302	4	0.04
80	0.32	565	490	180	1214	7	0.044
100	0.28	678	450	190	1124	11	0.05

SP – Superplasticizer, AGG – Aggregate

DISCUSSION OF TEST RESULTS

The properties of the high performance concretes produced are summarised in Tables 7 and 8.

Compressive strength

Compressive strength tests on cubes at 7, 28 and 42 days showed that the rate of development of strength of SCC with blended cement was similar to that for control specimen. The compressive strengths of high strength SCC with blended cement containing rice husk ash and Portland cement was somewhat higher than the control specimen. It was observed that compressive strengths kept increasing, as it can be seen from 42 day strengths, as due to low w/c ratios, water is required from external sources for hydration of cement which keeps progressing with time. High strength SCC with blended cement containing rice husk ash and Portland cement was observed to develop 80 to 85% of its 28 day characteristic strength in 7 days. The complete section of high strength SCC specimen including the aggregate and the paste, tends to reach failure simultaneously, typical of high strength concretes.

Flexural strength

From the values given in Table 7, it can be seen that the flexural strength of high strength SCC with blended cement are higher by 8 to 10% as compared to control specimen.

Table 7   Properties of high strength SCC

W/C RATIO	MIXES	STRENGTH, N/mm^2				
		cube, 7 days	cube, 28 days	cube, 42 days	cylinder	flexural
0.36	Control	54	63	66	53.5	6.4
	Blended SCC	56	70	74	58	7.3
0.32	Control	73	82	85	66	7.9
	Blended SCC	74	88	94	74	9
0.28	Control	92	103	106	86	9.8
	Blended SCC	91	108	114	90	11.4

Stress/strain behaviour

The general form of the stress/strain characteristics of high strength SCC with blended cement were similar to that for control specimen as shown in Figure 2. All the curves were observed to be virtually linear up to the point of failure, except for the initial small portion, typical of high strength concretes. Higher moduli of elasticity were observed for high strength SCC with blended cements.

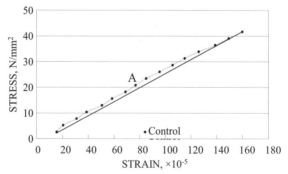

Figure 2   Idealised stress – strain curves
Control – 100% Portland cement, A – High strength SCC with blended cement

Static modulus of elasticity

Static modulus of elasticity was observed to be around 38,000 to 40,000 N/mm^2 for high strength SCC with blended cement containing rice husk ash and Portland cement as compared to 36,800 to 38,000 N/mm^2 for high strength concrete containing Portland cement only i.e. about 5% higher.

Dynamic modulus of elasticity

The average dynamic modulus of elasticity for concrete with blended cement containing 50% rice husk ash + 50% Portland cement was observed to be higher by about 4% than the control.

Ultrasonic pulse velocity

Average pulse velocity across high strength SCC with blended cement was observed to be 5.3 km/s as compared to an average velocity of 4.82 km/s for control mixes, i.e. 10 to 12% higher than the control mixes which is due to better quality, higher density and reduced voids in the high strength SCC.

Density of hardened concrete

The average saturated and oven-dried densities for high strength SCC with blended cement containing rice husk ash + Portland cement were 2578 and 2449 kg/m^3 respectively, as compared to control mixes which were 2480 and 2461 kg/m^3, respectively. Hence the saturated and dry densities of high strength SCC with blended cement containing rice husk ash and Portland cement are about 4% higher than the control mixes due to better hydration and packing of finer materials in high strength SCC. In the presence of higher content of cementitious material and the low w/c ratios, most of the unhydrated cementitious material acts as filler to densify the concrete, whilst the hydration process continues over longer duration.

Initial surface absorption test (ISAT)

Results of ISAT are given in Table 8. Initial surface absorption for high strength SCC with blended cement containing 50% rice husk ash + 50% Portland cement was observed to be lower as compared to the control. The values are compared with the guidelines given by the Concrete Society Technical Report No. 31.

Table 8   Properties of high strength SCC

W/C RATIO	MIXES	ISAT, ml/m^2·s	MODULUS, N/mm^2		PULSE VELOCITY, km/s
			elastic	dynamic	
0.36	Control	0.23	36,872	51,378.2	4.79
	Blended SCC	0.16	38,397	56,615.8	5.21
0.32	Control	0.20	37,198	54,563.1	4.82
	Blended SCC	0.11	39,461	59,194.7	5.32
0.28	Control	0.17	38,218	56,684.7	4.84
	Blended SCC	0.10	40,623	57,954.8	5.40

Sulphate and chloride resistance

For HCL solution, the weight loss for control was 8% as compared to 4% for high strength SCC with blended cement containing 50% replacement of cement with rice husk ash. Similarly for H_2SO_4 solution, the weight loss for control was 6% as compared to 2% for 50% replacement of cement with rice husk ash. Therefore, the performance of high strength SCC with blended cements was two to three times better in acidic environment and three to four times better in sulphate environment as compared to concrete with ordinary Portland cement control mixes. It is mainly due to the negligible amounts of $Ca(OH)_2$ present in the products of hydration of blended cements, lower permeabilities and stable compounds formed due to secondary chemical actions by the silica content of the rice husk ashes.

Shrinkage

Shrinkage was observed to be similar for almost all specimens. No appreciable difference in shrinkage of specimen cast from high strength SCC with blended cements and control mixes were observed for 90 days.

CONCLUSIONS

High strength SCC specimen with blended cements developed satisfactory compressive strengths in high strength range of 60 to 80 N/mm^2, about 10% higher flexural strength, 3% higher static moduli of elasticity with values up to 40,000 N/mm^2, similarly higher values for dynamic moduli, about 10% higher pulse velocities, 4% higher density, very low permeabilities, similar shrinkage and two to three times improved sulphate and acid resistance as compared to control specimen. Better strengths and improved durability of such high strength SCC is likely to make it a more acceptable material for major construction projects. It will also help in consuming large volumes of agricultural wastes like rice husk ash thereby reducing its disposal problems along with resulting into cheaper cements with stronger and durable characteristics.

REFERENCES

1. GOODIER C I, Development of Self Compacting Concrete, Proceedings of the Institution of Civil Engineers Structures & Buildings 156, November 2003, pp. 405–414.

2. LOO Y C, NIMITYONGSKUL P AND KARASUDHI P, Economical Rice Husk Ash Concrete, CIB Journal, August, 1984.

3. KIBRIYA T AND BAIG N, High performance concrete with blended cements using agrowastes, Proceedings of McMat 2005, Joint ASCE/ASME/SES Conference on Mechanics and Materials, June 1 – 3, 2005, Baton Rouge, Louisiana, USA, 2005.

4. KIBRIYA T AND KHAN S, High performance concrete pavements using agrowaste blended cement, Proceedings of McMat 2005, Joint ASCE/ASME/SES Conference on Mechanics and Materials, June 1 – 3, 2005, Baton Rouge, Louisiana, USA, 2005.

5. KIBRIYA T AND KHAN S, Performance of ecologically friendly waste in high strength concrete, IV Regional Conference on Civil Engineering Technology, Joint ASCE/ESIE Conference, 7 – 9 June, 2005, Cairo, Egypt, 2005.

6. KIBRIYA T AND BAIG N, Agricultural wastes in construction, Blended cements containing rice straw ash, IV Regional Conference on Civil Engineering Technology, Joint ASCE/ESIE Conference, 7 – 9 June, 2005, Cairo, Egypt, 2005.

7. O'SHEA M D, Design of Circular Thin-Walled Concrete Filled Steel Tubes. Journal of Structural Engineering, ASCE, November 2000.

8. KNOWLS R B AND PARK R, Strength of concrete filled steel tubular columns, ASCE, J. Struct. Div., Vol. 95, No. 12, 1969, pp. 2565–2587.

9. KIBRIYA T, Performance of concrete filled steel tubes under uni-axial compression, IV Regional Conference on Civil Engineering Technology, Joint ASCE/ESIE Conference, Cairo, Egypt, 2005.

10. GEORGIOS G AND LAM D, Axial capacity of circular concrete-filled tube columns, Journal of Constructional Steel Research, Vol. 60, 2004, pp. 1049–1068.

11. TEYCHENNE D C, FRANKLIN R E AND ERNTROY H C, 1988, Design of normal concrete mixes, (Replacement to Road Note 4), Department of Environment - Transport and Road Research Laboratory, London.

SMART DYNAMIC CONCRETE: AN INNOVATIVE APPROACH FOR THE CONSTRUCTION INDUSTRY

R Khurana   R Magarotto
S Moro   J Roncero

BASF Construction Chemicals

Italy

ABSTRACT. This article introduces the Smart Dynamic Construction concept, which allows the ready-mix industry to reach increased energy efficiency, higher concrete durability and better construction process economy in order to reduce CO_2 emissions, also saving time and money. This concept is suitable for upgrading the consistence classes of S4 and S5 concretes to a higher performance level, with self-compacting characteristics and the same easiness of production as standard concrete. In fact, it creates a new generation of concrete which combines the benefits of traditionally vibrated concrete with those of self-compacting concrete and is easy to produce and robust in everyday use.

Keywords: Self-compacting concrete, Smart dynamic concrete, Superplasticizer, Compressive strength, Sustainability, Viscosity modifying agent.

Rabinder Khurana is the Technology Director for Business Unit Admixtures System Europe of BASF Construction Chemicals. He received his postgraduate diploma in reinforced concrete structures from Milan Polytechnic, Milan, Italy. He has more than 30 years of experience in concrete technology and admixtures applications.

Roberta Magarotto, graduated in Industrial Chemistry, Venice University, Italy, 1993. She has been working in the field of concrete admixtures, focusing on the development of new polymeric superplasticizers. Since 1997 responsible of Admixture Department of R&D at BASF Construction Chemicals.

Sandro Moro, graduated in Sciences and Technologies of Materials, Venice University, Italy, 2003. Since 2001 he is Head of Technological Laboratory at BASF Construction Chemicals, he has been working in the fields of concrete admixtures development and of concrete technology development.

Joana Roncero is head of the R&D Center in Barcelona of the Business Unit Admixtures Systems Europe of BASF Construction Chemicals. She is author or co-author of more than 40 papers in symposiums or journals. Her research interests include chemical admixture development and utilization, concrete technology, self-compacting concrete and high performance concrete.

INTRODUCTION

Smart Dynamic Construction is an innovative approach addressed to the construction industry that leads to an effective cost reduction and improves the durability of the structures. With this technology, concretes of consistence class S4 and S5 are transformed into Smart Dynamic Concrete (SDC) that has self-compacting properties. The total amount of the "fines" (material passing the 0.125 mm sieve) is less than 380 kg/m^3. The composition of the SDC is not very different from the normal vibrated concrete that is used daily in construction sites around the world.

The use of RheoMATRIX, an innovative viscosity modifying admixture (VMA) contributes to impart self-compacting properties to the fresh concrete without the addition of extra fillers normally required for the powder type self-compacting concrete. This eliminates the materials and handling costs of the extra "fines".

Self-compacting concrete normally used in the construction industry is not an easy material to produce and place. Besides the initial cost of the mixture there are other costs such as quality control of the components that makes this material less cost effective. A small variation in the moisture content of the aggregates may produce big variations in the rheological properties of the fresh self-compacting concrete. All these factors limit the use of SCC to a very small portion of the total ready mixed concrete produced in the European Union (less than 1% of the 360 million cubic meters produced: ERMCO 2006 Statistics).

In the SDC, the new VMA is incorporated and this allows to produce mixes that are more stable and robust towards the daily variations of materials in the plant, thus reducing the time required for the quality control. Besides, the elimination or reduction of extra fines makes this concrete more cost effective compared to normal concrete, while maintaining the positives characteristics of self-compaction.

This technology has been evaluated in several European countries, and in this article three examples of its application are reported, the first in Italy, the second in Turkey and the third in Germany.

The first one regards the construction of a new building including the foundations, floor slabs, columns and walls. Before the job was started several tests were carried out in the laboratory to design and evaluate the best SDC mix. Then preliminary tests were conducted in the ready mix plant. The construction involved a production of 800 m^3 of concrete.

In the second case in Turkey, instead of preliminary tests in the laboratory, SDC mixes were directly developed and experimented in the ready-mixed concrete plant that had central mixing equipment. On the site the contractor normally placed concrete of S3 consistence class by means of a pump.

The last one regards the test in Germany. First a mix was developed in the laboratory and the fresh properties of SDC were evaluated and compared to a normal concrete having F5 consistency class and $R_{ck} = 30$ N/mm^2, that is normally produced by the ready mixed concrete supplier. After the laboratory tests, 24 m^3 of SDC were produced in the central mixing plant and transported to the site in 8 m^3 truck loads and poured into the foundation slab of a villa being constructed. The fresh concrete was tested for evaluating the self-compaction properties.

Mix Design of SDC

The most important characteristic of SDC is its mix design. It can be designed according to the specifications of the project engineer, because the amount of cement and water are similar to those of normal concrete mixes produced daily. The good self-compacting properties of SDC are due to the specific PCE superplasticiser, the new VMA and to the quantity and quality of the mortar in the resulting concrete. The SDC has a higher quantity of material passing 4mm sieve as compared to normal vibrated concrete. This allows the aggregates to be transported easily during the placing of concrete, while the risk of bleeding and segregation is eliminated, thanks to the innovative VMA.

During the tests, three different superplasticizer based PCE (PCE1, PCE 2 and PCE 3) available in the market have been used.

LABORATORY AND FIELD TESTS

Laboratory Tests in Italy

The binders used were cement type CEM II/A-LL 32.5 R (according to EN 197-1) and fly ash available in the plant, while the aggregates were composed of a natural sand 0/4, a coarse aggregate of 8/16 size and a gravel of 16/25, according to EN 12620.

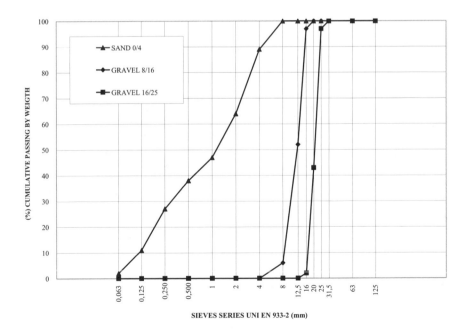

Figure 1   The grading of the aggregates, according to EN 933-1

In the SDC mix design it is important to evaluate the quality and quantity of mortar. In normal SCC the quantity of fines is more than 500 kg/m³. This amount is necessary for the self-compacting properties and stability. The stability of SDC is guaranteed by using a bigger quantity of mortar and by the new VMA.

The reference mix is a concrete having a S4 consistency class and XC2 exposition class (R_{ck} = 30 N/mm²) according to EN 206-1 standard.

The durability class determines the quantity of cement and the water to cement ratio.

The dosage of PCE 1 required to achieve the desired high fluidity inevitably leads to bleeding and segregation. The use of the innovative VMA is therefore necessary to balance the action of the superplasticizer and to obtain a homogeneous and stable concrete.

Table 1   The reference mix and the SDC mix

	C30 – S4	SDC
Sand 0/4 mm (kg/m³)	992	1090
Gravel 8/16 mm (kg/m3)	396	820
Gravel 16/25 mm (kg/m³)	596	
CEM II/A-LL 32,5 R (kg/m³)	280	300
Fly Ash (kg/m³)	30	30
Water (kg/m³)	165	180
Superplasticiser (PCE 1) (kg/m³)	2,6	3,2
Viscosity Modifying Agent (VMA) (kg/m³)		0,4

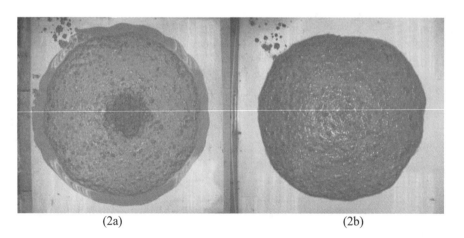

(2a)   (2b)

Figure 2   Slump flow before (a) and after (b) the VMA

The Slump Flow and V-Funnel time have been measured accordingly respectively to prEN 12350 – 8 and prEN 12350-9): the values measured were 68 cm and 11 seconds.

150 mm cubes were prepared for evaluating the compressive strength at 1, 7 and 28 days; Table 2 shows the results of the tests. The laboratory results indicate that the mixture composition may be tested in the ready mix plant.

Table 2   Compressive strengths

COMPRESSIVE STRENGTHS (N/mm^2)		
1 day	7 days	28 days
10,5	26,4	34,1

Field Tests in Italy

10 m^3 batches of SDC were produced by dry batching and mixing directly in the truck mixer. PCE 1 and VMA were introduced with the mix water, so that the admixtures were well distributed within the bulk of the mix.The properties of the fresh mix were measured. The slump flow was 68 cm and the V-Funnel time was 9 seconds. SDC was then transported for one hour at about 30°C to the job site. Here, again the slump flow was measured and the value was the same as before. For every 100 m^3, four 150 mm cubes were prepared to evaluate the compressive strength at 7 and 28 days; the average of the results is reported in Table 3:

Table 3   Compressive strengths

COMPRESSIVE STRENGTHS (N/mm^2)	
7 days	28 days
27,8	35,2

There is a very little difference between the fresh and hardened properties of SDC measured in the laboratory and in the field. This indicates that a correct study of the mix design reduces the risk of unexpected results in the plant. The concrete was placed by means of a pump and without any vibration. 800m^3 of SDC have been mixed, transported and pumped for the construction of the foundations, columns, slabs and walls that constituted the four floors building.

Figure 3   The foundation slab

Laboratory Tests in Turkey

In these tests only CEM II/B-S 42.5 R has been used without the aid of extra fines like fly ash or limestone filler. The aggregates were natural sand 0/2, crushed sand 0/5, coarse aggregate 5/12 and 12/22.

The following graph shows the grading curves of aggregates.

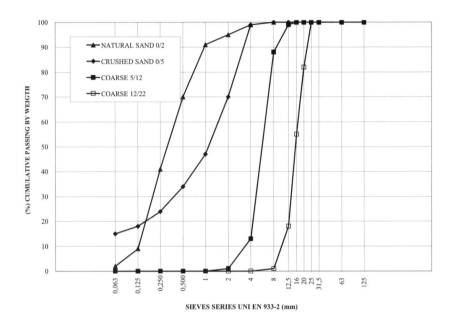

Figure 4   The grading curves of aggregates

The reference mix was a concrete with Rck = 25 N/mm^2 and S3 consistency class. Table 4 shows the comparison between the normal and SDC concrete mixtures:

In the reference mix a BNS (betanaphtalene sulfonate) – based admixture was used, while in the SDC a high range water reduction admixture based on polycarboxylate PCE 2 and the new viscosity modified agent (VMA) have been used.

The quantity of cement and water are the same, so the physical-mechanics properties of the concrete should be the same.

Field Tests in Turkey

8 m^3 of SDC were produced by mixing 2 m^3 batches in a pre-mixer and then loaded in a concrete truck mixer. The initial Slump Flow measured at the plant was 71 cm. The concrete was then transported for more than an hour and a half at 38°C to the job site where again the Slump Flow was measured. It was 60 cm.

Table 4   Comparison between the SDC composition and the reference mix

	C25 / S3	SDC
Natural Sand (kg/m^3)	266	371
Crushed Sand (kg/m^3)	578	658
Coarse 5/12 (kg/m^3)	497	371
Coarse 12/22 (kg/m^3)	478	430
CEM II/B-S 42,5 (kg/m^3)	330	330
Water (kg/m^3)	185	185
BNS (kg/m^3)	4,3	
Superplasticizer (PCE 2) (kg/m^3)		4,0
Viscosity Modifying Agent (VMA) (kg/m^3)		0,7
Slump Flow (cm)		65
Compressive Strength at 7 days (N/mm^2)	19,9	20,5
Compressive Strength at 28 days (N/mm^2)	28,9	29,7

Although the concrete did not have a high fluidity, it was still possible to place by pumping and without vibration. The quantity of reinforcement and the distance between the bars in the structure are classified as Rank 2 according to the "Recommendations for SCC" of the Japanese Society of Civil Engineers. Even then the concrete flowed easily and smoothly between the reinforcements.

Some pictures show the job site and the foundation:

Figure 5   Reinforcement of foundation   Figure 6   Job site

Laboratory Tests in Germany

The binders used were cement type CEM III/A 42.5 N (according to EN 197-1) and fly ash available in the plant, while the aggregates were composed of a crushed sand 0/2, a coarse aggregate 2/8 and 8/16. The grading curve of the aggregates, in according to EN 933-1, is reported in Figure 7.

The best mixture proportion has been studied and tested in the laboratory and the composition adopted is reported in Table 5 compared to a F5 concrete with the same compressive strengths.

Table 5   The mix composition

	C30 – F5	SDC
Sand 0/2 mm (kg/m³)	753	774
Coarse 2/8 mm (kg/m³)	368	172
Coarse 8/16 mm (kg/m³)	630	774
CEM III/A 42,5 N (kg/m³)	280	300
Fly Ash (kg/m³)	120	80
Water (kg/m³)	175	185
Superplasticiser (PCE 3) (kg/m³)	3,5	4,5
Viscosity Modifying Agent (VMA) (kg/m³)		0,6

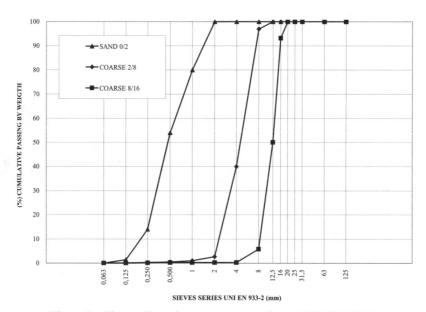

Figure 7   The grading of aggregates according to UNI EN 933-1

The Slump Flow and t_{500} time was measured according to prEN 12350 − 8: the values obtained were 64 cm and 2 seconds. The workability retention was evaluated by measuring the Slump Flow and t_{500} time after 1 hour; the values obtained were 64 cm and 3 seconds respectively.

The VMA doesn't influence the setting and hardening time. Only the superplasticiser can influence these parameters.

Table 6   The fresh and hardened properties of concretes

	C30 − F5	SDC
Initial Slump Flow (cm)		64
Initial t_{500} (s)		2
Slump Flow after 1h (cm)		64
t_{500} after 1h (s)		3
Compressive Strength at 7 days (N/mm^2)	27,5	29,2
Compressive Strength at 28 days (N/mm^2)	34,8	35,9

Field Tests in Germany

24 m^3 of SDC have been produced in the ready mix plant, filling three concrete truck mixers with 8m^3 each. In the plant the fresh properties were evaluated. The Slump Flow was 69 cm and t_{500} time was 2 seconds. The concrete was transported for 30 minutes and was poured by a chute into the reinforced foundation. The distance among the bars placed the reinforcement in Rank 3 of the "Recommendation for Self Compacting Concrete" of JSCE. Figure 8 below shows the placing of the concrete.

Figure 8   Placing of Smart Dynamic Concrete

CONCLUSIONS

All the field tests conducted had a positive result as it was possible to produce a concrete that has the compressive strengths required, but with the characteristics of self-compaction.

The use of the Smart Dynamic Concrete allows the reduction of the construction time, compared to the normal concrete.

According to ERMCO 2006 Statistic, 85% of ready mix concrete is produced to satisfy the R25 - R30 Strength Class. Therefore, SDC may be used in all structures that has these mechanical properties. Generally for such structures the amount of reinforcement is not very dense (< 150 kg/m^3).

Although SDC has a quantity of fines below 0.125 mm lower than traditional SCC, its self-compactability is possible thanks to the particular mix design, the use of a specific superplasticiser and the new viscosity modifying admixture.

The reduction of fines makes the concrete less cohesive than a traditional SCC, so that the placing and finishing becomes easier and faster.

Another important issue of SDC is the surface appearance of the hardened concrete. It is normally better than the conventional vibrated concrete and similar to the one of the powder type self-compacting concrete.

REFERENCES

1. RECOMMENDATION FOR SELF-COMPACTING CONCRETE – 1999. Japan Society of Civil Engineers, Tokyo.

2. EUROPEAN READY MIXED CONCRETE INDUSTRY STATISTICS - 2006. ERMCO - European Ready Mixed Concrete Association, Brussels, Belgium.

3. KHURANA R, MAGAROTTO R AND MORO S. Smart dynamic concrete: an innovative step towards a rationalized construction process for the ready mix concrete industry", Concrete Plant International (to be printed).

4. CORRADI M, KHURANA R AND MAGAROTTO R. Low fines content self-compacting concrete. 5th International RILEM Symposium on Self-Compacting Concrete. Vol. 2, 2007. pp. 839-844.

5. RONCERO J, CORRADI M AND KHURANA R. New admixture-system for low-fines self-compacting. 5th International RILEM Symposium on Self-Compacting Concrete. Vol. 2, 2007. pp. 875-880.

CRACKING BEHAVIOUR AT BENDING OF REINFORCED
HIGH-STRENGTH / HIGH PERFORMANCE CONCRETE BEAMS

C Magureanu

C-L Letia

Technical University of Cluj Napoca

Romania

ABSTRACT. This paper summarizes the research findings of the characteristics of high strength concrete for flexural cracks of reinforced concrete girders. Eleven HSC beams with different percentages of ρ (reinforcement ratio) were cast and incrementally loaded under bending. During the test, the strain on the concrete in compression, the tension zone, the tension bars, the crack widths and also the deflection at different points of the span length were measured up to failure. Based on the obtained results, the serviceability and crack width of the HSC members were reviewed in detail. A comparison between theoretical and experimental results is also reported here.

Keywords: High strength concrete, Crack width, Ultimate moments.

Professor C Magureanu works at the Technical University of Cluj Napoca, Romania. She specializes in high strength concrete, reinforced and prestressed concrete and also in ferrocement.

Engineer C-L Letia is a PhD student at the Technical University of Cluj Napoca, Romania.

INTRODUCTION

The use of high strength concrete (HSC) has become a common practice in the last ten years with strengths ranging up to 60 N/mm². High strength concrete not only implies increased strength, but also increased durability which corresponds to a reduced maintenance need of high strength concrete structures with longer lifetime.

Cracking behaviour is very important with regard to quality or durability requirement. HSC will improve the behaviour that means will diminish crack spacing and crack width under imposed loads.

EXPERIMENTAL PROGRAM

The experimental program was based on bending tests of eleven beams (three beams FT and two for each I beam). All beams had a width of 125 mm, height of 250 mm and span length $L_0 = 3000$ mm. The concrete strength at the day of testing was about 90 N/mm² (C80/90). The physical and mechanical characteristics of reinforcement are listed in Table 1.

Table 1   Mechanical characteristics of reinforcement

Type of reinforcement	Longitudinal	Transversal
Nominal diameter, mm	12, 14, 16	6
Yield strength, f_{ym}, MPa	320	210

All the beams are provided with high bond steel longitudinal reinforcement. The properties of specimens are detailed in Table 2 (FT and I are symbols of the beams). The loading schemes are presented in Figure 1.

Table 2   Parameters for bending tests

Test specimens	FT	I			
	5	1	2	3	4
$\rho_l = A_{sl}/b_w \cdot d$, %	2.06	2.59	3.03	3.40	3.83
$\rho_w = A_{sw}/b_w \cdot s$, %			0.152		

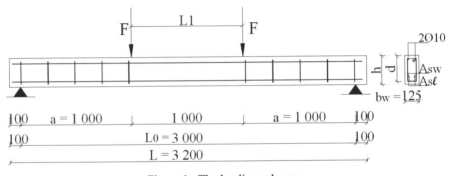

Figure 1   The loading schemes

All the beams were tested using a hydraulic testing system and loaded with two equal concentrated loads, F. The distance between the two concentrated loads was kept equal to $L_1 = 1000$ mm. Both ends of the beam were free to rotate and translate under load. At each load increment, the mid-span deflection and all strain reading were recorded and the developing crack patterns marked at the beam surface. The maximum compressive strain of concrete was recorded at the mid-span by strain gauges glued to one side of the beam in the tensile zone. Beams were submitted to monotonously increased loading until failure. The loading was applied at 10% of the calculated failure force.

TEST RESULTS AND DISCUSSION

The beams were all designed to fail in flexure. All beams exhibited vertical flexural cracks in the constant moment region before final failure of the beams due to crashing of concrete. Figure 2 shows the crack propagation under load.

Figure 2   Crack propagation under load

The experimental ultimate moment $M_{u(exp)}$ is the moment when ultimate load is reached during testing.

Cracking Moment

The cracking moment M_{cr} is usually estimated using the modulus of rupture [4] as:

$$M_{cr} = \frac{f_r \cdot I_g}{y_t} \tag{1}$$

where: f_r – the modulus of rupture
y_t – the distance from the neutral axis of the extreme tensile fibre of the beam
I_g – is the moment of inertia of the gross concrete section

The experimental cracked moment of inertia ($I_{cr,exp}$) based on the elastic deformation theory [5] is obtained by considering:

$$I_{cr.exp} = \frac{F_y \cdot a\left(3L_0^2 - 4a^2\right)}{48 \cdot E_c \cdot \Delta exp} \tag{2}$$

where: F_y – the load that causes yielding in the steel reinforcement
    a – the shear arm
    L_0 – the clear span of the beam
    Δ_{exp} – the measured deflection midspan

The traditional theoretical definition of I_{cr} based on the cracked transformed section [4] can be given for beams with single reinforcement:

$$I_{cr} = \frac{bc^2}{3} + n \cdot A_s(d-c) \tag{3}$$

where: $n = E_s/E_c$
    c – neutral axis depth
    b – beam width
    d – effective depth
    A_s – area of the tensile reinforcement

The result of the theoretical and experimental moment of inertia of cracked section are presented in Table 3.

Table 3   Theoretical and experimental cracked moment of inertia

Beam	$I_{cr} \times 10^6$, mm^4	$I_{cr.exp} \times 10^6$, mm^4	$I_{cr.exp} / I_{cr}$
FT5	54.21	32.30	0.59
I1	53.45	36.16	0.67
I2	69.37	54.35	0.78
I3	63.34	46.06	0.72
I4	70.52	69.61	0.98

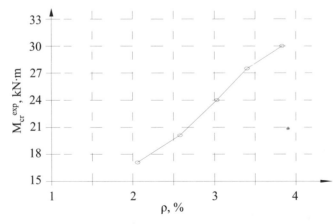

Figure 3 M_{cr}^{exp} vs ρ

By increasing the percentage of reinforcement ρ, the value of I_{cr} is increased. Table 4 presents the experimental and theoretical ultimate cracking M_{cr} in accordance with ACI 318 [2], Eurocode 2 [1] and the Romanian standard STAS 10107/0-90 [6] and ultimate moments (M_u) of the test specimens.

Table 4   Experiment and theoretical values of M_{cr} and M_u

| Beam | M_u, kN·m | | M_{cr}, kN·m | | | |
	exp.	theor.	exp.	theor. ACI 318	theor. EC2	Romanian standard
FT5	56.75	30.13	7.00	20.60	6.00	3.55
I1	50.50	36.60	10.00	24.52	6.11	3.71
I2	67.25	42.76	10.00	29.86	5.99	4.21
I3	75.00	47.04	12.00	31.87	5.99	4.73
I4	78.50	51.97	16.00	30.78	6.18	5.55

The cracking of a section occurs when the tension in the fibre reaches the value of the modulus of rupture. According to the Manual CEB 1990 and Eurocode 2 [1], this is the moment at which member cracks are completely independent of the reinforcement ratio, because most of the reinforced concrete members are cracked from bending in their working life (cf. the effect of the tensile stress induced by early shrinkage on the concrete).

The results obtained show that the experimental moments are higher than the theoretical values according to the Manual CEB 1990, Eurocode 2 [1] and Romanians standard [6].

By increasing the percentage of ρ, the values of M_{cr}^{exp} are increased.

The experimental cracking moments for crack width w = 0.1 mm and the ultimate moments are presented in Table 5. Therefore, the bending moment for a crack width of w = 0.1 is approximately 40% of the failure moment.

Table 5   Experimental values of M_{cr} at w = 0.1 mm and M_u

Beam	ρ, %	$M_{cr(w=0.1mm)}^{exp}$ kN·m	M_u^{exp} kN·m	$\dfrac{M_{cr(w=0.1mm)}^{exp}}{M_u^{exp}}$
FT5	2.06	17.00	56.75	0.44
I1	2.59	20.00	50.50	0.40
I2	3.03	25.00	67.25	0.37
I3	3.40	27.50	75.00	0.37
I4	3.83	30.00	78.50	0.38

CONCLUSIONS

By increasing the amount of ρ, the cracking moments are increased.

The value of $I_{cr(exp)}$ is lower then the values of $I_{cr(th)}$.

The experimental moments (M_{cr} and M_u) are higher than the theoretical values (Eurocode 2, ACI 318 and Romanian standard STAS 10107/0-90). The closest calculated values to the experimental values are those obtained in accordance with Eurocode 2.

The reinforcement is most effective in crack width reduction and with increasing reinforcement amount the crack width decreases.

REFERENCES

1. EUROCODE 2: Design of concrete structures – Part 1: General rules and rules for buildings, 2004.

2. ACI COMMITTEE 363, Review of ACI Code for Possible Revisions for High Strength Concrete (ACI 362 R- 92), American Concrete Institute, Detroit, 1992.

3. ACI COMMITTEE 318, Building Code Requirements for Structural Concrete, ACI 318, 2005.

4. MAGHSOUDI A A AND AKBAEZADEH BENGAR H, Effect ρ' on Ductility of HSC Members Under Bending, ACI-SP 228-26, 2005.

5. BALIUC R, Particularities of the Structural Behaviour of Reinforced High Strength Concrete Slabs, A thesis submitted in fulfilment of the requirements for the degree of Doctor of Philosophy, November, 2004.

6. STAS 10107/0-90 – Constructii Civile si Industriale. Calculul si alcatuirea elementelor structurale din beton, beton armat si beton precomprimat.

7. MAGUREANU C, Reinforced Steel - Concrete Bond of High Strength Concrete, Proc. of the 1st first "fib" Congress, 2002, Vol. 7, pp. 241-246.

8. MAGUREANU C AND HEGHES B, Experimental study on ductility reinforced concrete beams using high- strength concrete, Proc. of the 2nd International "fib" Congress, Naples, 2006.

THE BEHAVIOUR OF CONCRETE IN TENSION SURROUNDING TENSILE REINFORCEMENT

A Beeby

University of Leeds

R Scott

University of Durham

United Kingdom

ABSTRACT. The paper summarises a major research project on the behaviour of the concrete immediately surrounding the reinforcing bars in tension members. This region continues to carry some stress even after cracking, resulting in significant stiffening of the member (commonly referred to as tension stiffening). For convenience, both experimental and analytical work has been carried out members axially reinforced by a single bar. The experimental work has provided detailed information on the behaviour of the tension members and, in particular, the distribution of strains in the reinforcement. The results suggest that, between the formation of any two cracks, the elements behave elastically. Equations are derived for the prediction of both crack widths and deformations. Further analytical work is being carried out using Finite Elements and the results are in general agreement with the experimental findings. This work suggests an alternative model of tension zone behaviour to that commonly accepted at present.

Keywords: Reinforced concrete, Tension zones, Short term behaviour, Long term behaviour, Design.

Professor A W Beeby, Emeritus Professor, School of Civil Engineering, The University of Leeds. Professor Beeby has 40 years experience of research into the behaviour of reinforced concrete structures, particularly under service load conditions. He is a member of UK and European code committees and is a Fellow of the Royal Academy of Engineering.

Dr R H Scott, Reader in the School of Engineering at Durham University. A particular interests over the past 20 years has been the instrumentation of structures to measure strain and bond stress distributions. He is a chartered civil and structural engineer and a member of the American Concrete Institute.

INTRODUCTION

The interaction between reinforcement in tension and the concrete immediately surrounding it is fundamental to the behaviour of reinforced concrete. If concrete bonds poorly with the reinforcement then the bar will simply slip and there will be no transfer of stress between concrete and steel. If the concrete were to bond uniformly and firmly to the reinforcement then the concrete would disintegrate as soon as the ultimate tensile strain capacity of the concrete was exceeded and, again, reinforced concrete would not work. Fortunately, reinforced concrete does work! Steel and concrete thus interact in some way that will permit transfer of stress across the interface but in a manner which will result in the tensile strain being accommodated in discrete cracks. The most commonly assumed mechanism is that limited slip occurs between the reinforcing bar and the concrete accompanied by a shear or bond stress, the magnitude of which is related to the slip.

At the Dundee conference in 2002 we presented a paper setting out some preliminary results from a major research project on long term tension stiffening [1]. This was carried out mainly at the Universities of Durham and Leeds and was funded by two EPSRC grants and industrial support. Tension stiffening is a term used for the stiffening effect on the tension zone of the concrete immediately surrounding the reinforcement. This project is now complete. The quantity and quality of data obtained, particularly from the tests at Durham, have permitted a study of the steel-concrete interaction which has extended well beyond the original objectives of the project and has resulted in a major clarification of the nature of the interaction. Much of the work has taken place after completion of the original grants and all has now been published in a series of papers dealing with various aspects of the phenomena [2-9]. The objective of this paper is to bring together the results and to set out a possible new model of behaviour under service loading. Many of the ideas are not totally original but seem to have been largely ignored in the development of current theories of service load behaviour. The major practical aspects of behaviour which are dependant on the interaction of steel and concrete are cracking and deformation. A model is proposed which deals with both these phenomena.

THE RESEARCH PROGRAMME

Since the objective of the research was to study behaviour of the concrete in tension immediately surrounding the reinforcement, it was logical to use specimens which permitted this to be investigated in the most unambiguous way possible. For this reason, axially reinforced prismatic specimens subjected to pure tension were chosen. The specimens used were 120×120 mm in cross section and 1200 mm long. These were reinforced with single bars of either 12, 16 or 20 mm diameter. Three concrete strengths were used: nominally 30, 70 and 120 N/mm^2.

Two basic types of test were carried out. At Leeds long term tests were carried out on sets of 3 nominally identical specimens in simple rigs where the load was applied by hydraulic jacks which had to be adjusted at intervals to maintain a constant load. 9 rigs were built so that 3 sets of prisms could be tested simultaneously. The object of these tests was to obtain a large quantity of data on the load-deformation-time response. Instrumentation consisted of a load cell and demountable mechanical strain gauges (DEMEC gauges) of 200 mm gauge length applied on consecutive gauge lengths over the central 1000 mm on all four faces of the prisms to enable the average longitudinal strain to be measured. At Durham, the objective was to obtain much more detailed data for a smaller number of specimens. Two rigs were built which applied load weights through a system of levers. Pairs of specimens were tested; one loaded in stages with the load being held constant for up to 40 days at each stage and the other loaded at one time to the maximum load which was then held constant for the full period of the test. In these tests, the reinforcement was instrumented with internal ERS gauges at 15 mm centres using a method developed by Scott and Gill at Durham [10]. DEMEC gauges were also used as a check and to ensure consistency with the Leeds tests. The overall time under load varied from 48 to 104 days.

In addition to the prism tests, a series of 8 long-term simply supported slab tests were carried out at Leeds to attempt to establish the applicability of the tension test results to flexure.

The test programme described above gave data on both the short term and long term behaviour of tension zones. Detailed additional short term behaviour was made available from the series of 15 tests reported in [10]. This gave data for a much wider range of variables than were covered by the current tests. Fuller details of the test programme can be obtained from [2, or 3].

SHORT TERM BEHAVIOUR

Figure 1 shows the average strain in the reinforcement as a function of axial load for a typical axially reinforced specimen. The Figure shows how each crack causes an instantaneous increase in the reinforcement strain. After formation of a crack there is a linear increase in strain with increase in load until the next crack forms. It will be seen from Figure 1 that the linear sections, extrapolated back pass close to the origin, showing that behaviour is effectively elastic. At a higher load, there are two cases where sudden increases in deformation occur which are not associated with the formation of new surface cracks. It is suggested that these may be related to the formation or extension of internal cracks. It can be shown that the formation of each crack results in an approximately constant step increase in deformation.

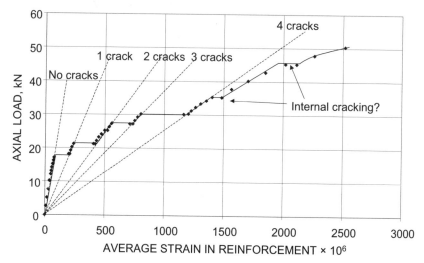

Figure 1   Average reinforcement strain as a function of load for a 100 × 100 mm specimen reinforced with a 12 mm bar (from [10])

Clearly, the spacing of cracks reduces with increase in loading and it has been found that in practically all circumstances for all types of test a linear relationship exists between the average crack spacing and 1/(average strain). A typical example is shown in Figure 2 which has been drawn from results from three nominally identical specimens tested by Jaccoud and Farra [11]. A final point will be illustrated from the short term results; this is the variation of stress along a bar in the region of a crack. This is illustrated in Figure 3 where the variation of stress between two cracks is shown for two levels of load for a 100 x 100 mm specimen reinforced with a 20 mm bar (from [10]). The figure shows that the variation in stress is close to linear between the crack and a point roughly mid-way between the cracks, suggesting a uniform bond stress along the bar. Also, the rate of change of steel stress is greater for the higher level of load, indicating an increase in bond stress. Often, this increase in bond stress is more or less proportional to the load, though in this case the increase is less than proportional, indicating an effective reduction in stiffness of the system.

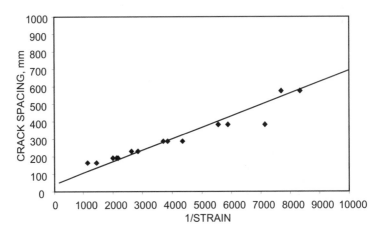

Figure 2   Crack spacing as a function of (1/strain) for 100×100 mm specimens reinforced with 20 mm bar (data from [11])

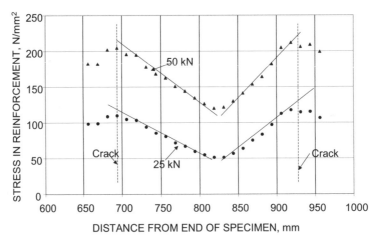

Figure 3   Variation of steel stress in region of a crack

LONG TERM BEHAVIOUR

Figure 4 shows the behaviour of two nominally identical specimens, one loaded in three stages with the load held for 15 days at each stage and one loaded up to the maximum load in a single operation and then held for 48 days. The specimens were 120 × 120 mm and were reinforced with a 20 mm bar. Several points may be observed. Firstly, the final total deformation is nearly the same for the two specimens and therefore seems largely independent of the load history. Secondly, the long term increment in deformation is relatively small compared with the short term increment when the load is applied. Thirdly, though this is rather difficult to see clearly, the total long term increment for the specimen loaded in one operation is substantially smaller than that for the specimen loaded in stages while the short term increment is larger.

Figure 5 shows the change in deformation during 15 days under a load of 57 kN plotted against log time. The total long – term increment in strain is 136×10^{-6}. Of this, 90×10^{-6} occur in two sudden events, the later of which was clearly associated with the formation of a new surface crack while the earlier, and smaller, increment is assumed to be related to some form of internal failure. It was found from all the long term tests that the long term increase in strain was substantially completed within a period of about 20 days from loading. A general summary of the long term tests carried out at Durham is given in Table 1. It will be seen that the points made in relation to Figures 4 and 5 are general to all the tests.

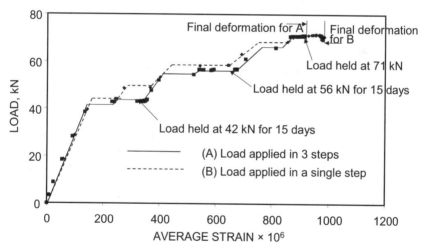

Figure 4   Load-strain history of 120 × 120 mm specimens under long term loading

DEVELOPMENT OF SIMPLE PREDICTION FORMULAE

The use of the results to develop a fuller, and more practically useful, understanding will be carried out in two stages: in the first stage, the results will be used to develop formulae for the prediction of crack widths and axial strains. In the second stage, a discussion will be presented of the mechanisms involved culminating in the proposal of a complete model for the prediction of tension zone behaviour.

Figure 6 shows conditions in the region of a crack. Within a distance of $\pm S_o$ of a crack, the stress in the reinforcement increases from a strain ε_o at S_o from the crack to ε_{s2} at the crack while the strain in the concrete decreases from ε_o to zero over the same distance. The maximum value of ε_o corresponds to conditions just before the formation of a crack. At distances greater than S_o from a crack, the strain is unaffected by the crack. The strain has been assumed to vary linearly both in the concrete and the reinforcement over the length S_o. As discussed in the previous section, this appears close to the experimental results. Inspection of Figure 6 will show that the total extension of the specimen is given by the extension of the specimen before formation of the first crack plus the shaded area in Figure 6(b) multiplied by the number of cracks.

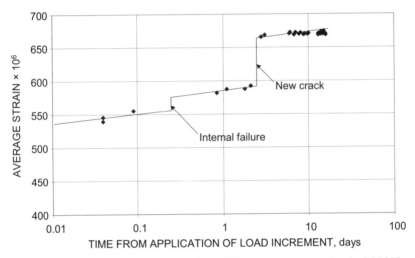

Figure 5   Load increment as a function of time under constant load of 56 kN

Table 1   Summary of results from long term tests carried out at Durham

BAR SIZE mm	CUBE STRENGTH N/mm^2	LOADING	STRAIN $\times 10^6$					
			Total	Short term	Long term	Sudden long term	Gradual long term	Estimated creep
16	24	stages	1499	1245	254	94	160	62
		single	1550	1507	43	10	34	28
20	34	stages	1540	1172	196	43	153	44
		single	1590	1492	47	16	31	30
16	69	stages	1735	1236	368	107	201	28
		single	1607	1442	98	22	76	8
20	89	stages	1022	826	295	80	215	34
		single	1097	1050	89	15	74	10
16	124	stages	1106	811	499	298	201	40
		single	1109	1020	165	53	112	18
20	118	stages	912	618	294	181	113	30
		single	976	901	75	0	75	18

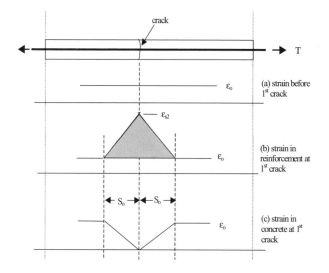

Figure 6   Strain conditions in the region of a crack where the crack spacing exceeds $2S_o$

Thus:

$$\varepsilon_{sm} = \varepsilon_{s2}(\alpha_e\rho + S_on/L)/(1 + \alpha_e\rho) \text{ or } \varepsilon_{sm} = \varepsilon_{s2}(\alpha_e\rho + S_o/S)/(1 + \alpha_e\rho) \tag{1}$$

where:

$L =$	length of the specimen
$n =$	number of cracks
$S =$	crack spacing under the loading considered $= L/n$
$S_o =$	length within which the strain is influenced by a crack
$\alpha_e =$	modular ratio $= E_s/E_c$
$\varepsilon_o =$	strain in the specimen away from a crack $= T/E_cA_c$
$\varepsilon_{sm} =$	average strain over the length of the specimen
$\varepsilon_{s2} =$	strain in the reinforcement at a crack $= T/E_sA_s$

Equation 1 strictly only applies when the crack spacing is greater than $2S_o$ however, an extensive experimental study shows that it fits the data well over all practical values of crack spacing.

The crack width can be derived similarly on the assumption that the width is equal to the integral of the shaded area in Figure 6(b) plus the shortening of the concrete over the distance $2S_o$, which can be obtained from Figure 6(c). This leads to the relationship:

$$w_{max} = S_o\varepsilon_{s2} \tag{2}$$

where w_{max} is the maximum crack width.

A fuller derivation of these equations is set out in [5].

Equation 1 requires knowledge of the crack spacing under the loading considered. The relationship demonstrated in Figure 2 provides a means of achieving this and a detailed study of some 130 specimens for which results are published in [11] led to values of coefficients defining the relationship between crack spacing and 1/strain. Unexpectedly, these coefficients turned out to be significantly influenced by the cement type. Studies of data from various sources show that S_o can be taken as proportional to the cover for the type of specimen considered.

From a practical point of view, the treatment of the long term deformation of tension zones is straightforward. It is clear that a stable long term deformation is reached in a few days from loading. The significant practical problem is the possibility of the deflection occurring after application of partitions and causing damage. Since the members are likely to have been loaded well before this, any increment in deflection due to loss of tension stiffening will have already occurred before finishes and partitions are in a state to be damaged. Thus the short term tension stiffening should be ignored in design. This is discussed in more detail in [3]. As will be discussed further below, the long term tension stiffening is simply calculated on the assumption that the tensile strength of the concrete is about 70% of the short term value.

MECHANISMS OF TENSION STIFFENING AND CRACKING

There are several aspects of behaviour which have been studied in more detail in order to develop a clearer model of tension stiffening behaviour, both in the short and long term. Three of these will be introduced below: the relationship between crack spacing and loading; the accommodation of high strains near the bar-concrete interface and the mechanisms of long term loss of tension stiffening.

The Relationship between Crack Spacing and Loading

Figure 2 shows the linear relationship which exists between crack spacing and 1/strain. No explanation has been offered so-far for this behaviour and so, though it was first noted in the 1960s, it has remained an empirical finding. This issue has now been explored further by means of Monte-Carlo studies of cracking behaviour.

Cracks do not form all at once; the first crack forms at the weakest section along the member; the second crack forms at the next weakest section, provided it is not too close to the first section where the stress has been reduced by the first crack. This process continues with each crack forming at a slightly higher load than the previous crack. If a frequency distribution of tensile strengths is assumed and each section along the member is ascribed a tensile strength from the chosen frequency distribution, then the development of the crack pattern can be followed. This process is then repeated many times to obtain a reliable picture of the result. Three slightly different models were investigated. These were:

Model 1. In this model, a distance S_o was defined and no crack was permitted to form within $\pm S_o$ of an existing crack.

Model 2. In this model, the tensile strengths within $\pm S_o$ of a crack were increased by a factor S_o/x where x is the distance from the crack. This is equivalent to assuming that the stress induced by the load increases linearly from zero at the crack to a value unaffected by the crack at S_o from the crack. The whole length of the member is searched to find the weakest section to establish the location of the next crack. Cracks can, therefore, occur within S_o of a previous crack if there happens to be a weak section within this region.

Model 3. In this model, the strengths are increased within $\pm S_o$ as in Model 2 but S_o is increased as a function of load to model a constant bond strength.

Only Model 2 gave results which corresponded to actual behaviour. Typical results from Model 2 are illustrated in Figure 7.

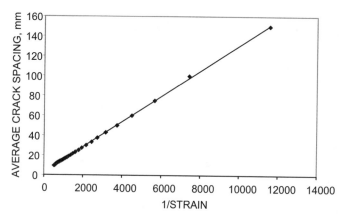

Figure 7   Relationship between crack spacing and 1/strain predicted by Model 2

Model 2 has two significant consequences: firstly it is implicit in the model that the bond stress increases with increase in load so that S_0 remains constant and secondly, that the concept of a stabilised crack spacing is not valid. Fuller details of this work are given in [2].

Accommodation of Interface Strains

The data from the strain gauges along the bars show that the strain in the reinforcement exceeds the strain capacity of the concrete over almost the whole of the length S_0 from the load which first forms the cracks. Thus, there must be some form of internal failure in the region of the interface at all loads above the cracking load. Two mechanisms have been proposed to accommodate this strain incompatibility: slip between the bar and the surrounding concrete or the development of internal cracking. Slip has been the mechanism most commonly invoked in the development of crack prediction formulae but it has considerable difficulty in adequately modelling behaviour (see, for example, [8]). Internal cracking was first proposed by Goto [12] but has also been demonstrated by Otsuka and Ozaka [13]. Analysis of internal cracking as a possible mechanism has not as yet been fully carried out though the possibilities are discussed in some detail in [5]. Recently, work has been carried out by Beeby and Forth [9] using Finite Element analysis to study the internal cracking mechanism. Analyses were carried out using axially symmetric elements to model a cylindrical specimen with a central bar subjected to tension. The structure modelled is sketched in Figure 8. The internal cracks have been assumed to decrease in height linearly with distance from the crack and slope towards the free end of the specimen at an angle of 60°. This pattern of internal cracks follows from arguments set out in [9] and there is some experimental justification in the crack patterns given in [13]. The right-hand end of the specimen is taken as an axis of symmetry and corresponds to mid-spacing between cracks. The crack width is given as twice the deformation of the point A relative to point B. Analyses have been carried out for a range of covers with the internal cracks scaled proportionally to the cover.

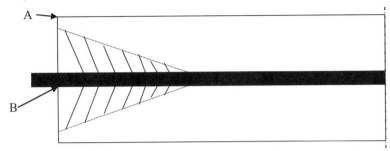

Figure 8   Axi-symmetric internal cracking model

Space does not permit a full presentation of the results of this study but Figure 9 shows the calculated variation of surface crack width as a function of cover. This shows a close to linear variation of crack width with cover. The results also show roughly linear distributions of steel strain which are similar to those obtained experimentally and a shape of the crack between the bar surface and the concrete surface which are qualitatively similar to those measured by various authors (see [8] for a discussion of crack profile). Exact comparison of the calculated crack widths with experimental values is difficult since few tests have been carried out on circular section specimens however, Figure 9 shows a crack width of 0.055 mm for a steel stress of 100 N/mm^2 and a cover of 40 mm while the maximum crack widths obtained for square section specimens with the same cover by Farra and Jaccoud [11] varied from 0.05 to 0.10 mm. Further analytical work is in progress to study the variation in crack width between bars in tension specimens with multiple bars.

Mechanisms of Long Term Decay of Tension Stiffening

The experimental results given in Table 1 show that, in the case of specimens loaded in several stages, about half of the long term deformation was related to sudden events, mostly the formation of new surface cracks. Far fewer cracks occurred during long term loading where the load was applied in a single operation. Interestingly, the total number of cracks in both sets of specimens were generally the same; the difference being that more cracks formed during the loading when the load was applied in a single operation. The same observation is true for the gradual changes in deformation. Some calculations of the likely effects of creep have been carried out. These were done by assessing the average stress in the concrete during each period under load. This strongly suggests that creep does not play a major part in the long term deformations of tension zones. The total creep was then calculated using the creep formulae in Eurocode 2 [14] but assuming that the creep coefficients in tension were twice those in compression. The calculated values are tabulated in Table 1 and it will be seen that the possible expected contribution of creep is small compared with the contribution from other causes. This is not unreasonable when the low tensile average stress present is considered.

One reasonably well established fact about tensile behaviour of concrete is that the tensile strength reduces when the concrete is subjected to permanent tensile stress. Data published by the CEB, in Bulletin d'Information 235 [15], suggest that the strength reduces to approximately 70% of the short term value in a period of about 24 hours. Other data suggests a somewhat longer period but the period remains short compared with the period required for creep to develop. It is suggested that the long term reduction in tension stiffening is mainly simply the result of this reduction in tensile strength with time. This leads to the formation of new cracks and the gradual extension of existing cracks, particularly internal cracks.

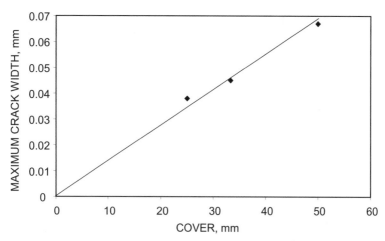

Figure 9   Calculated crack width at a steel stress of 100 N/mm² as a function of cover

SUMMARY AND CONCLUSIONS

It is shown that the deformation and cracking behaviour of axially reinforced tension members may be modelled on the assumptions:

(a) Concrete is elastic in tension up to its tensile strength at which stress it fails in a brittle manner

(b) Under service loads, there is no slip at the bar-concrete interface at the transverse ribs

(c) Internal cracks form from the ribs in the manner indicated in Figure 9.

(d) Long term effects can be modelled by assuming a tensile strength of 70% of the short term value.

It may be that assumption (c) could be dispensed with and that assumptions (a) and (b) would load to cracking of the form assumed in (c) but we have not yet tested this.

REFERENCES

1. BEEBY A W, ANTONOPOULOS A AND SCOTT R H, Preliminary Results for Long-Term Tension Stiffening Effects in R.C. Members, Int. Congress on Challenges of Concrete Construction (Concrete Floors and Slabs), Dundee, September 2002, pp. 57-67.

2. BEEBY A W AND SCOTT R H, Insights into the Cracking and Tension Stiffening Behaviour of Reinforced Concrete Tension Members Revealed by Computer Modelling, Magazine of Concrete Research, Vol. 56, No. 3, April 2004, pp. 179-190.

3. SCOTT R H AND BEEBY A W, Long Term Tension Stiffening Effects in Concrete, Proceedings of the American Concrete Institute (Structural Journal), Vol. 102, No. 1, January-February 2005, pp. 31-39.

4. BEEBY A W, SCOTT R H AND JONES A E K, Revised Code provisions for Long Term Deflection Calculations, Proc. ICE (Structures and Buildings), Vol. 158, Issue SB1, February 2005, pp. 71-75.

5. BEEBY A W AND SCOTT R H, Cracking and Deformation of Axially Reinforced Members Subjected to Pure Tension, Magazine of Concrete Research, Vol. 57, No. 10, December 2005, pp. 611-621.

6. SCOTT R H AND BEEBY A W, New Code Provisions for Long Term Deflection Calculations, Special Publication SP225 of the American Concrete Institute, 2005, pp. 119-132.

7. BEEBY A W AND SCOTT R H, Mechanisms of long term decay of tension stiffening, Magazine of Concrete Research, Vol. 58, No. 5, June 2006, pp. 255-266.

8. BEEBY A W, The influence of the parameter ϕ/ρ_{eff} on crack widths, Structural Concrete, Vol. 5, No. 2, 2004, pp. 71-83.

9. BEEBY A W AND FORTH J P, Finite element study of the interaction of steel in tension and the surrounding concrete, (in preparation).

10. SCOTT R H AND GILL P A T, Short term distributions of strain and bond stress along tension reinforcement, The Structural Engineer, Vol. 65B, No. 2, June 1987, pp. 39-48.

11. FARRA B AND JACCOUD J P, Influence du beton et de l'armature sur la fissuration des structures en beton – Rapport des essais de tirants sous deformation imposee de court duree, Departement du Genie Civil, Ecole Polytechnique Federale de Lausanne, Publication No. 140, November 1993.

12. GOTO Y, Cracks formed in concrete around deformed tension bars, Journal of the American Concrete Institute, 1972, Vol. 68, No. 4. pp. 244-251.

13. OTSUKA K AND OZAKA Y, Group effects on anchorage strength of deformed bars embedded in massive concrete block. Proceedings of International Conference on Bond in Concrete – Form Research to Practice, Riga Technical University, Riga, October 1992, Vol. 1, pp. 138-147.

14. BRITISH STANDARDS INSTITUTION, BS EN1992, Eurocode 2: Design of concrete structures, Part 1: general rules and rules for buildings, BSI 2004.

15. COMITE EURO-INTERNATIONAL DU BETON, Serviceability Models – behaviour and modelling in serviceability limit states including repeated and sustained loads – progress Report, CEB Bulletin d' Information 235, Lausanne, 1997.

PARAMETRIC STUDIES ON VARIABILITY OF FRACTURE PROPERTIES OF CONCRETE

G Appa Rao

Indian Institute of Technology Madras

P Sudhakar Rao

Government of Andhra Pradesh

India

ABSTRACT. Experimental studies on the behaviour and fracture properties of concrete have been reported. In this paper influence of various parameters such as SF, w/b ratio and compressive strength of concrete on variability of fracture properties has been reported. Fracture energy and crack tip opening displacement as functions of other influencing parameters have been studied. In concrete mixes with SF, the Abram's law did not comply with the generalized law. The highest compressive strength values are obtained with w/b ratio 0.4 at 10% SF. However, the highest tensile strength was observed with 15% SF. Fracture energy of concrete decreases as the w/b ratio increases from 0.40 to 0.60 in all the concrete mixes. The fracture energy of SF concrete has been observed to be higher than that of the plain concrete. Material characteristic length of concrete is higher for low compressive strength concrete. Higher values of characteristic length were observed with w/b ratio 0.4. The characetristic length decreases with increase in the w/b ratio in concretes containing silica fume.

Keywords: Fracture energy, Strength, Silica fume, W/B ratio, Ductility.

Dr G Appa Rao is a faculty of Civil Engineering, Indian Institute of Technology, Madras, Chennai-600 036, India. He has been awarded with Alexander von Humboldt Fellowship. He worked as an Associate Professor at Sri Venkateswara University, India. He obtained his graduation and doctoral degrees from Indian Institute of Science, Bangalore, India. He works in structural materials, fracture mechanics of reinforced concrete structures, seismic behaviour of beam-column joints and prestressed concrete. He has more 15 years of teaching and research experience. He has more than 70 publications in his credit.

Mr P Sudhakar Rao is an Executive Engineer in the Irrigation Department, Govt. of Andhra Pradesh, India. He has more than 20 years of service as an Engineer. He obtained his undergraduate and post graduate degrees from Sri Venkateswara University, Tirupati, India

.

INTRODUCTION

Fracture mechanics is a method of characterizing the fracture criterion in terms of applied stress, flaw size and fracture toughness in the design of structures. Concrete is a composite consisting of cement, fine aggregate, coarse aggregate and water. Cement hydrates when it is mixed with water. During fluid state the matrix fills the gap between aggregate and bonds them together to form hardened concrete. Both the matrix and the aggregate are bonded together at the interface. Due to bleeding or segregation, cracks form at the interface between the matrix and the aggregate. The interfacial transition zone between the matrix and aggregate often has more voids and is weaker compared to the bulk cement matrix. The properties of concrete are greatly influenced by the interfacial transition zone in concrete. Many internal flaws and cracks exist in concrete prior to loading. The mechanical behavior of concrete subjected to different loading conditions is governed by the initiation and propagation of these internal cracks and flaws during loading. For a quasi-brittle material like concrete, a substantial non linearity exists before the maximum stress. At some point before the peak stress, micro cracks begin to localize into a macro crack that critically propagates at the peak stress. When a concrete structure with a quasi-brittle crack subjected to loading results in an energy release rate G_q at the tip of effective quasi-brittle crack. In a quasi-brittle material, the total energy dissipated can be divided in to two forms; one is used to fracturing of material to create two surfaces, G_{1c} which is knwon as material surface energy and second form is to overcome cohesive pressure $\sigma(w)$ in separating the two surfaces, G_σ [3]. Though the value of G_{1c}, may be evaluated using LEFM, it is ignored as the other component is significantly higher.

REVIEW OF LITERATURE

Silica fume (SF) concrete produces higher compressive strength and dynamic modulus of elasticity under water curing. It is possible to develop HPC incorporating SF at higher water-binder ratio with proper curing [1]. The mechanism by which SF improves the performance of concrete are both physical and chemical. The physical effect arises from the filling up of pores. The chemical effect is due to the pozzolanic reaction. The open channels in SF concrete are filled by pozzolanic reaction products. This reduces the porosity and concrete becomes impermeable. Calcium hydroxide (CH) is one of the major phases formed in set portland cement. The formation of CH in hydrating Portland cement not only determines the percentage of reaction of hydration, but also influences its mechanical properties especially in the presence of SF [3]. When SF is added to cement in concrete, it acts both as an inert filler improving the physical structure and as a pozzolan, reacting chemically with CH. The pozzolanic reaction starts at the early ages in the presence of SF. The aggregate-cement paste transition layer is regarded as the most sensitive area with in the structure of concrete [4]. The structure of the transition zone and its properties are influenced by the type of constituents i.e. coarse aggregate, type of cement, admixtures and water-binder ratio. Porosity increases at the aggregate-cement paste interface caused by the higher water/binder ratios in this region. Since the interface between the aggregates and cement paste is the weakest link, the mechanical behavior of concrete is significantly affected by the properties of the interfacial zone. The effect of maximum size of aggregate on the fracture energy, G_F is important because the crack surface roughness induces aggregate interlock. Fracture energy, G_F is defined as the area under the load-deflection curve per unit fractured surface area. The fracture energy, G_F is one of the important material properties for the design of large concrete structures. Some studies on fracture behaviour of concrete have been reported by [5]. It has been reported that the characteristic length, l_{ch}, decreases as the concrete strength increases. The fracture surfaces are smooth and less tortuous in high-strength concrete (HSC) containing SF. Fracture energy decreases and the brittleness index increases significantly with the incorporation of large size aggregate. The type of fine aggregate has no influence on the fracture energy of concrete. The modulus of elasticity of concrete slightly increases as the maximum size of coarse aggregate increases. This is due to the addition of stiffer aggregate particles. In plain concrete, the characteristic length increases as the maximum size of coarse aggregate increases, which shows increased ductility, while concrete containing SF, the characteristic length decreases up to aggregate size of 12.5 mm, thereafter it increases.

The brittleness index is defined as the ratio of the elastic deformation energy, S_{II} to the irreversible deformation energy; S_I corresponding to the pre-peak point of the load-displacement at mid span curve [6]. The brittleness index is defined as $ß = S_{II}/S_I$, where $S_I =$ irreversible deformation energy due to damage and $S_{II} =$ elastic (reversible) deformation energy. When S_I becomes significantly large then the whole energy can be considered as irreversible due to deformation. RILEM [7] recommends a test method for the determination of G_F using simple three-point bend beam specimens. In the fictitious crack model proposed by Hillerborg [8], the fracture energy, G_F, tensile strength, f_t and the stress-CMOD relationship completely describe the fracture characteristics of concrete. It has been concluded that there is a tendency for G_F to increase when the maximum aggregate size increases form 8 to 20 mm.

Shah [1] introduced a material length Q, which is proportional to the size of FPZ in concrete. The value of Q is known as Brittleness Index

$$Brittlenes\ s\ Index\ =\ Q\ =\ \left[\frac{E\ CTOD_c}{K_{lc}^s}\right]^2$$

EXPERIMENTAL PROGRAMME

Materials

Ordinary Portland cement was used in this programme. The cement was tested for various properties. Normal consistency was 31%, initial and final setting times were 120 and 210 minutes respectively. Its specific gravity was 3.1 and its properties are given in Table 1. Silica fume (SF) has a silica content of 93.6%. Because of extreme fineness it is an efficient pozzolanic material, whose specific surface area is 20,000 m^2/kg. This property imparts high strength to concrete with enhanced qualities when mixed as a partial replacement of cement. The fine aggregate used in this programme was procured from a natural river bed. The fine aggregate fraction passing through 4.75 mm sieve and retained on 600 μm sieve was used for the preparation of test specimens. Its specific gravity was 2.78 and fineness modulus 2.82. Machine crushed granite aggregate of 20 mm size was used. The specific gravity was 2.80 and fineness modulus was 6.42. Potable water was used for the programme. The pH value was 7.76. The same water was used for mixing of concrete and curing of concrete cubes, cylinders and beams. HRWRA is a sulphonated naphthalene formaldehyde condensate having ability to disperse the finer particles and retarding properties. At a given w/b ratio, this dispersing action increases the fluidity of concrete, typically raising the slump. HRWRA of 0.01% of cement was thoroughly mixed with water before mixing in to concrete.

Table 1   Physical properties of Cement.

PROPERTY	RESULT
1. Normal Consistency	27.50
2. Seting Times	
(a) Initial (Minutes)	120
(b) Final (Minutes)	210
3. Specific Gravity	3.10
4. Compressive Strength, MPa	
(a) 3 Days	24.00
(b) 7 Days	34.50
(c) 28 Days	45.00

Table 2 Concrete mixes, w/b ratios and % SF.

SL.No	MIX	w/b RATIO	S.F %
1	M0400	0.40	0
2	M0410	0.40	10
3	M0415	0.40	15
4	M0500	0.50	0
5	M0510	0.50	10
6	M0515	0.50	15
7	M0600	0.60	0
8	M0610	0.60	10
9	M0615	0.60	15
10	M3510	0.35	10

Mix Proportions of Concrete

Mix proportions of cement concrete used in this programme are given in Table 2. The mix details are also given. The concrete mix contains 450 kg/m^3 cementitious material, 650 kg/m^3 fine aggregate and 1090 kg/m^3 coarse aggregate with 0.4, 0.5 and 0.6 w/b ratios with 0, 10 and 15% SF. In a mix the number following letter M indicates w/b ratio and another following number indicates % SF. For example in concrete mix M0410, the number 04 indicates 0.40 w/b ratio while 10 indicates 10% SF. Three w/b ratios 0.4, 0.5, and 0.6 along with three SF contents at 0, 10 and 15% as a cement replacement material by the weight of cement were adopted. All the constituents of the concrete were uniformly mixed and then the water mixed with superplasticizer was added to the concrete during dry mixing. The mixing was continued till a uniform and homogeneous mix was obtained

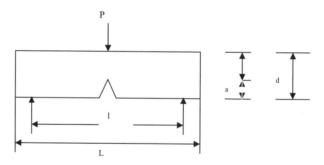

Figure 1   Typical three-point bend (TPB) specimen.

Where l = span of the beam (400 mm); P = load applied on the beam; L = length of the beam (500 mm), a = crack length, d = beam depth (100 mm) , b = thickness of the beam (100 mm).

Test Specimen Preparation

The test specimens for the determination of compressive strength of concrete were prepared using standard 100 mm size metallic cube moulds. The moulds were duly applied with oil on the inner surfaces. Concrete filled cubes were vibrated on power driven vibrating table for compacting the concrete. Metallic cylindrical moulds measuring 100 mm diameter and 200 mm height were used for the determination of split tensile strength. The test specimens for the determination of fracture parameters were prepared using standard metallic moulds of size (b x d x L) 100 x 100 x 500 mm. The specimen is shown in Figure 1. Edge notches were prepared by using 3 mm thick metal wedge plates over a depth of 50 mm from the edge. The metal wedge plates were properly fixed to the moulds and were set in their position before concreting. After 24 hrs, the specimens were demolded and cured for 28 days.

Testing of Specimens

The specimens were tested in a universal testing machine (UTM) to obtain compressive strength and split tensile strength of concrete. Compressive strength of concrete was determined at 3, 7, 28 and 90 days. Split tensile strength of concrete cylinders was obtained at 56 days after which the strength development was not significant. Testing of specimens for obtaining fracture parameters are done at 56 days. Dial gauges were used to measure the deflection. The three-point bend (TPB) beam specimens were tested under load control. It could be possible to obtain the post-peak response of the TPB under load control also with careful operation of the testing machine on small size TPB specimens. The load was increased at a constant rate up to the ultimate load. The point where the ultimate load is likely to attain in the specimen could be identified by the slow movement of the load indicator needle at which micro cracking has been initiated to widen. The micro cracking extension has been generally initiated at about 90 percent of the peak. This load reduction with the increased deflection could be interpreted as softening. Therefore, by a complete manipulation, it was possible to obtain softening by a less stiff UTM. Figure 1 shows a beam specimen. The whole process of loading the specimens could be completed in about 6 to 8 minutes. At every load increment of 200 N, readings of dial gauge for concrete beam deflection were noted.

RESULTS AND DISCUSSION

Variation of Compressive Strength

Abrams w/b ratio law has been described as the most important observation on the inverse proportionality between w/b ratio and strength of concrete. According to Abrams' generalization law, the compressive strength of plain concrete varies inversely with w/b ratio. The properties of cement-based materials are primarily affected by w/b ratio, chemical composition, microstructure and pore geometry of the cementitious materials, properties of aggregates, cement-aggregate ratio and properties of cement-aggregate interfacial zone.

Figures 2 and 3 show the compressive strength with SF and w/b ratio respectively. It has been observed that, at 3, 7 and 28 days the plain concrete without SF exhibited decrease in strength with water-binder ratio. This has been generally observed in Abram's law on compressive strength of concrete. But when tested at 90 days, the compressive strength attained higher value at w/b ratio 0.50 than 0.40 in plain concrete. At w/b ratio 0.40, the compressive strengths were observed to be higher with 10% SF at 3, 7, 28 and 90 days. Further it can be inferred from Table 3, that with 0 and 10% SF at w/b ratio 0.4, the 3 days compressive strength is the same. At 7, 28, and 90 days, it has been observed that the compressive strength is the highest at 10% SF. However, the compressive strength with 15% SF is lower than that of plain concrete. With w/b ratio 0.5, the compressive strength of plain concrete at 3 days is less than that of concrete with 10 and 15% SF.

With w/b ratio 0.6 also, the compressive strength with 10 and 15% SF is higher than that of plain concrete. However, the highest compressive strength values are obtained with w/b ratio 0.4 at 10 % SF.

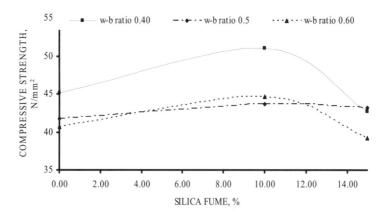

Figure 2   Compressive strength vs. Silica fume %

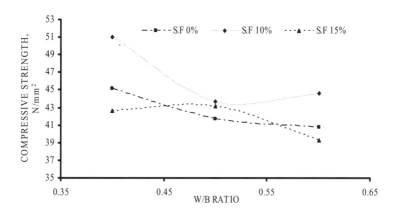

Figure 3   Compressive strength vs. w/b ratio

Table 3   Compressive strength and tensile strength of concretes

Mix	COMPRESSIVE STRENGTH (f_c) N/mm^2				f_t N/mm^2
	3 days	7 days	28 days	90 days	
M0400	34.33	36.29	45.13	52.00	2.81
M0410	33.35	43.16	51.01	56.00	3.59
M0415	20.60	31.39	37.00	38.33	2.34
M0500	25.5	35.32	41.69	43.50	2.03
M0510	27.46	36.29	43.65	46.33	2.81
M0515	28.45	30.41	43.16	42.67	2.81
M0600	22.56	32.37	40.71	42.70	2.65
M0610	34.33	35.32	44.64	47.75	2.50
M0615	21.58	20.60	39.24	41.60	2.18

Variation of Tensile Strength

The experimental values of tensile strength, f_t, are shown in Table 3. It has been observed that the maximum split tensile strength is 3.59 N/mm^2 with 10% SF at w/b ratio 0.4. The minimum value is 2.03 N/mm^2 with 0% SF at w/b ratio 0.5. The tensile strength with 10% SF are 3.59, 2.81 and 2.50 N/mm^2 respectively which are higher than that of plain concrete and with 15% SF. At this juncture, it is worth to mention that the Abram's generalized w/b ratio law cannot be applicable to concrete mixes containing SF. In other words, the variation of compressive strength of concrete containing SF did not comply with the generally observed Abrams' w/b ratio law in plain concrete. This is only an observation from the present study. It needs to be studied further for confirmation as it varies with the combination of various constituents and chemical composition of cementitious material.

Fracture Energy of Concrete

Fracture energy, G_F is one of the important material properties for the design of large concrete structures. The material fracture toughness G_F represents the energy absorbed per unit area of crack. In the fictitious crack model proposed by Hillerborg [8] the fracture energy G_F, tensile strength f_t and the stress-CMOD relationship completely describe the fracture characteristics of concrete. The fracture energy, G_F is defined as the area under the load-deflection curve per unit fractured surface area. Fracture energy of concrete observed in various concrete mixes is shown in Table 4. Figures 4 and 5 show the variation of fracture energy with w/b ratio and SF respectively in various concreet mixes. It has been observed that the fracture energy decreases as the w/b ratio increases from 0.40 to 0.60 with 0, 10 and 15% SF. From Figure 4, it has been observed that the fracture energy of SF concrete were higher than that of plain concrete with 10% SF and w/b ratios 0.4, 0.5 and 0.6. At 15% SF, G_F value decreases.

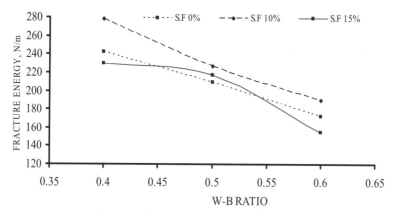

Figure 4   Fracture energy vs. w/b ratio

Fracture energy increases as the compressive strength increases. In HSC the total energy could be considered to be dissipated in two forms. The first form of energy seemed to be utilized in overcoming the surface forces of concrete (surface energy), while the second form of fracture energy seemed to be dissipated in overcoming the cohesive forces due to aggregate bridging, aggregate interlocking, friction forces and other mechanisms in FPZ. It has been reported [5] that interface between cement paste with SF and aggregate is very strong and concrete behaves like a composite material. The post-peak response appears to be steep with increase in the strength of concrete. The fracture energy of concrete varies between 155 to 280 N/m. The highest value of fracture energy is 280 N/m for concrete M0410 with w/b ratio 0.4 and with 10% SF. The lowest value is 151 N/m for concrete M0615 with w/b ratio 0.6 at 15% SF.

Table 4   Fracture energy, Characteristic length and BI.

Mix	E, GPa	G_F, N/m	l_{ch}, mm	BI, mm
M0400	32.15	240	300	92.38
M0410	34.18	280	294	89.70
M0415	31.25	227	302	93.63
M0500	30.90	206	282	94.15
M0510	31.61	224	295	93.12
M0515	31.44	210	283	93.37
M0600	30.53	170	242	94.69
M0610	31.97	192	244	92.62
M0615	29.97	153	228	95.53
M3510	30.16	175	241	95.25

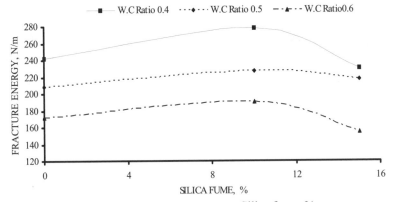

Figure 5   Fracture energy vs. Silica fume, %

Ductility of Concrete

The values of material characteristic length, "Q" shall be in the range of 12.5 and 50mm in cement paste, 50 to 150mm for mortar and 100 to 350 mm for concrete. The values of the factor have been furnished in Table 4, which vary with compressive strength of concrete for various w/b ratios and SF content. The brittleness index (BI) of concrete is higher for low compressive strength concretes. It can be inferred that the size of FPZ is large for concretes of low compressive strength. Hence for concretes of high compressive strength contain small FPZ. BI is observed to be varying in the range of 90-95 mm. The brittleness of concrete is indirectly measured by characteristic length, $l_{ch} = EG_F/f_t^2$.

In plain concrete, the characteristic length increases as the maximum size of coarse aggregate increases, which also indicates increase in the ductility [5].

In concrete containing SF, the characteristic length decreases up to aggregate size 12.5 mm, beyond which it increases. The characteristic length ranges between 228 and 300 mm. From Table 4 it has been observed that the highest and the lowest values of characteristic length are 300 and 228 mm in concrete mixes M0415 and M0615 respectively. The higher values of characteristic length were observed with w/b ratio 0.4 and lower values are observed in concrete mixes with increase in w/b ratio.

CONCLUSIONS

Based on the experimental studies, the following conclusions are drawn. Compressive strength of concrete containing SF did not comply with Abram's w/b ratio law. The highest compressive strength values are observed with w/b ratio 0.4 at 10% SF. The split tensile strength values with 10% SF are more than that of plain concrete and concrete with 15% SF. The highest value of split tensile strength is 3.59 N/mm^2. The fracture energy decreases as the w/b ratio increases from 0.40 to 0.60 at 0, 10 and 15% SF. However, the fracture energy of concrete with SF was observed to be higher. The brittleness index (B.I) of concrete is higher for low strength concrete and high strength concrete exhibited small fracture zone. The characteristic length of concrete decreases as the w/b ratio increases.

REFERENCES

1. SHAH S.P., STUART E.S., AND OUYANG, C., "Fracture Mechanics of Concrete: Applications of Fracture Mechanics to Concrete, Rock, and other Quasi-Brittle Mats", John Wiley and Sons, INC, 1994, New York.

2. ZAIN, M.F.M., SAFIUDDIN M., MAHMUD, H., "Development of high performance concrete using SF at relatively high water- binder ratios", 2000, Cem Con Res, 30 pp. 1501-1505.

3. ZELIC, J., RUSIC, D., VEZA, D., KRSTULEVIE, R., "The role of SF in the kinetics and mechanisms during the early stage of cement hydration", 2000, Cem Con Res 30, pp. 1655-1659.

4. PROKOPSKI, G., LANGIER, "Effect of water/cement ratio and SF addition on the Fracture toughness and morphology of fractured surfaces of gravel concretes.", 2000, Cem Con Res, 30, pp. 1427-1433.

5. APPA RAO, G., RAGHU PRASAD, .B.K., "Fracture energy and softening behavior of high strength concrete", 2002, Cem Con Res, 32, pp. 247-252.

6. TASDEMIR, C., TASDEMIR, M.A., LYDON, F.D., AND BARR, B.I.G., "Effects of SF and aggregate size on the brittleness of concrete", 1996, Cem Con Res, 26, pp. 63-68.

7. RILEM Committee on Fracture Mechanics of Concrete–Test methods, "Determination of Fracture Energy of Mortar and concrete by means of three point bend tests on notched beams", MS 8, (106), 1985, 285-290.

8. HILLERBORG, A., MODEER, M., AND PETERSSON, P.E., (1976), 'Analysis of Crack Formation and Crack Growth in Concrete by Means of Fracture Mechanics and Finite Elements", Cem and Conc Res, 6(6), pp.773-782.

LIMIT STATE ANALYSIS OF REINFORCED CONCRETE STRUCTURES BY ADDITIONAL FINITE ELEMENT METHOD

A Ermakova

Southern Ural State University

Russia

ABSTRACT. The paper considers the ways of solving some problems connected with the development of the Additional Finite Element Method (AFEM), which is a variant of the Finite Element Method (FEM) for analysis of reinforced concrete structures at limit state. AFEM is a combination of the three effective methods of structural design: finite element method, method of additional loads and limit state method. The problem is solved by means of ideal failure models (IFM) and additional design diagrams (ADD) which are composed of additional finite elements. Each additional finite element (AFE) describes a limit state reached by the main element. Examples for obtaining ideal failure models are also given.

Keywords: Additional finite element method, Limit state, Additional design diagram, Ideal failure model

Anna Ermakova, is Cand. Tech. Sc., Assistant Prof., Department of Building Structures of South Ural State University, Chelyabinsk, Russia

INTRODUCTION

Nowadays one of the important problems in the field of nonlinear analysis of reinforced concrete structures is the development of the variant of finite element method (FEM), intended for analysis of the structures at limit states. Additional Finite Element Method (AFEM) is this variant, the theses of which are given below.

ADDITIONAL FINITE ELEMENT METHOD

The Additional Finite Element Method originates from two well-known concepts: limit state of a structure and limit state of a finite element [1]. Hence from the very beginning it is necessary to determine these concepts in accordance with the logic of the finite element method. In the description of the structure at limit state, an ideal failure model may be used, which is the design diagram of this structure at the moment before its failure. The point in the initial design diagram of the structure under load gradually changes in accordance with nonlinear properties, which appear when the limit state is reached. In particular, the definite bonds between some finite elements are broken and finite elements with another nonlinear properties at limit state appear. As a results. the initial design diagram turns into an ideal failure model of the considered structure. The ideal failure model may be obtained by two ways. The first way is a step-by-step analysis of the structure under gradual increase of loading with an accompanying change of the initial design diagram. For example, some characteristics of the finite elements may change in it. The second way is a representation of a failure model well-known from the design of analogous structures or previously obtained test data. For example, the Codes for plane bending structures consider two failure models with transverse and oblique sections, respectively.

ADDITIONAL DESIGN DIAGRAM (ADD)

Then it is necessary to determine the method of description for the process of gradual transformation of an initial design diagram of the structure into a design diagram of this structure at limit state, i.e. into an ideal failure model. A variant of solving the problem is considered below.

Aim of the Additional Design Diagram

The use of the additional design diagram is proposed, which changes the properties of the initial design diagram comprising the linear finite elements, and turns into a design diagram of the considered structure at limit state, i.e. into an ideal failure model. The process is shown in Figure 1.

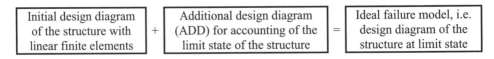

Figure 1   Scheme of action of the additional design diagram
for accounting of the limit state of the structure

The use of the additional design diagram for accounting of the limit state analysis of the structure, by means of the finite element method, is connected with the determination of the properties of the additional design diagram depending on the limit state of the considered structure.

Assembling the Additional Design Diagram (ADD)

The behaviour of a structure at limit state is an extreme case of its nonlinear behaviour. Therefore the action of the additional design diagram ought to ensure to take into account its any nonlinear properties attained as the limit state is reached. This means that the additional design diagram consists of some additional design diagrams, where each design diagram takes into account only one particular nonlinear property. The scheme of definition of the properties is shown in Figure 2.

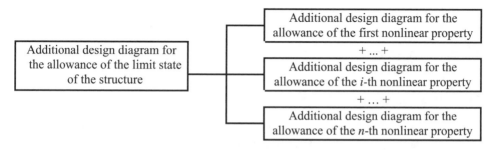

n – number of nonlinear properties considered in the design to the moment when the limit state is reached (i changes from 1 to n)

Figure 2   Additional design diagram for the allowance of the limit state of the structure

Such principle for the determination of the properties of additional design diagram opens some new possibilities for limit state analysis of structures by means of the finite element method as far as it allows the gradual change of properties of the initial design diagram consisting of linear finite elements. In particular, it clears the way for the method of additional loads whose efficiency has been already been proved.

Additional Finite Elements

Additional design diagrams consist of specially developed additional nonlinear finite elements [2] similar to the initial design diagram which is formed by linear finite elements. Additional finite elements change the properties of the main finite element in accordance with the stage of their behaviour. Moreover each main finite element is considered as a separate small structure which has its own limit state in behaviour at the given condition of loading. The limit state of the structure is reached when some finite elements (sometimes it may be the only finite element) reach the limit state. For example, the process of formation of additional design diagram when the initial design diagram consists of triangular deep-beam finite elements. Such finite elements are used for simulation of concrete in analysis of reinforced concrete structures in plane stress-strain state, where the distributed reinforcement is considered as bar finite elements or analogous triangular deep-beam finite elements.

The Codes [3] offer two types of concrete strength: compressive strength and tensile strength. Starting from the main strains appeared in a triangular concrete finite element, there are two types of analogous limit states in it. In this connection the behaviour of finite element at each limit state has different stages. Thus in compression, the finite element has two main stages, which reflect the singularity of concrete as a material. These are:

1) plastic behaviour up to the moment when the limit state of compression is reached;
2) collapse of the finite element after the compressive failure of concrete, as at this moment concrete loses its bearing capacity and ceases existing as integral material.

Under tension, when the limit state is reached, the finite element loses its bearing capacity in part as well as it may have some strains. This means that the finite element has four stages up to the moment of collapse:

1) plastic behaviour before cracking;
2) cracking and unloading in part;
3) partly reload up to the moment when the limit state of reloading is reached;
4) collapse.

The Additional Finite Element (AFE) must take into account all these stages depending on the particular step of the analysis.

Example for the Additional Design Diagram

Since the additional design diagram ought to take into account the character of behaviour of each finite element of the initial design diagram, the order of formation is determined by all stages of behaviour of the main finite element at each limit state, by means of corresponding additional finite elements (Figure 3). The example of the use of additional triangular deep-beam finite elements shows the way of realization of limit state analysis of reinforced concrete structures in a plane stress-strain state.

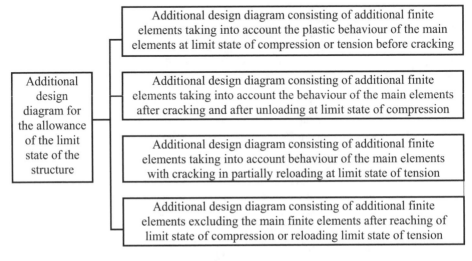

Figure 3   Example for the formation of the additional design diagram
consisting of additional finite elements

PROBLEM OF COMPUTER PROGRAMS

At the same time, the additional finite element method may be a theoretical basis for the creation of easily modified algorithms and programs for analysis of structures from another material at limit states. It is determined by the fact that the main sequence of solving the problem by means of the finite element method remains almost invariable. Simply, in this case, the additional operations added are: 1) composition of the design diagram from the main finite elements; 2) formation of additional design diagrams on the basis of an ideal failure model; 3) calculation of stiffness matrices of the main and additional finite elements; 4) composition of the general stiffness matrix of the system; 5) solution of an algebraic system of linear equations; 6) calculation of stresses, strains and reactions of the main and additional finite elements; 7) analysis of the reached stage of the stress-strain state.

This order of solving the problem allows transforming the existing programs for linear analysis into programs for analysis of limit states.

PROPERTIES OF THE ADDITIONAL FINITE ELEMENT METHOD

The developed additional finite element method presents a variant of the finite element method for limit state analysis and corresponds to follow requirements:

1. Additional finite element method has a generalized character and it is independent of the type and nature of accounting nonlinear properties of the structures exhibited up to the moment when the limit state is reached.
2. In spite of its generalized character it guarantees the flexible analysis and it has possibility to reflect the essential singularities of each type of nonlinearity, appeared when the limit state is reached.
3. It reflects the character change of the analyzed structure stresses due to accounting nonlinear property in the process of reaching the limit state.
4. It guarantees the possibility of change of strains due to the considered nonlinear property in the process of reaching of the limit state.
5. It guarantees the possibility of creation of algorithms and programs for nonlinear analysis of structures at limit states.
6. It guarantees the possibility of creation of auxiliary algorithms and programs for transformation of ordinary programs into programs for nonlinear analysis of the structure at limit state.

Use of additional finite element method for limit state analysis of reinforced concrete structures at limit state provides the opportunity in the field of rational design and investigation. Analysis of some spatial structures and structures at plane stress-strain state proved the efficiency of additional finite element method.

EXAMPLES OF IDEAL FAILURE MODELS

The solving of the problem of creation of additional finite element method as a variant of the finite element method for limit state analysis of reinforced concrete structures demands to introduce an ideal failure model of the considered structure [4]. This ideal model ought to correspond to the logic of finite element method as well as the limit state of the considered structure. An ideal failure model of the structure is the design diagram of the structure at limit state. It differs from the initial design diagram by next singularities: finite elements at limit state, breaks of some connections between another finite elements and additional changes connected with simulation of limit state of the considered structure.

Ideal Failure Model of the Shell

The choice of an ideal failure model of the shell of Chelyabinsk Shopping Centre is given later [5]. The application package «LIRA» was used for analysis. The design diagram was prepared with respect to the principles of this program and according to the library of finite elements [6]. The shell was simulated by means of two types of finite elements: "The bar of general attitude" and "The plane element of the shell of general attitude" (Figure 4). The first type was used for the simulation of the elements of index contour, columns and cross-beams; the second one simulated the plates of the roof cover. Since the applied loads are symmetrical only to the axis oriented from North to South, a half of the structure was considered in the analysis. In order to pass to the limit state analysis of the structure, it was necessary to determine the type of the limit state which corresponded to the type of deformation of the structure. It would allow choosing a failure model of the considered structure. The results of regular tests show that there were some cracks in the concrete elements of the contour from the South. This means that the given prestressing of concrete was insufficient for overcoming of unfavourable influence of temperature action. Thus the limit state of the structure is the state preceding the moment of the failure of the South index contour due to the tension under the action of increasing temperature. Therefore the ideal failure model of the shell is the initial design diagram where one of the bar elements of the South side of the index contour moved off. Such approach means that the limit state of the structure is determined by the only bar.

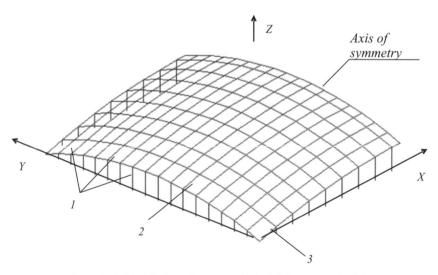

Figure 4   Initial design diagram and ideal failure model of the shell

Legend: 1 – FE "The bar of general attitude", 2 – FE "The plane element of the shell of general attitude", 3 – FE of index contour under the most unfavourable combinations of loads, which is moved off the initial design diagram for obtaining of an ideal failure model.

This element has the most unfavourable combinations of loads: rolling component of dead load, prestressing of concrete, action of temperature. The bar of general attitude with its own system of stresses was used for the simulation of elements of the index contour. Each stress has its own limit state which is determined by the properties of the material of the finite element, concrete in this a case. Concrete is characterized by two types of strength: tensile strength and compressive one. This fact determines the limit states of the finite element used in the problem, i.e. the bar of general attitude. Table 1 illustrates the process and reflects the peculiarities of behaviour of concrete in spatial frames.

Table 1   Limit states of concrete finite element «The bar of general attitude»

№	FORCE	LIMIT STATES	CHARACTER OF FAILURE
1	N – longitudinal force	1. Tensile stress reaches the value of tensile strength of concrete	Tension
		2. Compressive stress reaches the value of compressive strength	Compression
		3. Longitudinal compressive force reaches the ultimate value in statical stability design	Lost of stability
2	M_y – bending moment about axis y	1. Tensile bent stress in y direction reaches the value of tensile strength of concrete	Failure of tensile zone of the border of section
		2. Compressive bent stress in y direction reaches the value of compressive strength of concrete	Failure of compressive zone of the border of section
3	M_z – bending moment about axis z	1. Tensile bent stress in z direction reaches the value of tensile strength of concrete	Failure of tensile zone of perpendicular border of section
		2. Compressive bent stress in z direction reaches the value of tensile strength of concrete	Failure of compressive zone of perpendicular border of section
4	M_k – torsional moment about axis x	1. Torsional moment reaches the ultimate value in torsion with bent	Failure due to torsion with bending
5	Q_y – shear force in y direction	1. Shear force in y direction reaches the ultimate value in strength of oblique section in xoy plane	Failure of oblique section in xoy plane
		2. Tangential stress in xoy plane reaches the shear strength of concrete	Failure due to shear in xoy plane
6	Q_z – shear force in z direction	1. Shear force in z direction reaches the ultimate value in strength of oblique section in xoz plane	Failure of oblique section in xoz plane
		2. Tangential stress reaches the ultimate value of shear strength of concrete in xoz plane	Failure due to shear in xoz plane

Behaviour of a concrete bar finite element at each limit state mentioned above (see Table 1) demands a special theoretical research. In particular, since we may rarely come across the proper shear in reinforced concrete, shear limit state is unlikely to occur (see Table 1, articles 5.2 and 6.2). Out of all limit states of the bar element, only the axial tension (see Table 1, article 1) corresponds to the character of limit state of the shell, which was accepted for development of its ideal failure model.

Ideal Failure Models of the Beam

The ideal failure model may be set on the date of design of the analogous structures. For example, the Codes recommend two types of failure for bent beams: with transverse or oblique sections. This means that two ideal failure models ought to be considered in limit state design. In analysis of a beam with symmetrical loads and bar reinforcement, its initial design diagram consists of two types of finite elements: a triangular finite element of deep-beam and one connecting with the two nodes (Figure 5, a and b). The first type was used to simulate the bond between concrete and reinforcement. As the structure and the loading of the deep-beam are symmetrical, the design diagram includes one-half up to the axis of symmetry. It is necessary to insert the corresponding changes in the initial design diagram of the deep-beam responding to the character of the failure. It is accepted in the first failure model with transverse section that a dangerous transverse crack is formed in the middle of the beam along the axis of symmetry. In addition, the limit height of compressive zone x ought to be provided above the crack. The ideal failure model differs from the initial design diagram by the absence of some horizontal connections of concrete finite elements at the axis of symmetry. These connections occur only at the height of the compression zone of section x (Figure 6 a).

a) Initial design diagram

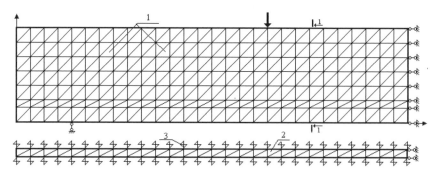

b) View 1-1

Finite elements:
1 – concrete FE;
2 – reinforcement FE;
3 – connecting FE.

Figure 5   Initial design diagram of the beam

a) Ideal failure model № 1 with transverse section

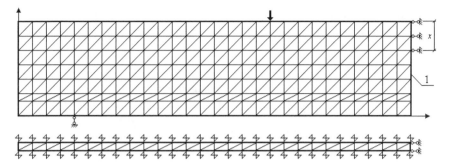

b) Ideal failure model № 2 with oblique section

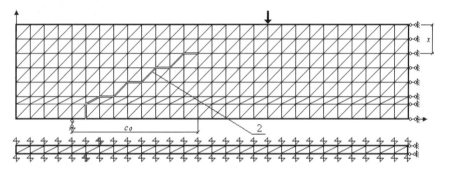

Figure 6   Ideal failure models of the beam 1 – transverse crack; 2 – oblique crack;
x – height of compressive zone; c_0 - plan view of the oblique crack

In the second case of failure with oblique section, another model may be used. It differs from the initial design diagram by distinguishing between concrete finite elements along the damage line with plan view c_0 and by introducing additional connecting elements at the points of intersection of crack and bar reinforcement. The height of compressive zone x (Figure 6 b) ought to be also provided in the simulation of oblique crack. The presence of two failure models in limit state design of the beam means that it is necessary to calculate each model.

Ideal Failure Models for FEM and AFEM

The considered examples of ideal failure models of the shell and bent beams have shown that their introduction enables the limit state design of reinforced concrete structures to be carried out by the finite element method. The additional finite element method can also be used. In the first example an additional finite element "The bar of general attitude" excludes the main finite element for obtaining an ideal failure model of the shell. In the second example an additional finite element "The triangular finite element of deep-beam" changes the main element for taking account of corresponding transverse or oblique cracks.

CONCLUSIONS

The additional finite element method by means of additional finite elements and ideal failure models enables carrying out limit state analyses of reinforced concrete structures by the well-known finite element method.

ACKNOWLEDGEMENT

The Department of Building Structures of the South Ural State University, where all research work has been and is being carried out, is acknowledged.

REFERENCES

1. ERMAKOVA A, Limit State Design of Reinforced Concrete Structures Using Complementary Finite Element Method, Proceedings of the Conference "Concrete and Reinforced Concrete: Ways of Development", September 5–9, Vol. 2, Moscow, 2005, pp. 386 – 391.

2. ERMAKOVA A, Limit State Analysis of Reinforced Concrete Structures by Finite Element Method, Application of Codes, Design and Regulations, Proceedings of the International Conference held at the University of Dundee, Scotland, UK on 6–7 July 2005, Thomas Telford Publishing, London, 2005, pp. 587–594.

3. SN AND P, 2.03.01-84, Reinforced Concrete Structures, State Administration of Construction, Moscow, 1985, 79 p.

4. ERMAKOVA A AND MAXIMOV J, Ideal Models of Failure for Limit State Analysis of Reinforced Concrete Structures by Finite Element Method, Proceedings of the Conference "Concrete and Reinforced Concrete: Ways of Development", September 5–9, Vol. 2, Moscow, 2005, pp. 392–397.

5. MAXIMOV J AND ERMAKOVA A, Necessity of Allowance for Daily Temperature Difference in Design and Operation of Long-Span Shells, Proceedings of the third International Conference on Concrete under Severe Conditions CONSEC'04, June 18-20 2001, Vancouver, BC, Canada, Vol. 1, pp. 1017–1023.

6. Methodical Recommendations on Listing of Results of the Work of Application Package for Computer-Aided Design of Reinforced Concrete Structures (PA CDZhBS) – Kiev, NIIASS, Gos. Stroy. USSR, 1987, 136 p.

TOWARDS ENHANCEMENT OF THE STATUS OF CONCRETE CONSTRUCTION

S Desai

BYL Group

UK

ABSTRACT. Concrete is widely used in all types of projects, to the extent of some two tonnes of concrete per person per year. It affords many advantages, e.g. economy of construction, low energy demand in making of concrete, fire resistance of concrete elements, saving of energy during the service life of the building afforded by inherent insulation properties of concrete, ability of concrete cladding to serve as load-carrying structural members, etc. However, such contribution of concrete construction could be enhanced further in the long term through research and development in various fields, including its performance in floors, precast cladding, precast concrete frames, etc. Additionally, it is widely recognised that there is a limit on the availability of natural minerals used for making cement and producing aggregates. Solutions to this problem are sought through research on binders, using industrial by-products to partially replace Portland cement, and on recycling of concrete aggregates. However, use of such alternative constituents is often hampered with the uncertainties associated with their chemical inertness and their adverse influence on quality of construction. This paper examines possibilities for employing advanced mineralogical assessment methods for this purpose. The paper also considers a potential for improving the status of concrete through investigation of its microstructure and properties using laboratory studies on porosity, permeability and chemistry of concrete before and after solidification. This could materialise with the development of suitable software and experience of the researchers in advanced methods using SEM (scanning electron microscope) mapping and confocal scanning laser microscopy (CSLM) techniques.

Keywords: Precast, Concrete, Recycled, Aggregate, Inertness, Microscopy.

Dr Satish Desai is a principal structural engineer with the BYL Group of consultants in Croydon, Surrey, UK. He worked until May 2000 as a principal civil engineer with Building Regulations Division of the Construction Directorate, Department of the Environment, Transport and Regions (as it was known at that time). His main research interest is in the fields of reinforced concrete design and collaboration between research in concrete technology, structural design and construction process for enhancing competitiveness and sustainability of concrete construction. He is Hon. Visiting Professor at the University of Surrey (UK) and a member of the Industrial Advisory Board at the Kingston University (UK).

INTRODUCTION

It is the responsibility of an engineer to choose construction material best suited for a building. The choice is often influenced by factors other than just the technical merits of the material, e.g. local availability of the material and its constituents. Concrete could often be the popular choice for common buildings, both for the superstructure and the substructure. However, it appears that tall building construction may be better served by steel superstructure, for example, providing speed of construction with standardized products. Designers of low-rise buildings may opt for timber superstructure, for the reasons of economy. For foundation construction, however, concrete would seem to be the common and obvious choice in most cases.

The pre-1992 version of the Approved Document Part A (structural safety, Building Regulations (England and Wales)) had mentioned concrete as a construction material suitable for use in pad foundations for low-rise residential buildings constructed with bricks and timber. After reading the old document, an inexperienced reader could easily have formed an opinion, perhaps mistakenly, that concrete is a low-tech material that can be obtained by mixing stones, sand and cement by hand, with no regard for the water content. The recent revision has revised the relevant paragraphs and mentioned appropriate grades of concrete.

It is encouraging that there are some examples of use of concrete in the superstructure of tall buildings and in construction of prestigious bridges. For common building construction in the UK, however, concrete is not the common choice of material. It seems that the situation could be improved and the status of concrete could be raised with research and development in the following fields:

1. Re-thinking in concrete construction to promote sustainable construction.
2. Promotion of precast concrete products
3. Development of methodology for safe use of industrial by-products and recycled concrete aggregate.

RE-THINKING IN CONCRETE CONSTRUCTION

Concrete Construction and Sustainability

At the 2002 UN World Summit on Sustainable Development, the participating governments recognised the importance of industry's contribution to sustainable development and pledged to enhance corporate environmental and social responsibility and accountability. Many industries in the UK are participating in voluntary reporting on sustainability under the auspices of "Global Reporting Initiative", a worldwide non-profit organisation that has a set of comprehensive sustainability disclosure standards. The UK Government has demonstrated its support to sustainable construction and sponsored many research projects, pertaining to concrete construction and concerning environmental issues, e.g. reduction in waste, use of industrial by-products, recycling of products of demolition, etc.

Sustainable construction represents a move away from the 19th century Industrial Revolution's linear processes, leading to inefficient conversion of energy and materials into a limited product with much waste, towards a 'cyclical' process for the 21st Century where much is recycled for reused [1]. This is a shift in thinking from linear, essentially 'physics' based processes, towards non-linear and complex 'biological' based systems, using the advantages of modern technology coupled with a major behavioural change of the users of natural resources. Such considerations produce the 4Rs cycle – Reduce, Refurbish, Reuse and Recycle – a cycle echoing biological processes.

If this cycle is adopted, it will benefit the planet by promotion of following considerations that readily apply to design and construction of buildings with concrete:

- Reduction in emission of greenhouse gases.
- Standardisation of formwork and temporary works designs to allow for reuse.
- Focus on design loads to be just adequate for proposed use (including seismic, typhoon and groundwater effects), i.e. without any over-design of structures.
- Assessment of implications of design life, remembering that design life is theoretical, service life is a reality.
- Choosing structural components capable of being reused or recycled elsewhere.
- Investigation of possibilities of structural determinacy to simplify fabrication and future reuse as against the economies from structural redundancy and continuity.
- Where appropriate, incorporation of recycled materials and reused elements into new build.
- Choosing the structural system with an eye to low initial cost and reasonable embodied energy.
- Consideration of structural adaptability for future use and adequacy for alternative uses.
- Keeping as-built documents to become part of requirements of ownership for efficient demolition.
- Use prefabrication to reduce waste on site and to recycle waste in manufacture at source.
- Design of joints for dismantling, demolition and reuse as well as ease of construction assembly.
- Demand on constructors to produce an environmental and waste management statement with tenders.

Current Initiatives in the Field of Sustainable Construction

The current thinking generally stems from the basic issue of sourcing and processing of materials. For steel, this process starts with extraction of iron ore and leads to manufacture of steel and its transportation for use in construction. Finally, subsequent reuse or recycling is achieved through the scrap recovery route. For concrete, there is an issue of sourcing the raw materials – aggregates, and limestone (for cement) in particular. In the UK, suitable sources are limited and often located in areas of outstanding natural beauty, making extraction of aggregate very detrimental to the environment. Furthermore, cement manufacture accounts for approximately 2% of UK carbon dioxide (CO_2) emission. Production of concrete could perhaps account for some 0.35% of the UK CO_2 emission, assuming an average of the commonly used cement content in general concrete construction.

All these circumstances have led to increasing interest in using recycled materials and cement replacements. A small number of projects have been developed using recycled aggregate concrete. At the same time, many other industrial by-products are being considered for recycling, e.g. glass, rubber tyres, incinerator ash, etc. However, there are difficulties associated with the supply of suitable material for reprocessing, potential reduction in quality and lack of detailed guidance. These factors tend to discourage general use of recycled aggregate. On the other hand, specifying cement replacements such as pulverised fuel ash (PFA) and ground granulated blast-furnace slag (GGBS) is something that can be seriously considered on many construction projects.

An urgent change is needed in the way our buildings are constructed and maintained, if we are to avoid the mistakes of the disastrous 1960s building boom. "Sustainable Buildings Task Group", a group of leading building experts, practitioners and academics, would address such issues [2]. This group would formulate a report and deliver it to the UK Government. This organisation is supported by many UK Government Departments, e.g. Office of the Deputy Prime Minister and the Department of Transport and Industries.

Reviewing the available technologies for environmental improvement, the Sustainable Buildings Task Group has concluded that a significant uplift in quality is both possible and affordable. Its report calls for the Government and the building industry to adopt a single national Code for Sustainable Building, which would raise the quality standards and reduce the unsustainable use of natural resources in the built environment. The Group would wish to see development of such a code that could afford a consistent approach towards making today's buildings fit for the future. The new Code would incorporate standards in energy and water efficiency and waste management that are above the minimum requirements of the Building Regulations. The Group recommends that Government should lead by example and adopt these standards for all new buildings in the Public Sector. The Group also calls for tighter regulations to be enforced, so that all new buildings would reduce energy and water consumption by 25%.

It is envisaged that 10% of materials used in the construction of new buildings would be recycled or reused and more sustainable sources of materials would be encouraged. Waste would be better managed during construction and buildings would be designed for households to dispose of waste more effectively. It is believed that the construction, development and house-building industries can subscribe much better to the sustainability agenda and that they need to be persuaded of the long-term benefits. The Code for Sustainable Buildings would provide them with a level playing field for quality buildings.

Recycled concrete aggregate (RCA)

Recycling concrete aggregate is important for reducing demand on primary sources of natural aggregate and for easing the pressure on sites for disposal of products of demolition. At present, RCA use is limited mostly to road construction and plain concrete. Rapid disposal of waste from demolished concrete structures has tended to be the preferred practice of the UK construction industry. Production of RCA for use in structural work would require careful and systematic processing, grading and testing of aggregate, essential for ensuring its acceptability in structural concrete. However, costs of such operations are considered to be too high at present.

There is enough evidence to show that RCA can be used to replace some 30% of the coarse aggregate in some grades of concrete meant for less demanding performance, e.g. concrete suitable for drives and parking for dwellings, external paving, foundations in non-aggressive soils, blinding, kerb bedding, etc. [3]. RCA can also be used in some concrete grades meant for internal reinforced concrete members. Further work may suggest that use of RCA concrete can be extended to higher grades of concrete. However, in such instances, a designer has to appreciate deficiencies of RCA concrete regarding workability and the problems presented in the form of harshness of the mix, together with porosity and inadequate durability of the hardened concrete. These problems could be resolved with careful mix design and addition of coarse pulverised fuel ash to the RCA concrete mix, provided that it is compatible with the other concrete constituents. It would be most inappropriate to increase water content and the cement content (to maintain the water/cement ratio) for the purpose of overcoming these difficulties. Such measures would jeopardize the basic principles of sustainability.

There are some well-recognised barriers to rapid increase in use of RCA. First, a quick disposal of the demolition waste is often preferred to a systematic processing and producing good quality recycled aggregates. This is because, in many cases, the cost of such activities, including the investment in the appropriate plant, is considered to be too high and economically unacceptable. However, introduction of special taxes, such as landfill and extraction taxes, can improve economic viability of such process and the price of good quality RCA could become comparable to that of natural aggregate.

Secondly, supply of recycled aggregate is sporadic by nature and it depends on a number of factors, including volume and type of demolition activities (which are linked to the economy, among other things) and recycling infrastructure. The latter is important, since it is customary for demolition contractors and concrete recyclers to crush the material only when they have sufficient proportion to justify operation of the crushing equipment.

Third obstacle is known to be the public perception of quality of RCA. Despite the fact that many governments and regulatory bodies throughout the world have developed specifications/ guidelines for the use of recycled aggregate, it is still perceived as "inferior". Quality control of the RCA could be difficult for that obtained from demolished concrete structures, on account of potential contaminations (wood, glass, etc.) and difficulties in identifying the source of original materials.

These aspects may make it difficult to check whether the RCA has any undesirable chemicals or inherent deterioration mechanisms like alkali-aggregate reaction. Such problems may not exist in RCA obtained from precast factory-reject elements. Therefore, it is of great importance to produce authoritative 'fit for purpose' specification for testing RCA. Under the present circumstances, RCA should not be used in any significantly important structural concrete elements, if there is any doubt about its chemical inertness. Furthermore, special design considerations may be required before indulging in any proposal to use RCA concrete in members sensitive to creep and shrinkage.

Other key barriers to wider acceptance of recycled aggregate include lack of experience and guidance supported by detailed case studies. Whilst all these factors are recognised and potential outlets, technologies and standards required for recycled aggregate usage are being researched and developed, a great deal of work is still needed before RCA can be marketed as a useful resource. Collectively, all these barriers confront the construction industry and these must be overcome before RCA use can be fully exploited in other value-added applications, such as general concrete structures.

Research at Kingston University

Kingston University has a long tradition of successful research in the field of design and construction of masonry. For example, a programme dedicated to development of new design approach for unreinforced concrete in domestic basements has been recently completed. The outcome of this research complements the work being undertaken into unreinforced masonry basements.

Over the last three years, a decision has been made to broaden and diversify the research base, resulting in further development of the Concrete and Masonry Research Group and the Sustainable Technology Research Centre of the Faculty of Engineering. The main aim is to address escalating costs of procuring natural aggregates and reduce the demand for landfill sites for disposal of construction and demolition waste. This development has led to investigation of feasibility of using re-cycled materials as aggregate in concrete, aided by procurement of two main sets of equipment:

- A specialist freeze-thaw chamber for use in the sustainability area of research.
- Carbonation tanks to accelerate the effect of carbonation on concrete.

Additionally, microscopical investigations into the chemical nature of concrete are undertaken as well as the usual wet, dry and non-destructive tests on concrete. Some concrete research projects are well under way at present, on the subject of feasibility of using recycled materials as aggregate in new concrete, as described below.

RCA concrete with Ordinary Portland Cement (Portland Cement Class 42.5)

This project has demonstrated the suitability of recycled aggregate for use in a range of concrete applications using Portland cement, through laboratory research and field trials. The findings have confirmed that the British Standard Institution (BSI) draft specifications are adequate and appropriate for their intended use, and that they provide suitable guidance for using re-cycled aggregates. Full-scale site trials confirmed that coarse Recycled Aggregate (RA) could be used successfully in a range of concrete applications. This investigation included characterisation of aggregates and development of suitable concrete mixes containing a range of blends of natural and coarse RA in trial mixes. Full-scale demonstrations were built with RA concrete and assessed, including a variety of structural and paving elements. These demonstrations are being monitored over the long term. The existing project output comprises technical papers and running of various workshops, including a technical skills workshop.

RCA concrete with binary cements containing industrial by-products

This project demonstrates the suitability of recycled aggregate in a range of applications using binary cements, thus enhancing the principle of sustainable construction further and combining the advantages of using recycled aggregate with the use of cement comprising industrial by-products. The binary or blended cements represent combination of Ordinary Portland Cement with PFA, Silica Fume, and GGBS. The findings of this project confirm the recommendations given in the new European Standards for concrete and aggregates, which allow the use of recycled aggregates and industrial by-products such as PFA, Silica fume and GGBS combined with Ordinary Portland Cement.

Other industrial by-products

Kingston University has developed good contacts with the industry and made significant progress in the field of using globules of various sizes with encapsulated crushed waste glass. Vehicle tyres are also cut very small and converted into aggregate that can be used in semi-flexible pavements, potentially for its use in children's playground.

PRECAST CONCRETE CONSTRUCTION

Earlier Situation

For a long time, the statutory guidance documents in UK, for example the Approved Documents of the Building Regulations (England and Wales), have given information predominantly related to masonry and timber construction. These materials have been used in construction of small buildings in the UK for some considerable time.

Such construction is often termed as "traditional", giving an impression that the statutory guidance would have some preference for the same. This is in fact not true and any system can be used to build houses, provided that the construction can comply with the principles of safeguarding health and safety of people in and around the buildings.

Potential Influence of Prefabrication on Status of Concrete Construction

The Egan Report [4] had identified importance of alternative construction techniques as an essential aspect of the way forward for meeting targets for the industry. These targets are summarised as follows:

- Reduction in capital costs and construction time   =   10% pa
- Reduction in defects   =   20% pa
- Reduction accidents   =   20% pa
- Increase in predictability of delivery of materials   =   20% pa
- Increase in turnover and profits   =   10% pa

Prefabrication and use of factory-produced concrete elements in building construction should lead to benefits that could serve the principles implied in the Egan report. For example, precast products made in advance and ready for incorporation in the work could save construction time and using precast slabs could reduce time spent in shuttering and scaffolding, thus affording economy of construction.

Use of precast products could lead to improved quality of concrete, gaining benefits of better quality control that is readily feasible in factory environment, right from the early stage of ensuring procurement of good quality constituents that are mutually compatible and consistently available for a range of products. Factory-made products should also benefit from superior mix design, better equipment for making and placing concrete and dimensional accuracy of the elements. Improvement in construction with good quality of finished products can be further enhanced with pre-planned integration of design and construction processes. At the same time, there is considerable potential for enhancement in quality and workmanship, a kind of "ramping up" effect, i.e. high technology products inducing extra demand on skills of the construction workers engaged in assembling the products.

Buildings made with large size precast concrete elements would have reduced concerns about the site-dependent aspects of quality control. For example, larger panels would require less site joints and provide better defence against the elements. Large precast panels perform two main functions, i.e. cladding and structural elements transferring the load to the ground. This would compare favourably with steel frames, which require cladding separately to the structural elements. Furthermore, the construction with precast panels could be inherently robust and alleviate foundation problems caused by nearness of trees and changes in moisture and volume in shrinkable clays, e.g. heave, shrinkage and differential settlement.

In addition to structural safety, precast construction could benefit fire safety, providing advantages of using non-combustible structural material. Similarly, resistance to moisture and driving rain could be improved with the use of less porous material, compared to that provided by masonry walls.

Future Initiatives

One of the problems with concrete construction has been inadequate training of operators. This may apply to joint making, a vital part of precast frame construction. Steel construction benefits from the work of trained and certificated operatives, e.g. welders. Concrete construction has not benefited from such requirements and anyone with marginal experience could work on site in the process of mixing and placing concrete.

There are initiatives in progress for training workers but the contractors may not necessarily give higher remunerations to workers with certificates and better training, thus reducing the incentive for acquiring training as such. It is feasible that precast concrete factories could make distinction between trained and non-trained workers and improve quality of construction through highly skilled workers.

For any change to the current practice and introduction of new system and materials, authoritative guidance should be readily available in the public domain. It is also essential that the precast industry should cast aside any short-term policies, whereby they would continue with their current products that are doing well, e.g. blocks and small span elements.

There should be intensive research and development, which could demonstrate that, with an adequate input of design and pre-planning, precast concrete elements can be used to build cost-effective buildings, which will comply with the Building Regulations and meet the expectations of the users. In the context of such development, it should be ensured that the earlier problems would not resurface in the new systems, e.g. inadequate attention to joint details and dimensional accuracy. With due attention to analyses and practical construction issues, it can be demonstrated that precast elements can provide aesthetic and energy-saving buildings, and meet the users' needs within the building enclosure, e.g., provision of services, wall-mounted items, etc.

Investigation into the beam and column connection could serve as a potentially beneficial initiative. Commonly prevalent connection details include angle cleats that are bolted and grouted in to provide transfer of loads from beams to columns. Such connections are treated as capable of transferring only shear and the building has to incorporate shear walls for the purpose of resisting transverse forces due to wind, unsymmetrical loading, constructional misalignment, etc

It can be shown that most connections are "semi-rigid" or capable of transferring some 60% of bending moments that are associated with the theoretically "fixed" connections. It should be possible to achieve economy of low and medium rise precast framed buildings by using frame action to resist horizontal forces, instead of depending on shear walls. Such construction could benefit from simplification in the building process, providing saving in time and materials that would lead ultimately to economy of construction.

ADVANCED METHODS OF ANALYSING CONCRETE CONSTITUENTS

General Nature of Microscopy Techniques

Microscopy provides an effective, convenient and non-destructive method for assessing quality of concrete, its composition and its chemistry as it sets and after solidification. It could become an ideal way for investigating a wide range of concrete construction issues, including its current condition and reasons for deterioration. Researchers at Kingston University aim to develop Micro-analytical and Micro-imaging facilities and plan to offer their expertise to develop enterprise activities in concrete testing and characterization of construction materials. It is also intended to investigate methods for assessment of various constituents of concrete and concrete construction techniques, centred on expertise at Kingston in Laser Raman Microspectroscopy, X-ray Diffractometry (XRD), X-ray Florescence Spectrometry (XRF), Scanning Electron Microscopy (SEM) and Electron Microprobe Analysis. Further investigations include hot and cold stage microscopy and laser-scanning confocal microscopy.

Examples of Use of Advanced Techniques at Kingston University

Delamination of concrete floors

An initial application of the advanced technique is planned to develop criteria to control quality of surface of concrete floors. There are questions associated with causes of delamination of concrete floors, a subject that has not received proper attention in the field of concrete technology research. This work is very important for the flooring industry and the clients, and it represents high quality of research, which will make use of modern and highly technological methods.

The methodology comprises laboratory study on porosity, permeability and chemistry of concrete before and after solidification. This is eminently feasible, with the expertise and experience of researchers at Kingston University in advanced methods using SEM mapping and confocal scanning laser microscopy (CSLM) techniques.

The research programme has received enthusiastic industry support, and benefit of expertise and experience from the industry. The output, a practical guidance document, will be based on studies and tests on various types of floors, followed by interim reports and website updates during the project. The final report should be very useful for the Concrete Society's working party looking at design and construction of concrete floors. On the whole, this work has the potential to benefit concrete industry with significant advancement of knowledge in the field of concrete floors.

Study of composition of constituents of concrete

Advanced techniques have been recently used to determine RCA composition, which should accord with BS 8500: Part 2 [5]. The results obtained were found to be well within the standard specified limits. RCA and natural aggregate size fractions were examined through XRF Spectrometry and XRD techniques to establish the chemical as well as mineralogical composition of 42.5 N Portland Cement used in concrete production. For these analyses, the samples were prepared by grinding and obtaining grading suitable for XRD and XRF tests.

This study included samples of commercially produced coarse RCA obtained from sources where quality control criteria were commonly adopted. The study has demonstrated that such RCA has chemical and mineralogical characteristics suitable for use in new concrete production. The results should help to reduce inhibition of some consulting engineers towards specifying this material in a range of concrete applications. Results also indicate that coarse RCA samples obtained from different sites should have no significant variation in their characteristics, provided that adequate quality control criteria are adopted during RCA production. This is important to aggregate suppliers and concrete producers, who wish to produce RCA and supply RCA concrete, and may wish to obtain recycled aggregate from different sources that observe good quality control procedures.

CONCLUSIONS

Concrete construction has to keep pace with developments in the construction industry during the current climate of competition and various initiatives leading to promotion of sustainable construction. All sections of the industry, including researchers, suppliers of materials and contractors should strive towards eradication of any impression whereby concrete is perceived as a low technology material, fit for its use only below ground and apparently lacking attributes of a product suitable for sustainable development. There are various initiatives dedicated to efficient, economical and sustainable concrete construction but their impact is not sufficiently visible so far. The writer would like to see speedier elevation of status of concrete construction with research and development in the fields of sustainable construction, use of precast concrete products, development of methodology for safe use of industrial by-products and recycled concrete aggregate, etc.

REFERENCES

1. DIXON M, Engineering buildings for a small planet, Towards construction without depletion, The Structural Engineer, Vol. 80, No. 3, 5 February 2002, pp. 35-43.

2. STRUCTURAL NEWS, The Structural Engineer, Vol. 82, No. 12, 15 June 2004, pp. 35

3. DESAI S B, Sustainable Development and Recycling of concrete aggregate, Proceedings of the International symposium "Sustainable Construction: Use of Recycled Concrete Aggregate" organised by the Concrete Technology Unit, University of Dundee, 11-12 November 1998, Thomas Telford, 1998, pp. 381-388.

4. Rethinking Construction – The Egan Report, The report of the Construction Task Force on the scope for improving the quality and efficiency of UK construction, published in July 1998.

5. LIMBACHIYA M C, MARROCCHINO E AND KOULOURIS A, Chemical– mineralogical characterisation of coarse recycled concrete aggregate, Waste Management, 2006, Available on www.sciencedirect.com.

LEARNING FROM HISTORY:
USING STAINLESS STEEL FOR THE FUTURE

D J Cochrane

Nickel Institute

United Kingdom

ABSTRACT. The United Kingdom is world renowned for its historical buildings, many of which are centuries old. Repair and maintenance is vital to sustaining the longevity of these structures and replacement of many corrosion prone components is key to realising this goal. St. Paul's Cathedral in London was one of the first major buildings to use stainless steel in the 'great restoration' of 1925-30. They used reinforcing bar, masonry fixings, and threaded rod, as an alternative to demolishing the dome. From the only inspection of the stainless steel, carried out in the mid 1990s, it was evident that the material was in the 'as installed' condition and eminently suited to the long-term preservation of major structures. The construction industry, however, has been slow to recognise the benefits of using stainless steel, particularly for reinforcing bar in concrete construction. The corrosion of carbon steel reinforcement has been most problematic, and costly, in highway bridges subject to the application of de-icing salts or sited in coastal environments. Durable construction can be achieved, cost-effectively, by intelligent placement of stainless steel reinforcement. This has been recognised by the UK Highways Agency in their design manual for roads and bridges, and by highways authorities in the USA, Canada, and Scandinavia who now use stainless steel reinforcement. Major new bridges such as the Western Corridor and Stonecutters Bridge in the Far East are current projects that incorporate stainless steel reinforcement, while the Lidingo river crossing railway bridge in Sweden is a notable example of repair for durability. These case histories will be used to illustrate the contribution of stainless steel to sustainable concrete construction.

Keywords: Corrosion, Cost-effectiveness, Highways authorities, Historic buildings, Road bridges, River crossing bridges, Stainless steel, Sustainable construction.

David J Cochrane served an apprenticeship in the aeronautical industry before joining the steel industry. A former technical marketing structural engineer with British Steel and the Steel Construction Institute, he specialises in the application of stainless steel in architecture, building and construction. He is the principal of Technical Publication Services and been a consultant to the Nickel Institute since 1987 for whom he has lectured world wide on stainless steel. He has served on a number of British Standards Institution committees and written numerous articles and publications on stainless steel.

INTRODUCTION

Discovered in Sheffield in 1913, the high corrosion resistance of stainless steel has become its most universally recognised property. It inspired the architects and engineers in the early 1920s to use it to save one of London's most famous buildings, St. Paul's Cathedral, from possible demolition as will be outlined in this paper. Since its discovery, the number of grades of stainless steel has since grown to over 100 as major industries have recognised the significant benefits from using stainless steel. The construction industry, however, has been slow to recognise the true potential of a high corrosion resisting material – and at a significant financial and environmental cost. For example, the corrosion of carbon steel reinforcement has been the principal cause of premature failure in many thousands of highway bridges necessitating costly repair and traffic delays, and creating disruption that could have been avoided by using stainless steel. Recognising the severity and cost of the corrosion problem, highways authorities have sought solutions to the corrosion problems in bridges over the last two decades and investigated many different materials and design solutions. Stainless steel has been proven in test [1] to have high resistance to the chloride ion, which is the principal cause of corrosion of carbon steel rebar. It has also been tested for any adverse galvanic reaction when in contact with [2] carbon steel rebar. The conclusion of feasibility studies in North America and Europe, is to use stainless steel reinforcement in the elements of the structure that are at high risk to corrosion and carbon steel where there is little or no risk. This solution has proven in practice to provide increased durability cost-effectively and has been adopted by highways authorities around the world. In Scandinavia it is now mandatory to use stainless steel reinforcement in bridge parapets – a recognised corrosion prone element.

This paper will show the contribution of stainless steel to sustainable concrete construction using repair case studies such as St. Paul's Cathedral in London, and the Lidingo railway river crossing bridge in Stockholm. The design guidance available for the use of stainless steel reinforcement will also be outlined with its application demonstrated in two of the world's major new bridges currently under construction.

CASE STUDY 1: ST. PAUL'S CATHEDRAL

Material selection is often one of the most difficult decisions the engineer has to make. When building St. Paul's Cathedral about three hundred years ago, Sir Christopher Wren did not have the luxury of selecting a strong corrosion resisting steel but recognised that the material available to them at that time, 'iron', was susceptible to oxidation. His instructions were *'in cramping of stones, no iron should lie within nine inches of the air if possible'*. Elaborate fixing methods such as setting the cramps in boiled lead were deployed to stave off the initiation of corrosion. Figure 1 shows the effect that corroding iron cramps has had on the stonework in the cathedral.

At every opportunity, stainless steel fixings have since been used to replace the old iron ones. In the early 1920s, St. Paul's was in need of major repair as the great dome was beginning to lozenge and the eight piers that support it were sinking at different rates. One option considered was to demolish the 65,000 tonne dome. But with the 'new' 'Staybrite' stainless steel available to them, they decided to use it in ingenious ways. The piers were reinforced with 165 tonnes of 'Staybrite' reinforcement. These were not the fluted shape as in current standards, but simply stamped (flattened) at 3 inch (75mm) intervals to provide a key with the concrete. (NB This is the earliest use of stainless steel reinforcement known to the author.)

Figure 1   Effect of corrosion on stonework in St.Paul's Cathedral.

The dome of St. Paul's is not a single dome but comprises of three domes; an outer lead covered one, an inner painted one, and between these is a third cone shaped dome supporting the lantern and cross. In order to stabilise the dome therefore, in addition to the pier reinforcement, the dome was encircled at two levels with a 'Staybrite' chain each formed from twenty two forged links joined with 4 inch (100 mm) diameter headed 'Staybrite' pins. The chains were encased in concrete but small areas are exposed to the elements. A further one hundred and forty four tie rods, also in 'Staybrite', were used to fasten the outer dome to the inner dome. These comprised: 48 tie rods, 4 inches in diameter and 38 feet six inches long, 48 at 2.5 inches diameter and 16 foot 6 inches long, and forty eight at 2.5 inches diameter and 18 foot 3 inches long. These were installed horizontally and at an incline around the dome.

This solution has proved to be very effective. No further spalling has since taken place, and no maintenance has been necessary to the 'Staybrite' components. The only close inspection of the stainless steel was carried out in 1993 by the author and a corrosion specialist from British Steel Stainless (now Outokumpu Stainless), who reported the stainless steel to be in the 'as installed' condition.

Figure 2   Assembling the links of the stainless steel chain at St.Paul's Cathedral

It is worth noting that other cathedrals have also used stainless steel in repair. In the late 1990s, Chester cathedral used stainless steel reinforcing bar in the repair of the top of the cathedral tower where the carbon steel reinforcement had corroded. Due to its height and position, a new 12 x 8 m ring beam was cast in-situ with pumped Lytag concrete and stainless steel reinforcement. In 2003, the new song school to the Cathedral was built entirely with stainless steel reinforcing bar. York Minster used specially turned stainless steel pins to join the large fourteen inch square fresh oak timbers in the rebuilding of the South Transept roof structure which was destroyed by fire in 1984. Stainless steel was used as it has the high corrosion resistance necessary to combat the aggressive sap in the fresh oak timbers.

CASE STUDY 2: REPAIR OF THE PIERS OF THE LIDINGO RIVER CROSSING RAILWAY BRIDGE

Built in 1971, the 1100 m long steel and concrete railway bridge in Stockholm required repair of all 24 supporting piers after only thirty years of service due to corrosion of the carbon steel reinforcement in the tidal zone. Carried out over a two year period, the €1.5 million repair involved constructing caissons around each pier and pumping out the water surrounding the pier to facilitate removal of the concrete cover by high pressure water jet. The bottom 2 m of concrete cover was removed and the carbon steel reinforcement repaired as necessary. A new outer layer of 16 mm diameter stainless steel reinforcement was wrapped around the pier and 65 mm of aerated concrete cover applied for frost protection as the river freezes over in winter periods. Austenitic stainless steel grade 1.4436 (316) was chosen for the repair to ensure that no further corrosion of the reinforcement will occur in the pier tidal zone for an estimated hundred years.

Figure 3   Lidingo river crossing railway bridge

Figure 4   New stainless steel reinforcement in the splash zone

CASE STUDY 3: ANGCHUANZHOU BRIDGE, HONG KONG
(STONECUTTERS BRIDGE)

Due to be completed in 2008, the Stonecutters Bridge, as it is referred to, will be the longest bridge of its type in the world. [3] With an overall length of 1596 m and a span of 1018 m, the bridge is at the entrance to the Kwai Chung container port in Hong Kong. The cable stayed bridge carries a twin deck suspended from two, 295 m high, single pole towers. The deck will be suspended 75 m above the water to allow for navigation clearance. The towers are concrete to a height of 175 m, (reinforced with stainless steel reinforcement in the splash zone), and a composite stainless steel and concrete construction above this level. Flanged sections fabricated from 20 mm thick duplex 2205 (1.4462) stainless steel plate clad the top 120 m of the towers. The sections are bolted together. About 40,000 300 mm long stainless steel studs are welded to the inside of the cladding to provide the bond to the concrete and transfer the load from the concrete to the stainless steel. The high design strength of the duplex stainless steel, 450 N/mm^2, and its high corrosion resistance in a marine environment were key to its selection. A matte non-directional finish was selected for the stainless steel for aesthetic reasons and to simplify fabrication. A significant advantage of the composite construction is that it stabilises the structure from unacceptable vibrations of the cable stays, and the stainless steel requires no maintenance.

In addition to the 2000 tonnes of duplex outlined, the concrete piers and the main tower splash zones will be constructed using about 2900 tonnes of stainless steel reinforcing bar up to 50 mm in diameter. This will provide for maximum durability and minimum maintenance of the structure over its service life of 120 years.

Figure 5   Stonecutters Bridge, Hong Kong
(artists impression)

CASE STUDY 4: SHENZEN WESTERN CORRIDOR BRIDGE

The second of two major bridge structures in the Far East currently under construction is the Shenzen Western Corridor Bridge. Both bridges incorporate stainless steel reinforcement to provide for maximum durability and minimal maintenance in this region where saline waters and high humidity levels create aggressively corrosive conditions. The 3.2 km long Shenzen bridge provides a major three lane highway across Deep Water Bay connecting Ngau Hom Shek and the Hong Kong Western Corridor. The stainless steel reinforcement was used in the splash zone of the supporting piers and select portions of the deck. Over 1250 tonnes of stainless steel reinforcing bar grades 1.4401 (316) and 1.4462 (duplex 2205) in diameters ranging from 16 to 40 mm were used in the construction.

Figure 6   Stainless steel rebar in the deck of the Western Corridor Bridge

DESIGN GUIDANCE AND STANDARDS

Since 2001, new Standards and design guidance on the use of stainless steel reinforcement has become available enabling new bridge structures, such as those illustrated in this paper, to be built for long term durability. A summary of the two major design documents is as follows.

BS6744:2001 STAINLESS STEEL BARS FOR THE REINFORCEMENT OF AND USE IN CONCRETE [4]

The revised standard contains a wider range of austenitic stainless steels than its predecessor and includes guidance on where each would be appropriate depending upon the expected service conditions. Stainless steel rebar is produced to two strength levels, 500 N/mm^2 and 650 N/mm^2, in the diameters shown in Table 1. Stainless steel reinforcing bar is produced in the material grades shown in Table 2 with the recommended service conditions.

Table 1   Strength and size range of stainless steel reinforcing bar

STRENGTH GRADE	NOMINAL SIZE, mm
500N/mm^2	3, 4, 5, 6, 7, 8, 10, 12, 14, 16, 20, 25, 35, 40, 50
650N/mm^2	3, 4, 5 ,6, 7, 8, 10, 12, 14, 16, 20, 25

Table 2   Material grades of stainless steel reinforcing bar & service conditions

GRADE	A	B	C	D
1.4301	1	1	5	3
1.4436	2	2	1	1
1.4429	2	2	1	1
1.4462	2	2	1	1
1.4529	4	4	4	4
1.4501	4	4	4	4

Key to Table 2

1 – Appropriate choice for corrosion resistance and cost

2 – Over specification of corrosion resistance for the application

3 – May be suitable in some instances: specialist advice should be obtained

4 – Grades suitable for specialist applications which should only be specified after consultation with corrosion specialists

5 – Unsuitable for the application

Key to service conditions

A – For structures or components with either a long design life, or which are inaccessible for future maintenance

B – For structures or components exposed to chloride contamination with no relaxation in durability design

C – Reinforcement bridge joints, or penetrating the concrete surface and also subject to chloride contamination (e.g. dowel bars or holding down bolts)

D – Structures subject to chloride contamination where reductions in normal durability requirements are proposed (e.g. reduced cover, concrete quality or omission of waterproofing treatment.)

HIGHWAYS AGENCY GUIDELINES BA84/02 [5]

The new guideline document BA84/02 was prepared for the Highways Agency [HA] by Arup R & D and incorporated in the HA Design Manual for Roads and Bridges. The agency recognises that stainless steel reinforcement can improve the durability of the structure, reduce maintenance, and minimise the costly effect of lane closures and traffic disruption during periods of maintenance/repair. As shown in Tables 4 and 5, BA84/02 recommends the use of stainless steel reinforcement according to the location of the bridge, and indicates where stainless steel should be used, and the appropriate grade of stainless steel.   The guidelines recognise the contents of BS6744: 2001, and that stainless steel reinforcing bar is issued with a CARES (a standards compliance certification body) certificate. The agency also accepts that stainless steel reinforcement can be used cost-effectively by siting the stainless steel in the high-risk elements of the structure and using carbon steel reinforcement in protected low risk areas.

The stainless steel content can vary, therefore, from a small percentage to 100% depending upon the bridge location. Due to the high corrosion resistance of stainless steel, the agency recognises that the rules developed to improve the durability of carbon steel reinforced structures can be relaxed when using stainless steel reinforcement. This relaxation, however, should not be taken to imply that the concrete quality and workmanship can be relaxed when using stainless steel, and it is important for these qualities to be maintained to realise the full durability benefits of the improved structure. The relaxation rules for using stainless steel are shown in Table 3.

Classification of Structures

The classification of structures and elements by the Highways Agency identifies the areas of potential corrosion risk and the potential disruption of carrying out future maintenance. The first classification identifies where stainless steel may not be appropriate.

Table 3   Relaxation rules for using stainless steel reinforcement

DESIGN CONDITION	RELAXATION
Cover	Cover for durability can be relaxed to 30mm where stainless steel is used irrespective of the concrete quality or exposure condition
Design crack width	Allowable crack width increased to 0.3mm
Silane treatment	Not required on elements with stainless steel

Table 4   Classification of structures and elements

CLASSIFICATION	RECOMMENDATION
U	New structures where stainless steel is not considered appropriate as they are unlikely to be exposed to high concentrations of chlorides. For structures in the UK motorway and local trunk road system this category is likely to include minor crossings such as footbridges, farm access crossings, buried structures and elements that are remote from the highway
Category C	New structures – stainless steel is appropriate for the partial replacement of reinforcement in selected substructure and superstructure elements exposed to chlorides from road De-icing salts. For example parapet edge beams, slab soffits, columns, and walls adjacent to the road.
Category B	New structures – stainless steel is appropriate for the partial replacement of reinforcement in substructure elements exposed to seawater. As for category C plus all parts of the substructure (columns, abutments, walls etc) within the tidal and splash zones.
Category A	New structures – stainless steel is appropriate for the complete replacement of reinforcement in the substructure, superstructure, and deck slab, except foundations or piles.

Selection of Steel Grade

With regard to BS6744:2001 and the HA BA84/02 guidelines, the advised grade of material is shown in Table 5 for a range of exposure conditions.

GALVANIC CORROSION ISSUES

Studies have shown [6,7] that cost-effective structures with enhanced durability can be achieved by the placement of stainless steel reinforcement in the elements of the structure at highest risk of corrosion, or in those elements of the structure which are inaccessible or expensive to repair. Carbon steel reinforcement would therefore be sited where chloride ion would be unlikely to reach during the design life. In new structures, and in the repair of damaged structures stainless steel reinforcement will be in direct contact with carbon steel reinforcement. The risk of a galvanic reaction has concerned civil engineers, however, this issue has been studied by practical test by authoritative bodies in Italy, Denmark, Sweden, USA, and Canada, and the results have conclusively shown that there is no detrimental reaction to the materials being in direct contact. On the contrary, it has been shown to have a beneficial effect in repair over coupling carbon steel with carbon steel. In a report on the studies by the National Research Council [8] of Canada for the Quebec Ministry of Transport et al, in 2004, techniques included cyclic voltammetry, linear polarisation, potential dynamic, AC impedance and galvanic coupling measurements. The results of galvanic coupling of corroding carbon steel (cs) and passive cs compared with corroding cs and stainless steel (ss), showed that the increased rate of corrosion of the corroding cs was significantly higher in the former. These results concur with the tests carried out by European research authorities. They all conclude that the use of stainless steel can significantly extend the service life of structures exposed to chlorides.

Table 5   Stainless steel grades and appropriate bridge applications

MATERIAL GRADE	EXPOSURE CONDITION
1.4301	Stainless steel embedded in concrete with normal exposure to chlorides in soffits, edge beams, diaphragm walls, joints and substructures
1.4301	As above but where design for durability requirements are relaxed in accordance with Table 3.7.
1.4436	As above but where additional relaxation of design for durability is required for specific reasons on a given structure or component i.e. omission of a waterproof membrane
1.4429 1.4436	Direct exposure to chlorides and chloride bearing waters, for example, dowel bars, holding down bolts and other components protruding from the concrete
1.4462 1.4429	Specific structural requirements for the use of higher strength reinforcement and suitable for all exposure conditions

CONCLUSIONS

The use of stainless steel in St. Paul's Cathedral has shown that sustainable construction can be realised in buildings with a very long design life. The infrastructure of highways and bridges also have a long design life and by using stainless steel as outlined in design guidance issued in the 21st century, sustainable structures can be built to last well into the next century. Corrosion of carbon steel reinforcement has left a legacy of bridges with an impaired life and an enormous repair bill.

The use of stainless steel has often been considered as too expensive, however, selective placement in high risk elements is now being adopted as it has been shown to be a cost-effective solution to building durable highway bridges as illustrated by the two major new structures in this presentation. Other countries such as Sweden and Eire have also adopted the design approach as they are designing for durability and minimal maintenance.

St. Paul's restoration shows that civil engineers are used to being at the forefront of innovation and the tradition continues with the new bridges illustrated in this paper.

REFERENCES

1. COX R.N., OLDFIELD J.W., The long term durability of austenitic stainless steel in concrete. Proceedings of the 4th International Symposium on Corrosion of Reinforcement construction, Robinson College, Cambridge, 1-4 July 1996.

2. L.BERTOLINI, M.GASTALDI, M.P.PEDEFERRI, P.PEDEFERRI., Effects of galvanic coupling between carbon steel and stainless steel reinforcement in concrete, International conference on corrosion and rehabilitation of reinforced concrete structures, Orlando, December 1998, Proceedings on CD-Rom.

3. 'Nickel', July 2005, Volume 20, Number 3. The Nickel Institute, Toronto.

4. BRITISH STANDARDS INSTITUTION, BS6744:2001 Stainless steel bars for the reinforcement of and use in concrete – Requirements and test methods.

5. THE HIGHWAYS AGENCY. BA84/02: Design manual for roads and bridges – Volume 1, Section 3, Part 15, Use of stainless steel reinforcement in highway structures.

6. KNUDSEN, A., JENSEN, F.M., KLINGHOFFER, O., AND SKOVSGAARD, T., Cost effective enhancement of durability of concrete structures by intelligent use of stainless steel reinforcement, International conference on corrosion and rehabilitation of reinforced concrete structures, December 1998, Orlando, Florida.

7. BRITISH STAINLESS STEEL ASSOCIATION, CD-Rom, Stainless steel reinforcement for concrete.

8. NATIONAL RESEARCH COUNCIL CANADA, Investigation of the effects of galvanic coupling between stainless steel and carbon steel reinforcements in concrete, report for Alberta Infrastructure, City of Ottawa, Quebec Ministry of Transport, Nickel Institute, Valbruna Canada Inc.

DEVELOPING WORLD: RESPONSIBILITIES IN CHANGING ENVIRONMENTS AND DEMANDS

CONCRETE CONSTRUCTION: SUSTAINABLE OPTIONS

S A Reddi
Gammon India
India

ABSTRACT. This paper deals with sustainable concrete construction in the context of the built environment being fundamental to a sustainable society. The built environment provides spaces for education, comfort, community, safety and science. Designers, concrete technologists, owners and contractors can make a difference in crating a more sustainable built environment. The paper highlights the influence of concrete in reducing adverse environmental impacts and use of industrial byproducts such as fly ash, GGBS, silica fume etc. Typical case histories highlighting the sustainable aspects in variety of projects- tall buildings, bridges, dams, sewage works etc. are included.

Keywords: Climate change, Impact on concrete raw materials, Energy consideration, Self compacting concrete, LEED certification and concrete, Codes, Specifications.

S A Reddi is a Chartered Civil Engineer and Deputy Managing Director (Retired) Gammon India Ltd. A Fellow of the Indian National Academy of Engineering, he is active in the development of cost effective High Strength Concrete and its applications.

INTRODUCTION

The world commission on Environment and Development defines Sustainable Development as "Development that meets the needs for present without compromising the ability of the future generations to meet their own needs". The built environment is fundamental to a sustainable society, as it provides spaces for education, comfort, community, safety and science. Designers, Engineers, Owners, Contractors and Manufacturers can make a difference in creating a more sustainable built environment.

The Fourth Report issued by the Intergovernmental Panel on Climate Change, has been named the top-most revelation of the year 2007. Unhindered emissions of carbon dioxide might lead to catastrophic changes including sea level rise, more severe weather, melting of glaciers, etc. The third assessment report of the same panel predicts an increase in surface temperature of $1.45°C$ over the period 1990–2100.

Concrete influences the impact on environment - quarrying of raw materials; the energy used in its production and consequent carbon dioxide emissions. Relevant industries are actively involved in reducing the environment impacts of the production of cement and concrete. There is accent on more efficient use of resources in concrete production including re-used materials and byproducts from industrial processes (Fly-ash, GGBS, Silica fume etc.).

Sustainability Options

- Sustainable cementitious materials
- Sustainable aggregates
- Sustainable concrete manufacture
- Sustainable designs and construction in concrete
- Minimum use of cementitious materials

Energy Efficient Raw Materials

An outstanding example is the construction of Sardar Sarovar Dam and concrete lined canal system in Western part of India (Figure 1). The project involves about 25 million m^3 of concrete, one of the largest concrete intensive projects in the World. Almost all the concrete is being manufactured utilizing the natural river gravel. Reliance on blasting and operation of crushers for production of aggregates has been eliminated.

Benefits of Concrete Frames in High Rise Building

Modern methods of construction aimed at range of products of techniques to improve efficiency in construction that includes off-site manufacturing of components, on-site fabrication and improved management methods. Ready-mixed concrete is now batched off-site and delivered in exact quantities to fill readymade formwork with minimal waste. Hundreds of Ready Mix Concrete (RMC) plants are now in operation all over India. The recent rises in reinforcement and steel prices have increased frame cost. However, in developing countries such as India, the difference between steel and concrete frame costs remains significant. Thus the majority of high rise structures in India are realized with concrete frames. The costs are further reduced by using high strength concrete (60 to 80 MPa) and Self Compacting Concrete in some cases. An outstanding example concerns the tallest building in Mumbai under construction (2008).

Figure 1   Sardar Sarovar Dam and 70 m bed width concrete lined canal system in Gujarat, Western India

Concrete and LEED (Leadership in Energy and Environmental Design)

Using concrete can facilitate the process of obtaining LEED Green Building certification. LEED is a point rating system devised by the US Green Building Council to encourage sustainable design. Use of sustainable material attracts one of the fine main credit categories. Using concrete can increase the number of points awarded in the LEED system. Above 20% of the credit goes towards Materials and Resources. Use of Self Compacting Concrete reduces site disturbance, earning additional credit points. Use of concrete instead of asphalt for construction of roads and pavements reflects solar radiation rather than absorbing the same. Concrete has a reflectance of about 0.35 as against 0.05 for asphalt

Buildings constructed in concrete or masonry possess thermal mass that helps moderate indoor temperature extremes and reduces peak heating and cooling loads. It can lead to a reduction in heating ventilating and air-conditioning equipment capacity. Such reduction can represent energy and construction cost savings.

Locally Available Materials: LEED credit supports the use of local materials and reduced transportation distances. Concrete usually qualifies since ready-mixed plants in India are generally within 20 km of a job site. Additional points are earned if the raw materials are extracted within 800 km. Cement and supplementary cementicious materials used for structures are also manufactured within 800 km. Four points are available through innovation credits.

Mumbai's Tallest Residential Tower

The Palais Royale tower design is based on innovative Green strategy. It also uses concrete with substantial cement replacement with fly-ash in the case of Self Compacting Concrete. Acute land shortages, together with demographic life style changes have dictated development of multi-storage residential structures. The Palais Royale residential tower under construction in Mumbai (2008) has a height of about 316 m (Figure 2).

The first apartment starts at 87 m above ground. All services (car parks, automated building management systems, back-up generator facilities, ventilation and air-conditioning, building security, air and water re-cycling facilities, etc.) are located below 87 m level.

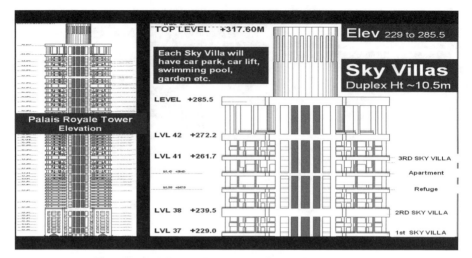

Figure 2   Palais Royale, 316 m Tall Building in Mumbai

Optimization of energy used and provision of all-round comfort include extensive use of natural light and ventilation as well as insulation. The building solar gains are controlled by active shading system while the natural air supply is sourced from the outdoors. Hot water for the residents is ensured by solar heating panels installed on the roof. Dependence on external source of power is partly reduced by installation of Wind Mill on the roof. The structure is located close to the sea, facilitates generation of about 2 MW power by the wind mill. Solar panels are expected to be active for nine months in a year, excluding the heavy fall periods during monsoon. The structural frame work is in High Strength reinforced concrete (40 MPa and 60 MPa) and Self Compacting Concrete (Grade 80 MPa). Ordinary Portland cement grade 53 with fly ash is used for making concrete to enhance the durable quality of concrete.

In the fresh state they enhance flowability and pumping qualities apart from reducing segregation, bleeding and plastic shrinkage. Concrete containing fly ash enables reduction in the cement content and therefore an increase in the building's "Green" ratings. This building is expected to be the first in the residential sector in India to obtain the international LEED rating.

US Green Building Council created the Leadership in Energy and Environment Designs (LEED) green building rating system. As many as 21 of the minimum 26 points needed for LEED certification can be earned through appropriate use of concrete. In Green Building, concrete helps in optimizing energy utilization, contains re-cycled materials, creates sustainable environment, is manufactured locally and builds durable structures. Since 9/11, focus on sustainability is even sharper all over the world. Very tall buildings have now shifted from using structural steel columns to either high strength concrete columns or concrete –steel composite columns, in order to provide better fire resistance.

Property prices in down-town Mumbai are one of the highest in the World (about US $10,000 per m^2.). As such, any saving in column size adds to the availability of saleable floor area. Hence the decision to go for 80 MPa concrete which is the highest grade permitted in the Indian Concrete Code (IS: 456 – 2000). The same code is quite conservative in design of reinforcement requirements and generally results in very close spacing of large diameter reinforcement bars leading to difficulties in proper placement and vibration of concrete. Hence it was decided to use self compacting concrete.

The tower is founded on a 3.5 m thick reinforced concrete raft. Self compacting concrete have been used. The space between the raft and the entrance level is utilized for two basements, housing the Building management system, utilities such as air-conditioning plants, cellars, stand by generators etc. A 7-level podium is followed by three amenities levels. All columns from the raft to the girder level are of 80 MPa grade of Self Compacting Concrete, against the normally adopted 40–60 MPa Standard Concrete.

The development of economical, cost effective proportioning of 80 MPa concrete mix by the author has resulted in very low cement content (350 to 400 kg/m^3) plus Fly-ash and Silica Fume. The concrete is pumped using high pressure single stage pump for the height of about 300 m. Reinforcement steel is of Fe500 grade, in pre-fabricated cages for columns, beams and slabs, interconnected by couplers.

Figure 3   Couplers for pre-fabricated reinforcement

The use of high strength Self Compacting Concrete and Fe500 grade reinforcement have contributed to sustainability:

- Cement content has been reduced by 25%
- 30% cement has been replaced by fly-ash
- Use of Self Compacting Concrete enables work to be carried out round the clock, as there is no disturbance to the residents in the surrounding areas (noise pollution created by the vibrators). Otherwise local regulations prohibit night work due to noise pollution
- Due to use of couplers, there will be a 20% reduction in the total tonnage of reinforcement.

The concrete columns and shear walls are jump-formed. The form panels for beams and slabs use a unique patented system, scratch proof and capable unlimited use, realizing smooth under-side not requiring further treatment or plastering.

The beams and slabs are in 60 MPa grade concrete. The mix is designed for High Early Strength to enable removal of formwork (with props left under) within 24 hours of concreting. A four day time cycle per floor is being aimed at. High Strength Concrete inhibits chloride ions from reaching reinforcement steel and consequent cracking and spalling of concrete.

The building has created a new environment performance benchmark for India's residential buildings.

Chenab Bridge at Aknoor near Jammu, North India

The main central span of 160 m across the Chenab river was dictated by the perennial, fast flowing river conditions where no foundation is practicable in the stream (Figures 4 and 5). The working season is short. 50 MPa grade concrete was used for the decking which has been constructed by symmetrical cantilevering out from each pier. Normally for such grades of concrete in India, about 450–500 kg of cement is used per m^3. The Indian Roads Congress specifies a minimum of 400 kg of cement. Special efforts where made to design a mix successfully with only the minimum cement content. Segmental construction was realized on a six day cycle, based on pre-stressing at 60 hours after concreting, with a minimum high early strength of 40 MPa. PC based admixtures were used.

Due to congestion created by the reinforcement and anchor studs of the special bearings, normal concreting was ruled out. Special Self Compacting Concrete with smaller size aggregates was poured to cover the area around the anchorages. The concrete was designed for 24 hour strength of about 40 MPa.

Figure 4   Chenab Bridge at Aknoor — Central Span under construction

Figure 5   Chenab Bridge at Aknor: Typical bearing

READY MIX CONCRETE

The concrete is manufactured by binding and mixing the constituents in large batches in ready-mixed concrete plants and hauled by truck mixer in to the construction site. These processes require energy and produce some waste (dust, unused concrete and washed water contaminated with concrete). Some of the waste may be, at-least partially, are claimed and re-used.

Embodied carbon dioxide of concrete mix for typical building structural concrete with OPC and Blended Cements are given in Table 1.

Table 1   Embodied carbon dioxide levels of concrete

GRADE OF CONCRETE	UNIT, kg carbon dioxide per m^3
40 OPC	372
40 OPC with 30% fly ash	317
40 OPC with 50% GGBS	236

Concrete is one of the more societal building materials when both the energy consumed during the manufacture and inherent properties in use are taken into account.

The concrete mix contains cement in varying quantities depending on the minimum specified in Codes and Specifications, the technology available and construction practices prevailing in the country. The Indian Code specifies minimum cement content of 300 to 360 kg/m^3 depending on environment exposures, higher than the EUROCODE. The Indian Roads Congress Specifications insist on minimum 400 kg/m^3. The construction practices are based on use of even higher cement content up to 540 kg/m^3 for higher grades of concrete. Apart from over conservative Codes and Specifications, the teaching of the subject concrete in the Universities leaves much to be desired and results in over utilization of cement with no corresponding benefits, but with problems of creep, shrinkage, crack etc.

Ready-mixed concrete plants are gaining popularity, particularly for large volume concrete. For the expansion of Delhi International Airport, two Nos. of 240 m^3 per hour batching plants manufactured in Bangalore are in use. The use of such large capacity batching plants results in lower unit power consumption.

Bandra Worli Sea Link in Mumbai

Bandra Worli Sea Link in Mumbai, 5.86 km long and 8 lanes under construction (2008) to decongest the city traffic (Figure 6): The sea link consists of predominantly 50 m spans pre-cast segmental construction and two cable stay bridges 600 m and 350 m long. The structure is founded on 1500 mm dia piles. The pylon is a diamond shaped four legged concrete structure. 60 MPa grade concrete mix contains only 320 kg cement plus 80 kg fly ash.

Figure 6   Bandra Worli sea link Mumbai (under construction – 2008)

Extradosed Viaduct Delhi

The extradosed viaduct is a railway bridge with a main span of 93 m, crossing five railway tracks. The deck cross section is U shaped. The extradosed cables are covered by a concrete beam which allows considering them as internal pre-stressing tendons. The bridge is first of its kind in the world erected using cantilever construction method. High strength concrete up to 75 MPa grade was utilized. The bridge is curved for a part of the distance. The deck consists of pre-cast U shaped segments with thicker webs to locate the extradosed cable anchorages (Figure 7).

Figure 7   Extradosed Viaduct in Delhi

Marine Outfalls at Bandra & Worli in Mumbai

The outfalls are constructed to dispose off waste water into the sea. The system consists of a 3.4 km long undersea tunnel, 3500 mm dia. located 30 m below sea level. The inlet is through a vertical shaft of 9 m diameter and 8 m inner diameter inlet shaft. Tunnel liner segments were pre-cast in 45 MPa concrete. Concrete mix with 70% GGBS (Ground Granulated Blast Furnace Slag) to satisfy the durability requirements, though it involved transportation of GGBS over a long distance (Figure 8).

Figure 8   Marine Outfalls at Bandra & Worli - Inlet Shaft

Codes and Specifications

The specification requirements should be appropriate. Currently there is a tendency to over-specify, over-generous floor ratings specified to provide flexibility for future use should be challenged and realistic floor ratings used instead.

Wood-free Formwork

In a typical patented system, any type of slab work can be realized with just one system. The same components are used for different applications and the number and position of props is determined by the system. Hence the most important requirements of a building are met with easily and flexibly for different ground plans, slab thickness, floor heights, concrete finish etc. The drop head permits early stripping and fast transport to the next floor level. Three days concreting cycles are feasible. It is planned to realize a four days cycle at Palais Royale.

The patented prop is a combination of steel inner tube and aluminium outer tube. The quick lowering system allows the stress in the prop to be released with one strike of a hammer. After stripping the prop automatically resets and locks in the original position. The slide carriage allows sliding back the formwork by 700 mm from the wall without a crane. Platform and formwork remain connected as one unit. The overall platform width of 2300 mm provides sufficient working space even with the slide back.

High Strength Concrete

High Strength Concretes are characterized by a high cement factors and very low w/c ratios. It is extremely difficult to obtain proper workability with such mixes and to retain the workability for a sufficiently long period of time. High dosage of water reducing agents then become a necessity. The resulting sticky mixes are equally difficult to place and compact fully and efficiently. There is a critical limit for the water content below which high admixtures dosages are unhelpful and undesirable.

Very low w/c ratios and very high admixture dosages lead to concrete to early cracking (15 minutes to six hours after placing). Such cracking is caused by dual effects of lack of bleeding and the inability of whatever bleed water is present to move up to the surface. Early occurrence of such cracking leads to wide cracks of one to three mm width whereas later age cracking results in greater number of hairline cracks.

High Early Strength

Nowadays there is emphasis on fast track construction based on faster cycle times, particularly in high rise buildings. High early strength requirement is also used for self compacting concrete. Cement manufacturers in India are achieving the results by finer grinding or control of particle size distribution. The chemical and physical effects of use of silica fume also contribute to high early strength.

Chemical admixtures are increasingly used to obtain high early strength, primarily by reducing the water content; their formulation can be tailored individual requirements; their dosages can also be tailored to local environmental conditions. However there are practical problems in implementation. Sometimes, it takes inordinately long time for the concrete to set.

Due to very low water content, the concrete tends to develop shrinkage cracks despite precautions. Addition of a shrinkage reducing admixture to the concrete is some times useful; such admixture helps reduce the surface tension of the pore solution. Alternatively reducing the fineness of cement is another route; however it is difficult to obtain cement with reduced fineness in India.

SUMMARY

The paper highlights developments in sustainable concrete construction in a developing country. Few real life cases are also described, highlighting new technologies and use of fly ash for high strength as well as self compacting concrete.

THE EFFECTS OF BOUNDARY CONDITIONS ON REINFORCED CONCRETE DEEP BEAMS

A Arabzadeh

Tarbiat Modares University

Iran

ABSTRACT. Reinforced concrete deep beams could be divided into four categories, Simply Supported, Fixed Ended, Continuous and Cantilever. Therefore, their end supports could have different effects on the behaviour of the beams. In this paper, 43 deep beams with span to depth ratio less than 3 and different reinforcement arrangement, boundary conditions are constructed and tested, and their results have been investigated. All the beams were loaded from the top mid centre, except four beams which were loaded unsymmetrical. The results indicate that the boundary condition have the most important effects on ultimate loads, modes of failure and defections. The crack formation and their patterns are almost the same. Further more, the ultimate loads of the beams with fixed ends are 1.2 to 6 times simply supported, and continuous deep beams, depends on amount of main bottom and top reinforcement used. Also in fixed ended deep beams, modes of failure and ultimate loads mostly depend on top main reinforcement. For theoretical study of the beams most of the available standards and published proposed methods are used. It can be seen from the results that none of the existing methods are suitable for the fixed ended deep beams, except the two proposed methods. But the method by ACI (1999) and CIRIA (1977) have given relatively acceptable results for simply supported deep beams.

Keywords: Deep Beam, Reinforce Concrete, Shear, Flexural, Failure.

Dr Abolfazl Arabzadeh, Associate Professor of Civil engineering, Tarbiat Modares University, Tehran, Iran

INTRODUCTION

Concrete has been used in construction of buildings, bridges, dams, silos, shear walls etc. for many years. An important element of these structures is the beam, being constructed with different properties and geometry. When the span/depth ratio is below 4, it is considered as a deep beam with its behaviour and analysis differing from those of ordinary shallow beams.

The behaviour of these beams is studied by many researchers [1-12], and many Codes of practice proposed different formula for prediction of ultimate load [13, 14, 15]. But none of the existing Codes of practice refer to deep beams with fixed ends condition, almost all the published papers and Codes of practice have given recommendation for simply supported deep beams and to lesser extend for continuous deep beams. The only published works on fixed ended deep beams are the works published by the author [4, 11, 12]. In this paper the effects of boundary condition on ultimate load and behaviour of deep beams are investigated and discussed.

To study the effects of boundary condition and other properties of deep beams, 18 fixed ended, 18 simply supported and 8 continuous deep beams with the same properties are constructed, and tested. Considering the geometry of these beams their strength controlled by shear and this is because before failure distribution of forces take place, and shear strength of these beams are mush higher than shear strength of shallow beams. Furthermore, stress-strain relationship is not linear, therefore flexural theory for ordinary beam is not suitable for these beams and especial analysis is required. Because of above differences between deep and shallow beams, many methods and Codes of practice have been proposed. The CIRIA Guide 2 [14] has given a method for prediction of ultimate load of simply supported deep beams and continuous deep beams. Recently ACI 2005 [13] and CAN (1984) [15] have proposed Strut and Tie method for the analysis of simply supported and continuous deep beams.

Experimental Work

To achieve the objective of the investigation, 48 deep beams with different properties and boundary conditions were constructed and tested, Table 1. The tests included a wide range of geometrical parameters and reinforcement content which were described in detail in previous papers [1-5, 11, 12], here, only the summarized data are presented in Table 2. The main reinforcement arrangement of the beams were such that, different modes of failure to happen.

Table 1   Dimensions of tested beams

BEAM SERIES AND NUMBER	GEOMETRY				
	Span (L), mm	Depth (h), mm	Width (b), mm	Span/depth ratio (L/h)	Slenderness ratio (h/L)
1B1 to 2B8	800	400	50	2	8
3B9 to 3B13	1680	600	60	2.8	10
4B14 to 4B18	1680	750	75	2.24	10

28 days after construction of each beam, they were tested with loading system shown in Figure 1. All the beams were loaded from top mid–centre, except beams, (4B16 and 4B18) which were loaded unsymmetrical. The concrete mix was based on concrete grade 40 N/mm^2 and for each beam was almost the same [1, 2, 4].

Table 2   Summarized test results

Beam Series and Number	REINFORCEMENT				BEAM						Ratio		
	No. and Type				SS: simply supported; C: continuous; FE: fixed ended								
	Main		Web		Mode of Failure			Ultimate Load P$_s$, kN			$\dfrac{P_s}{P_s}$	$\dfrac{P_c}{P_s}$	$\dfrac{P_f}{P_s}$
	Top	Bottom	Hour.	Veer.	SS	C	FE	SS	C	FE			
1B1	1R6	1T16	3R6	7R6	S	S	F+S	155	183.4	267	1.00	1.18	1.72
1B2	1T12	1T16	3R6	7R6	S	S	S	181	182.2	282	1.00	1.00	1.56
1B3	1T16	1T16	3R6	7R6	S	S	S	202	193.4	275	1.00	0.96	1.36
1B4	1T20	1T20	3R6	7R6	B	S	S	165	184.0	262	1.00	1.12	1.59
2B5	1R6	1R6	3R6	7R6	F	-	F+S	38	-	231	1.00	-	6.08
2B6	1T10	1R6	3R6	7R6	F	-	F+S	46	-	253	1.00	-	5.50
2B7	1T12	1T12	3R6	7R6	S	S	S	132	140.0	236	1.00	1.06	1.89
2B8	1T16	1R6	3R6	7R6	F	-	S	47	-	267	1.00	-	5.68
3B9	2T12	2T12	10R6	34R6	F	-	S	264	-	512	1.00	-	1.94
3B10	2T10	2T20	10R6	34R6	S	-	S	406	-	496	1.00	-	1.22
3B11	3T12	3T20	10R6	34R6	S	S	S	418	420.0	525	1.00	1.01	1.26
3B12	2T16	2T20	10R6	34R6	B	-	S	311	-	500	1.00	-	1.61
3B13	2T20	2T20	10R6	34R6		B	S	385	420.0	502	1.00	1.09	1.3
4B14	2R6	2R6	10R6	34R6	F	-	F+S	270	-	688	1.00	-	2.55
4B15	1T20	1T20	12R6	34R6	F	-	B	287	-	725	1.00	-	2.53
4B16	2T12	2T12	8R6	22R6	B	-	F+S	418	-	775	1.00	-	1.85
4B17	2T16	2T15	6R6	18R6	S	-	S	492	-	580	1.00	-	1.18
4B18	2T20	2T20	8R6	22R6	B	-	B	364	-	675	1.00	-	1.85

Notes: S = shear failure, F = flexural failure, F+S = flexural plus shear failure, B = bearing failure

a) Simply supported

b) Continuous

c) Fixed ended

Figure 1   Loading System of the beams

Analysis

The ultimate strength and modes of failure of the beams were obtained by using different Codes and proposed methods, and their results are compared with experimental results. The Codes of practice considered are ACI (1999 Revised 5), CIRIA Guide 2 (1977) CEB-FIP Model Codes (1978), CAN 23.3-M84 (1984) and proposed methods by different authors. All the above Codes and proposed methods are described in detail in pervious papers [1, 5, 6]. Those Codes and methods which have given reasonable results are presented in Table 3 and discussed here. In summary only the idealized failure planes together with assumed forces are shown in Figure 2.

It should be noted that the idealized failure planes have been carefully chosen from the experimental observation and assumptions made are:

1.   Failure plane take place between clear shear span,
2.   The web strength controlled by concrete

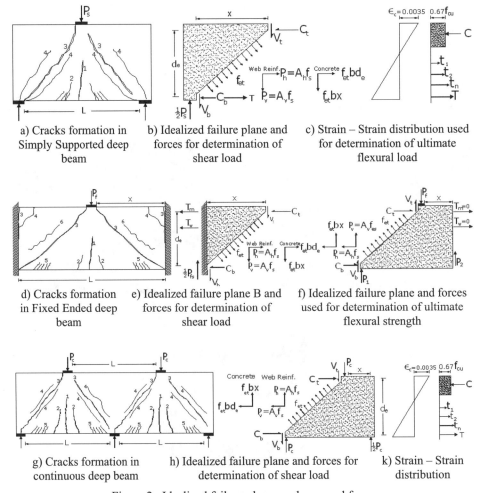

a) Cracks formation in Simply Supported deep beam

b) Idealized failure plane and forces for determination of shear load

c) Strain – Strain distribution used for determination of ultimate flexural load

d) Cracks formation in Fixed Ended deep beam

e) Idealized failure plane B and forces for determination of shear load

f) Idealized failure plane and forces used for determination of ultimate flexural strength

g) Cracks formation in continuous deep beam

h) Idealized failure plane and forces for determination of shear load

k) Strain – Strain distribution

Figure 2   Idealized failure planes and assumed forces

Table 3 Comparison of the ratio of Analysis and Test Results by different methods

Beam Series and Number	BEAM											
	Simply Supported				Fixed-Ended				Continuous			
	ACI	CIRIA	FEC	S.T.	ACI	CIRIA	FEC	S.T.	ACI	CIRIA	FEC	S.T.
1B1	0.82	0.75	0.97	1.08	0.97	0.48	0.85	1.00	0.38	0.67	0.95	0.85
1B2	0.74	0.64	0.88	0.93	0.92	0.46	0.96	0.95	0.39	0.65	0.92	1.01
1B3	0.65	0.58	0.76	0.83	0.94	0.47	1.00	0.98	0.41	0.64	0.88	0.97
1B4	0.83	0.86	1.02	01.13	0.99	0.56	1.10	0.97	0.41	0.64	0.95	0.98
2B5	0.45	0.36	1.12	0.46	0.76	0.12	0.74	0.92	-	-	-	-
2B6	0.35	0.28	0.72	0.40	0.71	0.11	0.70	0.88	-	-	-	-
2B7	0.60	0.52	0.83	0.75	0.76	0.42	0.74	0.90	0.46	0.65	1.00	0.96
2B8	0.36	0.29	0.92	0.41	0.67	0.1	0.84	0.82	-	-	-	-
3B9	0.48	0.48	0.72	0.75	0.54	0.39	0.75	1.03	-	-	-	-
3B10	0.64	0.53	0.50	0.75	0.65	0.56	1.04	1.02	-	-	-	-
3B11	0.73	0.68	0.56	0.60	0.51	0.44	0.91	0.98	0.73	0.64	1.00	1.04
3B12	0.89	0.75	0.67	1.04	0.53	0.46	0.95	1.00	-	-	-	-
3B13	0.79	0.74	0.62	0.96	0.52	0.45	0.95	1.00	0.73	0.64	1.05	1.04
4B14	0.12	0.11	0.53	0.24	0.48	0.07	0.81	0.99	-	-	-	-
4B15	0.12	0.1	0.48	0.22	0.41	0.07	0.89	1.00	-	-	-	-
4B16	0.42	0.38	0.62	0.51	0.48	0.23	0.67	1.00	-	-	-	-
4B17	0.53	0.49	0.64	0.66	0.63	0.53	1.20	0.99	-	-	-	-
4B18	0.77	0.87	0.80	0.66	0.51	0.44	0.86	1.00	-	-	-	-

Notes: FEC = force equilibrium concept; S.T. = Strut and Tie method

The ultimate shear strength (P_s) and flexural strength-(P_f) obtained from equilibrium of above forces, and following equations:

1. Simply Supported:

$$P_{ss} = 2[V_b + \frac{P_v}{2} + \frac{P_h d}{2} + f_{et}(\frac{x^2 + d^2}{2x}) + \frac{C_b d}{x}]$$

P_{fs} = Strain compatibility method is used [3, 5], Figure 2c.

2. Continuous:

$$P_{sc} = 2[V_b + V_t(1 - \frac{x}{L}) + f_{et} b(2x - \frac{x^2 - d^2}{L}) +$$

$$(2T_x + T_w + 2C_t - P_h)\frac{d}{L} + P_v(2 - \frac{x}{L}) + f_{et}b\frac{4de}{3}$$

P_{fc} = Strain compatibility method is used [2, 6], Figure 2k.

3. Fixed ended:

$$P_{sf} = 2[V_t(1 + \frac{d''}{d}) + f_{et} bx + P_v]$$

$$P_{ff} = V_t[2\frac{d''}{d'} + 2 - \frac{x}{L}] + P_V(2 - \frac{x}{L}) +$$

$$f_{et}b[2x - \frac{x^2}{L} - d^2] + (C_t - P_h)\frac{d}{L}$$

where

$$f_{it} = 0.3 f_c'^{2/3}, \qquad f_{et} = \frac{f_{cu}}{f_{cu}/f_{it} + (\frac{d}{x})^2},$$

$$f_s = \frac{E_s}{E_c} f_{et}, \quad P_v = A_v f_s, \quad P_h = A_h f_s$$

$C = 0.67 f_{cu} bd'; \; C_t = C \cos\alpha, \; V_t = C \sin\alpha$ and α is the angle between failure plane and soffit. It should be noted that the term in above equations are defined in Figure 2, and previous papers [1, 4, 5].

EXPERIMENTAL AND ANALYSIS RESULTS

The general behaviour and results may be summarized as follows:

a) The first crack to form was the flexural cracks at mid-span region. This crack labelled 1 and 2 in Figure 2a. The flexural cracking load was typically from 10% to 30% of ultimate load of simply supported and continuous deep beams and 17% to 42% of ultimate load of fixed ended deep beams Figure 2d. If the modes of failure happened to be flexural, these cracks continued up to point load, other wise, it was closed after formation of major diagonal cracks.

b) When the load increased, the diagonal cracks which were formed at mid-shear span were extended toward load bearing and point load [crack No.3 Figure 3]. These cracks were typically 0.5h to 0.9h long when first formed.

c) As load increased the number of diagonal cracks increased and struts formed [crack No. 4 Figure 2a].

d) As the load was further increased under load bearing (in fixed ended beams) and over the supports (in continuous and simply supported beams) start to spill. For those beams with small amount of top main reinforcement, 0.002bde, (in fixed ended beams 1B1, 2B5 and 2B6) flexural cracks (Figure 2d: crack labelled 3) occurred along the support, from top surface, and extended downwards. At the point of failure one of plane of failure take place along these cracks and other plane of failure along major diagonal cracks. This kind of failure for fixed ended beams called flexural plus shear failure.

e) From the experimental results four modes of failure were observed:

1- Pure flexural in simply supported deep beams, Figure 3a.

2- Flexural plus Shear in fixed ended deep beams, Figure 3b

3- Shear failure, Figure 4.

4- Local failure (Bearing), Figure 5.

Furthermore in fixed ended beams the modes of failure mainly depends on top main reinforcement, where as in continuous and simply supported beams both mode of failure and ultimate strength depends on bottom main reinforcement.

 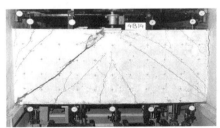

a) Simply supported   b) Fixed ended

Figure 3   Flexural and Flexural Plus shear failure of the beams

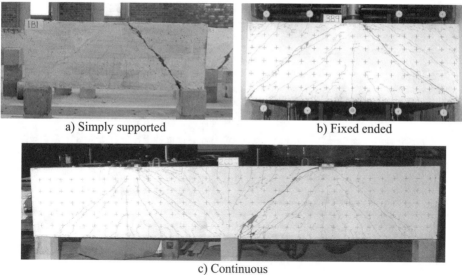

<table>
<tr><td>a) Simply supported</td><td>b) Fixed ended</td></tr>
</table>

c) Continuous

Figure 4   Shear failure of the beams

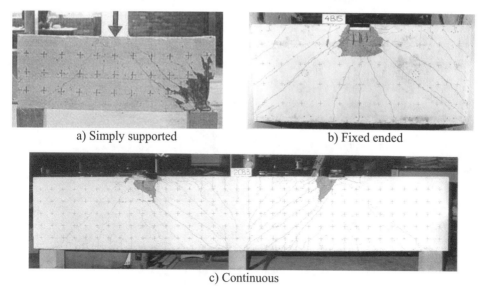

a) Simply supported   b) Fixed ended

c) Continuous

Figure 5   Local failure of the beams

f) Comparing the ultimate load of two simply supported beams, 2B5 (R6/R6) and 2B8 (T16/R6), it can be seen that they have almost the same ultimate load, where as the same beams with fixed ends have ultimate load, six times bigger, this high-light the effects of boundary conditions.

g) The comparison of results in Tables 1, 2 and Figure 6, indicate that not only the boundary conditions but also the dimensions, concrete strength, position of loading, amount of web reinforcement and amount of main reinforcement would affect the ultimate loads and modes of failure.

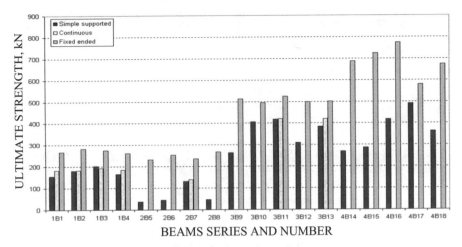

Figure 6 Ultimate strength of the tested beam

h) The maximum deflection occurs at mid–span, irrespective of boundary conditions, Figure 7, but the slop of deflection for fixed ended beam is much higher than simply supported and continuous deep beams. This is expected, because the fixed ends do not allow rotation of the end supports.

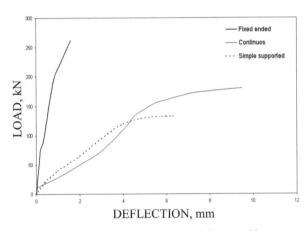

Figure 7   Deflection at mid-span of the tested beams

k) From a close examination of Table 3 it can be understand that proposed methods can predicate the ultimate loads of test beams and the results are accurate, safe, and consistent. All though Table 3 indicates that the Force Equilibrium method over estimated the ultimate strength of the fixed ended beam 4B17 (2T16/2T16), if the properties of this beam compared with beam 4B14 (2R6/2R6) it is expected that this beam, should carry more load. The lower ultimate strength could be due to the smaller web reinforcement in 4B17 compared to 4B14 and 4B15, this high lights the effect of the web reinforcements [4, 12].

CONCLUSIONS

From the experimental observations and analysis results the following conclusions are drawn:

1. Failure modes of the test beams were strongly dependent on boundary conditions top and bottom main reinforcement and the clear shear – span/depth ratio. When small amount (0.1 % to 0.4%) of top and bottom main reinforcement were used, the failure modes were flexure plus shear for fixed ended beams and flexural for simply supported beams. When the beams were reinforced moderately or heavily (0.568% to 1.6%) and had a clear span/depth ratio of 2 to 2.8, the failure mode was shear. When the clear shear span / depth ratio was 0.62 for main reinforcement, 1.1%, at the top and bottom, the failure mode was bearing.

2. The change of failure mode from shear to flexure did not affect the failure load in fixed ended beams, where as in simply supported beams this change has been significant. However, transition from shear or flexure plus shear failure to bearing failure was accompanied by a significant reduction in the ultimate load.

3. The major diagonal crack loads were considerably lower than the ultimate loads. These cracks usually formed from 35% to 70% of the ultimate load accompanied by a typical loud splitting sound for fixed ended beams. In simply supported and continuous beams at lower load (17% to 40% of ultimate load) these cracks formed. The propagation of these cracks happened in almost the same manner for all the beams.

4. The ultimate strengths of test beams were strongly influenced by:
   a) the amount of bottom reinforcement, in simply supported beams and continuous.
   b) the boundary condition for instance fixed end supports can increase the ultimate loads 1.2 to 6 times the simply support and continuous deep beams.

5. None of the existing methods, which were primarily developed for simply supported conditions, are applicable to the test beams in this investigation.

6. The proposed Force Equilibrium Concept and Truss analogy methods are capable of dealing with any combination of beam parameters including the unsymmetrical arrangement of load and boundary conditions. In majority of cases the experimental and predicted results were found to be in good agreement.

7. Perfect flexural failure of fixed-ended deep beams was highly unlikely. The theoretical analysis (Strain Compatibility Method) indicated that much higher loads than the test results were needed for a perfect flexural failure.

8. Shear deformations of the beams were significant, especially in the case of fixed ended deep beams.

REFERENCES

1.   ARABZADEH A, Analysis of some experimental results of simply supported deep beams using Truss Analogy method, Journal of Science and Technology Shiraz University, Vol. 25, No. 1, 1990. pp. 115-128.

2.   ARABZADEH A, Behavior of reinforced concrete continuous deep beams, Proceedings of conference Creating with concrete, Dundee, 6-10th Sept, 1999.

3. ARABZADEH A, Ultimate strength and failure modes of simply supported deep beams, 4th International conference of Civil Eng. Sharif University, Tehran, Iran, 1997.

4. ARABZADEH A, Truss Analogy method for analysis of fixed ended deep beams, Bahrain Conference, April, 1996.

5. ARABZADEH A, ADIBI M AND ERSHADI S, Analysis of simply supported reinforced concrete deep beams, International Journal of Engineering Science University of Science and Technology, pp. 99-110, 1999.

6. ARABZADEH A, Shear and flexural behavior of reinforced concrete continuous deep beams. Modarres Technical and Engineering, Scientific Research Journal, No. 8, 2002.

7. KONG F K, Reinforcement concrete deep beams, Blackie and Son Ltd, Glasgow and London, 1990, 288 pp.

8. KOTSOVOS M D, Design of reinforced concrete deep beams, The Structural Engineering, Vol. 66, No. 2, January, 1988, pp. 28-32.

9. RICKETTS D R AND MacGREGOR J G, Ultimate behavior of deep continuous beams, Structural Engineering Report, No. 10, Department of Civil Engineering, University of Albarta, Edmonton, 1985, 89 pp.

10. ROGOWSKY D M, MacGREGOR J G AND ONG S Y, Tests of reinforced concrete deep beams, Structural Engineering Report, No. 109, Department of Civil Engineering, University of Alberta, Edmonton, 1983, 116 pp.

11. SUBEDI N K AND ARABZADEH A, Reinforced concrete deep beams with fixed boundary conditions, Development in Structural Engineering, Proceedings of the Forth Rail Bridge, Vol. 2, Edited by Topping, B. H. V. 21-23 August, Centenary conference, Edinburgh, 1991, pp. 833-845.

12. SUBEDI N K AND ARABZADEH A, Some experimental results for reinforced concrete deep beams with fixed-end supports, Structural Engineering Review, Vol. 6, No. 2, 1994, pp. 105-128.

13. ACI (1999 revised 2005), Building code requirements for reinforced concrete (ACI 318-83), American Concrete Institute, Detroit, 111 pp.

14. CIRIA Guide 2, The design of deep beams in reinforced concrete, Ove Arup and partners, Construction Industry Research and Information Association, London, 1977, 131 pp.

15. CANADIAN STANDARDS ASSOCIATION, Design of concrete structures for buildings. CAN3-A23.3-M84, National Standard of Canada, Canadian Standards Association, Rexdale, 281 pp.

DEVELOPMENT OF COMPRESSIVE STRENGTH OF MORTARS WITH FLY ASH FROM CO-BURNING OF COAL AND BIOMASS

M Kosior-Kazberuk

M Lelusz

Białystok Technical University

Poland

ABSTRACT. It is now well-established that the incorporation of industrial by-products such as fly ash, slag and silica fume in concrete or mortar can significantly enhance their basic properties in both fresh and hardened states. These materials greatly improve the durability of concretes through control of high thermal gradients, pore refinement, depletion of cement alkalis, resistance to chloride and sulphate penetration and continued microstructure development through long-term hydration and pozzolanic reactions. The present paper concerns the mortars and concretes containing supplementary cementing material: fly ash from co-burning of coal and biomass. The co-burning of coal and biomass becomes more and more popular. Factors such as the origin of the coal and biomass as well as burning conditions, strongly affect the chemical and mineralogical compositions of fly ashes. The rate of strength increase of fly ash mortar is slower but is sustained for longer periods than the rate of increase of Portland cement concrete. Due to this fact a research program concerning relationships between the composition of mortar and concrete and their compressive strength development has been realized. Mortar specimens prepared with 25% of fly ash from co-burning of coal and wood biomass replacement were evaluated for their compressive as well as flexural strengths at 14, 28 and 90 days and the results were compared with those of Portland cement mortars without fly ash. Additionally, the test of compressive strength of concretes with fly ash was conducted after 2 and 28 days of curing. The results of a short program of other properties of fly ash concrete tests are also included.

Keywords: Concrete, Compressive strength, Fly ash, Biomass, Co-burning.

Dr M Kosior-Kazberuk, works at the Chair of Concrete and Masonry Structures, Dept. of Civil and Environmental Engineering at the Białystok Technical University (Poland), where she received her PhD degree. Member of the Polish Ceramics Society. Her research interests include durability of building materials and structures with emphasis on diagnostic methods.

Dr M Lelusz, works at the Chair of Building Engineering and Prefabrication, Dept. of Civil and Environmental Engineering at the Białystok Technical University (Poland), where she received her PhD degree. Member of the Polish Ceramics Society. Her research interests include concrete technology and properties of building materials, particularly their durability.

INTRODUCTION

The development and use of blended cement is growing rapidly in the construction industry mainly due to considerations of cost saving, energy saving, environmental protection and conservation of resources. Fly ash, the siliceous material obtained from different thermal power stations is now being considered as a cementitious ingredient of concrete. The use of fly ash in mortar and concrete, as a partial replacement of Portland cement, appears to constitute a very satisfactory outlet for this industrial by-product. The utilization of by-products as the partial replacement of cement has important economical, environmental and technical benefits such as the reduced amount of waste materials, cleaner environment, reduced energy requirement, durable service performance during service life and cost-effective structures. It was found that, in order to get resemblance in properties with ordinary Portland cement, fly ash needs special treatment like mechanical grinding, thermal activation, alkali activation, etc. [1].

It is generally accepted that sustainable development of the cement and construction industries can be achieved by maximization of the use of cementitious and pozzolanic by-products. The beneficial effects of incorporating these materials in concrete are widely discussed in the literature [2, 3] but there are not enough data available concerning the properties of concrete with fly ash from co-burning of coal and other fuels, e.g. biomass. The available data are not sufficient to draw statistically significant conclusions. Factors such as the origin of the coal and biomass, the fraction of biomass in fuel mixture as well as burning conditions, strongly affect fly ashes chemical and mineralogical compositions.

The co-burning of coal and biomass is connected with enforcing EU regulations concerning the application of renewable sources of energy. The increased demands on energy have been forcing us towards the more rationalized use of non-renewable sources of energy, together with the expansion of the renewable sources of energy, as well as the creation of ecologically and economically acceptable combination. The combustion of wood biomass and coal in thermal power stations gives a lot of benefits in comparison to the burning of coal only. The wider and more efficient utilization of that fuel affects the ecology as well as the economy. Wood biomass as an ecologically less-damaging fuel enables a considerable reduction of local pollution of the environment. The occurrence of that raw material will enable the supply of fuel from local sources yielding savings in coal transportation, especially to remote regions [4].

The present paper deals with the mortars and concrete containing supplementary cementing material: fly ash from co-burning of coal and biomass (wood chips). Because of the wide availability and low cost, fly ashes are the most commonly used in the manufacture of cement-based materials to improve their microstructure [2, 5]. The compressive strength developing behaviour of cement based materials containing fly ash widely differs from that without fly ash, depending on the method and amount of fly ash addition [1, 6].

The rate of strength increase of fly ash concrete is slower, but it is sustained for longer periods than the rate of the strength increase of Portland cement concrete. Due to this fact, a research programme concerning the relationships between composition of mortar and concrete and their compressive strength development was carried out. This paper is part of a research project on evaluating the various aspects of mortar as well as concrete with fly ash from co-burning of coal and biomass.

EXPERIMENTAL DETAILS

Materials

Fly ash from co-burning coal and biomass (FAB) was used for the analysis. The fuel mixture burned in a thermal power station consisted of 60% of coal and 40% of biomass from wood residues (wood chips). The density of fly ash used was 2360 kg/m³. The phase composition of the fly ash was determined on the basis of diffractional X-ray analysis (Figure 1).

The X-ray diffraction analysis was done using a Philips X'PERT PRO diffractometer, according to Debye-Scherrer-Hulle powder diffraction method. CuKα radiation was used, the measuring angle range was as wide as 5 to 70° 2Θ, the tube voltage and currant were 45 kV and 30 mA, respectively. The patterns obtained were compared with the PDF database (ICDD), which was a component of the diffractometer software system.

The X-ray pattern of fly ash tested was different from that of conventional fly ash from coal combustion. The material tested contained glassy and crystalline phases. The amorphous glass is the active part of fly ash. Two main crystalline phases were determined, β-quartz and hematite. The fly ash also contained magnetite and iron sulfite. There was no mullite in the fly ash tested. The lack of mullite, typical phase in conventional fly ash, can be explained by the relatively low temperature of burning.

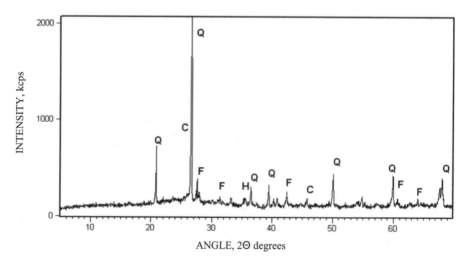

Figure 1   X-ray pattern of fly ash used
(Q – β-quartz, H – hematite, C – carbon, F – iron sulfite)

The test was carried out on specimens containing fly ash in various forms. Part of the fly ash was used in an unprocessed (as-received) form and another part was mechanically ground for 6 hours in a ball mill to the fine powder (< 0.063 mm). The mortar mixtures also contained fly ash, from which the particles greater than 0.125 mm were eliminated by sieving.

Portland cement CEM I 42.5, observing the requirements of EN 197-1, was used as binder. The physical properties and chemical analysis of cement used are given in Table 1.

Table 1   Selected physical, chemical and strength characteristics of Portland cement used

CHARACTERISTICS OF CEMENT USED	
Loss on ignition, %	1.37
Insoluble residue, %	0.19
SO_3, %	2.96
Cl^-, %	0.016
Initial setting time, minutes	106
Le Chaterlier, mm	0.0
f_c (at 2 days), N/mm^2	28.8
f_c (at 28 days), N/mm^2	54.6

The fine aggregate used was natural sand with a maximum diameter of 2 mm, the natural coarse aggregate maximum diameter was 8 mm. The aggregate mix consisted of 40% of the 0/2 mm fraction, 25% of the 2/4 mm fraction and 35% of the 4/8 mm fraction.

SPECIMEN PREPARATION AND TESTING

Compressive strength changes were tested on both mortars and concretes. The mortar mixture consisted of binder, water and fine aggregate in a proportion of 1:0.5:3. The tests were carried out on control specimens, containing Portland cement, as well as on specimens with fly ash replacing 25% of cement. FAB was used in three various forms: as-received fly ash, sieved fly ash (< 0.125 mm) and ground fly ash (< 0.063 mm).

Concrete specimens were prepared of mixtures with three different percentages of fly ash content: 5, 15 and 25% related to the cement mass (FAB/C). Unmodified control specimens were also tested. Part of fly ash in the mixture (40% for CEM I 42.5) was taken into account as binder and the remaining part – as filler, according to the standard EN 206-1:2003 [7]. The cement content in control concrete was 350 kg/m³. Two systems were used: as-received fly ash and ground fly ash. A water to binder ratio in the tested concretes was constant (w/b = 0.50). All mixtures were characterised by plastic consistence tests. The mixture proportions of all concrete specimens are summarized in Table 2.

Table 2   Mixture proportions for concrete specimens

FAB/C	CEMENT, kg/m³	FAB, kg/m³	WATER, kg/m³	AGGREGATE, kg/m³
0.0	350	0.0	175	1887
0.05	343	17.2	175	1873
0.15	330	49.5	175	1847
0.25	318	79.5	175	1823

Mortars and concrete were cast in prism shaped moulds (40 × 40 ×160 mm) and compacted by vibration. After 24 hours, the specimens were removed from the moulds and stored in water at a temperature of 18±2°C. The aim to apply the prism shaped specimens was to evaluate the compressive as well as the flexural strength. The mortar specimens were tested for compressive strength after 14, 28 and 90 days of curing, whereas the concrete specimens were investigated after 2 and 28 days. Every series consisted of six replicates.

Flexural strength was determined in a single point bending test and every series consisted of three replicates. The results of flexural strength tests are not comparable with those of the splitting tensile strength test, but they make it possible to evaluate the rate of strength increase. The strength test results were used to evaluate the pozzolanic activity of FAB.

RESULTS AND DISCUSSION

Strength Characteristics of Mortars

The variation of compressive strength and flexural strength of mortars with curing time are presented in Tables 3 and 4. The strength results make it possible to evaluate the rate of strength development. The changes in strength with age can be analysed in details on the basis of relative strength to the 28 day strength, as a percentage.

Table 3   Compressive strength development of mortars in relation to 28-day strengths

FORM OF FLY ASH	FAB/C	COMPRESSIVE STRENGTH after days, N/mm^2			STRENGTH relative to 28-day strength, %		
		14	28	90	f_{cm14}/f_{cm28}	f_{cm28}/f_{cm28}	f_{cm90}/f_{cm28}
–	0.0	37.9	42.0	57.9	90	100	138
As-received	0.25	23.1	23.3	35.6	99	100	153
Sieved < 0.125 mm	0.25	10.9	11.2	17.4	97	100	155
Ground < 0.063 mm	0.25	30.5	30.8	44.6	99	100	145

Table 4   Flexural strength development of mortars in relation to 28-day strengths

FORM OF FLY ASH	FAB/C	FLEXURAL STRENGTH after days, N/mm^2			STRENGTH relative to 28-day strength, %		
		14	28	90	f_{tm14}/f_{tm28}	f_{tm28}/f_{tm28}	f_{tm90}/f_{tm28}
–	0.0	8.0	9.0	11.0	88	100	122
As-received	0.25	5.5	5.5	9.9	100	100	180
Sieved < 0.125 mm	0.25	3.5	3.6	4.7	98	100	130
Ground < 0.063 mm	0.25	7.3	8.3	9.4	88	100	113

The observed results of the mortar strength test proved the FAB influence on the strength development at all ages. It was found, in general, that the rate of strength development was slower for mixtures containing FAB, at an early age, in comparison to that of control specimens. The rate of strength gain in mortars with FAB was significant between 28 and 90 days. The mixtures containing ground fly ash achieved the highest and the mixtures with sieved fly ash achieved the lowest 90-day compressive strength; however at the same time, the control mortar (FAB/C = 0.0) attained higher strength than the modified mortars.

The most crucial property of fly ash is its pozzolanic activity. The pozzolanic activity index was determined by comparing the compressive (f_{cm}) as well as flexural strength (f_{tm}) of mortar containing 75% of cement and 25% of fly ash with that of cement mortar without fly ash addition. The results of pozzolanic activity evaluation are presented in Tables 5 and 6.

Table 5   Results of pozzolanic activity of fly ash from co-burning, determined on the basis of the compressive strength of mortars, f_{cm}

POZZOLANIC ACTIVITY INDEX BASED ON f_{cm}	COMPOSITION OF FAB USED, %			EN-450 REQUIREMENT, %
	As-received	Sieved < 0.125 mm	Ground < 0.063 mm	
after 14 days	62	29	80	-
after 28 days	55	27	73	75
after 90 days	61	30	77	85

Table 6   Results of pozzolanic activity of fly ash from co-burning, determined on the basis of the flexural strength of mortars, f_{tm}

POZZOLANIC ACTIVITY INDEX BASED ON f_{tm}	COMPOSITION OF FAB USED, %			EN-450 REQUIREMENT, %
	As-received	Sieved < 0.125 mm	Ground < 0.063 mm	
after 14 days	70	44	91	-
after 28 days	59	40	92	75
after 90 days	90	43	85	85

Grinding of fly ash from co-burning resulted in significant improvement of its pozzolanic activity. The values of activity index were comparable with the requirements valid for fly ash from burning of coal only, according to the standard EN-450 [8]. FAB devoid of particles greater than 0.125 mm by sieving was characterized by poor pozzolanic activity.

Selected Properties of Hardened Concrete with FAB

Selected properties, characterising the concretes tested, are summarised in Table 7. The results obtained for control concrete as well as for concrete with as-received raw fly ash and ground fly ash were compared.

Table 7   Selected properties of concretes with FAB addition after 28 days of curing

PARAMETER	CONCRETE MIXTURE, FAB/C						
	Control	As received Fly Ash			Ground Fly Ash		
	0.0	0.05	0.15	0.25	0.05	0.15	0.25
Bulk density, kg/dm³	2.211	2.211	2.209	2.191	2.255	2.206	2.218
Compressive strength, MPa	63.7	57.7	59.8	56.6	64.0	66.1	64.1
Water absorption, %	5.10	5.18	5.26	5.37	5.20	5.37	5.38
Capillary porosity, %	11.27	11.45	11.62	11.77	11.72	11.85	11.94
Sorptivity, %	2.04	2.13	2.13	2.38	2.36	2.28	2.32

Fly ash addition, in the adopted range of FAB/C values, had no significant effect on bulk density. Concretes with ground fly ash achieved higher compressive strength after 28 days of curing than other concretes tested.

The fly ash addition caused a slight worsening of other properties depending on the concrete microstructure. The properties were determined after 28 days, but the hydration process, pozzolanic reactions and changes in microstructure continue for a longer period [2].

Compressive Strength of Concrete

The test results of compressive strength of concrete containing FAB, after 2 and 28 days of curing, are given in Figure 2.

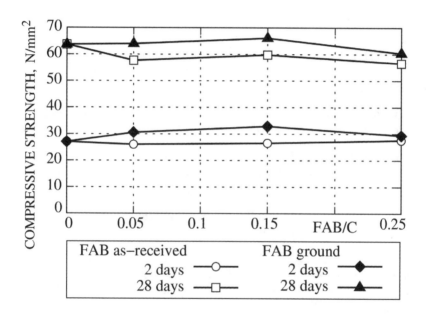

Figure 2   Compressive strength of concretes with FAB at different ages

After 2 days of curing, the concrete containing ground fly ash attained higher strength than concretes with unprocessed fly ash. The test results after 28 days showed that all mixes containing fly ash were able to develop a compressive strength comparable with the strength of the unmodified control concrete.

The laboratory tests also evaluated how much FAB could be added to concrete. FAB was added up to 25% by weight replacement of cement, although the best results were found at about 15% of FAB. At the considered range of FAB/C ratio, there were no harmful effects on workability and strength.

CONCLUSIONS

The tested fly ash from co-burning of coal and biomass can be considered as supplementary cementing material, although the biomass content in the fuel mixture was relatively high. The analysis of test results showed that the cement based material containing FAB can achieve applicable strength. The results generally indicated that the compressive strength of mortars and concretes tested increased with age. The slight decrease in strength with inclusion of fly ash was observed for mortars. Mortars with FAB showed a significant increase in compressive and flexural strengths after 28 days of curing. Concrete with FAB gained a compressive strength comparable with the strength of Portland cement concrete, particularly when ground FAB was used. Mechanical grinding of fly ash activated this material significantly. The experimental results obtained showed that the finely ground FAB can be used up to 25% of total cementitious material in order to produce acceptable strength of mortar or concrete.

REFERENCES

1. SARASWATHY V, MURALIDHARAN S, THANGAVEL K AND SRINIVASAN S, Influence of activated fly ash on corrosion-resistance and strength of concrete, Cement and Concrete Composites, Vol. 25, 2003, pp. 673-680.

2. PAPADAKIS V G AND TSIMAS S, Supplementary cementing materials in concrete, Part I: efficiency and design, Cement and Concrete Research, Vol. 32, 2002, pp. 1525-1532.

3. ANTIOHOS S, MAGANARI K AND TSIMAS S, Evaluation of blends of high and low calcium fly ashes for use as supplementary cementing materials, Cement and Concrete Composites, Vol. 27, 2005, pp. 349-356.

4. ILAVSKY J AND ORAVEC M, Utilisation of biomass in Slovakia, Ecological Engineering, Vol. 16, 2000, pp. 83-89.

5. SWAMY R N, Holistic design of concrete technology: The only route to durability and sustainability in construction, Proc. of 2nd International Symposium on Cement and Concrete Technology in the 2000s, Istanbul, 6-10 September 2000, pp. 58-72.

6. HWANG K, NOGUCHI T AND TOMOSAWA F, Prediction model of compressive strength development of fly-ash concrete, Cement and Concrete Research, Vol. 34, 2004, pp. 2269-2276.

7. EN 206-1: 2003 Concrete – Part 1: Specification, performance, production, and conformity.

8. EN 450: 1998 Fly ash for concrete – Definitions, requirements and quality control.

A COMPARISON OF MATERIAL PROPERTIES FOR NATURAL AND RECYCLED AGGREGATES

J J Bester

D Kruger

University of Johannesburg

South Africa

ABSTRACT. In most developed countries there is a determined effort to safeguard our diminishing natural resources by establishing and enhancing the use of recycled aggregates as a viable construction material. This is most certainly not the case in developing countries such as South Africa. Reasons for this lack of commitment are mainly as a result of misconceptions about its suitability for re-use in new construction ventures and the fact that recycled aggregate properties have not been explored in depth in these countries. In an attempt to make recycled aggregates more viable in South Africa and specifically in the province of Gauteng, it is believed that the first step needed to establish a commitment to using recycled aggregates, is to identify the important material properties required for producing durable concrete using recycled aggregates. Once it has been shown that these properties can be achieved and sustained with recycled aggregates, recycled aggregates will become a more viable option in South Africa. This paper presents the results of a study performed in which the properties required for durable concrete were investigated by comparing various recycled aggregates with natural properties.

Keywords: Recycled aggregates, Material properties.

Mr J J Bester, Technical lecturer, Department of Civil Engineering Science, University of Johannesburg, South Africa

Mr D Kruger, Senior lecturer, Department of Civil Engineering Science, University of Johannesburg, South Africa

INTRODUCTION

Worldwide our natural resources are diminishing. This has forced the construction community to search for alternatives. The sustainable recycling and re-use of construction and demolition waste has become an economically viable option in many parts of the world. In South Africa, however, this is not the case. The perception is that construction and demolition waste is inferior in quality and not for construction purposes.

NATURAL AGGREGATES AND THEIR PROPERTIES

The Oxford dictionary defines natural as: "existing in or produced by nature" and aggregate as: "hard substances (sand, gravel, broken stone etc.) mixed with cement to make concrete" [1]. In South Africa most of the aggregates used in concrete is derived from the blasting and crushing of solid rock.

Some years ago aggregates were treated as inert filler in concrete. This was due to the fact that aggregates do not enter into the complex chemical reactions with water. However, an increase in awareness of the role that aggregates play has put a focus on aggregates properties such as porosity, grading, water absorption, shape, surface texture, aggregate crushing value, elastic modulus and deleterious substances [2].

METHODS FOR CRUSHING SOLID ROCK

Various types of mechanical crushers have been developed over the years, and crushing processes have evolved along with it. Many crushing processes involve primary, secondary and sometimes even tertiary phases.

The different types of crushers include jaw-, gyratory-, cone and impact crushers, each of which has different advantages and disadvantages. This is determined by the properties of the material being crushed and the particle size and shape required.

It is quite common for jaw- or gyratory crushers to be used in the primary crushing process from the large raw feed. This is then followed by cone crushers or impact breakers to reduce the material to the final aggregate sizes. Impact crushing is useful for producing acceptable particle shapes from difficult material which would otherwise produce flaky or elongated particles [3].

WHY DO WE NEED ALTERNATIVES FOR NATURAL AGGREGATES?

The two major reasons driving the need for alternatives to natural aggregates are:
1. The social impacts of the quarrying and crushing processes
2. The environmental impacts of the quarrying and crushing processes

Social Impacts

Owing to the fact that personal use of aggregates by the public is limited, many individuals do not grasp the dependence of society on aggregates. Thus the public does not consider the mining of aggregate as necessary land use.

Furthermore, individuals has the "not in my back yard (nimby)" attitude towards the mining of aggregates. This is due to the noise pollution, dust pollution and vibrations emanating from the mining process.

Environmental Impacts

The operations are inevitably visually intrusive. This holds true not only during the active mining process, but also after operations has ceased. There is a hole left where mining operations were, and must be rehabilitated. The land use after operations has ceased is a prominent consideration for the successful rehabilitation of the area.

The pollution generated during operations from the crushing (dust) and washing (dust and fines) of aggregates must be considered. In addition there is also pollution from the loading machinery and vehicles that transport the aggregates from the operations [4].

RECYCLED AGGREGATES AND THEIR PROPERTIES

The Oxford dictionary defines recycle as "to convert (waste material) into a form in which it can be re-used". Globally it is considered that construction and demolition waste (C&D waste) is suitably materials for recycling. Unfortunately C&D waste (concrete, brickwork and masonry) is widely used as daily cover in landfills.

A recycling plant will receive C&D waste from various sources and in different sizes. The waste is crushed and screened giving a material that has unique properties. The quality of the end product may vary greatly, depending on the initial source and the composition.

INTERNATIONAL PERSPECTIVE

Significant attention is given to research and experimental work on recycled aggregates, especially coarse aggregates. There is a definite shift from using recycled materials for more that just fill material. Many countries have regulatory legislation in place that governs the use of recycled aggregates and it is more commonly used every year.

SOUTH AFRICAN PERSPECTIVE

Limited research and experimental work have been done on recycled aggregates. Most of the provinces in South Africa do not have recycling operations. In 1999 it was only KwaZulu Natal and the Western Cape that have sustainable recycling operations in progress. The recycled material is mostly used as backfill material and road building material. Extremely limited use of coarse aggregates in concrete has been noted [5, 6].

MATERIALS USED

Natural aggregates (13.2 mm and 19.0 mm) were supplied by Lafarge Aggregates & Readymix Division from 10 different stone quarries in Gauteng province. The following identification system was used:

N1	Zimbiwa dolomite	N2	Lyttleton dolomite
N3	Drift supersand granite	N4	Glen Douglas dolomite
N5	Rosslyn norite	N6	Laezonia norite
N7	Rossway granite	N8	Klipfontein granite
N9	Donkerhoek quartzite	N10	Willows quartzite

Recycled aggregates were manufactured in the laboratory using a jaw crusher. Three different specimens were used:

1. 25 MPa pump mix from a demolished floor slab
2. Concrete paving blocks
3. 28 day compressive strength cubes from a pre-cast manufacturing plant.

EXPERIMENTAL SET-UP

The experimental work was divided into two different stages: Stage 1 was aggregate characterization and stage 2 was strength development.

For stage 1 the natural and recycled aggregates was tested and compared against each other with the list of tests as set out below.

For stage 2 six different mixes of dimensions $150 \times 150 \times 150$mm were cast and cured in a curing bath at 22-25°C for 7, 14 and 28 days. At each of the 7, 14 and 28 days compressive strength tests were performed. The mixes were all made from 19.0 mm stone, had a w:c ratio of 0.65 and were mixed in the following proportions:

1. 100% natural dolomite stone
2. 30% recycled pump mix, 70% natural dolomite stone
3. 50% recycled pump mix, 50% natural dolomite stone
4. 100% recycled pump mix
5. 50% paving blocks, 50% natural dolomite stone
6. 50% recycled pre-cast stone, 50% natural dolomite stone

TESTS PERFORMED

Table 1 lists the tests that were performed during stage 1.

Table 1   Aggregate tests performed during stage 1

TESTS PERFORMED ON AGGREGATES	
Grading analysis	SABS SM 829 / SANS 201
Aggregate crushing value (ACV)	SABS SM 841 / SANS 5841
10% FACT – Dry	SABS SM 842 / SANS 5842
Water absorption (WA)	SABS SM 843 / SANS 5843
Relative density (RD)	SABS SM 844 / SANS 5844
Loose bulk density (LBD)	SABS SM 845 / SANS 5845
Compacted bulk density (CBD)	SABS SM 845 / SANS 5845
Voids content (V)	SABS SM 845 / SANS 5845
Flakiness index (FI)	SABS SM 847 / SANS 5847
Chemical composition	
X-ray diffraction & fluorescence analysis	

RESULTS

Stage 1

Grading

Figures 1-4 indicate the grading results for all the specimens. All the Figures indicate a grading envelope. Figure 1 and 3 give the lowest and highest value of the ten different natural aggregates that were tested.

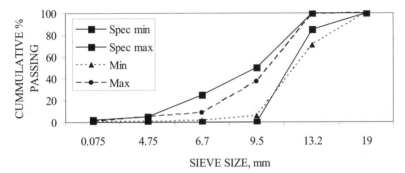

Figure 1   19.0 mm natural aggregate grading curve

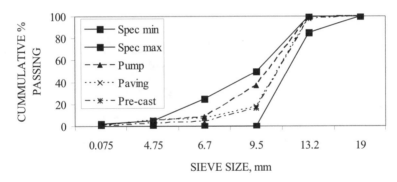

Figure 2   19.0 mm recycled aggregate grading curve

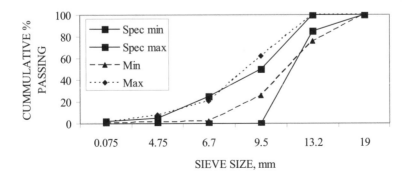

Figure 3   13.2 mm natural aggregate grading curve

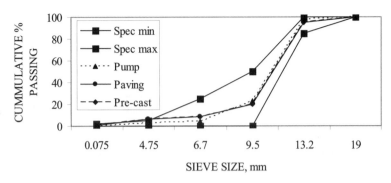

Figure 4   13.2 mm recycled aggregate grading curve

Material properties

Table 2 indicates the 13.2 mm aggregate characterization test results, whereas Table 3 indicates the 19.0 mm aggregate characterization test results.

Table 2   Results for 13.2 mm recycled aggregate

	13.2 mm RECYCLED AGGREGATE TESTS							
	FI, %	RD (SSD)	Water Absorption, %	Bulk Density, kg/m³		Voids, %	ACV (dry), %	10% FACT, kN
				CBD	LBD			
Specification	35	2.5 – 3	1.0	-	-		≤ 30	
N1	19.4	2.87	0.30	1635	1510	43	12.7	360
N2	17.5	2.86	0.20	1680	1530	41	11.9	365
N3	24.0	2.67	0.53	1440	1340	46	23.8	171
N4	19.8	2.86	0.20	1640	1520	43	12.8	340
N5	21.6	2.89	0.39	1610	1480	44	19.4	199
N6	23.3	2.94	0.36	1710	1560	42	7.0	540
N7	24.4	2.65	0.45	1420	1340	46	23.6	170
N8	24.8	2.63	0.51	1470	1360	44	27.6	-
N9	15.6	2.62	1.60	1500	1330	43	23.5	180
N10	29.2	2.64	0.66	1520	1400	42	26.1	
Pump mix	8.57	2.54	3.84	1311	1175	51	20.4	
Paving blocks	5.29	2.57	5.92	1128	1038	60	47.9	
Precast mix	13.03	2.64	1.67	1352	1239	53	21.2	

Table 3   Results for 19.0 mm aggregate

	19.0 mm RECYCLED AGGREGATE TESTS					
	FI, %	RD (SSD)	Water Absorption, %	Bulk Density, kg/m³ CBD	LBD	Voids, %
Specification	35	2.5 – 3	1.0	-	-	
N1	12.8	2.87	0.40	1620	1520	44
N2	10.9	2.86	0.18	1670	1530	42
N3	18.1	2.66	0.61	1470	1360	45
N4	15.8	2.86	0.22	1660	1510	42
N5	16.6	2.91	0.25	1630	1470	44
N6	13.8	2.93	0.25	1660	1550	43
N7	4.20	2.64	0.51	1480	1340	44
N8	17.2	2.64	0.58	1450	1380	45
N9	14.8	2.63	0.96	1480	1350	44
N10	18.4	2.64	0.42	1520	1370	43
Pump mix	7.15	2.57	4.65	1316	1184	52
Paving blocks	2.42	2.47	5.94	1090	943	60
Precast mix	9.98	2.64	1.95	1378	1275	50

Chemical composition

The loss on ignition of the three specimens is as follows:
- Paving blocks 32%
- Pre-cast mix   28.7%
- Pump mix   11.7%

The results of the XRF indicated:
1. A relatively low presence of SO_3, 1.03%, 1.28% and 0.69% was observed.
2. A low presence of Cl, 0.03%, 0.04% and 0.04% was observed.

Stage 2

The compressive strength results for the pump mix are shown in Figure 5. In Figure 6 the compressive strength results for the pre-cast concrete, pavers and natural aggregates are shown.

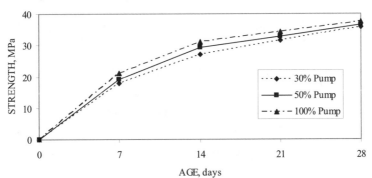

Figure 5   Compressive strength results for pumped concrete aggregates

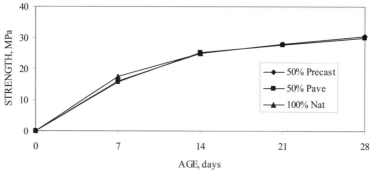

Figure 6   Compressive strength results for concrete aggregates

CONCLUSIONS

1. Coarse aggregate (19.0 mm and 13.2 mm) could be extract from the crushed samples to fit into a grading envelope.

2. For the 13.2 mm and 19.0 mm fraction the specification for FI and RD was easily satisfied.

3. The pump mix and the pre-cast mix easily satisfied the specification for ACV.

4. The WA for all the recycled materials was significantly higher than the specification. The mix designer will have to take not of this, but the recycled material could still be used in concrete with minor adjustments to the mix design.

5. The low presence of SO_3 and Cl indicates that there is a low risk to new concrete due to chemical impurities present in the recycled aggregates.

REFERENCES

1. HAWKINS J M, Oxford Dictionary, Oxford University Press, 1981.

2. MEHTA P K AND MONTEIRO P J M, Concrete - Structure, properties and materials, New Jersey, United States of America, 1993.

3. HEWLETT P C, Lea's chemistry of cement and concrete, 4th ed., Butterworth-Heinemann, United Kingdom, 2001.

4. UNITED STATES GOVERNMENT PRINTING OFFICE, Natural Aggregate-Building America's future, Denver, United States of America, 1993.

5. COUNCIL OF SCIENTIFIC AND INDUSTRIAL RESEARCH, Construction Waste Management, Pretoria, South Africa, 1999.

6. COUNCIL OF SCIENTIFIC AND INDUSTRIAL RESEARCH, Economics of recycling construction waste, Pretoria, South Africa, 2000.

TOWARDS A SUSTAINABLE MIX-DESIGN FOR CONCRETE

G Habert

N Roussel

Université Paris-Est LCPC

France

ABSTRACT. The building and construction sector is a major CO_2 producer and climate change perspectives urged to reduce CO_2 emissions. The impact of concrete buildings on environment is partly due to clinker, which is the main material used all over the world to produce cement and which releases a bit less than 1 ton of CO_2 per ton of clinker produced. In this study, we correlate CO_2 reduction possibilities with other concrete requirements such as mechanical strength or durability. Two different environmental options are considered and evaluated in this paper. The first one is the substitution of clinker by mineral additions in cement in order to reduce the environmental cost of the material for a given volume of concrete produced; the second one is the reduction of the concrete volume needed for a given construction process by enhancing the concrete performances. The impact on CO_2 emissions of a combination of these options is also roughly evaluated.

Keywords: Mix-design, Mechanical strength, Sustainability, Evaluation.

Dr G Habert is a researcher at the Université Paris-Est, LCPC, Paris, France.

Dr N Roussel, is a researcher at the Université Paris-Est, LCPC, Paris, France.

INTRODUCTION

The industrial sector is responsible for approximately 25% of global carbon dioxide (CO_2) emissions among which CO_2 emissions from cement plants represent no less than 5% of total anthropogenic emissions [1] despite the efforts of the cement industry. Recent studies on the Life Cycle Assessment for concrete structures show that 85% of the CO_2 emissions come from the cement [2]. Moreover, life Cycle Assessment for cement shows that 95% of the CO_2 is produced during the fabrication of the cement powder, compared to the low energy consumption of the transport of raw materials and finished products [3]. Therefore it seems obvious that the necessary effort in the building and construction sector in term of CO_2 reduction has to be made on the type and amount of cement used in concrete, at least as a first step.

In this study, two different environmental options are considered. The first one is the substitution of clinker by mineral additions in cement in order to reduce the environmental cost of the material for a given volume of material; the second one is the reduction of the concrete volume needed for a given construction process by enhancing the concrete performances. It has to be noted that, in this study, we limit our observations to the French context, but the conclusions drawn here could be easily extrapolated to other national concrete industries. Moreover, we will only focus in the following on the CO_2 emission aspect of a sustainable mix design.

ENVIRONMENTAL IMPACT OF CONCRETE

Limestone is the major raw material used in the production of cement. It is burnt at 1450°C to make clinker and is then blended with additives. The finished product is finely grounded to produce different types of cement.

Through the cement production process, around 0.706 ton of CO_2 is released for each ton of clinker produced. Its release is mainly due to the decarbonation of limestone (0,521 ton), and the use of coal and fossil fuels for heating (0.185 ton) [4-6]. These estimations come from the European cement industry where important investments have already been done to enhance the combustion efficiency of the cement kilns. In the United States of America, the production of one ton of clinker still release 0.935 ton of CO_2. In this study we will not develop these technologic improvement options, but a massive investment to enhance the efficiency of world cement kilns could lead to 25% of $C0_2$ reduction per ton of clinker produced.

Another way to reduce the greenhouse gas emission from cement production is to partially replace clinker. The substitution to clinker often used are mainly fly ash from coal-fired thermal power plants, slag from blast furnaces in the iron and steel industry and natural pozzolans. These additives contain large quantities of reactive SiO_2 and Al_2O_3, which produce cementitious materials in the presence of lime. In France, electricity is produced by nuclear power generation which has a very low impact in term of CO_2 emissions. As the addition of alternative material to clinker in the cement is essentially consuming electricity for grinding, we assume that no supplementary CO_2 emissions are released when replacing clinker by other mineral additions except for meta-kaolin where the heating of clays up to 700°C has to be taken into account. We consider in our approach fly ash and blast furnace slags as industrial wastes, which therefore do not release CO_2 to be produced. However, the energy consumption for clay heating can not be neglected but may be calculated from Gartner [4] and induces CO_2 emissions from 0.088 to 0.159 ton of CO_2 per ton of clay, for heating at 500°C or 900°C respectively. It is an approximation, especially for granulated blast furnace slags as the granulation and the vitrification are additional industrial processes used exclusively for the slag valorisation, which are not CO_2-neutral. However, these emissions are negligible in comparison to clinker production emissions [7].

Finally, it can be kept in mind that the amount of CO_2 released during the production of 1 kg of clinker is of the order of 0.7 kg and the additional cost of most additives can be neglected compared to this figure.

MATERIAL SUBSTITUTION OPTIONS

One option in order to reduce CO_2 emissions is to substitute a large part of the clinker by the mineral additions described above. Depending on their physical properties (grading curve, average size…) or their chemical nature and properties, mineral additions will have either a filler comportment (*i.e.* they will fill the porosity of the material and thus enhance its elastic modulus and its mechanical strength) or a binding capacity (*i.e.* they will react with water or with clinker hydration products in order to form stable hydrates). Mineral additions can also be activated before incorporation into cement in order to enhance their binding properties. Such activation treatments may be mechanical, chemical or thermal. As an example, we consider here the effect of thermal treatments as they have more impacts on CO_2 emissions than the others. Figure 1 shows that it is possible to reduce the environmental impacts while maintaining quality requirement up to an *optimum* for most of the substitution options. Treated natural pozzolan allows for instance for a reduction of 40% of the CO_2 emission without any impact on the mechanical strength of the material. The substitution where the strength is linearly reducing while increasing the substitution content, can be considered as a substitution material with no additional properties that takes up some volume in the material for a rather low economical and environmental cost. For such a material, it therefore depends on the concrete designer to choose between a strength requirement and a concrete environmental impact as there does not exist any optimal amount. It has however to be kept in mind that the level of knowledge in this field is still at the state of empirical exploration and that no tool exist allowing for the *a priori* evaluation of the potential of a given mineral addition and of the existence of these *optima*.

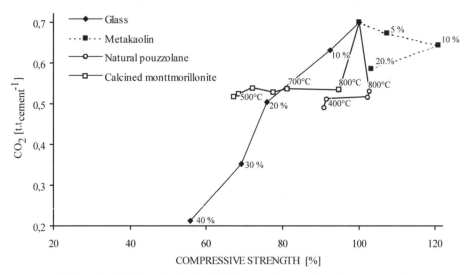

Figure 1   Relationship between CO_2 emission per ton of cement and
the associated concrete strength

In Figure 1, for Glass and Metakaolin, percentage of substitution are indicated. For Natural pozzolans and calcined Montmorillonite, temperatures of the thermal treatment are indicated. Data comes from [8-11].The same type of presentation have been used to deal with durability requirements, but is not developed in this study. It leads to the same conclusion, that there are an optimum in substitution proportions where durability performances are maintained while environmental impacts are reduced.

The cement substitution by alternative mineral additions is therefore a powerful option process to reduce the CO_2 emission from concrete production. The type of presentation we have developed here, where the environmental cost is compared to quality requirements has the potential ability to permit to the mix-design engineers to balance the society demand in terms of environment with the technical building requirements. With these diagrams, it seems obvious that, whatever the type of substitution, there are *optima* (here of the order of a few tens of percent) above which concrete performances may be strongly deteriorated. However, the existence of these limits demonstrates that the cement substitution with our current level of knowledge can not be the exclusive solution to answer at the same time to the needed reduction of CO_2 emissions to meet the Kyoto goals and to the predicted increase in cement demand due to the economic growth [12]. Additional knowledge in the concrete design field concrete will therefore be strongly needed in a close future. It can however be estimated that the today level of knowledge should allow for a 40% reduction of CO_2 emission per cube metre of concrete compared to a pure clinker based material. In France, it is already rather common to reach at least 25% CO_2 emission reduction by using fly ashes.

MATERIAL PERFORMANCE OPTIONS

Instead of reducing the amount of CO_2 per ton of concrete produced, another option is to reduce the total quantity of concrete produced. In this study, we have tried to roughly predict the material volume reduction that could occur by enhancing the concrete mechanical strength. However, increasing the mechanical strength without any of the substitution described above can be most of the time associated with an increase in the CO_2 emission per cube metre of material as it is shown in Figure 3 where mechanical strength of various concretes is plotted as a function of their CO_2 emissions per cube metre [13-30]. CO_2 emissions have been calculated here by considering exclusively the cement contribution as detailed in the section 2. We assumed that each ton of CEM I produces 0.7 tons of CO_2 and that any mineral addition and all aggregates can be considered as CO_2 neutral compared to clinker. Figure 3 shows that, as a first approximation, the usual way to increase strength resistance is to increase the cement content (and the CO_2 emissions) in concrete up to 70 MPa, where compressive strength becomes strongly dependant on the quality of the granular skeleton [19]. It can be worth noting that, if all other parameters are kept constant (number of granular classes, nature of the cement…), then the Ferret relation [19] which predicts that mechanical strength is proportional to the power 2 of the cement amount per cube metre can also be derived to predict that mechanical strength is proportional to the power 2 of the CO_2 emissions per cube metre.

$$f_c \approx \left(CO_2^{m^3} \right)^2 \tag{1}$$

The question we tackle here is the following: by increasing the mechanical strength, we increase on one hand the CO_2 emissions per cube metre of concrete produced but, on the other hand, we could decrease the amount of concrete needed to build a given structural element. There are however many types of structural elements in the construction industry and the problem is complex to deal with . We will only develop here a very simple and quasi dimensional approach for standard cases of structural elements. It has moreover to be kept in mind that structural considerations does not always give access to the minimum size of a given element as acoustic, thermal or fire safety aspects may lead to higher minimum sizes. We will however in the following focus on four simplified cases which are representatives of many structural elements in construction industry. We will moreover neglect the presence of steel bars and consider the material as homogeneous.

We will limit here our study to horizontal element. In the first case, we consider an horizontal element of thickness h which only carries itself. The other dimensions of this element such as width and span are considered as given in this study. This type of element is representative of beams in housing construction, for which external load is small compared to the weight of the element itself. This element is thus only submitted to its own weight and to a flexural torque M proportional to h. The maximum stress can be written as $\sigma \approx Mh/I$ where I is proportional to h^3. As a consequence, $\sigma \approx 1/h$. In other words, for a given width and span, the maximum stress in the element is inversely proportional to its thickness and thus to its total volume. This total volume writes $V^{Total} = CO_2^{Total}/CO_2^{m^3}$ where CO_2^{Total} is the total CO_2 emission involved in the building of this element and $CO_2^{m^3}$ is the CO_2 emission per cube metre of concrete. We have moreover shown above that the compressive strength of a given cementitious materials is more or less proportional to the power 2 of its CO_2 production per cube metre (Eq. (1)). It could of course be objected that tensile strength is playing a stronger role that compressive strength but, in the frame of the simple approach proposed here, it could be answered that only proportionality relations are written and that, at this level of simplification, tensile strength could be considered as roughly proportional to compressive strength. Finally, the above derivations lead to the following relation:

$$CO_2^{Total} \approx 1/CO_2^{m^3}$$

(2)

This means that, because of the strength increase of the cementitious material and of the element volume reduction, the total CO_2 production CO_2^{Total} decreases when the CO_2 production per cubic metre $CO_2^{m^3}$ increases in the case of horizontal elements in housing construction. As a consequence and if, as a first approximation, considerations other than structural aspects are neglected, high mechanical performances concretes are the most environmental friendly materials for this type of structural element.

In the second case, we consider an horizontal element of thickness h which carries an external load. This type of element is representative of beams in bridges, for which external load is large compared to the weight of the element itself. If we carry the same analysis as above, we obtain that total CO_2 production CO_2^{Total} does not depend on mechanical strength of the concrete. This means that the strength increase of the cementitious material and the element volume reduction more or less compensates the increase in CO_2 production per cube metre. As a consequence, and if, as above, considerations other than structural aspects are neglected, high mechanical performances concretes does not bring anything from an environmental point of view for this type of structural element.

Same studies for vertical elements have been performed, but are not developed in this study. They lead to the same conclusions. High mechanical performances concretes are the most environmental friendly materials for vertical element which carries an external load. Vertical elements only carrying themselves are indeed rare in the construction industry.

It should therefore be kept in mind that higher mechanical performances, on the whole, could reduce total CO_2 emissions following roughly Eq. (1). In France, it is rather common to use concretes displaying 25/30 MPa compressive strength for housing while using concretes displaying 50/60 MPa compressive strength for bridges. By doubling these values and of course making the needed structural design changes, it can thus be estimated using the very simplified rough approach developed here (Eq. (2)) and using Eq. (1) that CO_2 emissions could be reduced by 30%. It can also be extrapolated that using ultra high performance concretes (mechanical strength of the order of 120 MPa) could lead in housing constructions to reductions of the order of 50%. Other materials than clinker based concrete should then be used to fulfil the other requirements such as fire resistance, acoustic or thermal behaviour as all these other materials (plaster, bricks, wood…) have all lower CO_2 emissions than concrete and can be easily dedicated to these functionalities. Strongly composite walls and slabs could fulfil all the technical requirements with a reduced total environmental impact.

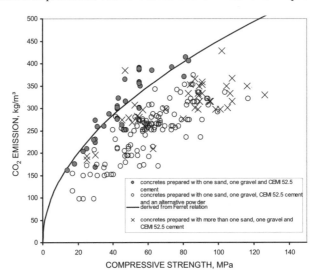

Figure 3 CO_2 Emissions per cubic metre of concrete as a function of compressive strength resistance after 28 days of curing. Data comes from [13-30]

CONCLUSIONS

In this study, two different environmental options for sustainable concrete mix design were considered and evaluated. The first one is the substitution of clinker by mineral additions in cement in order to reduce the environmental cost of the material for a given volume of concrete produced. The second one is the reduction of the concrete volume needed for a given construction process by enhancing the concrete performances. It has been estimated that, in France, the CO_2 emissions could be reduced by 15% by increasing the level of substitution in concrete. It has also been estimated that the second option could lead to reduction of the order of 30%. But it has to be kept in mind that, as it can be seen in Fig .3, it is possible to combine cement substitution and mechanical strength increase. From the present results and the observation of the nowadays French practice, this could lead to CO_2 emissions reduction of the order of 40% (15% for substitution and 30% for mechanical strength increase). This would still not be sufficient to reach the "*factor four*" objectives as it is edited in the European climate action plan, which aim to reduce by four 1990 emissions in 2050, but this represents what could be achieved with the actual level of knowledge in concrete industry. This will have to be improved by an additional effort in research and development.

ACKNOWLEDGMENTS

This work was funded by the LCPC and the French ministry for ecology and sustainable development. The authors wish to thank L. D'Aloia and A. Pavoine for supervising the operation 11L062 on sustainable concrete. Comments on drafts on the manuscript by A. Haricot are acknowledged.

REFERENCES

1. HENDRICKS, C.A., WORRELL, E., PRICE, L., MARTIN, N. Emission reduction of greenhouse gases from the cement industry, 4th International Conference on Greenhouse Gas Control Technologies, Interlaken, Austria, IEA GHG R&D Program, 1998.

2. PARROTT, L. Cement, concrete and sustainability, a report on the progress of the UK cement and concrete industry towards sustainability. British Cement Association, 2002.

3. TELLER, PH., DENIS, S., RENZONI, R., GERMAIN, A., DELAISSE, PH., D'INVERNO, H.. Use of LCI for the decision-making of a Belgian cement producer: a common methodology for accounting CO_2 emissions related to the cement life cycle. 8th LCA Case Studies Symposium SETAC-Europe, 2000.

4. GARTNER, E. Industrially interesting approaches to "low-CO_2" cements. Cement and Concrete Research, Vol. 34, 2004, pp 1489-1498.

5. SFIC, SYNDICAT FRANÇAIS DE L'INDUSTRIE CIMENTIÈRE. Activity report 2000. 2001.

6. CEMBUREAU. Main Characteristics of the Cement Industry. Key facts, 2005.

7. ECOLE DES MINES DE DOUAI. Guide technique régional relatif à la valorisation des laitiers de haut-fourneau, 2005, 21pp.

8. HABERT, G., CHOUPAY, N., MONTEL, J.M., GUILLAUME, D., ESCADEILLAS, G. Effects of the secondary minerals of the natural pozzolans on their pozzolanic activity. Cement and concrete research, Accepted.

9. HABERT, G., CHOUPAY, N., ESCADEILLAS, G., GUILLAUME, D., MONTEL, J.M. Clay content of argillites: Influence on cement based mortars. Applied clay science, submitted.

10. SHAYAN, A., XU, A. Value-added utilisation of waste glass in concrete. Cement and concrete research, Vol. 34, 2004, pp 81-89.

11. COURARD, L., DARIMONT, A., SCHOUTERDEN, M., FERAUCHE, F., WILLEM, X., DEGEIMBRE, R. Durability of mortars modified with metakaolin. Cement and concrete research, 33, 2003.

12. HABERT, G., CHEN, C., ROSSI, P. Perspective assessment of CO_2 emission reduction from the cement industry by using a nomogram of production growth, thermal heating and decarbonation processes. Journal of Environmental Management, Submitted.

13. REINHARDT, H.W., WUSTHOLZ, T., About the influence of the content and composition of the aggregates on the rheological behaviour of self-compacting concrete, Materials and Structures, vol. 39, 2006, pp. 683-693.

14. LEE, H.K., LEE, K.M., KIM, Y.H., YIM, H., BAE, D.B., Ultrasonic in-situ monitoring of setting process of high-performance concrete, Cement and Concrete Research, vol. 34, 2004, pp. 631-640.

15. ERDEM, T.K., TURANLI, L., ERDOGAN, T.Y., Setting time: An important criterion to determine the length of the delay period before steam curing of concrete, Cement and Concrete Research, vol. 33, 2003, pp. 741-745.

16. ROUSSEL, N, STAQUET, S., D'ALOIA SCHWARZENTRUBER, L., LE ROY, R., TOUTLEMONDE, F., SCC casting prediction for the realization of prototype VHPC-precambered composite beams, Materials and Structures, Vol. 40, 2007, pp. 877-887.

17. NUNES, S., FIGUEIRAS, H., OLIVEIRA, P.M., COUTINHO, J.S., FIGUEIRAS, J., A methodology to assess robustness of SCC mixtures, Cement and Concrete Research, vol. 36, 2006, pp. 2115-2122.

18. CHOPIN, D., DE LARRARD, F., CAZACLIU, B., Why do HPC and SCC require a longer mixing time?, Cement and Concrete Research, vol. 34, 2004, pp. 2237-2243.

19. DE LARRARD, F., Concrete mixture proportioning, E & FN Spon, London, 1999.

20. BOEL, V., AUDENAERT, K., DE SHUTTER, G., HEIRMAN, G., VANDEWALLE, L., DESMET, B., VANTOMME, J., Transport properties of self compacting concrete with limestone filler of fly ash, Materials and Structures, vol. 40, 2007, pp. 507-516.

21. CHARRON, J.P., DENARIE, E., BRUHWILER, E., Permeability of ultra high performance fiber reinforced concretes (UHPFRC) under high stresses, Materials and Structures, vol. 40, 2007, pp. 269-277.

22. PAPADAKIS, V.G., TSIMAS, S. Supplementary cementing materials in concrete. Part I: efficiency and design. Cement and concrete research. Vol. 32, 2002, pp 1525-1532.

23. VU, D.D., STROEVEN, P., BUI, V.B. Strength and durability aspects of calcined kaolin-blended Portland cement mortar and concrete. Cement and concrete composites. Vol. 23, 2001, pp 471-478.

24. RODRIGUEZ-CAMACHO, R.E. AND URIBE-AFIF, R. Importance of using the natural pozzolans on concrete durability. Cement and concrete research. Vol. 32, 2002, pp 1851-1858.

25. GASTALDINI, A.L.G., ISAIA, G.C., GOMES, N.S., SPERB, J.E.K. Chloride penetration and carbonation in concrete with rice husk ash and chemical activators. Cement and concrete composites, Vol. 29, 2007, pp 176-180.

26. POON, C.S., KOU, S.C., LAM, L. Compressive strength, chloride diffusivity and pore structure of high performance metakaolin and silica fume concrete. Construction and building materials, Vol. 20, 2006, pp 858-865.

27. MOULIN, E., BLANC, P., SORRENTINO, D. Influence of key cement chemical parametres on the properties of metakaolin blended cements. Cement and concrete composites. Vol. 23, 2001, pp 463-469.

28. KOCKAL, N.U. AND TURKER, F. Effect of environmental conditions on the properties of concretes with different cement types. Construction and Building materials. 2006.

29. WONG, H.S. AND RAZAK, A.H. Efficiency of calcined kaolin and silica fume as cement replacement material for strength performance. Cement and concrete research, Vol. 35, 2005, pp. 696-702.

30. DE LARRARD, F., BELLOC, A., BOULAY, C., KAPLAN, D., RENWEZ, S., SEDRAN, T. Formulation de référence Projet National: Bétons à Hautes Performances, BHP 2000. 1996.

DURABILITY OF GEOPOLYMER GBFS CONCRETE

R Mejía de Gutiérrez   **S A Bernal López**

E Rodríguez Martínez   **S Delvasto Arjona**   **J Maldonado Villa**

Universidad del Valle

Colombia

F Puertas Maroto

Eduardo Torroja Institute

Spain

ABSTRACT: This paper presents the properties of alkaline activated concrete mixes prepared using a Colombian granulated blast furnace slag (GBFS). The alkaline activator was a mix of sodium silicate and sodium hydroxide. These solutions were blended providing modulus (SiO_2/Na_2O ratio) equal to 1.3 and 2.4. Siliceous gravel and river sand were used as aggregates. The properties of Alkali activated slag concretes (AASC) are compared to those of Portland cement concrete (OPCC) with the same proportion of binder (400 kg/m^3 of concrete) and water to binder ratio in the range of 0.38 to 0.40. The tests carried out were mechanical behaviour, total and capillary absorption, chloride permeability and carbonation rate. Compressive strength of the AAS concretes at 28 days is better than those of a Portland cement concrete with equivalent water / cement ratio. The highest strength of AASC (70 MPa at 28 curing days) was obtained with a modulus of 2.4. The results obtained showed that AASC have lower total absorption (around the half) and higher capillary absorption than the reference concrete prepared with OPC. According to ASTM C1202, the chloride permeability of all the AAS concretes was lower than that of the reference specimens. Higher depth of carbonation (up to five times) was found than in OPC concrete; therefore it was proved that there exists a strong dependence between AAS concrete strength and its carbonation performance.

Keywords: Geopolymeric concrete, Alkali-activated concrete, New binders, Granulated blast furnace slag, Mechanical properties, Chloride permeability, Carbonation rate.

Dra Mejía de Gutiérrez Ruby. Universidad del Valle, Director of Composite Materials Group (GMC), Cali, Colombia.

Eng Bernal López Susan Andrea. Universidad del Valle, PhD. Student in Engineering. Composite Materials Group, Cali, Colombia.

Eng Rodríguez Martínez Erich. Universidad del Valle, Research Assistant in the Composite Materials Group, Cali, Colombia.

Dra Puertas Maroto Francisca. Eduardo Torroja Institute (CSIC), Madrid, Spain, Scientific Researcher.

Dr Delvasto Arjona Silvio. Universidad del Valle. Titular Professor of Materials Engineering, GMC, Cali, Colombia.

Eng Maldonado Villa Jorge. Universidad del Valle, PhD. Student in Engineering, GMC.

INTRODUCTION

Concrete, based mainly on Portland cement clinker, is the most widely-used material on the world after water. The annual production of cement is of the order of 1.8 billion tons. About 3 billion tons of natural resources per year [1] are needed in the production of Portland cement (PC). During its production, about one ton of carbon dioxide is released into the environment in addition to emissions sulphur dioxide (SO_2) and nitrogen dioxide (NO_2). The concrete industry is also a largest consumer of natural resources (8 billion tons per year of natural aggregates) and it is estimated that cement and concrete together consume 600 to 700 billion gallons of water per year [1]. Additionally, the production of PC is an energy intensive process (about 3700 kJ/kg of cement) [2]. In order to protect the environment it is necessary to develop new materials which present fewer durability problems and are less susceptible to different types of internal and external chemical attack.

The study of the microstructure and the chemistry of hydration of Portland cement, the main hydraulic binder since the 19th century, have led to the development of novel cements with environmental sustainability [3-5]. The replacement of cement by active nanopowders or supplementary cementing materials, such as ground granulated blast furnace slag (GBFS), silica fume (SF), rice husk ash (RHA), metakaolin (MK) or fly ash (FA) are examples of those new binders and constitutes a significant contribution to the eco-efficiency of the global economy [5-8]. Other materials are those called chemically bonded ceramics (CBC) [4], the macrodefect-free cements (MDF) and densified systems containing homogeneously arranged ultrafine particles (DSP) [3-5]. The development of new binders, alternative to traditional and blended cements and concretes, obtained by the alkaline activation or geopolymerization of different industrial by-products (blast furnace slag and/or fly ashes), is an ongoing study and research topic of the scientific community [9-13]. The production of alkali-activated slag (AAS), which is a binder based on 100% of ground granulated blast furnace slag (GGBS) added with an alkaline activator, presents comparative ecological advantages with respect to that of Portland cement, such as the utilization of an industrial by-product, low energy consumption and low greenhouse gas emissions (CO_2, SO_2, NO_x, etc.). These materials open new opportunities for development of high-performance structural materials.

The basic principles of alkaline activation of slags have been known since the 1940's [14], although its application as a binder in the construction industry started in Ukraine between 1960 and 1964 [15]. These materials, alkali-activated slag and geopolymeric concretes are being investigated by numerous authors [16-22].

The purpose of this research was to study the durability properties of concrete mixes made with a Colombian GGBS activated with waterglass ($Na_2SiO_3.nH_2O+NaOH$) instead of Portland cement. The performance of alkali activated slag concretes (AASC) is compared to that of Portland cement concretes (OPCC) with the same proportion of binder.

EXPERIMENTAL PROCEDURE

Materials

Colombian granulated blast-furnace slag (GBFS) from the factory *Acerías Paz del Río*, first producer of steel in the country, was used. The basicity coefficient (K_b = $CaO+MgO/SiO_2+Al_2O_3$) and the quality coefficient ($CaO+MgO+Al_2O_3/SiO_2+TiO_2$) based on the chemical composition (Table 1) were 1.01 and 1.92, respectively. Its specific gravity and Blaine fineness was 2900 kg/m³ and 399 m²/kg, respectively. Its X-ray diffraction pattern is given in Figure 1; only quartz and calcite are observed as crystalline phases. The particle size range, evaluated with laser granulometry, was of 0.1–74 μm, with D_{50} of 5 μm. An optimum range of fineness of 450-650 m²/kg for acid and neutral slags is recommend by Wang et al. [23], taking workability and strength into account.

An ordinary Portland Cement (OPC) with specific surface (Blaine) of 363 m²/kg and specific gravity of 3030 kg/m³ was used as reference. It is noted that this material, produced in Colombian southwest, contains 30% of limestone addition approximately.

Table 1 Chemical Composition of the GBFS

CHEMICAL COMPOSITION, %	
Loss on ignition	2.08
SiO_2	31.08
Al_2O_3	13.98
Fe_2O_3	3.09
CaO	43.92
MgO	1.79
SO_3	0.66

AAS concretes using two types of solutions as activators were prepared:

- *Water glass*: A commercial Sodium Silicate solution ($Na_2SiO_3 \cdot nH_2O$) with 32.4% of SiO_2, 13.5% of Na_2O and 54.1% of H_2O, and density of 1520 kg/m³. The modulus in solution ($M_s = SiO_2/Na_2O$) was 2.4.

- *Water glass* prepared blended liquid: sodium silicate and sodium hydroxide (50% w/v water solution) ($Na_2SiO_3.nH_2O + NaOH$) in order to obtain a modulus, M_s, equal to 1.3.

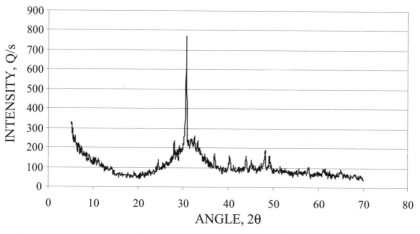

Figure 1 XRD spectra of GGBS

Crushed gravel and river sand were used as coarse and fine aggregates in the manufacture of concrete, respectively. The coarse aggregate was of 19 mm maximum size, specific gravity of 2790 kg/m^3 and absorption of 1.23%. The values of specific gravity, absorption and fineness modulus of sand were 2450 kg/m^3, 3.75% and 2.57, respectively. The grading of each aggregate is shown in Table 2.

Table 2   Coarse and Fine Aggregate Particle Size Distribution

PARTICLE SIZE DISTRIBUTION, % retained accumulated		
Sieve N°	Coarse Aggregate	Fine Aggregate
1 (25.4 mm)	0.52	-
¾ (19.0 mm)	9.85	-
½ (12.7 mm)	56.67	-
3/8 (9.5 mm)	86.01	-
4 (4.75 mm)	97.27	4.29
8 (2.36 mm)	-	8.38
16 (1.18 mm)	-	21.69
30 (600 μm)	-	42.40
50 (300 μm)	-	83.17
100 (150 μm)	-	96.88

Concrete Mixes

The mix design for AASC concrete specimens was made in accordance of ACI recommendations, with a slag content of 400 kg/m^3 as binder. The mixtures were proportioned to achieve initial slump flow consistency ranged from 70 to 100 mm with total water / (slag + activator) ratio of 0.38 to 0.42. This relation is expressed only with the solid portion of the activator. The alkaline activator, *waterglass* was incorporated at 5% and 4% of Na$_2$O by mass in proportion to the slag content, respectively.

Additional concrete mixes with ordinary Portland cement (OPCC) as binder were prepared as reference using 400 kg of cement and water / cement of 0.4. The proportions of coarse and fine aggregate were 55% and 45%, respectively. The AASC mixture codes are GC-SS (M$_s$: 2.40) and GC-W (M$_s$: 1.3). The OPCC specimens were cured underwater and the AASC were kept in moisture chamber (90% RH) at environmental temperature (25±5 °C).

Experimental tests

Cylindrical concrete specimens (76.2 mm in diameter) of each mix were cast and cured until test ages. The concretes were characterized by the following tests: compressive strength (ASTM C109), total absorption and porosity (ASTM C642), capillary absorption [24], and chloride permeability (ASTM C1202).

Additionally, concrete specimens were exposed to natural environment and climatic chamber of CO$_2$ in order to determine the susceptibility to carbonation. The properties of AASC are contrasted with those of Portland cement concrete.

RESULTS AND DISCUSSION

Compressive Strength Behaviour

Compressive strength at the ages of 14 and 28 days was determined. The results of this property are shown in Figure 2. Each value is an average of three specimens. Concrete GC-SS reported higher strength than the other mixes at the studied ages. GC-SS at 28 days of curing developed a compressive strength of 69 N/mm^2 (69 MPa), 59% and 142% higher than that obtained with code GC-W and the equivalent control mixture (OPCC), respectively. The 14 days strength of AASC concrete is around 86% of the 28 days strength while the OPCC was reported to be 75%. It is noted that its contribution to the compressive strength is especially significant at early ages.

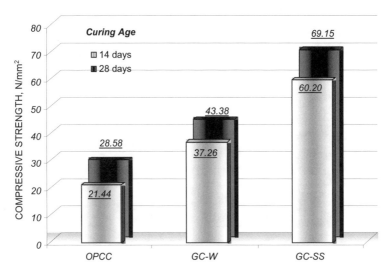

Figure 2   Compressive strength of concrete

Several reports have indicated that the strength development of AASC depends on the activator type and concentration, and in general the sodium silicate-based activator was found to have the best performance on strength development [25-27]. It is explained by the structure and composition of the calcium silicate hydrate, which is the main hydration product of alkaline activation of slag [17, 26-30]. The formation of abundant cross linked structures is considered responsible for the mechanical properties. The highest strength and the difference between the alkaline activated concretes, observed in the present study, are mainly due to the nature of alkaline activator and its higher modulus. Wang *et al.* [23] studied the effect of several factors on the strength of AASC and found an optimum range of M_s of 0.7 to 1.5 as a function of the type of slag. In the present work the optimum was 2.4.

Water Absorption Tests

The ingress of various ions from the environment and their movement through building materials are responsible for the deterioration of structures. By this reason, the control of the permeability of concrete plays an important role in providing resistance to aggressive environments. Water absorption was evaluated by ASTM C642 "Standard Test Method for

Density, Absorption, and Voids in Hardened Concrete" and by the test of capillarity. The results of the application of procedures established on specimens at 28 and 60 days curing are presented in Figure 3 and Table 3.

The results obtained in the first test with GC-SS at 28 days showed a reduction of 67% in the absorption and porosity when compared with OPC concretes. These results are in accordance with the higher compressive strength obtained for this concrete. Total absorption and porosity values lower than 3 and 10%, respectively, are considered acceptable as parameters of compaction and durability. In general, the concrete specimens evaluated at 60 days curing comply with this criterion.

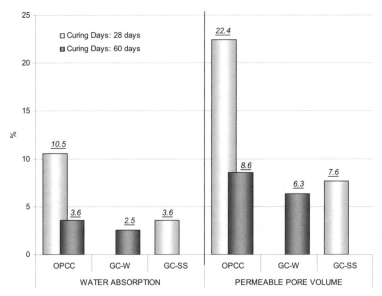

Figure 3   Water absorption and total porosity

Table 3   Capillary coefficient and water resistance (penetration test)

CAPILLARY TEST			
Concrete mix	Curing age, days	k_C, kg/m^{2}s$^{1/2}$	m, s/m^2
OPCC	60 60 28	0.0155	4.3 E07
GC-W		0.0077	9.5 E07
GC-SS		0.0009	7.4 E07

The capillary suction technique was carried out using cylindrical concrete specimens with 76.2 mm in diameter and 50 mm in height. A coating was applied on the curved surface of the samples to make them impermeable. The samples were conditioned at 60°C for 48 hours before the test. A substantial increment of the resistance to the water penetration (m) and a reduction of the capillary Index (k_C) in the order up to 50% in AASC concretes were observed (Table 3) when compared to that of the reference concrete. The observed behaviour is comparable with that found in blended cement concretes including high reactivity pozzolans. These results agree with those reported by Shi [31].

Chloride Permeability

The resistance to the penetration of chloride ions was measured following the ASTM C1202 "Standard Test Method for Electrical Indication of Concrete's ability to resist Chloride Penetration" as the charge passed through the concrete under the application of an external electrical field (60 V) during a period of six hours. The test was carried out on cylindrical specimens of 50 mm thick after curing for 60 days. This test, called rapid chloride permeability test (RCPT), is essentially a measurement of electrical conductivity which depends on both the pore structure and the chemistry of pore solution. The results are shown in Figure 4.

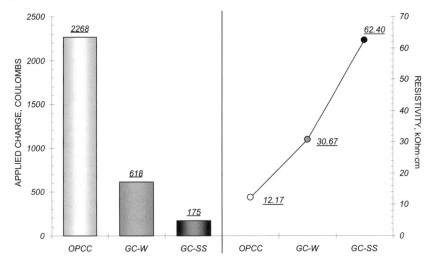

Figure 4   Electrical indication of the ability of concrete to resist chloride ion penetration (ASTM C1202) and concrete resistivity

According to ASTM C1202, a concrete with a charge less than 1000 coulombs can be considered as a material with high resistance to the penetration of chloride ions. AASC concretes activated with water glass and silicate gave a value of 618 and 175 coulombs, respectively. These values are up to 13 times lower than that of the reference concrete. It is noted that GC-SS shows the best performance in this test.

For a given specimen size and applied voltage in this test, the recorded initial current can be regarded as a representative of electrical conductivity. Based in this concept, the value of electrical resistivity was calculated. AASC specimens exhibited an electrical resistivity of approximately five times superior than that of the one obtained with the OPCC specimens. This result is accounted to its lower porosity.

Carbonation Performance

The effect of CO_2 on Portland cement concretes is well known, but fewer studies have been realized on AAS concretes. Carbonation is a chemical reaction between CO_2 and some hydrated cement compounds, mainly calcium hydroxide, although this can affect other hydrated compounds such as CSH. The rate of carbonation depends on the CO_2 concentration, the relative humidity and the temperature of the environment. This reaction modifies the microstructure of the concrete, but its final effect is a function of the carbonation degree. The reaction reduces the alkalinity of the pore solution and the steel becomes susceptible to corrosion too. This reaction requires a certain amount of water in the pores since CO_2 must dissolve and enter in contact with reactive species. The reaction rate depends on the relative humidity of the pore, and increases when the pores are partially filled with water (between 50 and 80% in Portland cement concretes). The natural carbonation process is very slowly and requires several months or years in order to obtain results, mainly due to the low concentration of carbonic gas, CO_2, ($\approx 0.03\%$) in the atmosphere. Therefore it is recommended to do accelerated tests. This process uses a carbonation (or climatic) chamber where the variables can be controlled.

The carbonation test was carried out with concrete specimens of 76.2 mm diameter after 60 days curing. Since there is no accepted standard test for the carbonation of concrete, two kinds of carbonation tests were performed: exposure in a natural environment ([CO_2] = 0.04%±0.01) during 150 days and in the accelerated carbonation chamber (or climatic chamber) for 350 hours (Figure 5). The temperature and relative humidity in the carbonation chamber were controlled at 20°C and 65%, respectively. The CO_2 concentration was 3%±0.3. At the age of measurement, carbonation depths were evaluated. The samples were split and the depths of carbonation were measured by spraying phenolphthalein solution on the broken specimen surface. The non-carbonation part becomes purple and the carbonation portion is colourless. The test results on the carbonation depth of concrete samples are shown in Figure 6.

In general, the carbonation is more intense in AASC concretes. A relationship between the carbonation depth and the compressive strength of concrete was observed. The carbonation of the GC-S concrete with a high initial strength (69 MPa) was 50% lower compared to that of concrete GC-W with 43 MPa. This observation is in agreement with the findings of Atis [32] and Pu et al. [33]. According to published papers, there exist a few articles about the carbonation in natural environment of geoconcretes and alkali-activated slag concretes (AASC) [21, 34-36]. Some authors have concluded that carbonation is more intense in AASC, mainly due to the action of CO_2 on the calcium silicate hydrated gel (CSH) [21, 35].

Puertas et al. [21] analyzed the behaviour of waterglass (SiO_2/Na_2O= 1.18) and NaOH activated slag mortars in an accelerated carbonation test using 100% CO_2 and 43% relative humidity and conclude that the nature of alkali activator affect the carbonation rate. This behaviour is attributed to the differences in the CSH gel formed. On the contrary, Xu and Pu [37] have reported lower carbonation rates in AAS than in Portland cement pastes. The latter authors sustained that this was a consequence of its low permeability.

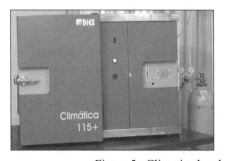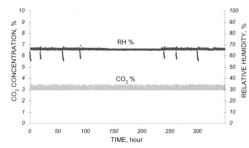

Figure 5   Climatic chamber of CO_2 and control parameters

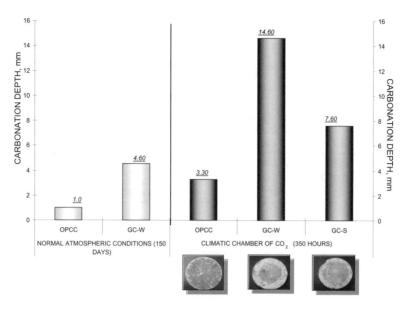

Figure 6   Natural and accelerated carbonation depths

Several studies have proposed equations in which the main factors affecting the inclusion of CO_2 into concrete were taken into account but, in general, the transportation mechanism of carbon dioxide through concrete and its movement rate are based on Fick's law. The classical model of carbonation presented by Equation 1 is also based on the definition of diffusion.

$$K_C = \frac{X}{\sqrt{t}} \tag{1}$$

Where K_C represents the carbonation coefficient (mm/year$^{1/2}$), X is the depth of carbonation (mm), and t is the period of exposure (year). This expression is used by numerous researchers [38-41]. Mathematical models have been developed [39] to predict the CO_2 diffusion coefficient in the natural environment based on the accelerated tests, and the rate has been related with the CO_2 concentration using Eq. 2:

$$\frac{K_c}{K_N} = \sqrt{\frac{C}{N}} \tag{2}$$

where K_N and K_C represent the carbonation coefficient to natural and accelerated conditions, respectively, C is the concentration of CO_2 used in the accelerated test (3.0%) and N the concentration of CO_2 of the natural environment (0.04%). The carbonation coefficients are shown in Table 4. According with these results, the natural coefficient calculated using Equation 2 was approximately 20% higher than those obtained in the experimental test.

The carbonation depth of OPCC and GC-W in the accelerated test was approximately three fold compared to those of natural environment at the conditions used in the experimental work, and the relationship between the two carbonation processes was $K_C \approx 10.3\ K_N$. Based on these results, it is possible to estimate $K_{N\ (experimental)}$ of 3.51 mm/year$^{1/2}$ for GC-SS.

AASC represent a higher carbonation rate than OPCC, however, it is noted that with a GC-SS concrete cover of 30 mm the corrosion of reinforcement would not take place for the first seventy years of service life in a natural environment with 0.04% CO_2. These results are in accordance with previous reports [36]. The probable reason for the higher rate of carbonation can be the larger scope of microcracking in these concretes [21], which could be reduced with a good initial curing period before exposing the concrete to carbonation [34].

Table 4   Carbonation coefficients of concretes

Concrete Mix	CARBONATION CO-EFFICIENT		
	Carbonation co-efficient in a CO_2 Climatic Chamber [CO_2] = 3% (mm/year$^{1/2}$)	Carbonation co-efficient with [CO_2] = 0.04% (mm/year$^{1/2}$)	
		Experimental (Eq. 1)	Calculated (Eq. 2)
OPCC	16.51	1.58	1.91
GC-W	73.04	7.18	8.43
GC-SS	38.02	-	4.39

CONCLUSIONS

Based on the results of this research, the following conclusions can be drawn:

- Strength and permeability properties of the concrete types used in this study are in the order: GC-SS, GC-W, OPCC, where the best activator was *Water glass* with SiO_2/Na_2O ratio of 2.4.
- AASC concretes present lower resistance to carbonation than OPC concrete, but this susceptibility is related to the compressive strength. In this sense, it is possible to produce AASC with low potential susceptibility to carbonation.
- It was proved that there is a relation between carbonation depths of concrete exposed in natural and accelerated environments at the conditions used. This relation can be expressed by $K_C \approx 10.3 \; K_N$ at the test conditions used in the study.
- In general, the results of the experimental work carried out until now confirm the excellent durability of AASC concretes.

ACKNOWLEDGEMENTS

The authors wish to acknowledge the support of Colciencias, Universidad del Valle (Cali, Colombia) and Excellence Center of Novel Materials (CENM).

REFERENCES

1. SARKAR S AND ROUMAIN J C, New Cements for sustainability. International Symposium. Theme 1, Modern Cement Manufacture, 2003, pp. 45-57.
2. GARTNER E, Industrially Interesting Approaches to Low CO_2 Cements. Cement and Concrete Research, Vol. 34, 2004, pp. 1489-1498.
3. ROY D M AND SILSBEE M R, Novel Cements and Cement Products for applications in the 21st Century, Symposium Concrete Technology, Past, Present and future, ACI, SP 144-18, 1994, pp. 349-382.
4. ROY D M, Advances in Cements/Chemically Bonded Ceramics, Ceramics toward the 21st Century, Centennial International Symposium, Ceramics Society, Japan, 1991. pp. 535-551.

5. MORANVILLE-REGOURD M, Portland Cement-Based Binders Cements for the Next Millennium, Modern Concrete Materials, 1999, pp. 87-100.

6. TAILING B AND KRIVENKO P, Blast Furnace Slag - The Ultimate Binder, Chapter 5, Waste Materials Used in Concrete Manufacturing, William Andrew Publishing Noyes, 1997.

7. VAN LOO W, The production and use of cement in a Sustainable Context. Proceedings of the International Symposium - Symposium I, Role of Cement Science in Sustainable Development, 2003, pp. 485-494.

8. ROY D M, SCHEETZ B E AND TIKALSKY P J, Major changes in cements current status and the effects on a sustainable future, Role of Concrete in Sustainable Development - Proceedings of the International Symposium, 2003, pp. 521-536.

9. GLUKHOVSKY V D, ZAITSWEV Y AND PAKHONOW V, Slag-Alkaline Cements and Concrete Structures, Properties, Technological and Economical Aspects of the Use, Silicates Ind., Vol. 10, 1983, pp. 197-200.

10. ROY D M AND SILSBEE M R, Alkali Activated Materials, An Overview, Mat. Res. Soc. Symp. Proc., 245, 1992, pp. 153-164.

11. WANG S D et al., Alkali Activated Slag Cement and Concrete: A Review of its Properties and Problems, Advances In Cement Research, Vol. 27, 1995, pp. 93-102.

12. PUERTAS F, Cementos de Escoria Activados Alcalinamente: Situación Actual y Perspectivas de Futuro, Materiales de Construcción, Vol. 45 (239), 1995, pp. 53-64.

13. SHI C A AND JUESHI QIAN B, High Performance Cementing Materials from Industrial Slags - A Review, Resources, Conservation and Recycling, Vol. 29, 2000, pp. 195–207.

14. PURDON A O, The action of alkalis on blast-furnace slag, J. Soc. Chem. Ind. Trans. Commun., Vol. 59, 1940, pp. 191–202.

15. GLUKHOVSKY V D AND PAKHOMOV V A, Slag-Alkali Cements and Concretes, Budivelnik Publishers, Russia, 1978.

16. CHENG T W AND CHIU J P, Fire-resistant geopolymer produced by granulated blast furnace slag, Minerals Engineering, Vol. 16, 2003, pp. 205-210.

17. FERNÁNDEZ A AND PUERTAS F, Effect of Activator Mix on the Hydration and Strenght Behavior of Alkali-Activated Slags Cements, Advances In Cement Research, Vol. 15, No. 3, 2003, pp. 129-136.

18. FERNÁNDEZ A AND PUERTAS F, Setting of Alkali-Activated Slag Cement. Influence of Activator Nature, Advances in Cement Research, Vol. 13, No. 3, 2001, pp. 115-121.

19. MEJÍA DE GUTIÉRREZ R, BERNAL S AND RODRÍGUEZ E, Non Conventional Concrete Based on GGBS, The 20th International Conference on Solid Waste Technology and Management, Chester, PA, 2005, pp. 259-267.

20. MEJÍA DE GUTIÉRREZ R, MALDONADO J, DELVASTO S, PUERTAS F AND FERNANDEZ A, Durability of Mortars made with Alkali Activated Slag, 11th International Congress on the Chemistry of Cement, ICCC. South Africa, 2003, pp. 1005-1012.

21. PUERTAS F, PALACIOS M AND VAZQUEZ T, Carbonation process of alcali-activated slag mortars. Journal of Materials Science, Vol. 41, 2006, pp. 3071-3082

22. CHANG J J, A study on the setting characteristics of sodium silicate-activated slag pastes, Cement and Concrete Research, Vol. 33, No. 7, 2003, pp. 1-7.

23. WANG S, SCRIVENER K AND PRATT P L, Factors Affecting the Strength of Alkali-Activated Slag, Cement and Concrete Research, Vol. 24, No. 6, 1994, pp. 1033-1043.

24. FAGERLUND G, On the Capillarity of Concrete, Nordic Concrete Research, Vol. 1, No. 6, 1982, pp. 20.

25. WANG S D AND SCRIVENER K L, Hydration products of alkali activated slag cement, Cement and Concrete Research, Vol. 25, 1992, pp. 561-571.

26. FERNÁNDEZ-JIMÉNEZ A, PALOMO J G AND PUERTAS F, Alkali-activated slag mortars mechanical strength behaviour, Cement and Concrete Research, Vol. 29, No. 8, 1999, pp. 1313-1321.

27. BROUGH A R AND ATKINSON A, Sodium silicate-based, alkali-activated slag mortars, Part I, Strength, hydration and microstructure, Cement and Concrete Research, Vol. 32, 2002, pp. 865-879.

28. FERNÁNDEZ A AND PUERTAS F, Structure of Calcium silicate hydrates formed in alkaline activated slag: Influence of the type of alkaline activator, J. Am. Ceram. Soc., Vol. 86, No. 8, 2003, pp. 1389-1394.

29. WANG S D, Alkali activated slag: hydration Process and development of microstructure, Advances in Cement Research, Vol. 12, No. 4, 2000, pp. 163-172.

30. SHI C, On the state and role of alkali's during the activation of alkali-activated slag cement, Proc. 11th International Congress on chemistry of Cement, South Africa, 2003, pp. 2097-2107.

31. SHI C, Strength, pore structure and permeability of alkali activated slag mortars, Cement and Concrete Research, Vol. 26, No. 12, 1996. pp. 1789-1799.

32. ATIS C D, Accelerated Carbonation and testing of concrete made with fly ash, Construction and Building Materials, Vol. 17, 2003, pp. 147-152.

33. PU X C, GAN C C, WANG S D AND YANG C H, Summary Reports of Research on Alkali-Activated Slag Cement and Concrete, Vol. 1–6, Chongqing Institute of Architecture and Engineering, Chongqing, 1988.

34. CRIADO M, PALOMO A AND FERNÁNDEZ-JIMÉNEZ A, Alkali activation of fly ashes, Part 1: Effect of curing conditions on the carbonation of the reaction products, Fuel, Vol. 84, 2005, pp. 2048–2054.

35. BAKHAREV T, SANJAYAN J G AND CHENG Y-B, Resistance of alkali-activated slag concrete to carbonation, Cement and Concrete Research, Vol. 31, 2001, pp. 1277–1283.

36. BYFORS K, KLINGSTEDT G, LEHTONEN V, PYY H AND ROMBEN L, Durability of concrete made with alkali-activated slag, Proceedings 3rd CANMET/ACI Inter. Conf., Trondheim, Norway, Vol. 2, 1989, pp. 1429–1466.

37. XU B AND PU X, Second International Conference Alkaline Cements and Concretes, 1999, pp. 101.

38. STEFFENS A, Modeling carbonation for corrosion risk prediction of concrete structures, Cement and Concrete Research, Vol. 32, 2002, pp. 935-941.

39. CASTRO A, FERREIRA R, LOPES A M, CASCUDO O AND CARASEK H, Relationship between results of Accelerated and Natural Carbonation in Various Concretes, International RILEM Conference on the use of Recycled Materials in Buildings and Structures, España, 2004, pp. 988-997.

40. KHUNTHONGKEAW J, TANGTERMSIRIKUL S AND LEELAWAT T, A study on carbonation depth prediction for fly ash concrete, Construction and Building Materials, Vol. 20, No. 9, 2006, pp. 744-753.

41. ROY S K, POH K B, NORTHWOOD D O, Durability of concrete-accelerated carbonation and weathering studies, Building and Environment, Vol. 34, 1999, pp. 597-606.

DYNAMIC BEHAVIOUR OF FERROCEMENT PLATES

Y B I Shaheen

A A Hamada

M A El-Sissy

Menoufia University

Egypt

Abstract. Thin ferrocement plates reinforced with various types of expanded metal meshes were developed with high strength, crack resistance , high ductility and energy absorption properties, which could be used for dynamic applications. Five designation series of plates were cast and tested under four different loading conditions. The dynamic responses such as: frequency, mode shape and damping factor were extensively investigated using FFT analyzer. The experimental analysis and finite element technique were utilized to study the effect of network configurations and boundary fixations on dynamic characteristics. In addition, the investigated composite plates were tested in the high frequency range (up to120 kHz) through ultrasonic attenuation technique. For this purpose, an experimental setup was designed and constructed to measure dynamic elastic modulus, phase velocity and damping attenuation. The effect of mesh-layer debonding on the dynamic characteristics (natural frequency and damping ratio) was investigated. Damage was detected using vibration measurements and identified by comparing signals in higher frequency ranges before and after damage.

Keywords: Modal testing; Finite element; Ferrocement composite; Ultrasonic test.

Yousry Bayoumy Ibrahim Shaheen is professor of Strength and Testing of Materials, Department of Civil engineering, Faculty of Engineering at Menoufiya University, Shebin El-Kom, Egypt.

Dr A A Hamada is associate professor**,** Department of Production Engineering and Machine Design, Faculty of Engineering at Menoufiya University, Shebin El-Kom, Egypt.

M A El-Sissy is graduate student, Department of Civil engineering, Faculty of Engineering at Menoufiya University, Shebin El-Kom, Egypt.

INTRODUCTION

Ferrocement is a type of thin wall reinforced concrete commonly constructed of hydraulic cement mortar reinforced with closely spaced layers of continuous and relatively small size wire mesh [1]. In its role as a thin reinforced concrete product and as a laminated cement-based composite, ferrocement has found itself in numerous applications both in new structures and repair and rehabilitation of existing structures. Compared with the conventional reinforced concrete, ferrocement is reinforced in two directions; therefore, it has homogenous-isotropic properties in two directions. Benefiting from its usually high reinforcement ratio, ferrocement generally has a high tensile strength and a high modulus of rupture. In addition, because the specific surface of reinforcement of ferrocement is one to two orders of magnitude higher than that of reinforced concrete, larger bond forces develop with the matrix resulting in average crack spacing and width more than one order of magnitude smaller than in conventional reinforced concrete [2-3]. Other appealing features of ferrocement include ease of prefabrication and low cost in maintenance and repair. Based on the aforementioned advantages, the typical applications of ferrocement include water tanks, boats, housing wall panel, roof, formwork and sunscreen [4-6].

The renaissance of ferrocement in recent two decades has led to the ACI design guideline "Guide for the Design, Construction, and Repair of Ferrocement" [7], and publications such as "Ferro-cement Design, Techniques, and Application" [8] and "Ferrocement and Laminated Cementitious Composites" [9], which provide comprehensive understanding and detailed design method of contemporary ferrocement. However, the rapid development in reinforcing meshes and matrix design requires continuous research to characterize the new material and improve the overall performance of ferrocement. Thus far steel meshes have been the primary mesh reinforcement for ferrocement, but recently fiber reinforced plastic (FRP) meshes were introduced in ferrocement as an promising alternative to steel meshes [10-14]. Compared with steel, FRP materials possess some remarkable features such as lightweight, high tensile strength and inherent corrosion resistance. However, unlike steel that has an elastic-plastic stress-strain relationship, FRP materials behave elastically up to failure, thus do not yield and lack ductility.

Investigation of the dynamic behavior of ferrocement composites plates in the literature is rarely available. However, limited studies were carried out on reinforced concrete structures subjected to dynamic loads (e.g. bridge) to characterize their dynamic behavior for the purpose of fault diagnosis. Here are some examples:

Salawu [15] conducted full-scale forced-vibration tests before and after structural repairs on a multi span reinforced concrete highway bridge. The tests were conducted to study any correlation between repair works and changes in the dynamic characteristics of the bridge. Comparison of the mode shapes before and after repairs using modal analysis procedures was found to give an indication of the repair. The bridge response was measured using accelerometers and the modal parameters were extracted from the frequency response function. The result of this study showed that damping ratio could not be used as an indicator for damage.

Koh and Ray [16] used mode shapes and natural frequencies for model updating method. The finite element model updating process modifies parameters in the global stiffness or mass matrix to reproduce the measured modal data. Thus, local perturbation of parameters in the global stiffness or mass matrix indicates damage location.

Richardson [17] focused on the determination of the functional relationship between variations in the mass, stiffness, damping and the variations in the model properties of the structure. This function could be in a simple form in case of small changes to detect, locate and quantify structural faults by monitoring frequency and damping only. The complete sensitivity function for mass stiffness and damping also the validity of the stiffness sensitivity for small changes were verified using a 3 DOF numerical example [18].

In the present work, modal testing is performed using accelerometers and data acquisition system to measure structure dynamic response. Collected data are used through some signal processing analysis to extract the dynamic parameters. On the other hand, the theoretical models are fine tuned to simulate the existing structure. These tuned fine models are used to form a data base for structure dynamic behavior under different boundary conditions. This research covers the application of two different techniques, namely, mechanical excitation and ultrasonic to characterize the dynamic behavior of the investigated composite plates made from ferrocement. The scope of research covers the numerical simulation and experimental verification.

MODAL ANALYSIS USING THE FINITE ELEMENT METHOD

A typical composite ferrocement plates of dimensions (150 × 150 × 10) mm with various boundary conditions, C-F-F-F, C-S-F-F, C-C-F-F and C-C-C-C along the edges of plate are modeled using the finite element method where C clamped, S simply supported, F free. The equivalent elastic modulus and density of ferrocement composite are computed also, different open mesh, various volume fraction and boundary conditions are employed. A mesh of 20 x 20 elements, eight node brick elements are utilized in the analysis and as shown in Fig.1. The stiffness matrix of the element can be then formulated as [19]:

$$[K] = t \iint [B]^T [D] [B] \, dx \, dy \tag{1}$$

Where t is the thickness of ferrocement plate, [B] is strain matrix and [D] is the elasticity matrix of ferrocement plate, which can be computed according to Ref. [20]. Consequently, the mass matrix of element can be formulated [19] as:

$$[M] = \rho \iint [N]^T [N] \, dx \, dy. \tag{2}$$

Where ρ is the density of the equivalent composite plate with various mesh type, [N] is the matrix of shape function.

The eigen-frequency can be then evaluated from the solution of the characteristic equation for composite plate given by:

$$\left\| [K] - \lambda [M] \right\| [V] = [0] \tag{3}$$

The eigen values and mode shapes are computed using the F.M soft ware package ANSYS (Version 5.4).

Initially, the plates were modeled in order to get a first estimation of the undamped natural frequencies and mode shapes utilizing finite elements type SOLID 65. The material properties were then entered in the program, and the constraint imposed to simulate a type of fixation. The numerical results using FEM were computed for different mesh types and boundary fixations and are listed in Table 1.

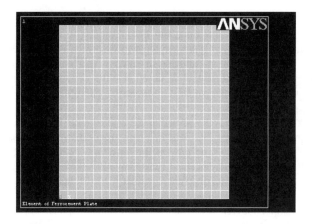

Figure 1   Finite element model for ferrocement plate

Table 1   Values of the first five frequencies in Hz for ferrocement plates under four different boundary conditions (Finite element and experimental results)

BOUNDARY CONDITIONS	F (C—F, F)		F (C—S, F)		F (C—C, F)		C (C—C, C)	
PLATE CONFIGURATIONS	F.E	Ex.	F.E.	Ex.	F.E.	Ex.	F.E.	Ex.
A	9.32	8.5	40	38	79.2	77	154.6	151
	80.90	76	150	148	232	230	526.7	524
	191.81	189	227.5	224	447.8	444	696.7	694
	264.3	262	349	346	504.9	501	1021.2	1018
	393.73	388	454.5	451	571.4	568	1328	1325
B	9.84	9	44	41	85.3	82	157.27	153
	83.5	79	147.9	144	229.1	226	530	527
	192.3	192	226.3	223	442	440	687	683
	278.1	275	347.8	346	498.4	494	1007	1003
	396.1	393	450.6	447	583.9	580	1339	1336
C	10.5	12	46.32	51	92.6	104	170	176
	88.07	86	151	148	235.7	233	536.9	532
	194.8	193	235.7	232	454.8	451	707	704
	320.7	318	382.9	380	513	510	1037	1034
	416	415	472	469	580	578	1351	1355
D	12.3	16	52.86	58	105	110	184.4	187
	94.8	92	156.9	153	243	240	554	552
	200.8	197	243	240	469	466	729	726
	339	336	400	397	529	525	1089	1085
	429	427	490	487	598	594	1389	1385
Plain	10.07	7	42.9	38	83	79	153	148
	72.63	68	147.6	146	228.8	224	519	518
	180	176	226.5	223	439.5	436	686	684
	262	258	346	342	490.8	488	1006	1004
	385	379	450	446	564	561	1318	1314

EXPERIMENTAL PROCEDURE

Materials preparation

The experimental program was designed to investigate the effect of reinforced steel mesh configurations and boundary conditions on the dynamic behavior of ferrocement plates. Five different patterns of mesh reinforcements were used. Different materials were used to produce the plates (150 × 150 × 10 mm) including: mortar, steel meshes, silica fume, super-plasticizer, fly ash and polypropylene fibers.

Frequency Response Function

The experimental set up, used in the present work, was described in details before, where the specimen was located in a test rig and excited by an Impact hammer type '8202', which resembles an ordinary hammer but has a force transducer type '8200' built into its tip to register the force input. The hammer was used to excite the specimen at the mid-point position. The charge amplifier type '2635' was used to generate the signal from the hammer to the dual channel analyzer type '2034'. The vibration response was registered by a suitable piezoelectric accelerometer (type '4374' its weight '2.4 grams'). The vibration meter (type 2511) was utilized in connection with the accelerometer to generate the signal to the dual channel analyzer (type 2034). The frequency response spectrum was recorded and printed. A sample of the recorded responses is shown in Figure 2. The frequency and damping factor measurements for the fundamental frequency and associated damping factor were carried out for each specimen. The experimental results were taken as an average of five measurements of each. The damping factor (ξ) of a particular resonance was calculated from the width of the resonance peak in the magnitude of the (FRF). The experimental measurements of frequency, amplitude and damping factor are listed in Table 2.

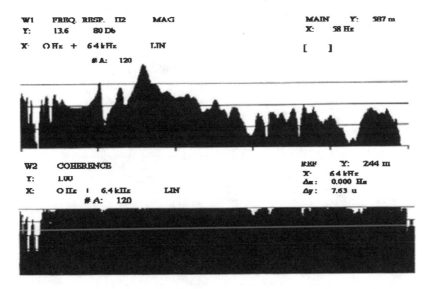

Figure 2   A sample of frequency response function and coherence function
for ferrocement plate (A-type, C-C-F-F).

Table 2   Values of fundamental frequency (Hz), amplitude (dB) and damping factor for ferrocement plates tested under four different boundary conditions* (experimental results).

BOUNDARY CONDITIONS		F C ☐ F / F	F C ☐ S / F	F C ☐ C / F	C C ☐ C / C
A	Frequency	8.5	38	77	151
	Amplitude	52	46	41	37
	Dam. Factor	0.19	0.17	0.12	0.08
B	Frequency	9	41	82	13
	Amplitude	48	42	36	33
	Dam. Factor	0.18	0.14	0.11	0.06
C	Frequency	12	51	104	176
	Amplitude	42	39	34	31
	Dam. Factor	0.15	0.09	0.07	0.02
D	Frequency	16	58	110	187
	Amplitude	38	34	32	30
	Dam. Factor	0.10	0.07	0.05	0.016
Plain	Frequency	7	38	79	148
	Amplitude	24	22	20	18
	Dam. Factor	0.04	0.015	0.012	0.008

Damage identification using structural dynamic analysis:

To study the effect of mesh–layer debonding on the vibration characteristics of ferrocement plates, the plates were provided with debonding lengths: 10, 20, 30, 40, 50 and 60 mm by inserting aluminum foils of various sizes on the upper surface of the reinforcing mesh. Figure 3 depicts the configuration of the experimental set-up for vibration testing of cracked plate. The fundamental frequency of cracked plate was recorded and compared with those for noncracked plate. The input parameters were crack length and plate type. The crack location was kept fixed at mid-line of the tested plate.

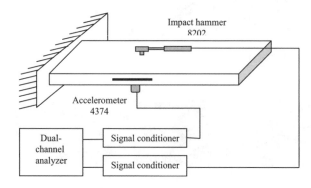

Figure 3   Configuration of experimental set-up for damage detection.

Ultrasonic measurement

Ultrasonic measurements for internal friction (Q^{-1}) and phase velocity (Cp) were based on the application of magnetostriction phenomena [22]. Figure 4, shows the basic system used for these measurements while the recorded signal echo is schematically shown Figure 5. The first part of the echo including the cross-over, is the direct return of the transmitted signal, whilst the second part, the decrement, is the exponential retransmission of the energy stored. The number of oscillations to the cross over is a function of the line (wire) cross-section and the properties of resonating material. The parameters shown in this figure are used to calculate, absolutely, the internal friction according to the following equation [22]:

$$\frac{Qc}{QM} = \frac{A0 + A\infty}{A0 - A\infty} = X \tag{4}$$

where Qc and Qm are coupling and material Q-values respectively.

and $$\frac{\Pi N}{Qc} = \frac{Ln[2/(1-X)]}{1-X} \tag{5}$$

Internal friction values were calculated with standard variation (±0.0019). The longitudinal resonant modes of vibration of each tested composite specimen were excited by cementing it with the remote end of the delay line (wire 1 m long and 1mm dia. made entirely from nickel) of the system. The corresponding resonance frequencies were detected.
For a specimen of length L(m), vibrating at its natural frequency f(Hz), the phase velocity C_p (m/sec) is related to this resonance frequency by [22].

$$C_p = 2L \, f / n \tag{6}$$

The most accurate dynamic Young's modulus (E_D) usually follows from determining ultrasonic phase velocity (C_p) as using the general relationship [23].

$$E_D = \rho C_p^{\,2} \tag{7}$$

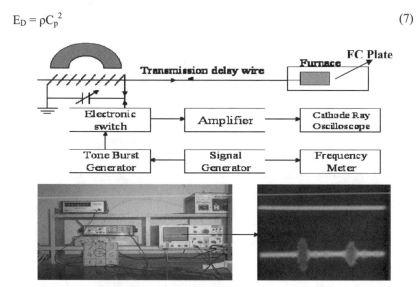

Figure 4   Set-up of ultrasonic measuring system and the resultant echo

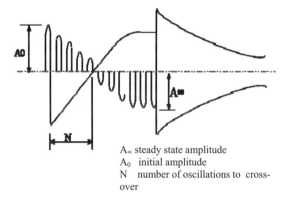

A∞   steady state amplitude
A₀   initial amplitude
N   number of oscillations to cross-
over

Figure 5   Schematic diagram of the resultant echo-pattern displayed
in the magnetostrictive delay-line system.

RESULTS AND DISCUSSION

The resonant frequencies, mode shapes and damping factors of square ferrocement composite plates have been measured and analyzed for different mesh configurations and boundary fixations. The measured and computed values of the frequencies are given in Table 1. Comparisons between the experimental and numerical results of the frequencies indicate good agreements. Table 2 shows the variation of fundamental frequencies, amplitude and damping factor for different mesh configurations at the same plate thickness. It can be seen that the damping factor of plate A is relatively high compared with the other. This is due to the maximum dissipated energy at this mesh reinforcement. Also, it can be noticed that the damping factor of plain plate is relatively low compared with the other due to the small stiffness value for this case of bulk material. As expected, the measured frequencies are inversely proportional to damping factor as shown in Figure 5.

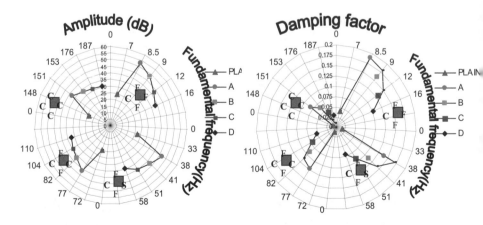

Figure 5   Variations of fundamental frequency with amplitude a), and with damping factor b),
for various boundary conditions and mesh configurations.

In general, the damping factor in composite materials is relatively high relative to bulk materials. It is difficult to control the value by variation of the mass and stiffness. From Fig. 5, it can be noticed that minimum values of the damping factor occur in the case of clamped [CCCC] plates with different types of mesh configuration. In all boundary conditions, it is observed that the damping factor is small for mesh type [D] compared with other reinforced type. This explained by the fact that mesh reinforcements are expected to increase the plate stiffness and result in less energy dissipation. In view of the state of fixation, it is observed that the effect of the degree of constraints is dominant on values of natural frequency and damping factor compared with the variation of the open mesh type. Figure 6 shows mode shape for a few selected cases of tested plates for different boundary conditions.

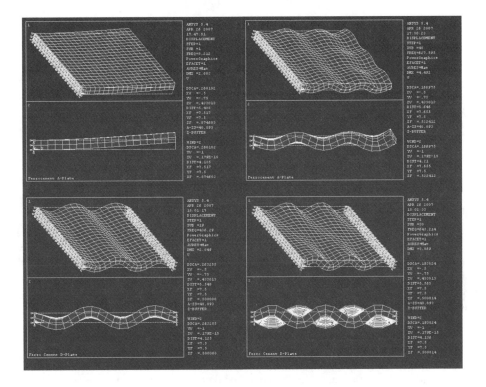

Figure 6   Mode shapes for a few selected cases of tested plates
for different boundary conditions.

Determination of the natural frequencies and mode shapes of a vibrating structure is an important aspect from the stand point of view of the structure dynamic behavior. The natural frequency gives information about resonance avoidance for certain loading conditions. Mode shape, on the other hand, gives indication about the vibration level at each position of the structure. One of the most important parameters from designer's point of view is the location of nodes and antinodes. The nodes are the positions at which the vibration vanishes, the maximum stresses induced at these nodes. While the antinodes are the positions at which maximum vibration level occurs.

In Table 3, the natural frequencies and damping ratios for the cracked D-plate under cantilever fixation are set out based on the dynamic measurements. A group of natural frequencies can be recognized that represent the dynamic characteristics of the noncracked plate; that the natural frequencies in this group decrease proportionally until a critical debonding extend ($\approx$ 20 mm) is reached. The damping ratio in this group shows an increase, while the natural frequency decreases. Beyond this critical debonding extent, another group of frequencies decreases which represents the dynamic characteristics of the damaged plate. The damping ratios in this group increase noticeably beyond the critical debonding extent.

Table 3   Natural frequencies and damping factor for the cracked D-plate under cantilever fixation

CRACK LENGTH, mm	N. FREQUENCY, Hz	DAMPING RATIO
Intact	96	0.09
10	93	0.10
20	83	0.14
30	70	0.18
40	53	0.39
50	29	0.56

Ultrasonic Results

The ultrasonic phase velocity propagating in the specimen (Cp), the dynamic elastic modules E_d as well as the internal friction (Q^{-1}) for the first two echoes are given in Fig. (7). It is clear that the ultrasonic wave propagating in plain specimen is attenuated by various factors mainly voids, lack of homogeneity, cracks etc. This explains the lower Cp-values for the dummy compared with the composites plates by almost 50%. Such attribution could explain the high (E_d) – values for composite compared with dummy by almost 70% .

Regarding internal friction values, Figure 7 shows that the composite plates have high Q^{-1} values compared with the dummy plates .

Let us try to offer an explanation for the improved damping for the composite plate based on arguments other than defectology. The composite plates consist of steel mesh sandwiches within more layer and so can be considered as a viscoelastic material, i.e. posses both elastic and viscous properties. Thus some of the energy stored in a viscoelastic system is recovered upon removal of the load, and the remainder is dissipated by the material in the form of heat, resistance,…. So when the composite plate is subjected to ultrasonic energy, the mesh layer begins to slide resulting in a shearing action compared to matrix material. The pattern does more than just bond the mesh to the matrix, it also provides the mechanisms that create the damping effects as these shear strains are converted to heat energy within viscoelastic material. The dependence of Q^{-1} values on steel mesh configuration is further observed from Figure 7.

From which, plate A has the higher Q^{-1} values compared with plate D. In ferrocement composites, two kinds of interfaces: strongly bonded interface and weakly bonded interface, can be identified. For a strongly bonded interface (case D), it is assumed that bonding between metal mesh is strong enough to resist sliding between matrix and reinforcement. When the interface is weakly bonded, sliding at the interface is more likely to occur. Interfaces between different phases in multi-component material system (the case of the present work), may offer various possibilities for vibration energy dissipation. Poorly bonded interfaces are generally expected to increase damping through friction or columbic sliding mechanisms.

The data shown in Figure 7, strongly support the idea that the stiffer materials (high Ed-values) could have a higher internal friction. In this regard, steel mesh configurations (volume fraction of fiber, open mesh area, wire diameter....) have a specific contribution to keep both stiffness and damping at high levels. However, to quantify the specific weight of these parameters on damping necessitates further experimentation. In our opinion, this step will be of special interest and will open a new important area for ferrocement composite structure in dynamic field.

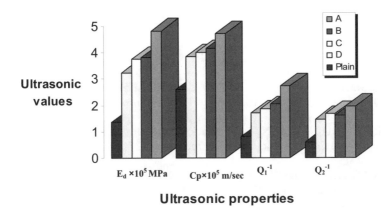

Figure 7   Experimental data from ultrasonic measurements

CONCLUSIONS

The dynamic behavior of ferrocement plates reinforced with various steel mesh configurations was investigated experimentally and numerically under different boundary conditions. The experimental techniques were employed, namely hammering excitation for law frequency ranges and ultrasonic attenuation for high frequency ranges (up to 120 kHz). The following conclusions were arrived at:

1. The dynamic characteristics of FC plates differ considering depending on mesh configu-rations and boundary conditions. There for FC structures may be tailored for specified modal parameters and nodes positions to satisfy certain operations conditions .

2. The numerical results from FEM indicate good agreement with those obtained from modal analysis. However, it is recommended to use finer mesh for the numerical FEM considering more nodes of vibration.

3. Damping of the investigated composites is relatively higher than those un-reinforced by almost five times. Also, the larger the open mesh area (type D-80%), the lower was damping capacity compared with type A (68%).

4. The positions of the nodes and antinodes are shifted for FC plates compared with the plain ones.

5. Damage identification based on FRF can accurately determine the extent of the damage from natural frequency and damping ratios .

6. In the high frequency range (ultrasonic data), the reinforced plates have higher stiffness and damping compared with the un-reinforced plates. In this regard, steel mesh configurations (volume fraction of fibers, open mesh area, wire diameter,..) have their specific contributions to keep both stiffness and damping at high levels. However, further experimentations one required to quantify the specific weight of those parameters on ultrasonic data.

.

REFERENCES

1. ACI COMMITTEE 549, "State-of-the-Art Report on Ferrocement", ACI 549-R97, in Manual of Concrete Practice, ACI, Detroit, 1997, 26pp.

2. SHAH, S.P AND NAAMAN, A.E., "Crack Control in Ferrocement and Its Comparison with Reinforced Concrete," Journal of Ferrocement, V. 8, No. 2, 1978, pp.67-80.

3. ARIF, P.M., AND KAUSHIK, M. A. "Mechanical Behavior of Ferrocement Composites", J. Mat. in Civ. Eng: V. 14, 2, 2002, pp. 156-163.

4. GUERRA, A.E., NAAMAN, A.E. AND SHAH, S.P., "Ferrocement Cylindrical Tanks: Cracking and Leakage Behavior," ACI Journal, Proceedings, V. 75, No. 1, Jan., 1978, pp. 22-30.

5. NIMITYONGSKUL, P., CHEN BOR-SHIUN AND KARASUDHI, P., Impact Resistance of Ferrocement. "Investigation on Metallic and Dielectric Solid Media, Boat Hulls," Journal of Ferrocement, V.10, No. 1, 1980, pp. 1-10.

6. KADIR, M.R.A., SAMAD, A.A.A., MUDA, Z.C., AND ABANG ABDULLAH, A.A., "Flexural Behavior of Composite Beams with Ferrocement Permanent Formwork," Journal of Ferrocement, V. 27, No. 3, 1997, pp. 209-214.

7. ACI COMMITTEE 549-1R-88, "Guide for Design Construction, and Repair of Ferrocement" ACI 549-1R-88 and 1R-93, in Manual of Concrete Practice, ACI, Detroit, 1993, 27pp.

8. BINGHAM, B., "Ferro-cement Design, Techniques, and Applications", Cornell Maritime Press, Cambridge, Maryland, 1974, 444pp.

9. NAAMAN, A., "Ferrocement and Laminated Cementitious Composites", Techno Press 3000, Ann Arbor, Michigan, 2000, 372pp.

10. NAAMAN, A.E., AND AL-SHANNAG, J. "Ferrocement with Fiber Reinforced Plastic Meshes: Preliminary Investigation," Proceedings of the Fifth International Symposium on Ferrocement, Manchester, England, September 1994. P. Nedwell and R.N. Swamy, Editors, E. and FN Spon, London.

11. GUERRERO, P. AND NAAMAN, A. E., "Bending Behavior of Hybrid Ferrocement Composites Reinforced with PVA Meshes and PVA Fibers," Ferrocement 6 - Lambot Symposium, Proceedings of Sixth International Symposium on Ferrocement, Naaman, A.E., Editor, University of Michigan, June 1998.

12. HAMMOUD, HASSEN "Study of Ferrocement Bolted Connection for Structural Applications" Ph. D Thesis, University of Michigan, 1993.

13. NAAMAN, A.E., AND CHANDRANGSU, K., "Bending Behavior of Laminated Cementitious Composites Reinforced with FRP Meshes," High Performance Fiber Reinforced Concrete Thin Sheet Products, Edited by A. Peled, S.P. Shah and N. Banthia, ACI SP-190, American Concrete Institute, Farmington Hills, 2000, pp. 97-116.

14. LOPEZ, M. AND NAAMAN, A. E., "Study of Shear Joints in Fiber Reinforced Plastic (FRP) Ferrocement Bolted Connections," Ferrocement 6 - Lambot Symposium, Proceedings of Sixth International Symposium on Ferrocement, Naaman, A.E., Editor, University of Michigan, June 1998.

15. SALAWU, W. "Bridge Assessment Using Forced-Vibration Testing', Journal of Structural Engineering, Vol.121, No.2, Feb., 1995, pp.161-172.

16. KOH B. AND RAY L., "Localization of Damage in Smart Structures Through Sensitivity Enhancing Feedback Control", Mechanical System and Signal Processing 1566, 2003, pp.1-19.

17. RICHARDSON M., "Determination of Modal Sensitivity Function for Location of Structural Faults", The 17th International Modal Analysis Conference, Kissimmee, FL, Feb. 1999.

18. ZIENKIEWICZ, O.C. AND TAYLOR, R. L., "Finite Element Method" 4th edn. McGraw-Hill, New York 1989.

19. KRAWCZUK, M. " A Rectangular Plate Finite Element With an Open Crack" J. of Computers and Structures Vol. 46, No.3, 1993, pp 487-493.

20. EPSTEIN, M. AND HUTTELMAIER, H.P., "A Finite Element Formulation for Multilayered and Thick Plates" J. of Composite Structure, Vol. 25, 1993, pp 645-650.

21. KHAFAGY A. H., "Acoutical and Mechanical Ph. D Thesis, university of London, 1985.

22. NESVIJSKI, E. G., "Some Aspects of Ultrasonic Testing of Composites", Composite Structures, V 48, 2000, pp 151-155.

23. MORENO E., CASTILLO, M., "Ultrasonic NDT in Ferrocement", Ultrasonic Symposium, IEEE, Atlanta, GA, USA, 2001, pp. 777-780.

ENVIRONMENTAL EFFECTS ON HEAT OF HYDRATION IN FOAMED CONCRETE

A Yerramala

M R Jones

University of Dundee

United Kingdom

ABSTRACT. Foamed concrete has a unique set of properties including flowability, self compactability, flexibility in producing range of densities, reduced cost and good thermal insulating and fire properties and it can be used in a wide range of construction applications. However, in large volume applications such as void filling, its high volume to surface area ratio and increased thermal insulation leads to significantly higher core temperatures due to heat of hydration. This paper describes a laboratory study carried out to examine the effect of constituent temperatures on heat of hydration and the peak temperatures resulting in the concrete mass. In addition, influence of ambient, pour volume and foamed concrete density on heat of hydration was discussed.

Key words: Foamed concrete, Effect of constituent and environmental temperature, Heat of hydration.

Amar Yerramala, is a PhD student in the Concrete Technology Unit, Division of Civil Engineering at the University of Dundee. His main interests are foamed concrete, use of recycled and secondary aggregates in concrete, concrete durability and ultra lightweight foamed concrete.

Rod Jones, is Dean of the School of Engineering, Physics and Mathematics and the Associate Director of the Concrete Technology Unit at the University of Dundee. A Chartered Civil Engineer, his research work concentrates on innovation in concrete construction.

INTRODUCTION

Foamed concrete can provide a sustainable fill material for many of the construction applications [1-3] such as high volume void backfills, bridge abutments, arch bridge infills, road sub-bases, floor and roof screeds, soil stabilisation, grouting tunnel walls, thermal and acoustic insulation and as a semi structural material [4, 5]. It has a unique set of properties such as flowability, self compactability, controlled low density and good thermal insulation that set it apart as a material to be considered in it's own right. In addition, with the use of foamed concrete, primary aggregate consumption can be minimised, as it does not require coarse aggregate and natural sand can be replaced either partially or fully with recycled and secondary aggregates, such as construction and demolition waste, conditioned fly ash, glass cullet, incinerated bottom ash and even crumbed rubber.

As noted earlier, good thermal insulating property (thermal conductivity between 0.15 and 0.21 W/mK at 600 to 900 kg/m^3 densities) of foamed concrete can result in high core concrete temperatures, due to heat of hydration, especially in high volume to surface area applications, such as fills. High temperature differentials between core and surface can cause plastic cracking and adversely effect performance of the concrete [6-8]. In addition, development and maintained high temperatures (>65°C) may lead to form delayed ettringite in the concrete [9] although there is no evidence that foamed concrete is affected by this and even if it did occur the bubble structure is capable for allowing phases to expand with cracking.

Thus, this paper reports a laboratory investigation carried out to examine the influence of constituents temperature on heat of hydration and feasibility to minimise peak temperatures of foamed concrete. In addition, influence of ambient, pour volume and foamed concrete plastic density on heat of hydration are presented.

EXPERIMENTAL PROGRAMME

Scope of the Work

The laboratory investigation considered the most common foamed concrete mixes used in large volume fill applications, viz flowing and self compacting 700 and 900 kg/m^3 plastic density foamed concrete. The aim of the study was to determine the effect of constituent temperatures, pour volume, ambient and density on heat of hydration of foamed concrete. The testing programme is given in Table 1. Temperature profiles of test specimens, subjected to near adiabatic conditions in an insulated hot-box, were measured.

Mix Constituent Proportions and Test Specimen Preparation

Portland cement confirming to BS EN 197-1 [10], sand as a fine aggregate having particle size <2 mm confirming to BS EN 12620 [11], water and commercially available protein surfactant was used to produce foamed concrete. The mix constituent proportions of the foamed concretes is summarised in Table 2. Rotary drum mixture (free-falling action) was used to produce foamed concrete. Firstly homogeneous base mix was obtained, and then designed pre-formed foam was produced and added to the base mix. The mixing was continued until uniform distribution of foam in base mix (approximately 1-2 minutes). The plastic density of the foamed concrete was then measured in accordance with BS EN 12350-6 [12] and values within ± 50 kg/m^3 of the target density accepted, which is typical tolerance used by industry [13]. If the density was higher, additional foam was produced and added incrementally until to reach the tolerance limit. Lower densities which were below acceptance range were discarded and repeated.

Table 1 Environmental and mix constituent temperatures.

TEST	TEMPERATURE, °C[*]			Number of pours
	Ambient	Cement	Water	
Test$_1$	20	50	15	Mix$_1$ single pour
Test$_2$	20	50	5	Mix$_1$ single pour
Test$_{11}$	20	50	15	Mix$_1$ 0-82.5mm pours
Test$_{12}$	20	50	15	Mix$_1$ 82.5-150mm pours
Test$_{21}$	20	50	5	Mix$_1$ 0-82.5mm pours
Test$_{22}$	20	50	5	Mix$_1$ 82.5-150mm pours
Test$_{23}$	35	50	5	Mix$_1$ single pour
Test$_3$	20	30	15	Mix$_1$ single pour
Test$_4$	20	30	5	Mix$_1$ single pour
Test$_5$	20	50	15	Mix$_2$ single pour

[*]Note: all aggregates were at 20$\pm$2°C, temperatures were to $\pm$2°C

Table 2 Mix constituent proportions of test foamed concrete.

MIX NO	DENSITY	CEMENT	WATER	SAND	FOAM
		kg/m^3			
1	700	315	181	204	32
2	900	252	176	474	28

Sampling of foamed concrete was carried out in accordance with BS EN 12350-1 [14]. The foamed concrete was then poured into the near-adiabatic hot-box within 5 min of completion of mixing. The top surface of the plastic cube was then covered with cling film and closed with insulated lid on top of the hot-box.

Test Procedures

The stability of the mixes was examined visually and by comparing designed and actual plastic densities. Temperature development due to heat of hydration of single pour and double pour was monitored by placing foamed concrete into 165 mm plastic cube, in insulated timber hot-box. In single pour a Type K thermocouple was placed in the centre of the cube. Whilst, for two pour testing, foamed concrete was poured in plastic cube in two layers with approximately 24 hr time interval. Two Type K thermocouples were placed in centre of each layer to monitor temperature development (See Figure 1). Temperature was recorded at 10 min intervals, and the data were used to determine peak temperature and maximum rate of temperature rise and decline. One sample was tested for each mix.

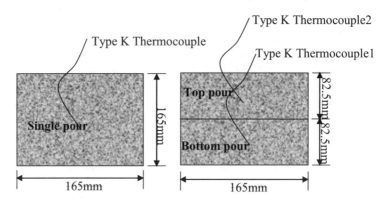

Figure 1   Schematic layout of concrete testing

Samples of the temperature profiles obtained from the foamed concretes subjected to near adiabatic heat cycles are shown in Figures 2 to 5, whereas, summary of the peak temperatures and time to reach these peak temperatures are shown in Table 3.

Table 3   Summary of the test results

TEST	PEAK TEMPERATURE, °C	TIME AT WHICH PEAK TEMPERATURE RECORDED, hr[*]
Test$_1$	55.2	10.6
Test$_2$	44.7	14.4
Test$_{11}$	37.3	11.0
	39.2	11.30
Test$_{12}$	41.8	11.33
Test$_{21}$	38.9	11.0
	38.0	14.0
Test$_{22}$	37.6	14.0
Test$_{23}$	59.4	11.47
Test$_3$	52.4	13.00
Test$_4$	49.3	14.00
Test$_5$	39.1	17.17

[*]After placing in the hot-box

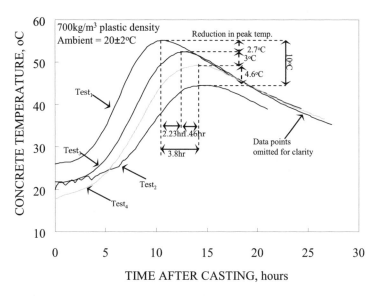

Figure 2   Effect of constituent temperature on peak temperatures reached in foamed concrete due to heat of hydration

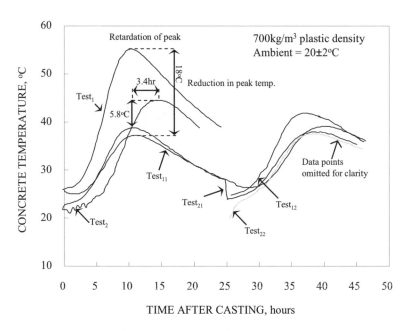

Figure 3   Effect of pour volume on peak temperatures reached in foamed concrete due to of heat of hydration

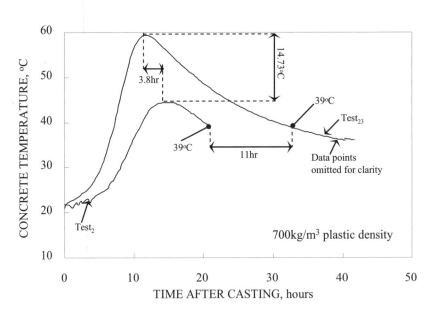

Figure 4   Effect of ambient temperature on peak temperatures reached in foamed concrete due to of hydration

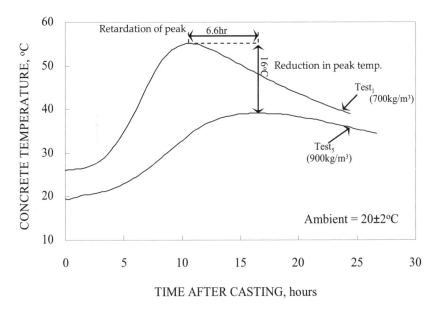

Figure 5   Effect of changing design plastic density on peak temperatures reached in foamed concrete due to heat of hydration

RESULTS

Effect of Mix Constituents Temperature

As can be seen in Table 1, Test 1 and Test 2 are the mix proportions made with same cement temperatures of 50°C and varied water temperatures of 15°C and 5°C respectively. Similarly, Test 3 and Test 4 are the mix proportions of same cement temperatures of 30°C and varied water temperatures of 15°C and 5°C respectively. The recorded temperature profiles of these tests are shown in Figure 2, from the figure it can be noted that the peak temperatures were 55.15 and 44.68°C for Test 1 and Test 2 and time taken to reach these temperatures was 10.6 and 14.4 hr respectively from test initiation. A delay of 3.8 hr was noted between them. Whereas, for Test 3 and Test 4 peak temperatures were 52.44 and 49.3°C, respective time to reach these temperatures were 13.0 and 14.0hr and retardation of 1.46 hr was noted.

When compared peak temperatures of Test 1 and Test 2, a reduction in peak temperatures of 10°C was observed, which was almost same as that of difference in water temperature. On the other hand, peak temperature difference of 3°C was observed between Test 3 and Test 4. From these observations it was noted that the change in water temperature was more effective at higher cement temperatures on heat of hydration of foamed concrete.

However, change in water temperature significantly influenced heat profiles, than change in cement temperature. Similar observations can be found from Figure 3 as can be seen in the figure, difference in peak temperatures was 2.7°C between Test$_1$ and Test 3, although, there was 20°C difference in cement temperature. Contrary to above observation, a decrease in cement temperature, resulted an increase in peak temperature to 4.6°C for Test 4 when compared to Test 2. As noted earlier Test 1 and Test 3 have same mix water temperature 15°C where as Test 2 and Test 4 have water temperature of 5°C.

Over all, the behaviour of heat of hydration due to change in constituent temperatures is not clear to the authors knowledge. However, it was believed that the temperature profiles of foamed concrete are combination of the factors [8]. To be specific, in addition to density and thermal insulating capacity, the significant difference in specific heats of individual mix constituents [8]. Specific heat of water is 4190 J/K.kg and specific heat of cement is 920 J/K.kg [15,16].

Effect of Pour Volume

For a given density, ambient and constituent temperatures effect of pour volume on heat of hydration of foamed concrete was examined by single and double pours. For mixes made with hot cement (50°C) and normal water (15°C), decrease in pour thickness from 165 to 82.5 mm showed a significant decrease in peak temperature (18°C) (see Figure 3). However, no big difference in time to reach peak temperatures was noticed. On the other hand, for mix proportions made with hot cement (50°C) and cold water (5°C), only a decrease of 5.8°C in peak temperatures was observed and retardation of 3.4 hr was recorded. From these observations, it can be noted that the influence of change in layer thickness on heat of hydration was significant at higher water temperature than at lower water temperature. Whilst, almost similar peak temperatures and time to reach these temperatures were observed for top and bottom layers, irrespective of water temperatures (see Figure 3).

Effect of Ambient Environmental Temperature

Temperature profiles obtained for mix proportions made with hot cement (50°C) and cold water (5°C), tested at two different ambient, one at 20±2 and the other at 35±2°C can be seen in Figure 4. From the figure it can be observed that more is the ambient more is the peak temperature. A difference in peak temperature for mixes tested between two ambient was noted as 14.73°C, which was almost equal to difference in ambient (15±2°C). A retardation of 3.8 hr was noted between the two peaks.

Effect of Plastic Density

It was believed that for given cement content, reduced density could result in increased temperatures [17], due to high insulating property. Given that, similar trends were observed in this study (see Figure 5). As can be seen in the figure, increase in density from 700 to 900 kg/m³, resulted in decrease of peak temperatures from 55.1°C to 39.1°C, a retardation of 6.6 hr between the peaks was noted.

Practical Implications

As observed in this study, in majority of the cases foamed concrete exceeds 20°C above ambient. Furthermore, such high temperatures are un-avoidable in foamed concrete due to its insulating property. However, based on the current study following suggestions are made to alleviate heat of hydration:

Decrease the mix constituent temperature: of the mix constituents, ie cement, aggregate, foam and water, the latter has the highest specific heat capacity and thus cooling it has the most significant effect of peak temperature. On the other hand, cooling the mix water is energy intensive and thus expensive for large pours.

Do not pour a fresh mix on an already hot substrate: the insulative effect of foamed concrete means that thinner pour depth allow heat to be more quickly dissipated to the surrounding environment, providing that the ambient conditions are themselves cool and have a low relative humidity, and peak temperatures reduced. On a practical basis, however, it may be impossible to wait for a substrate to cool before pouring a fresh layer on top, given that maximum pour depths are limited to 1000 mm and with lower density mixes (which tend to give higher temperatures.

Changing the ambient external temperature: as noted in b) dissipation of heat to the ambient environment reduces peak temperatures but in large enclosed void fills, such as mine reinstatements, the enclosed nature of the works may reduce heat loss from the concrete. In these cases cooling and drying the air layer adjacent to the pour will allow heat dissipation at a relatively low cost.

Increasing density: At lower densities thermal insulating capacity of foamed concrete is more, due to increased air content. Peak temperatures of foamed concrete can be reduced by increasing density and by decreasing cement content.

In terms of effectiveness *cooling the substrate>pour thickness>cooling the ambient environment>cooling the mix constituent temperature.*

However, more data is needed to develop exact trends and equations between the peak temperatures and other variables. It should also be noted that these tests were carried out using CEM 1 (Portland cement) and a potentially more effective, and indeed sustainable solution, is to use fly ash to replace around 30% of cement content, and the reader is referred to references 8 and 18-23.

CONCLUSIONS

- This study has shown that, at higher cement temperatures decrease in water temperature is more effective on heat of hydration than at lower cement temperatures.

- Reduction in cement temperature however, did not show any significant change in peak temperatures.

- Peak temperatures were decreased with the decrease in pour volume of foam concrete. However, change in peak temperatures was more significant at normal water temperatures (15±2°C) when compared to cold water (5±2°C). On the other hand no significant difference in peak temperatures was noticed between top and bottom layers, irrespective of water temperature.

- With the increase in ambient, heat of hydration also increased and it was almost equal to difference in ambient.

- As expected, the increased density from 700 to 900 kg/m^3 significantly decreased temperature development.

REFERENCES

1. JONES M.R., McCARTHY A. and DHIR R.K., Recycled and secondary aggregates in foamed concrete, WRAP, 2005, 67pp.

2. JONES M.R., L ZHENG, McCARTHY A., R.K. DHIR and YERRAMALA A., Increasing the use of foamed concrete incorporating recycled and secondary aggregates, WRAP, 2007, 94pp.

3. BUILDING RESEARCH ESTABLISHMENT, Concrete with no or minimal primary aggregate content, The MAGCON Pilot Study, in: P.J. Nixon (Ed.), Information Paper, 2004, 8 pp.

4. BRITISH CEMENT ASSOCIATION, Foam Concrete, a Dutch view, Reprint 1991, 6 pp.

5. BRITISH CEMENT ASSOCIATION, Foamed Concrete, Composition and Properties, 1994, 4 pp.

6. G. DE SCHUTTER and L. TAERWE, General hydration model for Portland cement and blast furnace slag cement. Cement and Concrete Research, Vol. 25, No. 3, 1995, pp 593–604.

7. C.D. ATIŞ, Heat evolution of high-volume fly ash concrete, Cement and Concrete Research, Vol. 32, No. 5, 2002, pp 751–756.

8. JONES M.R. and McCARTHY A., Heat of hydration in foamed concrete: Effect of mix constituents and plastic density, Cement and Concrete Research, Vol. 36, 2006, pp 1032-1041.

9. BRITISH CEMENT ASSOCIATION, in: C.D. LAWRENCE (Ed.), Laboratory studies of concrete expansion arising from delayed ettringite formation, BCA, 1993, 147 pp.

10. BRITISH STANDARDS INSTITUTION, BS EN 197-1: 2000, Cement Composition, specifications and conformity criteria for common cements.

11. BRITISH STANDARDS INSTITUTION, BS EN 12620: 2002, Aggregates for concrete.

12. BRITISH STANDARDS INSTITUTION, BS EN 12350-6:2000, Testing fresh concrete Density.

13. HIGHWAYS AGENCY and TRANSPORT RESEARCH LABORATORY, Application Guide AG39: Specification for Foamed Concrete 2001, 62 pp.

14. BRITISH STANDARDS INSTITUTION BS EN 12350-1: 2000, Testing fresh concrete Sampling.

15. PORTLAND CEMENT ASSOCIATION, Impact of hot cement on the concrete mix, http://www.cement.org/tech/cct_hot_cement.asp, Accessed on 3rd February 2007.

16. R.K. DHIR, K.A. PAINE, L. and ZHENG, Design data for use where low heat cements are used, Concrete Technology Unit, University of Dundee, 2004 DTI Research Contract No. 39/3/680 (CC2257), 171 pp.

17. INSTITUTION of STRUCTURAL ENGINEERS, Guide to the Structural Use of Lightweight Aggregate Concrete, IStructE and the Concrete Society, 1987, 56 pp.

18. JONES, M.R., McCARTHY, A., KHARIDU, S. and NICOl, L. Foamed Concrete – Development and applications. Concrete, Vol. 39 No. 8, August 2005 pp 41-43.

19. JONES, M R and McCARTHY, A. Utilising unprocessed low-lime coal fly ash in foamed concrete. Fuel. Vol. 84, Issue 11, August 2005, pp 1398-1409.

20. JONES, M.R. and GIANNAKOU, A. Preliminary views on the application of foamed concrete in structural sections using pulverized fuel ash as cement or fine aggregate. Magazine of Concrete Research. Vol. 57, No 1. Feb 2005 pp 21-31

21. JONES, M R and GIANNAKOU, A. Thermally Insulating Foundations and Ground Slabs Using Highly-Foamed Concrete. Journal of ASTM International, Vol. 1, No. 6 June 2004. (ISSN: 1546-962X)

22. JONES, M R and McCARTHY, A. Foamed Fly Ash Concrete. AshTech 2006. International Conference on Coal Fired Power Station Ash. (Editor L K A Sear) Birmingham 14 to 17 May 2006. (ISBN 0-9553490-0-1)

23. JONES, M R and McCARTHY, A. Behaviour and assessment of foamed concrete for fill and highway applications. Keynote Paper, Proceedings 6th International Congress: Global Construction: Use of Foamed Concrete. (Eds R K Dhir, M D Newlands and A McCarthy) Dundee pp 61-88. 5-7 July 2005. (ISBN 0 7277 3406 7)

WASTE WASH WATER IN PRECAST CONCRETE PRODUCTS

C Perlot

P Rougeau

Study and Research Center for the French Concrete Industry

France

ABSTRACT. For precast concrete industry, the optimisation of the processes regarding environmental impact is an important point. The making of precast concrete products induces waste water coming essentially from the cleaning plants. This water contains solid fines (cementitious particles, sand, filler) and organic substances (demoulding oil, mixer lubrication). Recycling waste appears to be a challenge: blending recycled water in concrete presents a positive alternative way to discharge placement. In this context, the feasibility of waste water recycling in substitution for a part of concrete mix water was studied. Then, the waste water characteristics were compared with the requirements of NF EN 1008 standard "Mixing water for concrete". The conformity regarding to directive limiting values was examined (chemical analyses and hydrocarbon oil index determination). To assess the suitability of water, the NF EN 1008 standard specifies controls on setting time and mechanical strength. From this point of view, waste water is suitable to be used as mixing water for concrete. Hence, the reuse of waste wash water containing solid particles and organic substances in concrete properties were discussed. A methodology to optimize the incorporation rate was investigated. Fresh properties, mechanical strength, microstructure and durability were compared in concretes incorporating different waste wash water rates. The main difficulty was set by the hydrocarbon oil contained: in contact with alkalis, the organic compounds acted as air entraining agent and led to an increase of air content in concrete, especially for the two lowest rates tested. Consequently, the concrete void structure was amplified thus mechanical and durability performances diminished. For the higher incorporation rate tested, concrete characteristics were similar to these of the reference concrete: the solid particles present in wash waste water balanced the lack of porosity created by the entrapped air brought by the foam.

Keywords: Waste wash water, Recycling, Mix water, Precast concrete, Mechanical strength, Durability.

Dr C Perlot, responsible of the Microstructural laboratory of the Material Division in the Study and Research Center for the French Concrete Industry.

Dr P Rougeau, head of the Material Division of the Study and Research Center for the French Concrete Industry.

INTRODUCTION

The management of the environmental protection strategy in concrete industry assumes the economical management of resources and materials used in production processes, the energy preservation and the prevention from producing waste materials. Regarding the above mentioned aspects, the recycling of waste water from precast concrete plant washings comes into sight as priority [1, 2]. There appears a potential for reuse of this waste water in concrete production, but investigation has to be carried out to establish the conditions of this recycling.

After the cleaning of the concrete plants (washout of the moulds, mixers and other devices), the water is collected in a sedimentation basin: the aggregates are separated, extracted and then reused in concrete. The rest of the waste wash water is pumped to a recycling station. The decantation process can involve a cone-filter or a succession of settling tanks, connected in series. The waste water is stored until reuse in concrete production. Before the injection in mixers, the waste wash water is homogenised by agitation. The waste wash water contains suspended fine solid particles, essentially hydrated cement particles, fine aggregates, fillers, and organic substances such as demoulding oil and mixer lubrication.

The reuse of the whole of the waste wash water, in contrast to using only the clarified waste wash water, presents the advantage to recycle fine particles. In this procedure, laborious pre-treatment of removing the sediments before discharge is saved.

The recycle of waste wash water in the production of concrete is regulated by the European standard NF EN 1008 "Mixing water for concrete – Specification for sampling, testing and assessing the suitability of water, including water recovered from processes in the concrete industry, as mixing water for concrete". This standard determines the suitability and the proportion of the waste wash water to be used as combined water with tap water for mixing water for concrete.

MATERIALS AND METHODS

Waste wash water sampling was performed in two different factories of precast concrete, designated by F or G manufacturing structural elements (e.g. beams) and floor systems (e.g. prestressed products), respectively. Water samples, about 50 l, were collected in plastic containers at the point of injection of the reuse water into the mixer, on different days, always during active stirring. These were designated by a letter (a, b, c, d…) after the reference of the factory (F or G).

WASTE WASH WATER CHARACTERIZATION

Before studying their incorporation in concrete, waste wash waters were characterised as given below.

Fine particles nature

The solid materials content in waste wash water was examined (Table 1). The dry material content represents the mass of particles in a litre of wash waste water after evaporation (samples were stored at 105°C for 10 days).

Table 1   Solid content of waste wash waters samples

FACTORY F							
	Fa	Fb	Fc	Fd	Ff	Mean value	Standard deviation
Solid content, g·l⁻¹	54	47	40	41	40	44	6

FACTORY G											
	Ga	Gb	Gc	Gd	Ge	Gf	Gg	Gh	Gi	Gj	Mean value	Standard deviation
Solid content, g·l⁻¹	2.7	2.7	2.5	3.1	3.3	12.7	26.4	34.6	35.4	42.1	16.5	16.2

X-ray diffraction

The mineralogical composition of the solid particles of the waste wash water was investigated by X-ray diffraction (Figure 1): calcium hydroxide (portlandite), hydrated calcium sulphate aluminates (ettringite) result from the partial hydration of cement grains; calcium carbonate in the form of calcite and silica in the form of quartz come from fine particles of aggregates. The presence of $CaCO_3$ can also be attributed to carbonation of portlandite if the solid particles were in contact with carbon dioxide from the atmosphere and dissolved in waste wash water. Gypsum is a component added to clinker to make cement. Thermogravimetric analyses reveal that the solid material from the recycled water contains cement grains which are either hydrated or not.

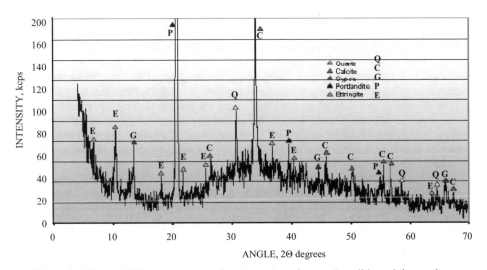

Figure 1   X-rays diffraction patterns from waste wash water's solid particles analyses

Particle size distribution

Preliminary granulometric analyses show that more than 90% of solid particles are smaller than 63 µm in diameter. Then, particle size distributions of the evaporation residues of the waste wash waters were determined by a laser granulometer. Figure 2 shows a slight difference between solid particles from F and G samples. This can be attributed to the production processes. Nevertheless, we can conclude that the solid particles of the waste wash water are very fine, less than 200 µm, while most of the particles are around 20 µm in diameter.

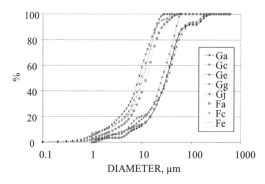

Figure 2   Particle size distributions in the waste wash water F and G

Chemical composition

Chemical composition of waste wash water was investigated. The Total Organic Carbon content (TOC) has been determined by IR method. Chemical analyses have been performed according to NF EN 1008 standard specifications: chloride, sulphate and alkali concentrations were determined according to the methods given in the standard NF EN 196-2 (Tables 2 and 3). The NF EN 1008 standard specifies a concentration of $NaO_{2eq} < 1500$ mg·l^{-1}. The chloride content is required to be < 500 mg·l^{-1} in the case of prestressed concrete or grout, < 1000 mg·l^{-1} for reinforced concrete and < 4500 mg·l^{-1} for non reinforced concrete.

The presence of organic compounds is due to the demoulding oil and mixer lubricants. The concentration is below the standard limit for water. For the waste wash water from factory F, the TOC content is more than ten times the values for factory G. This difference could be attributed to the specificities of the precast products process: in the case of prestressed products, a high volume of oil is used in the factory. From these first chemical tests, all the samples of waste wash water from the two factories were able to be used as mixing concrete water according to the NF EN 1008 standard.

Table 2   Chemical analyses of waste wash water samples from factory F

	Fa	Fb	Fc	Fd	Ff	MEAN	STANDARD DEVIATION
TOC, mg·l^{-1}	140	140	210	180	100	154	42
Concentration, mg·l^{-1}							
Na_2O_{eq}	20	18	16	15	14	17	2
Cl$^-$	23.5	21.0	21.4	18.7	21.7	21.3	2
SO_4^{2-}	< 5	< 5	< 5	< 5	< 5	< 5	-

Table 3 Chemical analyses of waste wash water from factory G

	Ga	Gb	Gc	Gd	Ge	Gf	Gg	Gh	Gi	Gj	MEAN VALUE	STANDARD DEVIATION
TOC, mg·l^{-1} Concentration, mg·l^{-1}	13.0	undo	12.7	undo	16.7	undo	10.1	25.4	24.7	20.0	17.5	6
Na$_2$O$_{eq}$	108	103	117	132	149	134	140	147	162	169	136	22
Cl$^-$	18.3	19.0	18.5	19.6	20.4	18.4	14.5	15.1	12.4	13.3	16.9	3
SO$_4^{2-}$	<5	<5	<5	<5	<5	<5	<5	<5	<5	<5	<5	-

Preliminary Tests

The NF EN 1008 standard recommends preliminary controls on water. If all the points checked are in accordance with the standard specifications, the water could be used as mixing water to concrete. The absence of grease, particles in suspension and humus material is verified. The odour and the colour of the samples have to be similar to those of drinking water. The water must not be acidic (pH ≥ 4). All the samples present a pH value around 13 due to the high amount of alkalis coming from cement grains. The examination reveals the presence of some detergents and oil which cause some foam on the surface of the water. Because one of the checked points not being conforming, the standard advocates setting time tests in addition to chemical tests.

Mortars Characterization

Once the waste wash waters have been recognized as suitable for concrete mixing from a chemical point of view, the NF EN 1008 standard advocates setting time and compressive strength tests on referring mortars. The values of these measurements have to be compared to those of the reference mortar made using demineralised water.

Reference mortar composition

The weight composition of reference mortars is set by the standard NF EN 196-1: 1 part of cement, ½ part of water and 3 parts of normalized sand (0/4 mm siliceous sand). The total water incorporated into these mortars is distilled water for the reference mortar or mixes of tap water and wash waste water in variables proportions (25%, 50%, 75% or 100% by mass of the total water).

Setting time

The influence of two parameters on setting time was observed: the influence of the sampling (from factory G) for a constant part of waste wash water (25% by the total mass of water, Figure 3) and the influence of the waste wash water content (Figure 4). The testing method is given in the standard NF EN 196-3. According to the standard NF EN 1008, the mix water can be used as mixing water for concrete if the initial setting time is greater than 1 hour and does not differ by more than 25% of that of the reference; and if the final setting time is shorter than 12 hours and does not differ by more than 25% of that of the reference. From this point of view, all the samples of wash waste water from factory G could be used as part of the mixing water for concretes (25% of the total mass of water) and the Fc sample could be used in all proportions between 25 and 100% of the total mass of water.

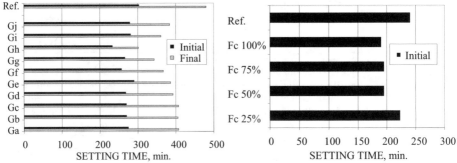

Figure 3   Influence of different waste wash
water sampling on the setting time
for 25% of substitution

Figure 4   Influence of waste wash water
content on initial setting time

Figures 3 and 4 show that the waste wash waters tend to decrease the initial setting time. This effect is more pronounced in the case of the final setting time. As referred to in the literature, this can be attributed to the presence of fine solid particles that accelerate the hardening of the cement pastes by developing nucleation sites [3, 4]. Moreover, the presence of calcium in wash waters and their high alkali contents tend to accelerate cement hydration [5], especially short-term hydration [6]. This result is confirmed by Figure 4. Beyond the rate of 25% of waste wash water, the initial setting time seems to be influenced just by the presence of reused water and not by its content: despite the rate of waste water increases, the setting times for 50, 75 and 100% are similar.

Air content and mechanical strength

As mechanical strength is partly influenced by the amount of entrapped air, these values were measured on three samples for each mortar according to the standard NF EN 196-1 (Figure 5). The entrapped air content increased as a function of the part of waste wash water used. This was due to the formation of foam in waste wash water that entrapped air in fresh concrete during mixing. This foam was generated by the saponification reaction which takes place in the presence of fats or oils and sodium or calcium hydroxides. Foaming tests showed that this production was enhanced in the waste wash waters because these contained organic compounds and were highly alkaline.

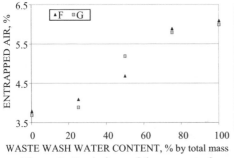

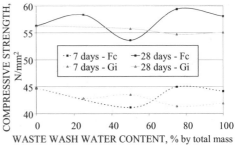

Figure 5   Evolutions of the amount of
entrapped air in mortars as a function of
waste wash water content

Figure 6   Evolutions of the mechanical
strength of mortars as a function of waste
wash water content

In spite of the fact that the entrapped air increased with waste wash water, the mechanical strength did not decrease in the same trend (Figure 6). For the mortars made of the water from factory G, the use of recycled water slightly decreased the mechanical strength: - 5% at 7 days, - 2% at 28 days for a total replacement of water. For the water from factory F, for an elevated content of waste wash water the mechanical strength was higher than for the reference one. This water had a large amount of solid particles. These mineral particles compensated the negative effect of the entrapped air increase by a filler effect and by the enhancement of cement hydration combined to those of chloride and alkalis present in waste wash water. Furthermore, Gagné et al. showed that the variations of mechanical strength have to be considered regarding the decrease of the W/B ratio with solid particles [7]. To verify these results, complementary tests were performed on concrete (see next section).

According to the standard NF EN 1008, the combined water (tap water + waste wash water) could be used only if mortars exhibited 7 and 28 days compressive strengths higher than 90% of the value exhibited by the samples prepared with distilled water. This criterion defines the maximal amount of wash waste water in the combined water. From these results and the NF EN 1008 standard, the different waste wash waters studied could be used as mixing water for concrete in different proportions ranging from 25 to 100%, depending to the factory. Also, the influence of waste wash waters on concrete properties were evaluated.

EFFECT OF INCORPORATION OF WASTE WASH WATER IN CONCRETE

Concrete Samples

For each factory (F or G), four concretes have been fabricated in addition to the reference sample, denoted "R", made with tap water. The influence of waste wash water content was evaluated with three concretes incorporating different levels of waste wash water (30, 42 and 54% of the mass of total water) that corresponded to a mass of solid particles in 1 m^3 of concrete (2, 3 and 4 kg) (Tables 2 and 3). To minimize the decrease of mechanical strength, an optimised mix was done (O) for which the W/C ratio was decreased by increasing the cement content and decreasing the efficient water content.

Table 4   Concrete samples mix proportions

COMPOSITION, kg·m^{-3}	R	F2 or G2	F3 or G3	F4 or G4	F4O or G4O
Cement CEM I 52.5 N			375		420
Aggregate 1 (sand 0/0.315 mm)			194		
Aggregate 2 (sand 0/5 mm)			565		
Aggregate 3 (sand 5/12 mm)			988		
Superplasticizer, % of cement mass			0.55		0.63
Efficient water			180		170
Tap water	180	127.5	105	83.4	63
Waste wash water	0	52.5	75	96.6	107
% of waste wash water	0	30	42	54	63
W/C			0.48		0.40
W/B = C + solid particles	0.480	0.477	0.476	0.474	0.406
Solid particles coming from waste wash water, kg·m^{-3}	0	2.1	3.0	3.9	4.3

The entrapped air in fresh concrete, just after the mixing, was measured using an aerometer. To simulate the precast processes, the concrete samples were cast and placed in an oven for 6 hours. The characteristics of the heat treatment were: 2 hours at 20°C, temperature increased to 65°C at around 0.5 min^{-1} for 1 hour, 3 hours at 65°C, then the oven was switched off and the samples slowly cooled to ambient temperature during a period of 2 hours. After demoulding, the samples were cured in a 20°C temperature room at RH>95%.

RESULTS AND DISCUSSION

Mechanical Strength

Compressive strength tests were performed at 8 hours, 24 hours and 28 days on three cubic samples (10 cm of length) at each time. The mean results are shown in Figure 7. The criterion of acceptance of the concrete mix is based on the 28 days compressive strengths that must be higher than 90% of the value exhibited by the reference sample. Then from the results, a replacement of the water by waste wash water could be made as far as the value of 42% by mass (F2, G2, F3 and G3). For the maximal value of water replacement tested (F4 and G4), the decrease of mechanical strength is just below the 90% of reference value. Meanwhile, an optimization of the mix proportions could be done and then a higher amount of recycled water could be incorporated (F4O and G4O).

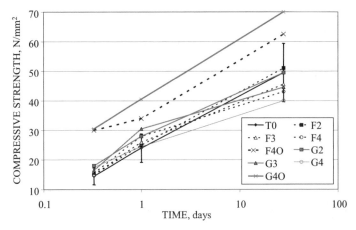

Figure 7 Compressive strength evolution with time for the concretes tested

Microstructure

The apparent density and the porosity accessible to water were determined according to AFPC-AFREM recommendations on four replicates for each concrete tested [8]. The incorporation of waste wash water did not seem to affect the porosity of the concretes: all values were in the same order of magnitude even if the entrapped air value in fresh concrete was different. It has been referred to in the literature [9] that solid particles of waste wash water have a filler effect and then can balance the increase of entrapped air consequent to the increase of the amount of recycled water and then foam.

Table 5 Concretes characterization results

	R	F2	G2	F3	G3	F4	G4	F4O	G4O
Entrapped air, %	1.5	2.4	2.5	3.1	2.6	3.2	4.1	1.9	2.6
Compressive strength at 28 days, N/mm²	49	51	49	45	44	43	40	62	70
Apparent density, kg·m⁻³	2180	2190	2170	2190	2180	2180	2180	2320	2400
Porosity, %	14.8	14.8	14.9	14.9	15.2	15.1	14.4	12.8	13.6

Microstructural investigation was performed with a Mercury Intrusion Porosimetry (MIP): this test allowed establishing the pores size distribution of the concretes.

The parameters measured during MIP tests are presented in Figure 8. The porosity range examined by MIP was about 3.8 nm to 68 µm. Three samples were tested from each concrete made with the water samples from the two factories at different rates of incorporation. The same trend as presented in Figure 8 was observed. The appearance of a supplementary peak centred around 10-50 nm for the samples using waste wash water confirmed that the incorporation of waste wash water introduces supplementary porosity as compared with the reference sample. But, as the capillary porosity, described by the intense peak around 100 nm, of the concrete using the highest rate of waste wash water was smaller than for the others, it can be supposed that beyond a certain rate of fines solid content, the filler effect and the slight decrease of the W/B ratio induced a decrease of this porosity. It is corroborated by the evolution of the average pore radius (Table 6).

As foreseen, the optimization of the W/B ratio (F4O sample) resulted in a significant decrease of the capillary porosity. This reduction of concrete capillary porosity was evoked by Sandrolini et al. [9] who also state that this improves the durability of concrete.

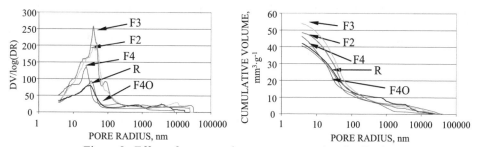

Figure 8 Effect of waste wash water on pores size distribution

Table 6 MIP results

	R	F2	F3	F4	F4O
Average pore radius, nm	38	39	37	38	32

Durability

The durability of the concretes was assessed by accelerated leaching tests [10]. The samples were immersed in an ammonium nitrate solution ($[NH_4NO_3] = 1$ mol·l^{-1}) for 104 days for F samples and 75 days for G samples. To preserve the aggressive potential of the leaching medium, the solution was renewed each time the pH reached a value of 8.3. At the end of the immersion, samples were sliced and sprayed with a colour indicator (phenolphthalein) which revealed the sound core and the peripherical degraded zone. The degraded depth was measured using a videomicroscope (Table 7).

Table 7   Results of accelerated leaching tests

	Factory F					Factory G				
	R	F2	F3	F4	F4O	R	G2	G3	G4	G4O
Degradation duration, days			104					75		
Degraded depth, mm	8.8	9.1	9.0	8.5	8.4	7.7	8.3	8.1	7.4	7.1
Degradation kinetics, mm/d$^{1/2}$	0.86	0.88	0.87	0.83	0.82	0.89	0.95	0.93	0.86	0.82

The degradation kinetics was linked to two effects: the open porosity of the material that promotes the penetration of aggressive agents and the amount of calcium hydroxide that controls the degradation front progression. In previous works [11] it has been shown that the incorporation of waste wash water could have a slight negative effect on the resistance on chemical degradation. According to these authors, it is caused by the addition of calcium hydroxide – the phase more susceptible for dissolution in concrete – present in the solid particles of the waste water. This could explain the behaviour against degradation for concretes incorporating medium quantity of waste wash water, but for the F4 and G4 samples, the resistance to aggressive environment increased by the presence of solid particles. There, the main factor influencing the degradation seemed to be the porosity: degradation kinetic ranges versus this parameter. Thus, the incorporation of a high rate of waste wash water was possible considering an adaptation to the cement and water proportions (decrease of the W/B ratio) to maintain suitable resistance to chemical attacks (case of F4O and G4O) or for high amount of waste wash water [12].

IMPACT ON PRECAST PRODUCTS

This study showed that the waste wash waters tested could be used as water for concrete mixing from a standardization point of view (NF EN 1008 specifications). However, their compositions (e.g. chemical elementary composition and solid particles content) were subject to change as a function of the factory process, the nature of the products, the formulation of the concrete, and the external conditions. It could be borne in mind that this point set a problem for the reuse of waste wash water in precast products because these need an accurate regularity of their characteristics. To test the influence of the variations of water composition, the entrapped air and mechanical strength of eight mortars made from different water samples were tested. Figure 9 shows the variations of mechanical strength correlated with entrapped air content. As it can be seen, the variation of entrapped air is important for the eight different waste wash waters (from 2.8 to 4.7%) but it does not induce important strength variations: the values measured were very similar considering the standard deviation (black bars) and the accuracy of the measurements.

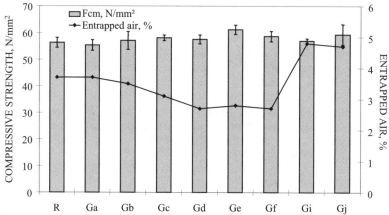

Figure 9   Variation of mechanical strength correlated with entrapped air content

Subsequently, waste wash water could be used for the manufacture of concrete elements while assuming a good regularity of the mechanical strength of the products. Various scenarios of recycling in concrete production could be developed and adapted to the manufacturing processes [13]. The study led by Gagne and Blanchard confirmed our results [7]. It established the feasibility of the incorporation of waste wash water by following the characteristics of concrete in factories and comparing these results with laboratory tests.

CONCLUSIONS

The chemical and mechanical tests revealed that the waste wash water tested in this study could be used as mix water for concrete, according to the NF EN 1008 standard in different proportions ranging from 25 to 100 % of the total mass of water, regarding the water characteristics depending on the factory processes. From setting time tests, the waste wash waters reduced the initial setting time. It has been reviewed that the chemical aspects linked to the high alkali and chloride contents (and then the basic pH imposed) and the physical effect (nucleation sites) of waste wash water accelerated cement hydration. The reuse of waste water in concrete increased the amount of entrapped air in concrete but did not significantly decrease the mechanical strength, the porosity and the durability behaviour. From the literature, this evolution can be attributed to the filler effect of the solid particles and the decrease of the W/B ratio due to fine solids contents in waste water.

However, these arguments are open to discussion: the amount of alkalis brought by the waste wash waters is more than ten times less than the amount contained in the cement. In the same way, the additional solid particles from waste wash water are minimal as far as the fine aggregates from concrete are considered, and the variation of W/B ratio is very small. Separately these effects appear to be powerless, but in combination they can be significant.

Besides, another parameter or compound of waste wash water could play a role in cement hydration or in porosity fulfilment. More investigations have to be led on this focus. As a conclusion, the recycling of waste water in precast industry appears to be a suitable technology preserving the environment without affecting product properties.

REFERENCES

1. BORGER J, CARRASQUILLO R L AND FOWLER D W, Use of recycled wash water and returned plastic concrete in the production of fresh concrete, Advanced Cement Based Materials, Vol. 1, Issue 6, 1994, pp. 267-274.

2. CHINI A R, MUSZYNSKI L C, BERGIN M AND ELLIS B S Reuse of wastewater generated at concrete plants in Florida in the production of fresh concrete, Magazine of Concrete Research, Vol. 53, Issue 5, 2001, pp. 311-319.

3. SU N AND LIU F-S, Effect of wash water and underground water on properties of concrete, Cement and Concrete Research, Vol. 32, 2002, pp. 777-782.

4. CYR M, LAWRENCE P AND RINGOT E, Mineral admixtures in mortars. Quantification of the physical effect of inert materials on short term hydration, Cement and Concrete Research, Vol. 35, 2005, pp. 719-730.

5. BROWN P, HARNER C AND PROSEN E, The effect of inorganic salts on tricalcium silicate hydration, Cement and Concrete Research, Vol. 16, 1986, pp. 17-22.

6. LAWRENCE P, CYR M AND RINGOT E, Mineral admixtures in mortars: Quantification of the physical effect of inert materials on short term hydration, Cement and Concrete Research, Vol. 33, 2003, pp. 1939-1947.

7. GAGNE R AND BLANCHARD S, Influence de l'eau recyclée traitée sur les propriétés des bétons, 7th Journées scientifiques (RF)²B, Toulouse, France, 2006, pp. 32-46.

8. AFPC-AFREM, Détermination de la masse volumique apparente et de la porosité accessible à l'eau. Méthodes recommandées pour la mesure des grandeurs associées à la durabilité, Compte rendu des journées techniques AFPC-AFREM, Toulouse, France, 1997, pp. 121-124.

9. SANDROLINI F AND FRANZONI E, Waste wash water recycling in ready-mixed plants, Cement and Concrete Research, 2001, Vol. 31, pp. 485-489.

10. PERLOT C, Influence de la décalcification de matériaux cimentaires sur les propriétés de transfert: application au stockage profond de déchets radioactifs (in French), PhD thesis, UPS Toulouse, 2005, 287 p.

11. CHATVEERA B, LERTWATTANARUK P AND MALUK N, Effect of sludge water from ready-mixed concrete plant on properties and durability of concrete, Cement and Concrete Composites, 2006, Vol. 28, pp. 441-450.

12. YAGUE A, VALLS S, VAZQUEZ E AND ALBAREDA F, Durability of concrete with addition of dry sludge from waste water treatment plants, Cement and Concrete Research, 2005, Vol. 35, pp. 1064-1073.

13. KUOSA H, Recycling of concrete, sludge and water in the concrete industry, BIBM 2005 proceedings, 2005.

GLOBAL CONCRETE COMMUNITY FORCASTING

M Ahmad Vand

H Hoornahad

F G Moghaddam

ConREC

Iran

ABSTRACT. There have been twenty-one different civilizations since the emergence of human settlement, of which the western society, as a great historical cultural unity, is one. There has been a complex variety of interweaving elements, be it the difference of religion, the rise and fall of political empires, or the succession of socio-economic, educational and technological systems. It is not just the task of the historian or sociologist or philosophers, concrete technologist also need to devise an intelligible unit if study to identify the structure features common to diverse institutions and more enduring and consistent patterns of change in our organizations within the international community. Such an analytic approach, however, risks flattening out what may be distinctive and meaningful in history of a particular society or organization and generation. Thus, one can take the evolution of some particular organizations such as ACI, ASTM, ISO and trace their rich yet idiosyncratic fate on the basis of their history, the character of their members, their "National, Regional and Global Will" and the so on. Yet it is also obvious that while its history may be individual, every society or institution shares elements with other communities-science, knowledge, technology, culture, economy and religion that cut across the particular organization or community of a people and influences them in specific ways. American economy is like, yet unlike, Japanese economy; one's purpose dictates the focus of attention.

Keywords: Technology paradigm shift, Theoretical knowledge, Forecasting, Knowledge Communities, Universities.

M Ahmad Vand is Director of the Concrete Research & Education Centre (ConREC) in Tehran. He is the author and co-author of numerous papers on concrete technology and cement chemistry.

H Hoornahad is a research civil engineer and in charge of the research department of ConREC. He is currently working in the field of concrete technology and has published several papers in this area.

F G Moghaddam is a research civil engineer and in charge of the education department of ConREC. He is the author of some papers in the area of concrete technology.

INTRODUCTION

This is an article in concrete community forecasting in the world. However, can one predict the future? The question is misleading. One cannot, if only for the logical reason that there is no such thing as "The Future". To use a term in this way is to reify it, to assume that such an entity is a reality. The word future is a relational term. One can only discuss the future of something. This article deals with the future of concrete community as a whole in the Global Post-Industrial societies.

It is a new school of thought that believes; it is the theoretical knowledge as the new strategical resource of innovations in technology controlled by intellectual technology. Forecasting differs from prediction. Though the distinction is arbitrary, it has to be established. Predictions usually deal with events-who will win a challenge, specification of a new invention; they center on decisions. Yet such predictions, while possible, cannot be formalized, i.e. made subject to rules.

The prediction of events is inherently difficult. Events are the intersection of sub system vectors (interests, forces, pressures, and so on). While one can to some extend assess the strength of the vectors individually, one would need a "Concrete Physics" to predict the exact cross points where decisions and forces combine not only to make up the events but, more importantly, their outcome. Prediction; therefore is a function largely of the detailed inside knowledge and judgment that come from long involvement with the situation as the creativities, versatility, knowledge, and wisdom prevailed in the life activities involving concrete service time.

Engineers and Concrete Technologist must rely on the application of a knowledge advancing system. Its processing system begins from bit, bytes, data, information, transparency and through understanding comes knowledge, and creativities, by having liberty to select the choices and assess the optimum application of innovations with a controlling tool of the new intellectual technology, which the concrete knowledge will be flourishing within the system of concrete technology, promoting sustainable development.

VISION 20/20

Here, have taken "Global Knowledge Community" as the intelligible unity of study. Knowledge based society is a concept that embraces the experiences of a half dozen different advanced and developed countries. This is primarily a change in the social structure, and its consequences will vary in societies with different political and cultural configurations. Yet as a social form it will be a major feature of the twenty-first century, in the social structures of the most advanced countries. There is differentiation of vision from other position in this manner that this proposal of going beyond control is relative to the growth of each individual. In other words, proper management is definitely preferred to disorder, when one is going up the steps of development and evolution. No social system organization or national society has a patent on the future, and in the next 20 to 30 years, we will see the integration of new knowledge communities with emphasize in the social and structural. In addition, its consequences will vary in societies with different political and cultural configurations; every flower has its own distinct sent in the global prairie. Yet as a social form it will be a major feature of the twenty first century, in the social structures of the most countries and any knowledge community or institutions.

MISSION

The concept of knowledge based Society is a view from the twenty-first century. It is an attempt, methodologically, to use a new kind of conceptual analysis, as a way that organized around the main resource of theoretical knowledge as the principle and knowledge communities as main structures, as a way of "Ordering" the complexity number of possible perspectives about macro-historical change.

It is an effort, to identify the substantive character of structural changes in the economy, technology, culture and the new and decisive role of theoretical knowledge in determining technology innovation and the direction of changes. It is a venture into the future of ACI as a knowledge society and need for change to become the leader in Global Concrete Community. "It is time to realize we are a part of an international community. We can no longer afford to have a protectionist attitude and attempt to shut out the rest of the world [1]".

CONCRETE COMMUNITY

We are discussing the Global Concrete Community and particularly ACI social holistic system forecasting. However, can one concrete technologist predict the future? One cannot, if for the logical reason that there is no such thing as "The Future". To use a term in this way is to reify it, to assume that such an entity is a reality. The word future is a relational term. One can only discuss the future of something. Like the future of concrete knowledge community, i.e. ACI concrete community. Forecasting is possible where there are regularities and recurrences of phenomena, or persisting trends whose direction, if not exact trajectory, can be plotted with statistical time-series or be formulated as historical tendencies and thus, probabilities and an array of possible projections. More importantly, at the crucial points the trends become subject to choice and are conscious interventions by people with power and decision to accelerate, swerve or deflect the trends is a policy intervention which may create a turning point in the history of an institution such as ACI.

To put it a different way; forecasting is possible only where one can assume a high degree of rationality on the part of peoples who influence events-the recognition of costs and constrains, the common acceptance or definition of the rules of the game, the agreement to follow the rules, the willingness to be consistent. Thus, even when there is conflict, the situation can be mediated through bargaining and trade offs, if one knows each party's schedule of acceptable costs and order of priorities. However, in any human situations- and particularly in politics-privileges and prejudices are at stake, and the degree of rationality or consistency is low. The degree to which crucial political decisions are carried out often depends on qualities of leadership and strength of will; such as aspects of personality are not easy to calculate, especially in crises.

PAST

In the last 100 years, the western society has worked out in the politics and social structure of industrial society. Looking ahead to the next decades, one sees that the desire for greater participation in the decision making of organizations that control individual lives in Universities, Business firms, Churches, Mosques, NGOs, Associations, Institutes and the increasing technical requirements of knowledge professionalism.

There are sure guide as to which new ideas, values or processes are genuine turning points in our history. We turn to change in social frame works, i.e., concrete communities which are the structures of the major institutions that order the lives of individuals in a society; the distribution of persons by occupation, the education of the young, as such. The change from a rural to an urban society, from an agricultural to an industrial economy, from a federalized to a centralized political state, is major changes in social frameworks. Because such frameworks are structural, they do not allow us to specify the exact details of a future set of social arrangements. When such changes are under way, they allow us not to predict the future but to identify an agenda of questions before the society and institution and have to be solved. It is this agenda that itself can be forecast for our concrete community.

PARADIGM

The knowledge based society and the new forms of its communities and institutions that are the subject of our discussion, which is a forecasting about a change in the social framework of industrial societies, and consequently the knowledge based communities, institutions and none profit organization such as ACI and its role in the local, regional and global community arena. There has been discussion that; it is the Theoretical Knowledge as the new strategically resources of innovations in technology and increase research, development and progress. However, can a concrete technologist predict the future service life of our institution?

Forecasting differs from prediction. Though the distinction is arbitrary, it has to be established. Predictions usually deal with events-who will win a challenge, specification of a new invention; they center on decisions. Yet such predictions, while possible, cannot be formalized, i.e. made subject to rules. The prediction of events is inherently difficult. Events are intersects of sub system vectors (interests, forces, pressures, and so on). While one can to some extend assess the strength of the vectors individually, one would need a "Physics" to predict the exact cross points where decisions and vector forces combine not only to make up the events but, more importantly, its outcome. Prediction is a function largely of the detailed inside knowledge and judgment that come from long involvement with the situation as the versatility, information, transparency, understanding, knowledge, wisdom and results of action prevailed in the life activities. Some of us live in the post-industrial societies (knowledge-based societies), the invention and the application of intelligent tools will decrease and has been to some extend, the technological necessity of using human beings.

We are witnessing the same thing that the first tolls making by Homo sapiens. "We are fortunate that we live in a very exciting time. With the development of the Internet as a communication tool, we can stay in touch with anyone virtually anywhere in the world [1]". We are at the dawn of a new civilization that will essentially be different from the existing one in the industrial societies.

PARADOX

The fundamental difference between the past and present activities is the shift in the centrality of human activities from the work and production-related activities to more creative, spiritual and knowledge activities. The wealth in these societies is generated, as the result of knowledge, versatility, creativity, knowledge and wisdom, and the gains precedence over wealth generated by industrial work. Acknowledging the inadequacy of the scientific paradigm and emphasize on the need to go beyond that. Not limiting ourselves to the economic and political sectors of life. The future technologist thinking issues range from human values and interpersonal relationships of individuals to the economics and innovation of technologies and the spiritual dimensions of life. That is, progress and justice are not just evaluated on a one-dimensional sliding scale of economic efficiency anymore.

PARADIGM SHIFT

A host of other factors ranging from environmental and biological imperatives to aesthetic and spiritual values also affect and define our concepts of progress and justice. Now, it is time to form new global institutions in the areas of collaborative research, judiciary, legislature, language and education, combined with strong local presence.

The above does not mean the invitation for aggression in the name of internationalism. Formation of global society is a voluntary conscious act rather than a coercive and aggression of leaders against lagers. If some nations similar to rigid families choose to enter this change later than others, we can only regret for their mistake rather than to impose the change on them.

Democracy has become more and more synonymous with the United Europe and the American ideal of individual's right to the pursuit of happiness. The relationship of different national organizations with global bodies will map the fate of all human rights, freedom, and justice, in the knowledge communities in the post-industrial societies of our time. "My advice is to all of you is to accept the idea that we get in a world community [1]."

GLOBAL INSTITUTE

The idea of moving towards a world community is going into a United ACI Regional Chapters Community with a unique constitution guide lines is a beginning of this great historical convergence into A Global Knowledge Society. The central principle for ACI regional chapter's constitution must be the recognition of the individual's right to the pursuit of happiness, within every institution: Nations, Corporations, NGOs, Nonprofit organizations, Knowledge organizations, Family, and other social bodies.

All human institutions such as family, schools, nations, religious institutions, professional, knowledge institute, associations, corporations, media, special interest groups, etc. have been created to respond to some particular human needs. Some of these institutions such as ACI will evolve, some will vanish, some will transform. One needs definitely understand the function, viability, and value of each one, individually, before deciding on whether an institution is a barrier to progress or can be shift and help progress. The pursuit of happiness for individuals will not be achieved by nagging all institutions as evil, and self-growth does not automatically "make things to fall in the right place!"

The success of the new civilizations and ACI as the Global Concrete Knowledge Community is not guaranteed. A worldwide economic disaster, environmental deterioration, nuclear war, or reversals such as revolutions, can put an end to humanity. Tyranny, poverty, menace of war and disease, injustice of all kinds, are surrounding us at this historic time.

The organization of change such as ACI Regional Chapters Union may also take different forms and any progressive organization will be one of the many international endeavors to help building a Global Knowledge Community in the post-industrial global world. I hope that the New Regional Chapters progressive organizations are created in a way, to function vertically and horizontally, to incorporate the reality of our future vision 20/20 in ourselves, elaborating on philosophical vision, we need to take a closer look at the future thinking of ACI.

The future thinking encompasses all aspects of society and is not confined to science, technology, politics, religion, science, or psychology. I define a future vision 20/20 of our regions as a singular conceptual category to mean the disenchantment with the philosophical paradigms of the industrial society in different parts of life and the search to go beyond them for our young engineers to come. Suggestion of the control paradigm shift is what the view as the ideal for humanity as a whole. It describes the alternative paradigm as an autonomous synchronicity or a mutual collaboration. How can all be whirling together with nature, with each other, different families, races, nations without giving up our identity? In sum, practical knowledge to be superseded by reflective wisdom, and the spiritual side of life rather than the mechanical side, to take the major portion of living hours, which has its own implications as regards to the issues of economic compensation and social justice.

The above paradigm shift is the reason that the thinkers differentiation of changing ourselves and changing the world (the emphasis being on the latter), which made sense in the industrial civilization, should now be superseded. It has believed that using the new paradigm in our relationships with our children, spouse, or friends, is as important as finding alternative organizational plans at work and our community in the global society. In fact, at the present this paradigm, in contrast to the industrial paradigm, is being more defined at the micro/individual level than at the grand/social scale. The reason it is that this view is challenging the whole history of humanity as a whole and not just one particular civilization in contrast to, i.e. the industrial civilization that was challenging only another civilization, i.e., the mediaeval society.

There are two changes have occurred. One has been the systematic joining of science to invention, principally through the organization of research and development efforts that has been driving our researchers to a new horizon by investigating in new projects such as Self Compacting Concrete that our organization has taken upon. The second, more recent changes has been the effort to "Chart the Sea" of technology by creating which will lay out the future area of development and which will allow industry, or capital possibilities, needs, and products. This new fusion of science with innovation, technological growth, is one of the understandings of the knowledge society. The science-based technologies and industries have a great advantage in achieving major advances in products and processes. Our research aimed at cracking open a through direct attention.

The role of "Research and Development" as a component of scientific and economic activity is another project that needs to addressed some other time. The bulk of technological forecasting today is done without explicit use of special techniques. The need for formal techniques was not felt until a few years ago. While the beginning of systematic technological forecasting can be situated at ConREC since 1995, the existence of a more widespread interest in special techniques for latest innovation in self consolidating concrete, mix proportional design of the aggregates system, first made itself with fore runners already experimenting in the late 1990. Now, a noticeable interest is developing in more elaborate multi-level techniques and integrated models that are amenable to computer programming as well. The fact that, at the existing moment, the engineering possibilities of moving beyond the present do not exist is, in and of itself, no barrier it is assumed that the engineering break through will occur.

If this vision is correct, which is correct the heart of the knowledge economy post-industrial society is a class that is primarily a professional class. As with any status group, the boundaries of definition are fluid and often indistinct, yet certain core elements are obvious. A profession is a learned (i.e. scholarly) activity, and thus involved formal training means to be within of a broad intellectual context. To be within the profession means to be certified, formally or professionally. And a profession embodies a norm of social responsiveness. The idea of a profession implies an idea of competence and authority, technical and moral, and that the professional will assume a hieratic place in the society.

CONCLUSION

To support the progress of innovation in concrete and new technologies such as space technologies, biotechnologies, Stem Cell, robotics, telecommunications, artificial intelligence, nanotechnology, etc. In addition, to champion research programs on the perplexity of social justice, the future of various human institutions, and the political and spiritual issues. "The professional of the world get along just fine. Maybe we should try to teach our governments to do the same [1]." We need to promote different organizations, institutions and publications that discuss these Global, Regional and Local collaborative activities under a new banner; Think Globally Act Local effectively. It is really a small world.

ACKNOWLEDGEMENTS

Dr. Luke Snell and Dr. S. Shaw valuable contributions and cooperation greatly appreciated on assisting our organization on this matter.

REFERENCES

1 CAGLY, J.R. It Really is a Small World. V. 27, NO.11, November 2005.

2 AN AUSTRALIAN FRAME WORK. Measuring a Knowledge-Based Economy and Society. Australian Bureau of Statistics. 2002.

3 SKYRME, D. The Global Knowledge Economy and Its Implication for Markets. 2002.

4 SMITH, K. What is the "Knowledge Economy"? Knowledge Intensity and Distributed Knowledge Bases. 2002.

5 BLOMSTRM, M., KOKKO, A. AND SJHOLM, F. Growth and Innovation Policies for a Knowledge Economy. Experiences from Finland, Sweden & Singapore. 2002.

6 UNDP. Human Development Report. 2001.

7   BARTELMUS, P. Quantitative Aspects of Sustainable Development. 1997.

8   BOEDIONO & MCMAHON. Education and the Economy: the Eternal Efficiency of Education. 1992.

9   BOSSEL, H. Indicators for Sustainable Development. Theory, Theod'Applications. Canada: IISD. 1999.

10 ERNST & YOUNG. Environmental & Social and Economic Impacts of the Tangguh LNG project. 2004.

11 INDICATORS OF SUSTAINABLE DEVELOPMENT. UN. Department of Economic and Social Affairs Division for Sustainable Development. 2004.

12 KUBBELDAM, LF.B. International Yearbook of Education. Unesco. 1994.

13 MEDIWAKL, W. Social Effects of the Victoria Dam Prefect. 2000.

14 PELLAUD, FRANCINE. The Difficulties of Presenting Sustainable Development in an Exhibition. Switzerland: LDES, Geneva. 2000.

15 RAVETZ, J. Integrated Economic Evaluation for Sustainable Development. England, Manchester University. 2000.

16 SOCIAL AND CULTURAL ASPECTS OF SUSTAINABLE DEVELOPMENT. 2000.

17 WHITELEGG, KATY. Organizing Research for Sustainable Development: An Assessment of National Research Programmers. Proceedings of the Berlin Conference. 2002.

18 COOPEY, R. AND UTTELY, M. Defence Science & Technology, Adjusting to Change. 1993.

19 KEMAL PAK, N. What is the Role of the S&T Parks in Knowledge Based Economy. Proceeding of the XVI IASP Conference, Istanbul, Turkey, August 1999.

20 LOWE, P. The Management of Technology, Perception and Opportunities. Chapman & Hall, 1995.

21 LAZKANO, M. Technology Parks & Economic Growth at a Regional Level. Proceeding of the xviii, IASP Conference, Bilbao, June 2001.

22 GARDNER, P.L. Science Parks as Gateways for International Technology Transfer in Developing Economics. Proceeding of the XVI , IASP Conference, Istanbul, Turkey, August 1999.

23 TUNCER, D. Transformation from Free zones to Techno Pole. Proceeding of the XVI, IASP Conference, Istanbul, Turkey, August 1999.

CLOSING
PAPER

GLOBAL SUSTAINABILITY:
TRENDS, DEVELOPMENTS AND ATTITUDES

P C Hewlett

John Doyle Group and British Board of Agrément

United Kingdom

ABSTRACT. Global development requires global available basic resources. For construction, these are needed for either domestic or industrial use. That, in the main, means concrete, steel, wood, masonry and water and linked to these is energy use and provision. These commodities, whilst readily available in many countries, should be used sensibly if their on-going use is to remain sustainable. Concrete in some form is a global commodity and is likely to remain so. The reasons for such status are explored. In those countries that do not have ready access to all of these materials, alternatives have to be sought. As a consequence, complimentary technologies develop and these have to be recognised. At the root of sustainable issues is the provision and consumption of energy, be it for production use or the running of buildings. Materials use cannot be divorced from design aspects and that ultimately reflects land use and peoples' quality of life expectations. These matters interact one with the other in technically complex ways, having societal outcomes. Conclusions are drawn and a prognosis for concrete's future presented.

Keywords: Sustainability, Emissions, Climate change, Waste, Resource, Bio/Geo-mimicry, Prognosis.

Professor P C Hewlett is a visiting industrial professor to the Concrete Technology Unit, University of Dundee. He is currently Principal Consultant to the British Board of Agrément, Group Technical Advisor to John Doyle Group plc and Chairman of the Editorial Board of Magazine of Concrete Research. He was awarded the UK Concrete Society's gold medal in November 2006.

INTRODUCTION

The issue of sustainability, coupled with environmental concern has made a mark in just about every aspect of human existence in the last decade or so. If by sustainable we mean 'giving support to' or 'keeping alive by tacit approval' we cannot pass responsibility for current concerns to others. It is our future that we are creating and ultimate responsibility is both individual and collective. In this regard, the Dundee 2007 congress focuses on construction and building activities and in particular concrete. These are the subject of my paper and it has to be seen in the context of Event 1 'Role for Concrete in Global Development' and it's supporting four Themes. Sustainability in this regard has to combine the pragmatic with the idealistic. A widely used definition of sustainability comes from the so-called Bruntland Report (refer 1987 World Commission on Environmental Development) and states 'sustainable development has to ensure that it meets the needs of the present without compromising the ability of future generations to meet their own needs'.

Such a definition implies all current needs are known and we also have some measure of what future generations may need. Since I am addressing sustainability in a global context and from a concrete perspective, I think we need to acknowledge global habits, cultures and aspirations. Selected options are likely to be more local – like a piece of jigsaw. The emerging response (picture) will be the totalising of individual parts and considering the four Themes in Event One, the suggested paradigm is very appropriate.

Issues of global concern as they relate to concrete are:

1.   Alleged climate change resulting from associated emissions
2.   Land and water pollution
3.   Energy conservation
4.   Consumption of finite resources

There are others but they tend to be sub-sets of these four. For instance, energy conservation impinges on design and life-style demands. Minimising waste reflects 1, 2 and 4 above.

On the issue of energy consumption it is sometimes difficult to build up a quantitative case because information varies. However the picture for the UK as outlined in the Energy White Paper 2003 has a commitment to cut CO_2 by 60% by 2050 with 20% of UK energy coming from renewable sources by 2020.

It is suggested that 80% of energy used is not optimised. In other words wastage is an issue. In addressing this issue it should be noted that the earth receives in one hour more energy than it requires for one year – an efficiency rating of about 0.01%. Not so much a lack of energy as having the means of using effectively what is available.

We need to use less energy as well as creating more effective ways of providing it and concrete has a net positive role to play.

When considering environmental concerns, there are three scales, namely, local, not so local and global. As far as gaseous emissions are concerned, they are exported and imported on a global scale – we are all affected to some extent. A rise in global temperatures is everybody's problem. Local relates to land fill and land abuse and not so local to ground water and river pollution. Local matters can and perhaps should be dealt with locally, reflecting local needs, demands and capabilities. Not so local and global require international controls and responses such as eco-debits and credits. Dialogue may have got that far but proven commitment falls somewhat short and how are acceptable limits to be set and with what veracity [1].

It is very clear that innovation and invention in the field of construction and with concrete in particular, is still running at a high level. The underlying research and development can be repetitive and individualistic, reflecting local competitive motivations rather than global needs. Striking a balance can be very difficult but I question whether the present way of identifying and prioritising issues, selecting topics and funding them is sustainable. There will always be conflicting priorities, national preferences and market competition and so priorities are not always concerned with global consequences. Over 2000 years ago, Aristotle said 'For that which is common to the greatest number has the least care bestowed upon it. Everyone thinks chiefly on his own, hardly aware of the common interest, and only when he is himself concerned as an individual'.

Such a sentiment is true today and such disregard of collective responsibility could become very serious.

Concrete in some form or other has been part of human social and commercial development for some 8000 years. Over this period it has become a global commodity, reflecting local materials, practices and needs. It has become established in technically advanced countries and can accommodate wide performance demands, as well as challenges from alternative materials such as steel, wood, glass, plastics and natural masonry. Concrete's availability and adaptability will remain dominant. What will change are, as a result of sustainability concerns, attitudes to its production and use and how it meets technical, functional, aesthetic and environmental concerns.

Concrete is a fascinating material. It is capable of an infinite variety of forms and as a commodity it can offer added value. Structures that we take for granted, for instance, roads, bridges, tunnels, sea defences and medium/high-rise buildings would be difficult to imagine without concrete.

This evolutionary process has been visually catalogued [2, 3] and it is apparent that jumps in concrete's development have been, until recently, rather fewer than might be imagined. For instance:

- Roman attainments
- Invention of Portland cement
- Advent of reinforced concrete
- Pre and post tensioning

However, in the last 50 years or so, many variations have come into play. For instance;

- Hydraulically active alternatives to Portland cement
- Non-ferrous reinforcement (organic and metallic)
- Admixtures
- Self-compacting concrete together with ultra high-strength concrete
- Surface treatments
- Non-cementitious mortars and concretes
- Pigments

These options result in a variety of forms that can offer added value.

In this regard, concrete is here to stay, there being no engineering alternative. The component cement, on a price per unit volume basis, is far cheaper than steel and aluminium [4]. Its use in developing and industrialising countries will increase, raising concern over its contribution to global warming by way of CO_2 emissions during cement manufacture, as well as its energy consumption, added to that resulting from the winning and transportation of aggregates. Seeking a balance is difficult and in some economies it may not be practical to adopt newer technologies aimed at reducing both energy demands and contaminating emissions.

However, there are concerns over sustainability issues and those relating to protection of the environment and these need to be addressed in some form. Nixon's paper given at the Dundee 2002 Congress [5] deals with sustainable concrete construction and is both forthright, to the point and very relevant. He makes some suggestions as to what we should be concerned about and trying to achieve,

- Adaptable buildings
- Minimising waste
- Design for de-construction
- Low energy cements
- Reduced energy in use by using concrete intelligently

This Congress allows us to exchange viewpoints and the papers in Event 1 try to bring together the four Themes underwriting the role of concrete in global developments.

The four Themes in Event 1 address these issues in some detail. Let us consider a selection of the papers given (where mentioned the first author of the paper and the country of origin are identified).

THEME 1 – CEMENTS AND THEIR MIXTURES: FUTURE DIRECTION AND PERFORMANCE

It is timely to note that the lead being given by the UK's cement manufacturing industry in an endeavour to address sustainability and environmental issues. In a recent report by the British Cement Association [6] the following achievements are recorded:

1. In 2005 the UK cement industry used 268,000 tonnes of waste fuel, saving some 225,000 tonnes of coal. It is reassuring that recent Spanish work indicates no harmful retained leachables from such fuel use [7]. In total, 1 million tonnes of waste were used as raw materials and kiln fuels.

2. Massive capital investment in production plants (15 in total) totalling approximately £100 million replacing old kiln technology with new, reductions in CO_2 and particulate emissions and improvements to ensure compliance with the European Union's Waste Incineration Directive

3. In 1998 some 289,207 tonnes of cement kiln dust went to landfill but in 2005 that had reduced to 67,682 tonnes as a result of returning the dust to the production line.

4. An 18.5% reduction in CO_2 emissions since 1998.

If such trends were echoed around the world a sustainable environmental impact would be made. After all, cement manufactures contribute approximately 2-6% to global CO_2 emissions.

Some 23 papers were submitted under this theme, eight on cement, five on additives, eight on admixtures and two across the boundaries. Geographically, the emphasis was in the Northern Hemisphere but spread east/west with a focus in Europe but notable submissions from the Far East such as India and China.

Considerable effort to replace at least in part Portland cement by using additions (Barbosa et al, Brazil). Of particular interest was the paper by Pavlenko (Russia) using high calcium fly ash and burnt sand from abrasives works coupled with furnace bottom ash to produce what is described as a cementless binder with strengths up to 60 MPa.

The prospect of zero shrinkage concrete (Saje and co-author, Slovenia) resulting from low w:c ratios and lower alite cements.

Examples where local conditions dominate choice were the use of abundant volcanic tuff (Brahma, Algeria) resulting in reduced costs but less performance, although the latter was considered adequate.

Likewise, the use of rice husk ash (Kothandaraman et al, India) to make more durable concrete by optimising the conditions to produce more amorphous material.

The paper by Justnes et al (Norway/Sweden) on additional grinding allowed low reactivity pfa to be used at significant cement replacement levels resulting in energy and CO_2 savings.

It is evident that chemical admixtures are now an integral part of good concrete practice and indeed the only means of achieving self-compacting concrete as well as high strength/ultra high-strength concretes. This is exemplified by the paper of Collepardi et al (Italy) using an acrylic polycarboxylate superplasticiser in conjunction with a polysaccharide viscosity modifier to yield optimum flow conditions.

We have available a wide spectrum of hydraulically active additions and admixtures to help make performance based optimised concretes with the out-turn of greater efficiency, lower associated emissions and better durability. Such a trend is sustainable.

As far as the UK is concerned whilst admixture use has grown substantially in the last decade we still lag behind countries such as Sweden, Italy, USA and Japan.

In explaining the UK deficit we have to acknowledge according to Dransfield [8] the on-going specifying of low slump concrete (50-75 mm) that reflects entrenched attitudes that fail to take advantage of more workable concretes.

Current usage levels show plasticisers and superplasticisers dominate with the introduction of polycarboxylate ether materials (PCEs). This category of admixture can be tailored to meet a range of performance requirements. The advent of self-compacting concrete could not have been achieved without admixtures [9, 10].

The search for improvement in the materials to make concrete and the concrete itself is global. Cost containment, sustainability, environmental and functional performance are the drivers.

Self-compacting concrete and high/ultra high strength Portland cement concretes are finding their places, particularly in industrialised countries. In developing countries sustainability reflects necessity and what is practical as much as environmental awareness and wellbeing.

THEME 2 – ENERGY AND RESOURCES: WHERE NEXT?

With the exception of the paper from South Africa (Jooste and co-author) the majority were from the Northern Hemisphere but this time with a non-European emphasis, e.g. Malaysia, Iran, China, India, Japan, Hong Kong and Italy.

The main emphasis was on materials and the connection with available resource. Materials disparagingly described as waste become value added. Familiar alternative materials such as glass cullet, rice husk ash, demolition waste, sit alongside quite radical prospects as high density polyethylene thermo-plastic waste (HDPE) being used as conventional aggregate replacement (Gavela et al, Greece), alkali activated phosphate slag ash (Fang et al, China) and electric-arc furnace slag replacing coarse aggregate (Shavarebi et al, Malaysia) that represent local initiatives. Such developments are innovative, technological in nature being problem solving but not advancing basic understanding. As such they may not be widely adopted and so the contribution to sustainability is limited.

It is clear that self-compacting concrete is being widely adopted and can offer a contribution to low energy concrete use (Ramasamy and co-author, India) (Fraaij and co-author, Netherlands) (Kumar and co-author, India) and others.

Adoption of new technology depends on easing the transfer from laboratory to practical application. It is difficult for a designer, consultant, specifier and main contractor to break away from convention and use such materials as incinerator ash, sewage sludge, reclaimed asphalt pavement and the like without off-setting the perceived risk by substantial benefits. Such attitudes are short-sighted but understandable. Somehow concern for the greater good needs to be rewarded.

Performance based specifications [11] could assist adoption coupled with certification and compliance auditing. This could alleviate the concerns of specifying engineers.

When considering recycled and replacement materials the concrete needs to be fit for its intended use but it is sometimes better than it needs to be [12]. Again performance specifications should help.

They should address long-term properties as well as short-term. A means of making durability assessments or rather service life predictions [13] depends upon knowing about and quantifying degradation mechanisms. Without such knowledge local innovation may remain local and potential wider benefit denied.

Within this theme the key note paper Damtoft and co-author, (Denmark) emphasises the benefits of concrete's thermal mass to reduce the running costs of buildings. The latter far exceed the energy demands to build to begin with because that is a fixed one off cost. Running costs extend to the life of the building.

THEME 3: ARCHITECTURE AND ENGINEERING: APPROPRIATE DESIGN

Two sub-themes were addressed namely what is sustainable design and minimising the environmental impact of buildings. Of the ten papers presented, nine were concerned with sustainable design and that by Desai (UK) addressed wide ranging issues linking materials options with design requirements highlighting the issues of retained quality and longer-term retention of function particularly with respect to recycled and unfamiliar materials use. It is persuasive but would have been more so had there been more information on cost/benefit. Sustainability has to offer a business solution unless regulation in some form imposes compliance and change. It could be that regulation and financial incentives are the only ways of changing habits.

A number of design options were presented. For instance, steel tube encased concrete used as columns (Kibriya, Pakistan). The concrete allowed as much as 25% cement replacement by rice husk ash and the composite achieved 350-430% increase in load carrying capacity. A combination of materials and concept.

A similar combination was described by Gopala Rao (India) in developing random rubble masoncrete for massive engineering structures. Design innovation can result in better performance and to be adopted has to address performance retention. Ermakova (Russia) presents an additional finite element method dealing with limit state analysis. Such methods are dependant upon sound input data and that in turn requires mechanistic understanding of what is occurring. This aspect was well documented in the paper by Kovler et al (Israel, Denmark and Russia) who addressed the issue of autogenous shrinkage when using low w:c ratio concretes. They adopted a logical and mechanistic approach using various water supply/retention techniques (internal curing). For instance, pre-soaked lightweight aggregate, super water absorbent polymers and water soluble chemicals reducing rates of evaporation.

Beeby and co-authors (UK) dealt with the behaviour of reinforced concrete in the tension zone. A fundamental issue governing elastic/brittle behaviours. Such knowledge underwrites confident design. In a similar vein, namely interfacial issues, Appa Rao and co-authors (India) rationalise the use of silica fume in high performance concrete and the contribution of the brittle index. Abram's Law did not seem to apply to such micro silica mixes. Desai, (UK) gave a wide ranging over view of concrete's ongoing role highlighting added performance from floors, pre-cast cladding and concrete frames. To maintain concrete's edge we have to change conventional process thinking from a linear approach to a cyclical one. Design should be adequate says Desai. Defining adequacy over 50-100 years can be variable and pre-supposes we know how design and materials will behave over such a period. Some built in redundancy can be a reassuring benefit – but how much?

Regulations can assist change and Desai makes mention of the UK's Code for Sustainable Building covering energy, waste management and water efficiency that may impose change particularly in public works. In this regard Government can act as a sustainability champion.

THEME 4 – DEVELOPING WORLD: RESPONSIBILITIES IN CHANGING ENVIRONMENTS AND DEMANDS

Whilst under this theme only five papers were presented they offer interesting possibilities. Firstly, that by Mejia de Gutierrez et al/Columbia) concerning alkali activated ggbs. Concretes using 100% ggbs and using no Portland cement at all had physical characteristics better than its

OPC counterpart with the exception of carbonation. Since slag from steel manufacture will be made regardless because steel is a primary commodity cement replacement could have a massive global impact. The authors remind us that 1.8 billion tonnes of cement are made globally each year resulting in a similar amount of carbon dioxide into the atmosphere. However, each tonne of cement also consumes 3530 kJ of energy. A potential double saving. Such prospects have to be taken seriously taking business as well as idealistic viewpoints. It is tantalising to think of the chemical extension of this technology to pfa although attempts to date at ambient temperature hydration have not been very successful. I suspect this is a topic that could/should be dealt with by a global perspective.

In a similar way the co-burning of biomass (wood chips) (Kosior-Kazberuk and co-author, Poland) with coal and incorporating the resulting fly ash in concrete has shown benefits. The prospect offered by biomass should be exploited since it offers the prospect of neutral CO_2 balance.

One of the concerns raised about recycling is the quality and consistency that can be maintained. This was certainly the case when incorporating waste water into concrete at pre-casting plants (Perlot and co-author, France) not withstanding standards (NFEN108) covering the same. For something like water the economic case would be dominant in water endowed countries less so for arid regions where water is precious.

The issue of water is also addressed in the paper by Huhammadiev (Tajikistan) and highlights problems of water collection and retention.

The final paper of Theme 4 endeavours to look ahead to concrete's future in what is termed a global post-industrial society. The bedrock would seem to be knowledge and its application but taking a global view. The means of sharing knowledge exists by way of the internet but there also has to be a will to share. Collaborative research of global relevance – sustainability and environmental wellbeing would seem to be contenders.

At the 2005 Congress I made mention of nanotechnologies [14] and the role it might play in making more efficient concretes. At that time nanotechnology was a curiosity and not seen in serious contention as far as cement and concrete are concerned [15]. It would seem that is now changing which perhaps is not so surprising because materials such as micronised silica fume (approximately 0.1 microns or 100 Nm diameter) whilst outside the nano-scale range is quite close. Likewise admixtures and water itself are within range and can undergo molecular transformations (chemistry!) resulting in interactions with the very fine suspended particulate cementing materials giving rise to increased hydration and greater fluidity. What Skalny et al [16] referred to as micro-structural engineering.

Therefore it is a pleasure to note the advent of Nanocem – a new European research network [17, 18].

This initiative is supported by academic and industrial partners gathered from Europe. The emphasis is on fundamental research but will endeavour to relate nano-scale processes (physical chemistry!) with macro-scale outturns.

Sustainability is on the agenda of objectives together with usability and multi-functionality.

Core projects will address,

- hydrate assemblies of CSH
- pore structures
- organo-alumina reactions
- hydration of blended cements

In addition to the core projects there are 21 partner contributed projects. One of these entitled 'Nano scale studies on the durability of cement based materials' contributed by the Universitat Polytechnica de Catalunia (Spain) that may have a direct bearing on the so-called coefficients relating to estimating life spans of cementitious materials. If that is the case this could be a very important advance.

So what lessons are to be learnt? What topics should have priority and can we map out a sustainable future for concrete as a main provider of infrastructure and global human security and wellbeing?

CONCLUSIONS

1   Sustainability is now firmly on the political/global agenda and we need to seek options of global significance either by way of many but collective actions and/or generic changes. Some form of centralised organisation would be better placed and aid dissemination and manage change.

2   Concrete based on Portland cement in some form will remain as the benchmark but with increasing cement replacement by waste/recycled based materials. In that regard concrete has to be construction's sustainable option.

3   Innovation is at a high level reflecting a search for alternatives to Portland cement, aggregates and reinforcements. Such a quest may give local respite but have limited application as well being quite generic allowing others to adopt sustainable practices.

4   The ultimate driver to adoption of change is likely to be functional and /or legal. Many quoted examples do not address the issue of cost and unless some alternatives are mandatory they are unlikely to be adopted. Such an approach can be dispassionate and cynical and regulations/legislation may have to be used.

5   Improvements in cement manufacture will continue by way of reducing emissions and improving thermal efficiency. Such a trend should be encouraged perhaps by way of tax concessions and capital subsidies. However, to be substantial long-term it has to make good business sense.

6   Efficient and attractive building designs. Concrete offers real opportunities and I am sure will be exploited. However, more regard has to be taken of concrete's retained appearance particularly in external and exposed situations.

7   Will sustainability catalyse new technologies and alleviate pollution and profligate energy use. Are technologies such as large scale CO_2 sequestration (bio/geomimicry) feasible both in treatable volume and rate of chemical conversion terms? There are some who think so [19]. It is issues such as these that require global as well as local responses. Do we have an international network capable of exercising such action?

8   In coming back to the title of my paper. Sustainability and all that is associated with it is an ongoing trend that requires national and international commitments. A case can be made for more science to underwrite proposed technological changes. Such an approach would give confidence and also support second generation developments.

Developments resulting from and allied to these trends are many and varied. Underlining the four Themes of Event 1 there were 28 topics. If we include the other five Events we add another 88 topics. The potential for development is huge and much is occurring but often only of local relevance. What should have or not have priority? Should we avoid duplication of effort or not? From the cauldron of development do we distil the winners?

As to attitudes towards sustainability they are variable particularly at the political level. Emotion sometimes exceeds objective data. Unfortunately such doubts can lead to complacency. If sustainability can generate new technologies and these in turn can improve effectiveness and reduce pollution whilst making good business sense, it surely can't be resisted even if history shows (and I doubt that) it was all unnecessary.

REFERENCES

1.   HARRISON J, The TecEco Times Issue 62, 5th Oct. 2006. p. 7.
2.   STANLEY C C, Highlights in the History of Concrete, ISBN 07210 1156X (C and CA) 1979. 44 p.
3.   Concrete Through The Ages, ISBN 07210 15476 (British Cement Association), 1999. 37 p.
4.   BIRCHALL J D, HOWARD A J AND KENDALL K K, Chemistry in Britain, Dec. 1982, 860 p.
5.   NIXON P J, More Sustainable Construction: The Role of Concrete, Proceedings of International Congress: Challenges to Concrete Construction, Conference 2 – Sustainable Concrete Construction, Univ. of Dundee, Scotland, 9-11 Sept. 2002. pp. 1-12.
6.   Performance – A Corporate Responsibility Report from the UK Cement Industry, British Cement Association, July 2006, 19 p.
7.   HIDALGO A, ALSONSO C, SANJUAN M A, GUEDD E AND ANDRADE C, Construction Materials, Thomas Telford Journals, Vol. 159, Issue 2, 2006, pp. 85-92.
8.   DRANSFIELD, D. Admixture Current Practise – Part 1 Concrete, Sept 2006m pp. 35-36.
9.   CAA Website – www.admixtures.org.uk
10. Relevant Standards BSEN 934 Part 2 (2001), Part 3 (2003), Part 4 (2001).
11. DHIR R K, DYER T D AND PAINE K, Appropriate Use of Sustainable Construction Materials Concrete. Oct. 2006, pp. 20-24.
12. NEVILLE A M. Concrete: Neville's Insights and Issues ESBN 07277 34687, Thomas Telford, 2006, 314 p.
13. ISO 15686-2:(2001) Building and Constructed Assets – Service Life Planning, Part 1 Service Life Prediction Procedures.
14. HEWLETT P C, Young Researchers Forum: Proceedings of an Intl. Conf. Dundee, 7 July 2005, pp. 1-20.
15. ROUVRAY D, Is the Future Nano? Chemistry in Britain, Dec. 2000, pp. 46-47.
16. SKALNY J AND YOUNG J F, Mechanisms of Portland Cement Hydration 7th International Congress on Chemistry of Cement 1, sub-theme 11-1 Paris 1980, 45 p.
17. GARTNER E, Nanowork for Cement Giants; Materials World, Sept. 2006, pp. 26-28.
18. www.nanocem.org
19. HARRISON J, The TecEco Times, issue 61, 5 Oct 2006, 13 pp.

INDEX OF AUTHORS

SUBJECT INDEX

This index has been compiled from the keywords assigned to the papers, edited and extended as appropriate. The page references are to the first page of the relevant paper.